Sources and Studies
in the History of Mathematics and Physical Sciences

Editorial Board
J.Z. Buchwald   J. Lützen   G.J. Toomer

Advisory Board
P.J. Davis   T. Hawkins
A.E. Shapiro   D. Whiteside

Springer
New York
Berlin
Heidelberg
Hong Kong
London
Milan
Paris
Tokyo

Sources and Studies in the
History of Mathematics and Physical Sciences

K. Andersen
Brook Taylor's Work on Linear Perspective

H.J.M. Bos
Redefining Geometrical Exactness: Descartes' Transformation of the Early Modern Concept of Construction

J. Cannon/S. Dostrovsky
The Evolution of Dynamics: Vibration Theory from 1687 to 1742

B. Chandler/W. Magnus
The History of Combinatorial Group Theory

A.I. Dale
A History of Inverse Probability: From Thomas Bayes to Karl Pearson, Second Edition

A.I. Dale
Most Honourable Remembrance: The Life and Work of Thomas Bayes

A.I. Dale
Pierre-Simon Laplace, Philosophical Essay on Probabilities, Translated from the fifth French edition of 1825, with Notes by the Translator

P.J. Federico
Descartes on Polyhedra: A Study of the *De Solidorum Elementis*

B.R. Goldstein
The Astronomy of Levi ben Gerson (1288–1344)

H.H. Goldstine
A History of Numerical Analysis from the 16th Through the 19th Century

H.H. Goldstine
A History of the Calculus of Variations from the 17th Through the 19th Century

G. Graßhoff
The History of Ptolemy's Star Catalogue

A.W. Grootendorst
Jan de Witt's *Elementa Curvarum Linearum, Liber Primus*

T. Hawkins
Emergence of the Theory of Lie Groups: An Essay in the History of Mathematics 1869–1926

A. Hermann/K. von Meyenn/V.F. Weisskopf (Eds.)
Wolfgang Pauli: Scientific Correspondence I: 1919–1929

Continued after Index

Andrew I. Dale

Most Honourable Remembrance

The Life and Work of Thomas Bayes

With 29 Illustrations

Springer

Andrew I. Dale
Department of Mathematical Statistics
University of Natal
Durban 4001, Natal
South Africa
dale@nu.ac.za

Sources and Studies Editor:
Gerald J. Toomer
2800 South Ocean Boulevard, 21F
Boca Raton, FL 33432
USA

Library of Congress Cataloging-in-Publication Data
Dale, Andrew I.
  Most honourable remembrance : the life and work of Thomas Bayes / Andrew I. Dale.
    p. cm.—(Studies and sources in the history of mathematics and physical sciences)
  Includes bibliographical references and index.
  ISBN 0-387-00499-8 (alk. paper)
  1. Bayes, Thomas, d. 1761. 2. Statisticians—Great Britain—Biography. I. Title. II. Sources and studies in the history of mathematics and physical sciences.
  QA276.157 .B39D35   2003
  519.5'092—dc21
  [B]   2003042474

ISBN 0-387-00499-8   Printed on acid-free paper.

© 2003 Springer-Verlag New York, Inc.
All rights reserved. This work may not be translated or copied in whole or in part without the written permission of the publisher (Springer-Verlag New York, Inc., 175 Fifth Avenue, New York, NY 10010, USA), except for brief excerpts in connection with reviews or scholarly analysis. Use in connection with any form of information storage and retrieval, electronic adaptation, computer software, or by similar or dissimilar methodology now known or hereafter developed is forbidden.
The use in this publication of trade names, trademarks, service marks, and similar terms, even if they are not identified as such, is not to be taken as an expression of opinion as to whether or not they are subject to proprietary rights.

Printed in the United States of America.

9 8 7 6 5 4 3 2 1   SPIN 10913000

Typesetting: Pages created by the author using a Springer T$_E$X macro package.

www.springer-ny.com

Springer-Verlag   New York Berlin Heidelberg
A member of BertelsmannSpringer Science+Business Media GmbH

To F.J.H.

PROLOCUTION

Si quelqu'vn me blasme de remarquer ces petites particularitez, ie luy répondray, que Plutarque témoigne qu'il auoit receu quelque sorte de déplaisir d'vne chose aussi legere: quand il se plaint de ceux qui n'auoient pas laissé par écrit les noms des meres de Nicias, de Demosthene, de Formion, de Thrasybule, & de Theramines, renommez personnages contemporains de Socrate: & au contraire il monstre sçauoir bon gré à Platon & à Antisthene, parce que le premier auoit écrit le nom du Precepteur d'Alcibiade, & l'autre, n'auoit pas dédaigné de nomer sa nourrice. La negligence des anciens Escriuains a este si grande, qu'elle est cause que sept villes ont esté en dispute pour la naissance d'Homere: chacune se donnant cet honneur d'auoir esté la mere nourrice du plus excellent Poëte de la Grece; Ce qui m'a fait toûiours penser, que c'est vn grand defaut à ceux qui se meslent d'écrire les Vies des Hommes illustres, de laisser en arriere quelques particularitez, qu'on n'eust peut-estre pas considerées dans le temps qu'ils ont vescu, mais qui servient infailliblement estimées en vn autre ciecle.

Hilarion de Coste.
La Vie du R. P. Marin Mersenne.

There are many passages in this book where I have been at some pains to resist the temptation of troubling my readers with my own deductions and conclusions; preferring that they should judge for themselves, from such premises as I have laid before them.

Charles Dickens.
American Notes.

Preface

> *I find it impossible to write a preface to this work, without discovering a little of the enthusiasm which I have contracted from an attention to it.*
>
> Joseph Priestley.
> The History and Present State of Electricity.

It is generally considered bad form in writing, unless on matters autobiographic, to make unbridled use of the perpendicular pronoun. The reader of the present book, however, may well wonder why one would want to study the life and works of Thomas Bayes, 'this strangely neglected topic'[1], and it is only by a reluctant use of the first person singular on the part of the author that this legitimate question can be answered.

It was in the late 1960s that my interest in various aspects of subjective probability was awakened by some of the papers of I.J. ('Jack') Good, and this was followed by the reading of works such as Harold Jeffreys's *Theory of Probability*. In many of these the (apparently simple) result known as Bayes's Theorem played a pivotal rôle, and it struck me that it might be interesting to find out a bit more about Thomas Bayes himself. In trying to satisfy this curiosity in spasmodic periods over many years I discovered that little information seemed to be available. Writings by John D. Holland in the 1960s shed some light on the matter, but it was only in periods of sabbatical leave that I was able to undertake the intensive (and extensive) archival research needed to flesh out the shadowy figure after whom one of the major branches of modern statistics is named.

Where, for instance, was Bayes educated? The answer to this question presented itself by chance to me when I was visiting the University of Chicago. While waiting for some rare volumes to be retrieved from the depths of the John Crerar Library I happened to pick up a catalogue of manuscripts in the Edinburgh University Library in which there appeared details of a certificate recording the admission of Thomas Bayes as a student. Visits to Edinburgh and the generous co-operation of the library staff resulted in the finding of a number of references to Bayes in the archives of the University.

Similar serendipity led to my finding of the proof, written mainly in Latin, of one of the rules in Bayes's *Essay towards solving a Problem in the Doctrine of Chances* in a Notebook attributed to him and held in the Equitable Life Assurance Society in England. This not only reinforced the opinion that the Notebook was indeed written in Bayes's hand, but also shed light on matters that Bayes had thought worth recording.

Once the question of Bayes's education had been solved, or at least answered to the best of my ability, it seemed necessary to investigate his family and life in Tunbridge Wells, where he spent many years as Presbyterian minister. Although some information on these matters was available from (fairly) modern writings, it seemed expedient rather to consult writers as near to Bayes as possible, and in doing so, of course, I was merely emulating Edward Gibbon, who, in the preface to the fourth volume of the quarto edition of his *Decline and Fall of the Roman Empire*, stressed that he had made it his earnest endeavour always to consult the original sources. Perhaps Gibbon, in his turn, was following the example set by Watson in his edition of *The Works of Horace*:

> Whoever attempts, at this distance of time, to write upon a *Classic*, ought not only to read over what the most approved commentators have written upon his *works*, but to compare the same carefully with the text itself; otherwise he is in danger of being misled: for many of them, instead of illustrating, only darken their author, and palpably mistake his meaning.
> [1792, p. v]

There are three eighteenth-century savants who are known to modern statisticians and historians of statistics for single contributions[2]: Thomas Bayes, Roger Boscovich[3] and Thomas Simpson[4]. Bayes is known for his Theorem, published in his posthumous *Essay* in 1764, Simpson for his work in 1755 (extended in 1757) on error distribution, and Boscovich for the posing of a question in 1757 (and its solution in 1760) on the fitting of a straight line to observational data[5]. Boscovich and Simpson have both been the subject of biographies[6], and it seems not inappropriate that Bayes be similarly commemorated, particularly as the tercentenary of his birth is upon us.

In drawing to a conclusion let me say a brief word about the presentation of material here. As regards the method of citation, works such as Bayes's *Essay towards solving a Problem in the Doctrine of Chances*, published in the *Philosophical Transactions* in 1764 in the volume for 1763, are cited as 'Bayes [1763]'. In the reprinting of Bayes's tracts, **[n]** indicates the end of page **n** in the original text. If a word in the original is broken across two pages, **[n]** appears here at the end of the word, with broken formulae being similarly treated. Footnotes are given in the usual way, irrespective of how they may be positioned (e.g., broken over two pages) in the original. Things that appear in crotchets, [...], are my corrections or additions, even in exponents in formulae. In giving Bayes's tracts I have allowed myself the liberty of a light editing. In the *Essay*, for example, what appear in the original as 'a remuch mistaken' and 'innature' are given here as 'are much mistaken' and 'in nature', and '$(a+b)^{b+q}$' in the first paragraph of Proposition 7 is reprinted as '$(a+b)^{p+q}$'. On page **[400]** the last occurrence of $2Ea^p b^q$ appears in the original as $2Ea_p b^q$, and the second Σ on page **[416]** appeared originally as E. No changes are more serious than these, and all would easily be made by the reader of the original texts.

Not being an expert in eighteenth-century history, I must bring to this study a certain naïveté, a quality that I hope will be seen by the reader as touching rather than irritating and obtrusive. There are no doubt portions of this work that could be better handled by a historian; but one of the results of approaching a topic in ignorance is the pleasure of discovery[7], and it is with the hope of communicating some of the pleasures I have met with in this journey through Bayes and other coves that certain passages — for example the remarks on the Great Plague and the descriptions of Tunbridge Wells — are presented.

I hope, though, that the reader will not view this work in the light in which Nicodemus Boffin, 'the minion of fortune and the worm of the hour', viewed his new mansion, viz.[8]

> He could not but feel that, like an eminently aristocratic family cheese, it was much too large for his wants, and bred an infinite amount of parasites.

Rather, as George Berkely put it in *Siris: a Chain of Philosophical Reflexions and Inquiries*:

> The displeasure of some readers may perhaps be incurred by surprising them into certain reflections and inquiries for which they have no curiosity. But perhaps some others may be pleased to find a dry subject varied by digressions, traced through remote inferences, and carried into ancient times. [¶350]

Durban, Natal, Andrew I. Dale
South Africa

Acknowledgments

I am grateful to the following for granting permission for quotation from the works mentioned: The Mathematical Association, from the paper published in *The Mathematical Gazette* by G.J. Lidstone in 1941; The Johns Hopkins University Press, from A.W.F. Edwards's *Likelihood* published in 1992; the American Mathematical Society, from (i) the paper published in *The Bulletin of the American Mathematical Society* by F.H. Murray in 1930, (ii) G. Boole's *A Treatise on Differential Equations* (1959, 5th edition) and (iii) F. Cajori's *A History of Mathematics* (1991, 5th edition); the Royal Statistical Society, from the papers published in the *Journal of the Royal Statistical Society, Series A* (i) by S.M. Stigler in 1982 and (ii) by D.J. Bartholomew (comment by D.A. Gillies) in 1988; Springer-Verlag, Heidelberg, from (i) A.N. Kolmogoroff's *Grundbegriffe der Wahrscheinlichkeitsrechnung* of 1933, (ii) O.B. Sheynin's 'R.J. Boscovich's work on probability' in *Archive for History of Exact Sciences* (1973) 9: 306–324 ©Springer-Verlag and (iii) my 'A newly-discovered result of Thomas Bayes' in *Archive for History of Exact Sciences* (1986) 35: 101–113 ©Springer-Verlag; Barnes & Noble, and, separately, Palgrave Publishers Ltd, from N. Page's *Dr Johnson. Interviews and Recollections* (Macmillan) of 1987; Hodder and Stoughton, from R.W. Dale's *A History of English Congregationalism* of 1907; T. & T. Clark, from the *Encyclopædia of Religion and Ethics*, edited by J. Hastings, of 1971; the Inter-Varsity Press, from *The New Bible Commentary*, edited by F. Davidson; C. Klincksieck et Cie, from H. Laboucheix's *Richard Price: théoricien de la révolution américaine; le philosophe et la sociologue; le pamphlétaire et l'orateur*; the Wesley Historical Society, from F.F. Bretherton's paper published in 1916 in the *Proceedings of the Wesley Historical Society*; the Associated University Presses (Golden Cockerel Press), from T. Hailperin's *Sentential Probability Logic. Origins, Development, Current Status, and Technical Applications*; the International Statistical Institute, from G. Shafer's 1985 paper in the *International Statistical Review* and F. Yates's paper in the *Revue de l'Institut International de Statistique* (1962); ITPS Ltd, for the passages (i) from

A. Browning's *English Historical Documents 1660–1714*, vol. VIII of *English Historical Documents* (General Editor D.C. Douglas) published by Eyre & Spottiswoode (Routledge) in 1953, (ii) from S. Körner's *The Philosophy of Mathematics: an introductory essay* (1960, Routledge) and (iii) from G.H. von Wright's *A Treatise on Induction and Probability* (Littlefield, Adams & Co.) of 1960; the Epworth Press, from W.B. Whitaker's *The Eighteenth Century English Sunday. A Study of Sunday Observance from 1677 to 1837* (1940); Faber and Faber, from M. Barton's *Tunbridge Wells* (1937) and S. Spender's *World within World* (1991); S.E. Black, from her *Bunhill Fields: the Great Dissenters' Burial Ground* (1990); Kegan Paul International, from M. Black's *The Nature of Mathematics: A Critical Survey* (1933); Oxford University Press, from (i) M.G. Kendall's 'Ronald Aylmer Fisher, 1890–1962', *Biometrika* (1963), 50: 1–15, (ii) R.L. Plackett's 'Studies in the History of Probability and Statistics VII. The principle of the arithmetic mean', *Biometrika* (1958), 45: 130–135 and (iii) J. Wishart's 'On the approximate quadrature of certain skew curves, with an account of the researches of Thomas Bayes', *Biometrika* (1927), 19: 1–38 (the citations are made by permission of the Biometrika Trustees). I regret that I have been unable to trace the authors or estates of these three items, and I thank them for their work; Edward Arnold, from A. O'Hagan's *Kendall's Advanced Theory of Statistics, Volume 2B. Bayesian Inference* (Charles Griffin); Academic Press, from my article in *Historia Mathematica* in 1991, and from the articles published in the same journal by G.C. Smith in 1980, vol. 7: 379–388 and R. Weinstock in 1998, vol. 25: 281–289; The Institute of Mathematical Statistics, from G. Shafer's article in *The Annals of Statistics* (1982) 10: 1075–1089; Carus Publishing Company (Open Court Publishing Company), for quotations from F. Cajori's *A History of the Conceptions of Limits and Fluxions in Great Britain from Newton to Woodhouse* of 1919; the Société Française d'Histoire des Sciences et des Techniques, from J.P. Clero's *Thomas Bayes: Essai en vue de resoudre un probleme de la doctrine des chances*, published in *Cahiers d'Histoire et de Philosophie des Sciences* No. 18 (1988); Random House, Inc., from F.R.B. Godolphin's *The Latin Poets* (1949) and from O. Johns's *Asphalt and Other Poems* (1917); Adolf M. Hakkert, from A.D. Leeman's *Orationis Ratio* of 1986; Palgrave Publishers, from J.M. Keynes's *A Treatise on Probability* of 1921 (Macmillan & Co. Ltd); the University of California Press, from (i) Andrew Motte's translation, *Sir Isaac Newton's Mathematical Principles of Natural Philosophy and his System of the World* as revised by Florian Cajori, of Newton's *Philosophiæ Naturalis Principia Mathematica* Copyright 1934 by The Regents of the University of California, Renewal copyright 1962 by The

Regents of the University of California and (ii) L.J. Savage's 'The foundations of statistics reconsidered' in the *Proceedings of the Fourth Berkeley Symposium on Mathematical Statistics and Probability*, (ed. J. Neyman), vol. I, pp. 578–586, Copyright ©1961 by The Regents of the University of California; David Higham Associates Limited, from A.W.M. Bryant's *Restoration England* published by Collins in 1960; P. Horwich, for the material from his *Probability and Evidence* (1982), published by Cambridge University Press.

The quotations from A. Hald's *A History of Probability and Statistics and Their Applications before 1750* ©1990 by John Wiley & Sons, Inc. and his *A History of Mathematical Statistics from 1750 to 1930* ©1998 by John Wiley & Sons, Inc. are reprinted by permission of John Wiley & Sons, Inc. The material from B. de Finetti's *Theory of Probability*, vols 1 & 2 (1974/5), © John Wiley & Sons Ltd, is reproduced with permission. The material from Lorraine J. Daston's *Classical Probability in the Enlightenment*, Copyright ©1988 by Princeton University Press, is reprinted by permission of Princeton University Press. The quotations from (i) R.C. Jeffrey's *The Logic of Decision*, ©1965, 1983 by The University of Chicago, (ii) D.M. Jesseph's *Berkeley's Philosophy of Mathematics*, ©1993 by The University of Chicago and (iii) E. Nagel's *Principles of the Theory of Probability*, copyright by The University of Chicago and printed in the *International Encyclopedia of Unified Science* (Editor-in-chief Otto Neurath), Volume 1, Number 6, are all reprinted by permission of the publishers, The University of Chicago Press, as is (i) J.B. Bond's paper published in *Isis* (1921), No. 11, vol. IV, 295–323 and (ii) P. Kitcher's paper published in *Isis* (1973), 64: 33–49. The passage from Popper's *The Logic of Scientific Discovery* (Hutchinson, 1968) is reprinted with the permission of Mrs Melitta Mew, executrix of Sir Karl Popper's estate. The material from Hans Reichenbach's *The Theory of Probability. An Inquiry into the Logical and Mathematical Foundations of the Calculus of Probablity* (1971), published by the University of California Press, is reprinted by permission of Dr M. Reichenbach.

Quotations of Samuel Pepys are reproduced from *The Diary of Samuel Pepys* edited by Robert Latham and William Matthews (Copyright © The Master, Fellows and Scholars of Magdalene College, Cambridge, Robert Latham and William Matthews 1983) by permission of PFD on behalf of The Master, Fellows and Scholars of Magdalene College, Cambridge, the Estates of Robert Latham and William Matthews. The numerous citations from manuscripts relating to Thomas Bayes's connexion with Edinburgh University are made with permission of Edinburgh University Library.

Extracts from Stephen M. Stigler's *Statistics on the Table: The History of Statistical Concepts and Methods*, Cambridge Mass., are reprinted by permission of the publisher Harvard University Press, Copyright ©1999 by the President and Fellows of Harvard College. The extracts from (i) J.E. Force's *William Whiston: Honest Newtonian* (1985), (ii) N. Guicciardini's *The development of Newtonian calculus in Britain 1700–1800* (1989), (iii) P. Walmsley's *The Rhetoric of Berkeley's Philosophy* (1990), (iv) E.G.R. Taylor's *The Mathematical Practitioners of Hanoverian England 1714–1840* (1966), (v) I. Hacking's *The Emergence of Probability* (1975), (vi) I. Parker's *Dissenting Academies in England, Their Rise and Progress and their Place among the Educational Systems of the Country* (1914) and (vii) H. Jeffreys's *Scientific Inference* (1973) are reprinted with the permission of Cambridge University Press. The extracts from T.L. Fine's *Theories of Probability: an examination of foundations* (1973) are reprinted by permission of the publisher, Academic Press. The material from R.W. Home's paper in *Notes and Records of the Royal Society* (1974–1975), 29: 81–90 is reprinted with the permission of the author and the Royal Society. The extract from *The Foundations of Scientific Inference*, by Wesley C. Salmon, ©1967, is reprinted by permission of the University of Pittsburgh Press. The quotations from (i) volumes III, IX & XXII of the *Dictionary of National Biography* (1917), edited by L. Stephen & S. Lee, © Oxford University Press, (ii) A.C. Fraser's *The Works of George Berkeley, D.D.* (1901, Clarendon Press), (iii) R.A. Fisher's *Statistical Methods, Experimental Design, and Scientific Inference* (1995, edited by J.H. Bennett), a re-issue of *Statistical Methods for Research Workers, The Design of Experiments*, and *Statistical Methods and Scientific Inference* ©1990 The University of Adelaide, (iv) G.B. Hill's edition of *Boswell's Life of Johnson* (1934, Clarendon Press), (v) H. Jeffreys's *Theory of Probability* © Oxford University Press 1961 (Clarendon Press), (vi) J.R. Lucas's *The Concept of Probability* (1970, Clarendon Press) and (vii) D.O. Thomas's *The Honest Mind. The Thought and Work of Richard Price* © Oxford University Press (1977, Clarendon Press) are all made by permission of Oxford University Press.

Permission is granted by the Director of Dr Williams's Library on behalf of the Trustees to quote (i) from a manuscript (1731–1732), copied by S. Palmer, entitled *A View of the Dissenting Intrest in London of the Presbyterian and Independent Denominations from the year 1695 to the 25 of December 1731. With a postscript of the present state of the Baptists* and (ii) from a note by W.T. Whiteley appearing in J.G. White's *The Churches & Chapels of Old London, with A Short Account of those who have Min-*

istered in Them (printed for private circulation in 1901), such permission being granted subject to the following declaration: (a) that the Trustees have allowed access to the manuscript but are not responsible for the selection made, and (b) that the author, both for himself and his publisher, waives whatever copyright he may possess in the extracts made, so far as the exercise of that right might debar other scholars from using and publishing the same material and from working for that purpose on the same manuscripts. Quotations from the 1974 edition of Richard Price's *A Review of the Principal Questions in Morals* are made by permission of the editor, D.D. Raphael, the quotations from pp. xvi, xvii, xxiv, and xxxii being from the editor's introduction and that from p. 74 being from the text of Price's essay itself. The extracts from Bayes's Notebook are given by permission of The Equitable Life Assurance Society, in whose possession the Notebook is. Transcripts of Bayes's letters to John Canton, and of other manuscripts by him in the possession of the Royal Society, are reprinted by permission of the President and Council of the Royal Society.

The Register of Burials in Bunhill Fields (RG4/3982) and the wills of Joshua (PROB 11/746) and Thomas Bayes (PROB 11/865) are housed in the Greater London Record Office in London, England. L. & B. Bailey's *History of Non-conformity in Tunbridge Wells* (1970) is in the Tunbridge Wells Library. The Stanhope of Chevening papers are housed in the Centre for Kentish Studies, Maidstone, Kent: I am indebted to the Research Archivist, Mrs L. Richardson, for her help in accessing the manuscripts and for permission to cite them here. W. Besant's *London in the Eighteenth Century* and *London City*, published by Adam & Charles Black, are now out of copyright, as is Besant and G.E. Mitton's *The Fascination of London. Holborn and Bloomsbury*. The material quoted from R. Blanchard's paper in *Restoration and Eighteenth-Century Literature. Essays in Honor of Alan Dugald McKillop* (The University of Chicago Press), is not under copyright. References to the list of subscribers to John Ward's *Lives of the Professors of Gresham College*, the letters of Ward to Thomas Bayes and Skinner Smith, John Rippon's lists of epitaphs in Bunhill Fields and *Tunbrigalia* are from The British Library's Add. Mss. 6207.f.1, 6222.c9, 28513–28523 and 11631.e.73(1), and are made with acknowledgments to the Library. Thomas Bayes's Election Certificate is copyright by The Royal Society, and is reproduced with permission. The material ©1979 from *The Science of Music in Britain, 1714–1830* by J.C. Kassler is reproduced by permission of Routledge, Inc., part of The Taylor & Francis Group. I regret that, despite serious endeavours, I have not managed to contact the author, whose work is hereby acknowledged.

Four sketches are reprinted by permission of The Brtish Library (with shelfmarks), viz. (i) the portrait of Joshua Bayes, from W. Wilson's *The*

History and Antiquities of Dissenting Churches [489.b.12, 13], (ii) Leather Lane Chapel, from G.W. Thornbury's *Old and New London*, vol. 2 [HLR 942.1], (iii) the Parade at Tunbridge Wells, from J. Phippen's *Colbran's New Guide for Tunbridge Wells* [796.e.14] and (iv) a map of Tunbridge Wells, from J. Britton's *Descriptive Sketches of Tunbridge Wells* [G3926].

Despite earnest endeavours, my attempts at tracing the current copyright holders (to whom my sincere acknowledgements are due) of the following works have been unsuccessful: (i) R. Bayne's edition of Joseph Butler's *The Analogy of Religion, Natural and Revealed*, published by J.M. Dent & Sons Ltd. in 1906; (ii) R. Nevill's *London Clubs their histories and treasures*, published by Chatto & Windus in 1911; (iii) A. Armitage's *Edmond Halley*, published by Thomas Nelson and Sons Limited in 1966; (iv) A.A. Luce & T.E. Jessop's *The Works of George Berkeley Bishop of Cloyne*, published by Thomas Nelson and Sons Ltd in 1948; (v) P.F. Mottelay's *Bibliographical History of Electricity and Magnetism chronologically arranged* (1922), published by Charles Griffin; (vi) D.D. Roller's *The De Magnete of William Gilbert* (1959), published by Menno Hertzberger; (vii) W.G. Bell's *The Great Plague in London in 1665* (1951), published by The Bodley Head (Random House); (viii) R. Lynd's *The Pleasures of Ignorance* (1928), published by Methuen & Co. Ltd; (ix) J.H. Wilson's *The Court Wits of the Restoration* (1967), published by Frank Cass & Co. Ltd; (x) C.B. Boyer's *The History of the Calculus and its Conceptual Development* (1949), Dover Publications; (xi) R. von Mises's *Probability, Statistics and Truth* (1981), Dover Publications; (xii) H. Poincaré's *la Science et l'Hypothèse* (1903), published by Flammarion; (xiii) T. Carlyle's *Scottish and Other Miscellanies* (1915), W. Cobbett's *Rural Rides* (1912), J. Stow's *A Survay of London* (1912) and J. Butler's *The Analogy of Religion* (1906), all published by J.M. Dent & Sons, Ltd (The Orion Publishing Group); (xiv) C. Eisenhart's 'Boscovich and the combination of observations', pp. 200–212 in L.L. Whyte's *Roger Joseph Boscovich, S.J., F.R.S., 1711–1787; Studies of his Life and Work on the 250th Anniversary of his Birth* (1961), L. Hogben's *Statistical Theory. The Relationship of Probability, Credibility and Error* (1957), B. Lillywhite's *London Coffee Houses* (1963), M.E. Ogborn's *Equitable Assurances. The Story of Life Assurance in the Experience of the Equitable Life Assurance Society 1762–1962* (1962), B. Russell's *History of Western Philosophy* (1961), all published by Allen & Unwin Ltd; (xv) F. Mosteller and D.L. Wallace's *Applied Bayesian and Classical Inference. The Case of The Federalist Papers* (1984), published by Springer-Verlag;

(xvi) B. de Finetti's 'La Prevision: ses lois logiques, ses sources subjectives', pp. 1–68 in *Annales de l'Institut Henri Poincaré* (1937), published by Gauthier-Villars; (xvii) P. Walley's *Statistical Reasoning with Imprecise Probabilities* (1991), published by Chapman and Hall; (xviii) F.M. Clarke's *Thomas Simpson and his Times* (1929), published by Columbia University Press; (xix) A.N. Whitehead's *Adventures of Ideas* (1967), published by The Free Press; (xx) D.T. Whiteside's *The Mathematical Works of Isaac Newton, vol. I* (1964), published by the Johnson Reprint Company Ltd; (xxi) H.H. Horne's *The Philosophy of Education. Being the foundations of education with the related natural and mental sciences* (1916), published by Macmillan Publishing Company, Inc.; (xxii) I. Tweddle's *James Stirling 'This about series and such things'* (1988), Scottish Academic Press Ltd; (xxiii) C.L.S. Linnell's *The Diaries of Thomas Wilson, D.D. 1731–1737 and 1750* (1964), published by the Society for Promoting Christian Knowledge; (xxiv) H. McLachlan's *English Education under the Test Acts* (1931), published by Manchester University Press; (xxv) T.J.I'A. Bromwich's *An Introduction to the Theory of Infinite Series* (1931), published by Macmillan; (xxvi) T. Huxley's *Science and Education. Essays* (1905) and *Science and Christian Tradition. Essays* (1909), published by Macmillan; (xxvii) H.E. Kyburg's *Probability and Inductive Logic* (1970), published by Macmillan.

My thanks are due to the staff at many institutions where I have spent periods of sabbatical leave during which much of this research was undertaken: in particular, the staff of University College, London; Nicholas Bingham and Rüdiger Kiesel, then of Birbeck College; and Stephen Stigler, of the University of Chicago. I also thank H.W. Johnson, of the Equitable Life Assurance Society, for providing me with a copy of Bayes's Notebook.

This tribute would be incomplete without mention of my indebtedness to Jo Currie, for her help in finding documents relating to Bayes in the archives of the Edinburgh University Library, to Roger Farthing, for sharing his expert knowledge of Tunbridge Wells with me and for the Jeffery Family Tree reproduced here with his permission, and to Jackie Sylaides, for her typing of much of the original manuscript. I also acknowledge the help of my colleagues Hugh Murrell, for his checking of some numerical results using *Mathematica*®, and Frank Sokolic and Brennan Walsh for the professionally produced diagrams. Finally I thank the editorial staff of Springer-Verlag New York, Inc., for their assistance.

Contents

Preface	ix
Acknowledgments	xiii
1 Introduction	**1**
2 A Bayesian Genealogy	**6**
2.1 Introduction	6
2.2 The early ancestors	7
2.3 Joshua Bayes	12
2.4 Appendix 2.1	33
3 Thomas Bayes: A life	**37**
3.1 Introduction	37
3.2 Thomas Bayes's education	38
3.3 Thomas Bayes and Tunbridge Wells	58
3.4 The faded Bayes	96
3.5 Appendix 3.1	99
4 Divine Benevolence	**101**
4.1 Introduction	101
4.2 The Tract	104
4.3 Commentary	140
4.4 Conclusion	181
5 An Introduction to the Doctrine of Fluxions	**182**
5.1 Introduction	182
5.2 The Tract	192
5.3 Commentary	219
5.4 Conclusion	249
5.5 Appendix 5.1	250

6	**On a Semi-convergent Series**	**253**
	6.1 Introduction	253
	6.2 The Tract	254
	6.3 Commentary	255
7	**The Essay on Chances**	**258**
	7.1 Introduction	258
	7.2 The Tract	269
	7.3 Commentary	298
	7.4 Editions of the *Essay*	334
8	**The Supplement to the Essay**	**336**
	8.1 Introduction	336
	8.2 The Tract	338
	8.3 Commentary	353
9	**Letters from John Ward**	**370**
	9.1 Introduction	370
	9.2 Letters to Bayes and Skinner Smith	371
	9.3 Translation of Letters	372
	9.3.1 The letter to Bayes	372
	9.3.2 The letter to Skinner Smith	373
	9.4 Commentary	374
10	**Miscellaneous Items**	**384**
	10.1 Introduction	384
	10.2 A Letter from Bayes to Canton	385
	10.3 Commentary on the letter to Canton	386
	10.4 Item on Electricity	395
	10.5 Commentary on the Item on Electricity	399
	10.6 Papers in the Stanhope Collection	412
11	**The Notebook**	**420**
	11.1 Introduction	420
	11.2 Attribution of the notebook	420
	11.3 The shorthand	421
	11.4 General plan of the notebook	423
	11.5 Mathematics	423
	11.5.1 Probability	423
	11.5.2 Trigonometry	429
	11.5.3 Geometry	434
	11.5.4 Solution of equations	436
	11.5.5 Series	440
	11.5.6 The differential method	451
	11.5.7 Numbers	454
	11.5.8 Miscellaneous mathematics	456

11.6	Natural philosophy	459
	11.6.1 Electricity	459
	11.6.2 On the weight of a body	464
	11.6.3 Optics	465
	11.6.4 Harmony	467
11.7	Celestial mechanics	471
11.8	Miscellaneous matters	480
	11.8.1 Of the pyramid measured by Greaves	480
	11.8.2 Of weights and measures	481
	11.8.3 Extract from a dissertation upon cubits by Sir Isaac Newton	482
	11.8.4 Aulay Macaulay's shorthand	486
	11.8.5 Estimate of the national debt upon 31 Dec. 1749	486
	11.8.6 A shorthand verse	486
	11.8.7 Warburton's syllogism	487
	11.8.8 Lettres Provinciales	487
	11.8.9 A Prescription	490
11.9	Appendix 11.1	491
11.10	Appendix 11.2	496
11.11	Appendix 11.3	497
11.12	Appendix 11.4	498

12 Memento Mori — 499

12.1	Introduction	499
12.2	A small old book	510
12.3	Register of Burials in Bunhill Fields	511
12.4	Inscriptions on vault	512
	12.4.1 Inscriptions from Rippon	512
	12.4.2 Further inscriptions from Rippon	513
	12.4.3 Inscriptions on present vault	514
	12.4.4 Inscriptions from Jones	516
12.5	Bayesian wills	516
12.6	Speldhurst memorials	524

Notes — 526

Bibliography — 595

Index — 645

1

Introduction

> *Suave est ex magno tollere accervo.*
>
> Quintus Horatius Flaccus.
> Satires i. i. 51.

In the introduction to the 1886 edition of Francis Bacon's *Essays* Storr and Gibson note that 'The guiding principles of Bacon's philosophy are utility and progress' [p. xi]. Few scientists would deny the impact that Bacon's writings had on guiding, if not directing, the development of science: few today would accept the slavish following of the precepts advanced in those writings.

There certainly were scientists who followed the Lord Chancellor's advice: in his autobiographical sketch Charles Darwin relates[1] that he 'worked on true Baconian principles, and without any theory collected facts on a wholesale scale', though a little later he writes

> I have steadily endeavoured to keep my mind free so as to give up any hypothesis, however much beloved (*and I cannot resist forming one on every subject*), as soon as facts are shown to be opposed to it. [F. Darwin, 1887, vol. I, pp. 103–104]

(emphasis added), a procedure that is perhaps not strictly Baconian. Darwin had perhaps adopted his earlier method of procedure from John Stevens Henslow, with whom he used to have long walks when at Cambridge, and of whom he wrote[2] 'His strongest taste was to draw conclusions from long-continued minute observations.'

There may still be situations in which Baconian experimentation is applicable; in general, however, its ignoring the critical function of an experiment makes it unsuitable for the advancement of science as at present understood. The situation was well summarized by Laplace, who wrote[3]

> Il [i.e., Bacon] a donné, pour la recherche de la vérité, le précepte et non l'exemple. Mais, en insistant avec toute la force de la rai-

son et de l'éloquence sur la nécessité d'abandonner les subtilités insignifiantes de l'école, pour se livrer aux observations et aux expériences, et, en indiquant la vraie méthode de s'élever aux causes générales des phénomènes, ce grand philosophe a contribué aux progrès immenses que l'esprit humain a faits dans le beau siècle où il a terminé se carrière. [1820, p. clvi]

In his *History of Western Philosophy* Bertrand Russell, although noting the somewhat unsatisfactory nature of Bacon's philosophy, wrote

[Bacon] has permanent importance as the founder of modern inductive method and the pioneer in the attempt at logical systematization of scientific procedure. [1961, p. 526]

Although Bacon emphasized the importance of induction as opposed to deduction, his method may be faulted in lacking a sufficient emphasis on hypotheses, such hypotheses usually being entertained before data are collected. The rôle of deduction in science was underplayed, and as deduction is often mathematical, this led Bacon to underestimate the importance of mathematics in scientific investigation[4].

I do not even attempt to argue that modern science is carried out in a Bayesian rather than Baconian manner: I only state that probability plays a large rôle in scientific reasoning, that classical methods of statistical inference may sometimes be inadequate, and that the Bayesian approach provides a viable alternative to older methods of scientific reasoning.

I do not know the effect that Bacon's writings had, or still have, on religious truths; but it would not be inappropriate to say something about them here. Storr and Gibson note that Bacon demanded that

information about things spiritual should be drawn fresh from the fountain-head, the Bible, not from "convenient cisterns," *i.e.* from cut and dried dogmatic systems, the work of men only. ... Bacon thus virtually accepted the principles of the Reformation as developed or explained by Puritanism.
[1886, pp. xxx, xxxi]

And although scientists are denied the strength or authority of Bacon's source, there can be little doubt of the importance of their consulting their own fountain-heads, at least from time to time. It is with this in mind that the examination of Bayes's *An Essay towards solving a Problem in the Doctrine of Chances* is undertaken here.

University courses in statistics are often designed in such a way that Bayesian methods and applications are appendages rather than substantive matter, being examined after much discussion of the more classical methods (Fisher, Neyman–Pearson). This is perhaps not altogether a bad thing, because knowledge of such classical techniques and results provides a starting point for the search for similar things in a Bayesian setting, the path perhaps following the classical route (the *Via Appia*?) in an analogical manner. This leads us back to Bacon — *à ces lards*, one is tempted, by analogy, to say — for we read in Storr and Gibson's introduction

> Bacon's preference lies in the direction of allegorical and analogical treatment, as having more elasticity, and being more adaptable to the wants of each generation. [1886, p. xxx]

Although the pertinence of allegory to Bayesian Statistics is uncertain (unless we try to work in a somewhat tenuous connexion between Sheridan and R.A. Fisher's Problem of the Nile ['As headstrong as an allegory on the banks of the Nile'[5] — a description that would not be totally misapplied to Fisher himself]), the relevance of analogy is undeniable. Indeed the connexion between analogy and probability was made by Laplace in his *Essai philosophique sur les probabilités* as[6] 'L'analogie est fondée sur la probabilité que les choses semblables ont des causes du même genre et produisent les mêmes effets'.

There are of course those who find little, if anything, appealing or satisfying in Bayesian methods. These are they who, if we say with Jeremiah,

> Stand ye in the old ways, and see, and ask for the old paths, where *is* the good way, and walk therein, and ye shall find rest for your souls [*Jeremiah* vi. 16]

will reply 'We will not walk *therein*.' They might even protest more strongly that 'Surely our fathers have inherited lies, vanity, and *things* wherein *there is* no profit' [*Jeremiah* xvi. 19].

Before one can exhort people to stand in the old ways or to ask for the old paths, however, it is necessary that the existence and accessibility of these ways and paths be determined. Indeed from time to time one is amazed at the knowledge acquired by members of the human race over the centuries of its development that has become overgrown and perhaps even lost to sight. Aristotle, Sir Thomas Browne informs us in his *Pseudodoxia Epidemica*, wondered 'why Sneezing from noon unto midnight was good, but from night to noon unlucky.' Like Sir Thomas Browne himself, one may wonder what song the Sirens sang, or what name Achilles assumed when he hid himself among women[7] — or even, more mundanely, Who put the 'Bop' in the bop. And although this knowledge has disappeared from the collective memory of mankind (if indeed it was ever there), it is certain that the entertaining of these or similar questions is in itself a source of some delight, or, as Lynd has it, 'The great pleasure of ignorance is, after all, the pleasure of asking questions' [1928, pp. 6–7]. One is perhaps surprised at the knowledge that has, seemingly by chance, been rescued almost from oblivion (one thinks immediately of much of the music of the Baroque period), and one notes de Morgan saying in the passage from his *Essay on Probabilities and their Application to Life Contingencies and Insurance Offices* from which the title of our work comes, that at that time (1838) the name of Thomas Bayes was 'now almost forgotten'. Indeed, one might almost have heard a spectral voice exclaiming with John Arbuthnot[8]

> What am I now? how produced? and for what end?
> Whence drew I being? to what period tend?
> Am I th' abandon'd orphan of blind chance,
> Dropt by wild atoms in disorder'd dance?

Fortunately, as Horace has it[9], *Dignum laude virum Musa vetat mori*, and de Morgan's jogging of the collective memory on this and other occasions, and the investigations of Ramsey, Wrinch, and Jeffreys in the early part of the twentieth century, followed later by the studies of de Finetti, Good, Lindley, and Savage, resulted in the preservation of Bayes's name.

Yet although Bayes's name may be saved from obscurity, little is known of his life; and his works, few in number, have not been given the attention I think they deserve. It is to redress this situation that this book is offered.

In our more biographical chapters we provide some details of the Bayes family, noting the importance of a number of its members in seventeenth- and eighteenth-century Nonconformity. We also provide some description of Tunbridge Wells, where Thomas spent much of his life as Presbyterian minister. Where he received his early schooling is unknown, though there is the possibility that he attended an academy at which John Ward was a tutor. A letter from Ward to Thomas is preserved in the British Library, and it is given here in our ninth chapter. There is certainly no doubt, however, that Thomas later studied at Edinburgh University, and details of his time there are also provided.

In considering Bayes's work we of course pay close attention to the *Essay*, a tract that, like Bacon's philosophy, could also be commended for its contributions to 'utility and progress'. Here the result is found that has formed the basis of modern Bayesian Statistics and that has come to play an important rôle in (non-Baconian) scientific practice in general. For once the *Essay* had been seen as providing a way of arguing from 'effects' to 'causes' (or 'hypotheses'), its relevance to scientific reasoning became patent[10]. We also find in the *Essay* and in the subsequent paper entitled *A Demonstration of the Second Rule in the Essay towards the Solution of a Problem in the Doctrine of Chances* less well known but nonetheless important matters: an evaluation of the beta-integral, the derivation of (a series expansion for) the Normal distribution as the limit of a beta posterior distribution, and the approximation of a skew beta-function by one that is symmetric, the latter having the same maximum and points of inflexion as the former — in which one might well see an early instance of the approximation of the posterior by another distribution by the matching of appropriate derivatives.

When we consider Bayes's other work, we find that there are a number of writings that warrant attention. These include a tract entitled *Divine Benevolence*, in which God's attributes are examined, and a mathematical work, *An Introduction to the Doctrine of Fluxions*, written in response to George Berkeley's *The Analyst*. In the first of these Bayes is seen as a writer whose views on theological matters are to be taken seriously, and in the second he appears as one knowledgeable in the fluxionary calculus.

Any work on someone's life should contain details of his death, and our essay is concluded with a chapter on Bunhill Fields, the nonconformist cemetery in London in which the Bayes family vault lies.

An author of a biographical work, or even of one that contains a fair proportion of biographical matter, must be constantly aware lest a justifiable and rational admiration for his subject degenerate into irrational hero-worship or, even less satisfactorily, into a *chronique scandaleuse*. There is always the danger, as Hogben [1957, p. 133] has noted, that, in attempting to draw illations from older work to modern research, one will read more into the original than the author of the former ever intended, and attribute deeper and perhaps more extensive results to one's subject than he meant. This is indeed the case with Bayes's *Essay* and the supplementary *A Demonstration of the Second Rule in the Essay*. Ignored to a large extent for many years before their relevance was realized in the early years of the twentieth century, these papers have been the source of considerable inspiration for modern statistics, realizing the sentiment expressed by T. B. Aldrich in his *At the Funeral of a Minor Poet*,

> 'Tis said the seeds wrapped up among the balms
> And hieroglyphics of Egyptian kings
> Hold strange vitality, and, planted, grow
> After the lapse of thrice a thousand years.

While taking cognisance of the importance of Bayes's work in probability, one must therefore constantly take care that that cognisance is no more than what is merited: Bayes's design in the *Essay* was the solving of what we would today view as a problem in inverse probability and nothing more.

We have tried to show in this book, with what measure of success only the reader can judge, that it is not Bayes's *Essay* alone, important though it of course is, that is of interest. The study of his life and works shows him as one who, despite (or perhaps because of?) his calling, not only had a deep interest in natural philosophy and mathematics, but also was able to comment in a logical and meaningful manner on these subjects. He is certainly worthy of 'most honourable remembrance'.

2
A Bayesian Genealogy

He shall bring back the faded bays.

Orrick Johns.
Second Avenue.

2.1 Introduction

In his review of John Dawson's *Logical Dilemmas: the life and work of Kurt Gödel*, Craig Bach states that a biographical study should be more than a mere chronological compilation of facts and events: the *importance* of such matters should also be considered. Indeed, in the early part of the eighteenth century biography was fourfold in aspect: anecdote, chronicle, moral tale, and compilation of letters, sayings, and excerpta from the works of him whose life was being studied[1]. Here three of these forms are to be found: the moral aspects are lacking, except inasmuch as any clerical record will naturally reflect the calling of those concerned.

The biographer must steer a course between chronological fact and biographical fiction, being careful also, I would suggest, to avoid both faction and friction. His task should be the determination of what details should be recorded and how much description the various facts should be allowed. He should try to avoid the finical; and although we attempt to do so here, the reader will find that a certain lack of uniformity in the recording of dates and events by earlier writers leads us from time to time to the recording of more 'facts' than seem necessary, but whose mutual adversation makes reconciliation at this distance from the events almost impossible. Following, however, the practice of nineteenth-century writers on the theory of errors, who took as fundamental the assumption[2]

$$\text{observation} = \text{truth} + \text{error},$$

we try, like good scientists (or at least agriculturalists), to winnow out the latent wheat of truth from the patent chaff of error[3], recording our findings as truthfully as possible.

In his *Proposals for printing by subscription ...* of 1803 Rippon wrote

> However difficult it be to appreciate the comparative value of the Publications which owe their celebrity to novelty, to genius, or to taste, it is of general admission, that few or none of the endless variety, which perennially court or charm an enlightened country, are of greater utility than those of correct BIOGRAPHY. The early days, the career, and the exit of wise men, especially of those who have been pre-eminently holy and good, awake our attention, arrest our curiosity, and interest all our intellectual and moral powers.

It is hoped that the reader of the present essay will concur in finding the following reference from the Advertisement to an Introduction to the Duke of Buckingham's seventeenth-century play *The Rehearsal* relevant.

> [The Reader] will find many obscurities removed; and numerous references recovered; far more of both than could reasonably be expected, considering that no assistance could be had but what is fetched from books, and that all personal information has been long since swallowed up in the gulph of time. It must however be acknowledged that our inquiries have not always been successful.

2.2 The early ancestors

In the *Familiæ Minorum Gentium* of 1895 Clay notes that the only family surnamed 'Bayes' was that in Sheffield. To some extent this makes our biographical task easier, in that we may be pretty sure that any reference we find to 'Bayes' will be to members of this family[4]. However, this invaluable genealogy contains little more than bare bones, and we therefore try, without forced feeding, to add a little flesh to the skeleton.

One should note, though, that many Bayes's (not Bayesians!) are listed in the *International Genealogical Index*. Although most appear to have a Sheffield connexion, Clay's genealogy seems sufficient — and I am not sure how accurate some of the details in the *Index* are.

The first member[5] of the family listed by Clay is one Hugh Bayes. Little is given about him, only that he married one Rose (of unknown maiden name) and was buried in Sheffield on the 7th of June 1628, Rose being buried there on the 2nd of September 1633.

Hugh and Rose had two children: Rose and Richard. Of the former, all we are told is that she married James Hurst in 1622; of the latter, we read that he was baptized on the 14th November 1596, and married Frances Howsley on the 9th February $162\frac{4}{5}$.

This Richard was clearly a person of some importance in Sheffield, for Clay records that he was Master of the Company of Cutlers of Hallamshire[6] in 1643. There is a further reference to him by Leader, viz.[7]

> And with money recd of Richard Bayze towards the walling betwixt the workhouse crofte and his close 0 12 0.
> [1879, p. 100]

One child, John, resulted from the marriage of Richard and Frances. Like his father, John became a cutler, being admitted to his freedom in the Company of Cutlers in 1654. Leader records that

> John, son of Richard Bayze, of Sheffield, cutler, was admitted apprentice 1656, according to the books of the Sheffield Cutlers' Company. John Bayes was assessed at two hearths, 1665.
> [1897, p. 189]

And further, in an entry for the year 1668, Leader notes

> And with money recd of John Bayes & Rich Parramour overseers of ye poore 0 12 0. [1879, pp. 106–107]

John married Sarah Newbold of Green Hill in the parish of Norton on the 4th of September 1654, there being three children of this marriage.

Frances having died, Richard married Alice Chapman on the 16th July 1628, seven children issuing from the union. The first son, also Richard, was short-lived: baptized on the 27th September 1629, he was buried two days later. The second son, Samuel, baptized on the 31st of January $163\frac{5}{6}$, was admitted sizar[8] at Trinity College, Cambridge, in April 1652, matriculating in the same year. He became a Scholar in $165\frac{5}{6}$, the B.A. following in 1656, and he was then appointed Minister at Beauchief Abbey, Derbyshire[9].

In 1662, after the restoration of Charles II, the anti-Puritan parliament passed the Act of Uniformity[10]. This Act, following a similar act passed in the reign of Elizabeth, was intended

> to regulate all public worship in England, and to secure perfect uniformity in the public religious services of all Englishmen. It excluded every kind of public worship not provided for in the Prayer-Book. [Dale, 1907, p. 414]

In terms of the Act every parson, vicar, or minister of any kind, had, before the Feast of St Bartholomew[11], to read morning and evening prayers in his church and make the following declaration before his congregation:

> I, $\mathcal{A}\,\mathcal{B}$, do here declare my unfeigned Assent and Consent to all and everything contained and prescribed in and by the book entitled The Book of Common Prayer and Administration of the

> Sacraments, and other Rites and Ceremonies of the Church, according to the Use of the Church of England, together with the Psalter or Psalms of David, pointed as they are to be sung or said in Churches; and the Form and Manner of Making, Ordaining, and Consecrating of Bishops, Priests, and Deacons.
> [Dale, 1907, pp. 414–415]

Any clergyman who failed to obey this instruction and to make this declaration was to lose all his spiritual promotions (bishopric, living, etc.).

The day on which the Act came into effect (24th August 1662) was perhaps deliberately chosen: church tithes were commonly due at Michaelmas, and hence those who refused to conform would lose the tithes they had earned by nearly a year's work[12].

Thus ejected, Samuel became Vicar of Grendon St Mary, a parish on the right bank of the River Nene, in the union of Wellingborough, hundred of Wymmersley, S division[13] of the county of Northampton, five and a half miles (S. by W.) from Wellingborough. In 1831 Lewis described the living as follows.

> The living is a discharged vicarage, in the archdeaconry of Northampton, and diocese of Peterborough, rated in the king's books at £8, endowed with £400 private benefaction, £400 royal bounty, and £400 parliamentary grant, and in the patronage of the Master and Fellows of Trinity College, Cambridge.
> [vol. II, p. 261]

I do not know the size of the parish at that time: in 1849, however, when the seventh edition of his *Topographical Dictionary of England* appeared, Lewis noted that the parish contained 595 inhabitants[14], and consisted of 1649a. 2r. 18p. of a rich and fertile soil[15].

Here, in Grendon St Mary, Samuel probably stayed for some time — perhaps until 1665, when the Five Mile Act (or the Act for restraining Nonconformists from inhabiting in Corporations) was passed by Parliament, sitting at Oxford while the Plague swept London.

This Act required all in holy orders, or who pretended to be in holy orders, and who had not taken the oaths required by the Act of Uniformity, to nuncupate[16]:

> I, $\mathcal{A}\,\mathcal{B}$, do swear, That it is not lawful upon any Pretence whatsoever, to take Arms against the King; and that I do abhor that traitorous Position of taking Arms by his Authority against his Person, or against those that are commissionated by him, in Pursuance of such Commissions, and that I will not at any Time endeavour any alteration of Government, either in Church or State. [Dale, 1907, p. 430]

A Nonconformist minister who refused to make this declaration was forbidden, except when on a journey, to come within five miles of any city, corporate town, or parliamentary borough, or, according to Dale [1907, p. 430] 'of any parish, town or place in which he had formerly been "parson, vicar, curate, stipendiary, or lecturer[17]," or had conducted any Nonconformist service'. The penalty for violation was severe: a fine of forty pounds.[18]

Once again thus ejected, Samuel moved, Turner recording

> A new built Meetting house att Sankey in Lancash by Presbyterien Sept^t 5th [1911, vol. I, p. 556]
>
> Licence to Sam̃: Bayes of Sankeÿ in Lancash: to be a Pr: Teachr. Sep^t 5th 1672. [1911, vol. I, p. 518]
>
> Sankey. (1) Samuel Bayes (t^r) (Cal. iii, 35), ej. from Grendon, Northants. (2) New Meeting House (m. pl.). [vol. II, p. 677]

Incidentally, according to Dale [1907, p. 416], schoolmasters and other teachers were licensed (at a cost of twelve pence) by the Bishop to teach: the amount may sound trifling, but one should bear in mind that it cost the same amount annually, and at about the same time, to license a coffeehouse in London[19] (a pound of coffee cost six shillings in 1696, the same as six ounces of snuff and three times the price of a pound of tobacco — see Browning [1953, p. 472]).

Turner also records

> Licence to Sam: Buze to be a Pr.[20] Teacher in his howse in Manchester. Sep^t 5th 1672. [1911, vol. I, p. 556]
>
> The howse of Sam: Buze at Manchester. Pr. [1911, vol. I, p. 518]
>
> Manchester. Samuel Buze (t^r) [1911, vol. II, p. 678]

and a question that immediately presents itself is: is 'Buze' the same as 'Bayes'? If so, why was the licensing recorded twice — and if not, on what authority is Samuel Bayes often fixed (mistakenly, I believe) in Manchester rather than in Sankey[21]? (The distance between the two is some twenty miles as the crow flies.) That authority might be Matthews, who, in his revision of Calamy's work, wrote of Samuel

> Licensed (P.), as of Sankey, Lancs., 5 Sep. 1672; also, as Buze, at his house, Manchester. [1934, p. 40]

(Here 'P.' stands for 'Presbyterian'.) The situation is complicated, however, by the fact that Nightingale [1890–1893, vol. 5, p. 85] records 'Buze' as 'Bure', with the date of the Manchester licence as 15th June 1672.

2.2. The early ancestors

According to Clay [1895] Samuel died in Manchester c.1681. One might perhaps go a little further and put his death at late 1681 or 1682, for it is noted in the Town Records of Sheffield that

> August 15th 1682. It is to be remembered that Mr. Samuell Baies did by his will bequeath unto the Poore of the Town of Sheffield Ten Poundes to be paid by his brother Joshua his executor, which said Ten pounds the said Joshua paid to Lionell Revell Town Collector the day and yeare above said.
> [Leader, 1897, p. 221]

Joshua, the third and youngest son of Richard and Alice, was baptized on the 6th of May 1638, and married Sarah Pearson[22] on the 28th May 1667. Like his father and half-brother John, he too became a cutler, becoming Master of the Company of Cutlers of Hallamshire in 1679. Leader has recorded several references to Joshua (many of these being to him as Town Collector), among which we mention the following.

> Joshua Bayes, cutler, his wife, one child, and one servant, were taxed £2. 16s in the poll-tax of 1692. He was a prominent man among the Nonconformists, and Master Cutler in 1679; Town Collector 1683–84. [1897, p. 213]

(It is not stated who this 'one child' was: probably Elizabeth, who would then have been some ten years old.)

> [In 1690] The Arrears unreceived are as follows ... Joshua Bayes 7li. 10s. [1897, p. 248]

> Also [in 1692] the acomptant craves an allowance of 50s. abated Joshua Baies of the rent, for damage by calling in his money in order to the purchase of Stansall farme. £39. 14. 8. [A footnote, however, notes that 'The Town trustees have no property called "Stansall Farme." The entry is not very intelligible.']
> [1897, p. 250].

> 1st of February, 1702. Whereas Mr. Thomas Diston, ye Capitall Burgess, is lately dead, and this day it is agreede yt Mr. Joshua Bayes shall serve out ye remaind' of his yeare; delivered him at the same time a key, with the Purse, the Bonds, and accounts.
> [1879, p. 144]

And at the end of Lionell Revell's accounts we have

> there remains in the Accomptants hands to balance the Accompt £132. 2s. $8\frac{1}{2}d$. which was paid to Joshua Baies att pass-

ing this Accompt who is chosen Collector for the ensuing year and soe the Accompt stands clear. [Leader, 1897, p. 224]

Joshua and Sarah had seven children. The eldest daughter, Ruth (baptized 9th July 1673), married Elias Wordsworth (who became a Town Burgess in place of Joshua on the latter's death); the third daughter, Elizabeth (born 31st May 1682), married John de la Rose, Minister of the Nether Chapel in Sheffield[23]. (We find these surnames occurring in the Wills of some members of the Bayes family, and attention is therefore drawn to them here.) The eldest son, John (baptized 31st March 1669), married Sophia Barrington, and the next son, Joshua (baptized 10th February $167\frac{0}{1}$), married Ann(e) Carpenter. A portrait of Joshua is shown in Figure 2.1. Boyd's *Citizens of London*[24] lists

1695 1732 Bayes Jn DR Sophia Barrington

Here 'DR' stands for 'Draper'.

Joshua *père* married the widow Martha Taylor in 1686, after Sarah's death. He was buried on the 28th of August 1703. His relict married Thomas Booth on the 2nd February 1707, and died on the 16th January 1711.

2.3 Joshua Bayes

Following the Nonconformist tradition of his family, perhaps more enthusiastically, and perhaps (at least initially) more under familial persuasion than divine guidance, Joshua *fils* entered Richard Frankland's academy[25] at Rathmell, in Yorkshire, on the 15th November 1686, where he pursued his studies 'with singular advantage' [Wilson, 1814, vol. IV, p. 396].

Frankland, described by Bogue and Bennett as 'an eminent dissenting[26] tutor, who taught university learning' [1808, vol. I, p. 225], started an academy at Rathmell in 1669 after he had been ejected from his living. 'Him,' writes Gordon [1902, p. 5], 'must we ever revere as the Founder in this Country of the Nonconformist Academy.' This academy, the first in the north of England for Nonconformists and certainly the most important there[27], was a victim of the Five Mile Act. It moved several times, eventually ending up again in Rathmell. In the almost thirty years of its existence (the academy closed in 1698) 303 students passed through its doors.

Such dissenting *academies* should not be confused with the dissenting *schools* of that period: the former were schools of university standing (and indeed were rivals of the universities)[28], whereas the latter were charity

FIGURE 2.1 Joshua Bayes.

foundations[29]. Many of these academies, which 'gave an education in advance of their day' [Parker, 1914, p. 53], in a sense rose out of the universities, being founded by men educated at the latter but ejected by the Act of Uniformity or the Five Mile Act[30]. Boys were admitted to the academies at the same age as they would have been admitted to the universities. In those days youths could matriculate at the University of Oxford at the age of sixteen, provided that they subscribed to the Thirty-Nine Articles (a subscription not required of Cambridge undergraduates) and took the oaths of supremacy and of obedience to the University's statutes: matriculation was possible over the age of twelve and under sixteen merely by subscription to the Thirty-Nine Articles, the taking of the oaths being required only at the end of the sixteenth year[31]. A number of these early academies were founded by ejected ministers, and the Act of Uniformity and the Five Mile Act resulted in many of these seminaries, like Frankland's, being migratory. For the Act of Uniformity was binding not on the clergy alone: every holder of any ecclesiastical office, every master, fellow, tutor of any college, every professor or reader in the universities of either Oxford or Cambridge, every schoolmaster keeping either a private or a public school, and every person instructing youth in any house or private family had to make the following declaration.

> I, $\mathcal{A}$ $\mathcal{B}$, do declare, That it is not lawful, upon any Pretence whatsoever, to take Arms against the King: ... and that I will conform to the Liturgy of the Church of England as it is now by Law established. [Dale, 1907, p. 415]

The dissenting academies did not profess to grant degrees, though, as Gordon notes,

> had they done so, I suspect that a degree at Rathmell in the seventeenth century, or one at Daventry in the eighteenth, would have meant a good deal more than a contemporary degree either at Oxford or Cambridge, if measured, not by its value for merely social purposes, but by its worth as an index of the intellectual stimulus promoted by careful and enlightened study. [1902, p. 6]

Although the training provided at the dissenting academies was primarily for those intending to enter the Church, it was by no means solely directed at those having a vocation. Gordon in fact records

> It [i.e., Rathmell] was no clerical seminary, either in design or in fact ... while the Academy was Nonconformist, its *alumni* were not asked to commit themselves, either actually or implicitly, to the Nonconformist position.

The immediate work of the Nonconformist Academies was to fit and equip men for public duty, not in the ministry alone, but in all the professions; it was to make them thinkers — not closing their minds with fixed opinions, but opening their intelligences, and giving them an impetus towards the acquirement of further knowledge; it was to make them workers for the good of their kind, to train them for the application of knowledge in all the departments of life. Far more was this their aim than to make Nonconformists. [1902, pp. 7, 14]

To give an idea of the subjects taught at the academies of this period[32] (generally with only one tutor), we mention that at Newington Green instruction was given in Latin, Greek, Hebrew, Logic, Mathematics, and Science [Parker, 1914, p. 63] (note the absence of instruction in English[33]).

James Clegg, a student at Rathmell in 1695, records that the day began at 7 a.m. (as Pippa passed?) with prayers and breakfast. Then followed lectures in Logic, Metaphysics, Somatology, Pneumatology, and Natural Philosophy. Dinner was followed by private reading or recreation, with prayers at 6 p.m. followed by supper and Students' Discussion in their own rooms. On Thursdays theses and public disputations were appointed by the tutor, and on Saturdays, from 5 to 6 p.m., there was Analysis, i.e., methodical and critical dissertation on some verses of a psalm or chapter.

Describing the unexpected benefits brought by intolerance to dissent Taylor writes

> intolerance had brought about another unexpected benefit — the dissenters, denied the universities, the church, and effectively the grammar schools, were obliged to found and staff their own teaching establishments. And this enabled them to break away from the age-old purely classical tradition of education and introduce new subjects into the curriculum. Without abandoning Latin, the boys at dissenting schools learned modern languages, practical mathematics, and natural philosophy. [1966, pp. 7–8]

Joshua was ordained in London on the 22nd of June 1694 along with six other candidates for the ministry (Joseph Bennett, Thomas Reynolds, Edmund Calamy, Joseph Hill, William King, and Ebenezer Bradshaw)[34]. This ordination was of particular importance: held at Dr Samuel Annesley's meeting-house[35], Bishops-gate Within, near Little Saint Helens, it was the first public ceremony of its kind to be held in the City after the passing of the Act of Uniformity. Calamy's description of the taxing nature of the proceedings — taxing not only on the candidates and the clergy, but also on the congregation, present the whole time[36] — is given as follows[37].

> The manner of that day's proceeding was this. First, Dr. Annesly began with prayer; then Mr. Alsop preached, from 1 *Pet.* v. 1, 2, 3. Then Mr. Williams prayed, and made a discourse concerning the nature of Ordination. Then he mentioned the names of the persons to be ordained, read their several testimonials, that were signed by such ministers as were well acquainted with them, and took notice what places they were severally employed in as preachers. Then he called for Mr. Bennett's confession of faith, put the usual questions to him out of the Directory of the Westminster Assembly[38], and prayed over his head. Then Mr. Thomas Kentish did the same by Mr. Reynolds,; Dr. Annesley did the like by me; Mr. Alsop, by Mr. Hill and Mr. King; Mr. Stretton by Mr. Bradshaw; and Mr. Williams again by Mr. Bayes. After all, Mr. Sylvester concluded with a solemn charge, a psalm, and prayer. The whole took up all the day, from before ten to past six o'clock.
>
> Before our being thus ordained, we were strictly examined, both in Philosophy and Divinity, and made and defended a Thesis each of us, upon a theological question, being warmly opposed by the several ministers present.
>
> [1830, vol. I, pp. 349–350]

The theological theses were debated in Latin, Joshua's topic[39] being 'An Deus sit Essentiâ suâ omnipresens?' *Aff.*

That such an examination was almost necessary before ordination may be seen by noting that, in 1690, the Presbyterian and Independent ministers in London made a serious attempt to unite the two denominations. One of the requirements of the doctrine drawn up was that, before ordination, the pastor-elect should satisfy the pastors of neighbouring churches that he had the necessary qualifications for the pastorate.

> That they may be sent forth with *Solemn Approbation* and *Prayer*; which we judge needful, *that no doubt may remain concerning their being called to the work*, and for the preventing (as much as in us lieth) ignorant and rash intruders.
>
> [Dale, 1907, p. 477]

The attempt at some sort of unification seems to have been unsuccessful. Calamy relates

> In December this year [1692], after much pains taken, certain "Doctrinal Articles" of religion were fixed upon, which were agreed to by the Dissenting ministers that had been contending with each other, and subscribed and published to the world, under the title of "the Agreement in Doctrine among the Dissenting Ministers in London," by which it was hoped farther

differences might have been prevented. But a right healing spirit was wanting. Opposite weekly meetings were kept up, and some seemed desirous to be thought to differ from their brethren, whether they really did so or no; or at least fancied that they did so, more than they did in reality; and this had ill effects and consequences. [1830, vol. I, pp. 326–327]

It was probably soon after his ordination that Joshua moved to Box Lane, Bovingdon, for we find Urwick recording

> The first Trust-Deed [for the Nonconformist chapel there] is dated 1697, and speaks of the chapel as "lately erected," indicating that the present building was erected soon after the passing of the Toleration Act. The deed is signed by the proprietors, Thomas Lomax and Mary Lomax, his wife ... They transfer the chapel to twelve trustees*....
> *Joshua Bayes, clerk, ... [1884, p. 389]

We also find Joshua's name as certifying the chapel as a place of religious worship in Midsummer, 1702, and Urwick records further that

> After his ordination Joshua Bayes came to Box Lane, and remained as pastor of the church (which was Presbyterian or Independent according to the terms of Agreement of the United brethren) for about eleven years. [1884, p. 390]

Describing the Box Lane Chapel, Urwick writes

> The site upon which BOX LANE CHAPEL stands must long ago have been used by the Romans for sepulchral if not for sacred purposes. For in the year 1837, upon the digging of a grave in the chapel yard, there was brought to light what remained of a massive oaken chest, in which with other relics had been deposited two large rudely-constructed glass vases, one of them globular, and the other square, each of which contains to this day the calcined bones originally put within them, and which may be seen in the British Museum.
> Upon the passing of the Conventicle and Five Mile Acts, the Nonconformists were sorely pressed, and the position of their earlier chapels witnesses to the necessity for retreat. This accounts for the erection of a chapel in such an obscure place as Box Lane then was, and the provision made for the escape of the preacher by the doorway still to be seen at the back of the pulpit. [1884, pp. 388–389, 389]

In his *Topographical Dictionary of England* Lewis describes Bovingdon[40] as follows.

Bovingdon, a chapelry, in the parish and union of Hemel-Hempstead, hundred of Dacorum, county of Hertford, 4 miles (S.W.) from Hemel-Hempstead; containing 1072 inhabitants. ... There is a place of worship ... for Independents in Box Lane.
[1849, vol. I, p. 318]

Boyd's *Citizens of London*, however, quite clearly gives Joshua Bayes's dates as 1701 and 1746, and it would seem from similar entries that these were the years during which Joshua was actually in the City. The distance from Bovingdon (or Bovington, according to Bowen & Kitchin's map of 1777) is some twenty-five miles: would this be near enough for Joshua to be regarded as being in London? I doubt it.

In 1707, however, Joshua took up a post as assistant to John Sheffield at St Thomas's, Southwark, where he remained until 1723[41] [Evans, 1897]. (Some sources state that Joshua succeeded Edmund Batson[42] as assistant, though Evans gives the latter's dates at St Thomas's as 1697 to 1726. Evans does remark, however, on the difficulty of giving the precise location and dates of ministers of that period.) In Anon [1731–1732] Joshua's dates at Southwark are given as 1703 and 'in or about' 1724. Wilson records that this chapel in Southwark

> was built in the year 1703, for Mr. John Sheffield, and is a large square structure, with three galleries, substantially built, and capable of seating a numerous congregation.
> [1814, vol. IV, p. 294]

He also lists Joshua Bayes as assistant to Sheffield from 1707 to 1723. During that time, since his services were required only in the morning, Joshua also took up an assistantship to Christopher Taylor at the Chapel at 26 Leather Lane, opposite Baldwin's Gardens[43] [Lockie, 1813], in Hatton Garden, becoming the full-time minister there on the latter's death. Here he stayed until 1746, during which time, feeling the weight of advancing years, he took on first John Cornish and afterwards his own son Thomas as assistant.

An anonymously recorded description of the chapel and congregation, Anon [1731–1732], runs as follows.

> This meeting house is about 15 square of building, with 3 Galleries. In 1695, Mr. Buris was minister to this people, but not living many yeares after that time Mr. Christopher Taylor was chosen Pastor in his room. he was accounted a G^{t}man [i.e., Gentleman] of a bold spirit & a good preacher & about 1714 Mr. Bayes was chosen to assist him. Mr. Taylor dying about 1724 Mr. Bayes succeeded as pastor, & since that time Mr.

2.3. Joshua Bayes

Bayes Junr was chosen to assist his father. This congregation was never large. but were a people generally of substance. It does not certainly appear what difference there is between the congregation in 1695 & the present, tho it is apprehended to be somewhat less. Mr. Bayes is a judicious serious and exact preacher and his composures appear to be laboured. He is of a good temper & well esteemed by his brethren. [p. 35]

It appears, however, that Joshua was not a very popular preacher, for one reads further in the same source that his congregation declined from 1695 to 1731: presumably this refers to the Leather Lane chapel.

Figure 2.2 shows the position of the Leather Lane Chapel, and Figures 2.3 and 2.4 show views of the lane. It is interesting to note that, according to the *Minute Books of the Body of Protestant Dissenting Ministers of the Three Denominations in and about the cities of London and Westminster*, Joshua's woning was situate in Little Kirby-Street, a street described by Besant and Mitton as 'a broad street, and in times past ... a place of residence for well-to-do people' [1903, p. 67][44]. The propinquity of the stable to the chapel leads one to suspect that the odour of sanctity was sometimes commingled with an equine effluvium, though the latter was no doubt no less salubrious to the flesh than the former was to the spirit. A precedent to this close proximity had in fact been set in the late sixteenth century: the steeple, choir, and side aisles to the choir of the church of the Augustinian Friars in Broad Street, London, were used by Sir William Powlet, Lord Treasurer of England, for household storage of coal, corn, etc. Sir William's son, the Marquis of Winchester, sold the monuments of those buried there, together with the paving stones and anything else he could lay his hands on, for £100 in all, and established stables in their room[45].

It would perhaps also be of interest to note that, after the Restoration, Leather Lane became a venue for the mercurial treatments required by those who had experienced Cyprian pleasures[46]. Indeed, at the time of the Great Plague this area seems to have been a haunt of harlots: it was believed by many that venereal disease provided immunity to the Plague, and a contemporary report from Boghurst's *Loimographia* seems to suggest that there was some truth in this belief[47]:

> Of all the common hackney prostitutes of Luten Lane, Dog Yard, Cross Lane, Baldwin's Gardens, Hatton Garden, and other places; the common Cryers of Oranges, Oysters, fruite, etc.; all the impudent, drunken, drabbing Bayles and fellowes and many others of the Rouge Route, there is but few missing.
> [1666, p. 96]

Bearing in mind the conditions under which, and the area in which, the Leather Lane Presbyterians worshipped, the reader of Restoration diatribes

20   2. A Bayesian Genealogy

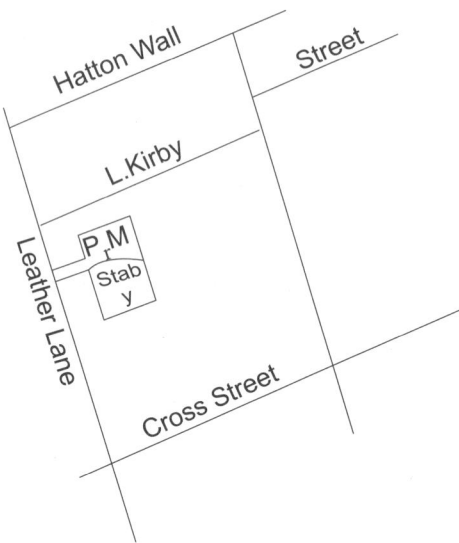

FIGURE 2.2 The site of the Leather Lane Chapel.

FIGURE 2.3 Leather Lane today.

FIGURE 2.4 Leather Lane in the nineteenth century.

may well be reminded of the title of one of Samuel Parker's works of 1669, in which the words *the Mischiefs and Inconveniences of Toleration* occur!

How Leather Lane got its name is uncertain, though there seems little doubt that it had no connexion with merchandise (as was the case, on the other hand, with Milk Street and Bread Street). Marryat & Broadbent suggest that the name is a corruption of the mediaeval 'Lither Lane', from an Anglo-Saxon word meaning *mud*, or *slough*. Rawlings, citing one 'Mr. Bonner', gives the form 'le Vrune Lane' of 1234 and 'Louerone Lane' in 1306 and 1456. She also quotes the suggestion of a Dr Bradley, identifying the original as the old French *leveroun* (greyhound), but notes too that *leuron* meant 'a pleasure-seeking young man', which ties in nicely with one of our earlier remarks. Fairfield follows Rawlings in accepting the thirteenth-century form of the name, but suggests that it came from a local merchant, whose name might have been a form of the Old English 'Leofrun'. And finally, Bebbington gives the earliest form as 'Le Vrunelàne', suggesting that it comes from the Flemish word *Vroon* meaning a *Soke* or *Manor* (the Flemish are known to have lived in this area at that time). The lane would then have been one leading to the Soke of Portpool, a pool near the port or gate to the city.

As we have intimated, Leather Lane was perhaps not the most salubrious of streets. In his 1720 edition of Stow's *Surveys of the Cities of London and Westminster* John Strype wrote

> *Perpole Lane* comes out of Gray's Inn lane, and falls into Leather lane; a Street of no great Account, being old Buildings, and none of the best inhabited, nor much frequented. Here are these Places of Name: *White Horse Inn*, but indifferent, and chiefly for Stablings. ...
> The East Side of this [i.e., Leather] Lane is best built, having all Brick Houses; and behind them several Yards for Stables and Coach Houses, which belong to the Houses in Hatton Street. [1720, Book 3, p. 255]

Commenting on Strype's description some two centuries later Besant & Mitton wrote

> Even in his day he reviles it [i.e., Leather Lane] as of no reputation, and this character it retains. It is one of the open streets of London, lined with barrows and coster stalls, and abounding in low public houses. The White Hart, the King's Head, and the Nag's Head, are mentioned by Strype, and these names survive amid immeasurable others. [1903, p. 69]

Similar remarks had been made somewhat earlier by Wheatley:

Leather Lane. The lane traverses a very poor neighbourhood; it is much infested with thieves, beggars, and Italian organ-grinders; and is in itself narrow and dirty, and lined with stalls and barrows of itinerant dealers in fish, bacon, vegetables, plasterers or image shops, and old clothes; a decidedly unsavoury and unattractive locality. [1891, vol. II, p. 378]

Even today (2000) the street is used as a market, though the nature of the wares offered has changed considerably. Greengrocers and mercers still have their stalls, and fish-and-chips shops, together with other providers of bodily sustenance (both liquid and solid), vie for the passenger's support. The winder of the mechanical barrel-organ has been replaced by the purveyor of the compact disc, and one sometimes gets the impression (piously hoped to be wrong) that the monkey on the former has merely been replaced by a similar simian on the back of the latter.

The idea that Roman Catholicism was a threat to Nonconformity led to no inconsiderable effort on the part of the dissenting clergy in the eighteenth century to expose the perceived dangers in this sect. The Merchants' Lecture, originally delivered at the Pinners' Hall, Broad Street, was aimed at the defence of the Protestant religion against Romanism, Socinianism,[48] and infidelity [Dale, 1907, p. 474]. Doctrinal disputes led to the venue for the lecture being changed, this being noted by Pike as follows.

From a union of Independent and Presbyterian Ministers in 1672, sprang the Merchants' Lecture, the first preachers of which were men with characters closely approaching the apostolic standard. ... Unfortunately, the tie which united this illustrious galaxy was sufficiently weak to be broken by the discussion of some doctrinal tenets. ... In a few years, when some of the original lecturers lay in their graves, disputes again arose to breed division between the two denominations [Independent and Presbyterian], so that a rival lectureship was instituted at Salters' Hall. These Tuesday services grew into real attractions, and hearers from the outskirts were wont to return at mid-day to their suburban homes, well repaid for a toilsome walk by having listened to some distinguished preacher. [1870, p. 67]

More specifically, the Salters' Hall lectureship was set up after Dr Daniel Williams had given a Merchants' Lecture on something he viewed as heresy. Most of the Presbyterians sided with Williams, and most of the Independents took the part of Dr Tobias Crisp and his son, whose theological writings on Antinomianism Williams had attacked.

Adjoining Salter's Hall, situated at Number 10 on the west side of St Swithin's Lane, south of New Court and in Walbrook Ward, was a chapel used by the Presbyterian and other religious bodies. It was originally erected during the reign of William III. The services were discontinued in 1820, the materials being sold and a new hall built in 1823–1827[49]. I do not know whether the lectures were given in the chapel or the hall itself.

On the death of Edmund Calamy (senior[50]) in 1732, the Merchants' Lecture at Salters' Hall fell vacant, and Joshua was elected to the post, one which he 'supported ... with great respectability for several years' [Wilson, 1814, vol. IV, p. 398]. In company with a number of other ministers, Joshua took part in a course of sermons against Popery, some of these later being printed.

'Respectability' is clearly an attribute that Wilson readily associated with Joshua[51], for he writes also that the latter was

> a worthy respectable man, of the old Protestant principles ... a minister of long standing and great respectability among the Presbyterians in the last century [1814, vol. IV, pp. 312, 396]

and he describes his pastoral work as follows.

> Mr. Bayes was a man of good learning and abilities; a judicious, serious and exact preacher; and his composures for the pulpit exhibited marks of great labour. In his religious sentiments he was a moderate Calvinist; but possessed an enlarged charity towards those who differed from him. His temper was mild and amiable; his carriage free and unassuming; and he was much esteemed by his brethren of different denominations. Though his congregation was not large, it consisted chiefly of persons of substance, who contributed largely to his support, & collected a considerable sum annually for the Presbyterian fund.
> [1814, vol. IV, p. 399]

Among the characteristics of dissenting clergy of that period were piety and learning, as Besant notes:

> [The Nonconformists] enjoyed the teaching of many remarkable men of great piety and profound learning and singular eloquence; they formed a compact body of men and women who maintained a pure and simple form of Christianity; they taught, in an extremely coarse age, the virtues of decency in conversation and in morals; in a drunken age, the virtue of moderation; in a frivolous age, the seriousness of life. [1902, p. 171]

James [1867, p. 670] describes the congregation at Leather Lane as 'mainly tradesmen'; I suppose there is no reason for one to see any contradiction between this description and that given above[52]. Writing of the decline of Dissent as a political force in London, as in the country, in the eighteenth century, Besant says

> The Dissenters generally belonged to the class of tradesmen and the better sort of working-men; the better families who belonged to the Nonconformist bodies in the seventeenth century dropped out with few exceptions in the eighteenth. [1902, p. 169]

In Palmer's list of the doctrinal positions of the London ministers in 1730, forty-four were found to be Presbyterian, nineteen of these being Calvinist, thirteen Arminian, and twelve Baxterian (Dale [1907, p. 543] interprets 'Baxterian' as *Broad Church*[53]). In Anon [1731–1732], 'Calvinists', among whom Joshua Bayes is counted, are described as 'such as agree with the Assembly's Catechism': Thomas Bayes is described in the same source as an Arminian,

> or such as are far gone that way, by which are meant such as are against particular election and redemption, original sin at least the Imputation of it, for the power of man's will in opposition to efficatious Grace, and for Justification by sincere obedience in the room of Christ's righteousness &c. [Anon, p. 89]

Joshua left behind four published sermons, viz. 'The Church of Rome's doctrine and practise with relation to the worship of God in an unknown tongue examined in a sermon' (1735), 'A funeral sermon occasioned by the death of Mr. J. Cornish, preached Dec. 10, 1727' (1728), 'A funeral sermon occasioned by the death of the Rev. C. Taylor' (1723), and 'A sermon preach'd to the Societies for the Reformation of manners, at Salters Hall, July 1, 1723' (1723). Dr Williams's Library has a copy of 'A sermon preached to the Dissenting ministers of the Three Denominations at the Library in Red Cross Street. Nov. 23. 1723'.

Joshua was also one of a number of clergymen engaged to complete Matthew Henry's 'Commentary on the Bible', his particular contribution being a commentary on Galations[54].

That Joshua was of no little prominence[55] in dissenting circles is evidenced not only by his participation in the Salters' Hall Lectures, but also by the records preserved in the *Minute Books of the Body of Protestant Dissenting Ministers of the Three Denominations in and about the cities of London and Westminster*. We find him elected to the general committee, together with Calamy, as the Presbyterian representative on the 7th of November 1727. He is listed as being present at various committee meetings, some held at North's and Hamlin's Coffee-houses, and was chairman of the committee on a number of occasions, the last such recorded occasion being a meeting at Dr Williams's Library on the 26th of December 1743.

In view of the repeated mention of 'coffee-houses' as meeting places of the tri-denominational committee, and on our bearing in mind present-day establishments of this kind, it might not come amiss to say something on the matter here[56].

Latham & Matthews [1983, vol. X, pp. 70–72] record that coffee-houses were introduced into Oxford in 1650 by one Jacob, a Jew, who moved to London two years later and set up shop in Holborn. A footnote to the first Number of *The Tatler*, however, records things rather differently: 'In 1652 an English Turkey Merchant brought home with him a Greek servant, who first opened a house for making and selling Coffee.' When coffee was first drunk in England is uncertain; Robinson relates that

> Nathaniel Conopius, a Cretan, who lived for some years at Oxford and was expelled during the troubles of 1648, was known to have made coffee for his own use while he continued at Balliol College. [1893, p. 71]

Whatever the origins of these establishments might have been, by May 1663 there were apparently eighty-two of them in the capital. As we have already mentioned, an annual licence cost $1s$, and sureties for good behaviour had to be provided. Such a coffee-house was but simply appointed: the cost of a dish of coffee (and tea or chocolate) was $1d$, tables and stools being provided for the customers. Newspapers were also available, and sometimes pamphlets were sold and advertisements displayed. By the reign of Queen Anne some 500 coffee-houses whose names are still known existed.

The conversation varied from one such venue to another:

> In those about the Temple, legal matters formed the principal subject of discussion. On the other hand, at Daniel's, the Welsh coffee-house in Fleet-Street, it was mostly of births, pedigrees and descents; Child's and the Chapter, upon glebes, tithes, advowsons, rectories, and lectureships; North's, undue elections, false pollings, scrutinies, and the like; Hamlin's, infant baptism, lay ordination, free-will, election, and reprobation.
> [Nevill, 1911, p. 9]

Joseph Addison, in his Letter No. 403 in *The Spectator*, relates that

> At a Coffee-house near the *Temple*, I found a couple of young Gentlemen engaged very smartly in a dispute on the Succession to the *Spanish* Monarchy.... I afterwards entered a By-Coffee-house that stood at the upper End of a narrow Lane, where I met with a Nonjuror[57], engaged very warmly with a Laceman who was the great Support of a neighbouring Conventicle.

The subjects discussed at some coffee-houses were even more varied than those instanced above: in Letter 454 for Monday the 11th of August 1712, Steele relates

> [I] passed the Evening at *Will*'s in attending the Discourses of several Sets of People, who relieved each other within my Hearing on the Subjects of Cards, Dice, Love, Learning and Politicks.

Matters discussed at coffee-houses were not always frivolous, however. Steele writes [Letter No. 49]

> It is very natural for a Man, who is not turned for Mirthful Meetings of Men, or Assemblies of the fair Sex, to delight in that sort of Conversation which we find in Coffee-houses. ... The Coffee-house is the Place of Rendezvous to all that live near it, who are thus turned to relish calm and ordinary Life.

Child's coffee-house, situated near Doctors' Commons, the Royal Society (then at Gresham College), and the College of Physicians, drew its patrons mostly from the clergy and the professional classes, mainly of the Tory persuasion. St James's was mostly Whig. The Grecian, chiefly frequented by lawyers, drew the learned for discussion of scholarship and philosophy. Jonathan's was haunted by the merchant and stock-jobbing set, their more fashionable customers going to Garraway's[58].

Mathematical matters were also discussed at these venues. Abraham de Moivre's association with Slaughter's Coffee-house is well known, and Henry Newman, in his letter to Richard Steele of the 10th August 1713(?)[59], wrote

> I thank you for y^r Kindness to Mr Whiston as it is a Charitie not only to him but to the Publick in putting him upon an Amusement which may divert him from those studies that have made him so obnoxious to the reproach of Good Men. I gave him notice immediately of yor favour & suppose he will wait upon you for y^r Commands. I only beg leave to suggest one thing to you when he does, because it will come with more authority from you than perhaps any man in ye Kingdom beside, and y^t is that you will be pleas'd to conjure him Silence upon all Topicks foreign to the Mathematicks in his Conversation or Lectures at ye Coffeehouse. He has an Itch to be venting his Notions about Baptism & the Arrian Doctrine but yor authority can restrain him at least whilst he is under your Guardianship.
> [Blanchard, 1963, p. 292]

Whiston also gave 'many astronomical lectures at Mr. *Button's* coffee-house, near Covent-Garden, to the agreeable entertainment of a good number of curious persons' [*Memoirs*, pp. 257–258]. These lectures were publicised in an advertisement to Richard Steele's Letter of Saturday the 9th of January 1714 in *The Englishman* as follows[60].

> Beginning January 11, 1713–1714, a course of Philosophical
> Lectures on Mechanics, Hydrostatics, Pneumatics, Opticks, ...
> This course of experiments is to be performed by Mr. William
> Whiston and Francis Hauksbee, the nephew of the late Mr.
> Hauksbee.

Such was the popularity of these lectures that the venue was later changed 'to a larger Room close by at Mr. Dale's, an upholsterer over the Corner of the nearest Piazza in Covent Garden', as reported in an advertisement to Steele's Letter of Tuesday the 26th January 1714.

In Stigler [1999, p. 232] the following passage from Thomas Sydserf's *Tarugo's Wiles; or, The Coffee-House. A Comedy*[61] of 1668, recording the conversation between two customers, is given.

> 'I'm told Sir, that Coffee inspires a man in the Mathematicks.'
> 'So far as it keeps one from sleep, which you know is the ready
> way to distract consequently the improvement of the Mathematicks.' [Act III]

Button's coffee-house was also the source of more frivolous exchanges, for we find a letter written 'By a Sparkish Pamphleteer of Button's Coffee-House' (John Arbuthnot) to Jonathan Swift, in which the former thanks the Dean 'in the name of the Publick, for your continuing to be useful, notwithstanding you are a dignified Churchman' [1770, vol. II, p. 115].

Lillywhite finds the subjects of discussion at Hamlin's coffee-house situated in 'Swithins Alley near the Royal Exchange' [1963, No. 532], better known as Sweeting's Alley, No. 5 Sweeting's Rents, 'odd', and says 'The subjects seem out of place near the Royal Exchange, but none-the-less may be true'. This coffee-house was consumed by fire on the 10th November 1759. From the records of the dissenting committee it would appear that matters more directly pertinent to Nonconformity were also discussed at North's Coffee House in King's Street, near the Guildhall.

In the first Letter in *The Tatler* of 1709 Richard Steele writes:

> All accounts of Gallantry, Pleasure, and Entertainment, shall
> be under the article of White's Chocolate-house; Poetry, under
> that of Will's Coffee-house; Learning, under the title of Grecian;
> Foreign and Domestic News, you will have from Saint James's
> Coffee-house.

Will's Coffee-house, in Russell Street, had been a favourite rendezvous of the Restoration Wits, but by this time — 1711 — its popularity had declined. It is perhaps worth noting that the word *Wit*, as used here, has not the meaning that we associate with it today: writing of the Wits at the Court of Charles II, Wilson, citing an eighteeth-century source[62], says

The name was as loose as the morals of the assemblage. A Wit was anyone from wild, malicious Harry Killigrew or George Bridges ('created a Wit for hard drinking') to George Villiers, Duke of Buckingham, the last splendid playboy of the fading Renaissance, or William Wycherley, the finest dramatic genius of the Restoration Court. A Wit was not necessarily one skilled at jest or repartee, nor need he be a poet, playwright, or maker of libels and lampoons. In the Restoration meaning of the term, a Wit was simply anyone who pretended to intellectuality, and (especially if he were a lord) he was often taken at his own evaluation. [1967, p. 5]

In his Note on Steele's Letter No. 54 in *The Spectator* Smith points out that 'Any distinction between *Coffee-house* and *Chocolate-house*, in respect of their names, must be more or less doubtful', and he cites Pepys's diary 'To a Coffee-house to drink Jocolatte[63], very good'. [24th Nov. 1664]

The uplifting, if not improving, nature of the conversation at these establishments was not appreciated by everyone, however. Hazlitt, in an essay of 1826, spoke of 'that lowest temple of fame, a coffee-shop', and a few years later, in a passage on the development of moral qualities, Quetelet wrote

L'ardeur avec laquelle on se porte au jeu, le nombre des faillites, la fréquentation des cabarets et des mauvais lieux, l'ivrognerie et plusieurs autres circonstances donneraient encore, au besoin, des élémens utiles pour l'appréciation du défaut d'ordre et de l'imprévoyance. [1835, pp. 134–135]

Very often the talk was of matters that were, or could perhaps be viewed as, dangerous to the State. Bryant notes

So popular did the coffee-houses become and so seditious the conversation therein, that the Government was forced to issue proclamations against them and subject them to the same licensing restrictions as their rivals the ale-houses. [1960, p. 40]

Thus in 1675 the following proclamation was issued.

CHARLES R.

Whereas it is most apparent that the multitude of coffee-houses of late years set up and kept within this kingdom, the dominion of Wales and the town of Berwick-upon-Tweed, and the great resort of idle and disaffected persons to them, have produced very evil and dangerous effects, as well for that many tradesmen and others do therein misspend much of their time, which might and probably would otherwise be employed in and about their lawful callings and affairs, but also for that in such

> houses, and by occasion of the meetings of such persons therein, divers false, malicious and scandalous reports are devised and spread abroad, to the defamation of his Majesty's government and to the disturbance of the peace and quiet of the realm, his Majesty has thought it fit and necessary that the said coffee-houses be for the future put down and suppressed, and doth (with the advice of his Privy Council) by this his royal proclamation strictly charge and command all manner of persons that they or any of them do not presume, from and after the 10th day of January next ensuing, to keep any public coffee-house, or to utter or sell by retail in his, her or their house or houses (to be spent or consumed within the same) any coffee, chocolate, sherbet or tea, as they will answer the contrary at their utmost perils. [Browning, 1953, pp. 482–483]

As might be expected, this decree was not received with universal approval. Thornbury relates that

> on a petition of the merchants and retailers of coffee, permission was granted to keep the coffee-houses open for six months, under an admonition that the masters of them should prevent all scandalous papers, books, and libels from being read in them, and hinder every person from declaring, uttering, or divulging all manner of false and scandalous reports against Government or the ministers thereof. The absurdity of constituting every maker of a cup of coffee a censor of the press was too great even for those days: the proclamation was laughed at, and no more was heard of the suppression of coffee-houses.
> [1887–1893, vol. I, p. 533]

After this somewhat protracted coffee-break, let us not rest any further on our laurels but return to our Bayes. As has already been mentioned, Joshua married Anne Carpenter. They had seven children: Thomas, John, Samuel, Nathaniel, Mary, Anne, and Rebecca. Of the eldest son, Thomas, we have more to say in the next chapter; for the moment let us devote a few paragraphs to the others.

Before doing so, however, we might perhaps note that although we have been able to give baptismal dates for some members of the Bayes family already mentioned, such information is lacking for many of them. Baptismal records would have been kept in parish churches, but Nonconformists would rather have had such sacraments performed by their own clergy. In his *Life* Edmund Calamy relates

> As to myself, I was baptized by my own father, soon after my birth, and trained up under his ministry, as well as his paternal

instruction; so that it cannot be said of me, as of several others, that I left the Established Church, because I was never joined to it, either by myself or my immediate parents.
[1830, vol. I, p. 72]

It could well have been the case that Joshua exercised a similar prerogative in the case of his children.

That this was not an abnormal custom is supported by MacKinnon's recording that, in 1734, a parish register had the entry 'Robert son of Robert Mercer j^r baptized in the Presbyterian way' [1930, p. 36]; but even though it was not unusual for dissenters to conduct their own rite, it did not meet with universal approval. Richard Steele writes, in his Letter No. 39 of Saturday the 2nd of January, 1714, in *The Englishman,*

> There is no Day set when Baptism is to be administred, consequently no Law is broken if the Administration be delay'd till a Priest can be had. The Church therefore did very wisely to make the Administration of that Sacrament a Part of the Ministerial Office; and since the Church has done so, it has a Right to punish Delinquents in that kind, as well as in any other. He offends therefore against the Laws of the Church who is baptized, or causes his Children to be baptized by Persons unordained; and nothing can in the least excuse it, but the most instant and urgent Necessity.

Let us make some general remarks about the children of Joshua and Anne. Boyd's *Roll of the Draper's Company of London* [1934, p. 9] lists

                Bayes   Jn   A1683   V1727   Nath   A1737   V1750

where 'Jn' and 'Nath' stand for John and Nathaniel, brother and son, respectively, of Joshua. Here 'A' denotes the year when the boy was bound apprentice (in many cases this happened in the fifteenth, sixteenth, or seventeenth year of age), and 'V' indicates that the draper was then alive.

Joshua's son John, who died on the 11th October 1743, aged 38, is mentioned with his father in the list of subscribers to John Ward's *Lives of the Professors of Gresham College* of 1740. The subscription is noted in MS 6207.f.1 in the British Library, a manuscript containing the names of the subscribers to this volume, as

           John Bayes of Lincoln's Inne Esquire   0–10–0

           The Rev. Mr Joshua Bayes   0–10–0

Samuel, who became a draper, married Theodosia Collier on the 13th of August 1747 in the Mercers' Hall Chapel, Cheapside[64]. His obituary in *The Gentleman's Magazine and Historical Chronicle* for 1789 ran as follows.

> Oct. 11. At Clapham, Sam. Bayes, esq. formerly an eminent linen-draper in London, son of the Rev. Mr. Sam [sic] Bayes, an eminent dissenting minister. His lady died a few weeks before him. [vol. 59, p. 961]

(The same source gives Theodosia's death as 'Aug. 16. At Brighthelmstone, Mrs. Bayes, wife of Sam. Bayes esq. of Clapham' [vol. 59, p. 955], but the Bunhill Fields memorial gives the date as the 22nd of September.) The draper's business is listed in *Kent's Directory* for 1759 as 'Bayes and Watts', in Cheapside[65]. Nathaniel, of Snow-hill, was a grocer (as was, in the same parish, Thomas Cromwell, great-grandson of the Protector — see Wheatley [1891, III, p. 260]). Mary died unmarried, Anne married Thomas West, and Rebecca married Thomas Cotton. We have more to say on Thomas Bayes's siblings, in particular about their variously recorded death dates, in Chapter 12.

Wallis & Wallis [1986] have recorded subscriptions made by divers mathematicians and mathematical writers to various books. In the list of subscribers to Samuel Say's *Poems on Several Occasions; and Two Critical Essays* of 1745 we find

> Rev. Mr Bayes, two books.   Mr. Samuel Bayes,

though whether the first of these entries refers to Joshua or to Thomas (as Wallis & Wallis suggest) I cannot say. However, Thomas certainly subscribed to James Foster's *Discourses on all the Principal Branches of Natural Religion and Social Virtue*[66] of 1752 and to Henry Pemberton's *A View of Sir Isaac Newton's Philosophy* of 1728[67].

Let us briefly say something about these three authors. Samuel Say (1676–1743) was a dissenting minister and an intimate friend of Isaac Watts. In 1734 he accepted a call to a congregation in Long Ditch (now Princes Street), Westminster, no minister having been there since the death of Edmund Calamy in 1732. The *Dictionary of National Biography* describes his poems as 'youthful rubbish'. James Foster (1697–1753), also a divine, was well known for the eloquence of his preaching. Pope immortalised him in his *Epilogue to the Satires*, Dialogue I, in the words

> Let modest FOSTER, if he will, excel
> Ten Metropolitans in preaching well. [131–132]

In 1728 he became Sunday-evening lecturer at Old Jewry, where Richard Price later ministered, and in 1744 he became pastor of the independent church at Pinners' Hall[68]. Henry Pemberton (1694–1771), although trained as a physician, devoted most of his time to mathematics. He was a great popularizer of Newton's work, and was in fact employed to superintend the third edition of the *Principia*, in the preface to which Newton described him as 'vir harum rerum peritissimus' (a man most knowledgeable in these matters). He became professor of physic at Gresham College in 1728[69].

Joshua Bayes died on the 24th April 1746 in the fifth-third year of his ministry. In view of his close connexion with the dissenting community it is surprising to read in the *Minute Books of the Body of Protestant Dissenting Ministers of the Three Denominations in and about the cities of London and Westminster*,

> North's Coffee House. Jan. 5th 1748–9. Mr. May reported that Mr. Bayes is dead.

Why this delay in reporting the death of one of the committee's most faithful and longest-serving members? One cannot say: presumably the phrase was not a Presbyterian equivalent of 'Queen Anne is dead.'

2.4 Appendix 2.1

The doctrinal differences between the Presbyterians, Calvinists, Arminians, and Baxterians were by no means insignificant in the seventeenth and eighteenth centuries, however much apathy and the passage of time may subsequently have blurred them. It is perhaps thus not inadvisable to say a word or two about them here.

According to Bailey & Bailey, English Presbyterianism was first offered to Edward VI as a form of church government. The movement, which 'found its first clear expression under the Marian exile', was suppressed under Elizabeth, 'who was convinced of the incompatibility of Presbyterianism with absolute monarchy.' After the Civil War English Puritanism began to show its pro-Presbyterian tendencies, the strong reaction to which resulted in the Clarendon Code and Presbyterian suppression (Bailey & Bailey suggest that as many as one thousand five hundred of the two thousand ministers who were ejected could have been Presbyterian).

The effect of the reformation initiated by John Calvin (1509–1564) upon Christianity is well known. A further widening of the Calvinistic attitude was advocated by Richard Baxter (1615–1691). Lewis writes

> The key to Baxter's doctrinal position is to be found in the fact that he always endeavoured to avoid the falsehood of extremes, and to find truth and harmony in the golden mean.
> [Hastings, 1971, vol. 2, p. 440]
> Those who shared Baxter's spirit of broad catholicity were called, often in reproach, 'Baxterians.' They never formed a sect or

even a school, but were men of independent minds who struck out paths for themselves, and in accordance with his principles distinguished between the essentials and non-essentials. 'In things necessary, unity; in things doubtful, liberty; in all things, charity.' [op. cit., 1971, vol. 2, p. 441]

Arminianism was founded by James Arminius (or Jacobus Arminius or Jakob Harmensen or Van Herman) as 'a revolt against certain aspects of Calvinism' [Hastings, 1971, vol. 1, p. 807]. Platt describes it as follows.

> As an active criticism of Calvinism it is based upon two positions — the restless and dominant demand for equity in the Divine procedure, on the one hand, and such a reference to the constitution of man's nature as will harmonize with the obvious facts of his history and experience, on the other. It sought to construct a system which should be dominantly ethical and human throughout. It contended, therefore, that moral principles and laws consistently condition the manward activities of the Divine will, and set human limits to the Divine action. The Calvinistic conception of justice was based altogether on the supremacy or rights of God; Arminianism so construed justice as to place over against these the rights of man. ... The Arminians were the fathers of toleration.
> [Hastings, 1971, vol. 1, pp. 811–812, 813]

The Anabaptists, or Katabaptists, as they were sometimes called, were the most radical movement of the Reformation. Generally they accepted the views on God set down in the Nicene and Apostles' Creeds, and as held in common by the Roman Catholic and Protestant doctrines. The freedom of the will and complete moral responsibility were paramount, as M'Glothlin wrote:

> The Anabaptists maintained the right of the individual to interpret Scripture for himself; and some of them, at least, asserted the superior authority and sanctity of the N[ew] T[estament] over the O[ld] T[estament] as the fuller, clearer revelation of God, thus approximating to the modern view of a progressive revelation. [Hastings, 1971, vol. 1, p. 410]

The name was determined from the beliefs on baptism:

> The purity of the Church was to be secured by the baptism of believers only, and preserved by the exercise of strict discipline. ... Infant baptism was regarded as without warrant in Scripture. ... In its stead they practised believers' baptism, administering the rite to those who had been baptized in infancy, thus

winning the name 'Anabaptists' — re-baptizers — from their opponents; but the name and its implication they earnestly repudiated, declaring that so-called baptism in infancy was no baptism, and claiming for themselves the name of 'brethren' or 'disciples'. [Hastings, 1971, vol. 1, p. 410]

Originating with Johannes Agricola (1492–1566), a coadjutor of Luther, Antinomianism is described as follows by J. MacBride Sterrett.

> It is the counterpart of modern political anarchism, being directed towards the destruction of the Moral Law of the O[ld] T[estament] in the interest of the new freedom of Christians and the testimony of the spirit. Antinomianism, as John Wesley defined it, is the doctrine that 'makes void the Law through faith.' Christians are free from the Law. The Law primarily referred to was the law of Moses. Agricola denied that Christians owed subjection to any part of this Law, even to the Decalogue.
> [Hastings, 1971, vol. 1, pp. 581–582]

The guiding, if not leading, spirit behind Socinianism was Fausto Paolo Sozzini (1539–1604), though he was heavily influenced by his uncle Lelio Francesco Maria (?) Sozzini. Possessing a Christian basis, Socinianism appealed to the intellectual and rational in man. In Clow's words:

> It asserted the necessity and the fact of a divine revelation, maintained that the Scriptures are its authoritative record, and declared that the reason — the moral and religious nature — is the sole and final arbiter of truth.
> [Hastings, 1971, vol. 11, p. 651]

The 'authority, sufficiency, and perspicuity' of the New Testament were postulated, the Old Testament being found to have no authority in itself. Tradition, creeds, and the authority of the Church were set aside. Further,

> such doctrines as predestination, original sin, the necessity of imposing a penalty or of adhering to any doctrine of forgiveness in relation to law, are swept away.
> [Hastings, 1971, vol. 11, p. 652]

The Socinians' views on Christ tended perhaps more to the anti-Trinitarian than the Unitarian:

> Jesus was truly a mortal man while He lived on earth, but at the same time the only-begotten Son of God, being conceived of the Holy Spirit and born of a virgin. ... By His resurrection He was begotten a second time and became, like God, immortal.
> [Hastings, 1971, vol. 11, p. 652]

Nonjurors are described by Charles Gaskoin in the following words.

> The Nonjurors were clergy and laymen who, though no Romanists, scrupled to take the oaths of allegiance and abjuration imposed, particularly on office holders, under William III. and George I. The term, strictly including some Scottish Covenanters, is commonly restricted to Episcopalians, but extended to persons who, though exempted from the oaths as unofficial, attached themselves to the Nonjurors proper. Those rejecting only the oaths of 1701 and 1714 are sometimes distinguished as Non-Abjurors. The English Nonjurors ... were often important individually as men of the highest character and of great learning. Corporately, they stood on the one hand for the doctrines of passive obedience and non-resistance, on the other for the right of the Church to independence in spiritual matters. [Hastings, 1967, vol. 9, p. 394]

Like the nice distinctions between the Arminians, the Calvinists, and the Baxterians, the differences between Dissenters and Nonconformists, although blurred today, were of no little importance in the eighteenth century. Adeney summarizes these latter differences as follows.

> While 'heresy' stands for opposition to ecclesiastically settled orthodoxy, 'schism' for separation from the communion of the society claiming to be the one true Church, and 'dissent' for divergence from the beliefs and doctrines maintained by the national settlement, Nonconformity consists in not carrying out the requirements of an 'Act of Uniformity,' which is a law of the State. [Hastings, 1967, vol. 9. p. 382]

3

Thomas Bayes: a life

> *I thought (quite rightly) that I would never know enough about anyone to write a biography.*
>
> Stephen Spender.
> *World within World.*

3.1 Introduction

In his *Portrait of Isaac Newton* Frank Manuel justified his examination of the early Lincolnshire years of his subject by suggesting that such consideration was warranted by an acquaintance with the adult savant[1]. Would that such were the case here! Our knowledge of the mature Thomas Bayes is largely restricted to one posthumously published essay on chance, and this is hardly sufficient for the attempting of a psychological study of the youth. Whereas contemporary opinions on Newton were so varied that Manuel was indeed justified in trying to find possible grounds for Newton's later development in a study of his youth, pertinent observations by Thomas's coevals are exceedingly sparse, and we must perforce rely for information about his early years, as we show, on genealogical records and the odd passing remark. I should not feel comfortable in attempting to amplify these few records and remarks into a fuller account, one that by its very nature would tend to the fictitious, or even fanciful. Hence our sketch here is necessarily somewhat drier, but no less accurate, I trust, than the appreciator of biography might desire — or even be entitled to expect.

In one of his lectures 'Read in the University of Edinburgh over 24 years' Hugh Blair said

> Biography, or the Writing of Lives, is a very useful kind of Composition; less formal and stately than History; but to the bulk of Readers, perhaps, no less instructive; as it affords them the opportunity of seeing the characters and tempers, the virtues

> and failings of eminent men fully displayed; and admits them into a more thorough and intimate acquaintance with such persons, than History generally allows. For a Writer of Lives may descend, with propriety, to minute circumstances, and familiar incidents. It is expected of him, that he is to give the private, as well as the public life, of the person whose actions he records; nay, it is from private life, from familiar, domestic, and seemingly trivial occurrences, that we often receive most light into the real character. [1784, p. 342]

Similar sentiments were expressed somewhat later by Thomas Carlyle, who, in his 1828 essay on Robert Burns, wrote

> if an individual is really of consequence enough to have his life and character recorded for public remembrance, we have always been of opinion that the public ought to be made acquainted with all the inward springs and relations of his character.
> [Carlyle, 1915b, p. 4]

Were the author of the present study fully to accept this viewpoint he would have more hesitation in offering this work to the public than he has. It is only because of the importance that Bayes's *Essay towards solving a Problem in the Doctrine of Chances* has come to hold that it is felt that such a memorial is necessary despite the lack of information about its subject's character.

In the Duke of Buckingham's burlesque, *The Rehearsal* (more of which anon), the chief character, one *Bayes*, says 'As, now, this next Scene some perhaps will say, It is not very necessary to the Plot', and the reader of the present work may perhaps view this as a fitting comment on some parts of this chapter. But the quotation continues 'I grant it', and with this I agree. Despite this, however, I feel that the less-biographical passages shed some light on Thomas Bayes's time and the milieu in which he lived and worked.

3.2 Thomas Bayes's education

Where Thomas Bayes was born is unknown, though it may be deduced from the previous chapter that the birth occurred in Hertfordshire. The year of his birth is also uncertain: it is usually derived by subtraction of his age at death (fifty-nine), from the year of his death, given on the present family vault in the Bunhill Fields Burial Ground as 1761 (we have more to say on this death date in Chapter 12). If he was born before the 25th of March (New Year's Day in the old English calendar), the year of birth could be 1701 (Old Style) or 1702 (New Style). Since it is known that he died on the seventh of April, one finds on subtracting the eleven days 'lost' by the calendar reform of 1752 that his death was on the 27th March, Old Style. Subtraction of his age at death then shows that he was born either on or shortly after New Year's Day 1702 (o.s.), or in 1701 (o.s.)[2].

Where Thomas was initially educated is similarly uncertain. Some sources[3] say he was 'privately educated', but others[4] believe 'he received a liberal education[5] for the ministry' — a requirement one might have thought necessary (but not sufficient), though Bogue & Bennett in fact say

> The necessities of the church may render it proper that men should be ministers, who have not enjoyed the advantages of an academical, or even a liberal education. [1809, vol. II, p. 7]

The idea of what constitutes a *liberal education* has changed considerably over the years, and I am not sure whether the term would be covered by what we today would term *liberal arts*. In the Middle Ages the seven topics covered by the latter term were grammar, logic, rhetoric, mathematics, geometry, astronomy, and music[6]. Alexander Grant writes of 'the seven subjects of the time-honoured Arts curriculum — Latin, Greek, Mathematics, Natural Philosophy, Logic, and Moral Philosophy and Rhetoric' [1884, vol. II, p. 115]. In the sixteenth century the Jesuits drew up a scheme for the education of boys: the lower classes read grammar, humanity, and rhetoric, and in the higher classes 'Aristotle's Logic, Physic, Metaphysic, and Ethics, with Euclid and the use of the globes, formed the staple of liberal education' [Parker, 1867, pp. 42–43] (Latin, with rather less Greek, was studied in every class).

The emphasis switched somewhat in the nineteenth century: for instance, in France under Napoleon, by 'liberal education' was meant 'the intimate union of literature and science' [Parker, 1867, p. 71], whereas in Austria

> Professor Bonitz, who was employed to reorganize the Gymnasia, defined it as the aim of liberal school education to impart a higher general culture, making such substantial use of classical literature, as to lay the foundation for University studies. [Parker, 1867, p. 63]

In 1868 Thomas Huxley gave a lecture entitled *A liberal education; and where to find it* to the South London Working Men's College, in which he said[7]

> That man, I think, has had a liberal education who has been so trained in youth that his body is the ready servant of his will, and does with ease and pleasure all the work that, as a mechanism, it is capable of; whose intellect is a clear, cold,

logic engine, with all its parts of equal strength, and in smooth working order; ready, like a steam engine, to be turned to any kind of work, and spin the gossamers as well as forge the anchors of the mind; whose mind is stored with a knowledge of the great and fundamental truths of Nature and of the laws of her operations; one who, no stunted ascetic, is full of life and fire, but whose passions are trained to come to heel by a vigorous will, the servant of a tender conscience; who has learned to love all beauty, whether of Nature or of art, to hate all vileness, and to respect others as himself.

Such an one, and no other, I conceive, has had a liberal education; for he is, as completely as a man can be, in harmony with Nature. He will make the best of her; and she of him. They will get on together rarely: she as his ever beneficent mother; he as her mouthpiece, her conscious self, her minister and interpreter. [Huxley, 1905, p. 86]

Somewhat more specifically Sidgwick [1867, p. 87] writes

taking the term [liberal education] in its ordinary sense, and applying it to those who are able to defer the period of professional study till at least the close of boyhood, a liberal education has for its object to impart the highest culture, to lead youths to the most full, vigorous, and harmonious exercise, according to the best ideal attainable, of their active, cognitive, and æsthetic faculties. What this ideal, this culture may be, is not easy to determine; but when we have determined it, and analysed it into its component parts, a natural education is evidently that which gives the rudiments of these parts in whatever order is found the best; which familiarises a boy with the same facts that it will be afterwards important for him to know; makes him imbibe the same ideas that are afterwards to form the furniture of his mind; imparts to him the same accomplishments and dexterities that he will afterwards desire to possess. An artificial education is one which, in order that man may ultimately know one thing, teaches him another, which gives the rudiments of some learning or accomplishment, that the man in the maturity of his culture will be content to forget.

In the nineteenth century John Locke spoke of an 'ingenuous' education, a term that Garforth (see Locke [1964, p. 60]) equates to 'liberal' and that Locke interpreted as the subordination of Inclination to Reason. In the early twentieth century Horne, describing a liberal education as 'rather an attitude of mind than a knowledge of courses' [1916, p. 147], somewhat chauvinistically wrote

> The Grecian liberal education was for the free man, implying
> the existence of the slave class; the English liberal education
> is still for the gentleman, implying the existence of the labor-
> ing caste; the American liberal education is for man as man,
> without qualifications, implying the equal rights of all to free
> self-realization. [pp. 244–245]

Under the umbrella of liberal education, Horne would include physics, biology, psychology, mathematics, grammar, literature, language, and history. Not only does a liberal study differ from a professional one in being pursued for its own sake rather than for the sake of its application, it also possess a non-specialized quality[8].

During the time of Gregory the Great the liberal arts were 'to be studied so far, as by their aid revealed truth is profoundly understood' [Parker, 1867, p. 9]. Somewhat later the moral and political sciences were seen to be of importance in a liberal education.

In his *A Discourse Concerning the Vegetation of Plants* of 1661 Sir Kenelm Digby wrote 'Enough, if not too much is said of these Curiosities by way of digression, and to entertain you (Noble Auditors) with pleasing variety; Let us come back to our Plant' [p. 72], and with this in mind let us return to our main topic.

Although we can say nothing definite about Thomas's early education, it is possible that he might have attended a dissenting academy in London. There remains, as we show in a later chapter, a letter written by John Ward to Thomas in 1720, and Ward (of whom we say more later on) is known to have been a schoolmaster[9] in Tenter Alley. We know also that John Eames, later one of the sponsors of Thomas's election to the Royal Society, was assistant tutor in classics and science at the Fund Academy in Tenter Alley. This migratory academy (Pinner, Moorgate) seems, according to the list of the chief dissenting academies given in Parker [1914, pp. 137–143], to have been the only one convenient to the Bayes house in London. The evidence for Thomas's attendance there is indeed scanty, but it is at least possible that Thomas was a student of both Eames and Ward[10].

Of Eames and the Academy D.O. Thomas has this to say:

> it was through him that Tenter Alley acquired its reputation
> as the leading Dissenting Academy in the development of sci-
> entific education. He taught divinity, the classics, mathemat-
> ics, anatomy, and natural and moral philosophy. His lectures in
> applied mathematics included mechanics, statics, hydrostatics,
> and optics, and at least some of these were delivered in Latin.
> [1977, p. 11]

Although, as we have already mentioned, dissenting academies were not solely concerned with religious studies, the exploration of the relation between man and God was of great importance in these institutions, as Thomas relates,

> The main purpose of education according to Eames and Densham, as it was to most of the Dissenters of the age, was to determine and clarify man's relation to God, and all studies were seen to be ancillary to this purpose. The enthusiasm for the development of natural science must be understood in this light; as R.T. Jenkins notes[11], in several of the Academies the study of mathematics was conceived as an apologia; science would serve theology by demonstrating the order upon which God founded the universe; it would fulfil the aim embodied in Boyle's motto 'Ex rerum causis supremam noscere causam' [to discover the Supreme Cause from the causes of things]. ... Later in the century more emphasis may have come to be placed upon the utilitarian justification of science and scientific education, but at Tenter Alley they still retained the inspiration of an older view which is embodied in such works as Ray's *Wisdom of God manifested in the Works of Creation*, Derham's *Physico-Theology*, and Nieuwentijdt's *Religious Philosopher*.
> [1977, p. 12]

Of Eames's exceptional abilities[12] McLachlan has the following to say.

> Appointed by the Congregational Fund Board to the academy afterwards at Hoxton, he was the only layman ever placed in charge of an academy, and, unlike most other tutors, published nothing[13]. He was eminent alike in classics and mathematics, attracted to his lectures, despite a lack of oratorical gifts, some of the most promising pupils of other academies, and after his death his lectures continued to be used in manuscript by tutors of academies other than his own. [1931, p. 18]

Commenting further on Eames's 'lack of oratorical gifts', Dale writes

> Eames, though distinguished as a scholar, was disabled for the ministry by a defect in the organs of speech, and by a pronunciation that was 'harsh, uncouth, and disagreeable.' He once attempted to preach, but broke down, and never repeated the experiment. [1907, p. 501]

As a final comment on this worthy man, one who was described by Jeremy [1885, p. 43] as being 'of a candid and liberal disposition and a friend of free enquiry', let us note the following remarks by Bogue & Bennett.

3.2. Thomas Bayes's education

> [Eames] was a native of London, studied the learned languages at Merchant Taylor's School, and afterwards received an academical education for the ministry; but extreme diffidence and a defect in the powers of elocution deterred him from preaching more than one sermon. His talents, however, were not lost in inutility. He employed them with great diligence and benefit in the instruction of youth ... His very superior attainments in the branches of science which he taught, entitle him to more than common praise, which it would be the more unjust to withhold, because excessive timidity and bashfulness veiled them so as almost to conceal his extraordinary talents. He was intimately acquainted with sir Isaac Newton, and it is said, assisted him on some occasions. [1810, vol. III, pp. 283–284]

Eames died suddenly on the 29th of June 1744, the inscription on his tombstone, in Bunhill Fields, reading simply 'The learned John Eames, F.R.S.'.

It is perhaps of some interest to note that Eames was not the only man whose awkwardness in locution led him away from the ministry to a profession in which he shone. Charles Lamb was prevented from becoming a clergyman by a stammer: Birrell writes in his introduction to Lamb [1894], 'His stutter saved him from the Universities', and Lamb himself wrote in the preface to the *Last Essays of Elia* 'The informal habit of his mind, joined to an inveterate impediment of speech, forbade him to be an orator'. One might well wonder whether Lamb's gentle humour would have sat comfortably under a cassock. Ainger [1905], while noting that school exhibitions to the universities were given under the tacit, or implied, condition that the recipients would proceed to holy orders, a calling in which Lamb's impediment would have been a hindrance, suggests that Lamb's not proceeding to a university might rather have been attributable to the poverty in the family home and the consequent need for the young Charles to earn some money.

What Thomas might have studied at the dissenting academy is uncertain, though the editors of Bacon [1886] draw attention to Whewell's definition of a *liberal education* as 'a little of everything and everything of some one thing' [p. 410]. Bogue & Bennett also note, in connexion with the training for the ministry,

> To mathematics and natural philosophy it has usually been judged proper to apply a portion of the student's time. As they tend to improve the mind, and peculiarly to exercise its powers, and call forth their energies, the general influence of both may be favourable to his future labours, and the hearers as well as the preacher experience their good effects.
> [1810, vol. III, p. 270]

It is interesting to note that, at the Attercliffe Academy (where Timothy Jolie, John de la Rose, and J. Wadsworth were tutors), mathematics was tabooed as 'tending to scepticism' [Gordon, 1902, p. 12]. Writing of the difference between the English and the French schools Taylor says

> In England, Eton set the example of a rigid adherence to the Classics, and Dr Johnson's narrow scheme for a grammar school may be remembered. But the French youth, while he studied the Classics, was grounded also in mathematics, and in the current natural philosophy (the Dissenters imitated this).
>
> [1966, p. 43]

Parker [1914, p. 55] states that training at such academies usually began with courses in Logic and Rhetoric, with students intending to enter the ministry going on to study Divinity, Greek, Hebrew, Jewish language and antiquities, Ethics, Natural Philosophy and Metaphysics, with lectures on the writing of sermons and pastoral care. Daniel Defoe, who attended the Academy at Newington Green, also studied French and Italian [Parker, 1914, p. 61]: if Thomas had a similar training it might account partly for the French passages in his Notebook (these passages are discussed in a later chapter).

In 1729 Philip Doddridge (1702–1751) was chosen to lead an academy at Market Harborough. An account of his life is given by Orton: we append some extracts on the conduct of the academy[14]:

> it was an established Law, that every Student should rise at *Six o'Clock* in the Summer, and *Seven* in the Winter. ... One of the first Things he expected from his *Pupils*, was to learn *Rich*'s Short-hand, which he wrote himself, and in which his Lectures were written; that they might transcribe them, make Extracts from the Books they read and consulted, with Ease and Speed, and save themselves many Hours in their future Compositions. Care was taken in the first Year of their Course, that they should retain and improve that Knowledge of *Greek* and *Latin*, which they had acquired at School, and gain such Knowledge of *Hebrew*, if they had not learned it before, that they might be able to read the *Old Testament* in its original Language. ... If any of his Pupils were deficient in their Knowledge of *Greek*, the Seniors, who were best skilled in it, were appointed to instruct them at other Times. Those of them, who chose it, were also taught *French*. He was more and more convinced, the longer he lived, of the great Importance of a *learned*, as well as a *pious* Education for the Ministry: And finding that some who came under his Care were not competently acquainted with *classical*

Knowledge, he formed a Scheme to assist Youths in their Preparations for academical Studies, who discovered a promising Genius and a serious Temper. ... Systems of *Logic, Rhetoric, Geography* and *Metaphysics* were read during the first Year of their Course ... To these were added Lectures on the Principles of *Geometry* and *Algebra*. These Studies taught them to keep their Attention fixed, to distinguish their Ideas with Accuracy and to dispose their Arguments in a clear, concise and convincing Manner.—After these Studies were finished, they were introduced to the Knowledge of *Trigonometry, Conic-sections* and *celestial Mechanics*. A System of natural and experimental *Philosophy*, comprehending *Mechanics, Statics, Hydrostatics, Optics, Pneumatics*, and *Astronomy*, was read to them; with References to the best Authors on these Subjects. ... Some other Articles were touched upon, especially *History, natural* and *civil*, as the Students proceeded in their Course, in order to enlarge their Understandings and give them venerable Ideas of the Works and Providence of GOD.— A distinct View of the *Anatomy* of the human Body was given them ... A large System of *Jewish Antiquities*, which their *Tutor* had drawn up, was read to them in the latter Years of their Course ... In this Branch of Science likewise, they were referred to the best Writers upon the Subject. *Lampe's Epitome of ecclesiastical History* was the Ground-work of a Series of Lectures upon that Subject; as was *Buddæi Compendium Historiæ Philosophicæ* of Lectures on the Doctrines of the ancient Philosophers in their various Sects.

But the chief Object of their Attention and Study, during three Years of their Course, was his *System of Divinity*, in the largest Extent of the Word; including what is most material in *Pneumatology* and *Ethics*. [1766, pp. 87–92]

Those who had entered on the Study of *Pneumatology* and *Ethics*, produced in their turns *Theses* on the several Subjects assigned them, which were mutually opposed and defended. Those who had finished *Ethics* delivered *Homilies*, (as they were called, to distinguish them from *Sermons*) on the natural and moral Perfections of GOD, and the several Branches of moral Virtue; while the *Senior-students* brought *Analyses* of Scripture, the *Schemes* of Sermons, and afterwards the Sermons themselves, which they submitted to the Examination and Correction of their *Tutor*. [ibid., p. 97]

If indeed Thomas did attend such a dissenting academy, then it is even more likely that he might have received some earlier general education, for Philip Doddridge refused in general to admit to his Academy boys who had not attended a grammar school and received a good knowledge of classics [Parker, 1914, p. 51], and I do not believe that Doddridge was alone in this requirement. Parker also notes that

> in 1695, the Independent or Congregational Fund Board was established (1) to assist poor ministers, (2) *to give young men who had already received a classical education*, the theological and other training preparatory to the Christian ministry.
> [1914, p. 54]

(emphasis added). Support for this Board was provided by the Presbyterians (£2,000 annually) and the Independents (£1,700). Among the distinguished men who were assisted by it were Samuel Wesley and Isaac Watts [Dale, 1907, p. 506]. Jeremy [1885] notes that contributions to the Fund were received from the Leather Lane congregation from 1690 to 1811.

Jeremy relates that the Presbyterian Fund originated in 1689.

> It was the joint enterprise of the leading Presbyterian and Independent Congregations of that day. It consisted of (1) Congregational collections, and (2) Individual subscriptions, amounting in the whole to about £2,000 per annum, which was dispensed as it arose. The Board of Managers consisted of (1) Elected representatives of the contributory congregations (always including their ministers), and (2) Individual contributories who subscribed not less than a specified sum per annum.
> [1885, p. ix]

Joshua Bayes (given by Jeremy as 'John') was a member of the board from 1712 to 1746, as were his successors, at the Leather Lane Chapel, Michael Pope (1746–1788), Edmund Butcher (1789–1797), and William Hughes (1798–1802). Hughes was the last minister of that congregation.

It is, however, certain that Thomas later studied at Edinburgh University. In the *Leges Bibliothecae Universitatis Edinensis. Names of Persons admitted to the Use of the Library*, detailing admissions for the year 1719 (Scottish rather than English style, presumably), one finds the following reference.

> Edinburgi Decimo-nono Februarij Admissi sunt hi duo Juvenes praes. D. Jacobo. Gregorio Math. P. Thomas Bayes. Anglus. John Horsley. Anglus.

This manuscript bears the signatures of those admitted: that of Thomas is markedly similar to that in the records of the Royal Society.

Thomas's name also appears in the Matriculation Album of the University under the following declaration.

Ego *A B Academiae Edinburgenae Discipulus*, quo me ad debita officia firmius astringam, sincere ac sancte promitto quod et *Syngrapha hâc meâ in per*petuum Testatum cupio, assistente Divina Gratia, mihi ante *Omnia cordi et curae* futurum verae Pietatis Studium atque Observantiam, Me etiam in Assuetis Academici curriculi Studijs Sedulum, & Legibus ac Disciplinae Academicae quamdiu in Curriculo illo permansero Obsequentem et Praeceptoribus Omnibus Morigerum memet praestiturum, Nec ullius Dissidii aut Tumultus Clam palamve vel Authorem vel Participem futurum et per reliquam vitam Academiam ipsam, Grato et benevolo Animo Prosecturum: Idque, omnibus officijs pro facultate mea & Occasione datâ Testarum.

Discipuli Domini Colini Drummond qui vigesimo-septimo die Februarij, MDCCXIX Subscripserunt.
Th. Bayes.

and further evidence may be found in the *List of Theologues in the College of Edinburgh since October 1711* (the date is obscure). It is stated there that Thomas entered both the College and the profession in 1720, having been recommended by 'Mr Bayes' (presumably his father Joshua). Other lists (perhaps of classes attended) give Thomas's name in the fifth section in 1720 and 1721. A further list, detailing the theological exercises to be delivered, mentions Thomas's name twice: on the 14th January 1721 he was to deliver the homily[15] on *Matthew* vii, vs 24–27, while on the 20th January 1722 he was to play the same part, the text now being *Matthew* xi, vs 29–30. Finally, his name appears in the list of theological students in the University from November 1709 onwards, as having been licensed but not ordained — i.e., a *probationer* in Church of Scotland terms. (Further references to Thomas in the records of Edinburgh University are given in Appendix 3.1; see also Dale [1990].)

How did Thomas, as well as Edmund Calamy (junior), John Horsley, Skinner Smith, Isaac Maddox, and Nathaniel Carpenter (a cousin-german of Thomas's on his mother's side), come to be students in Edinburgh[16]? As regards the cost of such study, it is possible that funds were made available for the purpose by the Congregational Fund Board (see the earlier quotation from Parker [1914]): indeed, the *Dictionary of National Biography*, Vol. XII, records that 'On an exhibition (1718–21) from the presbyterian fund, [Maddox] studied at Edinburgh University' — though one might note that the *List of Theologues*, while recording bursaries received by entering students, makes no mention of this award to Maddox.

Jeremy [1885, p. 12] provides a list of the 'Colleges and Academies at which, and the Tutors with whom, Students on the foundation of the Presbyterian Fund have been educated, together with Dates and the Number of such Students so far as can be ascertained', from which the following has been extracted.

    Foreign and Scotch Universities 1690–1754   59
    Rev. Richard Frankland, M.A., Yorkshire 1690–96   36
    Rev. John Eames, F.R.S., London   1719–25   3

English Nonconformists were of course excluded from the universities in their own country on doctrinal grounds. Jeremy [1885, p. 17] records that from 1690 to 1716 they tended to go to the universities of Utrecht, Leyden, and Halle, while from 1690 to 1754 they were to be found at Scottish universities, chiefly Glasgow and Edinburgh.

A further reason for the presence of English students in Edinburgh may perhaps be found in the attitude to dissent shown by William Carstares, principal of the College of James the Sixth[17] from 1703 to 1715.

William's father, the victim of Episcopalian persecution, spent several years in hiding separated from his family, and William himself, instrumental in the placing of William III on the throne, underwent excruciating torture at the hands of the executioner by the 'thumbkins' in Edinburgh. (After the Revolution the instrument was presented to Carstares by the Privy-council of Scotland[18].) Largely influenced, no doubt, by the treatment of the Nonconformists in Scotland, William busied himself in trying to stop the persecutions to which his predecessors had been exposed.

> Carstares first persuaded William III., who was wavering, to trust the Presbyterians of Scotland rather than the Episcopalians; and this produced the Revolution Settlement. And secondly, he succeeded in persuading the General Assembly to accept the Act of Union with England, which otherwise could not have been passed. [Grant, 1884, vol. II, p. 259]

(This Act of Union was described by John Arbuthnot in 1706 as 'a most precious Jewel indeed, and very well worth contending for'[19].)

Carstares's loyalty did not go unrewarded by William III or his successor Queen Anne: in 1708 the latter gave Carstares a bounty of £250 for the augmentation of the professors' salaries in the College. This was the year in which the Regents were turned into Professors, and such was Carstares's magnaminity that he shared the gift[20] between the professors of Humanity, Greek, Logic, Moral Philosophy, Natural Philosophy, Mathematics, and Hebrew, taking none of it for himself.

Incidentally, the royal altruism and magnaminity detailed above are reported somewhat differently by James Grant:

> William III. bestowed upon it [the College of Edinburgh] an annuity of £300 sterling, which cost him nothing, as it was paid

out of the bishops' rents in Scotland. Part of this was withdrawn by his successor Queen Anne, and thus a professor and fifteen students were lost to the university. [18–, vol. III, p. 26]

Alexander Grant relates that William III had intended to found twenty Bursaries in Theology, but that Anne, claiming that 'most of the Kirks now being supplied with learned and pious Ministers' [1884, vol. I, p. 232], decided to cut this number to five, the money thus saved being allotted as an endowment for 'a Professor of the Public Law and the Law of Nature and Nations' (loc. cit.).

Despite his political successes, and despite what Alexander Grant [1884, vol. II, pp. 259–260] has described as 'his power of influencing the minds of others', Carstares failed in one notable respect: namely, a scheme for bringing Nonconformists from England to Edinburgh for a university education. As part of his plan to form a Hall, affiliated to the College, for English students, he corresponded with a number of English dissenters, but to little avail. The failure of Carstares's plan is reported somewhat differently by M'Cormick:

> As his reputation had brought down many students from England, who complained of the want of proper accommodation in Edinburgh, he concerted a plan with his friends in that kingdom, which, if he had lived to carry it into execution, would probably have proved of great benefit to the college and city of Edinburgh. It was proposed, that a public contribution should be raised among the whole body of the dissenters in England, for the purpose of repairing the farbric [sic] of the college, so as to render it fit for accommodating all the English students who should resort thither. A public table was to be kept, at which they were to be entertained at a moderate expence. An English tutor, with proper assistants, was to be brought down, to have a particular inspection over the students, to preside at the common table, to assist them in their academical exercises, and to instruct them in such branches of education as were not taught in the university. By letters addressed to Mr Carstares from different parts of England, I find considerable sums were actually subscribed for these purposes some little time before his death [on the 28th December, 1715], which event overturned the whole project. [1774, pp. 70–71]

In 1709 Edmund Calamy took a trip to North Britain, attending the General Assembly in Edinburgh and making contact with a number of dissenters and University men. He found, though (according to the *Dictionary of National Biography*), that he 'relished the claret of his hosts more than

50   3. Thomas Bayes: a life

their ecclesiasticism' [vol. III] (Calamy does not say this in so many words in his *Life*, but he does comment frequently on the excellence of the French claret served in Scotland[21]). During this visit the principal of the College of James the Sixth made him an offer that he records as follows.

> Principal Carstaires, calling on me in the morning, told me, that at a meeting of the Masters of their college, (of which by the way I had not the least notice) it had been determined not to let me go from among them, without conferring a token of their respect, in an academical way. I told him, I was very thankful, (as I had good reason) for the many civilities already received, for which I was at a loss how to make them a suitable return. He said, they had agreed to present me with a Diploma, for a Doctorate, and begged my acceptance of it. My reply was, that if they would make me a Master of Arts, I should not at all demur, upon accepting it; but as for anything farther, I begged their excuse, and desired it might be waved, and that, for this reason, among others; that it would look like affectation, and a piece of singularity, for me to take the title of Doctor, when so many of my superiors went without it.
> [1830, vol. II, pp. 186–187]

Carstares[22] replied that he had heard that other Scottish universities intended to honour Calamy in a similar way and that Edinburgh wished to be first. Moreover, they would also send diplomas for the Doctorate to Daniel Williams and Joshua Oldfield (both dissenting English ministers of note). Calamy, who had already been honoured by the City on this trip by being made a free Burgess and Guild Brother of Edinburgh[23], therefore accepted Carstares's offer, and the degree of Doctor of Divinity[24] (*Doctoralis in S.S. Theologiâ*) was conferred on him on the 2nd of May (*VI nonas Maii*) in 1709. King's College, Aberdeen and Glasgow followed with similar distinctions on the 9th and the 17th (*16to Cal. Junias*) of May, respectively[25], with Aberdeen in fact conferring both magistral and doctoral degrees[26]. This is related somewhat differently by Alexander Grant, who says that the award of the D.D. by Edinburgh so irritated the members of Glasgow University that they refused to recognize it, and even opined that Edinburgh had no power to confer degrees. (This attitude on the part of the Glaswegians mirrored that shown in the mid-sixteenth century, when the City of Edinburgh and its Ministry wanted to found a university. Grant, citing Craufurd's *Memoirs*[27], notes that 'the three Universities of St. Andrews, Glasgow, and Old Aberdeen, by the power of the Bishops, still bearing some sway in the Kirk, and more in the State, did let their enterprize.' [1884, vol. I, p. 104].)

3.2. Thomas Bayes's education

It is perhaps doubtful to what extent Calamy supported Carstares's scheme for the drawing of English students to Scotland (though a number of English theologues were in fact recommended by Calamy to the college). During his visit to Scotland, and rehearsing the words of his Sovereign, he remarked on the number of theological students, writing in his *Life*

> They have too many small bursaries in their Colleges, which are temptations to the inhabitants to breed up for the ministry more than they are able to support and provide for, when they have gone through the course of their education.
> [1830, vol. II, p. 217]

The Queen and Calamy were in their turn but echoing the sentiment expressed in the second half of the seventeenth century by John Oldham, who, in his *A Satire. Addressed to a friend that is about to leave the university, and come abroad in the world*, wrote

> If you for orders and a gown design,
> Consider only this, dear friend of mine,
> The church is grown so overstocked of late,
> That if you walk abroad, you'll hardly meet
> More porters now than parsons in the street.
> At every corner they are forced to ply
> For jobs of hawkering divinity;
> And half the number of the sacred herd
> Are fain to stroll and wander unpreferred.

These trencher-chaplains, or *Mess-Johns*, as they were also called[28], were but indifferently treated by their wealthy patrons, often being invited to dinner merely to say grace before meat and to ask a blessing afterwards[29], being further expected to take a wife from the lowly female servants of the household[30]. As Burton has it in his *Anatomy of Melancholy* of 1621

> If he be a trencher Chaplain in a Gentleman's house, ... after some seven years' service, he may perchance have a Living to the halves, or some small Rectory with the mother of the maids at length, a poor kinswoman, or a crackt chambermaid, to have and to hold during the time of his life.
> [Part. I. Sect. II. Mem. III. Subs. XV.]

Isaac Bickerstaff's Letter 255 of the 25th November 1710 in *The Tatler* is devoted to this matter, and is an answer to the following 'letter' written to the *Censor of Great Britain*.

> I am at present under very great difficulties, which it is not in the power of any one, besides yourself, to redress. Whether or

no you shall think it a proper case to come before your court of honour, I cannot tell; but thus it is: I am chaplain to an honourable family, very regular at the hours of devotion, and, I hope, of an unblameable life; but for not offering to rise at the second course, I found my patron and his lady, very sullen and out of humour, though at first I did not know the reason of it. At length, when I happened to help myself to a jelly, the lady of the house, otherwise a devout woman, told me, that it did not become a man of my cloth to delight in such frivolous food: but as I still continued to sit out the last course, I was yesterday informed by the butler, that his lordship had no further occasion for my service.

In his reply Bickerstaff says 'I would fain ask these stiff-necked patrons, whether they would not take it ill of a chaplain, that in his grace after meat should return thanks for the whole entertainment with an exception to the dessert?'[31].

As we have already stated, Carstares corresponded with several English Nonconformists in an attempt to gain support for his scheme. Among these was Christopher Taylor, with whom, as we have seen, Joshua Bayes worked; and it was quite possibly as a result of this correspondence that Thomas Bayes (together with Skinner Smith, John Horsley, and Isaac Maddox) enrolled as students in the College. Indeed, in the College's *Library Accounts 1697–1765* we find these names listed under the heading 'supervenientes', that is,

> such as entered after the first year, either coming from other universities, or found upon examination qualified for being admitted at an advanced period of the course.
> [Dalzel, 1862, vol. II, p. 184]

Edmund Calamy (junior) might well have been in the same position, for in his father's *Life* we read

> Towards the latter end of this year [i.e., 1714], my eldest son, who had been trained up in grammar learning, at the school at Westminster, went to Edinburgh, to lay the first foundation of academical learning. [1830, vol. II, p. 307]

Indeed, one of the aims of the father's trip to Scotland in 1709 was to find out something about the universities there

> because my eldest son, bred at the Grammar School in Westminster[32], was in a little time to be sent to one place or other, in order to academical education [sic].
> [Calamy, 1830, vol. II, p. 145]

Even though the name of James Gregory appears on Thomas's admission certificate, there is unfortunately no record, at least in those documents currently accessible, of any mathematical studies. Thomas does however appear to have pursued logic (under Colin Drummond) and theology. There is also some slight evidence for a study of Greek, for the Rev. Richard Onely[33] A.M., late of Christ Church College, Cambridge, and Rector (or Vicar) of Speldhurst[34] and Ashurst [MacKinnon, 1902, p. 10–11], is reputed to have said that Thomas was the best Greek scholar he had ever met [Timpson, 1859][35].

The rules to which students were subject at the College of King James were strict: Dalzel notes that the seventh Article set down on the 10th of November 1668 said

> That the censors, in their respective classes, observe such as speak Scots, curse, swear, or have any obscene expressions, that the Regent may censure them according to the degree of their offence. [1862, vol. II, p. 198]

Some forty years later additional laws were introduced, the seventh (once again) running

> Students are obliged to discourse always in Latin ... Those who transgress, especially such as speak English within the college, are liable the first time in a penny; thereafter in twopence. [Dalzel, 1862, vol. II, p. 275]

The Latin requirement was something that Thomas might well have been used to, if our opinions on his earlier education are correct. For Parker [1914, pp. 78–79] states that students at John Jennings's Academy at Kibworth in Leicestershire, founded about 1715, were required to speak in Latin at specific times and in specific places, while the scriptures were read at family prayers from French, Hebrew, or Greek into English. Parker [op. cit. p. 92] also relates that most tutors of Jennings's time lectured in Latin. This practice later disappeared, as Gordon relates:

> Before his [i.e., Doddridge's] time, following the practice of the older Universities, all lectures were in Latin. Latin was the customary speech within the Academy walls; English being only permitted on stated occasions, *e.g.*, on Sunday evenings, when sermons were repeated. Doddridge changed all that ....
> [1902, p. 9]

In the sixteenth century the speaking of Latin in German schools was the norm. Parker records

54   3. Thomas Bayes: a life

> To gain colloquial readiness, all the boys speak Latin, even the obscure little Teutons in the dim regions of the lowest forms. The masters are forbidden to address them in German. The boys are severely chastised if they use their mother tongue. On the way to and from school, and in games, they are to speak only Latin, or Greek. A first fault may be pardoned, but contumacious use of the mother tongue is far too grave an offence[36]. [1867, p. 37]

The manuscript volume in the library of Edinburgh University that contains the list of theologues also contains a list of books. The range of topics covered seems too narrow for this to be a catalogue of the College Library, and it is possible that the works listed were for the particular use of the theologues. Most of the books (some in Latin, some in French, but most in English) are of a theological nature, but one finds in addition things like Wilson's *History of Great Britain*, Bacon's *Advancement of Learning*, Goodwin's *Roman Antiquities*, Low on Chirurgery, and the intriguing *Snake in the Grass*, a work we have identified as being written against the Quakers in 1696 by the Jacobite Charles Leslie (1650–1722)[37], the full title being *The Snake in the Grass; or Satan transformed into an Angel of Light*.

Only a few of these books seem at first sight to be mathematical in nature: they are

(i) *Alstedii Methodus Admirandorum Mathemat:* (quarto & octavo);

(ii) *Apollo mathematicus:* (octavo);

(iii) *Keckermanni systema mathem:* (octavo); and,

(iv) *Speedwells geometrical problems* (quarto).

The first is Johann Heinrich Alsted's *Methodus admirandorum mathematicorum; complectens novem libros Matheseos universae*, first published in 1613 with later editions in 1623 and 1641. The second, which, as it it turns out, is hardly mathematical, is Sir Edward Eizat's *Apollo mathematicus; or, The art of curing diseases by the mathematicks, according to the principles of Dr. Pitcairn*[38] of 1695. Although the *National Union Catalogue* describes this merely as a satire on works by Archibald Pitcairne, Stigler [1999, chap. 11], perhaps more charitably, finds in Eizat's sarcastic and witty (and, to my mind, distinctly humorous) tract a brilliant scientific attack. The third is Bartholomaeus Keckermann's *Systema compendiosum totius mathematices* of 1617, with a later edition in 1671. (Keckermann published a number of other 'Systema' during the early part of the seventeenth century: his *Systema logica* is also listed here.) The last work is probably John Speidell's *A geometricall extraction; or, A compendiovs collection of the chiefe and choyse problemes, collected out of the best, and latest writers*, first published in 1616, with a second edition appearing in 1657.

Those admitted to the use of the College Library were obliged to take the following oath[39].

> I $\mathcal{A}\,\mathcal{B}$. undersubscriband, Forasmeikle as I am Privileged to the accesse and use of the Bibliotheck of Edinburgh Colledge, at such times as are appointed, it be patent. Therefore I by this my present subscription bind & oblige me as I shall answer to the great GOD, that I shall neither steale, nor willingly blott, violat, cancell, or wrong any Book of the said Bibliotheck in whole or in part, and that I shall not move any Book out of the owne place without the consent and leave of the keeper. And that I shall according to my credite, place, & power be a favorer & friend of the Colledge of Edinburgh. So help me God. And farder I sweare as said is, that if it shall happen me rickeleslie to blot any Book, that I shall not conceale the same, but immediatelie show it to the keeper, that I may be censured for the skaith, aither in reponing a new Book, for it that is blotted, or otherwise to pay according to the discretion of him to whom the power belongs.
> And moreover I agree to these Lawes of the Bibilotheck Subsequent.

Here follow eleven *Leges Bibliothecae*, the translation of a sample of which follows[40].

1. No-one may enter the Library without the permission of the Librarian or leave without his knowing.
2. No-one may read unless admitted and duly sworn.
3. No-one may handle a book unless the Bibliothecary has handed it over.
4. No-one may carry off a book.
5. No-one may mark a book or dog-ear the pages.
7. No-one may read a book by a lamp, or bring it near a fire.
10. Let everyone read silently to himself and not interrupt the reading of others, and if it is necessary to speak to one's neighbour, let it be to whisper in his ear.

Some who entered the College at the same time as Thomas proceeded to the Master of Arts degree. In the *Library Accounts 1695–1746*, for instance, is an item, dated the 14th February 1723, detailing Isaac Maddox's application for this degree[41]. Further, on the same day,

John Horsley in the same circumstances (now a Preacher in London) had an Ample Diploma granted him for the degree of Master of Arts & the Person who appeared for him gave in a Guiney.

(The amount given in Maddox's case was only ten shillings sterling, which amount is also recorded as 6 — — . The conversion is done as follows: the Scottish *merk* was worth 13s. 4d. Scots or $13\frac{1}{3}d$. sterling. Thus ten shillings sterling was worth £6 Scots.)

A letter concerning the application for the degree is preserved in the Special Collections Department of the Library of Edinburgh University[42]. The text runs as follows.

> Nos Ingenŭi Adolescentes qŭi Nomina Sŭbsignamŭs Academiæ Edinbŭrgenæ Alŭmini, Agnoscentes nos ejŭsdem Academiæ Beneficio in Literis & honestis Disciplinis, ac præsertim in Pietate et Pŭrioris Religionis Professione Institutos & Edŭcatos esse, et Magisterij Titŭlo Donandos: Sancte coram Deo Cordium Scrutatore Spondemŭs nos in Pŭritate et veritate Religionis Christianæ ab omnibus Pontificiorum Erroribus Repŭrgatæ et in omni debitæ Gratitudinis officio erga eandem Academiam ad extremum vitæ halitum Perseveraturos. Nec-non Spondemŭs nos nunquam Commissuros ut Magisterij Titulo semel ornati, ita de Gradŭ Dejiciamŭr ŭt ad Baccalaŭreatŭm denuo Redeamus: In qŭarŭm fidem Chirographa nostra Apposŭimus Londini — *secundo* die *Martij* Anno Salŭtis Vicessimo tertio sŭprà Millesimŭm et Septingentesimŭm còram his Testibŭs.

> Gul. Hey A.M.   S. Ministerij
> Candidatus Testis   Isaac Maddox

> Thom. Bayes   Rev^{di} Josh. Bayes
> filius Testis   Johannes Horsley.

All that remains of the inscription on the other side of this letter is

>   For
> Maddox preacher of
> js[?] the care of Mrs Dunnells
> n green
> London

The *Dictionary of National Biography*[43] notes that, on 24th February,

> 'Johannes Horseley' and Isaac Maddox (*sic*), 'Angli præcones evangelici, academiæ olim alumni,' were 'nunc demum' admitted to the degree of M.A., the diploma being given on 9 March.

Neither John Horsley nor Madox (afterwards bishop of Worcester) seems to have held any dissenting pastorate.

The entry for Isaac Maddox in Volume XII of the same dictionary suggests that 'praeco' implies licensed but not ordained. The *List of Theologues*, however, gives an 'o' after Maddox's name, indicating that he was both licensed and ordained; Horsley was merely licensed.

The Master of Arts degree was perhaps less essential for those who intended to be Nonconformist ministers. Thomas himself did not take such a degree, a practice that was not uncommon at the time. Grant in fact notes notes that as 'after 1708 it was not the interest or concern of any Professor in the Arts Faculty ... to promote graduation ... the degree [of Master of Arts] rapidly fell into disregard' [1884, vol. I, p. 265].

There was, at this time, nothing in the University of Edinburgh worthy of being called a Theological Faculty. There had been a Professor of Theology (a title often assumed by the Principal) and a Professor of Divinity (charged with training graduates for the Ministry) since the early 1600s. The first Regius Professor of Ecclesiastical History was appointed in 1702, and thus things remained for almost a hundred and fifty years. (One might view the Professor of Hebrew as being in this Faculty.)

Worried by the lack of graduates the Senatus tried to make graduation in Arts a requirement for entry to the Ministry, resolving, in 1738,

> that the Professor of Divinity be enjoined that he shall receive no new Students of Divinity, nor consider them as scholars under his care, who cannot produce a certificate for having got the degree in Arts; and that such as are already listed students in Divinity shall have the degree gratis; and that the Rev. Professor of Divinity should advise such students to take the degree for a good example in this matter. [Grant, 1884, vol. I, p. 279]

This decree had little effect on the General Assembly of the Church of Scotland — this body perhaps recalling the words written by John Knox on his death-bed 'Above all things preserve the Kirk from the bondage of the Universities'.

Having completed his studies in Edinburgh, Thomas presumably returned south. His whereabouts for the next decade or so are uncertain, but thereafter we find him established as the pastor of the Presbyterian meeting-house in Tunbridge Wells, a town of which we now give some description[44].

3.3 Thomas Bayes and Tunbridge Wells

Situated some thirty-five miles from the Metropolis by road, and forty-six by rail, Tunbridge Wells (Motto: *Do well, doubt not*) nestles on and between two hills: Mount Ephraim and Mount Sion (see Figure 3.1). The spring whose presence is the reason for the town's existence, was discovered by the twenty-five-year-old Dudley, Lord North, in 1606. The chalybeate waters welled up in an area almost exactly in the meeting point of the Manors of Rusthall and South Frith, of the parishes of Speldhurst, Tonbridge, and Frant, and of the counties of Kent and Sussex. Lord North attributed the regaining of his health to the waters, and the fame of the wells soon spread, particularly after the visit by Queen Henrietta Maria, the wife of Charles I, in 1630 — a visit so noteworthy that the springs were called 'Queene Maries Wells' by some. Farthing [1990] records that two houses for the drinking of coffee and the smoking of pipes, with the necessary offices, were erected in 1636. Lodowick Rowzee, a medical doctor, writes in 1670 of this latter habit with qualified approval, saying

> Divers do take Tobacco after their Water, which I do not dislike, especially if they hold it a good While in their Mouths, before they puff it out. [1746, p. 327]

Others were perhaps less sympathetic to the combination of coffee and tobacco. Sir Henry Blunt wrote to Justice Walter Rumsey in the mid-seventeenth century

> Coffee and Tobacco have not the advantage of any pleasing taste wherewith to tempt and debauch our palat, as Wine and other such pernicious things have; for at first Tobacco is most horrid, and Cophie insipid. [Robinson, 1893, p. 61]

The town slowly grew, with a decline during the Civil War, until the visit of Charles II and his Court in 1663. Other royal visits followed (the future James II and his daughters Mary and Anne took the waters), and less royal personages (such as Moll Davis, Nell Gwyn, and Peg Hughes) were also to be seen there — though even these women had royal connexions!

One of the reasons for the visit by Queen Catherine, the wife of Charles II, was the desire for an heir[45]. Visits to watering-places with such ends in view were common: in his play *Epsom-Wells*[46] Thomas Shadwell relates the following exchange.

> *Cuff.* Others come hither to procure Conception.
> *Kick.* Ay Pox, that's not from the Waters, but something else that shall be nameless.

In the case of the Queen, however, the visits proved fruitless.

3.3. Thomas Bayes and Tunbridge Wells

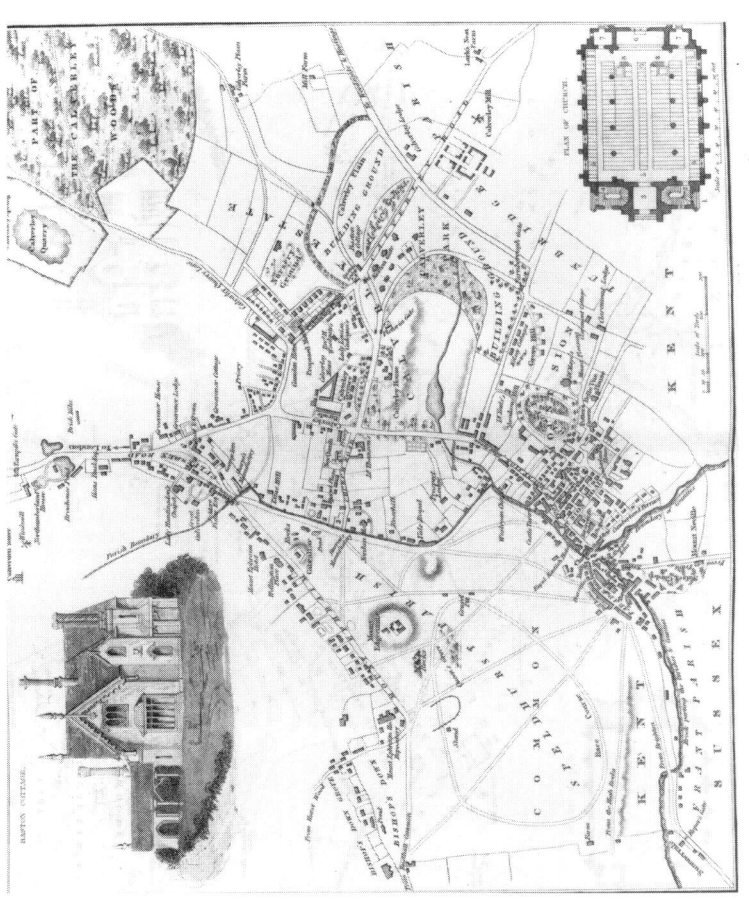

FIGURE 3.1 Plan of Tunbridge Wells.

The idea that the waters were useful in combatting infertility was even espoused by Rowzee, who wrote

> Yet must I not forget, in the Behalf of Women, to tell them, that there is Nothing better against Barrenness, and to make them fruitful, if other good and fitting Means, such as the several Causes shall requite, be joined with the Water. [1746, p. 326]

Even Richard Onely wrote that 'As they [the waters] are deobstruent and bracing, they are a very good remedy for sterility'. [1771, p. 7].

The eighteenth century was the time for satires and lampoons, many of which would have caused the vapours among the genteel Victorians, and some of which seem somewhat robust even in our more relaxed era. John Wilmot, Earl of Rochester, was a great writer of such verses, and in his satire *Tunbridge Wells*, first published in 1675, he imagines two women talking as follows.

> I pray, good Madam, if it may be thought,
> Noe rudenesse, what cause was it hither brought
> Your Ladyshipp? She soone replying, smild,
> Wee have a good Estate, but have noe Child;
> And I'm informed these Wells, will make a Barren
> Woman, as fruitfull as a Coney-Warren.

In 1698 the young Duke of Gloucester fell while playing soldiers on the Walks (see Figure 3.2); this prompted his mother, Princess (later Queen) Anne, to donate £100 for the paving of the Walks with pantiles. She was so incensed to discover, on her next visit, that no paving had yet been laid, that she never visited Tunbridge Wells again. (Calamy records in his *Life* [1830, vol. I, p. 420], that Gloucester died of a fever (like Sweet Molly Malone) on the 29th of July, 1699, aged eleven.) The effect of visits by Royalty to such spas should not be underestimated: Gaspey notes

> What George IV. effected for Brighton, by domiciliating himself and his suite there, and by making it one of the celebrities of the empire, Queen Anne accomplished for Tunbridge Wells, where she was a frequent resident anterior to her ascending the throne, and which she distinguished by many marks of favour.
> [1863, p. 9]

A contemporary description of the town was given by Anthony Hamilton in his memoirs of the Count Grammont:

> Tunbridge[47] is the same distance from London, that Fontainebleau is from Paris, and is, at the season, the general rendezvous of all the gay and handsome of both sexes. The company, though always numerous, is always select: since those who repair thither

for diversion, ever exceed the number of those who go thither for health, every thing there breathes mirth and pleasure: constraint is banished, familiarity is established upon the first acquaintance, and joy and pleasure are the sole sovereigns of the place.

The company are accommodated with lodgings, in little, clean, and convenient habitations, that lie straggling and separated from each other, a mile and a half, all around the Wells, where the company meet in the morning: this place consists of a long walk, shaded by spreading trees, under which they walk while they are drinking the waters: on one side of this walk is a long row of shops, plentifully stocked with all manner of toys, lace, gloves, stockings, and where there is raffling, as at Paris, in the Foire de Saint Germain: on the other side of the walk is the market: and, as it is the custom here for every person to buy their own provisions, care is taken that nothing offensive appears on the stalls. Here young, fair, fresh-coloured country girls, with clean linen, small straw hats, and neat shoes and stockings, sell game, vegetables, flowers, and fruit: here one may live as well as one pleases: here is, likewise, deep play, and no want of amorous intrigues. As soon as the evening comes, every one quits his little palace to assemble on the bowling-green, where, in the open air, those who choose dance upon a turf more soft and smooth than the finest carpet in the world. [Hamilton, 1859, pp. 267–268]

Not all accounts of this period were as flattering. In his satire *Tunbridge Wells* the Earl of Rochester wrote[48]

> At Five this Morne, when Phœbus raisd his head,
> From Thetis Lapp, I rais'd my self from Bed;
> And mounting Steed, I trotted to the Waters,
> The Rendevouz of Fooles, Buffoones, and Praters,
> Cuckolds, Whores, Cittizens, their Wives and Daughters.
> My Squeamish Stomach, I with Wine had brib'd,
> To undertake the Dose, it was prescrib'd.
> But turning head, a sudden cursed view,
> That innocent provision overthrew,
> And without drinking made me Purge, and Spew.

(He then went on to describe some of the visitors to the Wells.)

The rigours to which later visitors to the Wells were subjected were described in Thomas Wilson's diaries as follows.

> Wednesday 14th July 1736. Set out for Tunbridge [from London]. God grant we may return home safe. In Brown's Coach

FIGURE 3.2 The Walks at Tunbridge Wells.

and 4. Breakfast at Lewisham with Mr Lewis over against the
Rookery — dyed soon after of the Palsy — To Bromley. Farnborough. Dined at the Crown at Sevenoke. Pestered with Barbers, Butchers etc. from the Wells for our custom. Which is
called *touting* from a French word *tout* — one and all. From
Sevenoke to the Wells but an indifferent road. At Sevenoke
there are just beyond the town fine stone Buildings — Hospitals. Came to the Wells about 7 a clock. Plagued there again
with touters. Agreed for Lodging at £1. 5. 0. per week at one
Knights at the foot of Mount Sion. [Linnell, 1964, p. 160]

The accommodation at the Wells was not uniformly satisfactory: on Thursday the 22nd of July Wilson recorded

More company come in daily and the weather is very inviting.
All the inconvenience is that in a dry season it is prodigious
dusty and the houses are so thin and badly fitted up that they
are not at all made for wind or rainy weather. But a month ago
a sudden flood the Houses below the Hills were overflowed near
4 foot in the lower rooms. [Linnell, 1964, pp. 162–163]

A hundred years later the state of the Wells, particularly in the rainy season, was still a matter of concern. Lee writes

Nature having done so much in favour of the place, the inhabitants seem disposed to trust too exclusively to its scenic advantages, but little having been done towards improving the part
near the springs, which still remains in a very primitive condition. In wet weather there are no other means of recreation than
newspaper or novel reading; no public hall, or point of reunion
for the evenings when the days shorten; and, consequently, but
little association among visitors. Hence many who, on arriving,
are greatly pleased with the aspect of the place, after a few
days' sojourn become *ennuyés*, and hasten their departure.
[1856, p. 38]

The primitive nature of the area had been reflected a few years earlier by Britton, who wrote

In topographical language, it [i.e., Tunbridge Wells] must rank
as a hamlet, or a series of hamlets, connected with, and subordinate to, the mother parishes of Tunbridge, Speldhurst, and
Frant. [1832, p. 1]

It would seem that the hills on which the town was established only received their present names after steps had been taken to satisfy the spiritual (or spirituous) needs of the visitors and the inhabitants. The origin of the names *Mount Ephraim* and *Mount Sion* is given as follows by Gaspey.

> The place is reputed to owe to the Presbyterians and the Baptists, and also to an imaginary resemblance of its site to that of Jerusalem, the scriptural names by which two of its most commanding localities [i.e., Mounts Sion and Ephraim] are known. [1863, p. 7]

A less pious reason for Mount Sion's being thus named is given by Burr, who writes

> One of the first houses built upon this hill was an alehouse, to which the whimsical landlord, whose name was Mathews, affixed the sign of Mount-Sion, from which the name of the hill is derived. [1766, p. 103]

A similar report, though differing in some features, is given by Gaspey:

> *Mount Sion*, the other guardian hill of the Wells, faces Mount Ephraim and is entered from High street. Tradition ascribes the name to a Mr. Jordan, who erecting a house there, shewed his personal interest by calling it Mount Sion. Sheltered by trees from easterly winds, the lodging-houses find favour with invalids and visitors. There is a story (described by Britton as 'ludicrous and irreverend' [1832, p. 52]), told of a man called Okill (or O'Kill), who combined the avocation of a lodging-house keeper on this hill, with that of parish-clerk. Having an eye more to gain than to godliness, it was his practice when his apartments were vacant to intimate the same by giving out the Psalm[49] commencing:—
>
> "Mount Sion is a pleasant place."

[1863, p. 22]

The story about Okill is perhaps more accurately recorded in Sprange [1797, pp. 307–308], where, Gaspey's 'commencing' being eschewed, it is noted merely that this 'singular character' would, when his small lodging-house was empty, give out the psalm that *says* 'Mount Sion is a pleasant place'. (Britton [1832, p. 52] speaks of 'the following stanza from the Psalm ...'.)

The question of the identity of the original owner is resolved by Farthing [1990, No. 43], who notes that the first two landlords of the *Grove Tavern* were in fact Jordan and Mathews; and although differing as to the surname, Sprange [1797, p. 106] and Gaspey agree in describing the landlord as 'whimsical'.

With all the visitors to Tunbridge Wells lodging had to be provided, if both the visitors and the inhabitants were to be satisfied. In 1680 the first lodging-houses were erected. Speaking of such accommodation in general, Burr writes

> The lodging-houses are mostly situated on the hills before mentioned, and particularly on Mount-Sion, where there are a great many good houses, built in regular confusion, and so beautifully intermixed with trees and groves, that they cannot fail of having a most pleasing effect on a stranger. At a little distance, it bears the appearance of a town in the midst of woods, and conveys to the imagination the soothing idea of a rural romantic retirement, while it actually affords all the conveniences of a city life. [1766, pp. 102–103]

To give an idea of the accommodation to be found at these lodging-houses the following descriptions from Sprange's *List of Lodging-houses, their Situation, Names of Owners, and principal Rooms in each* are given.

> On the Side of Mount Ephraim, London Road. Mr Jeffery's, *at the gate* (with garden and other ground). 2 Parlours, 4 Chambers, Stables for 6. [1780, p. 6]

> On the top of Mount-Sion, near the Bowling-green and Great Grove. Rich. Jeffery, *late S. Jeffery* 5 Parlours, 4 Chambers, Stables for 12. [1780, p. 6]

> Those houses that have stabling belonging have likewise coach-houses in proportion; and every house contains garrets, kitchens, and under offices in proportion to the other rooms. [1780, p. 6]

> Mrs Jeffery, at the Gate, Jordan's Lane. (with Garden and other ground belonging. Every House contains Kitchens, and other Offices in proportion to the other Rooms). 2 Sitting Rooms, 4 Chambers, 4 Garrets, Stabling for 6, 1 Coach House.
> [1797, p. 317]

As we show in Chapter 12, Thomas Bayes demised

> to Sarah Jeffrey daughter of John Jeffrey living with her father at the corner of Jourdains Lane at or near Tonbridge Wells five Hundred pounds and my watch made by Ellicot and all my linnen and wearing apparell and Household Stuff.

Why this bequest should have been made is uncertain; but Roger Farthing[50] has made the reasonable suggestion (apparently at the prompting of Jean Mauldon, late of the Tunbridge Wells Library) that, as a bachelor, Thomas might well have lived at one of the lodging-houses. Jourdan's Lane was the old name for Church Road, and Farthing has identified the house mentioned in Thomas's will as now being 69 London Road. This house, which is known to have been a lodging-house (see the last quotation from Sprange given above), was probably built by John Jeffery (Sarah's grandfather) in the late seventeenth century. We have more to say on this matter in Chapter 12.

Incidentally, on his visit to Tunbridge Wells on Tuesday the 3rd of August 1736 Thomas Wilson wrote in his diary ' Jeffrey's very good lodgings' [Linnell, 1964, p. 166].

With the increase in the number of visitors and the growth of the town, accommodation had also to be provided for the meeting of the needs of the soul. In 1678 the requirements of the Established Church were met by the completion of the Chapel of King Charles the Martyr, built by public subscription[51]; and Nonconformist establishments followed soon thereafter. Thomson writes

> The history of Nonconformity in Tunbridge Wells is so curious as to deserve a passing notice. During the last century, congregations of each of the three denominations of Protestant dissenters existed in that place; and singularly enough every one of these congregations became extinct. [1871, p. 82]

Early Presbyterian meetings were held in secret. A church was apparently formed in 1688, with David Stott as minister from 1691 to 1700. He was followed by Humphrey Ditton[52], who was pastor from 1700 to his resignation in 1710.

Sunday observance was a matter of some importance to the Presbyterians (both of the English and of the Scottish varieties). Whitaker records

> The English of all sects, but particularly the Presbyterians, make profession of being very strict observers of the Sabbath Day. ... I have observed it particularly in the printed confession of persons that are hanged, ... Sabbath breaking is the crime the poor wretches begin with.
> Thomas Wright Hill, born c.1763, gave a description of a Presbyterian Sunday as follows: breakfast at 8, with extempore prayer by his father. Chapel at 9 until about 11. The service consisted of extempore prayer, the congregation standing; reading & explanation of a passage of scripture; a hymn or a psalm; a long prayer, sermon; hymns, prayer and blessing.
> [1940, pp. 32–33]

In further support of the view expressed in the first paragraph of this quotation, we find John Rippon writing, in his *Proposals for printing by subscription* ... of 1803, 'The four greatest vices of the day, Swearing, Sabbath Breaking, Intemperance, and Impurity' (to which he later added Popery).

3.3. Thomas Bayes and Tunbridge Wells

After the Restoration Puritanism sank into disrepute in England in general[53]. Even after the passing of the Act of Toleration in 1689 — an Act that 'exempted from attendance at church all persons taking an oath of allegiance to the new sovereigns [William and Mary] and making a declaration against Popery' [Hastings, 1971, vol. 3, p. 650] — the Presbyterians still seemed to be received with little favour. Sprange records

> The Presbyterians wanted to build them a conventicle[54] at this place [Rusthall ... about one mile from the Wells], when it was in the most flourishing state; but, though it was chiefly supported by that sect, the landlord refused to sell them a foot of land for that purpose, even at the most extravagant price, so inveterate was the hatred that Churchmen then bore to Presbyterians. [1817, p. 23]

Indeed, in an attack on a pamphlet written by Bishop Burnet in 1713, Jonathan Swift expressed the opinion that freethinkers and dissenters were of more danger to both Church and State than Roman Catholics (see Blanchard [1955, p. 416])[55].

One should bear in mind that the sentiment responsible for the passing of acts such as the Act of Uniformity or the Five Mile Act did not vanish with the passing of the Act of Toleration. In the reign of Queen Anne the Schism Bill was introduced into Parliament, a major point in which, at least as far as we are concerned, was the following.

> By virtue of this Act, Nonconformists teaching school were to be imprisoned three months. Each schoolmaster was to receive the sacrament, and take the oaths. If afterwards present at a Conventicle, he was incapacitated and to be imprisoned. He must teach only the Church Catechism. But offenders conforming, were recapacitated; and schools for reading, writing, and mathematics, after a warm debate, were excepted.
> [Calamy, 1830, vol. II, p. 283]

Not only were schoolmasters keeping private or public schools affected, but also 'every person instructing or teaching any youth in any house or private family as a tutor or schoolmaster' [Browning, 1953, p. 409]. Things were particularly difficult for Nonconformists, as Gordon relates:

> they [the authorities in Church and State] employed every effort in their power to debar Nonconformists from the exercise of the teaching profession, and to break up their schools. A bishop's licence was required in the case of every Teacher; the Teacher not so licensed was prosecuted, and subjected on conviction to fine and imprisonment. [1902, p. 4]

As always in such cases there were of course exceptions:

> Provided always, that this Act or anything therein contained shall not extend or be construed to extend to any tutor teaching or instructing youth in any college or hall within either of the universities of that part of Great Britain called England, nor to any tutor who shall be employed by any nobleman or noblewoman to teach his or her own children, grandchildren or great-grandchildren only in his or her family, provided such tutor so teaching in any nobleman or noblewoman's family do in every respect qualify himself according to this Act except only in that of taking a licence from the bishop.
> [Browning, 1953, p. 410]

The Section of the Act that exempted those teaching at schools that offered certain subjects ran as follows.

> Provided always, that this Act shall not extend or be construed to extend to any person who as a tutor or schoolmaster shall instruct youth in reading, writing, arithmetic or any part of mathematical learning only, so far as such mathematical learning relates to navigation or any mechanical art only, and so as such reading, writing, arithmetic or mathematical learning shall be taught in the English tongue only. [Browning, 1953, p. 410]

So distressed were the Dissenters by this Bill, that they proposed to wait upon Queen Anne, once it had passed both Houses, to ask her to reject it. Lord Sunderland told Calamy and some other Dissenters that the Queen had been behind this Bill from the start, and that any attempt to persuade her to veto it would be ridiculous. As chance would have it, however, the Queen died before approving the Bill, 'on the very day that the Schism Act was to have taken place' [Calamy, vol. II, p. 293]. The 'schemes of the enemies of the Hanover succession' [op. cit. pp. 293–294] were foiled, and the Dissenters were free to carry on as they had before.

Let us now return to Tunbridge Wells and its Presbyterians. The absence of a suitable meeting-house seems to have resulted in Ditton's having to preach under less than ideal conditions:

> Meetings were held in the ballroom of Mount Ephraim House[56] which was licensed for their use. On Sundays a temporary pulpit was affixed to the wainscot by iron hooks, which allowed it to be easily removed. During the rest of the week the room was used for dancing, card playing and other secular activities.
> [Bailey & Bailey, 1970, p. 43]

Ditton in turn was followed by John Archer, during whose ministry a plot of ground was obtained on Mount Sion and a chapel built. It is related

3.3. Thomas Bayes and Tunbridge Wells

(for example, by Bailey & Bailey) that the land for the chapel was bought by a Mr Jordan of Jordan Cottage, Jordan Lane, from the owner who was ignorant of the purpose to which the land would be put. It is sometimes said (see Bailey & Bailey once again) that a Quaker named Thomas Seal was the seller, but Farthing [1990, No. 42] suggests that the land, which contained a cottage and garden, was owned by the plumber Philip Seale.

It is generally thought that Thomas Bayes succeeded Archer on the latter's death, and although this is in some measure true, Bailey & Bailey report the succession somewhat differently:

> John Archer[57] died in 1733 but was succeeded by Thomas Bayes F.R.S. in 1730, who was a man of considerable literary attainment. It appears that Thomas Bayes left Tunbridge Wells in 1728 and returned in 1731. During this time he was at Leather Lane Presbyterian Church, London, where his father had been pastor. [p. 44]

(One should not be misled by the use of the pluperfect tense in the last sentence here into thinking that Joshua had retired from the Leather Lane Chapel by this time.)

This remark about Thomas's having been in Tunbridge Wells at an earlier stage than that generally received, although not mentioned by most sources, may well be true. However, it would appear that he had retained some connexion with London: for in the *Minute Books of the Body of Protestant Dissenting Ministers of the Three Denominations in and about the cities of London and Westminster* we find

> Oct. 3rd 1732. List of approved Ministers of the Presbyterian denomination.
>
> Mr. Bayes Senr ⎫
> ⎬ Leather Lane
> Mr. Bayes Junr ⎭

It would also appear that Thomas had not been permanently installed in Tunbridge Wells before 1728, for the *Minute Books* also record that on the 7th November 1727 John Evans presented a List of approved ministers of the Presbyterian denomination to the Committee of the Three Denominations, in which list we find 'Unfixed: Mr. Thomas Bayes, at Mrs. Deacle's'. James [1867, p. 669] notes that Thomas Baies [sic] was appointed to the Leather Lane Chapel in 1728.

The 'New Chapel' in Tunbridge Wells is generally reported to have been opened on the 1st August 1720, John Archer being the minister (see Figures 3.3 to 3.5). Timpson relates that, after the chapel's being opened, 'various ministers supplied the pulpit' [1859, p. 464], and it is possible that Thomas was one of these preachers, being called to the full-time charge on Archer's resignation or death.

Roger Farthing, to whom we are indebted for his painstaking research on Tunbridge Wells, is not completely convinced that the chapel was opened on the day mentioned above. He suggests that that date is quoted as the opening because of Archer's published sermon, but finds the published details to be 'very circumstantial'. He notes that the chapel is not to be found on a 1718 engraving, but is on a 1738 map.

By the early 1730s, then, Thomas had been elected minister of the Presbyterian Chapel, and in Tunbridge Wells he was to stay until his death in 1761. He was followed in the pastorate by William Johnson, after whose death the chapel had no permanent minister for some years, and was indeed only open during the visiting season, some five months of the year. Onely in fact notes that

> *Tunbridge Wells* season, as it is called, generally begins the latter end of *May*, or beginning of *June*, and lasts till *Michaelmas* or after; and is at its height about the middle of *August*.
> [1771, p. 6]

In connexion with the best time to visit the Wells, Rowzee writes[58]

> Concerning the Season of the Year, Summer is the fittest, when there is a settled warm and dry Weather, as in the dog-days especially.
>
> *Cum Canis arentes findet biulcus agros.*
>
> And the chiefest Months are *June, July, August,* and *September*; although the *Dutch*, who naturally love good Beer and Wine better than Water, use to have this rhyming Verse in their Months:
>
> *Mensibus in quibus R. non debes bibere* Water.
>
> [1746, p. 326]

Messrs Skinner, Hampson, and Gough followed Johnson as pastors, but Presbyterianism in Tunbridge Wells declined and the chapel was eventually almost deserted[59].

By the early years of the nineteenth century decay had in fact set in among the Nonconformist communities in Tunbridge Wells in general. Amsinck[60] writes

> Whiston, in his Memoirs (A.D. 1748) testifies to the respectability of the anabaptist congregation[61], under their minister Mr. Copper[62]. This was continued, in a dwindling condition, till it

3.3. Thomas Bayes and Tunbridge Wells

FIGURE 3.3 Mount Sion Chapel.

FIGURE 3.4 The Mount Sion Chapel in decay.

FIGURE 3.5 The Mount Sion Chapel today.

actually expired, with their late venerable and truly Christian pastor Mr. Joseph Haines. Those, who remember the Rev. Mr. Johnson, will attest to the respectability of the presbyterian congregation under his ministry. The independents have ceased about twenty years; and their meeting-house is converted into the lodging, now called SYDENHAM HOUSE. ... The result of a few years is curious. Like Pharoah's lean kine, the lean schismatics, thus nobly protected, have actually devoured each of the other dissenting establishments. The independents are no more; not a vestige of them remaining; the deserted meeting-house of the baptists merely proclaims the spot, where formerly their congregations assembled; whilst that of the presbyterians, is unblushingly given up, by a trust formerly deemed respectable, to another separation of the methodists. [1810, p. 25]

Phippen relates the closure of the Presbyterian chapel as follows[63].

Mount Sion chapel was eventually closed in the year 1814, and thus it remained year after year in a decaying and dilapidated condition, exhibiting a melancholy proof of the desertion of this once crowded sanctuary, and the decay of presbyterian vigour and piety. [1840, p. 98]

But the decline among the Presbyterians was followed by an upsurge, albeit brief, among the Independents. The latter (also known as Congregationalists) took over the Mount Sion Chapel in 1830, by which time only the walls and roof remained[64]. The meeting-house was entirely renovated, Phippen describing the new establishment as follows.

A gallery was erected, the lower part was fitted up with pews — having had before (with the exception of a table pew in front of the pulpit) only forms with backs attached to them. The old vestry was taken down; and a school-room, capable of containing seventy children, was built. The expenses incurred by the alterations and repairs amounted to upwards of £700. The chapel is a plain substantial building, nearly square — 40 feet by 34. It is distinguished by no architectural attractions, but its interior is comfortable and commodious. It is capable of seating 450 persons; allowing 400 for the chapel itself, and 50 for the adjoining school-room, which is separated from the chapel by sliding shutters. Of these sittings 170 are free. During the week the school-room is used for a Female School, on the plan of the British and Foreign School Society; the average attendance at which is 60 girls. [1840, pp. 98, 100]

74   3. Thomas Bayes: a life

(Farthing [1990] notes that, in this context, a 'British' school was one intended for Nonconformists, whereas a 'National' one was for Anglicans. The school attached to the chapel continued until 1875.)

On the vacating of the building by the Independents in 1848, the Primitive Methodists moved in in 1854 for three years, the Friends taking over in 1887. In 1875 the chapel became a secular edifice, as it still is today. (A copy of one of the sale notices is provided as Figure 3.6.)

The history of the chapel was succintly summarized by Colbran as follows.

> Sion Chapel.— This Chapel is situated on Little Mount Sion. It was built by voluntary subscriptions, and opened for divine service on 1st of August, 1720, for the use of the presbyterians. It was closed in the year 1814, and re-opened on the 8th of July, 1830, by the Independents. It is now used as a place of worship by the Wesleyan Reformers. [c.1852, pp. 34–35]

The state of Presbyterianism in Tunbridge Wells was summed up by Thomson as follows.

> Their [i.e., the Presbyterians'] meeting-house which was built in 1720, still remains, and after having been used by the Independents until their new place of worship was erected, has for some time been devoted to the instruction of the young, in connection with the British School Society. A somewhat distinguished series of ministers occupied the pulpit of the Presbyterian meeting. Humphrey Ditton, an eminent mathematician, and the author of a celebrated argumentative treatise on the Resurrection, was the pastor from about 1700 to about 1710, when delicate health, occasioned by devoted application to his duties, led him to resign his pastorate, and give himself to learned pursuits. He was succeeded by John Archer, who published some masterly sermons, and who died in 1733, as appears from a laudatory epitaph in Speldhurst Church. ... The next minister was Thomas Bayes, whose pastorate lasted from 1731 to 1752, and who was a Fellow of the Royal Society, and a man of very great acquirements in scholarship and in science. William Johnston, a Master of Arts, and an instructor of youth, was Presbyterian minister from 1752 to 1776; his epitaph also was to be read in Speldhurst Church. The following ministers were men less known. It is a testimony to the respect entertained for the Presbyterian clergyman of those days, that a subscription-book for his benefit lay on the table of the libraries, beside that for the maintenance of his Episcopalian neighbour. ... It seems that the introduction of Methodism into Tunbridge Wells, in

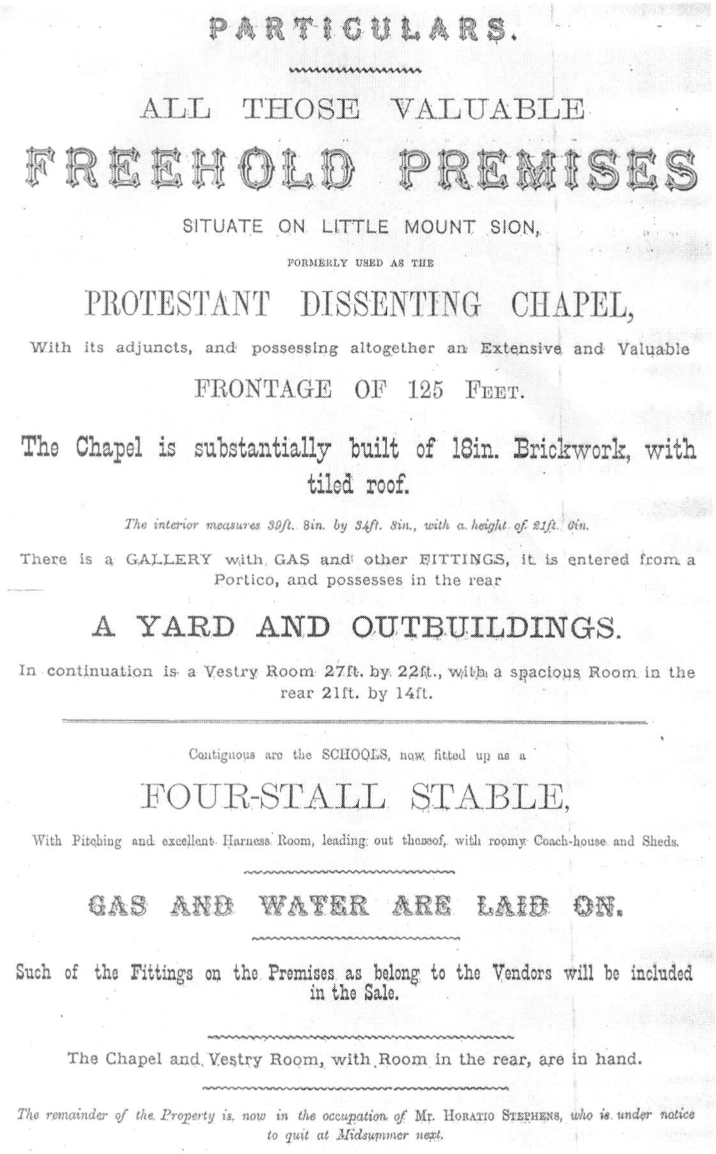

PARTICULARS.

ALL THOSE VALUABLE
FREEHOLD PREMISES

SITUATE ON LITTLE MOUNT SION,

FORMERLY USED AS THE

PROTESTANT DISSENTING CHAPEL,

With its adjuncts, and possessing altogether an Extensive and Valuable

FRONTAGE OF 125 FEET.

The Chapel is substantially built of 18in. Brickwork, with tiled roof.

The interior measures 39ft. 8in. by 34ft. 8in., with a height of 21ft. 0in.

There is a GALLERY with GAS and other FITTINGS, it is entered from a Portico, and possesses in the rear

A YARD AND OUTBUILDINGS.

In continuation is a Vestry Room 27ft. by 22ft., with a spacious Room in the rear 21ft. by 14ft.

Contiguous are the SCHOOLS, now fitted up as a

FOUR-STALL STABLE,

With Pitching and excellent Harness Room, leading out thereof, with roomy Coach-house and Sheds.

GAS AND WATER ARE LAID ON.

Such of the Fittings on the Premises as belong to the Vendors will be included in the Sale.

The Chapel and Vestry Room, with Room in the rear, are in hand.

The remainder of the Property is now in the occupation of Mr. HORATIO STEPHENS, who is under notice to quit at Midsummer next.

FIGURE 3.6 Sale Notice.

its two forms — Calvinistic Methodism under the auspices of Lady Huntingdon, and Arminian Methodism by the followers of John Wesley — had some effect upon the decline and extinction of the older dissenting congregations. The warmer and more emotional preaching, thus introduced, competed successfully with the intellectual, scholarly, and quiet teaching of the regular Nonconformists. [1871, pp. 83–84]

Tunbridge Wells, like other watering places, was frequented not only for its supposed help in promoting conception (as we have already mentioned in connexion with the visit of Queen Catherine), but also for its curative and restorative powers. Britton relates 'I do not know any place which is more free from local diseases than Tunbridge Wells' [1832, p. 75], and Burr, somewhat more expansively, writes

> On the little hills of Mount-Ephraim and Mount-Sion, it is remarkable that a gentle fragrant breeze unceasingly prevails, through all the summer months; which, in the hottest weather, generally keeps them mild and temperate. And it is acknowledged by every author who has occasionally mentioned the place, as well as by those who have professedly wrote on the subject, that this air is extremely benign, pure, and wholesome. [1766, p. 68]

There were, however, those on whose health the waters seemed to have little beneficial effect. One such was the novelist Samuel Richardson, who wrote to one Dr Young in a letter dated January 1758,

> I have often been at Bath; but remember not that I received benefit from the waters. The late worthy Dr. Hartley once whispered to me that I must not expect any.
> [Barbauld, 1804, vol. II, p. 46]

We shall find in Bayes's Notebook a prescription for rheumatism. Of this complaint, as evinced in Tunbridge Wells, Britton writes

> Chronic rheumatism has been said to be frequent here. I cannot say that I have observed such frequency; neither have I often seen cases of acute rheumatism. These diseases do, however, attack people, as in other places, during cold moist weather. [1832, p. 75]

Although Tunbridge Wells owed its nascence to the revivifying powers of its waters (its vivifying powers being less certain), its closeness to London led to its being seen as a holiday resort. It should not be thought, however, that the only pleasures to be found there in the early days were religion and taking the waters (the combination of which is perhaps typified by the tale of the Pool of Bethesda, though total immersion and divine intervention seemed unnecessary in the case of the Wells).

After a visit to the well, and the obligatory taking of the waters, in a state of undress, the company returned to their various lodging-houses to dress (Rowzee records that 'the Morning, when the Sun is an Hour, more or less, high, is the fittest Time to drink the Water' [1746, p. 327]). At 10 a.m. some went to the coffee-house and some to church, the company appearing together on the Walks after prayers. Conversation, promenading, gaming (or gambling), and the consumption of tea followed until 2 o'clock. The exercise taken after the drinking of the waters was not to be too strenuous, however: Rowzee writes

> Moderate Exercise after it is very available, but I utterly dislike it, if it be too violent, as Running, Leaping, and Jumping, as some in Wantonness use to do. [1746, p. 327]

The afternoons were devoted to excursions or games of bowls, with balls (at which minuets and country dances were executed) or cards in the evenings. With the thought of Augustus Carp's activities[65] in mind, one might be tempted to ask (if the appropriate verb may be coined) *Did Bayes Bourée?*. On his visit to the Wells in 1748 Samuel Richardson wrote to Miss Susanna Highmore as follows.

> Tunbridge, in high season, a place devoted to amusement. — Time entirely at command, though not hanging heavy; impossible indeed it should. — Vehicles, whether four-wheeled or four-legged, at will; riding, a choice. [Barbauld, 1804, vol. II, p. 204]

Indeed, the actual conduct was most accurately mirrored by Dickens in his description of Bath in *The Pickwick Papers*.

Whether a man like Richardson should ever actually have visited Tunbridge Wells is doubted by Barton, who, referring to the novelist, writes

> A man with a soul above scandal should never have gone to Tunbridge Wells — where everyone in sight suggested some anecdote of a quarrel, indiscretion or eccentricity, and where the coffee houses and taverns reeked with tittle-tattle from morning till night. [1937, p. 242]

On the other hand, in his *Tunbridge Wells Guide* of 1786 Sprange wrote

> The Circulating Library, and the Coffee House, ... are places where the social virtues reign triumphant over prejudice and prepossession.... Here divines and philosphers, deists and christians, wigs and tories, Scotch and English, debate without anger,

dispute with politeness, and judge with candour; while every one has an opportunity to display the excellency of his taste, the depth of his erudition, and the greatness of his capacity, in all kinds of polite literature, and in every branch of human knowledge. ... The bookseller's shop has indeed an advantage over the coffee-house, because there the ladies are admitted. [pp. 103, 104]

Sometimes, however, the frivolous activities just mentioned were overtaken by more sober pursuits, as Knipe relates:

About this year [1668] a pleasure-garden called "Fishponds" was opened on low ground behind Manor Lodge ... The place, it is reported, was for some years carried on with "decency and strict decorum"; but after a time it became the scene of such excessive impropriety that it was found necessary to close it. [In later years it was re-opened as a tea-garden[66], but did not meet with much success.] [1916, p. 8]

The 'excessive impropriety' is also remarked on by Wilson, who writes of 'the mad, bawdy riots at Epsom and Tunbridge' [1967, p. 199][67].

When it came to watering-places, the *genius loci* was the Master of Ceremonies. In Bayes's time it was Richard Nash, a regular annual visitor — though on each occasion for only a short time — from 1735 to 1761, arriving in time to supervise the preparations for the season and leaving after the first ball-night[68]. At this time certain principles were in train[69]: firstly, any person could talk to any other person in a public place, and secondly, everyone should in general do the same thing at the same time in a public place. The general procedure was described by Gaspey as follows.

In 1735, Beau Nash, the most famed of "Ladies' Men," appeared upon the scene, and, as Master of the Ceremonies, founded a code of laws for the government of the place, as irrevocable as those of the Medes and the Persians. Here, as at Bath, he assumed the title of "King," and the chief decree enforced by this despot of the *beau monde* was that every visitant should live in public. "The lodging houses were merely places of accommodation for eating and sleeping. The whole of the intermediate time of their temporary inhabitants was spent on the walks, in the assembly-rooms, in pleasurable excursions, or at chapel. Thus every hour of the day had its allotted occupation, the whole was regularly digested into system; and from the nobleman of the first rank to the meanest visitor, all were compelled to obey, and to yield to the established customs." [quoted from Amsinck] Nash realised no inconsiderable means by the regulated fees which he claimed from the gaming-tables, which, notwithstanding the source of revenue they yielded, he restrained within moderate bounds. [1863, pp. 10–11]

Rules for the company were recommended by the master of ceremonies: some of these, as listed by Sprange, were as follows.

> I. That there be two balls every week during the season, on Tuesdays at the Upper Rooms; on Fridays at the Lower Rooms; each to begin at seven, and end at eleven ...
> IV. That there be a card-assembly every Monday, Wednesday, Thursday, and Saturday, at each of the rooms alternately.
> V. That on Sunday Evenings, the Upper Rooms be opened for public tea-drinking: — Admission for that evening, 1s. each, tea included.
> VIII. It is humbly requested of all persons to subscribe to the rooms, to enable the renters of them to defray the many necessary and heavy expenses attending them.
> No hazard, or any unlawful game to be allowed in the public rooms, nor cards on Sunday Evening. [1797, pp. 120–122]

The 1786 edition of Sprange's *Guide* contained the rule that the dance evenings were 'to begin with Minuets, and then Country Dances' [p. 98].

Nash and his cronies did not spend all their time in riotous amusement, however. Their participation in the daily religious exercises has been recorded by Barton as follows.

> Punctually at ten o'clock every morning, Beau Nash and his silk-coated company of water drinkers could be seen mustering at the little red brick chapel near the spring. ... Contemptuous of any simulation of piety as savouring of Dissent — the whining in the conventicle on Mount Sion was really most illbred — these fashionable worshippers looked about them during the service with cheerful irreverence, made audible remarks to their neighbours, picked their teeth, combed their wigs, and exchanged nods, smiles and winks with their friends in the opposite pews. [1937, p. 213]

With the demise of Beau Nash[70] and the reigns of the seven Masters of Ceremonies following him, the enthusiasm with which the routine was conducted at the Wells declined, with Brighton, under the influence of the Prince of Wales, growing in popularity. The town[71] nevertheless grew.

Not everyone was equally enamoured of spas: on Saturday the 30th of September, 1826, William Cobbett, in his *Rural Ride from Ryall, in Worcestershire, to Burghclere, in Hampshire*, writing of 'one of the devouring WENS; namely CHELTENHAM', described a watering place[72] as

a place to which East India plunderers, West India floggers, English tax-gorgers, together with gluttons, drunkards, and debauchees of all descriptions, female as well as male, resort, at the suggestion of silently laughing quacks, in the hope of getting rid of the bodily consequences of their manifold sins and iniquities. ... To places like this come all that is knavish and all that is foolish and all that is base; gamesters, pickpockets, and harlots; young wife-hunters in search of rich and ugly and old women, and young husband-hunters in search of rich and wrinkled or half-rotten men, the formerly resolutely bent, be the means what they may, to give the latter heirs to their lands and tenements. These things are notorious.
[1912, vol. II, pp. 126–127]

A similar description of some of those drawn to Tunbridge Wells had been given by Isaac Bickerstaff in *The Tatler*, No. 47, in 1709:

It happened that, when he [i.e., Sir Taffety Trippet] first set up for a fortune-hunter, he chose Tunbridge for the scene of action, where were at that time two sisters upon the same design. The knight believed of course the elder must be the better prize; and consequently makes all his sail that way. People that want sense do always in an egregious manner want modesty, which made our hero triumph in making his amour as public as was possible. The adored lady was no less vain of his public addresses. ... the night before the nuptials, so universally approved, the younger sister, envious of the good fortune even of her sister, who had been present at most of their interviews, and had equal taste for the charms of a fop, ... unable to see so rich a prize pass by her, discovered to sir Taffety, that a coquet air, much tongue, and three suits, was all the portion of his mistress. His love vanished that moment, himself and equipage the next morning.

The copy of *Tunbrigalia, or, the Tunbridge Miscellany* for 1740 in The British Library has pasted in it an extract from a letter from Tunbridge Wells written on the 30th of July, 1773, from which the following is taken.

This place is extremely full, but the company consists of an odd olio of old maids, lively widows, polluted batchelors, Jews, parsons, and some few nobility, who visit the Wells to wash away sin and iniquity, or rather to cleanse their leprous habits of body, in consequence of irregularity, incontinence, and the excess of every sensual appetite; for unless they are invited here for these salutary purposes, it is past my judgment to conceive for what they are wheeled down here.

3.3. Thomas Bayes and Tunbridge Wells

A description of the company given by John Arbuthnot in a letter to Lady Suffolk in 1731 (i.e., at about the time of the start of Bayes's ministry) ran

> The company consists chiefly of *bon-vivants* with decayed stomachs, green-sickness virgins, unfrutiful or miscarrying wives. [Melville, 1912, p. 173]

A general description of the way in which things were done in Tunbridge Wells was provided by Onely:

> People of the greatest title, rank, and dignity; people of every learned profession, and of every religious and political persuasion; people of every degree, condition, and occupation of life, (if well dressed, and well behaved) meet amicably here together: some for the benefit of the water, and air; some for a little relaxation from study and business, and others for the pleasure of social and polished life. The morning is passed in an undress; in drinking the waters, in private or publick breakfastings, which are sometimes given by one of the company, in attending prayers at the chapel, in social converse on the parade, at the coffee-house, in the public rooms, or bookseller's shop; in raffling for, cheapening, and buying goods, at the milleners, turners, and other shops; billiards, cotillion dances, private concerts, cards, or some other adventitious and extraordinary curiosity and novelty; a painter, a musician, a juggler, a fire-eater, or a philosopher, &c. After dinner, all go dressed to the parade again, and the rooms, to tea, in private parties, or in public. — At night to the ball or assembly, and sometimes to a play. The ball nights are, *Tuesdays* and *Fridays*; and assemblies and cards every other night, except *Sundays*. The outward amusements are, cricketting, horse-races, and other diversions; such as walking, riding, and airing in carriages on the contiguous heaths and commons, or excursions to some of the most remarkable places in the environs of the Wells.
> [1771, pp. 10–11]

Incidentally, MacKinnon seems to have disapproved slightly of Onely's attachment to Tunbridge Wells, writing

> In contrast to his predecessor [James Kearsley], we find him a frequentor and admirer of "Life at the Wells." He writes in 1771 of the season there: "The appearance of the Company when assembled together is quite beautiful and noble; in the day walking along the Parade like a walking parterre, and at night, in the Rooms like a galaxy of stars in a bright nocturnal sky." Did Speldhurst often see him, one wonders? [1930, p. 95]

There were a number of distinguished visitors to Tunbridge Wells during Thomas's time there — Thomas Wilson, who visited in 1736, recorded, among the People of Quality, the presence of the Duke and Duchess of Norfolk, the Duke and Duchess of Richmond, a number of Marquesses, and lesser Lords and Ladies (including Lord Stanhope and his sister Lady Lucy), members of parliament, one Mr Debouverie — 'A Gentleman of Vast Estate' — and a Mrs Titchbourne, 'Dresser to the Queen'[73]. The grandeur of the occasion must almost have rivalled that seen by the Jackdaw at the Court of the Cardinal Lord Archbishop of Rheims!

The Prime Minister himself, Robert Walpole, was there to visit his sick mistress, Wilson recording this event as follows.

> Friday 23, This evening it is confidently reported that Sir R.W. came here to see Miss Skerrett ... They say he has a daughter by this lady 15 years old which he keeps a coach and 6 for and lives in great splendour in St James Square.
> Saturday 24. Miss Brett, daughter of Col. Brett by Macclesfield, came to the Wells. She is supposed Mistress of the late King. ... She behaves here very discreetly. People avoid her company when the same people are proud of being seen with Miss Skerrett. The Miss of the reigning Minister is to be feared when that of a Dead Lyon may safely be neglected. [Linnell, 1964, p. 163]

Among other visitors who contributed to the feasts of reason and flows of soul were Dr Robert Smith[74], Master of Trinity College, Cambridge, in 1745, and David Hume in 1756. G. Gregory Smith, in his Notes to Letter 154 in *The Spectator*, records that Tunbridge Wells and Bath were regarded as the high points in the scale of modishness.

The salubrity of the atmosphere also drew visiting clergymen, their services being offered to the congregations of the Wells.

> Of these, the most regular in his visits every season was William Whiston, but his discourses, in which astronomy and theology were bewilderingly blended, did little to bring his erring flock to a more devout frame of mind. [Barton, 1937, p. 216]

Thomas Wilson was more disparaging of Whiston's character[75], noting

> Oct. 1737. Sat. 8. The Queen once told the Bishop [of Durham] that Whiston was a very impertinent man though she had never talked to him about her sentiments in Religion. If I say so Whiston is a very false man for he has often asserted the contrary in publick. His Imprudence indeed may not be excuseable.
> [Linnell, 1964, pp. 213–214]

Barton has also commented on the frankness shown by Whiston in his dealings not only with the common man but also with the Queen:

Yet irreverence was one of the many human failings that sorely tried his [i.e., Whiston's] temper, and drew from him the frank rebukes for which he was famed. From Queen Caroline to the very frequenters of coffee houses who fled at his approach, no one was safe from the bitterness of his tongue. [1937, p. 216]

Elwig, however, was a little more appreciative of Whiston's visits, writing

> William Whiston, Scientist and Translator of Josephus, is in Tunbridge Wells in 1748. He interests the company by shewing them eclipses, lectures on the phenomenon of the stars and gives them his views on the millenium. Formerly of the Anglican Communion he now holds Anabaptist views and speaks of the respectability of the minister, a Mr Copper, of Mount Ephraim. [1941, p. 8]

Samuel Richardson also remarked on one of Whiston's visits, writing in a letter to Miss Sarah Westcomb,[76]

> Another extraordinary old man we have had here, but of a very different turn — the noted Mr. Whiston, showing eclipses, and explaining other phænomena of the stars, and preaching the millenium, and anabaptism (for he is now, it seems, of that persuasion) to gay people, who, if they have white teeth, hear him with open mouths, though perhaps shut hearts; and after his lecture is over, not a bit the wiser, run from him.

It might perhaps be worth noting that, in addition to an interest in the subjects above mentioned, Whiston not only co-operated with Humphrey Ditton on a scheme for the determination of longitude[77], but also presented, in his *New Theory of the Earth* of 1696, the view that, before the Deluge, the year was composed of 360 days, a view that was used as corroboration for some of the opinions expressed in Velikovsky's *Worlds in Collision*.

On one occasion Whiston visited Thomas Bayes, recording their meeting as follows.

> *Memorandum.* That on *August* the 24th this Year 1746, being *Lord's* Day, and St. *Bartholomew's* Day, I breakfasted at Mr. *Bay's*, a dissenting Minister at *Tunbridge Wells*, and a successor, tho' not immediate to Mr. *Humphrey Ditton*, and like him a very good Mathematician also[78]. I told him that I had just then come to a resolution to go out always from the public worship of the Church of *England*, whenever the Reader of Common Prayer read the *Athanasian* Creed; which I esteemed a publick cursing the Christians: As I expected it might be read at the Chapel that very Day, it being one of the 13 Days in the Year,

when the Rubrick appoints it to be read. Accordingly I told him that I had fully resolved to go out of the Chapel that very Day, if the Minister of the Place began to read it. He told me, that Dr. Dowding the Minister[79], who was then a perfect Stranger to me, had omitted it on a Christmas-Day, and so he imagined he did not use to read it. This proved to be true, so I had no Opportunity afforded me then to shew my Detestation of that Monstrous Creed; Yet have I since put in Practice that Resolution, and did so the first Time at Lincolns Inn Chapel on St. Simon and St. Jude's Day October 28, 1746, when Mr. Rawlins began to read it, and I then went out and came in again when it was over, as I always resolved to do afterwards.

Why should Whiston have visited Bayes — or rather, perhaps, why should the former have found the breakfast worth recording in his *Memoirs*? The reasons are perhaps threefold: firstly, both were clergymen with common scientific interests. Secondly, Humphrey Ditton, one of Thomas's predecessors, was a friend of and scientific collaborator with Whiston. And thirdly, William and Thomas may well have been acquainted for a number of years before this visit. For after his removal for Arianism from the Lucasian chair at Cambridge in 1710, Whiston, aged forty-three, moved to London, and established his home in Cross Street, very near the Bayes family home in Little Kirby Street[80] (see Figure 2.2). Here he established a Society for Promoting Primitive Christianity, whose weekly meetings drew Christians of all persuasions. Among those attending was Benjamin Hoadly, on whose writings we later find Thomas commenting. It is, I am sure, possible that Thomas, although no doubt too young at that time to be admitted to these meetings, might well have been acquainted with Whiston.

Taylor in fact notes that Whiston's having to move from Cambridge was not a complete disaster:

> Both Whiston (who had been expelled from Cambridge) and Stirling were fortunate alike in the fact that they came to London when the fashion for 'philosophy' was at its height. The official religious and political intolerance of the times did not prevent them from finding patrons and employment in the capital city. [1966, p. 7]

It has also been suggested (see Force [1985, p. 20]), and this perhaps sheds some light on an earlier quotation taken from Wilson's Diaries, that Queen Caroline might sometimes have attended these meetings incognita. This is hinted at in Whiston's *Memoirs*, where Whiston records that in 1736 he was told that:

> the late duke of *Somerset*, a great *Athanasian*, once forbad his chaplain to read the *Athanasian* creed, (which I imagined was

occasioned by a suggestion from the queen; to whom I had complained, that altho' she was queen, that creed was not yet laid aside). [I, p. 299]

Force relates that

> The queen enjoyed Whiston's plainspokenness and accepted his reproofs in good humour. She gave him a stipend of forty pounds "clear" a year, which was continued by George II after her death. [1985, p. 20]

If these royal visits to Whiston's meetings actually took place, then, bearing in mind our earlier remarks on the Leather Lane neighbourhood, we might say that the Three Estates of the Realm met the men of three letters in this area.

Whiston appears in the famous print recording *The remarkable characters who were at Tunbridge Wells with Richardson in 1748 from a drawing in his possession with references in his own writing*, reprinted here as Figure 3.7. Here Whiston is pictured on the extreme right, walking away from the crowd. Perhaps this was fortuitous, or perhaps the artist intended to depict Whiston's independence and intellectual opposition to the work and thought of many of his contemporaries. It has also been suggested (see Rothstein & Weinbrot [1976], as reported in Force [1985, p. 167]), that Whiston may be seen as the model for Goldsmith's Vicar of Wakefield.

Other characters depicted in this painting by the fan-painter Thomas Loggan (or, sometimes, Logan), originally dwarf to the Prince and Princess of Wales, are one Dr Johnson and his wife, Colley Cibber, Mr Garrick, Mr Pitt (Earl of Chatham), and The Baron (a German gamester). The presence of a gamester — or, less charitably, a gambler — is not surprising: Farthing [1990] records that gambling was uncontrolled until 1738 — although Nash insisted that it be restricted to the public rooms — yet in that year Parliament (following a precedent certainly extant in Justinian's time — see Arbuthnot [1770, vol. II, p. 258]) forbade the playing of games such as basset, pharoah, and hazard[81]. There then arose a new game, reputed to have been invented in Tunbridge Wells, called 'E & O', a kind of roulette based on even and odd numbers.

It is not clear whether the 'Dr Johnson' of this print was the great lexicographer. In the fifth edition of his *The Life of Samuel Johnson, LL.D.* of 1807, Malone[82] records that 'For the sake of relaxation from his literary labours, and probably also for Mrs. Johnson's health, he this summer [i.e., 1748] visited Tunbridge Wells' [1807, p. 43]. There seems, however, to be no mention of such a visit in the *Life* itself[83]; further, in Hill's edition of *Boswell's Life of Johnson* we read that 'There can be no doubt that the figure marked by Richardson as Dr. Johnson is not our Samuel Johnson, who

86   3. Thomas Bayes: a life

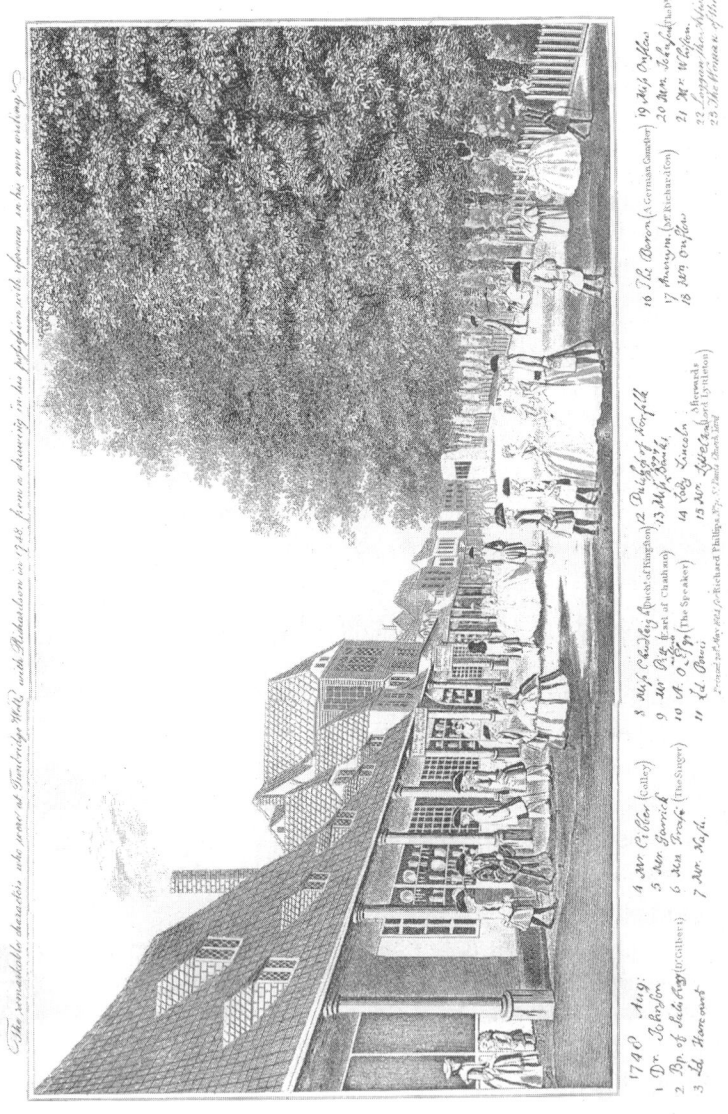

FIGURE 3.7 Characters at Tunbridge Wells.

did not receive a doctor's degree till more than four years after Richardson's death.' [1934, vol. I, p. 190]. *Ursa major* in fact received an honorary LL.D. (*Doctoratûs in utroque Jure*) from Trinity College, Dublin, in 1765 and an honorary D.C.L. (*Doctorem in Jure Civili*) from the University of Oxford ten years later[84].

Margaret Barton has suggested in her *Tunbridge Wells* of 1937 that the person in question was Dr James Johnson, later Bishop of Rochester, the 'Mrs Johnson' being his unmarried sister, who in her later years became one of the first to reside on Mount Ephraim.

One further candidate, though perhaps not one who can be supported as strongly as Barton's, is the Dr Johnson who was a medical man in Tunbridge Wells at this time. In Thomas Wilson's diaries, for the year 1750, there appears the following entry.

> My wife's expenses in my absence besides the £20 I left with her.
> To Dr Johnson ... 4. 4. 0
> Mr Dowding ... 1. 1. 0
> Sept. 15. Dr Johnson ... 1. 1. 0
> Sept. 15. Sent for Dr. J. [Wilson's wife being ill] who found her in a pretty high fever and full of stitches. Gave her Nervous Drafts with Senna Treacle. [Linnell, 1964, pp. 239, 250]

Among the distinguished visitors to Tunbridge Wells during Thomas's time there were some people from the East, whose visit was recorded by Phippen as follows.

> During the life of Mr. Bayes, an occurrence took place which is worthy of record. Three natives of the East Indies, persons of rank and distinction, came to England for the purpose of obtaining instruction in English literature. Amongst other places, they visited Tunbridge Wells, and were introduced to Mr. Bayes, who felt great pleasure in furnishing them with much useful and valuable information[85]. In the course of his instructions, he endeavoured to explain to them the severity of our winters, the falls of snow, and the intensity of the frosts, which they did not appear to comprehend. To illustrate in part what he had stated, Mr. Bayes procured a piece of ice from an ice-house, and shewed them into what a solid mass water could be condensed by the frost — adding that such was the intense cold of some winters, that carriages might pass over ponds and even rivers of water thus frozen, without danger. To substantiate his assertion, he melted a piece of the ice by the fire, proving that it was only water congealed. "No," said the eldest of them, "It is the work

of Art! — we cannot believe it to be any thing else, but we will write it down, and name it when we get home." [1840, p. 97]

(Phippen also notes that 'Mr. Bayes was a man of considerable literary attainments' [p. 96], though whether this is advanced in support of the second sentence in the above quotation or merely as a statement is uncertain.) Now Fraser writes of the visit to London by Persian Princes in 1835 and 1836,

> The author conceived that some account of the residence in England of the first Persian — in truth, the first *Asiatic* Princes, who ever visited this country — was due to the British public whose guests these Princes were. [1838, p. 5]

It is thus unclear to me where Bayes's visitors came from, though I suppose it is possible that Fraser erred in regarding the princes of whom he wrote as the first visitors of their kind. Perhaps Thomas's visitors, although 'persons of rank and distinction', were not princes.

The anecdote of visitors to Europe from lands in which the turning of water into ice was probably viewed with as much scepticism as that of the turning of water into wine, is one that has been recorded by a number of philosophers. For instance Hume, in his essay *Of Miracles, Part I*, writes

> The Indian[86] prince, who refused to believe the first relations concerning the effects of frost, reasoned justly.
> [Hume, 1894, p. 556]

It would appear, though, that this was not one of the visitors to Tunbridge Wells, for in a footnote Hume writes

> The inhabitants of Sumatra have always seen water fluid in their own climate, and the freezing of their rivers ought to be deemed a prodigy: But they never saw water in Muscovy during the winter; and therefore they cannot reasonably be positive what would there be the consequence. [Hume, loc. cit.]

Similar tales were related by visitors from Europe to more sultry climes: for example[87], John Locke, in his *Essay concerning Human Understanding*, writes

> a Dutch ambassador, who, entertaining the king of Siam with the particularities of Holland, which he was inquisitive after, amongst other things, told him, "that the water in his country would sometimes in cold weather be so hard that men walked upon it, and that it would bear an elephant if he were there." To which the king replied, "Hitherto I have believed the strange things you have told me, because I look upon you as a sober, fair man: but now I am sure you lie." [189-?, p. 557]

At least Bayes had the wit to supply precept by example! (the 'root and fruit', in Edgeworth's felicitous phrase[88]), and as Gordon wrote,

> Precept without Example avails little in making Proselytes. [1751, p. 143]

Perhaps Thomas had recalled the old Scottish paroemia[89],

> An Ounce of Mother Wit is worth a Pound of Clergy.

That Thomas should have had time to engage in pursuits other than those directly related to his calling is not surprising. In an editorial introduction to the *Account of the Life and Writings of the Rev. Dr. William Derham* we read

> The life of a country clergyman is in every respect more favourable to the cultivation of natural science, by experiment and observation, than any other professional employment. He has all the leisure that is requisite to philosophic researches; he can watch the success of his experiments from day to day, and institute long processes without interruption, or record his observations without chasm or discontinuation; and if he associates, as did our excellent Author, and all philosophers ought to do, the knowledge of Nature and the discovery of her laws, with the perfections and attributes of that wonderful *Being* who framed those laws and constantly superintends their operation and effect, he deviates not in any respect from his professional duties, but is, on the contrary, a most useful labourer in the vineyard of his heavenly Master. [Derham, 1798, pp. xxiv–xxv]

We have already noted Whiston's views of Thomas's mathematical abilities: Wilson in fact wrote

> Mr. Bayes was a man of considerable learning and judgement; of excellent natural abilities; and a good mathematician. His knowledge in the latter science is respectfully spoken of by Mr. Whiston, and procured his being elected a Fellow of the Royal Society. [1814, vol. 4, p. 402]

In the Restoration period and during the eighteenth century the word 'virtuosi' signified not only true men of science, but also dilettanti and 'mere' collectors of rarities[90]. *The Record of the Royal Society of London* records that

> The term [virtuosi], at first employed in rather a complimentary sense, came before very long, on account of the vagaries of these indiscriminate collectors of 'rareties,' to acquire a more or less contemptuous meaning. [1940, p. 42]

By many, all Fellows were tarred with the same brush: Shadwell ridiculed them in his comedy *The Virtuoso* of 1676, and similar disparaging remarks were made by Samuel Butler (*The Elephant in the moon*), Swift (*Voyage to Laputa*), and Pope (*The Dunciad*). Isaac Bickerstaff wrote in *The Tatler*,

> It is indeed wonderful to consider, that there should be a sort of learned men, who are wholly employed in gathering together the refuse of nature ... we must take this for a general rule, 'That whatever appears trivial or obscene in the common notions of the world, looks grave and philosophical in the eye of a virtuoso.' [Letter No. 216, 26th August 1710]

Recorded in this same Letter is the will of a virtuoso, here abridged:

> I, Nicholas Gimcrack, being in sound health of mind, but in great weakness of body, do by this my last will and testament bestow my worldly goods and chattels in manner following:
>
> *Imprimis*, To my dear wife,
>   One box of butterflies,
>   One drawer of shells,
>   A female skeleton,
>   A dried cockatrice.
>
> *Item*, To my daughter Elizabeth,
>   My receipt for preserving dead caterpillars,
>   As also my preparations of winter May-dew, and embryo-pickle.
>
> *Item*, To my learned and worthy friend doctor Johannes Elscrickius, professor in anatomy, and my associate in the studies of nature, as an eternal monument of my affection and friendship for him, I bequeath
>   My rat's testicles, and
>   Whale's pizzle,
> to him and his issue male; and in default of such issue in the said doctor Elscrickius, then to return to my executor and his heirs for ever.
>
> My eldest son John, having spoken disrespectfully of his little sister, whom I keep by me in spirits of wine, and in many other instances behaved himself undutifully towards me, I do disinherit, and wholly cut off from any part of this my personal estate, by giving him a single cockle-shell.
>
> To my second son Charles I give and bequeath all my flowers, plants, minerals, mosses, shells, pebbles, fossils, beetles, butterflies, caterpillars, grasshoppers, and vermin, not above specified; as also all my monsters, both wet and dry; making the

said Charles whole and sole executor of this my last will and testament: he paying, or causing to be paid, the aforesaid legacies within the space of six months after my decease. And I do hereby revoke all other wills whatsoever by me formerly made.

Matters, however, improved somewhat in the Royal Society during the eighteenth century.

Rousseau [1978, p. 155] records that in 1700 only 15 per cent of the members of the Royal Society were professional scientists, and it is possible that, even at the time of Thomas's election, many were virtuosi; the wording of his election certificate, however, suggests that he was not of this number. The presence of such people in the Society led to Steele's expressing the following opinion in the *Tatler*, No. 236, in 1709.

> There is no study more becoming a rational creature than that of Natural Philosophy; but, as several of our modern *virtuosi* manage it, their speculations do not so much tend to open and enlarge the mind, as to contract and fix it upon trifles. ...
>
> They seem to be in a confederacy against men of polite genius, noble thought, and diffusive learning; and choose into their assemblies such as have no pretence to wisdom, but want of wit; or to natural knowledge, but ignorance of every thing else. I have made observations in this matter so long, that when I meet with a young fellow that is an humble admirer of these sciences, but more dull than the rest of the company, I conclude him to be a Fellow of the Royal Society.

Although many of the Wits of the Restoration were Fellows, their counterparts in the reign of Queen Anne felt free to gird at them. Addison in fact wrote in 1711 in *The Spectator* No. 262,

> Among those Advantages, which the Publick may reap from this Paper, it is not the least, that it draws Men's Minds off from the Bitterness of Party, and furnishes them with Subjects of Discourse that may be treated without Warmth or Passion. This is said to have been the first Design of those Gentlemen who set on Foot the Royal Society; and had then a very good Effect, as it turned many of the greatest Genius's of that Age to the Disquisitions of natural Knowledge, who, if they had engaged in Politicks with the same Parts and Application, might have set their Country in a Flame. The Air-Pump, the Barometer, the Quadrant, and the like Inventions, were thrown out to those busy Spirits, as Tubs and Barrels are to a Whale, that he may let the Ship sail on without Disturbance, while he diverts himself with those innocent Amusements.

92   3. Thomas Bayes: a life

Although the Statutes of the Royal Society were originally drawn up in 1663, they underwent little change until 1752. In 1730, however, the following Statute (Stat. X of Cap. VI) was enacted.

> Every person to be elected a Fellow of the Royal Society, shall be propounded and recommended at a meeting of the Society by three or more Members; who shall then deliver to one of the Secretaries a paper, signed by themselves with their own names, specifying the name, addition, profession, occupation, and chief qualifications; the inventions, discoveries, works, writings, or other productions of the candidate for Election; as also notifying the usual place of his habitation.
>
> A fair copy of which paper, with the date of the day when delivered, shall be fixed up in the common meeting room of the Society at ten several ordinary meetings, before the said candidate shall be put to the ballot: Saving and excepting, that it shall be free for every one of his Majesty's subjects, who is a Peer or the Son of a Peer of Great Britain or Ireland, and for every one of his Majesty's Privy council of either of the said Kingdoms, and for every foreign Prince or Ambassador, to be propounded by any single person, and to be put to the ballot for Election on the same day, there being present a competent number for making Elections. [*Record*, 1940, p. 91]

Bellhouse [1999] has drawn our attention to the fact that Thomas attended a meeting of the Royal Society on the 25th of March 1742, as a guest of John Belchier. It is recorded in the Minute Books of that society that on the 8th of April of that year a certificate (see Figure 3.8) proposing Thomas's election to a fellowship was presented. He was put to the ballot and elected on the 4th of November, being admitted a week later. It is also recorded that he 'paid his admission Fee[91], and twenty guinea's in lieu of contributions', the latter being in essence a lifetime payment, in advance, of annual fees.

The reason for Thomas's election as a Fellow of the Royal Society is not clear today. His published works by that time were only two in number: *Divine Benevolence, or an attempt to prove that the Principal End of the Divine Providence and Government is the Happiness of his Creatures* of 1731, and *An Introduction to the Doctrine of Fluxions* of 1736 (both discussed in later chapters of this book). It is possible, on the face of it, that the latter of these two works was the sufficient cause of Thomas's election, though we later say something about Thomas's rôle in the mathematical community of his time.

Though his mathematical prowess may have won the approval of the Fellows of the Royal Society, Thomas's clerical abilities are less certain.

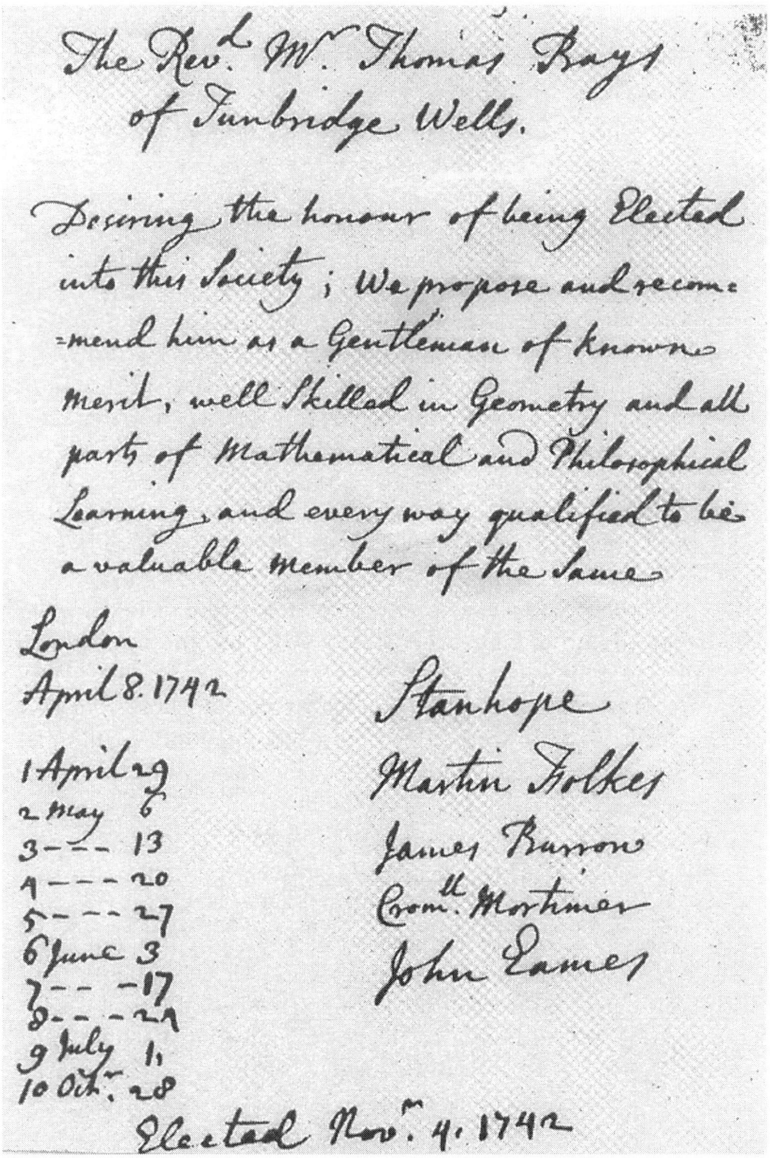

> The Rev.ᵈ Mr. Thomas Bayes
> of Tunbridge Wells.
>
> Desiring the honour of being Elected
> into this Society; We propose and recom-
> mend him as a Gentleman of known
> merit, well Skilled in Geometry and all
> parts of Mathematical and Philosophical
> Learning, and every way qualified to be
> a valuable member of the Same
>
> London
> April 8. 1742
>
> 1 April 29
> 2 May 6
> 3 — — 13
> 4 — — 20
> 5 — — 27
> 6 June 3
> 7 — — 17
> 8 — — 24
> 9 July 1
> 10 Oct.ʳ 28
>
> Stanhope
> Martin Folkes
> James Burrow
> Crom.ˡˡ Mortimer
> John Eames
>
> Elected Nov.ʳ 4. 1742

FIGURE 3.8 Certificate proposing Bayes's election to the Royal Society.

94   3. Thomas Bayes: a life

Timpson relates that that Thomas 'was not a popular preacher, nor evangelical in his doctrine' [1859, p. 464], though one must be careful of the interpretation to be placed on *popular* here. Blair writes

> One of the first qualities of preaching is to be popular; not in the sense of accommodation to the humours and prejudices of the people (which tends only to make a Preacher contemptible), but, in the true sense of the word, calculated to make an impression on the people; to strike and to seize their hearts.
> [1784, p. 264]

Whether the absence of popularity was of any significance or not, it appears that by 1749 Thomas wished to retire from his pastorate. Perhaps this was the reason for his permitting the Independents to use the Mount Sion Chapel, until an incident occurred that led to their having to obtain their own building. Timpson relates the events as follows.

> Having heard of a faithful dissenting minister at Goudhurst; they [Thomas Baker, Edward Jarret, an aged man, named Bunce, Robert Jenner] went one Lord's day, May 21st, 1749, to hear him. Delighted with his sermon, they conversed with him, and he informed them that the Rev. Mr. Jenkins would be ordained pastor of the Independent church at Maidstone on the following Wednesday. They went to that service, and became thus acquainted with the Rev. Mordecai Andrews[92], of London; and he came with Mr. Booth, for the season, to the Wells, where he engaged the Presbyterian chapel, from the Rev. Mr. Bayes, its minister. They enjoyed the gospel preached by ministers sent from London for nearly a year, until Easter Sunday, in 1750, when Mr. Bayes resumed his pulpit, disliking the doctrine of the Independents, and they again attended at the Established Church, for the sake of the Lord's Supper. [1859, p. 466]

In the Independents' Church Book, dated 1743 and entitled *A Brief Narration of the Work of Grace in raising and forming the Congregational Church of Christ at Tunbridge Wells, Kent*, this period is described as follows[93].

> All that summer 1749 we had supplies from London, Sabbath after Sabbath, 'twas indeed a summer to be remembered ... when our poor thirsty souls were made to drink of the Water of Life, and several were truly convicted and converted under these Ministrations.

In their *History of Non-conformity in Tunbridge Wells*, Bailey & Bailey date things rather differently:

> About 1746 Mr. Bayes decided that he would like to retire, but was loathe to do so as he wished the meetings to continue in the Chapel. The Independents arranged with him that speakers from London should take the services and it is recorded that this arrangement was in operation for the whole of the summer of 1747. All went well until Easter of 1748 when Rev. Bayes denied the pulpit to the visiting minister and forced the Independents to leave. Rev. Bayes presumably carried on until he handed over to the Rev. William Johnson M.A. in 1752. [1970, p. 44]

Yet other dates are given by Colligan, who writes

> In 1743, certain persons dissenting from the church of England, approached the Rev. Thomas Bayes for permission to use the meeting house. This was granted, but about seven years afterwards was withdrawn on account of their views. [1923, p. 224]

(Were these dissenting persons the Independents of the previous quotation?) In the light of these variously reported dates the most we can be sure of is that there was a time during which Thomas allowed the Independents to use the Presbyterian meeting-house for their services — and as Charles Dickens wrote[94],

> concerning the exact year there is no need to be precise.

Whatever the full reasons, real or supposititious, for Thomas's reclaiming his pulpit might have been, he seems to have continued as pastor until 1752, when he was succeeded by the Rev. William Johnson. The latter remained minister until his death in 1776, his remains being interred in Speldhurst cemetery, some three miles from Tunbridge Wells. During the incumbencies of subsequent ministers the congregation declined, the chapel being almost deserted by 1810. In 1814 it finally closed, and Presbyterianism vanished from Tunbridge Wells until 1938.

Thomas's refusal to allow the Independents further use of his pulpit was not without precedent or subsequent in Tunbridge Wells. In *Mr Perkins's Letter to Mr. Cornwell, And other Ministers at Tunbridge-Wells, Who denied him the Use of the Pulpit there* we read[95]

> Reverend Sir,
> I cannot blame you much for denying me the use of your Pulpit, since you did it not of your *own accord*, but by the Instigation of *other Clergy-men*, with whom you consulted, and who disavowed you, by telling you that *I should make some begging Business of it*. [Perkins, 1697, p. 3]

Further, citing the 'local record', Bretherton notes that

During this period [c.1808], and for some time after, the Wesleyans had the use of the chapel on Mount Sion on Sunday evenings, and also on week nights; as the Independents [I think this should certainly be Presbyterians][96] only used it on Sunday mornings and afternoons. Mr Gough, the schoolmaster of Southboro', was the minister, and the Wesleyans attended his ministry, till one Sunday he read a notice stating that the Wesleyans were not to have the use of the chapel any more.
[1916, p. 199]

After his retirement Thomas apparently remained in Tunbridge Wells until his death. His remains were taken to Founders' Hall, London, and he was interred in a family vault in the Bunhill Fields Burial Ground on the 15th of April, 1761.

3.4 The faded Bayes

One who is interested in Thomas Bayes will naturally consult appropriate seventeenth- and eighteenth-century sources. We have tried to include as many such pertinent references as possible here, hoping in some measure to rescue Thomas and his ancestors from oblivion, and perhaps also hoping, though less sanguinely, that a punning and pawkie reader will find the lines

> And on some far-off silent day
>   A thinker gazing on a hill,
> Shall cast his staff and horn away
>   And answer to your clamouring will;
>
> He shall bring back the faded bays,
>   The graces to their ancient rule,
> The harper to the market-place,
>   The genius nearer to the fool.

from Orrick Johns's *Second Avenue* to be applicable to the author. In this section we comment briefly on some other literary references to Bayes.

To the statistician Thomas's name is preserved, if not enshrined, in *Bayes's Theorem*: yet the eclectic and omniverous reader will have come across the saying 'Dead men may rise again, like Bayes's troops, or the savages in the Fantoccini'[97], and one may well wonder about its origin.

Indeed, to those who are more acquainted with seventeenth-century comedy than eighteenth-century clergy the name of Bayes will not be unknown. In 1670 John Dryden was appointed Poet Laureate and Historiographer to King Charles II. It was no doubt as a result of the bestowing of the laurel wreath that George Villiers, Duke of Buckingham[98], chose — in punning mood — to satirize Dryden under the name of *Bayes* in his burlesque *The Rehearsal*, first performed on the 7th December 1671. Here Bayes is a playwright, the last rehearsal of whose play is being depicted. In the play a ferocious battle is fought between foot-soldiers and hobby-horses, and, when all the actors are dead, the following exchange occurs between Bayes and Smith, one of those invited to watch this rehearsal.

> SMI. But, Mr. *Bayes*, how shall all these dead men go off? for I see none alive to help 'em.
> BAYES. Go off! why, as they came on; upon their legs: how should they go off? Why, do you think the people do not know they are dead? [Act V]

Thus arose the saying instanced earlier.

In John Wilmot, Earl of Rochester's, *Tunbridge Wells*[99], there occur the following lines[100].

> Listning I found the Cobb of all this Rabble
> Pert Bayes, with his Importance Comfortable:
> He being rais'd to an Arch-Deaconry,
> By trampling on Religion, Liberty;
> Was growne too great, and look'd too fat and Jolly,
> To be disturb'd with care, and Melancholly,
> Tho' Marvell has enough, expos'd his Folly.
> He dranke to carry off some old remaines,
> His Lazy dull distemper, left in's Veines:
> Let him drinke on, but 'tis not a whole Flood,
> Can give sufficient sweetnesse to his blood,
> To make his Nature, or his Manners good.
> Importance dranke too, tho' she'd beene noe Sinner
> To wash away some Dreggs, he had spewd in her.

Here *Bayes* refers to Samuel Parker (1640–1688), Archdeacon of Canterbury and later Bishop of Oxford. Parker wrote a number of works stressing the power of the King in ecclesiastical matters and against the toleration of dissent. Andrew Marvell, Member of Parliament for Hull and occasional writer, allowed these to pass without comment; but Parker's *Preface Shewing what grounds there are of Fears and Jealousies of Popery*, prefixed to Bishop Bramhall's *Vindication of himself and the Episcopal Clergy from the Presbyterian Charge of Popery* prompted him to write *The Rehearsal Transpros'd* in 1672 (a second part followed in 1673). Part of the aim of these marvellous works was to encourage the King in his support of the dissenters and to silence Parker. (Even some dissenters had attacked the Declaration of Indulgence, fearing that it might open the door to Popery.)

There were a number of replies to Marvell in which the name *Bayes* appeared; see the introduction to Marvell [1971].

The title of Marvell's work is taken from *The Rehearsal*, in which play the following exchange between Bayes and Johnson, who had been invited to view the last rehearsal of Bayes's play, takes place.

> ... I take a Book in my hand, either at home, or elsewhere, for that's all one, if there be any Wit in't, as there is no Book but has some, I Transverse it; that is, if it be Prose put it into Verse, (but that takes up some time) if it be Verse, put it into Prose.
> JOHNS: Methinks, Mr. *Bayes*, that putting Verse into Prose should be call'd Transprosing. [Act I, Scene I]

Citing various similarities between the main character in Villiers's play and the writer of the preface to Bramhall's tract[101], Marvell explains his reason for choosing this same name as follows.

> instead of Author, I may henceforth indifferently call him *Mr. Bayes* as oft as I shall see occasion. And that, first, because he hath no Name or at least will not own it, though he himself writes under the greatest security, and gives us the first Letters of other Mens Names before he asked them. Secondly, because he is I perceive a lover of Elegancy of Stile, and can endure no mans Tautologies but his own, and therefore I would not distaste him with too frequent repetition of one word. But chiefly, because *Mr. Bayes* and he do very much Symbolize[102]; in their understandings, in their expressions, in their humour, in their contempt and quarrelling of all others, though of their own Profession. Because, our Divine, the Author, manages his contest with the same prudence and civility, which the Players and Poets have practised of late in their several Divisions. And, lastly, because both their Talents do peculiarly lie in exposing and personating the Nonconformists. I would therefore give our Author a Name, the memory of which may perpetually excite him to the exercise and highest improvement of that Virtue. ...
> Besides, to say Mr. *Bayes* is more *civil* than to say *Villain* and *Caitiff*, though these indeed are more *tuant*. [1971, pp. 9, 10]

One might even try to draw some analogy here between Marvell's Bayes and our Thomas, for the latter certainly published some papers anonymously, and we later have occasion to comment on his use of style. But such analogy would be forced, and I would not like to try to pursue it.

3.5 Appendix 3.1

We are fortunate in that a number of College records for the eighteenth century are preserved in the archives of Edinburgh University. References there to Thomas Bayes, in addition to those given in the body of the text, in no particular order, run as follows (the references in crotchets are the shelf-marks of the university's Special Collections Department):

1. [Da]. *Matriculation Roll of the University of Edinburgh. Arts-Law-Divinity. Vol. 1, 1623–1774. Transcribed by Dr. Alexander Morgan, 1933–1934.* Here, under the heading 'Discipuli Domini Colini Drummond qui vigesimo-septimo die Februarii, MDCCXIX subscripserunt' we find the signature of Thomas Bayes. This list contains the names of 48 students of Logic.

2. [Da.1.38] *Library Accounts 1697–1765.* Here, on the 27th February 1719, we find an amount of £3–0–0 standing to Thomas's name — and the same amount to John Horsley, Isaac Maddox, and Skinner Smith.

3. [Dc.5.24^2]. In the *Commonplace Book of Professor Charles Mackie*, we find, on pp. 203–222, an *Alphabetical List of those who attended the Prelections on History and Roman Antiquitys from 1719 to 1744 Inclusive. Collected 1 July, 1746.* Here we have the entry

    Bayes (   ), Anglus. 1720,H. · 21,H.   3

    The import of the final '3' is uncertain.

4. *Lists of Students who attended the Divinity Hall in the University of Edinburgh, from 1709 to 1727. Copied from the MSS of the Revd. Mr. Hamilton then Professor of Divinity, etc.* Thomas's name appears in the list for 1720, followed by the letter '*l*', indicating that he was licensed (though not ordained).

5. *List of Theologues in the College of Edin[burgh] since Oct:1711. the 1st. columne contains their names, the 2d the year of their quūmvention, the 3d their entry to the profession, the 4th the names of those who recommend them to the professor, the 5th the bursaries any of them obtain, the 6th their countrey and the 7th the exegeses they had in the Hall.* Here we have

Tho.Bayes|1720|1720|Mr Bayes| — |London|E. Feb. 1721. E. Mar. 1722.

In a further entry in the same volume, in a list headed 'Societies', we find Thomas's name in group 5 in both 1720 and 1721. (These were perhaps classes or tutorial groups.) In the list of 'Prescribed Exegeses to be delivered' we have

1721. Jan. 14. Mr. Tho: Bayes. the Homily. Matth. 7.24, 25, 26, 27.

and

1722. Ja. 20. Mr Tho: Bayes. a homily. Matth. 11. 29, 30.

The final entry in this volume occurs in a list entitled '*The names of such as were students of Theology in the university of Edinburgh and have been licensed and ordained since Nov. 1709. Those with the letter .o. after their names are ordained, others licensed only*'. Here we find Thomas's name, but without an 'o' after it.

L'envoi

We often proceed farther than we at first intended, when we indulge ourselves in trifling liberties, which we think of no consequence; for though perhaps the heart takes no part at the beginning, it seldom fails to be engaged in the end.
[Hamilton, 1859, p. 173]

4

Divine Benevolence

> *The existence and character of the Deity, is, in every view, the most interesting of all human speculations.*
>
> *William Paley.*
> *Natural Theology.*

4.1 Introduction

Paralleling earlier discontent with the Church of Rome, the late seventeenth- and early eighteenth-century dissenters (or Nonconformists) grew disenchanted with the Established Church of England. The rise of Natural Philosophy led to an increased interest in Natural (as well as Revealed) Religion, and it was believed that the character of God could (and would) be revealed by a scientific study of His works. As a later exemplar of this we mention the following passage from William Paley's *Natural Theology*.

> The proof of the *divine goodness* rests upon two propositions: each, as we contend, capable of being made out by observations drawn from the appearances of nature.
> The first is, "that in a vast plurality of instances in which contrivance is perceived, the design of the contrivance is *beneficial*."
> The second, "that the Deity has superadded *pleasure* to animal sensations, beyond what was necessary for any other purpose, or when the purpose, so far as it was necessary, might have been effected by the operation of pain." [1825, vol. V, p. 316]

In his *The History of Statistics in the 17th and 18th Centuries* Karl Pearson remarks on the number of theologians, both Conformist and Nonconformist, who contributed to statistics. He instances William Derham (1657–1735), Johann Peter Süßmilch (1701–1767), and Richard Price (1723–1791), with William Whiston (1667–1752) and Humphrey Ditton (1675–1715), both of which names have occurred in our third chapter, being mentioned for their importance in the mathematical and religious post-Newtonian generation. We examine Thomas Bayes's contribution to the fluxionary debate in the next chapter: in this chapter we consider his only theological (or perhaps ethical[1]) tract, one in which God's fundamental attributes are explored[2].

Divine Benevolence, or, an attempt to prove that the principal end of the divine providence and government is the happiness of his creatures is often seen as a rebuttal to John Balguy's *Divine Rectitude* of 1730, and it was in turn followed in 1734 by Henry Grove's *Wisdom, the first Spring of Action in the Deity*. All these tracts were in fact published anonymously (at prices of 1s., 1s., and 1s. 6d., respectively).

Exhaustive of the exploration of the Divine Attributes though these tracts might seem to be, and exhausting though the reading of them today may in fact be, one should not be misled into thinking that these three works satisfied the need for writing on such matters at that time. We find further tracts by Balguy and Francis Hutcheson needful of comment in §4.3, and note for the moment only that the British Library has a work bound as *Theological Tracts* (478.b.20) that contains, in addition to *Divine Rectitude* and *Divine Benevolence*, the following Works[3].

1. William, Lord Archbishop of Dublin. *A Key to Divinity: or, a Philosophical Essay on Free-Will.* London: 1715.
2. Anon. *The Parent Disinherited by his Off-spring; or An Enquiry into the Descent of Penal Evil, With its Tendencies to subdue, and expel Moral Evil.* London: 1728.
3. Samuel Fancourt. *An Essay concerning Liberty, Grace, and Prescience.* London: 1729.
4. T. Melvil. *The Scripture Doctrine of Divine Grace: or, a few English Notes On a Book lately publish'd by Mr Nicholls, an Independent Teacher, Intituled, The Method of Divine Grace in the Salvation of the Fallen Man.* London: 1730.

Like Bayes, Balguy and Grove each sought a single principle (the *simplest* hypothesis?) to which God's moral principles could be ascribed — an implementation of Occam's Razor[4]? Although their conclusions were different (Bayes's fundamental attribute of *benevolence* being *rectitude* in Balguy's tract and *wisdom* in Grove's), Balguy and Grove both devoted some discussion to the place of benevolence in their schemes: we say more on this matter in §4.3.

In his article on *Benevolence* in the second volume of the *Encyclopædia of Religion and Ethics*, Kilpatrick discusses this quality under two main headings: I. As a quality of human character, and II. As a Divine attribute. Under the first of these he distinguishes the following aspects of benevolence as being most clearly indicated in the New Testament: χρηστότης (benevolence or benignity), ἀγαθωσύνη (active goodness or beneficence), εὐδοκία and εὔνοια (the gracious will that is the source of the benevolent deed, and the inner spring of conduct), φιλανθρωπία (philanthropy) and ἁπλότης (liberality). It seems that there was little, if any, attempt in the essays considered here to distinguish between *beneficence*, *benevolence*, and *benignity*.

In considering benevolence as a Divine attribute, Kilpatrick writes

> Consideration of the moral attributes of God can be fruitful only if two principles are borne in mind. (a) The attributes are not 'things' or 'forces.' ... They exist in the unity of the Divine character, and are partial manifestations of its inexhaustible wealth. (b) They are not to be viewed as given, in their truth and fullness, in man, and then applied to the Divine character as copies or reflexions of what they are in human nature. The moral attributes of God are to be learned, primarily, from the revelation which Christ has given, in His own person, of the Being whom He, alone among men, perfectly knew.
>
> It is important thus to connect the Divine benevolence with the aim of the Divine love. The Divine benevolence is part of the operation of the Divine love, which, from the beginning of creation, has been preparing a world in and to which God could fully reveal Himself. It signifies His goodwill toward all the creatures of His power, His determination to bless them according to their utmost capacity, and to bring them to the highest perfection of their nature.
>
> [Hastings, 1971, vol. II, pp. 477–478]

As indicated by the title of his tract, Bayes's concern is chiefly with God's benevolence, the human quality being considered only inasmuch as it may help us in our examination of the divine attributes: whether Bayes could be regarded as fulfilling the two principles laid down by Hastings emerges in due course. In §4.3 we (a) consider Bayes's comments on Balguy's work in some detail, (b) examine the tracts by Balguy and Grove as they relate particularly to Bayes's treatise, and (c) discuss the effect *Divine Benevolence* had on other writers on the Divine attributes.

In addition to the clearly Christian writings of the eighteenth century, the Age of Reason saw the start of the advocation of Deism. In his introduction to the 1906 edition of Joseph Butler's *The Analogy of Religion*, Bayne finds this starting point to be John Locke's *Reasonableness of Christianity* of 1695, subsequent advances being made in John Toland's *Christianity not mysterious* and Matthew Tindal's *Christianity as Old as the Creation*[5]. Butler's *Analogy* was a direct response to this movement, and we have something to say on this work also[6].

Divine Benevolence

Or, An ATTEMPT to prove that the

PRINCIPAL END

Of the DIVINE

PROVIDENCE and GOVERNMENT

IS THE

Happiness of his Creatures.

BEING

An ANSWER to a Pamphlet, entitled,

Divine Rectitude; or, An Inquiry concerning the Moral Perfections of the Deity.

WITH

A Refutation of the Notions therein advanced concerning Beauty and Order, the Reason of Punishment, and the Necessity of a State of Trial antecedent to perfect Happiness.

LONDON

Printed for JOHN NOON, at the *White-Hart* in *Cheapside*, near *Mercers-Chapel.* MDCCXXXI

Price One Shilling.

INTRODUCTION.

THERE cannot be a controversy of greater importance, than that which relates to the divine perfections. On our right apprehensions of these religion intirely depends: on these all our hopes of happiness are intirely built. For unless we know what God is, and entertain clear conceptions of his perfections, and particularly of his moral attributes, we shall neither know how to behave towards him, nor what we are to expect from him according as we do behave: We shall be apt to attribute to him those things that are real imperfections, and shall very probably be sadly perplexed in accounting for his various dispensations towards mankind. It is a work therefore of great service and use to mankind; to put them in a way of thinking more clearly concerning the perfections of God, and avoiding the mistakes we are apt to fall into relating to this point. **[3]**

 The author of a pamphlet entitled *Divine Rectitude*, &c. has made a laudable attempt of this kind, which I have observed to have been greatly commended by persons whose judgments I value, but to me, though there be many things in it which I could join in applauding, yet the general method of his reasonings appears rather to increase than to abate the perplexities of this subject, and (which seems to me a point of the greatest consequence) to render our expectations from the divine being very precarious and uncertain. This, I hope, will be esteemed a sufficient vindication of my attempting to reconsider this subject, and to make my remarks on the abovementioned dissertation; in which remarks, I hope, the author of that work will have no reason to complain of any designed misrepresentations of what he says, or of any treatment unsuitable to that excellent temper and good sense that discovers it self in his performances. **[4]**

Divine Benevolence.

Section I.

IT has been the common method of those who have treated of the moral perfection of God, to consider this as distinguished into several attributes; the principal of which are *goodness, justice* and *veracity*. And this has been done, I think, upon the justest ground, because the greater number of true propositions we know concerning God: provided each of these be clear in themselves, and really distinct from each other; the clearer notion we have of God himself, the better we are furnished for arguing concerning him, and may more easily judge what conduct to expect from him. This is certainly true, though all the moral perfections of God were allowed to be founded in one uniform principle of action. Suppose for instance, that this principle is moral rectitude, and that God is [5] good, and just, and true; because otherwise there would be a defect in this his moral rectitude. Yet it is certain, that I know more of God by knowing that he will ever act according to goodness, justice, and truth or veracity, than if I only knew that he would always act agreeable to the moral fitness of things, or as it is fit he should. And from the former I should at least more *readily* and *easily* know what to expect from him. It is told me in the word of God, *that Christ shall judge the world*; now, shall not I more readily discern the certain truth of this proposition, if I am assured that this is the declaration of him that cannot lie, than if I were only told, that this is said by him that always acts according to moral rectitude? For I can perhaps perceive no moral fitness in this, that Christ should be the judge of the world. 'Tis true, it may be said, but it is morally fit that God should not deceive and impose on his creatures; and therefore it follows from the moral rectitude of the divine being, that since he has said it, Christ will be the judge. I don't deny it. But yet it's plain, that before you can come to the conclusion, you are obliged to make the divine veracity a medium of your Proof. And the like examples might be easily given with respect to the other moral perfections of God. Wherefore the least inconveniency that can arise from abstracting from these, and resolving all into moral rectitude, [6] will be to make our arguments relating to this subject more tedious and laborious. If moral rectitude be indeed the foundation of all the moral perfections of God, the most proper method of procedure would be, from hence to deduce goodness, justice, and veracity, and the like, and then to go on in our enquiries concerning the divine dispensations; for till we have clear notions of these, and know whether they do, or do not belong to God,

4.2. The Tract 107

I am sure every one must acknowledge that nothing but mistakes and blunders can be expected in such enquiries; and therefore nothing can be more unreasonable than designedly to abstract from them, when we are employ'd in these disquisitions. But to come to the Author of the above-mentioned Dissertation, whom, for the future, to avoid multiplicity of words, I shall call our Author.

He tells us, *P.* 4. that "however we may divide and distinguish the moral "perfections of God, according to the different effects, dealings, and dis-"pensations resulting from them; yet in themselves they seem to be but "one and the same perfection variously exercised on different objects and "occasions, and in different cases and circumstances; and cannot therefore, "without error and inconvenience, be considered as distinct attributes." This "Perfection is that of God's determining himself by moral fitness, "or acting perpetually according [7] to the truth, nature, and reason of "things;" which is what, in other places, he calls Moral Rectitude. And, in *P.* 30. he tells us, that "his chief aim is to endeavour to shew the conve-"nience and advantage of considering the moral perfections of the Deity, "under the idea of rectitude, rather than in the mixed light of many distinct "attributes."

My design, on the other hand, is to show that there is no convenience in this method; but especially to take notice of those errors which he has fallen into, perhaps by adhering to it. As to the former part of this design, I have already, in some measure, prevented my self; but the consideration of what our author has advanced, in support of his opinion, will give occasion for some farther illustration of this point. He says the moral attributes of God are really one, and therefore they cannot without error and inconvenience be considered as distinct. If by their being but one he means only that they may be all comprehended under one general notion, as so many distinct species, I don't deny it; but that they are as such, really different, none, I think, can deny; and therefore it does not follow from the general notion in which they may agree, that they cannot, without error and inconvenience, be consider'd as distinct. Is it not plain that the ideas of goodness and veracity are distinct? May not a person be veracious, and yet wicked? [8] Where then can be the error of conceiving of these two attributes as distinct, when they really are so? If these two attributes really belong to God, then two distinct moral attributes belong to him; or else we must mean precisely the same thing, when we say of God that he cannot lie, as when we say that he is good to all, and his tender mercies are over all his works. And as there is no error in conceiving of the moral attributes of God as distinct, so neither is there any inconvenience in it, nor any occasion to have recourse to our Author's method to escape the perplexities to which the other exposes us, as he asserts, *Pag.* 6.

To support which, he represents it as common for men to bewilder themselves in searching for the boundaries between divine justice and goodness:

By the one of which, he says, is generally understood a *communication of blessings*; by the other, an *infliction of judgments and calamities*; and then adds, with a note of admiration, as if God were not equally righteous in both dispensations.

I can hardly think that any one did ever really entertain this notion of justice and goodness. To communicate blessings may in many cases arise from foolish fondness and partiality, and have nothing of real goodness in it; and to be sure the infliction of calamities, the causing of distress and misery may as well be supposed to arise **[9]** from malice and ill-nature, as from justice.

I can't therefore well understand what our author means by persons puzzling themselves in searching for the boundaries between goodness and justice, *i.e.* according to the definition he here gives of these words, between a communication of blessings, and infliction of judgments and calamities.

If any having entertained such an imperfect notion of justice and goodness have been perplexed in searching for the boundaries between them, this does not discover any inconvenience in considering justice and goodness as distinct, according to the proper signification of the words. By the goodness of God we ought, I think, to understand a *disposition to communicate happiness to his creatures in general*; so that the end of goodness is answered by every action that produces more happiness than misery. By justice we are to understand a *disposition to take care of the support of the cause of virtue and righteousness, by the distribution of proper rewards and punishments*: And, taking the words in this sense, there is so far from being any real, that there is not so much as the least seeming inconsistency between them; nor do these attributes, properly speaking, ever set bounds or limits to each other; for what justice requires, goodness cannot forbid, and what goodness demands, justice cannot oppose. 'Tis, indeed, usual enough **[10]** to represent these as opposite. Justice, we say, requires the punishment of the sinner, and goodness pleads for his pardon. But this is only a popular way of speaking; for justice requires the punishment of the sinner only as a means of preserving the reverence due to the divine laws; and where it is necessary for this end, goodness cannot oppose the punishment, because the general happiness of the creatures of God cannot be supported, but by maintaining the authority of his laws: and, on the other hand, where this end can be secured without the punishment of the guilty, justice does not absolutely require it. Indeed, if we consider justice as implying in it hatred and revenge against the sinner, or suppose that it obliges to punish the guilty, where no better end is answered by it, than merely his misery, which seems not to be very different from the opinion of our author, it would then be very difficult, if not impossible, to reconcile this with goodness; and the belief of two attributes in the divine being, really opposite to one another, might easily lead us into endless perplexities.

But, allowing the afore-mentioned to be the true ideas of divine goodness and justice, there can no difficulty arise from supposing them both to belong to God, and therefore, from hence, no necessity appears of having recourse to our author's method. But that I may once for all show, that as his method **[11]** may, very easily, lead us into mistakes; so it cannot possibly clear up any difficulty, or give a fair solution to any objection, unless where, by a round-about way, it at length coincides with the common method, consider the following instance.

Suppose a person were to object against what is said, 2 *Kings*, ch. xx. *ver.* 1. and should say, that it could not be the word of God, because it represents the prophet as telling *Hezekiah* that he should die of the sickness that was then upon him, whereas we find he recovered, and lived fifteen years afterwards: The difficulty of this objection lies plainly in this, that such an assertion being false, could not proceed from a being of perfect truth and veracity. Now is it any ways agreeable to the just rules of arguing in the solution of this difficulty to abstract from the particular consideration of the divine veracity, and to appeal to the moral fitness of things? Or what could we possibly gain by such a conduct, unless it were to impose on our selves or the objector? Suppose we could find an hundred reasons that should prove the fitness of the prophet's thus speaking to *Hezekiah*; yet nothing can really take off the force of the objection, but what shows that, in the place mentioned, there is nothing said inconsistent with the divine veracity; and the like may be said with respect to any other difficulties of this nature. If any phænomenon **[12]** be asserted, or at first view appear contrary to any one of the perfections of God, it is plain, I think, to any person that has the least notion of reasoning, that there can be no way of reconciling the proposed difficulty; but by showing either that the fact is falsely represented, that the perfection to which it is said to be contrary, is not really a perfection of God, or else that there is no inconsistency between them. But whilst we allow the perfection really to belong to God, and own the fact that seems contrary to it, to desire a person to abstract from that particular perfection, and to consider moral rectitude in general, looks more like a design to amuse than instruct; at least at first view it does not seem to be the conduct of a fair arguer.

I have all along hitherto gone upon the supposition that all the moral attributes of God might be comprehended under the idea of rectitude, or that the whole moral conduct of God is directed by a regard to the reason and fitness of things. But it seems plain to me, that in this description there is something wanting for the full and compleat explication of this point; for did we only know that God will act according to moral rectitude, which in plainer words is only that he will act, *as it is fit he should*, we should hence in no one particular be able to guess what conduct to expect from him, unless we **[13]** are also informed what it is in an action which renders it fit to be performed; nor without this can we possibly make any use of our author's general notion of rectitude: Now, as the desire of some end must be the motive to any action, so 'tis the nature of the end designed, and which the action is proper to effect that renders it good or bad, fit or unfit to be performed. When therefore we say, that God is in all his Actions governed by the reasons and fitness of things, we must, I think, mean, if we would understand ourselves, that he is moved to every action by a regard to some good and valuable end, and always chuses that way of conduct which is most proper to bring about the end designed. This seems to be the only notion we have of a wise and reasonable action, and *this end* ought to be taken notice of in the description of divine rectitude.

Thus, for instance, if you suppose with me, that the view by which the divine being is directed in all his actions*, is a regard to the greatest good or happiness of the universe, then the moral rectitude of God may be thus described, *viz.* That it is a disposition in him to promote the general happiness of the universe: Which definition would in effect be the same with that given by our author, if we **[14]** suppose that a tendency or fitness in things to produce happiness, or to prevent mischief, is that which constitutes the fitness of things; but whatever it be, whether a tendency to promote happiness, or order and beauty, or the like, that renders actions fit to be performed, this must be mentioned in the definition of divine rectitude, if we would have it of any use in our inquiries about the divine perfections or conduct; for only to know that God will act according to the fitness of things, will make us no wiser in relation to this matter, than a person's knowing that he ought to act according to the fitness of things, or, as it is fit he should, will lead him into a right moral conduct. The first thing therefore that we have to do before we can determine any thing concerning the moral perfections of God, or hope to solve the difficulties relating to the conduct of his providence, is to find out what it is that renders actions fit to be performed, or what is the end that a good and virtuous being, as such, is in pursuit of. And it is to be observed, that as we shall find this to be one and the same, or different and opposite, so it will appear that God is directed in his moral conduct by one uniform principle or not; for though we have got one word [Rectitude] to express the moral perfections of God by, as we might easily find another that should express both his natural and moral ones together; yet that perfection cannot be really one **[15]** and the same, that includes in it the desire of two opposite ends. If we suppose that God desires the happiness of his creatures, and at the same time the beauty and order of his works, if these two principles of action often clash and interfere, and set bounds to each other; this cannot be to represent him as governed by one uniform principle, tho' his love of order, and his love of his creatures, be both called moral rectitude: but if his moral rectitude be

---

* When I speak of the divine actions throughout this discourse, I always mean his actions towards his creatures, though that should not be particularly mentioned.

nothing else but a desire or inclination to preserve a constant regard to the general happiness of the universe in all his actions; then he is in this view represented as truly governed by one uniform principle; though we should afterwards find it very convenient to reckon justice, veracity, mercy, patience, &c. as particular species or branches of this rectitude. Our author therefore seems very far to have swerved from the main design he professes he had in view; which was to consider the moral perfection of God under one simple notion, rather than in the various and mixed light of many distinct attributes; when he supposes that the rectitude which he represents, as comprehending all moral perfections, really includes in it a regard to different and opposite ends; and consequently is only one word contrived to express several attributes, not only distinct from, but many times really opposite to one another. Had he, instead of thus comprehending **[16]** all the moral attributes of God under one word, shown how they are all derived from one uniform principle of action, or how they result from the supposition of the desire of one uniform end, he had acted much more agreeably to his own pretensions, and had explained a truth of very great importance in the consideration of the moral perfections of God; but he seems, on the contrary, to have studiously endeavoured to find out as many distinct reasons or ends of the divine conduct as possible he could, and then supposes that there be many more, of which we have not the least notion or conception. Now, can any one think that he represents the moral perfections of God in a distinct and unmixed light by including them all under the term Rectitude, when at the same time he tells us, that this rectitude implies in it a regard to different ends which do often interfere, but how far we know not; and a regard to various other reasons which we have no conception of? But lest this should be looked upon as misrepresentation, hear what he says himself. After having observed that it is commonly presumed that the communication of happiness was the sole end of God in creating the world, he says, *P.* 11. "I see no absurdity in supposing, that the creator "might have various ends and designs of which we have not the least con- "ception. But not to insist on this, it deserves to be considered, whether "within the compass of our own Ideas we may **[17]** not find some other in- "tention befitting the wisdom and rectitude of the supreme being. We seem "to have grounds for such a supposition both from reason and revelation." And again, having represented order and beauty as an end aimed at in the works of God distinct from the happiness of his creatures, he says, *P.* 14. "Where, and how far these ends do in fact interfere, is above the power of "man to determine. But it does not appear to me, that the order, beauty, "and harmony of the universe are meerly intended in subordination to the "welfare of creatures; on the contrary, I know not whether the latter be not "subordinate to the former, and limited by it."

I hardly need here observe what a melancholy aspect it would have upon the creatures of God were this supposition true, that God in creating, and in his acts of providence, has various ends and designs distinct from, and in some respect opposite to their happiness; for as of the importance of many of these ends, a regard to which may limit his regard to the happiness of his creatures, we know nothing, how can we be sure that they may not so far prevail as entirely to destroy it? We are sure, indeed, that God will act towards us according to the fitness of things but this fitness of things is very different from a fitness to produce happiness, and we must be acquainted with various other ends and designs of the divine being, before we can judge what it is fit [18] for him to do or not to do; and consequently before we can judge what conduct we are to expect from him. We are apt to flatter our selves that God designs our happiness, this is indeed allowed to be a right design, but it is not his whole design. He may perhaps have various other ends in view, ends for ought we know of much greater importance, and to which the happiness of his creatures must yield. Our author seems to suppose it impossible for us to determine how far a regard to the happiness of his creatures and the other views of the divine being do in fact interfere. I am sure then 'tis impossible that we should be certain that they will not so far interfere as wholly to destroy it; nor is it possible for us in any case to know what conduct we are to expect from God, when only one or two of those various ends that regulate his actions are known to us, and we suppose that there are various other ends, which may have as powerful influence, and may interfere with these. When therefore such melancholy consequences plainly arise from this notion, tho' it be no direct disproof of it; yet thus much will be gained by it, that I shall have every good-natured man so far on my side, as to wish it might prove false; and I may expect to be heard without prejudice, whilst disputing against an opinion, which, if true, would have such an ill aspect even upon good men. [19]

Section II.

This therefore I shall now set my self to do, and endeavour in opposition to it to prove these two points, 1. That God in his acts of creation and providence had a regard to the happiness of his creatures, and that he is really benevolent towards them. And, 2. That we have no reason to suppose that he is in his actions towards them influenced by any other principles; at least, by any other, that are not entirely coincident with, or perfectly subordinate to this. The former of these propositions I might omit the proof of, because it is not expresly denied by our author, and very plainly follows from some of his concessions; but as this is indeed a point of the highest importance, and ought to be proved before we come to consider the latter, I was unwilling wholly to pass it over. Now the only ways in which this point can be made out, are either from reason or experience: By reason, from the observations of the works of God, we justly conclude that he is a most wise, and powerful, and infinitely perfect being, and such a being can't be supposed any ways indigent, but must be perfectly happy, being liable to no infelicity from ignorance or weakness; and consequently he can't be influenced to do evil, or to lessen the happiness of his creatures from any interfering of their interests with his. Hence it is plain that the creatures of God have no reason [20] to fear any injurious treatment from him.

Thus far, I think, the case is very plain, but it is commonly apprehended that the goodness of God may from the foregoing principles be more strictly demonstrated after the following manner: The divine being, as infinitely wise, must be perfectly acquainted with the true nature and relation of things, must always know what is fit and proper to be done; and having no passions nor temptations to draw him aside from what is in it self reasonable and fit, must in every case act accordingly. But to make a good or innocent being miserable, is a thing in its own nature always unfit and unreasonable, and to communicate happiness to such kinds of beings, is always agreeable to reason; and therefore we may certainly conclude that God will conduct himself in this manner towards his creatures. I will not pretend to say that this argument is fallacious, but there is one part of it that to me is not so clear as I could wish. When it is asserted that to communicate happiness is a thing in it self fit and reasonable, I am at a loss to know what is the meaning of the expression. An action is, I think, then *reasonable*, when there is a good reason for the performing it. But if we thus understand it in the foregoing argument, there is something still wanting to compleat the demonstration; for it may then be justly ask'd, what is the reason why God should communicate happiness to the good and innocent? [21] Will you say that the reason for it is that such a procedure is agreeable, and the contrary opposite to the nature of things? If so, I should then ask to what things is such a proceedure agreeable to the nature of? Is is to that of the creatures? Is it agreeable to their nature as sensible beings? It is certainly pleasing to them as such; but this is entirely besides the question, and abstracting from this sense of the word, I don't see but that pain and misery is as agreeable to the nature of a sensible being, as pleasure and happiness. Is then such a proceedure agreeable to the nature of the divine being, and the contrary opposite to it? It certainly is if we suppose him benevolent; but this is begging the question: Or will you say it is agreeable to his nature, as a wise and intelligent being, without considering whether he is really benevolent or not? I should be extreamly pleased to see this proved, because nothing could then be a stronger demonstration of the divine goodness. But to do this, is at present beyond my skill. I don't find (I am sorry to say it) any necessary connexion between mere intelligence, though ever so great, and the love or approbation of kind and beneficent actions. I must therefore leave this proof to better hands, and content my self with showing in a more popular way, that God is really benevolent, without pretending to discover the reason why he is so. And to this purpose I would observe it as a matter of fact, that God has **[22]** made many beings capable of very great happiness, and that he has plainly made, if we only consider this world, very plentiful provision for their happiness. Now it is not conceivable with what view he has done this, if not with a design to promote their happiness. He must do it either with an intention to communicate happiness or misery to them, since we don't suppose that his felicity has any dependance on his creatures, or else with no view or design at all; for that action, by which the happiness or misery of no one being in the universe is promoted, is in effect the same as none at all. But we can't suppose that God designed the misery of his creatures; to suspect this, nothing in our minds, nothing in the appearances of things, gives us any reason: We must therefore conclude that it was with an intention to communicate happiness that the universe was created, and that the divine being is really most benevolent and kind to his creatures; and in this conclusion we shall be abundantly confirmed by observing the works of God. Take notice of them, and where you admire their curious and beautiful contrivance, you will always observe the marks of goodness as well as wisdom. Some particular appearances it may perhaps be hard to reconcile to our ideas of perfect goodness, but we shall ordinarily find that these are the effects of general laws, that in the main are useful and beneficial; and it is not to be expected but that we, who have but **[23]** very imperfect views of things, should sometimes meet with difficulties that we can't easily account for, though we shall never find any that really overthrow our notion of a perfectly good and benevolent Deity.

Besides this, it is plain that we are so formed as necessarily to approve of kind and beneficent actions, and to dislike a cruel and barbarous character. If this is a consequence of our being intelligent and rational creatures, God also is infinitely wise, and therefore infinitely good but if we cannot see this connexion between intelligence and goodness, or an approbation of what is kind and benevolent: yet our being thus formed, is a strong proof that our Creator is really good, since nothing can be a greater security to the general happiness than this; and a being that had no regard to the happiness of his creatures, would never have made the most sublime and satisfying pleasures the constant attendants of kind and beneficent actions. We have, I think, from such considerations as these, the justest reason to be assured that the divine being is really benevolent, or does intend the general happiness of his creatures. I put these as equivalent expressions, because I suppose he would not design their happiness, if he did not desire it, but should be full as well pleased to see them all *ruined* and *miserable*. But our author has nicely distinguished these two things: He freely owns that God intends the happiness of his creatures, **[24]** but seems very unwilling to allow him to be benevolent at all. He supposes he does good to them, not because he desires their happiness, but because such a conduct is agreeable to the reasons and natures of things. As if supposing the divine being perfectly indifferent to the happiness or misery of his creatures, there could be any reason why he should choose one rather than the other. But let us hear what our author has said himself on this head: "I cannot avoid thinking (says he, *P.* 9.) that "the divine goodness is very much misapprehended, when it is considered "as a physical propension, or disposition of nature analogous to those "affections and propensities which he has given us; such a disposition "would be so far from constituting moral goodness, that it would derogate "from it in proportion to its influence. To suppose in the Deity a benevolent "disposition necessary in it self and in its operations, is to suppose what is "utterly inconsistent with the perfections of the Creator, and the obliga- "tions of his creatures. And supposing it did not necessarily produce bene- "ficent actions, but to be consistent with freedom and choice; yet still, in "proportion to the extent of its influence, it would depreciate good actions, "and detract from the merit of the agent."

What would be the consequence of supposing such a benevolent disposition in the Deity, as would destroy the freedom of his actions I **[25]** am no ways concerned to mind, since all I contend for is, that God is necessarily good or benevolent, and this no more destroys his liberty, than his love of order and abhorrence of confusion, which is allowed by our author to bein the Deity, as unwilling as he is to own that he has any love to his creatures. Now against such a necessary benevolence he has said nothing, but that according to the extent of its influence, it would *depreciate good actions, and detract from the merit of the agent.* I can't but observe here, that by the same way of reasoning exactly he might prove that there is a contrary disposition in the divine being; for that would enhance the value of good actions, and exalt the merit of the agent; and his illustration of his argument from moral goodness in men, will as well suit this case as the other. If a person producing a certain quantity of beneficence, has the greater moral merit the less he is supposed influenced by benevolence; and if we suppose him entirely without any natural affection, his moral merit would still be increased: by parity of reason, 'tis certain it must follow, that to produce the same quantity of beneficence with a contrary natural disposition, must imply still a much higher degree of moral merit. But it is, I think, plain enough that there is no great force in such reasoning as this; for in what sense does the supposition of benevolence in the divine being depreciate his good actions? Does it **[26]** render them less fit to be performed? This certainly can't be said, consequently the actions are full as good in themselves, as if he were altogether destitute of benevolence. Or does the performance of them imply less perfection in him? This will probably be said, because the more he is supposed influenced by benevolence, the less he appears to regard the moral fitness of things, or moral rectitude. This indeed might be the consequence, if actions were not therefore fit to be performed, because they have a tendency to promote happiness;

or, which is the same thing, if benevolence were not the main part of divine rectitude, which seems to be a point extremely plain, and is very agreeable to the concessions of our author. He owns it is intrinsically right and fit for God to communicate happiness to his creatures; if so, to desire their happiness, must be a right affection, and consequently there would be a defect in the divine rectitude if he were not benevolent. But for what reason should I go about to show that the perfection of the divine being is no ways lessen'd by supposing benevolence to be the principle of his actions, when 'tis plain that we hardly have any other notion of a good and amiable action, but that it proceeds from this principle? If kindness or a good disposition be not the spring, no matter what the nature or consequence of the action be, however beneficial it may be to us, we like the being that produced it never [27] the better; we don't think our selves obliged to gratitude, or imagine him any ways the more perfect, as to his moral character on the account of it.

As to the other part of the objection against this truth, that it detracts from the divine merit to suppose him influenced by benevolence, I confess I really don't well understand it, and therefore can't give a particular answer to it. But if there be sense in saying that the divine being merits by any of his actions, kind and benevolent actions surely can't be the least meritorious. However, we may allow our author his own way of reasoning, and suppose that necessary benevolence in the divine being depreciates the value of his actions, and detracts from his merit as much as he pleases. But does it do this any otherwise than as it adds to perfection of his nature? If the more imperfect any being is, whether it be by want of benevolence or ability, though he should merit the more by the same good actions, does it yet follow for this reason that we are consulting the honour of any one by denying or lessening his natural perfections, in order to inhance the merit of his actions? In this case we may more admire the actions, but in the other shall more admire and love the being that was the author of them. If God have no benevolent disposition towards his creatures, 'tis indeed vastly admirable that he should be so constant in doing good to them; but still if they had just reason to apprehend [28] that all the good they receive from him was very far from being the effect of any real kindness to them, they could not, I think, according to the present constitution of our nature, be hence induced to love or gratitude. Take away the supposition of the divine benevolence, and he no longer appears amiable and lovely, he no longer remains the object of our trust and confidence; other perfections render him awful and great, but 'tis this alone renders him the delightful object of our contemplation and trust.

SECTION III.

This point therefore being, I hope, sufficiently clear, *viz.* That we ought to conceive of God as truly benevolent. I come now to consider whether we have any reason to imagine that he is actuated by any other principles distinct from this, or which are in any case opposite to it; and that he is not, will sufficiently be manifested, if it should appear that to a wise and rational being those things that have no relation to happiness, must necessarily be looked upon as indifferent. In order to make out which, let us suppose a person entirely unconcerned with any sensible being whatsoever, and see if you can observe any reason why he should at any time act contrary to his own happiness, or esteem any actions fit and proper to be performed that are not conducive to his own felicity. For it's plain, since we suppose him to be a solitary **[29]** being, that has no connexion with any other, no desire of their happiness or influence upon it, his condition can't be rendered better or worse, more or less eligible, but by that which increases or diminishes his own happiness; for what reason then can he chuse or approve of any action, but as it tends to this end? I don't deny but such a solitary being might have various inclinations and affections distinct from a calm regard to his own interest, some of which might engage him to do things whereby his own real happiness would be diminished. He might be led also by the constitution of his nature to approve of some actions as amiable and decent, without thinking of their tendency to happiness; but then, if he acted as a wise and rational being, he would reflect on these his inclinations, and affections, and determinations of his mind, and as far as he found that they were really prejudicial to him, so far he would be willing to alter and correct them: nor can any one possibly blame him for such a conduct, since being supposed to have no concern with any other sensible beings, he can't regard them; and on his own account, nothing can be eligible but what is conducive to his happiness. 'Tis easy to say that some actions are fit and reasonable in themselves, without any regard to happiness, on which account they ought to be performed; but really to conceive of any such fitness in the case of the being we have supposed is, I think, absolutely impossible. **[30]** For what is it can render an action fit to be performed, but the effects it tends to produce? and what can any effect be to such a solitary being, that is not an addition or diminution to his happiness? Does not every one at first view see, that to him it is to all intents and purposes the very same as none at all, and therefore must be utterly unworthy of his choice?

Besides, if this being could be reasonably influenced by any motives not entirely subordinate to a regard to his own happiness, these motives must have some weight even in opposition to it; otherwise their force will (if I may so speak) be infinitely small, or nothing: and from hence it would follow, that a person may have just reason to do those actions that will be hurtful to himself, when he knows no one being in the world can be the better for them. But if such a conduct can be fit and reasonable in it self, it would puzzle the wisest head to tell what can be absurd and foolish. And if it be allowed that there are some reasons which may justly sway with me to prejudice my self where none receives any advantage by it; since I can't be obliged to have a greater regard to another than my self, the same, or like reasons, may justly incline me to hurt him, where neither I, nor any one besides receives any advantage from what he suffers; which positions make morality, instead of consisting in following the dictates of a regular self-love and real benevolence, to be **[31]** an obedience to those reasons, which every sensible and benevolent mind must wish should often be disregarded.

Now, if it be allowed me that such a solitary being as has been supposed cannot reasonably act but from a view to his own happiness, it is plain that the same thing must hold true of all other beings with respect to that part of their conduct in which none besides themselves have any concern, *i.e.* which does good or harm to none besides themselves; and therefore all actions which produce neither happiness nor misery, either immediately or in their natural tendency, must be absolutely indifferent.

But surely this is a proposition that need not be enlarged upon, since an indifferent action, and one that does neither good nor harm, seem to be phrases that only differ in sound, when in sense they are exactly the same; and a person that can chuse to busy himself in such actions as these, seems to be something beneath the character of a fool. And yet the concession of this one principle will certainly go a great way towards confuting this notion of our author. For if actions doing neither good nor harm are indifferent, and even according to him it is *cæt. par.* reasonable and fit to produce happiness and to prevent misery, the rectitude or obliquity of actions must altogether consist in their tendency to produce happiness or misery. Those actions that have no good tendency, but only occasion **[32]** misery, must always be evil; and those only are really good in themselves, that produce more happiness than misery; the former therefore can never be chosen, but by a malicious or mistaken being.

And for this reason we may safely conclude, that the divine being is *by no consideration whatsoever* inclined to lessen the happiness of his creatures; but that whenever he inflicts evils on particular persons, or societies, it is for the sake of a greater or more general benefit: for since he himself can receive no advantage by hurting his creatures, since there can't be supposed to be any competition of interests between him and them; his method of conduct towards them must be regulated solely by a regard to their general happiness, if we suppose any good design in it at all.

Now such a notion of the divine conduct as this, renders him really most amiable, and lays all his creatures under the strongest obligations to love, gratitude, and obedience, and is the firmest foundation of our trust and dependence upon him: But to imagine that he would do infinitely more good to his creatures than he does, were it not for other motives, by which he is influenced in opposition to their happiness, destroys the glory of his goodness, and renders all our expectations from him perfectly uncertain.

SECTION IV.

After what has been said, I think, I might leave the controversy with any unprejudiced **[33]** person; but as it seems to me to be one of the greatest importance, I am unwilling to pass by any thing, that has been advanced by our author in favour of the opposite sentiment, without a particular answer. I shall therefore now set my self to consider the several *reasons* of the divine conduct, that he has mentioned, and represented as *entirely distinct* from a regard to the happiness of his creatures; and if I can plainly show that none of these can be such *distinct reasons* of his acting, it will follow at least that he has not proved his point, but has left the matter as he found it, and still it may be true for any thing he has said to the contrary, that happiness only is the end of God in his works.

His first instance is the regard which God has to beauty and order in his works. This our author represents, as an end which the divine being pursues, entirely distinct from, and in a great measure opposite to the creatures happiness. "It may safely be presumed, (says he, *P.* 14.) that the "Deity necessarily loves order, and abhors confusion. Were it possible that "deformity and disorder could be more conducive to the happiness of his "creatures than the contrary; even upon this supposition it seems not prob-"able that they would have been submitted to." And afterwards he tells us, that he apprehends these two ends a regard to happiness, and regard to order and beauty do sometimes interfere, and that the former is rather subordinate to the latter than contrarywise.

I am sensible it could not but appear very odd, should I, in opposition to this sentiment of **[34]** our author, assert, that God had no regard to order and beauty in his works; neither do I say that, I only deny that this was an end of his acting distinct from the happiness of his creatures, and on the contrary affirm, that it was a regard to their felicity which was the reason why he has made and disposes of all things in so orderly and beautiful a manner. It is very plain, that the beauty and regularity of the divine works does contribute to happiness; and it is impossible for any one to imagine that irregularity and confusion should answer this end as well as the contrary: This therefore is a plain reason why the beauty of any work is esteemed a recommendation of it; nor is it, I think, possible for us to find out any other excellency in the most beautiful object, besides its tendency to give pleasure in the contemplation, or to promote some other useful purpose. If there be, why are we not informed what that excellency is? To say that such objects are in themselves more perfect than others, is only to deliver your opinion that they are truly excellent. I also am of opinion that they are excellent; my reason for it is, that they give delight in the contemplation of them, or are some ways beneficial; and any further than they tend to these ends, can imagine no perfection, no excellency in them. Beauty in any object seems to be nothing else, but *such a relation, order and proportion of its parts, as renders the contemplation of it agreeable*: In the object it self therefore 'tis nothing else; but a certain relation, order, and proportion of its parts. Now, I appeal to any one, whether he can possibly conceive [35] of any objects being the more perfect meerly on the account of the order and proportion of its parts, any farther than as that order and proportion renders it more agreeable or more useful, *i.e.* any farther than that order and proportion contributes to happiness; and if this be true, 'tis plain that the divine being could never design beauty in his works, but in order to promote happiness. But let us fairly examine what our author has said to prove the contrary: And here his first argument is drawn from scripture, where God is said to have created all things *for his own glory*. By this I understand that in all his works he designed the illustration of his own perfections, that his creatures might discern and adore his infinite excellencies; but then he cannot be supposed to have done this for his own sake, for he cannot be profited by the praises and applauses of his creatures. For what then? undoubtedly for the happiness of his creatures themselves, whose glory and happiness it is to know, and love, and serve their great Creator. This phrase therefore thus interpreted is no objection against the opinion of those, who suppose that the end of all God's actions towards his creatures is their happiness. But our author has given a very different interpretation of it, and supposes that the glory, for which God created all things, is that glory which consists in the approbation of his own works, *i.e.* as I gather from what follows, in the approbation of his works and actions as perfect and beautiful. This, I confess, is a very odd interpretation of [36] the word; however, I shan't dispute it; nor do I deny that God did so make, and does so govern the world, that infinite wisdom approves of every part of his conduct, and of all his works as the most perfect, and if you please also as most beautiful; nor will he on any account do that which he can't approve of. But it still remains to be shown, that he can approve of any thing in his works that is an hindrance to his creatures happiness, or that he can approve of that as beautiful which is no ways beneficial; otherwise such places of scripture as describe God as acting for his own glory, that is, in our author's sense of them, as acting out of regard to his own approbation, do not in the least favour his sentiments. However, properly speaking, it seems plain, that God cannot make his own approbation a distinct end in his actions; for a wise being designs, acts, and approves of his own actions for the very same reasons. His own approbation cannot make his actions more fit to be performed, does not make him think them the more so (for that would be to suppose that he approves because he approves) and therefore cannot be any distinct reason of his acting. As for those other expressions which he has produced from scripture, that *God has made all things for himself*, and that *for his pleasure they are, and were created*, I might very well neglect to take any notice of them, since he has made no particular use of them, nor any ways endeavoured to show how they favour his hypothesis; especially as I apprehend, that any reader, who is not carried away with the first sound **[37]** of words, cannot possibly so far mistake them, as to imagine that they are at all to our author's purpose. The first of these expressions I find *Prov.* xvi. 4. *The Lord hath made all things for himself: yea, even the wicked for the day of evil*. The meaning of which verse is, I apprehend, very well expressed in bishop *Patrick*'s paraphrase of it. "The Lord disposes all things throughout the world to serve "such ends as he thinks fit to design, which they cannot refuse to comply "with. For if any men be so wicked as to oppose his will, he will not lose "their service; but when he brings a public calamity on a country, employ "them as the executioners of his wrath*". The other expression is in *Rev.* iv. 11. *Thou art worthy, O Lord, to receive glory, and honour, and power: for thou hast created all things, and for thy pleasure they are, and were created*. The meaning of which is, not that God made all things for his own pleasure, and entertainment, and delight; or, that his own pleasure was any end of his works of creation or providence: but that all things were made, and do exist *by his will*. This is plain in the original, the words of which are, Διὰ τὸ θέλημά σου εἰσὶ καὶ ἐκτίσθησαν.

It would be to abuse the reader's patience to say any thing more to show that these texts have no concern in the controversy before us: As for that other place, where it is said, that *God saw* **[38]** *every thing that he had made, and behold it was very good*, in which by this goodness in the works of God, our author says, that we must needs understand something more, than a tendency and conduciveness in every thing to the benefit of living creatures, I need only observe, that I apprehend he cannot prove that any thing farther is intended, and till that be done, I am not disposed to take it for granted. 'Tis certain things are most properly said to be good, because they are fitted to do good, or to promote happiness. Why then may we not suppose that this is the reason, for which God says of the things that he had made, that they were all very good?

But to leave the consideration of arguments drawn from scripture, by which kind of arguments, it is, I apprehend, impossible to decide such a controversy as this, let us proceed to consider what he has farther observed in confirmation of his hypothesis. And here the strength of his reasoning

---

* Or, if this interpretation be disliked, we may translate the words thus: The Lord hath made all things suited to each other; yea, even the wicked to the day of evil. This Dr. *Clarke* says is their proper rendring. *See his Sermons, Vol.* VII. *page* 315.

seems to be this: Order and beauty are real perfections in any work, and consequently the works of God could not have been so perfect, if they had not been so beautiful; wherefore, as we cannot but suppose that the works of God are most perfect, he must have had a regard to beauty as well as happiness in them. Here I don't deny the consequence, the only debate between us is, whether beauty be any inherent perfection in an object, and not merely a sentiment in our minds, arising from some particular order and proportion in the parts of the object we contemplate. To **[39]** prove that beauty is a real perfection, our author argues from matter of fact, in the following manner, *P.* 16. "If beauty and order are not real, but relative; "if they consist wholly in an arbitrary agreement between the object and "the sense, what means that wonderful apparatus, that boundless profu- "sion of art and skill, which we every where meet with among the Creator's "works? What means that curious contexture and elaborate arrangement "of parts? Why are they so nicely adjusted to each other, and all made "subservient to the grandeur and magnificence of the whole? According "to the present supposition, how shall we avoid looking on all this as mere "waste of workmanship? If there be no objective perfections, no real im- "provements hereby introduced, I am forced to conclude that a chaos would "have answered the purpose full as well." This is indeed a very heavy charge laid against those that deny the real intrinsic perfection of beautiful ob- jects: They represent, according to our author, all that curious art, that fine contrivance which is apparent in the works of God, as mere waste of workmanship, since a chaos would have answer'd the purpose full as well. But this is a charge which I would hope our author himself, after more cool deliberation, will not be willing to maintain. Will he say, that in all the curious contexture of the parts of the universe it was only beauty and not use that was intended? Or if he should, will he not be confuted almost in every instance? Was it usefulness or beauty **[40]** that was chiefly intended in the internal structure of animals and plants? Here 'tis evident that the former was chiefly designed; since in these we can see no beauty, but that of exquisite contrivance to answer a kind intention; and yet in this part of the divine workmanship, beyond any other, there appears the nicest art. Nor can there, I believe, be many instances given, where we can discern beauty in the works of God, and yet see no use in that contrivance which renders them beautiful. The ingenious author of the inquiry concerning beauty and virtue has observed, that that very quality in objects, which occasions their beauty, would render them more eligible, supposing we had no sense of beauty at all, *Treat.* 1. *Sect.* 8. *Par.* 2.

Now if this be really the case, there can be no reason to ask what occa- sion was there for such a profusion of beauty amidst the works of God, if there be no other excellency in beautiful objects besides their tendency to promote happiness? For the answer is very plain. The world is so framed, that every particular part of it is of use to promote happiness, and a sense of beauty was given still farther to advance this end. And it must be very absurd to say, that all the curious contexture and elaborate arrangement of the parts of the universe was upon this supposition mere waste of workmanship; since things were thus ordered, to promote the most useful purposes, and no less display the benignity than the wisdom of the Deity. Whether our sense of beauty be arbitrary or not, happiness alone required that a **[41]** regular world should be chosen, rather than a confused chaos; and therefore that the world is thus regular, can be no argument that our sense of beauty does not arise from the arbitrary constitution of the divine being. "But are not uniformity and variety real relations belonging to the objects "themselves? Are they not independent on us and our faculties? And would "they not be what they are, whether we perceived them or no?" *P.* 17. There is never any one denied this; yet hence our author seems to draw a very strange and surprising conclusion, when he asserts, that for this reason the author of the inquiry concerning beauty and virtue has fixed beauty on such a foundation, as is entirely inconsistent with his own hypothesis.

His hypothesis is, that those objects appear beautiful to us, in which there is *uniformity amidst variety*. He never imagined that uniformity and variety were themselves beauty, or that objects thus qualified must necessarily appear beautiful to every rational mind, either of which assertions seem inconsistent with his notions. But he asserts it only as matter of fact, that such objects, as are uniformly various, do appear beautiful to us. And this I suppose none who have examined the matter, but will acknowledge is pretty near the truth.

But what follows from hence? Does it follow because our perception of beauty arises from objects uniformly various, and uniformity and variety are real relations belonging to the objects themselves; that therefore beauty, which it is plain is neither uniformity nor variety, nor **[42]** any mixture of the two, is a real quality in objects, and not only a sensation in our minds, occasioned by somewhat real in objects themselves. Is it not possible that the sour taste of vinegar may arise from the acuteness of its particles? Now supposing this to be the real case, then the acuteness of the particles of vinegar are as much the foundation of its sourness, as uniformity and variety are the foundation of beauty. But did ever any man in his wits imagine that sourness is a real quality in vinegar? I cannot therefore but wonder that when Mr. *Hutcheson* (*Sect.* 2.) professed that it was his design to investigate what it is in objects that occasions their appearing beautiful to us, he should ever be taxed as contradicting his own hypothesis, by saying it is *uniformity amidst variety*.

The charge is no more just than if because a person should say that the intestine motion of the particles of bodies is the cause of heat, he should be represented as asserting, that heat is a real quality belonging to such bodies themselves.

Again our author says, (*Pag.* 21.) "It is not to be doubted, but that "the universe is most regular, most harmonious, and most beautiful in the "sight of the divine being. These beauties therefore must be real, absolute "and objective perfections; for whatever the Creator sees in his works or "ideas, must be actually in them. What they appear to him, that they are "precisely in themselves; either therefore the natural and moral world ap- "pear to the divine mind without beauty, or the **[43]** beauties he perceives "in them must be real and inherent." Here I can't but just take notice of the unfairness of this argument. Our author had before taken notice that order and beauty are therefore amiable in the sight of God, because they are real perfections; and had acknowledged that all that he had said concerning the Deities regard to order and beauty, was altogether groundless, but upon the supposition that these are real and inherent perfections in any object. From whence, I think, one may fairly conclude, that even he himself has no other reason to believe that God does discern any beauty in his works, or makes this a distinct end of his acting, but this, *that beauty is an inherent perfection in an object*. To what purpose then are the scales now turned, and that must be made a medium of proof, which before was the conclusion?

Before God could not produce disorder, and be the author of confusion; because beauty and order being real perfections in any work, must be perceived by him as such, and be amiable in his sight. *Now* he sees beauty and order in his works, and therefore they are real perfections in them; which last way of arguing is certainly very absurd, because it is impossible that we should know that God can discern any quality in an object, but by knowing first that it is a quality which does belong to it, unless it were by the help of a revelation. We can't know that the works of God appear beautiful to him, but by first knowing that they are beautiful in themselves, and must appear so to every one that **[44]** perceives them to be what they are; whilst this latter therefore is in debate, to take the former for granted is really begging the question that ought to be proved. I am sensible, that to doubt, whether the works of God appear beautiful to him or not, must appear strange to those that are used to conceive of beauty as a real distinct perfection in objects themselves. As to assert, that bodies appear without all colour to the divine mind, would for the same reason appear very surprizing to the vulgar; and he that can free himself of one of these prejudices, may as easily get rid of the other.

But I must here observe, for fear of being misunderstood, that I by no means doubt that the works of God are contrived in the most exact and regular manner to accomplish the ends he intended by them; the sum of which I take to be the happiness of his creatures. And this may, if you please, be called the beauty of his works; but without regarding an useful end and design, I think, there is no manner of reason to suppose, that any particular order and proportion of things appears to the divine mind more excellent and beautiful than another.

But I may here be asked, does confusion then and disorder in it self appear as amiable and lovely to him, as order, proportion and harmony? I answer, if we speak properly, nothing can appear to him confused and disorderly in it self, without a regard had to the end for which it may be designed. These are terms, that have an evident relation to the imperfection of our minds. Let an heap of stones be thrown together **[45]** without any design, and there will appear to us no order, no proportion in them; but still there is as real proportion in their sizes and distances, as if they had been ranged by the nicest hand, so as to make the most beautiful appearance to us. It is not any order and proportion of the parts of an object that renders it beautiful, as our author seems to imagine; for he says, *P.* 20. "The es-"sentials of beauty are order, symmetry and proportion; without these no "works of art are, or can be beautiful; and according to the degree wherein "those prevail, the beauty of these is greater or less." But there are only some particular orders, and some particular proportions of the parts of an object that appear agreeable and beautiful to us. Now, by what medium can we possibly prove that such an order, and such proportions in the parts of an object, which renders it most beautiful to us, must also make it appear so to the divine mind? especially when we often perceive beauty in an object, without so much as knowing what those proportions are which render it agreeable to us; nor till this be done, which has not been so much as attempted, can we have any reason to believe, that beauty, which in objects themselves is nothing else but a particular order and proportion of their parts, was any distinct end in the works of creation and providence.

On this single point the whole controversy seems to me to turn, and therefore I cannot but think it very well deserves a fresh consideration from our author. He says, the essentials of beauty are **[46]** order and proportion; but if I mistake not, the essence of deformity may as truly be asserted to be also order and proportion. In the same sense that one kind of order and proportion constitutes beauty, another kind of order and proportion seems to constitute deformity. Objects are not therefore ugly because they want order and proportion; for these as necessarily belong to all objects that have parts of the same kind, and that may be compared as to situation and quantity, as *figure does to all bodies. Divide any object into its several parts, and place them how you will, they will still be in some order or other; though perhaps in that order they may be of no use, and make no beautiful appearance. Again, enlarge, lessen, or change the parts of an object how you please, still every part must be in some proportion to the whole, and to every other part. 'Tis true indeed, that we usually say of objects that are deformed, that they want order and proportion; but the meaning of this is, only that they want that order and proportion that is

---

* Figure is nothing else, but the order and proportion of the external parts of a body.

necessary to their appearing beautiful to us. And the expression it self is no more proper, than when we call that body shapeless which is ill-shapen.

The reader will observe that I use the word proportion in that which is the common sense of it, as I apprehend, among all writers (unless we must except some of the mathematicians) to denote the comparative greatness of two quantities, in which sense we use the word when we **[47]** speak of the proportion between the circumference and the diameter of the circle; between the height of a building, and its length or breadth, and the like. But I thought it proper to observe, that some mathematicians using the word *ratio* instead of proportion in the foregoing sense, define *proportion* to be an *equality of ratios*; because in this sense of the word, it must be acknowledged that proportion has a considerable effect upon the beauty of an object; but not so much as that where there is the same proportions in this sense, or the same number of equal ratios, there must always be the same beauty. Take, for instance, two humane faces, the one perfectly beautiful, the other exactly like the former in all other respects, only let the eyes be too little, the nose too long, or the nostrils too wide; and I apprehend, that though these must needs differ in beauty, yet it will be impossible to show that in any sense there is more order or proportion in one than the other. So that it can't be justly said, that the essence of beauty is order and proportion; or that we have any reason to think that every being that perceives the same order and proportion in an object, must have the same sentiments of its beauty. When I say that order and proportion are not the essence of beauty, my meaning is, that it is not every order and proportion that constitutes beauty, but only some; and why these rather than others should make a beautiful appearance even to us, we can't assign any reason; and therefore can't possibly conclude, that they must raise the same sense of beauty in every other rational **[48]** mind that perceives them. But enough has been said on this head, at least to show that our author has given no sufficient proof of his assertion, that a regard to beauty and order was any distinct end of the divine being in his works.

Section V.

I proceed therefore now to another assertion of our author, which contradicts the opinion of those who suppose that God has no ends in his actions towards his creatures which are not intirely coincident with their happiness; which is this, (*P.* 52. *See also P.* 33, *&c.*) that "every instance, "every degree of sin, is a just ground for suitable punishment, whether the "ends and intentions of government require it or not; " and therefore, that when neither the reformation of the offender himself, nor the good of others is designed, the desert of sin it self is a sufficient inducement to the divine being to punish him in proportion to the greatness of his fault.

I should very readily own, that every instance, every degree of sin is a just ground for suitable punishment; because it is for the general happiness that every sin should be discouraged, and that all the rational creatures of God should be convinced that they can never offend against his righteous laws without injuring themselves. But if a person will abstract from this which is plain reason, and, I think, the only true reason of punishment; and will suppose the case, that no good is done or designed by **[49]** punishment, he certainly is obliged to assign some other reason for it. In order to do which, our author supposes that there was but one moral agent in the universe, and that he was obstinately and incorrigibly wicked; and on this supposition says (*P.* 34.) "It is not to be doubted, but he would feel the "effects of his Maker's displeasure: nay, that there are good grounds to "believe that his punishment would be proportioned to his crimes; and "that this is highly probable on many accounts, and if he mistakes not, "conformable to the doctrines both of reason and revelation."

I might here, I apprehend, very justly question the possibility of the fact supposed by our author, and therefore have still reason to doubt of his opinion, tho' it should be a just deduction from it. But I don't insist on this; and on the supposition of the reality of this case, I would willingly own that such a being might be punished, as far as this was necessary to prevent his future wickedness. But this will, I suppose, be hardly allowed to be any proper punishment; because it would be a less evil to the criminal himself, than his own wickedness would have been which is supposed to be prevented by it. However, 'tis plain this can't come up to our author's intention; for he supposes (or I strangely mistake his design) that punishment may and ought to be inflicted on a wicked person meerly on the account of the desert of sin, tho' no good were hereby derived to himself, or to any one else. **[50]**

To prove which he argues in the following manner, (*P.* 34.) "If we at-"tend to the principle of *divine rectitude*, we shall find the Deity ever act-"ing, and ever determined to act according to the *reason of the thing* and "the *right of the case*. If then it be right and reasonable in it self to punish "*wickedness*, as well as to reward *goodness*, he will assuredly do both."

Thus far we have no controversy, God will assuredly reward the virtuous, and punish the vitious; as far as it is reasonable, and fit he should. But what do we learn from hence? He goes on, "If *virtue* be necessarily amiable "in his sight, *vice* is necessarily odious; if there be *merit* in the one, which "naturally recommends it to his favour and esteem, there is equal *demerit* "in the other, which unavoidably excites his displeasure and indignation." This also is allowed, if I rightly understand the unusual phrases, that there is a merit in virtue that recommends it, and a demerit in sin that excites indignation against it. I suppose no more is intended by them, than what was as plainly said before, that virtue is necessarily amiable, and vice necessarily odious. But what is the proper consequence from this? Is it any thing more than this, that God will therefore take all proper methods to secure the cause of virtue, and to discountenance all sin and wickedness? Certainly it by no means follows, as our author would insinuate, that therefore it is always in it self morally fit, and agreeable to the rectitude and [51] sanctity of the divine nature, to punish the vitious and to reward the virtuous; even supposing the case, that such rewards and punishments had no tendency to answer the ends of the divine government, which are the promoting virtue and happiness. A love to *virtue* can't dispose a person to reward, but with a design to promote and encourage *virtue*. An hatred of *vice* can't dispose a person to punish, but with a design to discourage and discountenance *vice*. We may be very great enemies to vice, and yet not be enemies to vitious men, or inclined to hurt them; any further than an opposition to their vices may make it necessary. Otherwise no good christian could be a virtuous man, since with the greatest aversion to sin he is obliged to maintain the most real benevolence towards sinful men. And if this be commanded us as our constant duty, I can't see how we can assert that it is inconsistent with the divine rectitude to show mercy to sinners, and to make them experience the effects of his benevolence, where this can be done without prejudicing the cause of virtue and goodness. But "should "God communicate good, and finally dispense happiness to all deserving "or undeserving promiscuously; how could he be said upon this suppo- "sition to act according to the true nature, and reason of things? Virtue "and Vice are essentially opposite, and therefore it is not possible that the "same treatment should suit the votaries of each." (*P.* 35.) It is here readily acknowledged, that it would not be right to communicate [52] happiness equally to all sorts of persons without any difference, that it would not be right to give as great rewards to the evil as to the good. Such a procedure would not be treating persons according to their real character, nor the way most effectually to promote virtue and happiness. But what is all this to the point in view, which was to show, that without a regard to the deterring from sin, there is sufficient reason to punish bad men? Is there not as great a difference made between the virtuous and the vitious, as any wise and good being would desire; if the virtuous are always rewarded for what they do well, and the vitious are always punished, where their punishment is conducive to the general good and necessary to deter from sin? Is not the very reason why virtuous and vitious men should be treated differently, that virtue ought to be encouraged and vice discountenanced? And therefore where these ends are out of the question, there does not appear any reason from these opposite characters for an opposite treatment. But can we suppose that God would suffer an incorrigible sinner to go unpunished, when the quite contrary is deserved? Is not this essentially repugnant to the rectitude and purity of his nature?

I don't apprehend that God will suffer any impenitent sinner to go unpunished, but not meerly because this is deserved; for then no sin must go unpunished: but because his punishment is necessary to secure the respect due to the divine laws. And supposing this end out of **[53]** the question, no one can say, that meerly the deserving punishment is a sufficient reason for its being inflicted, that will allow any room for the exercise of pardoning mercy and grace.

As for what our author says, after having represented a most villainous character, which he desires us to reflect upon without the emotions of resentment, *viz.* that the most merciful man in the world would pronounce sentence upon such an offender, and give him up to condign punishment; it is not improbable that this would be true in fact, because the most merciful man in the world has his resentments; and it is impossible, but he should believe there would be good consequences from punishing such a person. But if we could suppose the contrary, that sparing him would certainly do no hurt, that punishing would do no good, I can by no means be of our author's opinion, that to spare would not be goodness, but weakness. For what weakness is it to spare, when all the reasons of punishment are supposed away? His reason for it, that to spare would be violating truth and nature, and acting contrary to the plainest reasons of things, is only a more fashionable way of saying that he is very positive, 'twould be exceeding wrong; for he does not tell us one truth that would be violated by it, nor give us one of those plainest reasons, which should persuade us to a contrary conduct. And I am satisfied I should be far from being singular if I should deliver my opinion on the opposite side of the question, in the same way with our author, **[54]** that to give pain to any sensible being where no good end is answer'd by it, nor any mischief prevented, is violating truth and nature, and acting contrary to the plainest reasons of things; 'tis certainly acting a mischievous and ill-natured part.

SECTION VI.

There is only one opinion more of our author, which, I think, it is needful to take notice of, as seeming too much to confine the divine goodness, and supported by no reason; and that is, "that it is not consistent with the "divine rectitude to bestow that supreme felicity, that indefectible state, "wherein consist the rewards of the righteous on any that have not first "gone through a state of probation," *P.* 46. Which is as much as to say, that God can't, consistently with his own attributes, create any being in such a degree of perfection, that he should be in no more danger of falling from his favour, than good men will be after death.

This proposition, I am sure, must appear at first view so inconsistent with our notions of the boundless goodness of God, that it can't be expected we should admit it without the plainest proof. 'Tis acknowledged by all, that God may bestow what favour he pleases on his creatures, antecedently to their doing any thing to deserve it. Why may he not then make creatures with the most pure and uncorrupt affections, and with such a large and distinct knowledge of the nature [55] of things, as to discern the connection between virtue and their own interest, as plainly as we know by experience the connection between eating and living? Now only allow that beings may be created thus perfect; and there will be no more danger of their falling, than there is that a man in his wits, for no reason at all, should thrust his hand into the midst of the fire, that is, there will be no danger of it at all. And if you deny this, how can you acknowledge that God may grant what favours he pleases to his creatures? Yea, if no beings were at first made thus perfect, what will become of the beauty of that scale of beings which our author speaks of, the highest of which was in danger of falling below the condition of the meanest sensible being, that is, into a state of misery? And is it not wrong in us to presume so far to limit the divine goodness, as to say that he can't bestow on any a favour that is perfect and compleat before they have deserv'd it, which it's plain no favour can be whilst there is any hazard of losing it?

But let us hear what our author says to justify this surprizing position. "Perfect happiness (he says, *P.* 41.) was and ought to be reserved for the "proper objects of God's love and favour; which none of his creatures could "be without virtuous qualifications and moral merit: and these imply a "state of probation." That as to confer the same happiness on the *faithful*, and on the *unfaithful* and *disobedient*, would be a manifest violation of truth [56] and nature; so "by parity of reason it was morally unfit to treat "those who had never been *tried*, and by consequence merited nothing, in "the same manner as if they had been tried and found faithful; because "such persons are in a moral sense *worthless*, and by consequence are in "a station as much below *merit* as it is above *demerit*." I have given, I apprehend, the full strength of our author's argument; and before I come to give an answer to it, I find it necessary to say somewhat in order to settle the meaning of the terms here made use of by him, on the ambiguity of which, if I mistake not, all the seeming strength of his reasoning intirely depends.

First then, let us enquire what we are to understand by *virtuous qualifications*; these cannot here signify meerly good dispositions, and such valuable qualifications of temper, as fit a person for the greatest serviceableness to himself and others. Even though a person have the strongest desires to promote the honour of the Creator, and the kindest affection towards his fellow creatures; if these arise from his original make, and are not acquired by himself with some difficulty, they are not according to our author virtuous qualifications; for these, he says, imply a state of probation, *i.e.* a state of difficulty and temptation. By virtuous qualifications therefore, I suppose, he must mean, what I would call, in opposition to those good dispositions which we receive more directly from our Creator, *acquired virtues*. **[57]**

The next word, that needs explication, is that of *moral merit*; by this he cannot mean only real worth and excellency; for that does not necessarily imply a state of probation. But by moral merit we must, I apprehend, understand a *right to a reward*, or that which gives such a right; in which sense of the word, I allow it can't be applied to any untried creatures. Those creatures, which by their original make are so constituted, that their desires and their duty always necessarily coincide, can't, I think, be said to have any claim to a reward: whereas those who are surrounded with difficulty and temptation, and who are obliged to deny themselves and submit to great inconveniencies that they may maintain their integrity, if notwithstanding this, they do behave uprightly, seem on this account to have an equitable claim to it, which they may deduce from this principle; that a wise and good God will certainly, in every instance, make it the interest of his creatures to behave virtuously. In this sense therefore creatures only in a state of probation can be said to merit.

I am sensible that to many it will seem absurd to say that a creature can any ways merit at the hands of God; and I confess, that in that sense in which men by showing kindness to each other merit a return of favour, it is absurd to say that any being whatsoever can merit at the divine hands. It seems also plain to me, that in particular the happiness that is promised to the righteous in the gospel, cannot be looked upon **[58]** as what they are capable of meriting by their obedience. Perfect and unalterable happiness is really of infinite value; and therefore cannot but appear as a reward greatly disproportionate to the temporary services of any creatures, and much more so to the imperfect obedience of men in this short life. So that if none must receive but according to their merit, which is the principle on which, if I mistake not, all our author's reasoning depends; mankind must for ever despair of a state of perfect and unalterable happiness. But if we take the word *merit* in the sense before mentioned, as signifying an equitable claim to a reward, *i.e.* to a reward proportional to the difficulty of performing, or the inconvenience that is submitted to in the performance of any virtuous action; in this sense, I think, a person may be said to merit, and thus it is plain he can merit only in a state of probation. And in analogy to this sense of the word must we understand that very harsh assertion of our author, that all beings that have not undergone a state of trial, are in *a moral sense worthless*, viz. that they have no proper claim to a reward. For in any other sense, nothing can be more strangely absurd, than to say that any creature is therefore in a moral sense worthless, because he was always possessed of so much wisdom and goodness as not to be in danger of offending, *i.e.* because he is most like the infinitely perfect being.

These things being observed, the forementioned **[59]** argument may be answered with very little difficulty. I readily allow, that perfect happiness will be given only to the proper objects of God's love, and favour, and esteem; but in order to these acquired virtues and moral merit, in the sense before explained, are not, I apprehend, absolutely necessary. For good dispositions and inclinations, a right temper and byass of mind, are of themselves a just foundation for love, favour, and esteem; though they don't seem to give a person a proper title to a reward. Nor does it follow, that because one has undergone the severest trials and overcome in them, that therefore he is a better and more valuable person than another that has not been thus tried; and consequently this latter may possibly have as much of the divine favour and esteem, as the former. We indeed can only value persons from their actions, because 'tis by this means only we form a judgment of their internal temper; but he, who immediately discerns the inward dispositions of the mind, can have no reason to wait till these be discovered by external actions, in order to know where his love and esteem should be fixed. Esteem, if I mistake not, ought always to be proportional to the good qualities of which a person is possessed, however these have been acquired; for *to have one in due esteem*, is to apprehend him to be what he really is in himself. And where a person is justly esteemed for his good dispositions of mind, it **[60]** can be no reason against granting him any favours which he is not likely to abuse, that he has no just claim to them as the reward of his actions. But, says our author, "if rewards are due to the righteous, and "punishments to the wicked; those who have never been tried can deserve "neither, and therefore cannot, agreeably to truth and rectitude, be treated "like either of the other." As this argument stands, the utmost that it can prove is, that an untried person ought neither to be rewarded nor punished; which is very different from that which he undertook to prove, that such an one ought not, through the undeserved goodness of God, to be made partaker of perfect happiness. But if this argument had been intended to have proved the point in view, it should have run in the following manner:

If perfect and unalterable happiness be due to the righteous as the reward of their services in a state of probation, and misery be the just portion of the wicked; those who have never been tried, and consequently have deserved neither, ought not to be treated like either of the other, *i.e.* ought not to be made either perfectly happy, or miserable. But in answer to this, I say, *perfect* happiness is not due to the righteous as the reward of their services; 'tis not on the merits of their own works, but on the gracious promise of God, through Christ, that their expectation of this is founded: and therefore, since they at last must not receive it, but as an undeserved **[61]** favour; why mayn't the same favour be conferred on others that are equally fitted for it, though supposed unequal to them in merit? Nor would it follow, if we could suppose a person capable of meriting perfect happiness; that therefore it should be bestowed on none, but those that do thus merit it; and that no room should be left for the exercise of undeserved favour. Because one has a just claim upon me, that is no reason why I may not freely give to another; even more than what the former can in justice demand.

'Tis a mighty weak way of arguing, to say that two persons ought not to be treated alike, that is, that they should not both receive the same happiness; because one has a particular claim to it, which the other has not: if one has merit, he will receive according to his merit, and when this is done, with regard to undeserved favours, he stands upon a level with the other creatures of God; and it is no absurdity to suppose, that it may in some cases at least seem good to his infinite wisdom to bestow these in the largest proportions on those that have not the greatest merit. Yea, as I think that none can deny, that God has a right to make a difference between the conditions of his creatures by his undeserved goodness, without any regard to their merits; so none can pretend to fix the bounds beyond which the difference hence arising ought not to proceed: and therefore we can't **[62]** possibly have any reason to assert, that God may not give some creatures greater happiness before they have merited it by any actual obedience in a state of trial; than what it is possible for others to gain by their merit.

But is not this to represent the divine being as acting in an arbitrary manner, and as having a particular fondness for some more than others without any reason at all? I answer, this no more follows from what I have said, than from the common opinion, that God has originally made creatures of different ranks and capacities for happiness. A man, for instance, has no more reason to complain of any arbitrary or partial proceeding in the divine being, though with all his boasted merit he should not be able to attain the happiness of angels; then a brute can complain that he is utterly uncapable even of that happiness, which men may enjoy in this life. But all that follows from what I have asserted is, that God has other reasons for bestowing happiness upon, and for diversifying the happiness of his creatures, besides a regard to their merit; which is what every one must allow that will not run contrary to the plainest matter of fact, as far as we are capable of judging concerning it. Besides this argument drawn from the divine equity, our author adds another taken from the natural connexion between virtue and happiness. Says he, supreme felicity is the peculiar portion of those that have gone through a state of probation, **[63]** "partly by God's ap-"pointment, and partly from the very nature of the thing; for since a state "of probation is necessary for the exercise and improvement of virtue, it is "by consequence necessary for the consummation and perfection of happi-"ness." That good or virtuous dispositions, and actions are necessary to the perfection of happiness is most certainly true: but I must confess, I can't imagine what our author intends by *virtue*; when he says that a state of probation, *i.e.* a state of temptation is necessary for the improvement, and even the exercise of it. Does the ignorance of men's minds, the irregularity of their affections, the external difficulties and temptations they meet with, which are what render the present a state of trial, give them any advantage for improvement in virtue? Or rather are not these very things hindrances to their progress in it? Are not these the reason that almost all men in this life are so very imperfect in virtue? But however this be, it is, I apprehend, certain, that but few even of good men do in this life arrive at that perfection in virtue, as without any farther improvement in it will fit them for a state of compleat happiness; so that if there can be no improvement after they have gone through this state of probation, I don't see how they can ever be capable of a state of the most perfect felicity. And if good men after death are capable of increasing in goodness and virtue, it can't be said that a state **[64]** of trial is necessary for this purpose; and therefore such a supposed necessity can be no solid objection against God's placing some creatures in a state of perfect happiness without obliging them first to go through the dangers, and temptations of a state of probation.

SECTION VII.

I have now gone through the several positions of our author, which seemed to me to contradict the notion of such as suppose that God has no other design in his dispensations towards his creatures but their happiness; and I hope, that what has been said sufficiently shows that he has not proved his point in a single instance. There is however one considerable objection still remaining against the opinion I have defended, which I have hitherto taken no notice of; because though adduced by our author in support of his notion of the divine Being's having a regard to the beauty and order of his works, as an end intirely distinct from the happiness of his creatures; it is not so properly a proof of his particular opinion with regard to beauty, as a general objection against his adversaries notion of happiness being the only end of God in his works; and for this reason I thought proper to defer the consideration of it to this place. The objection is, that if happiness had been the only end of God in his works, the greatest possible quantities **[65]** of it must have been produced at all times, and in all places; and it is thus urged by our author (*P.* 14.) "Had the production "of *happiness* been the sole end which the Creator had in view, it is "not, I think, to be doubted, but the utmost possibilities of it would have "been produced at all times and in all places. But as far as we are capable "of judging from the *phænomena* within our reach, this seems not to be the "case. That scale and subordination of beings beforementioned, may seem "to promote the *order* and harmony of the world more than the happiness "of its inhabitants. Had their several powers been more nearly equal, what "would have been the result? It seems probable that the latter of these ends "would have been advanced; and the former obstructed, if not destroyed." This is indeed at first view a very difficult objection; that small degree of happiness, that evil we see in the world, has been strongly urged against those that have believed that the world is governed by a wise, and powerful, and good God. But I hope it will not oblige us to let go this important truth; or incline us to imagine that the Deity is influenced by any principles of acting, that contradict his goodness.

In order to solve this objection, it will be proper first to enquire, whether the Divine Being was obliged by any of his perfections to create any beings at all. Indeed **[66]** after a universe of creatures is formed, his goodness seems to incline him to confer the greatest happiness upon it, of which it is capable; but it does not appear that even goodness it self obliged him to make those creatures. Creation is an argument of the goodness of God; because he would not have made any creatures at all, if he had not intended to communicate happiness to them; but I question whether in strict propriety it is always an instance of it. An act of goodness, one would think should require that the object of it should before exist; but as this argument may appear too subtle, and as I can't my self lay any great stress upon it, I only just hint at it, that the reader may further consider it, or pass it over, as he thinks fit. But I think we may more securely conclude, that God was not obliged by any of his perfections to create, at least not to create the universe in any particular degree of perfection, from the consequences that necessarily follow from such a supposition. Should we, for instance, suppose that the divine goodness obliged him to make a world, that there might be creatures to whom he might communicate happiness; the consequence of this seems plainly to be, that a universe must have been from eternity, in which there must have been the greatest possible happiness. Nor, if we should go upon our author's hypothesis, that the influence of the divine goodness is limited by his regard to the order and beauty of his works, **[67]** shall we be able to avoid the main part of this difficulty.

For if, notwithstanding a regard to order and beauty, the divine perfections did at any time oblige him to create; in every instant before that time there must have been the same obligation upon him, from his perfections, to have formed a most perfect and beautiful, though not a most happy world. It must follow also, even upon his hypothesis, that the world must be infinite as well as eternal, for no regard to beauty can set bounds to it; and that every creature must enjoy as much happiness as is consistent with the perfection of the whole; and that no more can possibly be created without destroying this perfection.

But as these consequences, particularly the last, are what I should by no means be willing to defend, I cannot entertain a favourable opinion of the principle from which they flow; and I think our author himself ought herein to agree with me, as he is in opinion for the* non eternity of the world. I know it is usually presumed that all the works of God are most

---

* *Letter to a Deist*, p. 7.

perfect; and thus they undoubtedly are, if by this be meant, that they are exactly agreeable to the highest goodness and wisdom, and have nothing in them that implies want of perfection in their author. But that the universe has such a degree and quantity of perfection, as that no addition **[68]** can possibly be made to it even by infinite wisdom and power, seems to me so far from being what we ought to suppose true; that I question whether the notion it self be not perfectly unintelligible. It is certain, that to speak of the greatest possible number, or the greatest possible triangle, is to use words without any meaning at all.

   A triangle may be made as large as you please, yet the largest possible cannot be; for such a one could not be a triangle, which is a surface bounded with three strait lines. In like manner we may, I apprehend, speak in relation to the happiness and perfection of the universe. God may make it as happy, and as perfect as he pleases, and may continually increase this in any proportion he thinks fit; but still I apprehend 'tis capable of this increase, without limits, and without end; and that to suppose the greatest possible quantity of happiness or perfection diffused through it, is to suppose that there is a certain fixed and determinate quantity of happiness and perfection, beyond which it is impossible, even for the power of God to proceed; which I must own seems to me absurd. So that to argue against the goodness of God, because there is not the greatest quantity of happiness and perfection in the universe, is to use an argument that can have no force, since, if put into form, one of the premises is unintelligible. I add moreover, that if it be a contradiction to **[69]** suppose that the universe should be absolutely most perfect, or most happy, so as to be incapable of further improvement, even from infinite wisdom and power; we can see no possible reason, why amongst the several degrees of perfection that might have been communicated, one should be chosen rather than another; for if one ought not to be chosen because there is another greater conceivable, for the same reason this ought also to be rejected, and so on for ever. And on this account we must be forced, I think, to acknowledge, that, for any thing we know to the contrary, God was not obliged to create the universe with any particular degree or quantity of perfection, or even to create it at all; and if this be true, 'tis very far from being a just consequence that God must be influenced by other principles that contradict his goodness, because we don't find the greatest possible happiness at all times and in all places. I have insisted something the longer on this point, because I take this to be the true answer to the objection beforementioned; and also because, if what has been said be true, it will make the solution much easier of many difficulties that are raised against the divine dispensations. If we conceive of the goodness of God as an unbounded inclination to create happiness, and consequently suppose he has made the world as happy and as perfect as he possibly could, there are undoubtedly abundance **[70]** of *phænomena*, the consistence of which with this supposition we cannot discern, and which we shall find some difficulty to perswade men are not incompatible with it: But if we only conceive of the divine goodness as a most kind affection towards his creatures, and as inclining him to confer upon that universe of creatures he has made the greatest happiness of which they are capable, still supposing that their original capacities for happiness were fixed by his will and pleasure, we shall find it much easier to satisfy our selves, that there is nothing in any of the appearances of providence contrary to the most perfect goodness of the divine nature.

But though, I think, the way I have already taken is really the proper way of answering our author's objection; yet, as I am sensible many will not readily come into it, I shall now make the most favourable supposition on his side, and see whether even then his argument will not admit of a fair answer. I will allow, that the consequence of supposing the Deity influenced by benevolence only, and no other principle contrary to it, is, that the universe must be created in the most perfect manner, and that the greatest possible quantity of happiness must be diffused through it; for even upon the supposition of the justness of this consequence, nothing that appears in the divine conduct towards his creatures, will oblige us to allow that the Deity is influenced by any other principle **[71]** of acting that contradicts or limits his benevolence. Our author supposes, that if the end of God in his works was only the good of his creatures, there must have been the greatest quantity of happiness communicated at all times, and in all places; and therefore that the universe must have been perfectly uniform, without any variety, and consequently without any beauty and order. For which reason he imagines that the world was formed in a different manner; but this is a consequence he can never possibly make good. All that can be imagined to follow from the most unbounded goodness, is, that the greatest quantity of happiness should be communicated to the universe; but it no ways appears that this must arise from the same happiness at all times, and in all places. Such a perfect uniformity may, for any thing we know to the contrary, be as inconsistent with the greatest happiness of the whole, as it is with its beauty. If the universe were to consist of one uniform sort of beings, however happy they might be, 'tis evident that they could not in some respects enjoy so great happiness as they might by variety; there would then be but one uniform object of contemplation, one uniform relation among all the creatures of the universe; which could never afford so great happiness as variety of these to intelligent and rational creatures, whose happiness is in a great measure derived from objects of contemplation, **[72]** and the relations they stand in to one another. Hence, I think, 'tis very plain, that from the greatest variety imaginable observed between the degrees of happiness bestowed on the several creatures of God, a defect of happiness in the whole cannot be inferred.

Besides this, it ought to be considered, that a most happy universe is so far from implying the same happiness in all creatures, and thereby contradicting that variety which is necessary to beauty; that the quite contrary is implyed in it: For the most happy universe is not one that consists of the greatest possible number of the most happy beings only; but one that consists of that, and the greatest possible number of beings next inferior to the first rank, and so downward, till we come to those that approach the nearest to insensible matter. This certainly must be allowed, unless it can be proved that the greatest possible number of the most happy beings having been created, no others can possibly be made. So that a most happy universe is so far from being unbeautifully uniform, that it must be most beautifully various; a most regular and orderly advance of perfection being made from insensible matter, *without breaking the scale of beings*, quite up to the highest possible rank. Strange therefore it is, that any, because he only sees the lowest part of this scale, should from hence infer a defect of happiness in the whole, and imagine a necessity of asserting another principle of acting in the Deity, **[73]** which in a great measure hinders those communications of goodness that would otherwise have been dispensed. To me it seems plain, that even upon this supposition we can have no reason to conclude, that the Deity has any other intention in his actions towards his creatures besides their good.

And I cannot but apprehend, that every person that wishes well to his fellow creatures, must be pleased to see there is so little reason to support an opinion, which if true would render all our expectations of happiness perfectly precarious and uncertain. To suppose that the almighty and infinitely wise governor of the world, who can do what he will with his creatures, has ends of acting any ways inconsistent with their general happiness, must needs be a very uneasy and uncomfortable reflection to every truly good and benevolent mind; especially as it is confessed, that we know not particularly what these ends are, and how far they may prevail. But to look upon God as the tender father of his creatures, as in all his actions towards them, not excepting those of judgment and vengeance, consulting the general happiness, is to represent him to our selves under a notion the most amiable and lovely that can be imagined. Separate infinite goodness from almighty power and perfect knowledge, and the idea is only awful and terrible; but that these are necessarily conjoined in the glorious **[74]** Author of our beings, renders him the object of our most delightful contemplation as well as of our highest reverence. That the governor of the world is the best as well as the greatest of all beings, is the support of a good man under all his troubles; this is his comfort and joy under all the evils he observes in this present state; this is his most powerful engagement to his duty, and gives him the greatest delight and satisfaction in acts of obedience and devotion. Yea, would sinners themselves be perswaded rightly to consider this attribute of the Divine Being; would they look upon God, though not as fond and partial, with respect to particular persons, to the general detriment of his creatures; yet as truly benevolent to them all, nothing could be more proper to make them ashamed of their disobedience, and thereby lead them to the truest repentance. And I may add, that in reason nothing can make them more afraid of continuing in a course of vice and wickedness; for whilst God is good the cause of virtue and goodness must be maintained and supported, which necessarily requires that sin should be punished, as well as that virtue should be honoured and rewarded.

ERRATA.

PAge 22. *line* 4. dele *to.* P. 43. *l.* 17. for *by* read *for.* P. 44. *l.* 28. for *can discern* r. *discerns.* P. 46. *l.* 20. for *renders* r. *render.* P. 48. *l.* 11. for *is* r. *are.*

FINIS.

4.3 Commentary

As has already been mentioned, *Divine Benevolence* was published anonymously, and although it has often been ascribed to Joshua Bayes, there can be no doubt that Thomas Bayes was indeed the author. For in a footnote to his *A Review of the Principal Questions in Morals*, Richard Price wrote

> The author [of *Divine Benevolence*] was Mr. BAYES, one of the most ingenious men I ever knew, and for many years the minister of a dissenting congregation at TUNBRIDGE WELLS[7].
> [1787, p. 429]

In his Introduction the young Bayes[8] stresses the importance of the correct apprehension of the divine perfections and attributes, and mentions the 'laudable' attempt made by Balguy in this respect, defective, at least in Bayes's eyes, though that may be.

In his first section Bayes considers the moral perfection of God as distinguished into several attributes, the principal ones being *goodness, justice*, and *veracity*[9]. On pages [5] to [7] he argues that the choice of *rectitude* as the fundamental principle results in 'tedious and laborious' arguments for the deduction of such attributes.

The first criticisms that Bayes makes of Balguy's work are based on the following passages from the latter's tract.

> However we may divide and distinguish God's Moral Attributes, according to the different Effects, Dealings and Dispensations resulting from them; yet in themselves they seem to be but one and the same Perfection variously exercised on different Objects and Occasions, and in different Cases and Circumstances; and cannot therefore, without Error and Inconvenience, be consider'd as distinct attributes.
> The Perfection I am speaking of, is that of God's determining himself by *Moral Fitness*, or acting perpetually according to the *Truth, Nature*, and *Reasons of Things*.
> By confining our Ideas of God's Moral Perfections to this Principle of *Rectitude*, we shall escape many Perplexities, if not many Errors. What more common, than for Men to bewilder themselves in searching for the Boundaries between Divine *Justice* and *Goodness*? By the one is generally understood a Communication of Blessings; and by the other, an Infliction of Judgments and Calamities. As if God were not equally righteous in both Dispensations! If doing that which is right, and reasonable, and fit toward his Creatures, be Moral Goodness in the Deity, as surely it must; then his Goodness is as conspicuous in one Case as in the other.

> My chief Aim is to endeavour to shew the Convenience and Advantage of considering the Moral Perfections of the Deity under the idea of Rectitude, rather than in the various and mixed Light of many distinct Attributes. [1730, pp. 4, 4, 6, 30]

Contra Balguy, Bayes argues not only that there is no convenience in this approach, but also that the moral attributes of God (especially goodness and veracity), though capable of being considered under one general notion, are really distinct.

Bayes then goes on to define the concepts of *goodness* and *justice* as they relate to God, between which terms he finds no inconsistency,

> for justice requires the punishment of the sinner only as a means of preserving the reverence due to the divine laws; and where it is necessary for this end, goodness cannot oppose the punishment, because the general happiness of the creatures of God cannot be supported, but by maintaining the authority of his laws: and, on the other hand, where this end can be secured without the punishment of the guilty, justice does not absolutely require it. [p. **[11]**]

As an example to illustrate the mistakes into which Balguy's method may lead us, Bayes considers the following scriptural passage.

> In those days was Hezekiah sick unto death. And the prophet Isaiah the son of Amoz came to him, and said unto him, Thus saith the LORD, Set thine house in order; for thou shalt die, and not live. [*2 Kings* xx. 1]

Since Hezekiah in fact recovered and lived for a further fifteen years[10], one who objected to this verse might well argue 'that such an assertion being false, could not proceed from a being of perfect truth and veracity'. Although Bayes argues that the difficulty is not resolvable under Balguy's scheme, it is perhaps unfortunate that he does not consider whether his own theory would allow its resolution. He notes, however, that such difficulties can be reconciled 'by showing either that the fact is falsely represented, that the perfection to which it is said to be contrary, is not really a perfection of God, or else that there is no inconsistency between them', and he argues that if the fact is accepted and the perfection really does belong to God, then 'to desire a person to abstract from that particular perfection, and to consider moral rectitude in general, looks more like a design to amuse than instruct'.

On page **[13]** Bayes defines God's acting according to moral rectitude as 'as it is fit he should [act]', which then requires our knowing what it is that renders an action fit to be performed:

Now, as the desire of some end must be the motive to any action, so 'tis the nature of the end designed, and which the action is proper to effect that renders it good or bad, fit or unfit to be performed. **[14]**

Thus *ends* should be taken account of in the description of divine rectitude. Supposing then that God is directed in all His actions by a 'regard to the greatest good or happiness of the universe', Bayes describes God's moral rectitude as 'a disposition in him to promote the general happiness of the universe'.

If we wish to determine the moral perfections of God, it is first necessary 'to find out what it is that renders actions fit to be performed, or what is the end that a good and virtuous being, as such, is in pursuit of', the coincidence of these two things then showing that God is indeed directed by one uniform principle. I cannot do better than quote Bayes on this:

> for though we have got one word [Rectitude] to express the moral perfections of God by, as we might easily find another that should express both his natural and moral ones together; yet that perfection cannot be really one and the same, that includes in it the desire of two opposite ends. If we suppose that God desires the happiness of his creatures, and at the same time the beauty and order of his works, if these two principles of action often clash and interfere, and set bounds to each other; this cannot be to represent him as governed by one uniform principle, tho' his love of order, and his love of his creatures, be both called moral rectitude: but if his moral rectitude be nothing else but a desire or inclination to preserve a constant regard to the general happiness of the universe in all his actions; then he is in this view represented as truly governed by one uniform principle; though we should afterwards find it very convenient to reckon justice, veracity, mercy, patience, *&c.* as particular species or branches of this rectitude. [pp. **[15–16]**]

After a fairly long passage criticising Balguy, Bayes records his observing that 'it is commonly presumed that the communication of happiness was the sole end of God in creating the world', the extract he has in mind possibly being the following.

> It is not to be doubted but the *Intention* of the Deity, in creating the World, was the *Production of Happiness*, or the Communication of Good. ... It seems equally manifest, that the *Reason, Motive,* or *Principle*, by which he was determined to execute this great Design, was the *Rectitude* of the Thing itself. [1730, pp. 8, 9]

Bayes then cites the following two passages from Balguy's work.

> I see no Absurdity in supposing, that the Creator might have various Ends and Designs of which we have not the least Conception. But not to insist upon this, it may deserve to be consider'd, whether, within the compass of our own Ideas, we may not find some other Intention befitting the Wisdom and Rectitude of the Supreme Being. We seem to have grounds for such a Supposition both from Reason and Revelation.
> Where and how far these Ends do in Fact interfere, is above the Power of Man to determine. I doubt not but God has been graciously pleased to reconcile, and make them co-incident, as far as was possible. But it does not appear to me, that the *Order, Beauty*, and *Harmony* of the Universe are meerly intended in Subordination to the *Welfare* of Creatures. On the contrary, I know not whether the latter be not subordinate to the former, and limited by it. [1730, pp. 11, 14]

The supposition expressed here, that God might have ends in view that are distinct from, or indeed opposed to, the happiness of His creatures[11], would strike a 'melancholy aspect', says Bayes, were it in fact true. He contrasts here 'the fitness of things' with 'a fitness to produce happiness', and notes the importance of knowing all God's ends and designs.

Price, incidentally, defines 'fitness' as follows,

> *Fitness* and *unfitness* most frequently denote the congruity or incongruity, aptitude or inaptitude of any means to accomplish an end. But when applied to actions, they generally signify the same with *right* and *wrong* [1787/1974, p. 104]

and William Paley[12] was yet more direct, writing in *The Principles of Moral and Political Philosophy*, 'The fitness of things, means their fitness to produce happiness' [1825, vol. IV, p. 38].

At the start of his second section Bayes lists two points that he proposes to prove, viz.

> 1. That God in his acts of creation and providence had a regard to the happiness of his creatures, and that he is really benevolent towards them. And, 2. That we have no reason to suppose that he is in his actions towards them influenced by any other principles; at least, by any other, that are not entirely coincident with, or perfectly subordinate to this. [p. **[20]**]

These, taken together, seem to provide what one might call an 'existence and uniqueness' proof for the principle of benevolence.

On page **[22]** he makes the somewhat depreciatory remark

> I don't find (I am sorry to say it) any necessary connexion between mere intelligence, though ever so great, and the love or approbation of kind and beneficent actions

(the emphasis perhaps being on *necessary*), a remark that might have some connexion with Balguy's

> Had all [i.e., creatures] been created *intelligent*, all endued with the Powers and Faculties of Angels; what may we suppose would have been the Consequence? In all probability an Encrease of *Happiness*, and a Destruction of *Order*. [1730, pp. 22, 23]

Not all agreed with Bayes on this point: for instance, Grove wrote

> I, in my turn, am heartily glad to hear him say, that he is *sorry* not to find this connexion, not only as his *good nature* breaks out in this expression; but as it is a confession, of which he himself was not aware, of the *intrinsick excellence* of beneficent actions. He is sorry not to discern a connexion between the highest degree of intelligence, and the approbation of kind and beneficent actions? Why so? but because he is inwardly conscious (as indeed every man must be who hath not debauch'd his reason, and much more one of our author's virtue and good sense) that such actions greatly deserve approbation. For if they don't deserve approbation, there is no cause for sorrow, that mere intelligence or reason would not lead any one to approve them. But, if they deserve love and approbation, as this author doth in a manner confess, certainly the same intelligence, which discovers the merit of such actions, as the highest intelligence cannot but do, must needs approve them, and delight in doing them, if there be no opposite principle to counteract it, which there is not in God. [1734, pp. 28–29]

Bayes remarks further that God 'has made many beings capable of very great happiness ... and ... he has plainly made ... very plentiful provision for their happiness' (p. **[23]**). What could His views have been in doing this? 'He must do it', writes Bayes, 'either with an intention to communicate happiness or misery to them, ... or else with no view or design at all' (p. **[23]**). He shows that God's communication of misery cannot be entertained, yet he does not explore the last of these options, and concludes therefore that the universe was designed with the appetence to communicate happiness — a conclusion that is confirmed by the observation of God's works. Paley also commented on the connexion between indifference and the happiness or misery as produced by actions in the fifth chapter 'The Divine Benevolence' of the first book of his *The Principles of Moral and Political Philosophy*, writing

> When God created the human species, either he wished their happiness, or he wished their misery, or he was indifferent and unconcerned about both. [1825, vol. IV, p. 45]

On page **[24]** Bayes writes 'If this is a consequence of our being intelligent and rational creatures, God also is infinitely wise, and therefore infinitely good', a passage that seems to suggest that infinite goodness is to be deduced from infinite wisdom — which would surely place infinite wisdom in a more fundamental position than infinite benevolence? Or if not, it surely implies some connexion — despite his earlier aphorism — between intelligence and the approbation of kind and beneficent actions, and indeed Bayes writes further

> but if we cannot see this connexion between intelligence and goodness, or an approbation of what is kind and benevolent: yet our being thus formed, is a strong proof that our Creator is really good, since nothing can be a greater security to the general happiness than this. [p. **[24]**]

Paley, incidentally, distinguishes in his *Natural Theology*[13] between *knowledge* and *wisdom* when applied to the Deity, the latter term 'always supposing action, and action directed by it' [1825, vol. V, p. 309].

The relationship between benevolence and wisdom was also discussed by Joseph Butler in his *The Analogy of Religion*[14]:

> Some men seem to think the only character of the Author of Nature to be that of simple absolute benevolence. This, considered as a principle of action and infinite in degree, is a disposition to produce the greatest possible happiness, without regard to persons' behaviour, otherwise than as such regard would produce higher degrees of it. And supposing this to be the only character of God, veracity and justice in him would be nothing but benevolence conducted by wisdom. Now surely this ought not to be asserted, unless it can be proved; for we should speak with cautious reverence upon such a subject. And whether it can be proved or no, is not the thing here to be inquired into. [Chapter 3; 1840(?), p. 53]

How the 'greatest possible happiness' is to be calculated is not discussed here; Francis Edgeworth, however, in his discussion of the Utilitarian Calculus in his perhaps underappreciated *Mathematical Psychics*, provides the following remarks.

> Of the Utilitarian Calculus ... the central conception is *Greatest Happiness*, the greatest possible sum-total of pleasure[15] summed through all time and over all sentience.

> *Greatest possible happiness* is the greatest possible integral of the differential 'Number of enjoyers × duration of enjoyment × degree thereof'. [1881, pp. vii, 57]

Such a definition would of course have been almost unthinkable in Bayes's time.

Although the preceding sentence is strictly correct, Hutcheson had attempted the incorporation of a mathematical calculus in the second part — *An Inquiry Concerning the Original of our Ideas of Virtue or Moral Good* — of his 1725 treatise. In Article XI of Section 3 we read

> To find a *universal Canon* to compute the *Morality* of any Actions, with all their Circumstances, when we judge of the Actions done by our selves, or by others, we must observe the following *Propositions*, or *Axioms*.
>
> 1. The *moral Importance* of any Character, or the *Quantity of publick Good* produc'd by him, is in a *compound Ratio* of his *Benevolence* and *Abilitys*: or (by substituting the inital Letters for the Words, as $M = $ *Moment of Good*, and $\mu = $ *Moment of Evil*) $M = B \times A$.
>
> 2. When in comparing the *Virtue* of two Actions, the *Abilitys* of the *Agents* are equal, the *Benevolence* is as the *Moment* of *publick Good*, produc'd by them in like Circumstances: or $B = M \times 1$.
>
> 3. When *Benevolence* in two *Agents* is equal, and other Circumstances alike, the *Moment* of *publick Good* is as the *Abilitys*: or $M = A \times 1$.
>
> 4. The *Virtue* then of *Agents*, or their *Benevolence*, is always *directly* as the *Moment* of *Good* produc'd in like Circumstances, and *inversly* as their *Abilitys*; or $B = \frac{M}{A}$.

(Similar 'Propositions' are given in Article XII in connexion with moral Evil rather than Good.) The discussion here, though, is clearly concerned with *human* rather than *divine* Benevolence, and we do not explore the matter further.

Further on, in Section 7 of the same Part, Hutcheson writes of the meaning of the generally held opinion that God's laws are 'just, and holy, and good',

> It must then first be suppos'd that there is something in Actions which is apprehended *absolutely good*; and this is *Benevolence*, or a Tendency to the *publick natural Happiness* of rational *Agents* ...
> Some tell us, "That the *Goodness* of the *divine Laws* consists in their Conformity to some *essential Rectitude* of his *Nature*."

But they must excuse us from assenting to this, till they make us understand the meaning of this Metaphor, *essential Rectitude*, and till we discern whether any thing more is meant by it than a *perfectly wise, uniform, impartial Benevolence.*

This view must clearly have been of encouragement (and perhaps even inspiration) to Bayes in the writing of his tract.

In the second edition (1726) Hutcheson made a number of Alterations and Additions. Particularly pertinent to our study is the following addition made to the first Treatise, Section V, at the end of Article XIV (one in which Hutcheson had mentioned the possibility of Design in the Universe).

One Objection further remains to be remov'd, *viz.* "That some imagine this Argument may hold better *à priori*, than *à posteriori*; that is, We have better Reason to believe, when we see a *Cause* about to act, without Knowledge, that he will not attain any given or desir'd *End*; than we have on the other hand to believe, when we see an *End* actually attained, that he acted with Knowledge: Thus, say they, when a Man is about to draw a Ticket in a *Lottery*, where there is but one *Prize* to a thousand *Blanks*, it is highly probable that he shall draw a Blank; but suppose we have seen him actually draw for himself the *Prize*, we have no ground to conclude, that he had *Knowledge* or *Art* to accomplish this End." But the Answer is obvious: In such Contrivances we generally have from the very Circumstances of the Lottery, very strong moral Arguments, which almost demonstrate that *Art* can have no place; so that a Probability of 1000 to 1, does not surmount those Arguments: but let the Probability be increased, and it will soon surmount all *moral* Arguments to the contrary. For instance, if we saw a Man ten times successively draw Prizes, in a Lottery, where there were but ten Prizes to ten thousand Blanks, I fancy few would question whether he us'd *Art* or not: Much less would we imagine it were Chance, if we saw a Man draw for his own Gain, successively, a hundred, or a thousand Prizes, from among a proportionably greater Number of Blanks. Now, in the Works of *Nature*, the Case is entirely different; we have not the least Argument against *Art* or *Design*. An intelligent Cause is surely, at least, as probable a Notion as *Chance, General Force, Conatus ad Motum*, or the *Clinamen Principiorum*, to account for any Effect whatsoever: And then all the *Regularity, Combinations, Similaritys of Species*, are so many Demonstrations that there was *Design* and *Intelligence* in the CAUSE of this Universe: whereas, in fair Lotterys, all *Art*, or *Wisdom* in drawing, is made, if not actually impossible, at least highly improbable.

Could this passage, in which, with hindsight, we may see prior and posterior odds, so have affected Bayes, even unconsciously, that it might have been a stimulus to his *Essay on Chances*?

Turning our attention back to Bayes's tract itself, we find here what is in effect a definition of God's *benevolence*, this being stated to be equivalent to the intending of the 'general happiness of his creatures' (p. **[25]**). In this respect Bayes differs from Balguy, who, while allowing that God intends the happiness of His creatures, seems reluctant to grant Him benevolence, supposing rather that God does good because 'such a conduct is agreeable to the reasons and natures of things' (p. **[25]**). Balguy's own words on this matter run as follows.

> I cannot avoid thinking that the Divine Goodness is very much misapprehended, when it is consider'd as a Physical *Propension*, or *Disposition* of nature analogous to those Affections[16] and Propensities which he has given us. Such a Disposition, as I apprehend, would be so far from constituting Moral Goodness, that it would derogate from it in proportion to its Influence. To suppose in the Deity a benevolent Disposition *necessary* in itself, and in its Operations, is to suppose what is utterly inconsistent both with the Perfections of the Creator, and the Obligations of his creatures. And supposing it did not necessarily produce beneficent Actions, but to be consistent with Freedom and Choice; yet still, according to the Extent of its Influence, it would depretiate good Actions, and detract from the Merit of the Agent. [1730, p. 9]

And further

> If we consider the *Goodness* of God as an unbounded Benevolence, prompting Him to produce all possible Happiness; we shall perhaps both expose ourselves to needless Difficulties, and entertain wrong Conceptions of Him. But if we consider Him as perpetually acting according to the true Reasons and Natures of Things, whatever they be; we may possibly escape such Difficulties, and at the same time form juster Notions of the Divine Nature and Conduct. According to the former of these Ideas, that unequal Distribution of Good which appears in the World, may be sufficient to puzzle, and put to silence, every Enquirer: according to the latter, it appears perfectly right, and agreeable to the Natures of Things. The Order of the Universe requires it; and this, if I mistake not, is a just Solution of the Difficulty. [1730, p. 23]

Price too was quite clear that divine benevolence was perhaps somewhat restricted, writing

> *Divine benevolence* is a disposition, not to make all indiscriminately happy in any possible way, but to make the *faithful*, the *pious*, and *upright* happy. [1787/1974, p. 251]

Whether Bayes's views can be taken as indicative of support for some form of eudæmonism is debatable: they might possibly be seen as a religious translation of Plato's thinking in which the final end of political development was the happiness of every member of the state inasfar as he was capable[17], God's intent being the happiness of His creatures, but I would not like to try to take this point further — nor to consider whether any suggestion of hedonism is to be found in the tract.

Bayes notes that benevolence is the main part of divine rectitude, writing

> This indeed might be the consequence, if actions were not therefore fit to be performed, because they have a tendency to promote happiness; or, which is the same thing, if benevolence were not the main part of divine rectitude, which seems to be a point extremely plain, and is very agreeable to the concessions of our author. [p. **[27]**]

Similar comments may be found in Price's *A Review of the Principal Questions in Morals*:

> in the divine intelligence, absolute rectitude is included; and that eternal, infinite power and reason are in essential conjunction with, and imply complete, moral excellence, and particularly perfect and boundless *Benevolence*.
> Happiness is the *end*, and the *only* end conceivable by us, of God's providence and government: But he pursues this end in subordination to rectitude, and by those methods only which rectitude requires.
> Absolute and eternal rectitude, (or a regard to what is in all cases most fit and righteous) is properly the ultimate principle of the divine conduct, and the sole guide of his power. In this GOODNESS is first and principally included. But GOODNESS and RECTITUDE, how far soever they may coincide, are far from being identical. The former results from the latter, and is but a part of it. [1787/1974, pp. 89, 250, 248]

And just as an earlier remark seemed to lead to the regarding of benevolence as a deduction from infinite wisdom, so this passage seems to show it as dependent on rectitude (which would surely suggest that the latter is more fundamental than benevolence).

Price was perhaps somewhat more explicit in dividing benevolence into two parts:

150   4. Divine Benevolence

> Benevolence, it has been shewn is of two kinds, *rational* and *instinctive*. *Rational benevolence* entirely coincides with rectitude, and the actions proceeding from it, with the actions proceeding from a regard to rectitude. And the same is to be said of all those affections and desires, which would arise in a nature as intelligent. ... But *instinctive benevolence* is no principle of virtue, nor are any actions flowing merely from it virtuous. As far as this influences, so far something else than reason and goodness influences, and so much I think is to be subtracted from the moral worth of any action or character.
> [1787/1974, p. 191]

Bayes also notes Balguy's suggestion that to suppose God influenced by benevolence is to detract from His merit, but confesses himself unable fully to comprehend, and therefore to answer, it. Without the assumption of divine benevolence, he concludes,

> [God] no longer appears amiable and lovely, he no longer remains the object of our trust and confidence; other perfections render him awful and great, but 'tis this alone renders him the delightful object of our contemplation and trust. [p. **[29]**]

We pass on next to the third section of the tract. Bayes begins by hoping that it is now sufficiently clear that one ought to conceive of God as truly benevolent, and proposes to examine here whether He might be actuated by any other principles, either distinct from, or opposed to, benevolence. He assumes that God is wise and rational, and demonstrates that He is indeed actuated by no other principle, since 'to a wise and rational being those things that have no relation to happiness, must necessarily be looked upon as indifferent' (p. **[29]**). The only thing that makes an action fit to be performed is the effects it produces. Here Bayes also defines *morality* as 'consisting in following the dictates of regular self-love and real benevolence', and notes too that 'all actions which produce neither happiness nor misery, either immediately or in their natural tendency, must be absolutely indifferent'. Thus 'the rectitude or obliquity of actions must altogether consist in their tendency to produce happiness or misery' (p. **[32]**). Incidentally, Price distinguishes between *self-love* and *benevolence* as

> SELF-LOVE leads us to desire and pursue *private*, and BENEVOLENCE, *public* happiness. [1787/1974, p. 69]

Concluding therefore that God is by no means inclined to lessen the happiness of His creatures, Bayes passes on to the question of evil, claiming that 'whenever [God] inflicts evils on particular persons, or societies, it is for the sake of a greater or more general benefit' (p. **[33]**). He also states that God is not prevented by motives other than benevolence from doing more good to His creatures than he does.

Balguy gave several reasons, distinct from a regard for His creatures' happiness, for God's conduct. The first of these, discussed in the fourth section of Bayes's tract, is concerned with the regard God has for *beauty* and *order* in his works, and is framed by Balguy as follows.

> It may safely be presumed, that the Deity necessarily loves Order, and abhors Confusion. Were it possible to suppose that Deformity and Disorder could be more conducive to the Happiness of His Creatures, than the contrary; even upon this Supposition it seems not probable that they would have been submitted to. [1730, p. 14]

Bayes does not deny that God has regard for beauty and order: rather, 'I only deny that this was an end of his acting distinct from the happiness of his creatures, and on the contrary affirm, that it was a regard to their felicity which was the reason why he has made and disposes of all things in so orderly and beautiful a manner' (p. **[35]**). Once again Price had something similar to say:

> Happiness is an object of essential and eternal value.... Beauty and order, which have been strangely said to be of equal, nay superior value, are chiefly to be regarded as subservient to this, and seem incapable of being proposed as proper *ends* of action. [1787/1974, p. 249]

Bayes further provides a definition of beauty:

> Beauty in any object seems to be nothing else, but *such a relation, order and proportion of its parts, as renders the contemplation of it agreeable*: In the object it self therefore 'tis nothing else; but a certain relation, order, and proportion of its parts. [p. **[35]**]

Balguy's first argument in support of his position is based on the scriptural comment that God created all things for his own glory:

> We find the Scripture referring every thing to the *Glory of God*. We are there expressly told, that Men, and by consequence other Beings, were *created by Him for His Glory*. That *He made all Things for Himself; and for His Pleasure they are and were created*. These, and the like Expressions, are sometimes strained up to very improper Heights; and at other times explained quite away. I am far from thinking, that either in the Creation or Government of the World, God proposed to Himself the poor Praises and Applauses of His Creatures. Sooner might an Angel

propose to himself the Homage of a Worm. Such an End as this was infinitely beneath the Divine Majesty. And when we are required to *glorify* Him, it cannot be upon this Account: but partly because it is absolutely *right* and *fit* in itself, and therefore *morally good*; and partly because it is essential to our *Happiness*, or *natural Good*. [1730, pp. 11–12]

'By this', says Bayes,

> I understand that in all his works he designed the illustration of his own perfections, that his creatures might discern and adore his infinite excellencies; but then he cannot be supposed to have done this for his own sake, for he cannot be profited by the praises and applauses of his creatures. For what then? undoubtedly for the happiness of his creatures themselves, whose glory and happiness it is to know, and love, and serve their great Creator. [p. **36**]]

This passage is thus seen to be no objection to Bayes's thesis. Balguy's interpretation, however, was different:

> Nevertheless, I can see no Reason why the Deity might not propose, and strictly intend, that *Glory*, which consists in *His own Approbation* of His Works and Actions. [1730, p. 12]

Bayes further notes that

> [God's] own approbation cannot make his actions more fit to be performed, does not make him think them the more so (for that would be to suppose that he approves because he approves) and therefore cannot be any distinct reason of his acting. [p. **37**]]

One finds something slightly similar to this in the *A Review of the Principal Questions in Morals*, where, in writing of the application of right and wrong, Price says

> Right and wrong when applied to actions which are commanded or forbidden by the will of God, or that produce good or harm, do not signify merely, that such actions are commanded or forbidden, or that they are useful or hurtful, but a *sentiment* concerning them and our consequent approbation or disapprobation of the performance of them. Were not this true, it would be palpably absurd in any case to ask, whether it is *right* to obey a command, or *wrong* to disobey it; and the propositions, *obeying a command is right*, or *producing happiness is right*, would be most trifling, as expressing no more than that obeying a command, is obeying a command, or producing happiness, is producing happiness. [1787/1974, pp. 16–17]

Expanding on Balguy's scriptural reference, Bayes cites *Proverbs* xvi. 4: 'The LORD hath made all *things* for himself: yea, even the wicked for the day of evil', a verse that is given in the *Revised Standard Version* as 'The LORD has made everything for its purpose, even the wicked for the day of trouble', and on which Davidson comments as follows.

> Jehovah made everything for His own purpose: and He will fulfil His purpose for the wicked when He brings about the day of evil for them. The phrase is compressed. There is no sense of an arbitrary predestination of men for this day of evil.
> [1954, p. 531]

Bayes finds the meaning of these words (on p. **[38]**) to be 'very well expressed' by the following commentary by Bishop Patrick[18].

> The Lord disposes all things throughout the world to serve such ends as he thinks fit to design, which they cannot refuse to comply with. For if men be so wicked as to oppose his will, he will not lose their service; but when he brings a public calamity on a country, employ them as the executioners of his wrath.
> [p. **[38]**]

In a footnote Bayes provides an alternative reading from Dr Samuel Clarke's *Sermons*[19] of 1731: the full relevant passage, from Volume VII, Sermon XIV, runs as follows.

> And if we render the words *more properly* and *more agreeably* to the original, not, *the Lord has made all things for himself,* but *the Lord has made all things suited* or *fitted to each other,*
> ... [p. 301]
> ... the Words of the Text in their other and more proper rendring, wherein the *latter* part of the words answer more strictly to the *former*; *The Lord has made all things suited to each other, yea even the wicked to the day of evil.* And in *this* Sense, the general doctrine they contain, is evident; that God has made every thing fitted for its proper end and purpose, and wisely contrived to answer its intention. [pp. 314–315]
> ... according to the nature of the *Jewish* language, *fitting the wicked to the day of evil,* signifies nothing more, than causing wickedness and punishment to be proportionable; that is, not that God *causes* wickedness to be at all, but that he causes punishment to be proportionable to the wickedness that he finds in Men. [p. 315]

Another passage of scripture quoted by Balguy is *Revelation* iv. 11:

> Thou art worthy, O Lord, to receive glory and honour and power: for thou hast created all things, and for thy pleasure they are and were created.

Once again Bayes's interpretation is supported by Davidson, who writes 'He willed the existence of all things. He has the right to deal with them in sovereign freedom' [1954, p. 1177]. Incidentally, note that Bayes quotes the Greek original here: is this perhaps supportive of Onely's comment, mentioned in an earlier chapter, of his excellence in this language?

Bayes summarily dismisses Balguy's discussion of his question

> Is nothing more meant by it [the Goodness], than the Tendency and Conduciveness of every Thing to the Benefit of Living Creatures? [1730, pp. 12–13]

Bayes now turns from the consideration of scripture to discuss what else Balguy has said in support of his hypothesis. While agreeing with Balguy that 'Order and beauty are real perfections in any work and consequently the works of God could not have been so perfect, if they had not been so beautiful' (p. **[39]**), — Balguy's words were 'It may safely be presumed, that the Deity necessarily loves *Order*, and abhors Confusion' ([1730, p. 14]) — and concluding that God must have had a regard to beauty as well as happiness in His works, Bayes finds his only possible point of disagreement with Balguy to be 'whether beauty be any inherent perfection in an object, and not merely a sentiment in our minds, arising from some particular order and proportion in the parts of the object we contemplate'. Balguy had claimed that beauty is a real perfection, arguing as follows.

> If Order and Beauty be not real, but relative; if they consist wholly in an arbitrary Agreement between the Objects and the Sense; what means that wonderful *Apparatus*, that boundless Profusion of Art and Skill, which we every where meet with among the Creator's Works? What means that curious Contexture, and elaborate Arrangement of Parts? Why are they so nicely adjusted to each other, and all made subservient to the Grandeur and Magnificence of the Whole? According to the present Supposition, how shall we avoid looking upon all this as meer Waste of Workmanship? If there be no objective Perfections, no real Improvements hereby introduced, I am forced to conclude, that a Chaos would have answered the purpose full as well. [1730, p. 16]

In discussing this opinion of Balguy's, Bayes instances Francis Hutcheson's *An Inquiry into the Original of our Ideas of Beauty and Virtue; in two treatises*, the pertinent passage — from *Treat. I. Sect. VIII. Pars 2 & 4* — running as follows.

Let it be here observ'd, that as far as we know of any of the great Bodys of the *Universe*, we see Forms and Motions really *Beautiful* to our Senses; and if we were plac'd in any *Planet*, the *apparent Courses* would still be *Regular* and *Uniform*, and consequently *Beautiful* to our Sense. Now this gives us no small Ground to imagine, that if the Senses of their Inhabitants are in the same manner adapted to their Habitations and the Objects occurring to their View, as ours are here, their Senses must be upon the same general Foundation with ours.

And possibly the DEITY could have formed us so as to have receiv'd no Pleasure from such Objects [a regular Form, Action, or Theorem], or connected Pleasure to those of a quite contrary Nature. We have a tolerable Presumption of this in the *Beautys* of various Animals; they give some small Pleasure indeed to every one who views them but then every one in its own Species seems vastly more delighted with their peculiar *Beautys*, than with the *Beautys* of a different Species, which seldom raise any desire but among Animals of the same Species with the one admir'd: this makes it probable, that the *Pleasure* is not the necessary Result of the *Form* it self, otherwise it would equally affect all Apprehensions in what Species soever. This present Constitution is much more adapted to preserve the *Regularity* of the *Universe*, and is probably not the Effect of *Necessity*, but *Choice* in the SUPREME AGENT, who constituted our Senses.

Turning his attention to Balguy's argument that states that

> For are not *Uniformity* and *Variety* real *Relations* belonging to the *Objects* themselves? Are they not independent on us, and our Faculties; and would they not be what they are, whether we perceived them or no? [1730, p. 17]

Bayes says that this cannot be denied, but that the author draws a very strange conclusion from it when he asserts

> The ingenious Author of the *Enquiry into the Original of our Ideas of Beauty and Virtue*, [i.e., Hutcheson] tho' he professedly maintains the contrary Opinion, yet has nevertheless fixed Beauty on such a Foundation, as seems to me entirely inconsistent with his own Notion. [1730, p. 17]

This notion, or 'hypothesis' as Bayes calls it, is that objects that appear beautiful to us exhibit *uniformity amidst variety*, a concept that Hutcheson introduced in his *Treat. I. Sect. VIII. Par. 5* as follows.

> Now from the whole we may conclude, "That supposing the DEITY so kind as to connect *sensible Pleasure* with certain Actions

or Contemplations, beside the *rational Advantage* perceivable in them, there is a great *moral Necessity*, from his *Goodness*, that the *internal Sense* of Men should be constituted as it is at present, so as to make *Uniformity amidst Variety* the Occasion of Pleasure." For were it not so, but on the contrary, if *irregular Objects, particular Truths*, and *Operations* pleased us, beside the endless Toil this would involve us in, there must arise a perpetual Dissatisfaction in all rational Agents with themselves; since *Reason* and *Interest* would lead us to simple *general Causes*, while a *contrary Sense* of *Beauty* would make us disapprove them: *Universal Theorems* would appear to our Understanding the best Means of increasing our Knowledge of what might be useful; while a *contrary Sense* would set us on the search after *singular Truths*: *Thought* and *Reflection* would recommend Objects with *Uniformity amidst Variety*, and yet this *perverse Instinct* would involve us in Labyrinths of *Confusion* and *Dissimilitude*. And hence we see "how suitable it is to the *sagacious Bounty* which we suppose in the DEITY, to constitute our *internal Senses* in the manner in which they are, by which Pleasure is join'd to the Contemplation of *those Objects*, which a finite *Mind* can best imprint and retain the Ideas of with the least Distraction; to *those Actions* which are most efficacious, and fruitful in useful Effects; and to *those Theorems* which most enlarge our Minds.

In his discussion of 'uniformity amidst variety' Price finds that regular objects are pleasing and preferred for the following reasons.

> *First*, They are more easily viewed and comprehended by our minds. ... Farther. Order and symmetry give objects their stability and strength, and subserviency to any valuable purpose. ... *Thirdly*. Regularity and order evidence art and design. [1787/1974, p. 65]

Bayes once again quotes a passage from Balguy's tract:

> It is not, I think, to be doubted, but it is most regular, most harmonious, and most beautiful in his sight. And such must the Plan of it have appeared, before the Creation, in his all-comprehensive Mind. These beauties therefore must be real, absolute and objective Perfections. For whatever the Creator sees in his Works, or Ideas, must actually be in them. What they appear to Him, that they precisely are in themselves. It is inconsistent with the Perfection of the Deity to ascribe any of His Perceptions to an accommodated Faculty, or to any thing analogous thereto. Either therefore the Natural and Moral World

appear without Beauty to the Divine Mind, or the Beauties which he perceives in them must be real and inherent
[1730, p. 21]

an argument that Bayes regards as 'unfair', for 'beauty is an inherent perfection in an object' (p. [**44**]).

Bayes pursues the question of beauty, but finds it does not lie in any specific order and proportion of the parts of an object, as Balguy seemed to imagine when he wrote

> Whatever Diversity there may be in Men's Tastes, Fancys or Perceptions; I presume the Essentials of Beauty are unconcerned therein. Without Order, Symmetry, and Proportion, no Works of Art are, or can be, beautiful; and according to the Degree wherein those prevail, the Beauty of these is greater or less. [1730, 20]

The whole controversy, Bayes states, appears to turn on the following point.

> Now, by what medium can we possibly prove that such an order, and such proportions in the parts of an object, which renders it most beautiful to us, must also make it appear so to the divine mind? especially when we often perceive beauty in an object, without so much as knowing what those proportions are which render it agreeable to us; nor till this be done, which has not been so much as attempted, can we have any reason to believe, that beauty, which in objects themselves is nothing else but a particular order and proportion of their parts, was any distinct end in the works of creation and providence. [p. [**46**]]

Pointing out the importance of order and proportion in the essentials of beauty, Bayes notes the sense in which he uses the latter term:

> I use the word proportion in that which is the common sense of it, as I apprehend, among all writers (unless we must except some of the mathematicians) to denote the comparative greatness of two quantities, in which sense we use the word when we speak of the proportion between the circumference and the diameter of the circle; between the height of a building, and its length or breadth, and the like. But I thought it proper to observe, that some mathematicians using the word *ratio* instead of proportion in the foregoing sense, define *proportion* to be an *equality of ratios*; because in this sense of the word, it must be acknowledged that proportion has a considerable effect upon the beauty of an object; but not so much as that where there is the same proportions in this sense, or the same number of equal ratios, there must always be the same beauty. [pp. [**47–48**]]

158   4. Divine Benevolence

In Section V Bayes considers Balguy's assertion contradicting the opinion of those who think that God has no ends in his actions towards his creatures that are not entirely coincident with their happiness. He cites Balguy:

> Every Instance, every Degree of Sin, is a just Ground for suitable Punishment, whether the Ends and Intentions of Government require it or no.
> But what shall we say of those Evils which are not only judicial, but strictly *penal*? When, the Measure of Men's Iniquities being filled up, they are actually under the Executions of Divine Vengeance. [1730, pp. 52, 33]

Although Bayes grants that every degree of sin is a just ground for suitable punishment, he asks what reason can be assigned for punishment if no good is done or designed by it. In attempting to answer this, Balguy considers what would happen were there but one moral agent in the universe, and were this agent 'obstinately and incorrigibly wicked' (p. **[50]**). Then

> It is not, I think, to be doubted, but he would feel the Effects of his Maker's Displeasure. Nay, there are good grounds to believe, that his Punishment would be proportioned to his Crimes[20]; and this is highly probable on many accounts, and is, if I mistake not, conformable to the Doctrines both of Reason and Revelation. [1730, p. 34]

Bayes now claims that Balguy supposes 'that punishment may and ought to be inflicted on a wicked person meerly on the account of the desert of sin, tho' no good were hereby derived to himself, or to any one else' (p. **[50]**), in support of which he cites the following passage from Balguy.

> If we attend to the Principle of *Divine Rectitude*, which comprehends all Moral Perfections, and I think exhibits them in a clear Light; we shall find the Deity ever acting, and ever determined to act, according to the *Reason of the Thing*, and the *Right of the Case*. If then it be *right* and *reasonable* in itself to punish *Wickedness* as well as to reward *Goodness*; He will assuredly do both. [1730, p. 51]

Bayes finds no difficulty with Balguy's argument so far: 'God will assuredly reward the virtuous, and punish the vitious; as far as it is reasonable, and fit he should' (p. **[51]**). However, Balguy continues,

> If *Virtue* be necesssarily *amiable* in his sight; *Vice* is necessarily odious. If there be *Merit* in the one, which naturally

recommends it to his Favour and Esteem; there is equal *Demerit* in the other, which unavoidably excites his Displeasure and Indignation [1730, p. 51]

and here Bayes finds nothing more said than that virtue is necessarily amiable and vice necessarily odious. As a consequence he deduces that God will take all proper methods to secure the cause of virtue, and to discountenance sin: he does not agree with Balguy's deduction that

Is it not then as *morally fit*, and as agreeable to the Rectitude and Sanctity of the Divine Nature, to punish the one, as to reward the other? [1730, p. 34]

Bayes does not find it inconsistent with divine rectitude to show mercy to sinners, though Balguy's passage

Should God communicate Good, and finally dispense Happiness to all his Creatures, deserving or undeserving, promiscuously and indiscriminately; how could He be said, upon this Supposition, to act according to the true Natures and Reasons of Things? *Virtue* and *Vice* are essentially opposite; and therefore it is not possible that the same Treatment should suit the Votaries of each. [1730, p. 35]

possibly contradicts this.

Turning his attention next to the question of punishment (the reason for which is 'to secure the respect necessary to the divine laws'), Bayes writes

Is there not as great a difference made between the virtuous and the vitious, as any wise and good being would desire; if the virtuous are always rewarded for what they do well, and the vitious are always punished, where their punishment is conducive to the general good and necessary to deter from sin? But can we suppose that God would suffer an incorrigible sinner to go unpunished, when the quite contrary is deserved? Is not this essentially repugnant to the rectitude and purity of his nature? [p. **[53]**]

A perhaps slightly stronger position was taken by Butler, who wrote[21]

Moral government consists, not barely in rewarding and punishing men for their actions, which the most tyrannical person may do: but in rewarding the righteous and punishing the wicked; in rendering to men according to their actions, considered as good or evil. And the perfection of moral government consists in doing this, with regard to all intelligent creatures, in an exact proportion to their personal merits or demerits.

> If a more distinct inquiry be made, whence it arises that virtue as such is often rewarded, and vice as such is punished, and this rule never inverted, it will be found to proceed, in part, immediately from the moral nature itself which God has given us; and also in part from his having given us, together with this nature, so great a power over each other's happiness and misery. [Chapter 3; 1840(?), pp. 53, 42]

In his discussion of the infliction of punishment Balguy considers the notion of a villainous character:

> When a virtuous Man hears or reads of a vile or villainous Character, he naturally resents, and is offended at it. And the more virtuous he is, and the viler the Character is, the higher his Indignation rises in proportion.
> Let us further consider the Sentiments of a virtuous Man in respect of a very base Character. He not only perceives it to be very odious in itself, but judges it immediately most worthy of *Punishment*. Let us suppose the Case of some Wretch, who has murdered his Father, or ruined his Country, or sacrificed the Lives of his Children to some mean Passion. I enquire not what Emotions of Resentment would rise in a good Man's Breast, upon the first hearing or reading of such Crimes. I only ask, how he would judge of them in cold Blood. Could he possibly desire or approve the Impunity of such a Criminal? It may be said, that if his Punishment were not necessary *in terrorem*, that a good Man would encline to grant him an Opportunity of Repentance and Amendment. Be it so. But if it be further supposed, that such a Criminal is altogether hardened and incorrigible, and that the longer he lives the more wicked he grows; the most merciful Man in the World, if it were in his power, would pass Sentence on him, and give him up to condign Punishment. To do otherwise, would be not Goodness but Weakness. It would be violating Truth and Nature, and acting counter to the plainest Reasons of Things. [1730, pp. 35–37]

Bayes cannot agree with Balguy on the matter of inflicting punishment when such punishment would do no good, writing[22]

> And I am satisfied I should be far from being singular if I should deliver my opinion on the opposite side of the question, in the same way with our author, that to give pain to any sensible being where no good end is answer'd by it, nor any mischief prevented, is violating truth and nature, and acting contrary to the plainest reasons of things; 'tis certainly acting a mischievous and ill-natured part. [pp. **54–55**]

Hutcheson also had something to say on the connexion between justice and benevolence:

> The *Justice* of the DEITY is only a Conception of his *universal impartial Benevolence*, as it shall influence him, if he gives any Laws, to attemper them to the *universal Good*, and inforce them with the most effectual Sanctions of *Rewards* and *Punishments*. [1725, *Treat. II. Sect. VII.*]

In his sixth section Bayes turns to the last opinion of Balguy's to be examined, one that seems to lead to the confining of the divine goodness:

> but that supreme Felicity, that indefectible State, wherein consist the Rewards of the Righteous, could not, I suppose, consistently with the *Divine Rectitude*, have been conferred on any untryed Creatures. [1730, p. 46]

Bayes finds this proposition, at least in the form in which he states it, to be, at first sight, 'so inconsistent with our notions of the boundless goodness of God, that it can't be expected we should admit it without the plainest proof' (p. **[55]**). Having argued against this position, Bayes cites Balguy's justification of it:

> Perfect Happiness was and ought to be reserved for the proper Objects of God's Love and Favour; which none of his Creatures could be, without virtuous Qualifications and Moral Merit: and these imply a State of Probation.
> To confer the same Happiness on the Unfaithful and Disobedient, would be a manifest Violation of Truth and Nature.
> By Parity of Reason it was morally unfit to treat those who had never been *tryed*, and by Consequence merited nothing, in the same manner as if they had been tryed and found faithful because such persons are in a moral sense *worthless* and by Consequence are in a Station as much below *Merit* as it is above *Demerit*.
> Those who have never been tryed, can deserve neither [i.e. rewards or punishments]; and therefore cannot, agreeably to Truth and Rectitude, be treated like either of the other. They are in a Moral Sense worthless; .... [1730, pp. 41–42]

Finding that the apparent strength of Balguy's argument lies in an ambiguity in some of his terms, Bayes begins his criticism of this argument by defining *virtuous qualifications*, which Balguy took as implying a state of probation, i.e., 'a state of difficulty and temptation'. 'By virtuous qualifications therefore,' writes Bayes, 'I suppose, he must mean, what I would call, in opposition to those good dispositions which we receive more directly from our Creator, *acquired virtues*' (p. **[57]**).

162   4. Divine Benevolence

The next term requiring discussion is *moral merit*, which Bayes defines as 'a *right to a reward*, or that which gives such a right' (p. **[58]**), this definition not being applicable to untried creatures. He concludes that, in his sense of the term, only creatures in a state of probation can be said to merit. He contrasts his approach and Balguy's as follows.

> So that if none must receive but according to their merit, which is the principle on which, if I mistake not, all our author's reasoning depends; mankind must for ever despair of a state of perfect and unalterable happiness. But if we take the word *merit* in the sense before mentioned, as signifying an equitable claim to a reward, *i.e.* to a reward proportional to the difficulty of performing, or the inconvenience that is submitted to in the performance of any virtuous action; in this sense, I think, a person may be said to merit, and thus it is plain he can merit only in a state of probation. [p. **[59]**]

Bayes finds Balguy's assertion 'that all beings that have not undergone a state of trial, are in *a moral sense worthless*, viz. that they have no proper claim to a reward' to be 'very harsh' (p. **[59]**).

Bayes now answers Balguy's argument, noting firstly that acquired virtues and moral merit are not absolutely necessary for the acquiring of God's love, favour, and esteem. Further, one who has undergone severe trials and overcome them is not necessarily a better and more valuable person than one who has not been thus tested. *Esteem*, Bayes states,

> ought always to be proportional to the good qualities of which a person is possessed, however these have been acquired; for *to have one in due esteem*, is to apprehend him to be what he really is in himself. [pp. **[60–61]**]

Balguy writes

> Are Rewards due to the Righteous, and Punishments to the Wicked? Those who have never been tryed, can deserve neither; and therefore cannot, agreeably to Truth and Rectitude, be treated like either of the other [1730, pp. 41–42]

an argument that Bayes interprets as saying that an untried person ought to be neither rewarded nor punished. Had this argument been intended to prove what Balguy initially set out to show, viz. 'that such an one ought not, through the undeserved goodness of God, to be made partaker of perfect happiness' (p. **[61]**), it should have run differently, as Bayes now shows. Balguy claimed that if happiness be due as a reward to the righteous in a state of probation, and misery to the wicked, then those who have not been tried ought to be treated differently to the others. To this Bayes responds

> *perfect* happiness is not due to the righteous as the reward of their services; 'tis not on the merits of their own works, but on the gracious promise of God, through Christ, that their expectation of this is founded: and therefore, since they at last must not receive it, but as an undeserved favour; why mayn't the same favour be conferred on others that are equally fitted for it, though supposed unequal to them in merit?
> Because one has a just claim upon me, that is no reason why I may not freely give to another; even more than what the former can in justice demand. ... 'Tis a mighty weak way of arguing, to say that two persons ought not to be treated alike, that is, that they should not both receive the same happiness; because one has a particular claim to it, which the other has not: ... .
> [pp. **[60–61, 62]**]

One is reminded of the parable of the labourers and the vineyard (see *Matthew* xx)[23]. Bayes notes that his argument does not in fact represent God as acting in an arbitrary manner:

> But all that follows from what I have asserted is, that God has other reasons for bestowing happiness upon, and for diversifying the happiness of his creatures, besides a regard to their merit.
> [p. **[63]**]

Balguy also adds an argument 'taken from the natural connexion between virtue and happiness' (p. **[63]**): speaking of the supreme felicity that those who have passed through a state of probation may expect, he says

> This is their [i.e., the '*untryed* Creatures'] peculiar Portion, partly by God's Appointment, and partly from the very nature of the thing For, in the last place, since a State of Probation is necessary for the Exercise and Improvement of Virtue; it is, by Consequence, necessary for the Consummation and Perfection of Happiness. [1730, p. 46]

Again Bayes has difficulty with Balguy's sentiment, this time with what the latter means by *virtue*.

Paley, in *The Principles of Moral and Political Philosophy*, states that some moralists have divided *virtue* into *benevolence, prudence, fortitude*, and *temperance*, whereas others distinguish only the first two [1825, vol. IV, pp. 28, 29]. Price, however, in the seventh chapter, *Of the Subject-matter of Virtue, or its principal Heads and Divisions*, of his *A Review of the Principal Questions in Morals*, takes exception to those writers who have found virtue to consist in benevolence, writing of the former quality:

164   4. Divine Benevolence

> *First*, To what particular course of action we give this name, or what are the chief *heads* of virtue.
> *Secondly*, What is the *principle* or *motive*, from which a virtuous agent, as such, acts.
> *Thirdly*, What is meant by the different *degrees* of virtue, in different actions and characters, and how we estimate them.
> There would be less occasion for the first of these enquiries, if several writers had not maintained, that the *whole* of virtue consists in BENEVOLENCE. [1787/1974, p. 131]

The point is further pursued in Butler's *Dissertation II: Of the Nature of Virtue*, appended to his *The Analogy of Religion*[24]:

> *Fifthly*, Without inquiring how far, and in what sense, virtue is resolvable into benevolence, and vice into want of it: it may be proper to observe that benevolence, and the want of it, singly considered, are in no sort the whole of virtue and vice. For if this were the case, in the review of one's own character or that of others, our moral understanding and moral sense would be indifferent to everything but the degrees in which benevolence prevailed and the degrees in which it was wanting. That is, we should neither approve of benevolence to some persons rather than to others, nor disapprove injustice and falsehood upon any other account, than merely as an overbalance of happiness was foreseen likely to be produced by the first and of misery by the second. [1840(?), p. 329]

Price[25] also considers whether God's moral attributes are reducible to benevolence:

> another Question of considerable importance relating to the Deity; I mean, the Question 'whether all his moral attributes are reducible to benevolence; or whether this includes the *whole* of his character?'
> It has been shewn, that the negative is true of inferior beings, and in general, that virtue is by no means reducible to benevolence. If the observations made to this purpose are just, the question now proposed is at once determined. Absolute and eternal rectitude, (or a regard to what is in all cases most fit and righteous) is properly the ultimate principle of the divine conduct, and the sole guide of his power. In this GOODNESS is first and principally included. But GOODNESS and RECTITUDE, how far soever they may coincide, are far from being identical. The former results from the latter, and is but a part of it. [1787/1974, pp. 247–248]

4.3. Commentary

We now come to the last section of the tract. Bayes begins by saying that at least so far, Balguy 'has not proved his point in a single instance' (p. [65]). One point, however,

> though adduced by our author in support of his notion of the divine Being's having a regard to the beauty and order of his works, as an end intirely distinct from the happiness of his creatures; it is not so properly a proof of his particular opinion with regard to beauty, as a general objection against his adversaries notion of happiness being the only end of God in his works
> [p. [65]]

remains to be discussed, viz. if happiness were the only end of God in His works, the greatest amount of it must have been produced at all times and in all places. This opinion was stated by Balguy as follows.

> Had the Production of Happiness been the Sole End which the Creator had in view, it is not, I think, to be doubted, but the utmost Possibilities of it would have been produced, at all Times and in all Places. But as far as we are capable of judging from the *Phænomena* within our Reach, this seems not to be the case. That Scale and Subordination of Beings before-mentioned, may seem to promote the *Order* and Harmony of the World, more than the Happiness of its Inhabitants. Had their several Powers and Perfections been more nearly equal, what would have been the Result? It seems probable that the latter of these Ends would have been advanced; and the former obstructed, if not destroyed. [1730, pp. 14–15]

Although admitting the importance and difficulty of this point, Bayes hopes that it will not lead us to think that God is influenced by any principles of acting that counteract His goodness.

In an attempt to answer the objection, Bayes first asks whether God 'was obliged by any of his perfections to create any beings at all' (p. [66]). Although His goodness seems to incline God to confer happiness on His creatures, goodness did not oblige Him to make His creatures. 'Creation is an argument of the goodness of God' (p. [67]). Bayes also notes that 'An act of goodness, one would think should require that the object of it should exist', but this argument he dismisses as appearing too subtle, and leaves it to his reader to consider it or not, as he pleases.

> But I think we may more securely conclude, that God was not obliged by any of his perfections to create, at least not to create the universe in any particular degree of perfection, from the consequences that necessarily follow from such a supposition.
> [p. [67]]

4. Divine Benevolence

Bayes concludes that it follows from Balguy's work that the world must be infinite as well as eternal,

> for no regard to beauty can set bounds to it; and that every creature must enjoy as much happiness as is consistent with the perfection of the whole; and that no more can possibly be created without destroying this perfection. [p. **[68]**].

These are consequences that Bayes is unwilling to accept. Noting that there is no such thing as 'the largest possible triangle', he argues by analogy that

> to argue against the goodness of God, because there is not the greatest quantity of happiness and perfection in the universe, is to use an argument that can have no force, since, if put into form, one of the premises is unintelligible. [p. **[69]**]

Proceeding from this argument, Bayes then reasons that

> God was not obliged to create the universe with any particular degree or quantity of perfection, or even to create it at all; and if this be true, 'tis very far from being a just consequence that God must be influenced by other principles that contradict his goodness, because we don't find the greatest possible happiness at all times and in all places. [p. **[70]**]

Although he believes that he has by now answered Balguy's objection, Bayes, to convince others, is prepared to make

> the most favourable supposition on his side, and see whether even then his argument will not admit of a fair answer. I will allow, that the consequence of supposing the Deity influenced by benevolence only, and no other principle contrary to it, is, that the universe must be created in the most perfect manner, and that the greatest possible quantity of happiness must be diffused through it; for even upon the supposition of the justness of this consequence, nothing that appears in the divine conduct towards his creatures, will oblige us to allow that the Deity is influenced by any other principle of acting that contradicts or limits his benevolence. [pp. **[71–72]**]

In refuting Balguy's theory Bayes notes that

> Besides this, it ought to be considered, that a most happy universe is so far from implying the same happiness in all creatures, and thereby contradicting that variety which is necessary to beauty; that the quite contrary is implyed in it: For the most happy universe is not one that consists of the greatest possible

number of the most happy beings only; but one that consists of that, and the greatest possible number of beings next inferior to the first rank, and so downward, till we come to those that approach the nearest to insensible matter. [p. **[73]**]

A most happy universe is then found to be most beautifully various, rather than most unbeautifully uniform, and Bayes completes the argument by claiming that 'we can have no reason to conclude, that the Deity has any other intention in his actions towards his creatures besides their good' (p. **[74]**).

Incidentally, a similar opinion was later expressed by Edgeworth, who wrote

> that in order to the greatest possible sum total of happiness, the more capable of pleasure shall take more means, more happiness. [1881, p. 10]

Bayes's concluding paragraph contains the following sentiment.

> Separate infinite goodness from almighty power and perfect knowledge, and the idea is only awful and terrible; but that these are necessarily conjoined in the glorious Author of our beings, renders him the object of our most delightful contemplation as well as of our highest reverence [pp. **[73–74]**]

and the essay is concluded with the following assertion, as to which no doubt is entertained;

> whilst God is good the cause of virtue and goodness must be maintained and supported, which necessarily requires that sin should be punished, as well as that virtue should be honoured and rewarded. [p. **[75]**]

Just as Bayes's tract was in a large measure a response to Balguy's, so Grove's *Wisdom, the first Spring of Action in the Deity* was to no small extent a commentary on, if not a criticism of, Bayes's work. We now cite a number of comments from Grove's treatise in which exception to Bayes's discussion is taken and reasons for Grove's inability to accept the arguments of *Divine Benevolence* are given[26]:

> "What is the reason" (saith an ingenious person in this way of thinking*) "why God should communicate happiness to the *good* and *innocent*? Will you say, that the reason for it is, that such a procedure is agreeable, and the contrary opposite to the nature of things? If so, I should then ask, to what things is such a procedure agreeable to the nature of? Is it to that of the

creatures? Is it agreeable to their nature as *sensible* beings? It is certainly pleasing to them as such; but this is entirely besides the question, and abstracting from this sense of the word, I don't see but that pain and misery is as agreeable to the nature of a sensible being, as pleasure and happiness." But, with this gentleman's leave, I would ask, how long a *sensible*, and an *innocent* being, have been equivalent terms? The question was concerning communicating happiness to a *good* and *innocent* being; and certainly, in the nature of the things, there is a greater agreement between *innocence* and *happiness*, than between *innocence* and *misery*, were it for no other reason but this, that the happiness of an innocent being, must, *in part*, arise out of his innocence or goodness itself, while his misery must be wholly *external*, and so (if not grounded in a *mistake*) not only suit but ill with his innocence, but, like a *negative quantity*, help to destroy the satisfaction that flow'd from the consciousness of it; for inward satisfaction may be so equally balanc'd with outward pain, as taken with that, to have no more value than *non-existence*, or so over-balanc'd as to be unspeakably worse than that. If any one saith, that he cannot see how happiness agrees better with innocence than misery does, I can only wish him a better eye-sight. Should this author, upon second thoughts, say, that tho there is no reason why God should communicate happiness to the good and innocent, yet there are plain reasons, why he should not inflict misery. I shall think it a considerable gain to have this one moral fitness granted me; and laying this for a foundation, make no doubt of raising a superstructure of many other moral fitnesses upon it. [pp. 27–28]
*Divine Benevolence, p. 21.

§5. The same author adds, a few lines after this: "I don't find (I am sorry to say it) any necessary connexion between mere intelligence, tho ever so great, and the love or approbation of kind and beneficent actions." And I, in my turn, am heartily glad to hear him say that he is *sorry* not to find this connexion, not only as his *good nature* breaks out in this expression; but as it is a confession, of which he himself was not aware, of the *intrinsick excellence* of beneficent actions. He is sorry not to discern a connexion between the highest degree of intelligence, and the approbation of kind and beneficent actions? Why so? but because he is inwardly conscious (as indeed every man must be who hath not debauch'd his reason, and much more one of our author's virtue and good sense) that such actions greatly deserve approbation. For if they don't deserve approbation, there is no cause for sorrow, that mere intelligence or reason would not lead any one to approve them. But, if they deserve love and approbation, as this author doth in a manner confess, certainly the same intelligence, which discovers the merit of such actions, as the highest intelligence cannot but do, must needs approve them, and delight in doing them, if there be no opposite principle to counteract it, which there is not in God. [pp. 28–29]

Second Corollary. *The divine rectitude is a complex term, including several ideas under it*; as, for instance, a *negative rectitude*, in opposition to every wrong inclination; *rectitude of judgment*, in opposition to all ignorance and mistake about *right* and *wrong*, whether in respect of the actions and operations of the Deity himself, or the actions and operations of his free creatures: *a rectitude of will*, denoting our invariable determination of the will, by a *right judgment*, in opposition to a will that is capable of being determin'd *without*, or *contrary* to, such a judgment: and, finally, a *rectitude of delight*, signifying that as some things are fit to be delighted in, others not, so God is delighted in that, and nothing else, which is a proper foundation and object of delight, and that his delight is always proportionable to the occasion, and the value of the object, in opposition to a satisfaction or delight that is *unreasonable*, because *misplac'd* as to the object, or *excessive* as to the degree. All these are comprehended in the *rectitude* of the *divine nature*; the inseparable effect of which is a *rectitude* in the *divine conduct and government*; by which rectitude, besides his never doing any thing that had better not be done, is farther meant his doing every thing that is fit and becoming him to do. [pp. 37–38]

Rather than admit of any original fitnesses in things, by the idea of which God determin'd himself, there are those who have recourse to a *natural benevolence*, prompted by which the Deity exerts his almighty power in producing the *greatest sum* of happiness that can possibly be. This greatest happiness of the whole system of rational beings taken together God *absolutely wills*, not because it is *fit*, but because his *inclinations oblige him* to it; and, accordingly, the *sum total* of happiness, let men and other free beings act how they please, will, in the event, be the greatest that infinite power and wisdom could possibly produce. Or (in the words of a late author*) "the greatest of which the universe of creatures which God hath made, is capable: still supposing that their *original capacities* for happiness were fixed by his *will* and pleasure." I shall not take advantage of this author's manner of expressing himself, when he saith, that the original capacities for happiness were fix'd by the will and pleasure of God, which, according to the propriety of language, should signify that the *very same beings* might have been created with greater or lesser capacities than those which God hath actually assign'd them; from which if true, it follows, that they were capable of greater or lesser *capacities of happiness*, that is, were originally capable of greater or lesser *degrees of happiness*: a capacity to receive a greater capacity of happiness, being, in effect, the same as a capacity of greater happiness; and, *consequently*, God bestows upon no being the utmost happiness of which he is capable. Letting this pass, I shall confine myself to the general notion, which, if some men are not mistaken, is such a glorious discovery as does at once dispel the darkness, wipe off every aspersion, and shew us the face of providence in its full beauty. Let us see whether it does so or no. [pp. 67–68]

*Divine Benevolence, p. 71.

§10. I imagine that in the preceding discourse I have overturn'd the very *foundation* of this *theory*; viz. the notion of *benevolent inclinations* in the Deity, of which his wisdom is not the *exciting cause* or reason, but merely the *servant* or *minister* to execute what they order. At present, without insisting upon that, I shall endeavour to demonstrate, that granting the existence of such a *natural benevolence*, it will by no means account for the origin of evil. For if all the works of creation and providence owe their birth to *mere benevolence*, without all regard to *moral fitness*; why is not every creature of God, that is capable of happiness, as happy as it is capable of being made? Why is there any such thing as misery in the world? Particularly, in the world of mankind? The answer, I apprehend, must be, that evil, or rather a liableness to evil, is the unavoidable consequence of something which the greatest happiness of man, or the entire system of rational beings, made necessary. But I very much doubt this is not so easily prov'd as said. Let them tell us what that is which, while man, or other beings of a higher order than man, cannot be happy without it, is yet the unhappy occasion of misery? [pp. 68–69]

"We have hardly any notion (saith one*) of a good and amiable action, but that it proceeds from this principle? If kindness or a good disposition be not the spring, no matter what the nature or consequence of the action be, however beneficial it may be to us, we like the being that produced it never the better; we don't think our selves obliged to gratitude, or imagine him any ways the more perfect, as to his moral character on the account

of it." Most surprizing news! that I should have no reason to be grateful to God, because he had his reason for being kind to me, tho' that reason was not borrowed from any merit of mine. [p. 92]
*Divine Benevolence, p. 27.

Grove summarizes his investigations as follows.

> *Upon the whole*, since we can have no reason to doubt of the truth of that notion which *best consults* the honour of the divine perfections, *best agrees* with the universal sense of mankind, and is *best adapted* to promote the cause of virtue and religion, and to answer the most difficult questions on the subject of *creation* and *providence*; I take leave to conclude, that wisdom (and not *arbitrary will* or *blind inclination*) is the *first spring of action in the Deity*. [p. 110]

Having considered the salient features of Bayes's essay as they refer particularly to Balguy's tract, and having noted the criticisms of the former as they were given by Grove, let us say something about the general contents of the three works. An accurate summary of the salient features of each was given by Philip Doddridge (1702–1751) in his 'A course of Lectures on the principal subjects in pneumatology, ethics and divinity: with reference to the most considerable authors on each subject', printed in Volume 4 of his *Works*, pp. 279–574, from which the following extended passage is taken.

> Lect. LVII. Of the Spring of Action in the Deity.
> §1. Schol. 6. It may not be improper here to take some notice of the celebrated controversy, between Mr. Balguy, Mr. Bayes and Mr. Grove, concerning the Spring of the divine actions....
> §2. Balguy maintains, that God always does that which is right and fit, and that all his moral attributes, viz. justice, truth, faithfulness, mercy, patience, &c. are but so many different modifications of *rectitude*[27].
> §3. He [Balguy] grants that the *communication of good* is *one* great and right end of the creator; but maintains that it is not the only end: he ultimately aims at *his own glory*, i.e. the complacental approbation of his own actions, arising from a consciousness of having inviolably preserved a due decorum, order and beauty in his works: and if ever the happiness of any particular creature, or of the whole system interfere with this, (as he thinks it sometimes may) it must so far give way to it.
> §5.7. To this Mr. Bayes objects, that to consider God first in general as doing all that is right, and then to deduce his particular moral attributes, as branches of this universal rectitude of

his nature, is going further about than is necessary, and leaves particular attributes entangled in just the same difficulty as before. But if it were otherwise, he says, that as nothing can be *fit* but what tends to promote happiness, the best idea we can entertain of the rectitude of God, is a disposition in him to promote the general happiness of the universe; and that we may as well consider all the other moral attributes as comprehended in this, and different modifications of it, as to consider them united in Balguy's view of rectitude; but with this advantage, that here we shall have something certain to depend upon; whereas it must throw the mind into perpetual perplexity, if (for aught we know) God may have some ends in his actions and dispensations, entirely different from and perhaps opposite to the happiness of his creatures.

§7. On the whole, he concludes that the divine benevolence is not to be stated, as "an unbounded inclination to communicate the highest degree of happiness," which is a contradiction, as it would be to suppose the greatest possible triangle actually described; but "as a kind affection towards his creatures, inclining him to confer upon that universe which he has made (and which he might have created or not, or have created with inferior or superior capacities for happiness) the greatest happiness of which it is capable." But if it be asked, why it was not made capable of more, he supposes that must be referred into the will and pleasure of God.

Lect. LVIII. Of the Spring of Action in the Deity; continued.

§1. Schol. 8. Mr. Grove refers all into the *wisdom* of God, which he says is "the knowledge that God has of what is fitting or unfit to be chosen in every imaginable circumstance[28];" and taking it for granted that he is under no wrong bias, concludes that he always chuses according to this fitness[29]. He adds, that nothing can be fit to be chosen by any being, but what has some reference to *happiness*, either that of the agent or some other; and that *beauty* and *order* are nothing any further than as they tend to communicate pleasure to percipient beings: therefore the end of God in the creation must be *happiness*; as to the degree and manner of obtaining it, suited to the faculties, dependencies and freedom of his rational creatures.

§2. As Bayes and others have maintained, that benevolence is a *kind inclination* or affection in God, Grove endeavours to prove, that properly speaking, there is no inclination in him; and maintains, that to suppose such an inclination as depends not on the previous act of the divine understanding, will be in effect imputing to him a blind and irrational propensity; and that nothing could be more dishonourable to the divine being, than universally to assign this reason for his conduct in any instance, "that he was inclined, or had a mind to do it." But he further maintains it, as probable at least, that there are no inclinations in God at all distinct from his actual volitions, but that the actings of the divine will are immediately and inseparably connected with those of his understanding: to suppose the contrary, he thinks would in effect be supposing, that reason would not be sufficient to determine the divine mind.

§4.9. From the survey we have taken of this controversy, it may be natural to make the following remarks.

§5.1. That each of these ingenious writers discover a pious temper, a concern for the honour of the divine being, and the advancement of virtue in the world.

§6.2. That they all acknowledge that God does always what is right and good: nay, when one thing is on the whole more fit than another, he invariably chuses it.

§7.3. That both Mr. Grove and Mr. Balguy acknowledge the *communication of happiness*, a noble and excellent end, which the deity in some measure has always in view; and which he prosecutes, so far as to bring happiness at least within the reach of all his rational creatures; never inflicting any evil upon them out of caprice, or without some just and important reason.

§8.4. That there is very little difference between the foundation of Grove's discourse, and that of Balguy's; *wisdom* in the former being so stated, that to be always governed by it coincides with the notion of *rectitude*, maintained by the latter.

§9.5. That Mr. Bayes himself does not assert, that it would have been impossible for God to have produced a greater sum of happiness; and by granting the contrary seems to overturn the foundation of those arguments, by which he attempts to prove, that God has made the creation as happy as its present capacity would admit.

§10.6. It seems that a virtuous mind may be as easy, in considering God as a being of universal *rectitude*, as if we were to consider him as a being of unbounded *benevolence*: nay it seems, that in some respects the former will have the advantage; as it is impossible for us confidently to say, what will be for the greatest happiness of the whole; but on the other hand, we may naturally conclude, that rectitude will on the whole incline God to treat the virtuous man in a more favourable manner than the wicked.

§11.7. That the scheme of universal benevolence in the highest sense seems evidently to imply *fatality*: for if all the sin and misery of the creatures were necessary to produce the greatest possible sum of happiness, and if the perfection of the divine nature determined him to produce this greatest sum, then sin and misery would be necessary; whereby the doctrine of liberty is destroyed, and such a seeming reflection thrown on the divine character, as few would be able to digest.

§12.8. It seems therefore on the whole best to keep to that in which we all agree, and freely acknowledge, there are depths in the divine counsels unfathomable to us; so that though we may justly believe God has his reasons for suffering evil to be produced, we cannot certainly determine what those reasons are; and when we go about particularly to explain them, we find it difficult, according to the different schemes we embrace, on the one hand to vindicate his goodness, or on the other his omnipotence.

Doddridge himself, incidentally, was also exercised by the problem of evil in the world, writing in Lecture LV 'Of God's Goodness',

§9.7. We have reason to believe that God is perfectly good.
§10. Schol. 1. The great objection to this, is the mixture of evil in the world, natural evil, i.e. pain, and moral evil, i.e. vice: and it is questioned, how far the existence and prevalence of it in so great a degree can be reconcileable with what has been said of the divine goodness, since God has already been proved an almighty being.
§11. Ans. 1. We cannot possibly judge as to the proportion there is between the quantity of happiness and misery in the creation, merely from what we observe in this part of it, which is our own abode.
§12.2. It is possible that there is no evil of any kind, from which a degree of good may not proceed, more than sufficient to counterbalance it.

The work started in *Divine Rectitude* was continued in the *Essay on Redemption*, in which Balguy defined his main topic as follows.

By the *Redemption* of Mankind, I understand in general, *Their Deliverance, or Release, from the Power and Punishment of Sin, by the meritorious Sufferings of Jesus Christ.* [1741, p. 6]

Again Balguy pursues his firmly held views on *rectitude* as the fundamental attribute of God:

the moral Perfection of the Deity is unquestionable, whether we call it Mercy, Benignity, or Rectitude: but to ascribe *Affections* to Him correspondent to ours, seems to me derogating from that Perfection.

[God] is pleased to direct all his Proceedings by the immutable Rule of Rectitude. Not only therefore his Esteem, but his *Benevolence* likewise is limited and governed by the Merits and Demerits of Moral Agents. [1741, pp. 40, 43]

Passing over Bayes's essay, Balguy cites Matthew Tindal's *Christianity as Old as the Creation; or, the Gospel, a republication of the religion of nature* of 1730, writing

For indeed he [i.e., Tindal] sometimes represents *Rectitude*, and sometimes *Benevolence*, as the chief Principle and Perfection of the Deity. To suppose them the same, is confounding the most distant ideas. [1741, p. 51]

Thus Balguy is adamant that rectitude and benevolence are distinct (at least when referred to God), although he sees the latter attribute as not inconsistent with his views on redemption, writing

Those Readers who prefer the Principle of *Benevolence*, will find it, if I mistake not, not only uninjured thereby, but directly consulted. Redemption, as above explained, appears an illustrious Instance of *Divine Benevolence*; and at the same Time eminently conducive to the Promotion of *Human*.
[1741, pp. 104–105]

We have occasion in a later chapter to comment on the possibility that Bayes and David Hartley were acquainted. Such a possibility perhaps gains credence when we note that Hartley, in writing of natural religion and man's duty and expectations, said:

I do not presume to give a complete treatise on any of these subjects; but only to borrow from the many excellent writings, which have been offered to the world on them, some of the principal evidences and deductions, and to accommodate them to the foregoing theory of the mind. [1834, Part II, Introduction]

Although on its own this passage does not seem particularly supportive of the suggested possible acquaintance, we find that the views expressed in Hartley's extensive treatment of the benevolence of God are markedly similar to those given in Bayes's treatise, and it is possible that this tract was one of the 'excellent writings' drawn on by Hartley.

In order that the reader may be spared the consulting of *Observations on Man, His Frame, His Duty, and His Expectations* (at least on this matter), I append some of Hartley's thoughts on benevolence, cited here in support of the suggestion proposed in the previous paragraph:

PROP. III.– *The Infinite Independent Being is endued with infinite Power and Knowledge.*

PROP. IV.– *God is infinitely Benevolent.*

As all the natural attributes of God may be comprehended under power and knowledge, so benevolence seems to comprehend all the moral ones. This proposition therefore, and the foregoing, contain the fundamentals of all that reason can discover to us concerning the divine nature and attributes.

Now, in inquiring into the evidences for the divine benevolence, I observe, first, That as we judge of the divine power and knowledge by their effects in the constitution of the visible world, so we must judge of the divine benevolence in the same way. Our arguments for it must be taken from the happiness, and tendencies thereto, that are observable in the sentient beings, which come under our notice.

Secondly, That the misery, to which we see sentient beings exposed, does not destroy the evidences for the divine benevolence, taken from happiness, unless we suppose the misery equal or superior to the happiness. A being who receives three degrees of happiness, and but one of misery, is indebted for two degrees of happiness to his Creator. Hence our inquiry into the divine benevolence is reduced to an inquiry into the balance of happiness or misery[30], conferred, or to be conferred, upon the whole system of sentient beings, and upon each individual of this great system. If there be reason to believe, that the happiness which each individual has received, or will receive, be greater than his misery, God will be benevolent to each being, and infinitely so to the whole infinite system of sentient beings; if the balance be infinitely in favour of each individual, God will be infinitely benevolent to each, and infinito-infinitely benevolent to the whole system.

Thirdly, Since the qualities of benevolence and malevolence are as opposite to one another, as happiness and misery, their effects, they cannot co-exist in the same simple unchangeable being. If therefore we can prove God to be benevolent, from the balance of happiness, malevolence must be entirely excluded.

Fourthly, Since God is infinite in power and knowledge, *i.e.* in his natural attributes, he must be infinite in the moral one also; *i.e.* he must be either infinitely benevolent, or infinitely malevolent. All arguments, therefore, which exclude infinite malevolence, prove the infinite benevolence of God.

Lastly, As there are some difficulties and perplexities which attend the proofs of the divine self-existence, power, and knowledge, so it is natural to expect, that others, equal, greater, or less, should attend the consideration of the divine benevolence.

Let us now come to the evidences for the divine benevolence, and its infinity.

First, then, It appears probable, that there is an over-balance of happiness to the sentient beings of this visible world, considered both generally and particularly.

Secondly, If we should lay down, that there is just as much misery as happiness in the world, (more can scarce be supposed by any one,) it will follow, that if the laws of benevolence were to take place in a greater degree than they do at present, misery would perpetually decrease, and happiness increase, till, at last, by the unlimited growth of benevolence, the state of mankind, in this world, would approach to a paradisiacal one. Now, this shews that our miseries are, in a great measure, owing to our want of benevolence, *i.e.* to our moral imperfections, and to that which, according to our present language, we do and must call *ourselves*. It is probable therefore, that, upon a more accurate examination and knowledge of this subject, we should find, that our miseries arose not only in great measure, but entirely from this source, from the imperfection of our benevolence, whilst all that is good comes immediately from God, who must therefore be deemed perfectly benevolent. ... It follows hence, that malevolence, and consequently misery, must ever decrease.

Fifthly, The whole analogy of nature leads us from the consideration of the infinite power and knowledge of God, and of his being the Creator of all things, to regard him as our father, protector, governor, and judge. We cannot therefore but immediately hope and expect from him benevolence, justice, equity, mercy, bounty, truth, and all possible moral perfections.

Seventhly, Supposing that every single thing is, other things remaining the same, the most conducive to happiness that it can be, then the real deficiencies that are found in respect of happiness, and which, at first sight, appear to arise from a proportional deficiency in the divine benevolence, may be equally ascribed to a deficiency in the divine power or knowledge.

Ninthly, ... infinite benevolence, which always appears to us under the idea of perfection and happiness, seems to be the immediate and necessary consequence of the natural attributes of infinite power and knowledge: since the wishing good to others, and the endeavouring to procure it for them, is, in us, generally attended with a pleasurable state of mind, we cannot but apply this observation to the divine nature, in the same manner that we do those made upon our own power and knowledge.

Tenthly ... benevolence must arise in all beings, other things being alike, in proportion to their experience of good and evil, and to their knowledge of causes and effects. One cannot doubt, therefore, but that infinite benevolence is inseparably connected with the supreme intelligence: ... the infinite benevolence of the Supreme Being is the same thing with his infinite perfection and happiness.

Twelfthly, It is probable, that many good reasons might be given, why the frame of our natures should be as it is at present, all consistent with, or even flowing from, the benevolence of the divine nature; and yet still that some supposition must be made, in which the same difficulty would again recur, only in a less degree. However, if we suppose this to be the case, the difficulty of reconciling evil with the goodness of God might be diminished without limits, in the same manner as mathematical quantitites are exhausted by the terms of an infinite series.

Thirteenthly ... Let us suppose then, that we may call that infinite benevolence, which makes either

1. Each individual infinitely happy always. Or,

2. Each individual always finitely happy, without any mixture of misery, and infinitely so in its progress through infinite time. Or,

3. Each individual infinitely happy, upon the balance, in its progress through infinite time, but with a mixture of misery. Or,

4. Each individual finitely happy in the course of its existence, whatever that be, but with a mixture of misery as before; and the universe infinitely happy upon the balance. Or,

5. Some individuals happy and some miserable upon the balance, finitely or infinitely, and yet so that there shall be an infinite overplus of happiness in the universe.

All possible notions of infinite benevolence may, I think, be reduced to some one of these five;

... we seem able, at present, to express the real appearance [of things], in the same way as mathematicians do ultimate ratios, to which quantitites ever tend, and never arrive, and in a language which bears a sufficient analogy to other expressions that are admitted. So that now (if we allow the third supposition) we may, in some sort, venture to maintain that which at first sight seemed not only contrary to obvious experience, but even impossible, *viz.* that all individuals are actually and always infinitely happy. And thus all difficulties relating to the divine attributes will be taken away; God will be infinitely powerful, knowing, and good, in the most absolute sense, if we consider things as they appear to him. And surely, in all vindications of the divine attributes, this ought to be the light in which we are to consider things.

We cannot doubt but the Judge and Father of all the world will conduct himself according to justice, mercy, and goodness. However, I desire to repeat once more, that we do not seem to have sufficient evidence to determine absolutely for any of the last three suppositions. We cannot indeed but wish for the third, both from self-interest[31] and benevolence; and its coincidence with the first and second, ... appears to be some presumption in favour of it.
[1834, Part II, Chap. I]

The reader will see in these passages from Hartley's book the repetition of several of the opinions displayed in Bayes's treatise; attention may be drawn to the following: the emphasis on God's benevolence at the expense of detailed consideration of His rectitude or wisdom, the arguments for such benevolence are found by Hartley to be derivable from the happiness observable in sentient beings, the presence of misery is not to be seen as destructive of the evidences for the divine benevolence, God is infinitely benevolent, and malevolence is excluded.

It may be deduced from parts of the preceding quotation that Hartley perhaps argues too strongly in favour of certain divine attributes on the basis of the display of similar characteristics in mankind (recall Kilpatrick's remarks mentioned in §4.1). However, he is careful to note the interpretation that should be placed upon such argument:

... the events of life, and the use of language, beget such trains of ideas and associations in us, as that we cannot but ascribe all morally good qualities, and all venerable and amiable appellations, to the Deity; at the same time that we perceive the meaning of our expressions not to be strictly the same, as when they are applied to men; but an analogical meaning, however a higher, more pure, and more perfect one. ... even the attributes of independency, omnipotence, omniscience, and infinite benevolence, though the most pure, exalted, and philosophical appellations, to which we can attain, fall infinitely short of the truth, of representing the Deity as he is, but are mere popular and anthropomorphical expressions. [1834, Part II, Chap. I]

There is one other writer whose work needs to be mentioned in connexion with *Divine Benevolence*, and that is Richard Price. Well known to the statistician as the communicator to the Royal Society of Bayes's *An Essay towards solving a Problem in the Doctrine of Chances*, Price devoted no inconsiderable time to writing on moral matters, and it is to some of these that we now turn our attention.

Like Bayes, Price ascribes justice and veracity to the Deity:

> *Justice* and *Veracity* are *right* as well as *goodness*, and must be ascribed to the Deity.—By *justice* here I mean principally *distributive justice*, impartiality and equity in determining the state of beings, and a constant regard to their different moral qualifications in all the communications of happiness to them. [1787/1974, p. 250]

However, in his *A Dissertation on the Being and Attributes of the Deity*, appended to the third edition of his *A Review of the Principal Questions in Morals*, Price goes perhaps even further than Balguy, Bayes, and Grove, ascribing to God not adjectives but identifying Him with nouns:

> [God] is therefore, WISDOM, rather than *wise*; and REASON, rather than *reasonable*. In like manner; he is ETERNITY, rather than *eternal*; IMMENSITY, rather than *immense*, and POWER, rather than *powerful*. In a word; he is not *benevolent* only, but *benevolence*; not absolutely *perfect* only, but absolute *perfection* itself; the root, the original (or to speak after Dr. *Clarke*, and perhaps less improperly) the *substratum* of all that is great and wise and good and excellent. [1787/1974, p. 290]

(William Hazlitt, in his 1823 essay 'On the old age of artists', writes that Richard Cosway's miniatures 'were not merely fashionable — they were fashion itself.')

The reasons for Price's general position, that there are no works of supererogation, are given by Thomas as follows.

> One of the reasons why this is so is that under some of the 'heads of virtue', as, for example, beneficence, it is acting from the principle of beneficence, and not any particular manifestation of it, that is obligatory. Where there is an obligation to give to someone, there may not be an obligation (though it may be right) to give to a particular one.
>
> The principal form of the doctrine to which Price was opposed — that there is but one test or principle of rectitude — was the view that virtue was founded in benevolence ... His main target appears to have been Hutcheson who has become widely celebrated as the originator of the utilitarian formula 'that *that action* is *best* which procures the *greatest happiness* for the *greatest numbers*'[32], but he also appears to have had in mind those who argued in a more theological context that God in His creation and providence is influenced by no other principle than a regard to the happiness of his creatures[33].
> [1977, pp. 69, 72]

4.4 Conclusion

I find it difficult, as one who is not theologically trained, to pass an authoritative judgement on the relative merits of the tracts by Balguy, Bayes, and Grove. The summary given by Doddridge, which we have already quoted, seems to be a fair reflexion of the contents of each, and it is unclear, at least to me, who, if any, emerged the victor in the 'celebrated controversy'. Perhaps Doddridge, finding some inadequacies in Bayes's work, would assign the laurels to the treatises by Balguy and Grove rather than to that of the more obvious recipient[34], though against this we must weigh the prominence given to benevolence in Hartley's book.

After all this perhaps the most we can say is that we are, if not wiser, at least better informed on God's attributes and the motives guiding Him — or, as Edward FitzGerald put it perhaps rather despondently in his translation of the *Rubáiyát of Omar Khayyám*,

> Myself when young did eagerly frequent
> Doctor and Saint, and heard great Argument
>   About it and about; but evermore
> Came out by the same Door as in I went.

5

An Introduction to the Doctrine of Fluxions

> *They have pretty sharp distinctions of the fluctuating and the permanent.*
>
> Charles Lamb.
> *The Essays of Elia: Popular Fallacies.*

5.1 Introduction

Although it is often a bad thing to rush into print, to delay in publishing can sometimes have a more serious effect. Thus it was that as a result of his reluctance to publish, Newton's own work on the fluxionary calculus was slow in being disseminated, it being known for a long time only to a few privileged correspondents and confidants[1]. Newton's first published work on this matter, the *Tractatus de quadratura curvarum*[2], appeared as an appendix to his *Opticks* in 1704, and *The Method of Fluxions and Infinite Series*, although written in 1671, was first printed[3] in 1736. It was thus left to others to expound the calculus in print, but despite such exposition, clarity on the underlying methods and logic was slow in being achieved. In her biography of Thomas Simpson, Clarke in fact writes

> Newton recognized three types of the calculus. In his *Principia* (1687) he made use of infinitely small quantities. His second method was that of fluxions ... Newton's third method, that of limits, ...[1929, p. 5]

Similarly, but perhaps more accurately, Kitcher notes

> the theory of fluxions yielded the heuristic methods of the calculus. Those methods were to be justified rigorously by the theory

of ultimate ratios. The theory of infinitesimals was to abbreviate the rigorous proof, and Newton thought he had shown the abbreviation to be permissible. Rather than competing for the same position, the three theories were designed for quite distinct tasks. [1973, pp. 33–34]

The situation was quite different with regard to Newton's work on optics and gravitational theory, however, the diffusion of which was such as to validate Alexander Pope's famous distich[4]

Nature and Nature's Laws lay hid in Night:
GOD said *Let Newton be!* and all was Light

whereas the status of the fluxionary calculus could perhaps be expressed in similarly pious vein as[5]

Though Alpha and Omega sternly spake,
Primes, Ultimates and Fluxions stayed Opaque.

It was, indeed, difficult to obtain a coherent idea of the fluxionary calculus from Newton's own publications. The solution of the inversely related problems of quadrature and tangency was approached by the development of two devices: the theory of moments and the theory of prime and ultimate ratios[6]. The *moment* of a fluent, or flowing quantity, was defined as its 'momentaneous synchronal increment' or the amount by which it was increased in an indefinitely small time period. The theory of prime and ultimate ratios, on the other hand, was concerned with the ratios between magnitudes as generated by motion. The prime ratios of nascent magnitudes were obtained as the magnitudes began to be generated, whereas the ultimate ratios of evanescent magnitudes were the ratios obtaining between magnitudes that diminished to nothing and vanished. For instance, in his 'An account of the book entituled *Commercium Epistolicum Collinii & aliorum, De Analysi promota*' of $171\frac{4}{5}$, Newton[7] presented two different explanations of the method of fluxions and moments: in the first of these, moments, divorced from their kinematic origin, are seen to be equivalent to Leibniz's differentials, and in the second approach one finds essentially a calculus of limits[8] (or at least a simulacrum thereof).

A number of attempts at the popularization and the exposition of the fluxionary calculus, many of these showing also the influence of Leibnizian methods, were therefore made by textbook writers in the early 1700s, the earliest of these perhaps being John Harris's *A New Short Treatise of Algebra* of 1702, some twenty pages of which were devoted to fluxions. It was, however, only in the 1730s that the previous limited demand for such knowledge grew[9], with consequent increase in the number of treatises on, and expositions of, the subject.

184   5. Introduction to the Doctrine of Fluxions

Because of its fundamental position in the history of textbooks on fluxions, Harris's work should perhaps receive some slight attention here. Facing the title page are the following words, 'All Kinds of Mathematicks Theorical [sic] and Practical, are Taught by the Author, at his House in *Amen-Corner*, near *Pater-Noster-Row*; where also Gentlemen may be Boarded.' Harris sets out his reasons for writing the book as follows.

> To the Reader. This small Tract of that admirable Science, *Algebra*, was written Primarily for the Use of my Auditors at the *Publick Mathematick Lecture*: Which is set up at the *Marine Coffee-house* in *Birchin-Lane* entirely for the Publick Good, by the Generous Charles Cox Esq; Member of Parliament, for the Burgh of *Southwark*. And because I have not yet found it done by any one in the *English* Tongue, I have given you a short Account of the Nature and Algorithm of *Fluxions*, and one or two Instances of their Use and Application: But in this I have been designedly as brief as possible I could, intending by it only to stir up the Readers Curiosity to peruse those excellent Treatises which I have there mentioned: And which when he hath done, I know I shall have his thanks for that little *Sketch* of Fluxions, which he will find there.

The section 'Of Fluxions' occupies pages 115 to 136. Harris defines the subject and its main concern as follows.

> By the Doctrine of Fluxions, then we are to understand the Arithmetick of the *Infinitely* small Increments or Decrements of *Indeterminate* or *variable Quantity*, or as some call them the *Moments* or *Infinitely small Differences* of such variable Quantities. [p. 115]
>
> The main Business of the Algorithm or Arithmetick of Fluxions, consists in these two Things.
> I. From the Flowing Quantity given, to find the Fluxion.
> II. From the Fluxion, to find the Flowing Quantity. [p. 118]

Controversy over the concepts and implementation of the fluxionary calculus abounded; and, as Charles Lamb said of the rogues of the world in the quotation given at the head of this chapter, each commentator was clear in his own mind[10] on the meaning of fluxions, permanents, prime and ultimate ratios, etc., though, as Pope said[11],

> 'Tis with our judgments as our watches, none
> Go just alike, yet each believes his own

with some commentators being describable by his further words[12]

> Some ne'er advance a Judgment of their own,
> But catch the spreading notion of the Town.

5.1. Introduction

Of course, the aim was always the sentiment expressed by Sir John Suckling in the Epilogue to his *Aglaura*:

> But as when an authentic watch is shown,
> Each man winds up and rectifies his own,
> So in our very judgments.

Among these controversialists, and by no means the least, was Bishop George Berkeley, who in 1734 published *The Analyst; or, a Discourse Addressed to an Infidel Mathematician*, in which he severely criticized the methods of the fluxionary and the differential calculi and the ontological status of the objects therein considered[13]. Many were the replies that this diatribe provoked[14], among which was the essay by Bayes that follows.

Although it was *The Analyst* that prompted Bayes's tract, Berkeley had in fact begun his criticism of the calculus some years before[15], with comments appearing in his *Philosophical Commentaries*[16], in the *Principles of Human Knowledge* and in an undated essay[17] 'Of Infinites'. Some of the notes in the *Philosphical Commentaries* are perhaps not such as one would expect from one who would later write on more abstruse matters[18]: for example, 'I say there are no incommensurables, no surds, I say the side of any square may be assign'd in numbers' and, no doubt as a consequence of this, 'One square cannot be double of another. Hence the Pythagoric Theorem is false.' And further 'Newton's fluxions needless. Anything below a M.[19] might serve for Leibnitz's Differential Calculus.' In some cases Berkeley's thoughts conflicted as the year during which the *Philosophical Commentaries* was compiled passed: thus we find

> Mem: upon all occasions to use the Utmost Modesty. to Confute the Mathematicians wth the utmost civility & respect. not to stile them Nihilarians &c.

and also 'I see no wit in any of them but Newton, The rest are meer triflers, meer Nihilarians'. It is, however, in *The Analyst* that the first full expression of Berkley's criticism of the fluxionary calculus is to be found.

The Analyst is both a theological and a mathematical work: as a theological one, it is intended to show the unjustness of certain criticisms of orthodox religion, and as a mathematical one its aim is to expose defects in the fluxionary calculus, Berkeley's thesis being that a freethinker who scorned (orthodox) religion because of its mysteries could not consistently and in conscience accept the tenets of the fluxionary calculus, the latter being at least as incomprehensible as any aspect of revealed religion. On this point the reader's attention may perhaps pertinently be drawn to a quotation from one of James Foster's sermons, printed in the volume for 1733, i, 175, viz. 'where mystery begins, religion ends'.

Berkeley's aims, then, were two: the theological aim was to show that this calculus was no less mysterious than Christianity, whereas the other, the threefold mathematical aim, was to show (a) that the postulation of infinitesimal quantities is needed for the fluxionary calculus, (b) that even the most fundamental proofs smack of fallacy and sophistry, and (c) that any successes that may be seen in the calculus may be attributed to compensatory errors, though perhaps Berkeley went too far in attributing 'absolute contradiction' rather than 'ultimate imperfect comprehensibility' to Newton's fluxions[20]. In his biographical account of Berkeley, Joseph Stock wrote as follows.

> The Bishop's chief objections to the doctrine of fluxions may be comprized under these two heads:
> I. That the object (viz. fluxions) was inconceivable.
> II. That the arguments, brought to prove the truth of the fundamental proposition, were fallacious and inconclusive.
> [1735(?), p. 75]

More than a century later Gibson framed Berkeley's mathematical contentions somewhat differently:

> (1) the conception of fluxions is unintelligible, inasmuch as they suppose the ratio of quantities that have no magnitude, for, in Berkeley's view, prime and ultimate ratios are such; (2) the demonstration of the value of a fluxion, say that of x^n, rests on the violation of an axiomatic canon of sound reasoning
> [1898, p. 15],

but essentially the various formulations are the same.

A severe critic of the methods used by the continental school, Robert Woodhouse, Lucasian Professor and later Plumian Professor of Astronomy and Experimental Philosophy in Cambridge University, expounded the differential method. Being unafraid of controversy himself, he viewed the dispute engendered by Berkeley's tract with enthusiasm, writing in his book *The Principles of Analytical Calculation* as follows.

> The name of Berkeley has occurred more than once in the preceding pages: and I cannot quit this part of my subject without commending the analyst and the subsequent pieces, as forming the most satisfactory controversial discussion in pure science, that ever yet appeared: into what perfection of perspicuity and of logical precision, the doctrine of fluxions may be advanced, is no subject of consideration: But, view the doctrine as Berkeley found it, and its defects in metaphysics and logic are clearly made out.

> If, for the purpose of habituating the mind to just reasoning, (and mental discipline is all the good the generality of students derive from the mathematics) I were to recommend a book, it should be the *Analyst*. Even those, who still regard the doctrine of fluxions as clearly and firmly established by their immortal inventor, may read it, not unprofitably, since, if it does not prove the cure of prejudice, it will be at least the punishment.*
>
> *The most diffuse and celebrated antagonists of Berkeley, are Maclaurin and Roberts[21], men of great knowledge and sagacity: but the prolixity of their reasonings confirm the notion, that the method they defend, is an incommodious one. The reason why Berkeley's ideas have not obtained a more general reception, seems to be this; unbiassed men, earnest lovers of truth, and moderately skilled in mathematics, read not the Analyst, because they imagined the discussion too deep for them; and professed mathematicians, in judging of an hostile tract, felt a zeal for the honor of their order, and a more reasonable affection for their favourite study. [1803, pp. xvii–xviii]

The *Analyst* opens with the following address to the 'infidel mathematician'.

> Though I am a stranger to your person, yet I am not, Sir, a stranger to the reputation you have acquired in that branch of learning which hath been your peculiar study; nor to the authority that you therefore assume in things foreign to your profession, nor to the abuse that you, and too many more of the like character, are known to make of such undue authority, to the misleading of unwary persons in matters of the highest concernment, and whereof your mathematical knowledge can by no means qualify you to be a competent judge. [§1]

It appears from this, as Gibson [1898, pp. 11–12] has noted, that Berkeley's aim in the writing of *The Analyst* was not so much to attack the conclusions reached by the new analysis, but rather to question the claims of the mathematicians *as such* to be regarded as authoritative in matters of religion.

We note too that Berkeley was quite firm in asserting that the infidel mathematician's opinion was to be taken *cum grano salis* when he was writing on matters other than those on which he was qualified to speak, as Sir Thomas Browne had stressed, in a wider context, some years before in the seventh chapter of the first book of his *Pseudodoxia Epidemica*: 'A testimony is of small validity if deduced from men out of their own profession'. One wonders whether in other circumstances Berkeley would have agreed with the statement made in *The Tatler*[22] when we see a man making profession of two different sciences, it is natural for us to believe he is no pretender in that which we are not judges of, when we find him skilful in that which we understand. [Letter No. 240, 21st October 1710]

The contents of *The Analyst* may be described briefly as follows[23]. The first two sections serve as a short introduction to the essay, the main argument against the calculus following in §§3–20, where Berkeley argues that the obscurity of the object of the calculus is matched by its unscientific principles and demonstrations. The rôle of compensatory errors in the success of the calculus is discussed in §§21–29, and the next eighteen sections are devoted to possible responses to this criticism and their rebuttal[24]. Sections 48–50 contain an indictment of the 'obscurity and incomprehensibility of [the infidel mathematician's] metaphysics' and the lack of scientificalness of the modern analytics, and the essay is concluded with a list of sixty-seven wide-ranging Queries[25]. We do not consider the contents of *The Analyst* in more detail here (except as they concern us in connexion with Bayes's essay); see Jesseph [1993, chap. 6] for a careful discussion.

Who the 'infidel mathematician' was to whom *The Analyst* is addressed is uncertain, though some clue can be found in Berkeley's 1735 tract entitled *A Defence of Free-Thinking in Mathematics*, a response to James Jurin's *Geometry no Friend to Infidelity*[26] (a work that was among the first responses to *The Analyst*). Here we find the following.

> I do not say, as you would represent me, that we have no better reason for our religion than you have for fluxions: but I say that an infidel, who believes the doctrine of fluxions, acts a very inconsistent part in pretending to reject the Christian religion because he cannot believe what he doth not comprehend; or because he cannot assent without evidence; or because he cannot submit his faith to authority. Whether there are such infidels, I submit to the judgment of the reader. For my own part I make no doubt of it, having seen some shrewd signs thereof myself, and having been very credibly informed thereof by others. ... [The late celebrated Mr. Addison] assured me that the infidelity of a certain noted mathematician, still living, was one principal reason assigned by a witty man of those times for his being an infidel. [§7]

This is all we find here, but the modern enquirer is usually referred for further details to Stock's *An Account of the Life of George Berkeley*, where we read[27]

> The occasion was this: Mr. Addison had given the bishop an account of their common friend Dr. Garth's behaviour in his last

illness, which was equally unpleasing to both those excellent advocates for revealed religion. For when Mr. Addison went to see the doctor, and began to discourse with him seriously about preparing for his approaching dissolution, the other made answer, "Surely, Addison, I have good reason not to believe those trifles, since my friend Dr. Halley, who has dealt so much in demonstration, has assured me, that the doctrines of Christianity are incomprehensible, and the religion itself an imposture." The bishop therefore took arms against this redoubtable dealer in demonstration, and addressed *The Analyst* to him, with a view of shewing, that mysteries in faith were unjustly objected to by mathematicians, who admitted much greater mysteries, and even falsehoods, in science, of which he endeavoured to prove that the doctrine of fluxions furnished an eminent example. [pp. xix–xx]

The referral is usually accompanied, however, by the mention of Stock's unreliability. It is further to be noted (see Jesseph [1993, p. 179]) that Garth and Addison both died in 1719, in January and June, respectively; at this time Berkeley was in Italy, and although he could have been informed of this incident by letter, no such letter survives. It also seems a long time for Berkeley to have waited to enter the lists against Halley, though he does say

> Of a long time I have suspected that these modern analytics were not scientifical, and gave some hints thereof to the public twenty-five years ago. [§50]

Incidentally, Garth's suggested irreligious views were reported somewhat differently by Pope in a letter to Jervas of the 12th of December 1718:

> But ill tongues, and worse hearts, have branded even his last moments, as wrongfully as they did his life, with irreligion. You must have heard many tales on this subject; but if ever there was a good Christian without knowing himself to be so, it was Dr. Garth. [1872, vol. VIII, p. 28]

In his Notes to this edition of Pope's *Letters*, Whitwell Elwin wrote that Garth was 'lax and epicurean', and that

> Any definition of a christian which was applicable to the life of Garth, would make manifest the inconsistency of the assertion that a better christian never existed.

Not all were as convinced as Stock, however, that the infidel mathematician was correctly identified as Halley. In 1844 Rigaud published a tract in which he defended Halley against this charge, writing 'no part of Halley's many writings has ever been brought forward in support of his alleged infidelity' [p. 10]. Rigaud notes that the Addison–Garth–Halley story was published by an anonymous writer in 1776, long after the bishop's death, and summarises his account with the words

> we have argued that the story connecting the publication of the Analyst with facts which would irrefragably convict Halley of infidelity, is deficient in authority, and inconsistent with other facts, as well as with the probabilities of the case. It appears that bishop Berkeley was not personally acquainted with Halley, and it is not even certain that Garth was so; and thus we are led to believe that Addison had merely repeated to the bishop an unfounded rumour, which they both, in common with many other men, believed. [p. 31]

We note that Rigaud's thesis is directed at Halley's alleged infidelity, and not at whether *The Analyst* was addressed to him. Indeed he writes

> When the Analyst was addressed to Halley under the ignominious distinction of "an Infidel Mathematician," it was not an incidental notice of a disappointed and injudicious man, nor the railing of acrimonious hostility: it was a regular argument drawn up by one of acute mind and charitable disposition; one likewise who was actuated by a sincere zeal for the truths of Christianity, and whose inference, (as far as affects the moral character of the individual,) cannot be parried, if his premises be admitted. These premises we now proceed to examine.
> [1844, p. 24]

In the absence of any other candidate, therefore, ignoring the adjective and not querying whether it should be understood in the sense of 'unbelieving' or of 'sceptical', we may assume that the 'infidel' mathematician is correctly identified[28]: the identification, however, is of little consequence to the interpretation of *The Analyst*.

Benjamin Robins's *A Discourse Concerning the Nature and Certainty of Sir Isaac Newton's Methods of Fluxions, and of Prime and Ultimate Ratios* of 1735 is described by Jesseph as 'an account of fluxions which is an original and important contribution to the foundations of the calculus ... a very good response to Berkley' [1993, pp. 259–260, 269][29]. Gibson views it as 'a masterpiece of its kind' [1898, p. 22].

Here Robins develops definitions and lemmata that are markedly similar to those of the modern theory of limits, distinguishing between the fluxionary doctrine and that of prime and ultimate ratios. The existence and uniqueness of limits and ratios of limits are established, and it is also shown that the limit of a ratio is the ratio of the limits. The theory does, however, differ in one important respect from that used today: Robins insists that a variable *never* attains its limit[30]. Whether a theory of limits can also be seen in Bayes's work emerges in due course[31].

It is perhaps worth noting, albeit briefly, that it was not only his work on fluxions that made Newton the subject of criticism. For example, in 1710 Johann Bernoulli objected to the proof, in Propositions XI–XIII along with Corollary I in Book One of the *Principia*, that conic-section orbit is implied by inverse-square force (see Weinstock [1998, pp. 284–285]). Weinstock argues that the fallacy exposed by Bernoulli is basically a violation of a logical principle (one that says in essence that a statement that one wishes to prove should not be assumed and used as part of the proof), a principle that he finds more flagrantly flouted in the proof offered by Newton for Proposition XV/Theorem XII in Book Two of the *Principia* (a result denoted by '2XV' by Weinstock)[32]. Perhaps echoing Berkeley's (tacit) opinion, Weinstock writes

> Newton's thought on the matter at hand [i.e., 2XV] is not accessible to us, but it is difficult to suppose that he was unaware of the logical principle violated in the argument he presented as proof of 2XV. Can it then be possible that the Cambridge professor was aware of his inability to prove 2XV and therefore presented an intricacy-infested counterfeit proof while entertaining the hopeful expectation that no one would detect the fallacy? Scholars in our time have pointed to items in the *Principia* that betray the aroma of swindle knowingly committed by the tome's author. [1998, p. 285.]

We now turn to *An Introduction to the Doctrine of Fluxions*, leaving further discussion of Bayes's response to Berkeley to the Commentary in §5.3. It suffices for the moment to note that this reply was concerned more with the logical analysis of the theory of prime and ultimate ratios than with the methods of the fluxionary calculus or moments[33], and it is with this in mind that Bayes's essay should be read. Indeed, such a response was directly to the matter of *The Analyst*, Berkeley having written

> I beg leave to repeat and insist, that I consider the geometrical analyst as a logician, *i.e.* so far forth as he reasons and argues; and his mathematical conclusions, not in themselves, but in their premises; not as true or false, useful or insignificant, but as derived from such principles, and by such inferences. [§20]

AN INTRODUCTION

TO THE

Doctrine of *Fluxions*,

And DEFENCE of the

MATHEMATICIANS

AGAINST THE

OBJECTIONS of the Author of the *ANALYST*, so far as they are designed to affect their general Methods of Reasoning.

L O N D O N:

Printed for J. NOON, **at the** *White-Hart,* near *Mercers-Chapel,* in *Cheapside.* MDCCXXXVI.

PREFACE

I HAVE *long ago thought that the first principles and rules of the method of* Fluxions *stood in need of a more full and distinct explanation and proof, than what they had received either from their first incomparable author, or any of his followers; and therefore was not at all displeased to find the method itself opposed with so much warmth by the ingenious author of the* Analyst; *and had it been his only design to bring this point to a fair issue, whether a demonstration by the method of Fluxions be truly scientific or not, I should have heartily applauded his conduct, and have thought he deserved the thanks even of the Mathematicians themselves. But the invidious light in which he has put this debate, by representing it as of consequence to the interests of religion, is, I think, truly unjustifiable, as well as highly imprudent. Among all wise and fair inquirers, 'tis beyond all contradiction plain, that religion can be no ways affected by the truth or falshood of the doctrine of Fluxions. And tho' prejudiced minds may be variously affected by it, yet it is not easy to be conceived what advantage this debate is likely to give to the cause of religion and virtue in general even among them. Whereas it is easy to guess of what disservice our author's representation of a controversy in which* [iii] *religion has no manner of concern, may be towards raising and inflaming the passions of weak men on both sides of the question: And I wish he had been pleased coolly to consider beforehand of what consequence the result of this dispute is likely to be to the cause of religion, among those for whose conviction his* Analyst *is chiefly designed. If he should not be able to make out his point, will not the blind followers of the Infidel Mathematicians be more confirmed in their errors than they were before? Will they not be more prejudiced against religion, and established in their esteem and veneration of their masters by a weak and fruitless attempt to depress their characters; and by finding that a zeal for it has occasioned so strong an attempt to wound the reputation of Sir* Isaac Newton *as a cautious and fair reasoner? And on the other hand, if our author should carry his point, and his proofs should be allowed, that the doctrine of Fluxions is an incomprehensible mystery, and that the most accurate Mathematicians have, one after another, imposed upon themselves in the most egregious manner, by false and inconclusive reasonings, what consequences can we suppose that such persons will draw from these premises? Our Author indeed would have them only from hence make this one conclusion, That their masters, the Mathematicians, are not to be depended upon when they speak against religion. But I believe it can't in reason be expected that they should stop here.*

If such men as Dr. Barrow, *Dr.* Clarke, &c. *and the incomparable Sir* Isaac Newton, *were capable of imagining that they saw with the greatest clearness and perspicuity, where they had nothing but absolute and incomprehensible darkness before them, what conclusion will persons, used to take their opinions from authority, be likely to make* [***iv***] *from these premises, but that all pretences to knowledge in religion, and every thing else, are only confidence and presumption?*

If they are taught that it is inconsistent for a person to reject the mysteries of religion, and yet believe the mystery of Fluxions, will they not know how to draw the opposite conclusion themselves, that it is inconsistent to reject the doctrine of Fluxions because mysterious, and yet receive the mysteries of religion? And* *when they are taught to think that a person may be* justly *said to have faith, because they give into what they can neither demonstrate nor* conceive; *if this give them a mean opinion of the Mathematicians, 'tis odds if it don't give them a mean opinion of faith itself. I am sure 'tis a very strange account of that which may justly be called faith: For without clear notions no man can believe any more than demonstrate.*

Considering these things, I can't help thinking it was highly wrong to bring religion at all into this controversy, which may inflame the dispute, but can hardly do any real service: Of which, to me, it is a very strong presumption, that every thing urged by the author of the Analyst *against* Infidels *in general, would have sounded full as well in the mouth of a Papist, if urged against those Mathematicians that don't believe the doctrine of Transubstantiation, as it would have been peculiarly in character for such a one to have made his chief attack upon a great enemy of all superstition and tyranny, and an hearty friend to the reasonable religion of Christians and Protestants. But enough of this. I shall now consider my subject as stript of all relation it has to religion, and merely as a matter of human science, and endeavour to shew that the method of Fluxions is founded upon clear and substantial principles.* [***v***]

---

* Defence of Freethinking, *p.* 62.

SECT. I.

IT cannot be doubted but that Sir *Isaac Newton* well understood the doctrine of which he was the original inventor; and his proofs of it are very far from being fallacious and deceitful, or their force hard to be understood by those that are used to these kinds of subjects.

BUT it is also very plain, that the question, which is the main dispute between our author and his adversaries, whether Mathematicians take the notion and certainty of the method of Fluxions implicitly from him or not, does not depend upon our being able to defend the exact accuracy of the demonstrations he has made use of, and the propriety of every phrase by which he has explained his notions upon this head. He always seems to have studied conciseness of expression, and to depend on the good sense and judgment of his reader. And on this account some of his demonstrations are not the most full and compleat that might be given, and must remain obscure to those who have no genius for the mathematical science, and can't find out those steps in a demonstration, which a writer often omits in confidence of the sagacity of his reader. In my opinion this is in some measure the case with respect to his proofs of **[7]** ^{34} the first principles of Fluxions, and therefore I don't wonder persons differ in their sentiments about them. But it is truly provoking to find that the greatest genius that ever appeared in the philosophical world, and one whom the lovers of knowledge must always think of with respect and gratitude, should be represented contrary to his known character, as craftily imposing on the world, in confidence of his own authority, and the obscurity of his subject. And therefore I would hope that the author of the *Analyst* did not design that severe reflection his words seem to carry with them, when he says, "Such reasoning as this, nothing but the obscurity of the subject could "have *encouraged* or *induced* the great author of the fluxionary method *to* "*have put upon* his followers; and nothing but an *implicit* deference to his "authority could have moved them to admit." To suspect Sir *Isaac Newton* of the mean design of seeking reputation among the ignorant, by venting unintelligible notions, and defending them by artful and cunning sophistry, is what I think no man is capable of doing. And therefore if the author of the *Analyst* does not think fit, for his own reputation, to revoke or explain the sentence just mentioned, it needs not a particular confutation. Nor do I propose particularly to follow him in all the objections he has made against Sir *Isaac*'s notions and demonstrations, being of opinion that the best way of answering him is to assist persons in understanding the subject itself; for if any one can do this, he will easily see there is little weight in what he has said against it. However, as I go along, I shall endeavour to obviate any thing that I think may create a difficulty; but my main **[8]** view is to settle the first principles on which the doctrine of Fluxions depends, and then to shew that, by just reasoning from them, the rules for finding the

Fluxions of equations, as delivered by Sir *Isaac*, do truly follow.

THE notion of Fluxions was originally gained by the observation of quantities being described by a continual motion; and the method of Fluxions was designed to do these two things. 1*st*, From the magnitude of a quantity continually changing being given, to find out the rate or velocity according to which the quantity itself continually increases or decreases. And, 2*dly*, From this latter continually given, to find the former.

WHERE you see the main thing taken for granted is this

POSTULATE.

Quantities may be supposed as continually changing, so as every distinct instant of time to be different from what they were before.

ILLUSTRATION.

SUCH quantities are the following, Time from a given hour, the distance of a body from a plane to or from which it moves, the amount of money lent out at interest, *&c*. However, it is not the business of the Mathematician to dispute whether quantities do in fact ever vary in the manner that is supposed, but only whether the notion of their so doing be intelligible; which being allowed, he has a right to take it for granted, and then to see what deductions he can make from that supposition. It is not the [9] business of a Mathematician to show that a strait line or circle can be drawn, but he tells you what he means by these; and if you understand him, you may proceed farther with him; and it wou'd not be to the purpose to object that there is no such thing in nature as a true strait line or perfect circle, for this is none of his concern: he is not inquiring how things are in matter of fact, but supposing things to be in a certain way, what are the consequences to be deduced from them; and all that is to be demanded of him is, that his suppositions are intelligible, and his inferences just from the suppositions he makes. In the case before us, Whether quantities in fact do ever vary in the manner before explained, or not, is nothing to the purpose, but whether the notion of their so doing be intelligible; and that it is so is plain, because tho' this should not be fact in any case, yet in a great many it seems to us so to be. When a stone falls to the ground, the line it describes seems continually to increase; nor can I avoid the same sentiment when I describe a circle or any other line upon a paper. And it may as well be pretended that the letters I now write consist of a number of distinct and separate dots, and no continued lines, as that each of them does not seem to be formed by a continued motion. Now if I really think I see quantities formed by a continued motion, it is plain that this sort of increase is not unintelligible, and therefore may be supposed by the Mathematician. The reader perhaps may here think that I intend to obviate an objection that even the author of the *Analyst* himself would never make; but I must own my suspicion of the contrary, because I think he can't allow of such sort of [10] increase, without giving up his cause, and allowing me a right to make this farther

POSTULATE II.

The notion of FLUXIONS *is intelligible.*

FOR if quantities may increase or decrease so as to be different every distinct instant of time from what they were before, it will follow that they must change at a certain rate either fixed or variable, it being impossible that a man shou'd conceive of a quantity continually changing, without knowing that it must either alter at the same rate always, or else sometimes faster, and at other times slower; *i.e.* he can't do this without knowing what Sir *Isaac Newton* means when he defines the Fluxion of a quantity to be the velocity or swiftness with which the quantity changes its magnitude: And the Fluxion of a quantity cannot be an unintelligible notion, when it necessarily arises from a plain and easy supposition. Nay, I am very well satisfied that our author himself must have a notion both of a first and second Fluxion, if he at all understands himself, when he supposes *a line described by the motion of a point continually accelerated.* At least I am sure 'tis as hard for me to conceive of the motion of a point, and the acceleration of that motion, as to form an idea of a first and second Fluxion. [11]

SECT. II.

DEFINITIONS.

1. A FLOWING quantity is one that continually increases or decreases, and in such a manner that some time is requisite to make any increment or decrement.

2. THE Fluxion of a flowing quantity is its rate or swiftness of increase or decrease.

3. THE change of a flowing quantity is the difference between the flowing quantity itself, and its value at a particular instant of time.

4. THE time in which a change is made, is the time the flowing quantity takes to alter from a prior to a subsequent value, whose difference is the change.

5. THAT change is said to vanish at a given instant, which is the difference between the flowing quantity before that instant and its value then; and that change is said to begin to arise at a given instant, which is the difference between the flowing quantity after that instant, and its value at that instant.

ILLUSTRATION.

LET y be a quantity continually increasing, A its value at a given instant; then $A - y$ will represent its change [taken from that instant] so long as y is less than A; and this change will vanish when y becomes equal to A; and by the [12] continued increase of y, there will then begin to arise another change which is ever afterwards equal to $y - A$.

Def. 6. IF there be two permanent quantities A and B, and two other flowing Quantities a and b, and the ratio of a to b be always, during a given time, that of the sum or difference of the first permanent quantity A, and another flowing quantity x to the sum or difference of the second permanent quantity B, and another flowing quantity y, and at the end of the given time all the flowing quantities vanish; then the ratio of the permanent quantities A and B, is the last or ultimate ratio of the vanishing quantities a and b; which I thus express, ult. $a : b :: A : B$. *i.e.* if $a : b :: A \mp x : B \mp y$ always, during the time $\mathcal{T}$, and at the end of that time, a, b, x, y all vanish; then ult. $a : b :: A : B$.

7. IF other things being supposed the same, a, b, x, y, be each equal to nothing at the beginning of the time $\mathcal{T}$, then the ratio of A to B is the first ratio of the nascent quantities a and b; which I thus express, $\text{p}^{\text{mo}} a : b :: A : B$.

Remark. THESE two definitions are in effect the same with those given by Sir *Isaac Newton*, and can't be disputed; for whether a and b, properly speaking, have any proportion as they arise, or vanish, yet A and B have; and that I am at liberty to call by what name I please.

Coroll. 1. THE ultimate ratio of two vanishing quantities, is a determinate ratio; *i.e.* if ult. $a : b :: A : B$, and ult. $a : b :: A : N$. B is $= N$. For since ult. $a : b :: A : B$; therefore, by *Def.* 6. $a : b :: A \mp x : B \mp y$ during a given time $\mathcal{T}$, at the end of which the flowing quantities a, b, x, y, all vanish; and by the same [13] *Defin.* $a : b :: A \mp v : N \mp z$, during a given time t, at the end of which the flowing quantities a, b, v, z, all vanish; and the end of t and $\mathcal{T}$ coincide, because that is the instant when a and b vanish. Taking therefore the shorter of these times t, during that $A \mp x : B \mp y :: A \mp v : N \mp z$. Now because x, y, z, v, change only as flowing quantities, and vanish together at the end of the time t, before the end of the time t they will be all less than any assignable quantity D [for suppose any of them never less than D before that instant, then a decrement as large as D must be made instantaneously, in order to its vanishing at that instant, which is contrary to the supposition of its changing only as a flowing quantity.] But since x, y, v, z, may be as small as you please, and yet the analogy $A \mp x : B \mp y :: A \mp v : N \mp z$ be preserved, it is plain that $A : B :: A : N$; and that therefore $B = N$. **Q.E.D.**

Coroll. 2. IN like manner it may be proved that the first ratio of two arising quantities is a given ratio.

AXIOM I.

THE sum and difference of two flowing, and the fourth proportional to three flowing quantities; and therefore a quantity any how made up of given and flowing quantities, must be itself a flowing or permanent quantity; *i.e.* some time is requisite to its receiving any increment or decrement, because a change in it must imply a change at least in some one of them.

Coroll. 3. IF $a : b :: A \mp x : B$ during the time $\mathcal{T}$; then if, at the end of the time $\mathcal{T}$, the flowing quantities a, b, x vanish, ult.$a : b :: A : B$; and if, at the beginning of $\mathcal{T}$, they are equal **[14]** to nothing, $\mathrm{p^{mo}}\, a : b :: A : B$. For because $a : b :: A \mp x : B$, therefore $a : b :: A + a \mp x : B + b$ during the time $\mathcal{T}$, and a and b, and $a \mp x$ are flowing quantities all vanishing together, or all arising together; and therefore, by *Defin.* 6 and 7, in the former case ult.$a : b :: A : B$, and in the latter $\mathrm{p^{mo}}\, a : b :: A : B$.

Coroll. 4. IF $a : b :: A : B$ always, a and b vanishing or arising together, then ult.$a : b :: A : B$; or $\mathrm{p^{mo}}\, a : b :: A : B$. This is proved as the foregoing.

Coroll. 5. IF the ratio of two quantities continually decreasing approach continually nearer and nearer to a given ratio, and by so doing at length come nearer to it than by any assignable difference, that given ratio is the last ratio of the flowing quantities when they vanish. Let the flowing quantities be a and b, the given ratio $A : B$. Then supposing $a : b :: A : B \mp x$ always, as a and b decrease, x will also continually decrease in the same manner as a flowing quantity, by $Ax.$ 1. and x may be as small as you please, because the ratio of a to b may approach nearer to that of A to B, than by any assignable difference. And therefore when a and b vanish, x must vanish with them; [for if it then be equal to D, it never before was less than D, since it was continually decreasing; whereas before it was smaller than any assignable quantity] and therefore, by *Coroll.* 3. ult.$a : b :: A : B$.

Coroll. 6. IF two flowing quantities a and b arise from nothing at a given instant, and the nearer the time be taken to that of their rise, so much the nearer is their ratio to that of A to B; and the time may be taken so small, that the **[15]** ratio of a to b shall differ from that of A to B, less than by any assignable difference, in this case, $\mathrm{p^{mo}}\, a : b :: A : B$. The Proof of this is similar to that of the foregoing.

Remark. IN these two last corollaries you have Sir *Isaac Newton*'s description of the ultimate ratio of vanishing quantities, and the* prime ratio

---

* We have not, as I know of, any direct definition of these prime ratio's, but we are left to form it from that most accurate definition of the ultimate ratio's of vanishing quantities; which we have at the latter end of *Schol. Lemm.* II. *Princ.*

200   5. Introduction to the Doctrine of Fluxions

of nascent or arising ones, which is in effect the same with that given in *Defin.* 6, 7. But I chose to make this little alteration for the greater expedition in practice. And it is to be noted, that tho' Sir *Isaac*, to explain what he means by the ultimate ratio of vanishing quantities, describes it as in *Coroll.* 4. yet in practice he seems evidently enough to make use of one similar to that of *Defin.* 6.

Coroll. 7. IF ult. $a : b :: A : B$, then ult. $a + b : b :: A + B : B$. For since ult. $a : b :: A : B$, by *Defin.* 6. a time $\mathcal{T}$ may be assumed; during which $a : b :: A \mp x : B \mp y$; and at the end of **[16]** which the flowing quantities a, b, x, y, all vanish; and therefore also during that time, $a + b : b :: A + B \mp x \mp y : B \mp y$; and at the end of that time also, the four quantities $a + b$, $b, x \mp y$, and y, which are flowing quantities, by *Axiom* 1. vanish: wherefore, by *Defin.* 6. ult. $a + b : b :: A + B : B$.

Coroll. 8. IF ult. $a : b :: A : B$, then ult. $a - b : b :: A - B : B$; A being greater than B.

Coroll. 9. IF ult. $a : b :: A : B$, and ult. $b : d :: B : D$; then also by equality, ult. $a : d :: A : D$.

Remark. THESE two last corollaries are demonstrated by the like contrivance as the sixth; and the like propositions are true, and in like manner to be proved concerning prime ratio's.

Axiom 2. A QUANTITY flows uniformly, or with a permanent Fluxion, when the changes made in it are always proportional to the times in which they are made.

Axiom 3. THE Fluxions of quantities uniformly flowing, are always in the proportion of their synchronal changes; and therefore, because this is a given ratio (by *Coroll.* 4. *Defin.* 5 and 6.) in their ultimate ratio when they vanish, or their prime ratio when they arise.

Axiom 4. IF thro' any time two quantities be generated, or changes be made in two quantities, one with an uniform Fluxion or rate of increase or decrease, and the other with a Fluxion continually increasing; then the ratio of the quantity generated, or the change made by the uniform or

---

and which is so plain, that I wonder how our author could help understanding it; which had he done, I am apt to think that all his *Analyst* says concerning the proportion of quantities vanishing with the quantities themselves, had never been heard: For according to this definition, we are not obliged to consider the last ratio as ever subsisting between the vanishing quantities themselves. But between other quantities it may subsist, not only after the vanishing quantities are quite destroyed, but before when they are as large as you please. And the reason we consider quantities as decreasing continually till they vanish, is not in order to make, but to find out, this last ratio. Sir *Isaac Newton* does indeed say that this last ratio is the ratio with which the quantities themselves vanish; but whether he herein speaks with the utmost propriety or not, is a mere nicety on which nothing at all depends.

permanent Fluxion to the quantity generated, or the change made by the continually increasing Fluxion, will be always less **[17]** than the ratio of the Fluxion of the former is to the Fluxion of the latter at the beginning of the time, and greater than the ratio which the Fluxions are in to one another at the end of the time.

Axiom 5. BUT if, other things being supposed the same, the latter Fluxion continually decrease, the ratio of the quantity generated, or the change made by the uniform or permanent Fluxion to the quantity generated, or the change made by the continually decreasing Fluxion, will be always greater than the ratio of the Fluxion of the former is to the Fluxion of the latter at the beginning of the time, and less than the ratio of the Fluxion of the former to the Fluxion of the latter at the end of the time. Both these axioms are plain consequences from this obvious truth, that if any thing increase with an accelerated velocity or swiftness, the increase made in a given time will be greater than if it had all along increased with the same swiftness it did at the beginning, and less than if it had all along increased with the swiftness it did at the end of it; and that the contrary will happen if the velocity be continually retarded. And the reader will please to take notice, that as the foregoing axioms are the sole principles on which the doctrine of Fluxions is built, so they were generally allowed, before this doctrine was ever thought of, in relation to the velocities of moving bodies, and the spaces described by their motion. The only thing new in this case is, that the idea of velocity is render'd more general, and, in order to prevent ambiguity, the word Fluxion has been substituted in the room of it: but still the same general notion is preserved; for as velocity, at least in its more **[18]** usual sense, signifies the degree of quickness with which a body changes its situation in respect of space, so the Fluxion of a quantity signifies the degree of quickness with which the quantity changes its magnitude. Why then is the notion of a Fluxion harder to be understood than that of a velocity? Why may not we make use of the same axioms in relation to one, which all the world allows in relation to the other? Surely some reason ought to be assigned of the difference between the two cases; or it is not fair to make those things peculiar objections against the notion of Fluxions, which equally affect the notion of velocities. And I am apt to think that the reader is mistaken, if he imagines that Sir *Isaac Newton* is peculiarly attacked in the objections of the *Analyst* on this head, and not the common sense of mankind both before and since his time. And what good reason can be assigned why the common notions of all mankind are represented as the peculiar blunders of Sir *Isaac* and his followers, it is not easy to determine. As to the second, *&c.* Fluxions, there may, I allow, seem to be some peculiar difficulty in them to persons that are not used to these subjects; but it seems to me great weakness to imagine that there can be any peculiar objections against them, because tho' the farther we go in the orders of Fluxions, our ideas become more and more complex, yet no new simple ideas are admitted. And when our author asserts, that in order to conceive of a second Fluxion, we must conceive of the velocity of a velocity, and that this is nonsense; he plainly appeals to the sound and not the sense of words. When velocity is considered only as an affection of motion, the velocity of a velocity is nonsense. But then this is not the notion of a second Fluxion; and if by inlarging the idea of velocity, you make it synonymous to the word *Fluxion*, then the velocity of a velocity, however oddly it may sound, is nothing but plain common sense. For the degree of quickness with which any quantity increases, may vary to greater or less with any [19] degree of quickness imaginable; and this last is called the second Fluxion of the quantity. Now if the author of the *Analyst* can shew the absurdity of this notion, he is welcome so to do; but let him not first dress it up in an ambiguous phrase, and then think to confute it by laughing at his own way of understanding that phrase.

PROP. I.

IF there be two flowing quantities, one of which flows with a permanent Fluxion, and the other with a continually increasing one, then the Fluxions of these quantities, at any given instant of time, will be in the last ratio of their synchronal changes vanishing at that instant, and in the prime ratio of their synchronal changes which begin to arise at that instant.

1. LET the given instant be the fix'd end of the variable time t, the determinate values of the two flowing quantities at the given instant A and B, their Fluxions then $\dot{y}$ and $\dot{z}$; the flowing quantities themselves y and z; the synchronal changes made in the time t in the quantities y, z, and the Fluxion of z, which continually increases, and I suppose now only as a flowing quantity, whilst the Fluxion of y remains always the same, call respectively $\acute{y}, \acute{z},$ and $\acute{\dot{z}}$. Then since $\dot{z}$ is the Fluxion of z at the given instant, or the end of t, and $\acute{\dot{z}}$ the change made in it during the time t, and this Fluxion continually increases; therefore $\dot{z}-\acute{\dot{z}}$ will always express the Fluxion of z at the beginning of t, whether t be longer or shorter; and consequently, by *Axiom* 4. the ratio of $\acute{y}$ to $\acute{z}$ will be [20] always less than that of $\dot{y}$ to $\dot{z}-\acute{\dot{z}}$, and greater than that of $\dot{y}$ to $\dot{z}$. If therefore $\acute{y} : \acute{z} :: \dot{y} : \dot{z} - x$ always, x will be always less than $\acute{\dot{z}}$. Now as t continually decreases, it is plain, by *Axiom* 1. that the quantities $\acute{y}, \acute{z}, x, \acute{\dot{z}}$ being always made up of given and flowing quantities, can themselves only vary as flowing quantities, and as such they must vanish, when t vanishes that is at the given instant: [$\acute{y}$ must then vanish, being always equal to the difference between A and y; and A and y are equal at the given instant: for the same reason $\acute{z}$ and $\acute{\dot{z}}$ vanish at the given instant, and x also must vanish, it being all along only a part

of $\overset{\prime}{\dot{z}}$] Wherefore since $\overset{\prime}{\dot{y}} : \overset{\prime}{\dot{z}} :: \dot{y} : \dot{z} - x$ always, and $\overset{\prime}{\dot{y}}, \overset{\prime}{\dot{z}}$, and x varying only as flowing quantities, vanish at the given instant, (by *Cor.* 3. *Def.* 6.) ult. $\overset{\prime}{\dot{y}} : \overset{\prime}{\dot{z}} :: \dot{y} : \dot{z}$.

2. AGAIN; let $\mathcal{T}$ denote any time that begins at the given instant, $\dot{y}$ and $\dot{\mathfrak{z}}$ the Fluxions of the quantities at the beginning of the time $\mathcal{T}$; $\overset{\prime}{\dot{y}}, \overset{\prime}{\dot{\mathfrak{z}}}$, and $\overset{\prime}{\dot{\mathfrak{z}}}$, synchronal changes of the quantities and of the Fluxion that increases, made in the time $\mathcal{T}$, and other things being supposed as before. Then by a like reasoning as the foregoing, it will appear, from *Axiom* 5, that if $\overset{\prime}{\dot{y}} : \overset{\prime}{\dot{\mathfrak{z}}} :: \dot{y} : \dot{\mathfrak{z}} + x$, that x is always less than $\overset{\prime}{\dot{\mathfrak{z}}}$; and by *Axiom* 1. and *remark* at the end of *Coroll.* 9. that $\mathrm{p^{mo}}\overset{\prime}{\dot{y}} : \overset{\prime}{\dot{\mathfrak{z}}} :: \dot{y} : \dot{\mathfrak{z}}$. [21]

3. IF the Fluxion of z at the beginning of the time $\mathcal{T}$, receives an instantaneous increment, which is not impossible, nor contrary to any thing supposed in the proposition, then z is to be looked upon as having two Fluxions at the given instant, one of which is that it would have had if no such alteration had been in it; and the other is that it would have had if the same alteration had been made in it before the given instant, and not then. And taking the times t and $\mathcal{T}$ so short, that during both there be no other instantaneous alteration made in the Fluxion of z, besides that at the given instant; and following the steps of the foregoing demonstrations, you will find that the ultimate ratio of the changes of y and z vanishing at the given instant, is the same as the ratio of their Fluxions, on the supposition that no alteration had been made in the Fluxion of z at the given instant; and that the prime ratio of the changes of y and z beginning to arise at the given instant, is the same as the ratio of their Fluxions, on the supposition that the alteration in the Fluxion of z had been made before the given instant: all which is manifest, since on the former supposition, the Fluxion of z thro' the time t, and on the latter thro' the time $\mathcal{T}$, varies only as a flowing quantity. Wherefore, *&c.* **W.W.D.**

P R O P. II.

IF two quantities flow, one uniformly, and the other with a decreasing Fluxion, or with a Fluxion increasing to a given instant, and then decreasing continually, or *vice versâ*; and even tho' this Fluxion should not always before or after the given instant increase, or always decrease, but for some time only immediately before it, shou'd continue in the same state of increase or decrease, and so likewise immediately after it; yet still the Fluxions of these quantities will be in the last ratio of their synchronal [22] changes vanishing at the given instant, and in the prime ratio of their synchronal changes which begin to arise at that instant; *i.e.* this is true in whatever manner one of these quantities flow. All which is to be made out by following the steps of the demonstration of the foregoing proposition, having assumed your times t and $\mathcal{T}$ so short, that during either of them there be no alteration of the Fluxion from a state of increase to that of decrease, or *vice versâ*.

PROP. III.

IF two quantities flow any how, their Fluxions, at a given instant of time, will be in the last ratio of their synchronal changes vanishing at that instant, and in the first ratio of synchronal changes then beginning to arise.

1. LET the given instant be the end of the time t, the flowing quantities z and x, an uniformly flowing quantity y; changes made in these quantities in the time t, $\overset{\prime}{z}, \overset{\prime}{x}, \overset{\prime}{y}$; their Fluxions at the end of t, $\dot{z}, \dot{x}, \dot{y}$. Then (by Prop. 2.) ult. $\overset{\prime}{z} : \overset{\prime}{y} :: \dot{z} : \dot{y}$, and by the same ult. $\overset{\prime}{y} : \overset{\prime}{x} :: \dot{y} : \dot{x}$; wherefore, by equality, ult. $\overset{\prime}{z} : \overset{\prime}{x} :: \dot{z} : \dot{x}$.

2. LET the given instant be the beginning of $\mathcal{T}$, the changes made in the time $\mathcal{T}$, $\overset{\prime}{Z}, \overset{\prime}{X}, \overset{\prime}{Y}$, and their Fluxions at the beginning of $\mathcal{T}$, $\dot{Z}, \dot{X}, \dot{Y}$. **[23]** Then also (by Prop. 2.) $\text{p}^{\text{mo}} \overset{\prime}{Z} : \overset{\prime}{Y} :: \dot{Z} : \dot{Y}$, and $\text{p}^{\text{mo}} \overset{\prime}{Y} : \overset{\prime}{X} :: \dot{Y} : \dot{X}$; and therefore, as before, $\text{p}^{\text{mo}} \overset{\prime}{Z} : \overset{\prime}{X} :: \dot{Z} : \dot{X}$.   **W.W.D.**

LEMMA I.

If a determinate ratio be never greater than the greatest, nor less than the least of two variable ratio's, it will not be greater than the greatest, nor less than the least of those ratio's when they are prime or ultimate; *i.e.* if $A : B$ is never greater than $x : y$, nor less than $v : z$, it will be a mean also between the ultimate or prime ratio of x to y, and v to z; or of the ultimate ratio of x to y, and the prime ratio of v to z, or *vice versâ*. All which is plain; for, by an evident consequence from the definitions of prime and ultimate ratio's, the quantities v, z, x, y, may be so small, as that their ratio's shall differ less from their prime and ultimate ratio's than by any assignable difference; and therefore if, $x : y$ being the greater of the two ratio's, $A : B$ should be greater than the ultimate ratio of x to y, let the difference between them be D. Then because $x : y$ may be nearer to the ultimate ratio of x to y, than by the difference D, the ratio of $A : B$ may be greater than that of x to y; but it is supposed to be never greater; which things are inconsistent: and therefore $A : B$ is never greater than the ultimate ratio of x to y. And in like manner every thing else asserted in this *Lemma* may be proved. **[24]**

Coroll. HENCE if the ultimate ratio of v to z, and the prime ratio of x to y be equal; and the ratio of A to B be never less than that of x to y, and never greater than that of v to z, then the ultimate ratio of v to z, and the prime ratio of x to y, will be each equal to that of A to B.

PROP. IV.

SUPPOSING that at the time you seek the Fluxions, there be no instantaneous change made in them; *i.e.* supposing the Fluxions you seek to be themselves flowing quantities, or permanent ones, then if you can take synchronal increments or decrements of each, made partly before and partly after a given instant, in such a manner, that however small they shall always be in a given ratio, that ratio will be the ratio of the Fluxions of the quantities at the given instant; *i.e.* if the flowing quantities be z and x; and if whilst z changes from $A - y$ to $A - y + a$, x changes from $B - v$ to $B - v + b$, a being greater than y, and b than v; and the ratio of a to b be a given ratio, then the ratio $a : b$ is the ratio of the Fluxions of z and x at the instant that $z = A$ and $x = B$, A and B being permanent quantities: For calling the Fluxions of z and x at that time $\dot{z}$ and $\dot{x}$, since v and y are* synchronal changes of z and x vanishing at the given instant, ult. $y : v :: \dot{z} : \dot{x}$; and because $a - y$ and $b - v$ are synchronal changes of the same quantities arising at the given instant, and there are no instantaneous [25] changes made in the Fluxions, therefore $\text{p}^{\text{mo}} a - y : b - v :: \dot{z} : \dot{x}$; but the ratio of a to b, as is evident, is never greater than the greatest, nor less than the least of the ratio's $y : v$, and $a - y : b - v$; and therefore (by *Coroll. Lemm.* preced.) the ratio $a : b$, which is a given ratio, is equal to $\dot{z} : \dot{x}$, which is the ult. ratio of $y : v$, and the prime ratio of $a - y : b - v$.

W.W.D.

Remark 1. IT must be owned that this proposition affords only a more indirect way of finding the Fluxions of quantities, and has this disadvantage above the foregoing, that it not only cannot be applied where there is an instantaneous change made in the Fluxions; (it then giving only a mean between the greatest and least Fluxion of the quantity whose Fluxion you seek) but also when the Fluxion of the quantity you seek, compared with a permanent Fluxion, changes from an increasing to a decreasing one, or *vice versâ*. However I thought proper to take notice of it, because in particular cases it affords the most elegant manner of demonstrating the propositions of Fluxions: and the way of demonstrating from this proposition, has at least some similitude to that which Sir *Isaac Newton* uses in *Lemm.* 2. Lib. II. *Princ.*

Remark 2. THE observation I made at the end of Prop. 1. That the Fluxion found out from the last ratio's and first ratio's may possibly be different at the same instant of time, cannot create any difficulty in prac-

---

*See *Defin.* 5.

tice, nor require that we should investigate them both **[26]** ways,* because if the Fluxions of any quantities as found out one way, appear to be flowing quantities, they must be so in reality, and therefore are always the same when found out either way. In what follows, therefore, I shall only consider Fluxions as found from the ultimate ratio's of the vanishing changes, but the reader will easily see that exactly the same conclusions would result just in the same manner, by arguing from the prime ratio's of the changes that begin to arise at any instant.

P R O P. V.

THE Fluxion of the sum of two quantities is, at the end of any given time, or always, equal to the sum of the Fluxions of each, if they both increase or decrease together. Let the two flowing quantities be z and y, their sum, by *Axiom* 1, is a flowing quantity, which call s; then is $z + y = s$ always; and calling the Fluxions of these three quantities at the end of any given time $\dot{z}, \dot{y}, \dot{s}$, and their synchronal changes vanishing then $\acute{z}, \acute{y}$, and $\acute{s}$. Then because z and y increase or decrease together, $\acute{z} + \acute{y} = \acute{s}$ always; therefore ult. $\acute{s} : \acute{z} + \acute{y} :: 1 : 1$; but (by Prop. 3.) ult. $\acute{z} : \acute{y} :: \dot{z} : \dot{y}$, and therefore **[27]** ult. $\acute{z} + \acute{y} : \acute{y} :: \dot{z} + \dot{y} : \dot{y}$; and therefore from the first and third analogies it follows that ult. $\acute{y} : \acute{s} :: \dot{y} : \dot{z} + \dot{y}$; but (by Prop. 3.) ult. $\acute{y} : \acute{s} :: \dot{y} : \dot{s}$; and therefore because the ultimate ratio of two vanishing quantities is a given ratio $\dot{y} : \dot{z} + \dot{y} :: \dot{y} : \dot{s}$, and $\dot{z} + \dot{y} = \dot{s}$; and consequently the Fluxion of s, the sum of two flowing quantities increasing or decreasing together, is always equal to the sum of the Fluxions of each.   **W.W.D.**

P R O P. VI.

IF an equation involve ever so many flowing quantities, turn them all on one side, and then instead of each quantity putting its Fluxion, you will have the equation expressing the relation of the Fluxions when all the quantities flow one way; but if not, change the sign before the Fluxions of those quantities which decrease, and this will give you the equation of the Fluxions.

1. SUPPOSE the quantities to flow all one way, and that the Fluxions of $y, z, \&c.$ are $\dot{y}, \dot{z}, \&c.$ then if $z - y + x = v + r - s$ always, it will follow that $\dot{z} - \dot{y} + \dot{x} - \dot{v} - \dot{r} + \dot{s} = 0$. For from the first equation $z + x + s = v + r + y$

---

* *i.e.* By considering the changes made in the flowing quantities as the differences between their values at the given instant and their former values, which differences vanish at the given instant; and then considering them as the differences between their values at the given instant, and their subsequent values, which differences begin to arise at the given instant. See *Defin.* 3 and 5.

always; and taking $S = z + x$, $\Sigma = S + s$, and $R = v + r$, and $\omega = R + y$ always, 'tis plain that S, Σ, R, ω, which are flowing quantities, by *Axiom* 1. must flow the same way with z, y, *&c.* and therefore (by **[28]** Prop. 5.) $\dot{S} = \dot{z} + \dot{x}$, $\dot{\Sigma} = \dot{S} + \dot{s}$, $\dot{R} = \dot{v} + \dot{r}$, and $\dot{\omega} = \dot{R} + \dot{y}$; and because $\Sigma = \omega$ always, therefore $\dot{\Sigma} = \dot{\omega}$, and $\dot{R} + \dot{y} = \dot{S} + \dot{s}$; that is, $\dot{v} + \dot{r} + \dot{y} = \dot{z} + \dot{x} + \dot{s}$, or $\dot{z} - \dot{y} + \dot{x} - \dot{v} - \dot{r} + \dot{s} = 0$. And the same way this part of the proposition is evident, if there had been ever so many flowing quantities in the equation.

2. SUPPOSE some of the quantities to decrease whilst others increase, and let the equation be $Z - Y + v - x = R + r + T$ always; I say then, that if the great letters denote increasing quantities, and the small ones decreasing ones, that $\dot{Z} - \dot{Y} - \dot{v} + \dot{X} - \dot{R} - \dot{T} + \dot{r} = 0$. For taking $V = 2D - v$, and $X = D - x$, and $P = D - r$ always, D being a determinate quantity large enough, it is evident that whilst v, x, r decrease, V, X, P must increase, and that $Y + R + T + V = Z + X + P$ always. Wherefore by the foregoing part of this proposition, $\dot{Y} + \dot{R} + \dot{T} + \dot{V} - \dot{Z} - \dot{X} - \dot{P} = 0$. But because $V = 2D - v$ always, the synchronal changes of V and v are always equal, since D is a standing quantity; and therefore $\dot{V} = \dot{v}$. And in like manner $\dot{X} = \dot{x}$, and $\dot{P} = \dot{r}$. Putting therefore $\dot{v}, \dot{x}, \dot{r}$ in the places of $\dot{V}, \dot{X}, \dot{P}$, it will follow that $\dot{Z} - \dot{Y} - \dot{v} + \dot{x} - \dot{R} - \dot{T} + \dot{r} = 0$. And in the same way the justness of the **[29]** rule might have been demonstrated (as is evident) had there been ever so many more quantities in the equation.

Coroll. THE same rule holds, tho' besides the flowing quantities there be determinate ones in the equation, supposing their Fluxions = 0, as is too evident to need any demonstration.

N.B. IN practice this change of the sign is neglected as useless, but then we consider the Fluxions of decreasing quantities as negative, which comes to the same thing.

P R O P. VII.

LET A be a determinate quantity, z and x flowing ones, their Fluxions $\dot{z}$ and $\dot{x}$; I say then, that if $Az = x$ always, $A\dot{z}$ will be $= \dot{x}$ always: for $z : x :: 1 : A$; and therefore the synchronal changes, and consequently the Fluxions of z and x, will be always in the constant ratio of 1 to A; *i.e.* $\dot{z} : \dot{x} :: 1 : A$, and consequently $A\dot{z} = \dot{x}$ always.   **W.W.D.**

P R O P. VIII.

LET x and y be two flowing quantities whose Fluxions are $\dot{x}$ and $\dot{y}$ always; I say then, that if $x^2 = y$ always, that $2x\dot{x} = \dot{y}$ always. Let A

be the value of x at a given instant, $\acute{x}$ and $\acute{y}$ synchronal changes of x and y vanishing at that instant, $\dot{X}$ and $\dot{Y}$ their Fluxions at the same instant; **[30]** then because $x^2 = y$ always, $2A\acute{x} \mp \acute{x}^2 = \acute{y}$ always, as is evident; and therefore $\acute{x} : \acute{y} :: 1 : 2A \mp \acute{x}$ always, and therefore ult. $\acute{x} : \acute{y} :: 1 : 2A$ (by *Coroll.* 3. *Defin.* 6.) but ult. $\acute{x} : \acute{y} :: \dot{X} : \dot{Y}$ (by Prop. 3.) $\dot{X}$ and $\dot{Y}$ being the Fluxions of x and y at the instant their synchronal changes $\acute{x}$ and $\acute{y}$ vanish; and therefore $\dot{X} : \dot{Y} :: 1 : 2A$, or $2A\dot{X} = \dot{Y}$. But at the given instant, when $\dot{x} = \dot{X}$, and $\dot{y} = \dot{Y}$, x is also $= A$, and therefore at that instant $2x\,\dot{x} = \dot{y}$. And because the same may be proved in the same manner at any other instant of time, therefore $2x\,\dot{x} = \dot{y}$ always. **W.W.D.**

Or thus:

WHILST x changes from $A - \frac{1}{2}\acute{x}$ to $A - \frac{1}{2}\acute{x} + \acute{x}$, x^2 or y changes from $A^2 - A\acute{x} + \frac{1}{4}\acute{x}^2$ to $A^2 - A\acute{x} + \frac{1}{4}\acute{x}^2 + 2A\acute{x}$; and therefore (by Prop. 4.) the ratio of $\acute{x}$ to $2A\acute{x}$, or 1 to $2A$, being a constant ratio, is the ratio of the Fluxions of x to x^2 or y at the instant that x becomes $= A$, and $x^2 = A^2$. **[31]**

P R O P. IX.

LET x, y, and z be three flowing quantities whose Fluxions are always $\dot{x}$, $\dot{y}$, and $\dot{z}$; I say then, that if $xy = z$ always, and x and y flow the same way, that $\dot{x}y + \dot{y}x = \dot{z}$. But if whilst one of them increases the other decreases, that $\dot{z}$ is equal to $\dot{x}y - \dot{y}x$, or $\dot{y}x - \dot{x}y$, according as z flows the same way with x or y. Let the sum of x and y be v, and its Fluxion $\dot{v}$; then because $x+y = v$ always, $v^2 = x^2 + 2yx + y^2$, i.e. $v^2 = x^2 + y^2 + 2z$ always. And if x and y flow the same way, so it is plain must v, v^2, y^2, x^2, and $2z$ do. Taking therefore the Fluxions (by Prop. 6 and 8.) $\dot{x} + \dot{y} = \dot{v}$, and $2v\dot{v} = 2x\,\dot{x} + 2y\,\dot{y} + 2\dot{z}$ always; but $2v\dot{v} = \overline{2x + 2y} \times \overline{\dot{x} + \dot{y}} = 2x\,\dot{x} + 2y\,\dot{y} + 2\dot{x}\,y + 2y\,\dot{x}$ always, and therefore $\dot{z} = \dot{x}y + \dot{y}x$ always.

AGAIN; if z and y flowing one way, x flows the contrary, assume a determinate quantity A large enough, and suppose $A - s = x$ always, the Fluxion of s being $\dot{s}$ always; then will $Ay - sy$ be $= xy$, and consequently $Ay - sy = z$ always. But here it is manifest, because $A - s = x$ always, that s flows the contrary way to x, and therefore the same way with y and z; and **[32]** therefore that Ay, sy, and z, and s and y all flow the same way, and the contrary way to x. And therefore from the two preceding equations collecting the Fluxions by the foregoing propositions, and the part of this that has been proved, $A\dot{y} - \dot{s}y - \dot{y}s = \dot{z}$, and $\dot{s} = \dot{x}$; and instead of s and $\dot{s}$ in the former of these, putting their values $A - x$ and $\dot{x}$, $A\dot{y} - \dot{x}y - \dot{y}A + \dot{y}x = \dot{y}x - \dot{x}y = \dot{z}$. And lastly, if z and x flow one way, and y the contrary, in the same manner you may prove that $\dot{x}y - \dot{y}x = \dot{z}$.

Coroll. HERE if the Fluxions of decreasing quantities are considered as negative ones, then in all cases the Fluxion of $xy = \dot{x}y + \dot{y}x$. And from hence the rule in Sir *Isaac Newton*, from the equation of the Fluents to find the equation of the Fluxions, is easily proved.

S E C T. III.

I HAVE now proved, I hope, beyond exception, that considering Fluxions as the velocities or rates with which quantities increase or decrease, that their proportions may be always found, if the last ratio of their synchronal changes vanishing at every given instant is known. **[33]** But I would here observe, that whatever false metaphysics there may be in the notion of Fluxions considered as velocities, it does not at all affect the general method; as Sir *Isaac Newton* himself has informed us, that instead of these Fluxions we may make use of any other finite quantities found out from the last ratio of the synchronal increments or decrements of Flowing Quantities; and therefore the Author of the *Analyst* hardly acted the part of a fair adversary, in making such a pother about the notion of Fluxions as incomprehensible, since if he had not been able to understand them, he might have made use of any other quantities he did understand in their stead. I don't say this as if I thought there were any difficulty in the notion of Fluxions thus considered; I am on the contrary very positive that no man can make any objection against a velocity of increase or decrease in general, that will not as strongly lie against a velocity of motion. If quantities may increase faster or slower, as well as bodies move faster or slower, there is no greater impropriety in saying that one quantity increases with a greater velocity than another, than in saying that one body moves with a greater velocity than another. But tho' I think there is no difficulty in the notion of Fluxions consider'd as velocities, yet in order to understand equations where Fluxions of different orders are jumbled together, it would be convenient to represent all Fluxions not as before, but as quantities of the same kind with their Fluents; and therefore I should chuse to do it, were I to write a treatise on this subject: And this may be done in the following manner. Take *Defin.* 1, and *Axiom* 2 for the two first definitions: Define also ultimate **[34]** ratio as before, and then proceed thus: *Defin.* 4. The Fluxion of an uniformly flowing quantity is the change made in it in a certain given time. *Defin.* 5. The Fluxion of a quantity any how flowing at any given instant, is a quantity found out by taking it to the Fluxion of an uniformly flowing quantity in the ultimate proportion of those synchronal changes which then vanish. And considering Fluxions in this light, all the uses might be made of them as are done under the foregoing notion. But I go on to what the *Analyst* gives me more occasion to take notice of. I might have proved from what went before, that the Fluxion of x is to that of x^n, as 1 to nx^{n-1}. But for the sake of justifying Sir *Isaac Newton*, I shall now take his method.

LET x increase uniformly, and 'tis proposed to find the Fluxion of x^n.

WHILST x by flowing becomes $x + o$, x^n will become $\overline{x+o}|^n$, that is, $x^n + n\,o\,x^{n-1} + \frac{n^2-n}{2} o^2\, x^{n-2} +$ &c. and the synchronal augments o and $n\,o\,x^{n-1} + \frac{n^2-n}{2} o^2\, x^{n-2} +$ &c. are to one another as $1 : nx^{n-1} + \frac{n^2-n}{2} o\,x^{n-2} +$ &c.

LET now these augments vanish, and their last ratio will be as $1 : nx^{n-1}$; and therefore the Fluxion of x is to the Fluxion x^n, as 1 is to nx^{n-1}.

W.W.F.

THIS our author says is no fair and conclusive reasoning, because when we suppose the "increments to vanish, we must suppose their proportions, "their expressions, and every thing else derived from the supposition of "their existence **[35]** to vanish with them." To this I answer, that our author himself must needs know thus much, *viz.* That the lesser the increment o is taken, the nearer the proportion of the increments of x and x^n will arrive to that of 1 to nx^{n-1}, and that by supposing the increment o continually to decrease, the ratio of these synchronal increments may be made to approach to it nearer than by any assignable difference, and can never come up with it before the time when the increments themselves vanish. And no more nor less than this does Sir *Isaac* mean (as he himself informs us) when he says, that the last ratio of the vanishing increments is that of 1 to nx^{n-1}. When therefore our author must own that to be true which Sir *Isaac* intends, what signifies it to dispute whether it be proper to speak of the proportion of the increments as still in being, when the quantities themselves vanish or become = o.

FOR tho', strictly speaking, it should be allowed that there is no last proportion of vanishing quantities, yet on this account no fair and candid reader wou'd find fault with Sir *Isaac Newton*, for he has so plainly described the proportion he calls by this name, as sufficiently to distinguish it from any other whatsoever: So that the amount of all objections against the justice of his method in finding out the last proportion of vanishing quantities, can arise to little more than this, that he has no right to call the proportion he finds out according to this method by that name, which sure must be egregious trifling. However, as on this head our author seems to talk with more than usual confidence of the advantage he has over his opponents, and **[36]** gives us what he says is the* amount of Sr *Isaac*'s reasoning in a truly ridiculous light, it will be proper to see on whom the laugh ought to fall, for I am sure somebody must here appear strangely ridiculous. His arguings and illustrations founded on† the *Lemma* he proposes as self-evident, I think I have no occasion to meddle with, because I readily allow whatever consequence he is pleased to draw from it, if it appears that Sir *Isaac*, in order to find the last ratio's proposed, was obliged to make two inconsistent suppositions. To confute which nothing more need be said than barely to relate the suppositions he did make.

1. THEN he supposes that x by increasing becomes $x+o$, and from hence he deduces the relation of the increments of x and x^n.

2. AGAIN; in order to find the last ratio of the increments vanishing, he‡ supposes o to decrease till it vanishes, or becomes equal to nothing. Besides these he makes no other suppositions, and these are evidently no more inconsistent and contradictory, than to suppose a man should first go up stairs, and then come down again. To suppose the increments to be something and nothing at the same time, is contradictory; but to suppose them first to exist, and then to vanish, is perfectly consistent; nor will the consequences drawn from the supposition of their prior existence, if just, be any ways affected by the supposition of their subsequent vanishing, because the truth of the latter supposition no **[37]** ways contradicts the truth of the former. To make this more plain, consider what is made out from each supposition: from the first that x has increased by o, this consequence is drawn, that the proportion between the increments of x and x^n, *so long as they exist*, may be expressed by that of 1 to $nx^{n-1} + \frac{n^2-n}{2} o x^{n-2}$, &c. if o always express the increment of x. And this consequence is no ways affected by supposing o continually to decrease, and at length to vanish. But from this last supposition we may gather, that the lesser o is, so much the nearer the ratio of 1 to nx^{n-1} comes to the ratio of the increments; and that by a continual diminution of o, it may come as near to it as you please, but can never equal it before o quite vanishes; and therefore this ratio, and no other whatsoever, agrees to the description which Sir *Isaac* has given of the ultimate ratio of the vanishing increments. His conclusion therefore comes out without the supposition of any thing inconsistent. And

---

* *Analyst*, p. 21.
† Id. p. 20.
‡ That what is here made the second supposition, is truly the meaning of *Evanescant jam augmenta illa* in Sir *Isaac Newton*, is very plain from the manner in which he supposes quantities to vanish throughout the *Introduction to the Quadratures*.

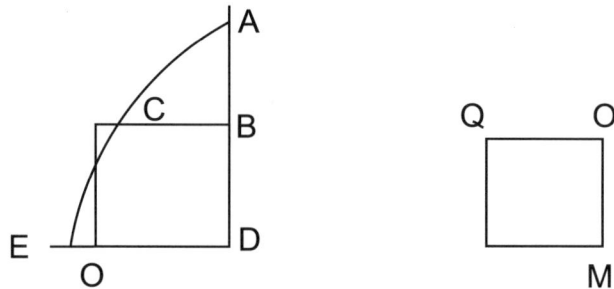

our Author, by his way of objecting, seems to make no distinction between two opposite things being done at the same time, and their being done at different ones; between a line's being drawn, and then rubbed out; and lines supposed to be drawn, and not drawn at all; as will appear still farther by considering his objection against the way of finding out the ordinate of a curve from the area and abscisse being perpetually given (which is a problem analogous to that of finding the Fluxion from the Fluent perpetually given, and) which I shall now represent somewhat more distinctly than he has done, and then consider his objection against it. **[38]**

LET the area $ABC = A$, the abscisse $AB = b$, the ordinate $BC = k$, all which are standing quantities as well as the square $QM = M$, and its side $QO = q$; and let any other ordinate, with its correspondent abscisse and adjoining area, as DE, AD, AED, be called respectively $z, x,$ and V. And the nature of the curve is such, that the area of the curve is to QM as the cube of the adjoining abscisse is to the cube of QO; that is, $A \times q^3 = M \times b^3$, and $V \times q^3 = M \times x^3$. From whence I find the area $BCED \times q^3 = M \times \overline{x^3 - b^3}$; and supposing $BCED =$ to the rectangle Bo, and calling Do, y, I find, instead of $BCED$, putting its value, and dividing by $x - b$, that $yq = x^2 + xb + b^2$; which equation I see is true, let x be of any magnitude either greater or less than b; from whence I conclude it must be true also when $x = b$. And because I find that y is always a mean between k and z, I conclude that when $x = b$, since then $z = k$, that then also $y = k$; and therefore that at that time the equation $yq = x^2 + bx + b^2$ must degenerate into this, $kq = 3b^2$, which gives me the relation of the ordinate and abscisse.

W.W.F.

NOW in this way of reasoning, says our Author, there is a direct fallacy; because, *first*, we are obliged to suppose that b and x are unequal, without which we could not proceed one step; and, in the *second* place, it is supposed that they are equal, which is a manifest inconsistency. But cannot so accurate a reasoner as our Author distinguish between different times? To suppose x and b equal and unequal at the same time, **[39]** would have been an inconsistency; but to suppose them first unequal, and afterwards to become equal, has not the shadow of difficulty in it. And there is nothing wonderful in it, that from the supposition of both these things happening one after another, I can deduce a conclusion which would not follow from either supposition singly taken; and that the conclusion I have drawn is just, will appear to any one that considers what is distinctly made out in each step of the demonstration.

BY taking x different from b, I find this equation, $yq = x^2 + xb + b^2$, and observe it holds true of whatever magnitude x be taken; the meaning of which may be thus expressed in words at length: There always is a line greater than one and less than the other of the two quantities k and z, so long as they remain different; which multiplied by q is always greater than one, and less than the other of the two quantities $3x^2$ and $3b^2$. Now this proposition is always true whatever becomes of x afterwards, and from it I infer that at the time when $x = b$, and $k = z$, kq must also be equal to $3b^2$. The force of which inference depends upon this plain axiom, That if so long as two quantities remain different, a third be always greater than one and less than the other of them, as soon as by a continued increase or decrease any two of these quantities become equal, all three of them must be equal. The truth of which, were there any manner of occasion for it, might easily be made out by reducing the contrary position to an absurdity.

IN what has been said, I have designedly taken notice only of those objections of the ingenious author of the *Analyst* which relate to the **[40]** doctrine of Fluxions; as for the differential method of *Leibnitz*, I do not undertake its defence, because I think the notion of Fluxions considered as velocities, or rather as finite quantities of the same kind with their Fluents, as before described, is much more easily conceivable, and frees from difficulty. Nay, I must confess there seems to me to be some objection against considering quantities as generated from moments. What moments, what the *principia jamjam nascentia finitarum quantitatum*, are in themselves, I own, I don't understand; I can't, I am sure, easily conceive what a quantity is before it comes to be of some bigness or other; and therefore moments considered as parts of the quantities whose moments they are, or as really fixed and determinate quantities of any kind, are beyond my comprehension, nor do I indeed think that Sir *Isaac Newton* himself did thus consider them. But when he says, for instance, that the moment of A^2 is equal to the moment of A into twice A, his meaning is only this, that if the increment of A be continually diminished, the proportion between that into A, and the correspondent increment of A^2 will approach towards, and at length come nearer to, a ratio of equality than by any assignable difference. And thus understanding him, the proportion of moments is a phrase easily intelligible, and his meaning very evident in every sentence where he uses it, altho', with me, you can't imagine what a moment in itself is. 'Tis true indeed, upon this supposition, it may not be agreeable to the exact accuracy of language, to speak of moments as if they were real intelligible quantities; but I am sure this is **[41]** all that can justly be objected against it, and that there is no false logic or metaphysics in it. Whether I have here spoke the sense of Sr *Isaac* himself, or not, I won't be positive, and must leave it to the judgment of the reader conversant in his works; but I think he has given us intimation sufficiently plain, that we may at least thus interpret him if we please, he being not at all solicitous under what notion we consider moments, whilst we take their proportions in that manner he directs us.

OUR Author may perhaps think himself not fairly used, that there has been no more notice taken of second and third *&c.* Fluxions in this treatise, but it is to no purpose to defend these before the objections against the first are given up; and if the first are well understood, there is no manner of difficulty in the other, except what arises from this consideration, that our ideas become somewhat more complex as we go on to the higher orders of Fluxions. However I shall mention one thing for the sake of young beginners in this science. Suppose $y^2 = x$, and y flows uniformly, then, according to the method of Fluxions, $2y\dot{y} = \dot{x}$, and $2\dot{y}^2 = \ddot{x}$, it may be inquired what is the meaning of this equation taking Fluxion in the notion of velocity, since upon that supposition first and second Fluxions are quantities of different kinds, and therefore have no proportion to one another? To this I answer, that the only meaning of the equation is, that if you take the **[42]** value of $2\dot{y}^2$ and $\ddot{x}$ at two different times, the $2\dot{y}^2$'s are in the same proportion to one another that the $\ddot{x}$'s are. And in like manner are equations always to be understood which involve quantities of different sorts; they don't express the proportion of heterogeneous quantities to one another, but of different values of homogeneous quantities among themselves. And from hence our Author may, if he pleases, receive an answer to the latter part of his thirty-first query; an odd query, surely, to be made by one that pretends to answer Sir *Isaac Newton*: But several of this kind are to be met with, which shew greater prejudice against the Mathematicians, than knowledge of the principles which they maintain. And I can't help observing, that tho' our Author professes the utmost caution as to what he admits as true, and a concern for the utmost accuracy in reasoning, yet neither the one nor the other appears when he is making his objections against the Mathematicians. Of this I should not have taken any notice, if his design had only been to correct some mistaken notions among the Mathematicians; but as his intention is evidently to run down their method of reasoning, and to represent them as Bigots and Enthusiasts in their own science, it is but a piece of justice to them to shew that their adversary is as inaccurate in his reasonings, and as incautious in his assertions, and in forming his accusations, as any of them can be. Of this the reader may have already observed some instances, and I shall beg his patience whilst I mention one or two more. **[43]**

1. HE urges against the Mathematicians with great injustice, I think, that they imagine there may be a proportion between nothings, and yet he himself argues from this principle, *Anal.* p. 18. "Points are undoubtedly "equal, as having no more magnitude one than another, a limit or point, "as such, having no magnitude."

2. EVERY reader of the *Analyst* will observe what stress he lays on this maxim, Inaccurate premises cannot infer an accurate conclusion; as if nothing more was necessary than the belief of this principle to confute his adversaries; whereas the inference he would draw from thence, that we therefore cannot, in any case, know the conclusion to be right, if we have argued from the supposition of the truth of what we know to be false, or don't know to be true, is what no rules of logic will justify: because it is possible for me, by virtue of false or inaccurate premises, to gain a conclusion, and at the same time to know that the error in the premises will not cause any error in the conclusion. Thus in Algebra the supposition of a negative quantity, or a quantity less than nothing, is an absurdity, but yet this is no hindrance to the evidence of those conclusions that according to the rules of art are derived from it. And again, to bring an instance more to our present purpose, I'll allow that the supposition of an infinitesimal is absurd, and also that the supposition that an infinitesimal added to a finite quantity does not increase its magnitude is absurd, when an infinitesimal is consider'd as a fix'd determinate **[44]** quantity; and yet, I say, in deducing the value of the subtangent of a Parabola from these suppositions, I may be sure that I make a right conclusion; because those quantities which I suppose to be infinitely small, if they are not so will be very small, and those I suppose to be equal will be very nearly so, and therefore the error in the conclusion can be but very small; *i.e.* the subtangent will be equal to twice the abscisse very nearly. But I am certain that this cannot possibly be demonstrated concerning any finite and determinate quantities, unless they are truly and exactly equal, if you don't fix upon some determined proportion or difference as the standard of what is nearly equal, but leave every person at liberty to chuse what he will. And from this principle I would observe, that the conclusions found by *Leibnitz*'s method of differences may be proved to be true, notwithstanding the error or inaccuracy of his principles: for supposing in his method that the mark of equality = does not signify truly but nearly equal in the sense before given, and then correct your conclusion by the forementioned axiom, and you may be sure it is just.

3. HE represents the Mathematicians as founding their reasonings on maxims shocking to good sense, as particularly when they take it for granted that a finite quantity divided by nothing is infinite, *Query* 16. I suppose our Author will not pretend to say here that he designed only to ask a civil and innocent question, but to charge the Mathematicians with proceeding on such absurd maxims as this; and if so, it is an hasty and ill-grounded charge; 'tis a **[45]** charge, I think, he himself can hardly imagine to be true. Does our Author take all Mathematicians to be fools; or did ever any fool imagine that a reasonable answer could be given to this question, How many nothings will fill a quart; and that the proper answer is, An infinite number? Yet this is only to say that a quart divided by nothing is infinite. 'Tis allowed that this is a rule in Algebra, *that finite divided by nothing is infinite*: but the meaning of it is only this, that if I inquire how big any quantity must be taken in order to answer any purpose, and the answer come out according to the rules of Algebra $\frac{1}{0}$ or $\frac{2}{0}$ *&c.* this is a sign that no finite quantity is large enough. And is there any thing shocking in this to the sense of any man, from the porter to the philosopher? The same kind of answer ought to be given in relation to other maxims of Algebra, which appear strange to those unaccustomed to the phrases, and really are in themselves senseless expressions, as that the sum or product of two impossible quantities may be possible. Nay this may be said of the terms multiply and divide, add and subtract, as used in Algebra. An Algebraist never scruples to subtract a greater quantity from a less; but if he really designs to do this, he may try till his heart akes before he will be able to accomplish it, or to know what he is about. All these phrases are therefore to be considered merely as terms of art to help the memory in an algebraic process, or signs of rules, which from other principles we must know to be just, and not from any idea we have of the things, which, according to the **[46]** common use of the words, an ignorant man may fancy they are intended to express. 'Tis pity, indeed, these things have not been more particularly explained by the writers of Algebra, the want of which may well make it abstruse and confounding to beginners. Yet for a person to make these objections against the art or those that understand it, is as if I should impute it to the Logicians as a fault in theirs, that they use the horrid terms *Barbara, Celarent,* &c. Nor do I think the use of these absurd expressions peculiar to Algebra, but they are frequently used where, thro' their being familiar, and the design of them easily intelligible, they are not thought to have any odd sound. Thus it is a maxim in Arithmetic that $2 \times 0 = 5 \times 0$, or in general any number multiplied by 0 is $= 0$. Now I say that if by this any thing else be understood than that this is a good rule to go by in managing figures in Arithmetic, I will so far venture the laugh of the Public as to declare I don't understand it. He that can take two nothings and find their sum, and then five nothings and find their sum equal to the former, seems to me to be in a fair way to be able to divide

by nothing, and find the quotient; and, with a little more pains, may prove that two is equal to five. To multiply by nothing, is as absurd as to divide by nothing; and to suppose we can do either, is to imagine nothing to be a real quantity or number: for *nihili nullæ sunt affectiones*. Here I suppose our Author agrees with me from his 40th *Query*; but no one will say that for this reason we ought to reject the maxim considered **[47]** as a rule of Arithmetic, that the product of any number by 0 is = 0.

4thly, *Query* 50. HE represents the disputes and controversies among Mathematicians as disparaging the evidence of their methods: and, *Query* 51. he represents Logics and Metaphysics as proper to open their eyes, and extricate them out of their difficulties. Now were ever two things thus put together? If the disputes of the professors of any science disparage the science itself, Logics and Metaphysics are much more disparaged than Mathematics; why therefore, if I am half blind, must I take for my guide one that can't see at all? And to say the truth, it can hardly be look'd upon as fair to represent Mathematics as disparaged by the disputes of its professors, which insinuates that in this way it is peculiarly disparaged, whereas the quite contrary is true, there being fewer disputes among Mathematicians, as such, than any other persons whatsoever. The disputes of Mathematicians about metaphysical and philosophical principles have nothing to do here; and take away these, hardly any remain.

Lastly, HIS frequent insinuations that Mathematicians don't care to be tried by the rules of good logic, and require indulgence for incorrect and false reasonings, and think that the truth of the premises are proved without more ado by the truth of the conclusion, are entirely groundless. They, as well as all other mortals, desire indulgence for inaccurate expressions, but none for false reasoning; and for **[48]** following the exactest rules of logic, they pride themselves in being the most perfect patterns.

To make a thing plain as a proposition in *Euclid*, is to give it the last degree of evidence.*Nor is it any objection to the justness of their reasoning, that an algebraical note is sometimes to be interpreted, at the end of the process, in a sense which cou'd not have been substituted in the beginning of it; since if quantities themselves are considered as continually changing, the sense of the mark which represents or expresses them must, in order to its doing so, continually change along with them. And they never imagine that any particular supposition can come under a general case which is inconsistent with the reasoning thereof, or any just reasoning whatsoever; tho' they have many times good reason to conclude that particular cases in a general theorem are true, tho' they could not be proved in the same manner with the theorem itself.

To conclude; as I would not be thought, by any thing I have said, to

---

* *Query* 43.

be an enemy to true Logic and sound Metaphysics; and on the contrary think the most general use of the Mathematics is to inure us to a just way of thinking and arguing; it is a proper inquiry, I imagine, for those who have the direction of the education of youth, in what manner mathematical studies may be so pursued as most surely to answer this end: upon which head the hints our Author gives, *Queries* 15, 38, 56, 57, deserve to be consider'd: for **[49]** so far as Mathematics do not tend to make men more sober and rational thinkers, wiser and better men, they are only to be considered as an amusement, which ought not to take us off from serious business.

$$\mathcal{FINIS}.$$

ERRATA.

Preface, pag. 5. *l.* 11.   *for* a person *r.* persons.
26. *l.* 23.   *for* propositions *r.* proportions.
32. *l.* 16.   *for* $2y\,\dot{x}$ *r.* $2\dot{y}\,x$.

5.3 Commentary

The first thing to notice in considering this essay is that the title page bears no author's name. Perhaps we can attribute it to Bayes on the authority of Augustus de Morgan, who wrote

> This very acute tract is anonymous, but it was always attributed to Bayes by the contemporaries who *write in* the names of authors; as I have seen in various copies: and it bears his name in other places. [1860, p. 9]

The accuracy of de Morgan (an ardent bibliophile) in such matters is almost beyond reproach, and we may take it that the authorship is correctly attributed.

Bayes begins his Preface by writing

> I have long ago thought that the first principles and rules of the method of Fluxions stood in need of a more full and distinct explanation and proof, than what they had received either from their first incomparable author, or any of his followers [p. *[iii]*].

Now Bayes was in his early thirties when this tract appeared, so the phrase 'long ago' perhaps indicates that it was in his late twenties that he became exposed to the fluxionary calculus. It is, I suppose, even possible that he had gained some knowledge of it during his time at the College of King James, when, as we have seen in an earlier chapter, he is known to have had some connexion with James (II) Gregory (though he is not known to have attended any of the latter's mathematical courses). Further, the phrase 'or any of his followers' suggests that Bayes was well acquainted with the then current works on the fluxionary calculus[35], and it is not impossible that he knew the writings of Humphrey Ditton, an earlier incumbent of the meeting-house at Tunbridge Wells (as we have already noted), and the writer in 1706 of an early careful study[36] of Newton's tract of 1704.

Note that it emerges from the Preface that Bayes, while applauding (if not lauding) Berkeley's attention to fluxionary methods, was less (than) enamoured by the bishop-elect's[37] incorporation of religious matters. Bayes notes that Berkeley represented this debate 'as of consequence to the interests of religion' (p. *[iii]*), and indeed mathematical, philosophical, and theological matters are commingled throughout *The Analyst* (see Jesseph [1993, p. 183]).

I think that Bayes's logic fails him slightly when he writes

> If they are taught that it is inconsistent for a person to reject the mysteries of religion, and yet believe the mystery of Fluxions,

220   5. Introduction to the Doctrine of Fluxions

> will they not know how to draw the opposite conclusion themselves, that it is inconsistent to reject the doctrine of Fluxions because mysterious, and yet receive the mysteries of religion? [p. [*v*]].

For surely $\bar{p} \wedge q$ may be inconsistent without the 'opposite conclusion', presumably $p \wedge \bar{q}$, also being inconsistent?

One might note too the curious conjunction 'Christians and Protestants': presumably one may be the former without being the latter, but not vice versa!

It is perhaps because of the last sentence of this preface, viz.

> I shall now consider my subject as stript of all relation it has to religion, and merely as a matter of human science, and endeavour to shew that the method of Fluxions is founded upon clear and substantial principles [p. [*v*]]

that Berkeley did not respond to Bayes in the manner in which he did to Jurin and Walton, for example[38]. Indeed, shortly after considering *Philalethes*'s response, Berkeley received Walton's *A Vindication of Sir Isaac Newton's Principles of Fluxions*, in answer to which he wrote

> As this Dublin professor gleans after the *Cantabrigian*, only endeavouring to translate a few passages from Sir Isaac Newton's *Principia*, and enlarge on a hint or two of *Philalethes*, he deserves no particular notice. [1735a, Appendix, §1]

Perhaps Berkeley's opinion was coloured by Walton's having written

> But he [i.e., Berkeley] ought to have read Sir *Isaac* with more Care and Attention than he seems to have done, before he set up to decide and dictate in Matters of this Nature; and he wou'd do well yet to read him with Attention. [1735, p. 13]

Note too that Bayes's reference to 'human science' is to Berkeley's sixty-second Query, which runs

> Whether mysteries may not with better right be allowed of in Divine Faith than in Human Science?

The reference 'Defence of Freethinking, p. 62' at the end of the Preface is of course to Berkeley's *A Defence of Free-thinking in Mathematics*, where, at the start of a long list of questions addressed to 'those learned gentlemen of Cambridge', we find the following.

> I desire to know whether those who can neither demonstrate nor conceive the principles of the modern analysis[39], and yet give in to it, may not be justly said to have faith, and be styled believers of mysteries? [§50]

There is not much to comment on in the first (general) section of the tract. Perhaps we might note only that it follows from the two postulates and the remarks following the second postulate (a postulate that, in the light of the remarks following its being given, may be viewed as redundant[40]) that Bayes purposively declines to discuss the exact meanings of the notions of continuous change and instantaneous velocity, though these were in fact matters specifically raised by Berkeley[41] (in fact, Bayes was among the many mathematicians who concentrated more on the mathematical aspects of *The Analyst* than on the metaphysical arguments that were the other reason for its having being written).

The second section of Bayes's tract begins with seven definitions and two axioms. The first, the second, and the fourth definitions seem unexceptionable, though it is perhaps not quite clear from the first definition whether Bayes would allow a 'flowing quantity' to have discontinuities, or whether he would require it to be continuous and either (a) monotone increasing or (b) monotone decreasing throughout its entire domain of definition (except perhaps at the end-points), or whether he would allow it to switch from increasing to decreasing or vice versa (later passages in the tract suggest that such switches would be acceptable). Bayes's first two definitions are given as follows in Newton's *Method of Fluxions and Infinite Series*.

> Now those quantities which I consider as gradually and indefinitely increasing, I shall hereafter call *Fluents*, or *flowing Quantities* ... And the velocities by which every Fluent is increased by its generating motion (which I may call *Fluxions*, or simply Velocities, or Celerities,) ... .[1737, p. 27]

The third and the fifth definitions also present a slight problem. It appears from other remarks that Bayes always works with differences (between magnitudes) taken as non-negative: for example, he writes

> in Algebra the supposition of a negative quantity, or a quantity less than nothing, is an absurdity, but yet this is no hindrance to the evidence of those conclusions that according to the rules of art are derived from it [p. **[44]**]

and further

> An Algebraist never scruples to subtract a greater quantity from a less; but if he really designs to do this, he may try till his heart akes before he will be able to accomplish it, or to know what he is about. [p. **[46]**]

Thus it would seem that the difference mentioned in Definition 3 should be regarded as the *absolute value* of the difference[42] between 'the flowing quantity itself, and its value at a particular instant of time', with a similar interpretation in Definition 5. This viewing of quantities, fluxions, and changes as non-negative, noted by Smith [1980, p. 381], complicates some of the following work. One might also note that Bayes's 'differences' are what we might call *increments* or *decrements*.

Further clarity on this point can be obtained from Emerson [1763], where the initial definitions lead to the deductions[43]:

(a) if w is increasing, $\Delta w \equiv w(t + \Delta t) - w(t)$, and

(b) if w is decreasing, $\Delta w \equiv w(t) - w(t + \Delta t)$,

so that in any case the change Δw is positive. Similarly, when it comes to interpreting fluxions as derivatives (with respect to time), we must think of $\dot{w}$ as 'connected' with $[w(t + \Delta t) - w(t)]/\Delta t$ if w is increasing, and with $[w(t) - w(t + \Delta t)]/\Delta t$ if w is decreasing. Thus $\dot{w}$ is always non-negative.

One might also note, in connexion with Definition 5, that a change in x vanishing at a given instant t^*, say, requires consideration, for $\varepsilon > 0$, of $x(t^* - \varepsilon) - x(t^*)$, whereas a change arising at the same instant involves $x(t^* + \varepsilon) - x(t^*)$ (compare the notion of left and right derivatives).

Bayes's sixth definition, described by Jesseph as 'a prolix definition which does little to overcome Berkeley's objections' [1993, p. 275], requires more attention (and similar remarks apply *mutatis mutandis* to Definition 7). We note firstly that Bayes requires only that the flowing quantities *vanish* at the end-points, but does not require that they should tend to zero as the end-points are approached. Further, it is unclear whether Bayes would allow $a : b$ to be a function of t (say), or whether this ratio should be a constant[44]. The following examples illustrate these points.

Example 1.

Let $a(\cdot)$ and $b(\cdot)$ be defined on $T = [0, 2]$ by

$$a(t) = \begin{cases} t, & 0 < t < 2 \\ 0, & t = 2 \end{cases} \quad ; \quad b(t) = \begin{cases} 2t, & 0 < t < 2 \\ 0, & t = 2. \end{cases}$$

If $A = 4$ and $B = 8$, we could take $x(t) = 2 - t$ and $y(t) = 2(2 - t)$. Then

$$a(2) = b(2) = x(2) = y(2) = 0,$$

and, for $t \neq 2$,

$$\frac{A + x(t)}{B + y(t)} = \frac{4 + (2 - t)}{8 + 2(2 - t)} = \frac{1}{2} = \frac{a(t)}{b(t)}.$$

Example 2.

Let $a(t) = 4(2-t)$, $b(t) = 3(2-t)$ on $T = [0, 2]$. If $A = 16$ and $B = 12$, then we may take $x(t) = 12(2-t)$ and $y(t) = 9(2-t)$ so that

$$a(2) = b(2) = x(2) = y(2) = 0,$$

and, for $t \neq 2$,

$$\frac{A + x(t)}{B + y(t)} = \frac{4[4 + 3(2-t)]}{3[4 + 3(2-t)]} = \frac{4}{3} = \frac{a(t)}{b(t)}.$$

Example 3.

Let $a(t) = 4(4-t^2)$, $b(t) = 2(2-t)$ on $T = [0,2]$. For $A = 8$ and $B = 1$ we may take $x(t) = 3(2-t)$ and $y(t) = 5(2-t)/(2t+4)$. Then

$$a(2) = b(2) = x(2) = y(2) = 0,$$

and, for $t \neq 2$,

$$\frac{A + x(t)}{B + y(t)} = \frac{8 + 3(2-t)}{1 + 5(2-t)/(2t+4)} = 2t + 4 = \frac{a(t)}{b(t)}.$$

In all these examples ult.$a : b :: A : B$.

One should also note that an 'ultimate ratio' is concerned with a ratio *of vanishing quantities*, unlike the modern theory of limits[45], in which infinitesimals do not arise[46] (provided, of course, that we except non-standard analysis); but one must also note that Bayes refuses to treat ultimate ratios as limiting values of ratios that are continually decreased. It is necessary, therefore, to be somewhat hesitant about interpreting Bayes's approach as a limiting theory[47].

In the remark following Definition 7 Newton's name is mentioned, but unfortunately no reference to a pertinent work is given. Now although prime and ultimate ratios certainly appear in the *Principia*, clear definitions (at least of the former) appear to be lacking, and Newton in fact 'abandoned [these ratios] in those places in which fluxions are alluded to' [de Morgan, 1852b, p. 322]. However, in the *Tractatus de quadratura curvarum* of 1704 we find the following[48].

> Fluxiones sunt quam proxime ut Fluentium augmenta æqualibus temporis particulis quam minimis genita, & ut accurate loquar, sunt in prima ratione augmentorum nascentium; exponi autem possunt per lineas quascunq; quæ sunt ipsis proportionales ... Eodem recidit si summantur fluxiones in ultima

> ratione partium evanescentium. [Fluxions are very nearly as the Augments of the Fluents, generated in equal, but infinitely small parts of Time; and to speak exactly, are in the *Prime Ratio* of the nascent Augments: but they may be expounded by any Lines that are proportional to 'em. ... 'Tis the same thing if the Fluxions be taken in the *ultimate Ratio* of the Evanescent Parts.]

And in the later *Account of the Commercium Epistolicum* we read

> We have no Ideas of infinitely little Quantities, and therefore Mr. *Newton* introduced Fluxions into his Method, that it might proceed by finite Quantities as much as possible. It is more Natural and Geometrical, because founded upon the *primæ quantitatum nascentium rationes*, which have a Being in Geometry, while *Indivisibles*, upon which the Differential Method is founded, have no Being either in Geometry or Nature. There are *rationes primæ quantitatum nascentium*, but not *quantitates primæ nascentes*. [Newton, $171\frac{4}{5}$, p. 205]

In his first two corollaries Bayes proves essentially the uniqueness of the prime and ultimate ratios. It is perhaps worth comparing his proofs with those used today to show the uniqueness of the left- and right-hand limits $\lim_{t \uparrow t_0} f(t)/g(t) = A/B$ and $\lim_{t \downarrow t_0} f(t)/g(t) = A/B$: the proofs are seen to be remarkably similar, provided that one does not interpret Bayes's 'vanishing' of flowing quantities too literally. Note also that a 'determinate' ratio is the same as a 'given' ratio.

It is perhaps also worth noting the first Lemma in Book I of Newton's *Principia*, which runs as follows[49].

> Quantitates, ut & quantitatum rationes, quæ ad æqualitatem dato tempore constanter tendunt & eo pacto propius ad invicem accedere possunt quam pro data quævis differentia; fiunt ultimo æquales. [Quantities, and the ratios of quantities, which in any finite time converge continually to equality, and before the end of that time approach nearer to each other than by any given difference, become ultimately equal.]

This lemma is interpreted by Benjamin Robins in his *A Discourse Concerning the Nature and Certainty of Sir Isaac Newton's Methods of Fluxions, and of Prime and Ultimate Ratios* as follows.

> In this method any fix'd quantity, which some varying quantity, by a continual augmentation or diminution, shall perpetually approach, but never pass, is considered as the quantity, to which the varying quantity will at last or ultimately become equal;

provided the varying quantity can be made in its approach to
the other to differ from it by less than by any quantity how
minute soever, that can be assigned.... Ratios also may so vary,
as to be confined after the same manner to some determined
limit, and such limit of any ratio is here considered as that
with which the varying ratio will ultimately coincide. From any
ratio's having such a limit, it does not follow, that the variable
quantities exhibiting that ratio have any final magnitude, or
even limit, which they cannot pass. [1761, §§95, 98, 99][50]

This interpretation, Gibson believes, disposes of a large part of Berkeley's objections (see Gibson [1898, p. 23]).

Axiom 1 is fairly straightforward. Presumably as 'a quantity any how made up of given and flowing quantities' Bayes would accept something of the form $f(t)/g(t)$, where $f(\cdot)$ and $g(\cdot)$ are polynomials in t, though I am not sure whether he would consider something like $(t+2)/2(t+2)$ to be a permanent quantity.

Passing over Corollaries 3 and 4, we come to the fifth corollary, and we note that Bayes's remark that the ratio of two continually decreasing quantities eventually comes 'nearer to it [i.e., a given ratio] than by any assignable difference' is once again reminiscent of modern ideas on limits (cf. Cauchy's work). The last part of this proof, viz. 'and therefore, by *Coroll.* 3. ult. $a:b::A:B$', perhaps actually follows *invertendo*, i.e., by writing $a:b::A:B+x$ as $b:a::B+x:A$, and then using Corollary 3.

In the Remark following Corollary 6 Bayes refers to Newton's *description* of the ultimate and prime ratios, his reference being to *Schol. Lemm. II. Princ.* Now this reference is somewhat confusing: neither the first nor the third edition of the *Principia* has a Scholium to Lemma II in Book I, the first such-named distinguished passage following Lemma X. Further, the Scholium to Lemma II in the second book of the *Principia*, though vastly differing in the first and third editions, deals in both with the priority and independence of the invention of the (differential) calculus by Newton and Leibniz. I suspect strongly that the numbering of the pertinent result in the footnote in Bayes's essay should actually be Arabic rather than Roman, and this suspicion is confirmed by the following passage from the Scholium to Lemma 11 in the first book of the (third edition of the) *Principia*[51]:

> Objectio est, quod quantitatum evanescentium nulla sit ultima proportio; quippe quæ, antequam evanuerunt, non est ultima, ubi evanuerunt, nulla est. Sed & eodem argumento æque contendi posset nullam esse corporis ad certum locum, ubi motus finiatur, pervenientis velocitatem ultimam: hanc enim, antequam corpus attingit locum, non esse ultimam, ubi attingit, nullam esse. Et responsio facilis est: Per velocitatem ul-

timam intelligi eam, qua corpus movetur neque antequam attingit locum ultimum & motus cessat, neque postea, sed tunc cum attingit; id est, illam ipsam velocitatem quacum corpus attingit locum ultimum & quacum motus cessat. Et similiter per ultimam rationem quantitatum evanescentium, intelligendam esse rationem quantitatum, non antequam evanescunt, non postea, sed quacum evanescunt. Pariter & ratio prima nascentium est ratio quacum nascuntur. Et summa prima & ultima est quacum esse (vel augeri aut minui) incipiunt & cessant. Extat limes quem velocitas in fine motus attingere potest, non autem transgredi. Hæc est velocitas ultima. Et par est ratio limitis quantitatum & proportionum omnium incipientium & cessantium. Cumque hic limes sit certus & definitus, problema est vere geometricum eundem determinare. Geometrica vero omnia in aliis geometricis determinandis ac demonstrandis legitime usurpantur.

... Ultimæ rationes illæ quibuscum quantitates evanescunt, revera non sunt rationes quantitatum ultimarum, sed limites ad quos quantitatum sine limite decrescentium rationes semper appropinquant; & quas propius assequi possunt quam pro data quavis differentia, nunquam vero transgredi, neque prius attingere quam quantitates diminuuntur in infinitum. Res clarius intelligetur in infinite magnis. Si quantitates duæ quarum data est differentia augeantur in infinitum, dabitur harum ultima ratio, nimirum ratio æqualitatis, nec tamen ideo dabuntur quantitates ultimæ seu maximæ quarum ista est ratio. In sequentibus, igitur siquando facili rerum conceptui consulens dixero quantitates quam minimas, vel evanescentes, vel ultimas; cave intelligas quantitates magnitudine determinatas, sed cogita semper diminuendas sine limite.

[Perhaps it may be objected, that there is no ultimate proportion of evanescent quantities; because the proportion, before the quantities have vanished, is not the ultimate, and when they are vanished, is none. But by the same argument it may be alleged that a body arriving at a certain place, and there stopping, has no ultimate velocity; because the velocity, before the body comes to the place, is not its ultimate velocity; when it has arrived, there is none. But the answer is easy; for by the ultimate velocity is meant that with which the body is moved, neither before it arrives at its last place and the motion ceases, nor after, but at the very instant it arrives; that is, that velocity with which the body arrives at its last place, and with which the motion ceases. And in like manner, by the ultimate ratio of evanescent quantities is to be understood the ratio of the quantities not before they vanish, nor afterwards, but with which they vanish. In like manner the first ratio of nascent quantities is that with which they begin to be. And the first or last sum is that with which they begin and cease to be (or to be augmented or diminished). There is a limit which the velocity at the end of the motion may attain, but not exceed. This is the ultimate velocity. And there is the like limit in all quantities and proportions that begin and cease to be. And since such limits are certain and definite, to determine the same is a problem strictly geometrical. But whatever is geometrical we may use in determining and demonstrating any other thing that is also geometrical.

... For those ultimate ratios with which quantities vanish are not truly the ratios of ultimate quantities, but limits towards which the ratios of quantities decreasing without limit do always converge; and to which they approach nearer than by any given difference, but never go beyond, nor in effect attain to, till the quantities are diminished *in infinitum*. This thing will appear more evident in quantities infinitely great. If two quantities, whose difference is given, be augmented *in infinitum*, the ultimate ratio of these quantities will be given, namely, the ratio of equality; but it does not from thence follow, that the ultimate or greatest quantities themselves, whose ratio that is, will be given. Therefore if in what follows, for the sake of being more easily understood, I should happen to mention quantities as least, or evanescent, or ultimate, you are not to suppose that quantities of any determinate magnitude are meant, but such as are conceived to be always diminished without end.]

Corollaries 7 to 9 are regarded by Smith as being 'of considerable interest' [1980, p. 382]. Once again they may be compared to modern limiting results, this time concerning sums, differences, and products. Notice that Bayes's penchant for non-negativity results in his requiring, in Corollary 8, that A be greater than B.

Axiom 2 may be symbolized as follows. Let $x(\cdot)$ be a fluent having a permanent fluxion; e.g., $\dot{x}(t) = k$, where k is a constant. Then $x(t) = A + kt$ and $\Delta x = x(t + \Delta t) - x(t) = k\,\Delta t$, or, in Bayes's terminology, the change in x is proportional to the time in which that change is made.

Denoting the 'synchronal changes' in the flowing quantities $z(\cdot)$ and $x(\cdot)$ by Δz and Δx, respectively, one may write Axiom 3 as $\Delta z : \Delta x :: \dot{z} : \dot{x}$, the cited references then ensuring that, for example, ult. $\Delta z : \Delta x :: \dot{z} : \dot{x}$.

This may be alternatively but somewhat crudely expressed in the form

$$\frac{z(t+\Delta t) - z(t)}{x(t+\Delta t) - x(t)} = \frac{z(t+\Delta t) - z(t)}{\Delta t} \bigg/ \frac{x(t+\Delta t) - x(t)}{\Delta t} \approx \frac{\dot{z}}{\dot{x}}.$$

Turning to Axiom 4, let us denote by $x(\cdot)$ and $y(\cdot)$ the fluents having uniform and continually increasing fluxions, respectively. Let[52] $T = [t_0, t_1]$. Then, for $t \in (t_0, t_1)$,

$$\Delta x \equiv x(t+\Delta t) - x(t) \approx \dot{x}(t)\,\Delta t$$

$$\Delta y \equiv y(t+\Delta t) - y(t) \approx \dot{y}(t)\,\Delta t.$$

Since $x(\cdot)$ has a uniform fluxion, $\dot{x}(t) = \dot{x}(t_0)$, and since $y(\cdot)$ has a continually increasing fluxion, $\dot{y}(t_0) < \dot{y}(t)$ for $t_0 < t$. Thus

$$\frac{\Delta x}{\Delta y} \approx \frac{\dot{x}(t)\,\Delta t}{\dot{y}(t)\,\Delta t} = \frac{\dot{x}(t_0)}{\dot{y}(t)} < \frac{\dot{x}(t_0)}{\dot{y}(t_0)},$$

and similarly, for $t < t_1$,

$$\frac{\Delta x}{\Delta y} > \frac{\dot{x}(t_1)}{\dot{y}(t_1)}.$$

Axiom 5 may be similarly handled. These axioms will be of considerable importance in the proofs of some subsequent results.

Having laid down his definitions and axioms and deduced some consequences of these, Bayes now sets down nine propositions in which various properties of fluxions are derived. The first three propositions are of the same kind: we therefore, after having considered Bayes's proof of the first in some detail, are able to dismiss the second and third fairly rapidly.

In the first part of Bayes's proof of the first proposition, then, let the time interval $t = [t_0, t^*]$, where t^* is the 'given instant'. Let $y(\cdot)$ denote the fluent flowing with a permanent fluxion, and $z(\cdot)$ the one flowing with a continually increasing fluxion. Further, let $y(t^*) = A$ and $z(t^*) = B$ where A and B are determinate (i.e., constant), with $\dot{y} \equiv \dot{y}(t^*)$, $\dot{z} \equiv \dot{z}(t^*)$, and $\acute{y} = y(t^*) - y(t_0) = \Delta y$, and where the changes $\acute{z}$ and $\acute{\dot{z}}$ are similarly defined. Then $\dot{z}(t_0) = \dot{z} - \acute{\dot{z}}$, and hence, by Axiom 4,

$$\frac{\dot{y}}{\dot{z}} < \frac{\acute{y}}{\acute{z}} < \frac{\dot{y}}{\dot{z} - \acute{\dot{z}}}.$$

If $\acute{y} : \acute{z} :: \dot{y} : \dot{z} - x$ always, then

$$\frac{\dot{y}}{\dot{z} - x} < \frac{\dot{y}}{\dot{z} - \acute{\dot{z}}},$$

and hence $\overset{\prime}{\overset{\cdot}{z}} > x$. Now $\overset{\prime}{y}$, being made up of flowing and given (i.e., constant) quantities, is itself a flowing quantity (see Axiom 1), and thus vanishes at the given instant t^*. Hence $\overset{\prime}{y} \equiv y(t^*) - y(t_0) = A - y(t_0) \to 0$ as $t_0 \uparrow t^*$. Similar remarks apply, as $t_0 \uparrow t^*$, to the vanishing of $\overset{\prime}{z}$ and $\overset{\prime}{\overset{\cdot}{z}}$, and x vanishes since it is part of $\overset{\prime}{\overset{\cdot}{z}}$. Since $\overset{\prime}{y} : \overset{\prime}{\overset{\cdot}{z}} :: \dot{y} : \dot{z} - x$, and since $\overset{\prime}{y}$, $\overset{\prime}{z}$, and x vary as flowing quantities that vanish at t^*, we find, on using Corollary 3 and on recalling the definitions of A and B, that ult.$\overset{\prime}{y} : \overset{\prime}{z} :: \dot{y} : \dot{z}$.

The proof of the second part of Proposition I is similar, the time interval with given instant t^* now being $T = [t^*, t_1]$ and a result involving prime ratios being obtained.

The third and final part of the proof deals with the case in which a fluxion, instead of being uniformly increasing or uniformly decreasing, 'receives an instantaneous increment' (and presumably an instantaneous decrement could be similarly handled): i.e., the fluxion has a discontinuity of the first kind at t^*. Bayes's idea is essentially to consider the two (short) time intervals $t = [t_0, t^* + \varepsilon]$ and $T = [t^* - \varepsilon, t_1]$, with $\varepsilon > 0$, and with no other 'alteration' than that at t^* occurring in the fluxion in t, and similarly in T. The method of proof is then similar to the methods used in the proofs of the first and second parts.

Let us now turn to Proposition II, and let us denote by $y(\cdot)$ the second fluent (i.e., the one that is not uniformly flowing). Let t^* denote the distinguished point. The behaviour of the fluxion of $y(\cdot)$ may then be described by any one of the following cases (see Figure 5.1 for pertinent illustrations).

(a) $\dot{y}$ is *constantly increasing*.

(b) $\dot{y}$ is *constantly decreasing*.

(c) (i) $\dot{y}$ is *increasing* before t^* and *decreasing* thereafter;
   (ii) $\dot{y}$ is *decreasing* before t^* and *increasing* thereafter;
   (iii) $\dot{y}$ is *decreasing* in an ε-neighbourhood of t_1^*, and *increasing* in an ε-neighbourhood of t_2^*;
   (iv) $\dot{y}$ is *increasing* in an ε-neighbourhood of t_1^*, and *decreasing* in an ε-neighbourhood of t_2^*.

It should be noted that the distinguished point is not permitted to be a point of inflexion of the fluent.

To prove the first part of Proposition III, let z and x be quantities that 'flow any how'. Bayes cunningly introduces a uniformly flowing quantity y, and applies Proposition II to the pairs (y, z) and (y, x) in turn, the ratios ult.$\overset{\prime}{z} : \overset{\prime}{y} :: \dot{z} : \dot{y}$ and ult.$\overset{\prime}{y} : \overset{\prime}{x} :: \dot{y} : \dot{x}$ being obtained. It then follows that

230   5. Introduction to the Doctrine of Fluxions

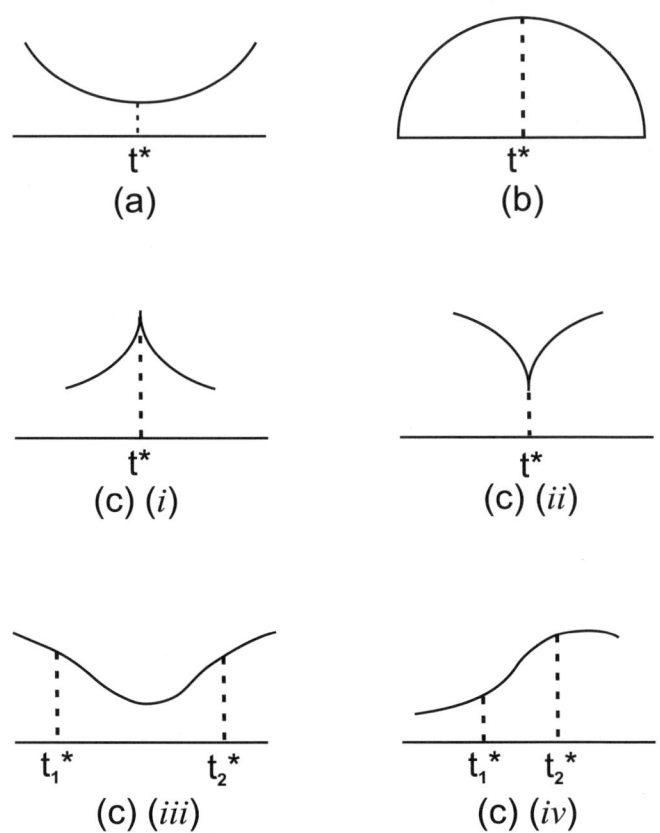

FIGURE 5.1 Curves with different fluxions.

ult. $\overset{\prime}{z} : \overset{\prime}{x} :: \dot{z} : \dot{x}$. A similar result is obtained in the second part of the proof for the prime ratio.

The assumptions of Lemma I may be stated as follows. Let $v : z$ and $x : y$ be variable ratios while $A : B$ is a constant ratio, and suppose further that $\min\{x : y, v : z\} \leq A : B \leq \max\{x : y, v : z\}$. Let these minimum and maximum ratios be $v : z$ and $x : y$, respectively. (Note the similarity between Bayes's statement that

> the quantities v, z, x, y, may be so small, as that their ratio's shall differ less from their prime and ultimate ratio's than by any assignable difference [p. [**24**]]

and the modern theory of limits.) Then it is asserted that

$$\left\{ \begin{array}{c} \text{ult.}\, v : z \\ \text{p}^{\text{mo}}\, v : z \end{array} \right\} \leq A : B \leq \left\{ \begin{array}{c} \text{ult.}\, x : y \\ \text{p}^{\text{mo}}\, x : y \end{array} \right\}, \tag{5.1}$$

where each of the embraced terms on the left-hand side may be taken with either of the embraced terms on the right-hand side.

Now for 'any assignable difference' ε we have $|x : y - \text{ult.}\, x : y| < \varepsilon$. If $A : B - \text{ult.}\, x : y = D > 0$, then, by taking ε to be D we would have

$$(A : B - \text{ult.}\, x : y) = D > |x : y - \text{ult.}\, x : y|,$$

whence $A : B > x : y$, contradicting the assumption that $A : B \leq x : y$. The remaining assertions can be similarly proved.

The Corollary to the Lemma is an obvious consequence of (5.1) above when $\text{ult.}\, v : z = \text{p}^{\text{mo}}\, x : y$ (Bayes now has $\min\{x : y, v : z\} = x : y$ and $\max\{x : y, v : z\} = v : z$).

The fourth proposition is one of the most difficult in the treatise, and there is some doubt about its correctness. Smith in fact wrote

> In Proposition 4 Bayes attempts more than he can handle with prime and ultimate ratios. [1980, p. 385]

I suspect, though, that something of the proof can be salvaged: in any case, this proposition is not crucial to most of the remaining work.

Before discussing the proof, however, we might note that in Bayes's opening words, viz.

> Supposing that at the time you seek the Fluxions, there be no instantaneous change made in them; *i.e.* supposing the Fluxions you seek to be themselves flowing quantities, or permanent ones, then if you can take synchronal increments or decrements of each ... [p. [**25**]]

232   5. Introduction to the Doctrine of Fluxions

it should be understood that *each* refers to the fluents rather than the fluxions (this becomes clear on reading Bayes's proof).

Briefly put, Bayes's argument runs as follows. Let z and x be the flowing quantities with flowing or permanent fluxions, and let $z(t^*) = A$ and $x(t^*) = B$ at the 'given instant' t^* (A and B are permanent quantities). Suppose that in a short interval $[t_0, t_1]$ containing t^*, z and x change from $A-y$ and $B-v$ (both values being taken on at t_0) to $A-y+a$ and $B-v+b$ (values taken on at t_1). Further, suppose that $a > y$ and $b > v$. If $a : b$ is a given ratio, then

> the ratio $a : b$ is the ratio of the Fluxions of z and x at the instant that $z = A$ and $x = B$, A and B being permanent quantities. [p. [25]]

Now v and y are synchronal changes in x and z, respectively, both vanishing at t^*, and hence, by Proposition III, ult. $y : v :: \dot{z} : \dot{x}$. Similarly, since $a - y$ and $b - v$ are synchronal changes arising at t^*, $\text{p}^{\text{mo}} a - y : b - v :: \dot{z} : \dot{x}$. But $\min\{y : v, a - y : b - v\} \leq a : b \leq \max\{y : v, a - y : b - v\}$, and hence, by the preceding Lemma,

$$a : b = \dot{z} : \dot{x} = \text{ult.} y : v = \text{p}^{\text{mo}} a - y : b - v.$$

The difficulty, of course, arises in connexion with the use of the ratio $a : b$ for the $A : B$ of the Lemma, a matter that we consider in the following examples.

Example 4.

Let $t^* = 0$ be the distinguished point and let $z(\cdot)$ and $x(\cdot)$ be defined by

$$z(t) = A + kt, \quad x(t) = B + ht, \quad \text{with } h > 0, \ k > 0.$$

On $[-\varepsilon, \varepsilon]$ we then have

$$z(-\varepsilon) = A - k\varepsilon, \quad x(-\varepsilon) = B - h\varepsilon$$
$$z(\varepsilon) = A + k\varepsilon = A - k\varepsilon + 2k\varepsilon$$
$$x(\varepsilon) = B + h\varepsilon = B - h\varepsilon + 2h\varepsilon.$$

Thus $y = k\varepsilon$, $a = 2k\varepsilon$, $v = h\varepsilon$, $b = 2h\varepsilon$, and y, a, v, and b all vanish at t^*. Further, $a > y$, $b > v$, and $a : b :: k : h$. Since

$$y : v :: k : h \text{ and } a - y : b - v :: k : h,$$

we clearly have

$$\dot{z}(0) : \dot{x}(0) :: \text{ult.} y : v = \text{p}^{\text{mo}} a - y : b - v :: k : h.$$

In the following example we consider possible forms of $z(\cdot)$ and $x(\cdot)$.

Example 5.

Let $t^* = 0$ be the distinguished point and let $z(\cdot)$ and $x(\cdot)$ be defined by

$$z(t) = A + \varphi(t), \quad x(t) = B + \psi(t),$$

with[53] $\varphi(0) = \psi(0) = 0$. Then $\dot{z}(t) = \dot{\varphi}(t)$ and $\dot{x}(t) = \dot{\psi}(t)$. For any $\varepsilon > 0$

$$z(-\varepsilon) = A + \varphi(-\varepsilon), \quad x(-\varepsilon) = B + \psi(-\varepsilon),$$

so that Bayes's y and v are given by $y = -\varphi(-\varepsilon)$ and $v = -\psi(-\varepsilon)$. Further,

$$\begin{aligned} z(\varepsilon) &= A + \varphi(\varepsilon) &= A + \varphi(-\varepsilon) + [\varphi(\varepsilon) - \varphi(-\varepsilon)] \\ x(\varepsilon) &= B + \psi(\varepsilon) &= B + \psi(-\varepsilon) + [\psi(\varepsilon) - \psi(-\varepsilon)], \end{aligned}$$

with $a = [\varphi(\varepsilon) - \varphi(-\varepsilon)]$ and $b = [\psi(\varepsilon) - \psi(-\varepsilon)]$. The ratio $a : b$ is then $[\varphi(\varepsilon) - \varphi(-\varepsilon)] : [\psi(\varepsilon) - \psi(-\varepsilon)]$. Now since y, v, $a-y$, and $b-v$ are synchronal changes vanishing at t^* (or, perhaps more correctly, since these functions tend to 0 as $\varepsilon \downarrow 0$),

$$\text{ult.}\, y : v :: \dot{z} : \dot{x}, \quad \text{p}^{\text{mo}}\, a - y : b - v :: \dot{z} : \dot{x};$$

and since

$$\min\{y : v, a - y : b - v\} < a : b < \max\{y : v, a - y : b - v\},$$

the given ratio $a : b$ satisfies

$$a : b = \dot{z} : \dot{x} = \text{ult.}\, y : v = \text{p}^{\text{mo}}\, a - y : b - v.$$

Now let us examine the ratio $a : b$ in more detail. Using L'Hospital's Rule (in a formal manner, if nothing else) we have

$$\lim_{\varepsilon \downarrow 0} \frac{[\varphi(\varepsilon) - \varphi(-\varepsilon)]}{[\psi(\varepsilon) - \psi(-\varepsilon)]} = \frac{\dot{\varphi}(0)}{\dot{\psi}(0)} = \frac{\dot{z}(0)}{\dot{x}(0)},$$

and so $a : b$ 'tends' to $\dot{z}(0) : \dot{x}(0)$. Similarly

$$\lim_{\varepsilon \downarrow 0} \frac{y}{v} = \lim_{\varepsilon \downarrow 0} \frac{-\varphi(-\varepsilon)}{-\psi(-\varepsilon)} = \frac{\dot{\varphi}(0)}{\dot{\psi}(0)} = \frac{\dot{z}(0)}{\dot{x}(0)},$$

and the ratio $a - y : b - v$ may be similarly treated.

As a particular case let us suppose that $\varphi(\cdot)$ and $\psi(\cdot)$ are defined by

$$\varphi(t) = \sum_{i=1}^{2n} \alpha_i t^i, \quad \psi(t) = \sum_{i=1}^{2m} \beta_i t^i,$$

with $m > n$ (a similar argument will apply if $m \leq n$ or if either of the upper limits of summation is odd). For given $\varepsilon > 0$ let $\rho(\cdot)$ be defined by

$$\rho(\varepsilon) = \frac{\varphi(\varepsilon) - \varphi(-\varepsilon)}{\psi(\varepsilon) - \psi(-\varepsilon)}.$$

Since this is to be a 'given' ratio, let us suppose that $\rho(\varepsilon) \equiv k$, where $0 < |k| < \infty$.

Now neither φ nor ψ can be an even function, for in such a case either the numerator or the denominator in $\rho(\varepsilon)$ would be zero. Further,

$$\varphi(\varepsilon) - \varphi(-\varepsilon) = \sum_{i=1}^{2n} \alpha_i [1 - (-1)^i] \varepsilon^i = 2 \sum_{i=1}^{n} \alpha_{2i-1} \varepsilon^{2i-1},$$

a similar expression obtaining for $\psi(\varepsilon) - \psi(-\varepsilon)$. Thus

$$\rho(\varepsilon) \equiv k \Leftrightarrow 2 \sum_{i=1}^{n} \alpha_{2i-1} \varepsilon^{2i-1} \equiv 2k \sum_{i=1}^{m} \beta_{2i-1} \varepsilon^{2i-1}$$

$$\equiv 2k \left[\sum_{i=1}^{n} \beta_{2i-1} \varepsilon^{2i-1} + \sum_{i=n+1}^{m} \beta_{2i-1} \varepsilon^{2i-1} \right].$$

Since this is an identity we must have

$$\alpha_{2i-1} = k \beta_{2i-1}, \ i \in \{1, 2, \ldots, n\}$$

$$\beta_{2i-1} = 0, \ i \in \{n+1, n+2, \ldots, m\}.$$

Thus

$$\varphi(t) = \sum_{i=1}^{n} \alpha_{2i-1} t^{2i-1} + \sum_{i=1}^{n} \alpha_{2i} t^{2i}$$

$$= k \sum_{i=1}^{n} \beta_{2i-1} t^{2i-1} + \sum_{i=1}^{n} \alpha_{2i} t^{2i}$$

$$= k \left[\psi(t) - \sum_{i=1}^{n} \beta_{2i} t^{2i} - \sum_{i=2n+1}^{2m} \beta_i t^i \right] + \sum_{i=1}^{n} \alpha_{2i} t^{2i}$$

$$= k \left[\psi(t) - \sum_{i=1}^{m} \beta_{2i} t^{2i} \right] + \sum_{i=1}^{n} \alpha_{2i} t^{2i}$$

$$= k \psi(t) + \sum_{i=1}^{m} \zeta_{2i} t^{2i},$$

where

$$\zeta_{2i} = \begin{cases} \alpha_{2i} - k\beta_{2i}, & i \in \{1, 2, \ldots, n\} \\ -k\beta_{2i}, & i \in \{n+1, n+2, \ldots, m\}. \end{cases}$$

Even more generally, for given k and ψ with $0 < |k| < \infty$ and ψ not an even function, we may define φ by

$$\varphi(t) = k\psi(t) + \chi(t^2).$$

It thus appears that Bayes's 'proof' is valid in certain circumstances.

The next three propositions are concerned with linearity properties of fluxions: in Proposition V Bayes shows that the fluxion of a sum of two fluents is the sum of the fluxions, provided that the fluents both increase or decrease together; in Proposition VI this last proviso is removed; and in Proposition VII the fluxion of Az is shown to be $A\dot{z}$ (with A being determinate). Although the fifth and seventh propositions seem fairly clear, there are a few points that require discussion in the sixth.

Let us then consider Proposition VI. The proof is given in two parts, in the first of which the fluents are all required to flow the same way (i.e., all increasing or all decreasing). Essentially the proof here consists of the repeated application of Proposition V. In the second part of the proof, however, where some quantities (or fluents) may increase while others decrease, Bayes's preference for non-negativity presents some problems. He considers the equation $Z - Y + v - x = R + r + T$, in which majuscules and minuscules denote increasing and decreasing quantities, respectively, and sets $V = 2D - v$, $X = D - x$, and $P = D - r$, where D is a 'large enough' determinate (or 'standing') quantity. Presumably things are written this way to ensure that, on the appropriate range of definition of the fluents, V, X, and P not only increase but are also positive, so that all terms in the expression $Y + R + T + V = Z + X + P$ are increasing. (For example, if $v = (1-t)^2$, then v is positive and decreasing on $(0,1)$. Thus $-v$ is increasing but negative on the same range, whereas $V = 2 - (1-t)^2$ is positive.) Presumably D is chosen so large that V, X, and P are all positive, with

$$Z - Y + v - x = R + r + T \iff Z - Y + (D-x) + (D-r) = R + T + (2D-v)$$

$$\iff Z + X + P = R + T + V + Y.$$

Bayes also notes that

> because $V = 2D - v$ always, the synchronal changes of V and v are always equal, since D is a standing quantity. [p. **[29]**]

This looks a bit odd, since the definition of V would seem to imply that $\Delta V = -\Delta v$, but we must recall Bayes's requirement that changes be non-negative.

In the corollary to this proposition Bayes remarks that determinate (i.e., constant) quantities have zero fluxions, and this remark is followed by his noting that the proof (of the proposition) is expedited by assuming that the fluxions of decreasing quantities are negative.

The proof of Proposition VII is a straightforward application of earlier results.

The eighth proposition states that the fluxion of x^2 is $2x\dot{x}$. Two proofs are given: the first seems correct, but the second needs some attention. Bayes begins the latter by saying

> Whilst x changes from $A - \frac{1}{2}\dot{x}'$ to $A - \frac{1}{2}\dot{x}' + \dot{x}'$, x^2 or y changes from $A^2 - A\dot{x}' + \frac{1}{4}x'^2$ to $A^2 - A\dot{x}' + \frac{1}{4}x'^2 + 2A\dot{x}'$. [p. [31]]

This allows him to conclude, by Proposition IV, that the ratio $\dot{x}' : 2A\dot{x}'$, the constant ratio $1 : 2A$, is the ratio of the fluxions of x to x^2 or y at the time at which x becomes A and y becomes A^2.

But suppose that instead of considering a change from $A - \frac{1}{2}\dot{x}'$ to $A - \frac{1}{2}\dot{x}' + \dot{x}'$ we had considered a change from $A - \alpha\dot{x}'$ to $A - \alpha\dot{x}' + \beta\dot{x}'$, where α and β are (positive) real numbers with $\beta > \alpha$. Then $(A - \alpha\dot{x}')^2$ becomes

$$[(A - \alpha\dot{x}') + \beta\dot{x}']^2 = (A - \alpha\dot{x}')^2 + 2(A - \alpha\dot{x}')\beta\dot{x}' + (\beta\dot{x}')^2,$$

and hence, by Proposition IV, the ratio to be considered would be $\beta\dot{x}' : 2(A - \alpha\dot{x}')\beta\dot{x}' + (\beta\dot{x}')^2$ or $1 : 2A + (\beta - 2\alpha)\dot{x}'$. Although this ratio certainly reduces to Bayes's constant ratio of $1 : 2A$ for $\alpha = 1/2$ and $\beta = 1$ (and more generally for $\beta = 2\alpha$), in general it is not a constant ratio, and Proposition IV is inapplicable.

One can also consider this problem using the approach presented in Example 4. Thus let $Z(t) = x(t)$ and $X(t) = [x(t)]^2$, with $A = x(0)$. Then, for any $\varepsilon > 0$,

$$\begin{aligned} Z(-\varepsilon) &= x(-\varepsilon) &&= x(0) + [x(-\varepsilon) - x(0)] \\ X(-\varepsilon) &= [x(-\varepsilon)]^2 &&= x(0) + [(x(-\varepsilon))^2 - x(0)] \\ Z(\varepsilon) &= x(\varepsilon) &&= x(0) + [x(-\varepsilon) - x(0)] + [x(\varepsilon) - x(-\varepsilon)] \\ X(\varepsilon) &= [x(\varepsilon)]^2 &&= x(0) + [(x(-\varepsilon))^2 - x(0)] + [(x(\varepsilon))^2 - (x(-\varepsilon))^2], \end{aligned}$$

and hence, as $\varepsilon \to 0$,

$$\frac{a}{b} \equiv \frac{x(-\varepsilon) - x(\varepsilon)}{[x(-\varepsilon)]^2 - [x(\varepsilon)]^2} \to \frac{1}{2x(0)} = \frac{1}{2A}.$$

The ninth, and last, proposition is concerned with the fluxion of the product of two flowing quantities[54]. Thus if x, y, and z are three flowing quantities with $xy = z$, then Bayes shows (a) that $\dot{z} = \dot{x}y + \dot{y}x$ when x and y flow the same way, (b) that $\dot{z} = \dot{x}y - \dot{y}x$ when z and x flow the same way and y flows the opposite way, and (c) that $\dot{z} = \dot{y}x - \dot{x}y$ when z and y flow the same way and x flows the opposite way (the negative signs in the expressions in (b) and (c) once again arise from the requirement that changes be positive). As a corollary Bayes notes that one will always have $\dot{z} = \dot{x}y + \dot{y}x$ if fluxions of decreasing quantities are taken to be negative.

We now come to the third section of the essay, in which, leaving his development of the fluxionary calculus, Bayes replies to some of Berkeley's criticisms of Newton's arguments. Pointing out that the general method may be based not on fluxions as velocities but on any other finite quantities given in terms of the last ratio of synchronal increments or decrements of fluents, Bayes asserts that this might indeed be a preferable approach when fluxions of different orders occur in the same expression. This leads him to suggest a way in which a treatise on fluxions might be written, one in which some of the definitions and axioms of the present paper would be needed. Let us give these fundamentals as Bayes has outlined them (the bracketed terms refer to Bayes's tract).

Defin. 1. (Definition 1). A Flowing quantity is one that continually increases or decreases, and in such a manner that some time is requisite to make any increment or decrement.

Defin. 2. (Axiom 2). A quantity flows uniformly, or with a permanent Fluxion, when the changes made in it are always proportional to the times in which they are made.

Defin. 3. (Definition 6). If there be two permanent quantities A and B, and two other flowing Quantities a and b, and the ratio of a to b be always, during a given time, that of the sum or difference of the first permanent quantity A, and another flowing quantity x to the sum or difference of the second permanent quantity B, and another flowing quantity y, and at the end of the given time all the flowing quantities vanish; then the ratio of the permanent quantities A and B, is the last or ultimate ratio of the vanishing quantities a and b; which I thus express, ult. $a : b :: A : B$. *i.e.* if $a : b :: A \mp x : B \mp y$ always, during the time $\mathcal{T}$, and at the end of that time, a, b, x, y all vanish; then ult. $a : b :: A : B$.

Defin. 4. The Fluxion of an uniformly flowing quantity is the change made in it in a certain given time.

Defin. 5. The Fluxion of a quantity any how flowing at any given instant, is a quantity found out by taking it to the Fluxion of an uniformly flowing quantity in the ultimate proportion of those synchronal changes which then vanish.

These definitions, Bayes then asserts, are sufficient to allow the development of the theory he has already given.

Bayes now proposes to show, following Newton, that the fluxions of x and x^n are in the ratio $1 : nx^{n-1}$ (this was a matter specifically raised by Berkeley). First he expands the binomial $(x+o)^n$, and then he concludes that the synchronal augments o and $n\,o\,x^{n-1} + \frac{n^2-n}{2} o^2\, x^{n-2} + \&c.$ are to one another as $1 : nx^{n-1} + \frac{n^2-n}{2} o\,x^{n-2} + \&c.$ So far there can be no disagreement. The difficulty arises in his now saying 'Let now these augments vanish', an assumption that is needed to conclude that the fluxions of x and x^n are in the ratio $1 : nx^{n-1}$. The problem is of course avoided in modern practice by the use of limits, an approach that was not open to Bayes, although some notion of a limiting argument is evinced in the text following this result in the essay (see p. [**36**]). With the acceptance of Bayes's and Newton's idea of 'the last proportion of vanishing quantities', one may well find the proof satisfactory, provided one does not actually permit o to vanish. It is perhaps worth drawing attention to the fact that, after having shown that the proportion between the increments of x and x^n, *so long as they exist*, is expressible as 1 to $nx^{n-1} + \frac{n^2-n}{2} o\,x^{n-2} + \&c.$, Bayes writes

> But from this last supposition [i.e., that o continually decreases and eventually vanishes] we may gather, that the lesser o is, so much the nearer the ratio of 1 to nx^{n-1} comes to the ratio of the increments; and that by a continual diminution of o, it may come as near to it as you please, but can never equal it before o quite vanishes; and therefore this ratio, and no other whatsoever, agrees to the description which Sir *Isaac* has given of the ultimate ratio of the vanishing increments. [p. [**38**]]

Walton was quite explicit, even if not acceptably so, in his discussion of moments:

> By Moments we may understand the nascent or evanescent Elements or Principles of finite Magnitudes, but not particles of any determinate Size, or Increments actually generated; for all such are Quantities themselves, generated of Moments.

> The Magnitudes of the momentaneous Increments or Decrements of Quantities are not regarded in the Method of Fluxions, but their first or last Proportions only; that is, the Proportions with which they begin or cease to exist: These are not their Proportions immediately before or after they begin or cease to exist, but the Proportions with which they begin to exist, or with which they vanish. [1735, pp. 6–7]

Two further references from *The Analyst* occur on p. **[36]**: the first is not easily identifiable: presumably it is to §§13 & 14, in which Berkeley discusses the finding of the fluxion of x^n in Newton's fashion, the reasoning being stated to be 'not fair or conclusive' [§13]. Valiant-for-truth though Bayes may be, however, he does not manage to conquer the Giant Despair raised by Berkeley: it is difficult to accept that if a conclusion is reached on the predication that a certain quantity is non-zero, that conclusion will remain valid if that same quantity is put equal to zero in the final result (though of course a limiting argument may well be invoked). In this connexion the following passage is worth noting.

> Nor is it any objection to the justness of their [i.e., the mathematicians'] reasoning, that an algebraical note is sometimes to be interpreted, at the end of the process, in a sense which cou'd not have been substituted in the beginning of it; since if quantities themselves are considered as continually changing, the sense of the mark which represents or expresses them must, in order to its doing so, continually change along with them. [p. **[49]**]

The second reference from *The Analyst* is to the following Lemma,

> If with a view to demonstrate any proposition, a certain point is supposed, by virtue of which certain other points are attained; and such supposed point be it self afterwards destroyed or rejected by a contrary supposition; in that case, all the other points attained thereby, and consequent thereupon, must also be destroyed and rejected, so as from thence forward to be no more supposed or applied in the demonstration [§12]

a result Berkeley claims to be 'so plain as to need no proof'[55].

In this same passage Bayes cites two assumptions made by Newton, the second of which he gives as follows.

> in order to find the last ratio of the increments vanishing, he [i.e., Newton] supposes o to decrease till it vanishes, or becomes equal to nothing [p. **[37]**]

and in a footnote he says that this is the true meaning of Newton's phrase *Evanescant jam augmenta illa*, one that is found in the derivation of the fluxion of x^n given in the Introduction to Newton's *Tractatus de quadratura curvarum*. Berkeley does not in fact quote this sentence in *The Analyst*: he chooses rather to follow the translation given in Harris [1710], writing

> Let now the increments vanish, and their last proportion will be 1 to nx^{n-1}. But it should seem that this reasoning is not fair or conclusive. For when it is said, let the increments vanish, *i.e.* let the increments be nothing, or let there be no increments, the former supposition that the increments were something, or that there were increments, is destroyed, and yet a consequence of that supposition, *i.e.* an expression got by virtue thereof, is retained. Which, by the foregoing lemma, is a false way of reasoning. Certainly when we suppose the increments to vanish, we must suppose their proportions, their expressions, and every thing else derived from the supposition of their existence to vanish with them. [§13]

The lemma referred to by Berkeley is that instanced above, writing of which Cajori said

> It is interesting to observe that no British mathematician of the eighteenth century acknowledged the soundness of Berkeley's *lemma* and its application. [1917, p. 148]

He goes on to attribute the first recognition of the *fallacia suppositionis* or the *shifting of the hypothesis* to Robert Woodhouse in 1803.

In his reply to *The Analyst* Jurin comments on Berkeley's interpretation of Newton's phrase as follows (we quote the passage in full, not only because of the importance of the sentence to the understanding of the fluxionary calculus, but also as an example of Jurin's response in general).

> *Evanescant jam augmenta illa*, let now the increments vanish, *i.e.* let the increments be nothing, or let there be no increments. Hold, Sir, I doubt we are not right here. I remember Sir *Isaac Newton* often uses the terms of *momenta nascentia* and *momenta evanescentia*. I think I have seen you likewise several times using the like terms of nascent and evanescent increments. Also, if I am not mistaken, both he and you consider a nascent, or evanescent moment, an increment or decrement, as the same quantity under different circumstances; sometimes as in the point of beginning to exist, and other times as in the point of ceasing to exist. From this methinks it should follow that the two expression subjoined, will be perfectly equivalent to each other.

> Nascantur *jam augmenta illa, & eorum ratio* prima *erit*
> Evanescant *jam augmenta illa, & eorum ratio* ultima *erit*

> The meaning of the first can possibly be no other than to consider the first proportion between the nascent augments, in the point of their beginning to exist. Must not therefore the meaning of the latter be to consider the last proportion between the evanescent augments, in the point of evanescence, or their ceasing to exist? Ought it not to be thus translated, Let the augments now become evanescent, let them be upon the point of evanescence? What then must we think of your interpretation, *Let the increments be nothing, let there be no increments?* Do not the words *ratio ultima* stare us in the face, and plainly tell us that though there is a last proportion of evanescent increments, yet there can be no proportion of increments which are nothing, of increments which do not exist? I believe, Sir, every thinking person will acquit Sir *Isaac Newton* of the gross oversight you ascribe to him, and will acknowledge that it is your self alone, who have been guilty of a most palpable, inexcusable, and unpardonable blunder. [pp. 56–58]

Gibson finds that Jurin has completely missed the point in his expounding of Newton's treatment of prime and ultimate ratios — Cajori finds Jurin's replies to Berkeley to be 'full of noisy rhetoric and giving little that was truly substantial' [1917, p. 146] — and he claims that Jurin interpreted that doctrine

> exactly as Berkeley had done, so that an ultimate ratio is not the *limit* of a varying ratio, but the *last value* of a ratio. Berkeley very properly maintains that there is no last value of the augments except zero, so that the phrase "the ratio with which they vanish," used by Newton himself, and so often repeated by his expounders, does not represent any mathematical operation, and so far from explaining anything, requires explanation. [1898, p. 20]

Incidentally, Gibson [1898, pp. 24, 27] also asserts that Robins's explanation of the terms *evanescens* and *nascens* is not only the most reasonable, but also the most consistent with Newton's writings.

Berkeley took up this point in his reply, *A Defence of Free-thinking in Mathematics*, writing

> ... the question between us is, whether I have rightly represented the sense of those words, *evanescant jam augmenta illa*, in rendering them, let the increments vanish, *i.e.* let the increments be nothing, or let there be no increments? ... I ...

affirm, the increments must be understood to be quite gone, and absolutely nothing at all. ... Further by *evanescant* must either be meant, let them (the increments) vanish and become nothing, in the obvious sense, or else let them become infinitely small. But that this latter is not Sir Isaac's sense is evident from his own words in the very same page, ... where he expressly saith, *volui ostendere quod in methodo fluxionum non opus sit figuras infinite parvas in geometriam introducere.* ... You raise a dust about evanescent augments, which may perhaps amuse and amaze your reader, but I am much mistaken if it ever instructs or enlightens him. [§33]

Nothing, I say, can be plainer to any impartial reader than that by the evanescence of augments in the above-cited passage, Sir Isaac means their being actually reduced to nothing. But, to put it out of all doubt that this is the truth, and to convince even you, who shew so little disposition to be convinced, I desire you to look into his *Analysis per Æquationes Infinitas* (p. 20), where, in his preparation for demonstrating the first rule for the squaring of simple curves, you will find that on a parallel occasion, speaking of an augment which is supposed to vanish, he interprets the word *evanescere* by *esse nihil*. Nothing can be plainer than this, which at once destroys your defence. [§34]

Bayes next turns his attention to the problem of finding the ordinate of a curve when the area and the abscissa are given, a problem that, he notes, is 'analogous to that of finding the Fluxion from the Fluent perpetually given' (p. [**38**]). (Notice in his proof that *Bo* and *Do* should be *BO* and *DO*, respectively.) We might frame his discussion as follows. Let $f(\cdot)$ be a 'nice' function on some interval $[a, b]$, say, with $f(a) = 0$, and let

$$A = \int_a^{x_0} f(x)\,dx.$$

If $A = x_0^3$, then

$$x_0^3 = \int_a^{x_0} f(x)\,dx \implies 3x_0^2 = f(x_0),$$

that is, the ordinate y_0 corresponding to the abscissa x_0 and the area $A = x_0^3$ is given by $y_0 = 3x_0^2$.

In connexion with this proof the following passages from Newton's *Tractatus de quadratura curvarum* may be of interest[56].

> PROB. IX. THEOR. VII. Æquantur Curvarum areæ inter se quarum Ordinatæ sunt reciproce ut fluxiones Abscissarum. [The

Areas of these Curves are equal to one another whose Ordinates are as the Fluxions of the Abscissa.] Nam contenta sub Ordinatis & fluxionibus Abscissarum erunt æqualia, & fluxiones arearum sunt ut hæc contenta. [For the Rectangles under the Ordinates and the Fluxions of the *Abscissæ* are equal, and the Fluxions of the *Areas* are as those Rectangles.]

COROL. I. Si assumatur relatio quævis inter Abscissas duarum Curvarum, & inde per Prop 1. quæratur relatio fluxionum Abscissarum, & ponantur Ordinatæ reciproce proportionales fluxionibus, inveniri possunt innumeræ Curvæ quarum areæ sibi mutuo æquales erunt. [If any Relation between the *Abscissæ* of two Curves be assumed, and thence (by *Prop.* 1.)[57] the Relation between the Fluxions of the *Abscissæ* be sought, and the Ordinates be supposed reciprocally proportionable to the Fluxions; then innumerable Curves may be found, whose *Areas* shall be mutually equal to one another.]

Bayes is of course correct in saying that there is no inconsistency in assuming that, at one time, x and b are unequal, and that, at another, x and b are, or become, equal. But he is wrong in supposing that the conclusion of a mathematical illation arrived at on the assumption that x and b are distinct necessarily holds when x and b are equal. His argument following this (p. [40]) seems to be what we might today write as follows.

If $\lim_{x \to x_0} f_1(x) = A = \lim_{x \to x_0} f_3(x)$, and if $f_1(x) < f_2(x) < f_3(x)$ for all x, then $\lim_{x \to x_0} f_2(x) = A$.

Bayes shares the thoughts of others when he writes

> What moments, what the *principia jamjam nascentia finitarum quantitatum*, are in themselves, I own, I don't understand
>
> [p. [41]]

and he goes on to note that, although moments on their own may be difficult to imagine, the *proportion* of moments is easily intelligible.

It is indeed difficult to get an idea of moments from Newton's own writings[58]: in the first edition of the *Principia* [1687] moments seem to be infinitely small quantities, Newton writing[59] in Book II, Lemma II,

> Cave tamen intellexeris particulas finitas. Momenta, quam primum finitæ sunt magnitudinis, desinunt esse momenta. Finiri enim repugnat aliquatenus perpetuo eorum incremento vel decremento. Intelligenda sunt principia jamjam nascentia finitarum

244   5. Introduction to the Doctrine of Fluxions

> magnitudinum. Neq; enim spectatur in hoc Lemmate magnitudo momentorum, sed prima nascentium proportio. [But take care not to look upon finite particles as such. Moments, as soon as they are of finite magnitude, cease to be moments. To be given finite bounds is in some measure contrary to their continuous increase or decrease. We are to conceive them as the just nascent principles of finite magnitudes. Nor do we in this Lemma regard the magnitude of the moments, but their first proportion, as nascent.]

In the third edition of 1726 this passage runs as follows[60].

> Cave tamen intellexeris particulas finitas. Particulæ finitæ non sunt momenta, sed quantitates ipsæ ex momentis genitæ. Intelligenda sunt principia jamjam nascentia finitarum magnitudinum. Neque enim spectatur in hoc lemmate magnitudo momentorum, sed prima nascentium proportio. [But take care not to look upon finite particles as such. Finite particles are not moments, but the very quantities generated by the moments. We are to conceive them as the just nascent principles of finite magnitudes.]

It seems then that Bayes is right in emphasizing the interpretation of moments via proportions.

Bayes's dismissal of second- and higher-order fluxions as being no more mysterious than those of first-order (pp. **[19, 42]**) echoed the earlier tracts by Jurin and Walton. Indeed, the avoidance of this issue by Walton prompted the following somewhat sarcastic remark from Berkeley.

> [Walton] discreetly avoids ... to say one syllable of second, third, or fourth fluxions, and of divers other points mentioned in the *Analyst*, about all which I observe in him a most prudent and profound silence. And yet he very modestly gives his reader to understand that he is able to clear up all difficulties and objections that have ever been made. [1735a, Appendix, §2]

On (p. **[42]**) Bayes cleverly notes that, when one deduces from $y^2 = x$ (where y is uniformly flowing) that $2\dot{y}^2 = \ddot{x}$, one should interpret this only as $2\dot{y}^2|_{t_0} : 2\dot{y}^2|_{t_1} :: \ddot{x}|_{t_0} : \ddot{x}|_{t_1}$, an interpretation that avoids dimensional difficulties (this perhaps suggests that the replacement of '::' by '=' should be cautiously executed). This remark is seen as an answer to part of Berkeley's thirty-first query, which runs as follows.

> Where there are no increments, whether there can be any *ratio* of increments? Whether nothings can be considered as proportional to real quantities? Or whether to talk of their proportions

be not to talk nonsense? Also in what sense we are to understand the proportion of a surface to a line, of an area to an ordinate? And whether species or numbers, though properly expressing quantities which are not homogeneous, may yet be said to express their proportion to each other?

One may perhaps see Bayes's carefulness on this point as an answer to the following comment in Berkeley's *Philosophical Commentaries*.

> Mathematicians seem not to speak clearly & coherently of Equality. They nowhere define w^t they mean by that word when apply'd to Lines.

Bayes also speaks here of *Bigots* and *Enthusiasts*, and although these words may be given their modern meanings, the *Oxford English Dictionary* also gives religious interpretations[61]. Such interpretations were in fact advanced by Robert Burton in 1621 in the third book of *The Anatomy of Melancholy*, where *Enthusiasm* is included in the discussion of religious melancholy, and where the consideration of the causes of such melancholy causes Burton to cite Polydore Virgil as holding

> that those prophesies, and Monks' revelations, Nun's dreams, which they suppose come from God, do proceed wholly *ab instinctu dæmonum*, by the Devil's means: and so those Enthusiasts, Anabaptists, pseudo-Prophets from the same cause.
> [Burton, Book III, Part. III. Sect. IV. Mem. I. Subs. II]

Perhaps such religious connexions were lurking in Bayes's mind in view of the general nature of Berkeley's tract.

In his further discussion of Berkeley's misrepresentation of the mathematicians Bayes remarks that

> we ... cannot, in any case, know the conclusion to be right, if we have argued from the supposition of the truth of what we know to be false, or don't know to be true [p. **[44]**]

and although there may be some truth in this, at least inasmuch as a conclusion deduced from false premises cannot be held to be true as a consequence of those premises, the strength of his further argument about the absurdity of algebraic arguments involving negative quantities has been somewhat weakened by developments subsequent to his time. It also appears from this passage that Bayes had some familiarity with Leibniz's method of differences. Note again here the adumbration of limit theory.

In his next paragraph Bayes refers to Berkeley's sixteenth query, viz.

> Whether certain maxims do not pass current among analysts which are shocking to good sense? And whether the common assumption that a finite quantity divided by nothing is infinite, be not of this number?

Bayes's allowing it to be a rule in Algebra that 'finite divided by nothing is infinite' (p. [**46**]) is again something that one would not wish to accept today, though it seems that this is to be understood as $\lim_{n \to 0}(1/n) = \infty$.

Bayes also quotes here the 'horrid' logical terms *Barbara* and *Celarent*. To illustrate the use and meaning of these terms, let us consider the following scheme.

All Xs are Ys	...A
No Xs are Ys	...E
Some Xs are Ys	...I
Some Xs are not Ys	...O

(the labels being given by the vowels in *affirm* and *nego*). Now consider the general syllogism

Premises. $\begin{cases} \text{All } Y\text{s are } X\text{s.} \\ \text{All } Z\text{s are } Y\text{s.} \end{cases}$

Conclusion. All Zs are Xs.

Here X is called the *major* term, Z the *minor* term, and Y the *middle* term. The *figure* of a syllogism consists in the situation of the middle term with respect to the terms of the conclusion. The possible figures are given by Boole [1847, p. 31] as follows.

1st. Fig.	2nd. Fig.	3rd. Fig.	4th. Fig.
YX	XY	YX	XY
ZY	ZY	YZ	YZ
ZX	ZX	ZX	ZX

Designation of the three propositions of a syllogism by the symbols A, E, I, or O is said to determine the *mood* of the syllogism. Thus the general syllogism given above belongs to the mood AAA in the first figure. A general mnemonic for the determination of all valid syllogisms is given by Boole (loc. cit.) as follows.

Fig. 1. bArbArA, cElArEnt, dArII, fErIO que prioris.

Fig. 2. cEsArE, cAmEstrEs, fEstInO, bArOkO, secundæ.

Fig. 3. Tertia dArAptI, dIsAmIs, dAtIsI, fElAptOn, bOkArdO,
fErIsO, habet: quarta insuper addit.

Fig. 4. brAmAntIp, cAmEnEs, dImArIs, fEsApO, frEsIsOn.

Thus our general syllogism is an example of *Barbara*. One cannot but agree with Bayes's adjective.

In connexion with the matter raised by Bayes in the following passage,

> He that can take two nothings and find their sum, and then five nothings and find their sum equal to the former, seems to me to be in a fair way to be able to divide by nothing and find the quotient [p. **[47]**]

one might recall the note

$$a + 500 \text{ nothings} = a + 50 \text{ nothings an innocent silly truth,}$$

from Berkeley's *Philosophical Commentaries*. Although modern algebraic custom forbids division by zero, it does not frown upon multiplication by the cipher; nor does it stop us from allowing 'nothing to be a real quantity or number' (p. **[47]**), though we must of course bear in mind that Bayes is not using the term 'real number' as we do today.

On this same page Bayes names Berkeley's fortieth query, viz.

> Whether it be not a general case or rule, that one and the same coefficient dividing equal products gives equal quotients? And yet whether such coefficient can be interpreted by 0 or nothing? Or whether any one will say that if the equation $2 \times 0 = 5 \times 0$, be divided by 0, the quotients on both sides are equal? Whether therefore a case may not be general with respect to all quantities and yet not extend to nothings, or include the case of nothing? And whether the bringing nothing under the notion of quantity may not have betrayed men into false reasoning?

(In Luce & Jessop's edition of *The Analyst* the '0' we have used here is given as '*o*'. The question is whether Berkeley is considering division and multiplication by nought or by an increment: Bayes takes the former.)

In his fourth point (p. **[48]**) Bayes refers to two more of Berkeley's queries:

> *Qu.* 50 Whether, ever since the recovery of mathematical learning, there have not been perpetual disputes and controversies among the mathematicians? And whether this doth not disparage the evidence of their methods?
>
> *Qu.* 51 Whether anything but metaphysics and logic can open the eyes of mathematicians and extricate them out of their difficulties?

And although it may be true that in the land of the blind the one-eyed is king[62], Bayes perhaps goes further in suggesting that he who is half-blind should not take the cecitous[63] as a guide.

Returning again to matters algebraic, Bayes notes Berkeley's forty-third query:

Whether an algebraist, fluxionist, geometrician, or demonstrator of any kind can expect indulgence for obscure principles or incorrect reasonings? And whether an algebraical note or species can at the end of a process be interpreted in a sense which could not have been substituted for it at the beginning? Or whether any particular supposition can come under a general case which doth not consist with the reasoning thereof?

Again these are answered in the negative.

Finally, in noting the importance of a mathematical education, Bayes instances as relevant the following of Berkeley's queries.

Qu. 15 Whether to decline examining the principles, and unravelling the methods used in mathematics would not shew a bigotry in mathematicians?

Qu. 38 Whether tedious calculations in algebra and fluxions be the likeliest method to improve the mind? And whether men's being accustomed to reason altogether about mathematical signs and figures doth not make them at a loss how to reason without them?

Qu. 56 Whether the corpuscularian, experimental, and mathematical philosophy, so much cultivated in the last age, hath not too much engrossed men's attention; some part whereof it might have usefully employed?

Qu. 57 Whether from this and other concurring causes the minds of speculative men have not been borne downward, to the debasing and stupifying of the higher faculties? And whether we may not hence account for that prevailing narrowness and bigotry among many who pass for men of science, their incapacity for things moral, intellectual, or theological, their proneness to measure all truths by sense and experience of animal life?

It must have been with difficulty that Bayes refrained from quoting what has perhaps become the most well-known phrase from *The Analyst*, viz. 'ghosts of departed quantities'. One might well see an echo of Berkeley's difficulty with these momentous shades in Hughes Mearns's lines

> As I was going up the stair
> I met a man who wasn't there.
> He wasn't there again today.
> I wish, I wish he'd stay away.

5.4 Conclusion

The approach taken by Bayes in the more mathematical parts of his essay is one that marks the author as having a logical mind keenly attuned to mathematical rigour. Postulates, definitions, axioms, and propositions succeed each other in logical order, and the general appearance of the work does not seem foreign to the modern eye. As an example, consider the following passage,

> it is not the business of the Mathematician to dispute whether quantities do in fact ever vary in the manner that is supposed, but only whether the notion of their so doing be intelligible.
>
> [p. [9]]

This is patently the sort of remark that could be uttered comfortably by a modern pure mathematician.

Although later mathematical work on limits and non-standard analysis would provide a rigorous and satisfactory theory for the calculus and overcome Berkeley's objections, the lack of such work made contemporaneous criticism of *The Analyst* awkward. The replies were in some cases, as that of Bayes, as mathematically sound as was possible with the tools available at that time, and any lack of conviction carried that may be seen today should not result in their summary dismissal[64].

Whether Bayes's reply to Berkeley can be regarded as a success is arguable. Thus Fraser, in the third volume of his 1901 edition of Berkeley's works, lists Bayes's tract among the 'more important of the relative publications' (p. 9) issued in response to *The Analyst*; Jesseph, on the other hand ([1993, chap. 7]), in considering the most important responses to Berkeley's treatise, mentions Bayes's tract, but finds [p. 259] that, in his view, it features neither among the best nor among the worst responses, and he concludes that

> We can grant that Bayes's *Introduction* is a serious work by an important mathematician, but it is hardly a conclusive answer to Berkeley. [Jesseph, 1993, p. 277]

Cajori, perhaps more cautiously, says 'The pamphlet [by Bayes] of 1736 represents a careful effort to present an unobjectionable foundation of fluxions' [1919, p. 157][65]. As we have seen, some of Bayes's arguments seem to foreshadow aspects of limit theory later to be used in the rigorization of the calculus; but presented as they were (and as they could then only possibly be), they fail to convince today, adequate though they might have been at the time.

The responses to *The Analyst* were indeed varied, and perhaps there is some validity in Elizabeth Montagu's opinion of Berkeley's philosophical works expressed in a letter[66] to Gilbert West of the 28th January 1753:

> they are some of them too subtile to be even the object of most peoples consideration. He has the hard fate of not convincing any one, tho he cannot be confuted; a judgment of his metaphysical works must be pass'd by superior intelligences, it falls not within the measure of 5 senses.

Berkeley's last word on the *Analyst* controversy was published in 1744 in his tract *Siris: a Chain of Philosophical Reflexions and Inquiries*[67], where we read in a footnote to §271

> Our judgment in these matters is not to be overborne by a presumed evidence of mathematical notions and reasonings, since it is plain the mathematicians of this age embrace obscure notions, and uncertain opinions, and are puzzled about them, contradicting each other and disputing like other men: witness their doctrine of Fluxions, about which, within these ten years, I have seen published about twenty tracts and dissertations, whose authors being utterly at variance, and inconsistent with each other, instruct bystanders what to think of their pretensions to evidence.

There can be no doubt as to the importance of *The Analyst* in drawing attention to flaws in the fluxionary (and differential) calculus. The exposure of such inadequacies and the various attempts to satisfy both them and their criticism led to considerable discussion in British mathematical circles during the eighteenth century, and to the later perfection of the calculus. Among these attempts Bayes's tract is surely of no little importance.

5.5 Appendix 5.1

In view of the importance of Robins's work, and for ease of comparison of his approach with that of Newton (and perhaps of Bayes), we give some extracts from Robins's *Mathematical Tracts* here. *The whole of the rest of this section, though not written as a quotation, is from this book.*

The reader will perceive, I am endeavouring to explain Sir Isaac Newton's expression Ratio ultima quantitatum evanescentium; and I have rendered the Latin participle evanescens, by the English one vanishing, and not by the word evanescent, which having the form of a noun adjective, does not so certainly imply that motion, which ought here to be kept carefully in mind. The quadrilaterals $ABCD$, $BEFC$ become vanishing quantities from the time we first ascribe to them this perpetual diminution; that is, from that time they are quantities going to vanish. And as during their diminution

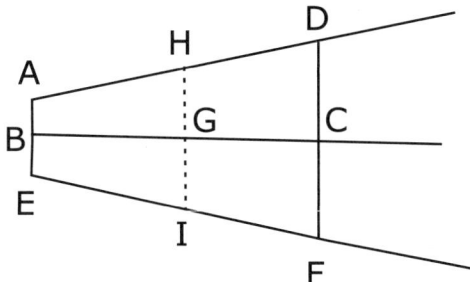

they have continually different proportions to each other; so the ratio between AB and BE is not to be called merely Ratio harum quantitatum evanescentium, but Ultima ratio. [§105, p. 52] Here I have attempted to explain in like manner the phrase Ratio prima quantitatum nascentium; but no English participle occurring to me, whereby to render the word nascens, I have been obliged to use circumlocution. ... we must not use the simple expression Ratio quantitatum nascentium; for by this we shall not specify any particular ratio; but to denote the ratio between AB and BE we must call it Ratio prima quadrilaterum nascentium. [§107, p. 53]

The other part of this method, which concerns varying ratios, may be put into the same form by defining ultimate ratios, as follows. [§123, p. 57]

If there be two quantities, that are (one or both) continually varying, either by being continually augmented, or continually diminished; though the proportion, they bear to each other, should by this means perpetually vary, but in such a manner, that it constantly approaches nearer and nearer to some determined proportion, and can also be brought at length in its approach nearer to this determined proportion than to any other, that can be assigned, but can never pass it: this determined proportion is then called the ultimate proportion, or the ultimate ratio of those varying quantities. [§124, p. 57]

To this definition of the sense, in which the term ultimate ratio, or ultimate proportion is to be understood, we must subjoin the following proposition: That all the ultimate ratios of the same varying ratio are the same with each other. [§125, p. 57]

The two definitions here set down, together with the general propositions annexed to them, comprehend the whole foundation of this method, we are now explaining. [§128, p. 58]

5. Introduction to the Doctrine of Fluxions

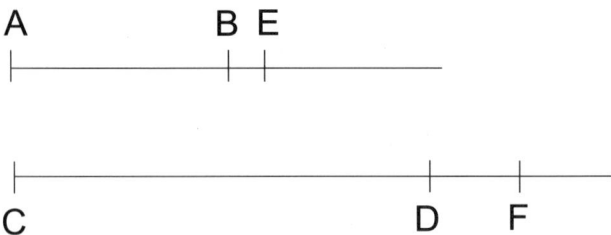

If now the augment BE be denoted by o, the augment DF will be denoted by

$$nx^{n-1}o + \frac{n \times \overline{n-1}}{2} \times x^{n-2}o^2 + \frac{n \times \overline{n-1} \times \overline{n-2}}{6} x^{n-3}o^3 + \&c.$$

And here it is obvious, that all the terms after the first taken together may be made less than any part whatever of the first, that shall be assigned. Consequently the proportion of the first term $nx^{n-1}o$ to the whole augment may be made to approach within any degree whatever of the proportion of equality; and therefore the ultimate proportion of

$$nx^{n-1}o + \frac{n \times \overline{n-1}}{2} \times x^{n-2}o^2 + \frac{n \times \overline{n-1} \times \overline{n-2}}{6} x^{n-3}o^3 + \&c.$$

to o, or of DF to BE, is that of $nx^{n-1}o$ only to o, or the proportion nx^{n-1} to 1. [§147, p. 70]

And now having justified Sir Isaac Newton's own demonstrations, we have not only shewn, that his doctrine of fluxions is an unerring guide in the solution of geometrical problems, but also that he himself had fully proved the certainty of this method. [§151, pp. 72–73]

... it must always be remembered, that the only use, which ought ever to be made of these momenta, is to compare them one with another, and for no other purpose than to determine the ultimate or prime proportion between the several increments or decrements, from whence they are deduced.

6
On a Semi-convergent Series

This about series.
James Stirling.

6.1 Introduction

In 1764 two articles by Bayes appeared in the same issue of the *Philosophical Transactions*. The second of these, *An Essay towards solving a Problem in the Doctrine of Chances*, is examined in the next chapter: the first, headed *A Letter from the late Reverend Mr. Thomas Bayes, F.R.S. to John Canton, M.A. and F.R.S.*, is the subject of consideration here.

In the half-century preceding the publication of this short note infinite series had been examined in a number of papers in the *Philosophical Transactions*, those of particular relevance[1] being by Abraham de Moivre [1714], James Dodson [1753], John Eames [1736], John Landen [1760], Thomas Simpson [1748], [1751], [1755], [1758], and Brook Taylor [1717]. Yet Bayes's 'Letter to Canton' was in a sense unique, addressing itself, as we show, to the question of the divergence of a certain series.

This is the only published work by Bayes on infinite series *per se*, although such series are used in his evaluation of the incomplete beta function (see Chapters 7 and 8). Other fragments dealing with this matter remain: manuscripts in the Stanhope collection in the Centre for Kentish Studies in Maidstone, and several pages in a Notebook (now the property of the Equitable Life Assurance Society) that has been attributed to Bayes: these are considered in later chapters. We show that Bayes explored the series for $\log z!$ in some detail in the Notebook[2], calculating sufficiently many coefficients in the series to become aware of its divergence.

XLIII. *A Letter from the late Reverend Mr.* Thomas Bayes, *F.R.S. to* John Canton, *M.A. and F.R.S.*

SIR,

Read Nov. 24, 1763.

IF the following observations do not seem to you to be too minute, I should esteem it as a favour, if you would please to communicate them to the Royal Society.

It has been asserted by some eminent mathematicians, that the sum of the logarithms of the numbers 1.2.3.4. &c. to z, is equal to $\frac{1}{2}\log.c + \overline{z + \frac{1}{2}} \times \log.z$ lessened by the series $z - \frac{1}{12z} + \frac{1}{360z^3} - \frac{1}{1260z^5} + \frac{1}{1680z^7} - \frac{1}{1188z^9} +$ &c. if c denote the circumference of a circle whose radius is unity. And it is true that this expression will very nearly approach to the value of that sum when z is large, and you take in only a proper number of the first terms of the foregoing series: but the whole series can never properly express any quantity at all; because after the 5th term the coefficients begin to increase, and they afterwards increase at a greater rate than what can be compensated by the increase of the powers of z, though z represent a number ever so large; as will be evident by considering the following manner in which the coefficients of that series may be formed. Take $a = \frac{1}{12}$, $5b = a^2$, $7c = 2ba$, $9d = 2ca + b^2$, $11e = 2da + 2cb$, $13f = 2ea + 2db + c^2$, $15g = 2fa + 2eb + 2dc$, and so on; then take A$= a$, B$= 2b$, C$= 2 \times 3 \times 4c$, D$= 2 \times 3 \times 4 \times 5 \times 6d$, E$= 2 \times 3 \times 4 \times 5 \times 6 \times 7 \times 8e$ and so on, and A, B, C, D, E, F, &c. will be the coefficients of the foregoing series: from whence it easily follows, that if any term in the series after the 3 first be called y, and its distance from the first term n, the next term immediately following will be greater than $\frac{n \times \overline{2n-1}}{6n+9} \times \frac{y}{z^2}$. Wherefore at length the subsequent terms of this series are greater than the preceding ones, and increase in infinitum, and therefore the whole series can have no ultimate value whatsoever.

Much less can that series have any ultimate value, which is deduced from it by taking $z = 1$, and is supposed to be equal to the logarithm of the square root of the periphery of a circle whose radius is unity; and what is said concerning the foregoing series is true, and appears to be so, much in the same manner, concerning the series for finding out the sum of the logarithms of the odd numbers 3.5.7. &c. ... z, and those that are given for finding out the sum of the infinite progressions, in which the several terms have the same numerator whilst their denominators are any certain power of numbers increasing in arithmetical proportion. But it is needless particularly to insist upon these, because one instance is sufficient to shew that those methods are not to be depended upon, from which a conclusion follows that is not exact.

6.3 Commentary

The fact that Bayes died in 1761, together with the heading and the opening lines, suggests that the letter was communicated to Canton only after Bayes's death, perhaps by Richard Price (who was responsible for the submission of the *Essay towards solving a Problem in the Doctrine of Chances*).

A series of the form $a_0 + a_1 x + a_2 x^2 + \cdots$ is said to be an *asymptotic series for* $f(x)$, near $x = 0$, if

$$f(x) = a_0 + a_1 x + \cdots + O(x^{n+1}),$$

for small x and each n. The series considered by Bayes is one of this type, and hence may be referred to as asymptotic. Since, however, the remainder of Bayes's series alternates in sign, I prefer to be more specific and to refer to it as *semi-convergent*[3].

For convenience, let us write the series under examination in the form

$$\sum_{1}^{z} \log k = \tfrac{1}{2} \log c + \left(z + \tfrac{1}{2}\right) \log z - \mathfrak{S}, \qquad (6.1)$$

where $c = 2\pi$ and

$$\mathfrak{S} = z - \frac{1}{12z} + \frac{1}{360z^3} - \frac{1}{1260z^5} + \frac{1}{1680z^7} - \frac{1}{1188z^9} + \text{etc.} \qquad (6.2)$$

The (absolute values of the) coefficients of the nth and $(n+1)$th terms on the right-hand side in (6.2) — call them y_n and y_{n+1} — satisfy[4]

$$y_{n+1}/y_n > (n-1)(2n-3)/(6n+3), \quad n \geq 3.$$

(This is given in a slightly different form by Bayes, but the equivalence of the two can be seen by replacing Bayes's n by $n - 1$.)

The Stirling–de Moivre series[5] 'for' $\log z!$ was of course well known (see Archibald [1926] and Tweedie [1922] for a discussion). However, it appears that neither de Moivre nor Stirling had appreciated the mere symbolism of the series: not having continued the series in (6.2) far enough, although the rule for continuance was known, they failed to spot the divergence (see Hardy [1949] for a discussion of such series). As W. Edwards Deming [1940] notes, Euler, some six years before the death of Bayes, had noted the divergence of the factorial series for $z = 1$ in §§157–159 of the second part of his *Institutiones calculi differentialis* of 1755. Here, from the expression

$$\sum_{1}^{x} \log k = x \log x - x + \frac{1}{2} \log x + \frac{A}{1.2x} - \frac{B}{3.4x^3} + \frac{C}{5.6x^5} - \frac{D}{7.8x^7} + \text{etc.} + \text{Const.}$$

where A, B, C, D are Bernoulli numbers, Euler concluded that the constant, found on setting $x = 1$, is given by

$$\text{Const.} = 1 - \frac{A}{1.2} + \frac{B}{3.4} - \frac{C}{5.6} + \frac{D}{7.8} - \text{etc.,}$$

'which series', he then noted, 'is, on account of excessive divergence, unsuitable for the obtaining of the value of [the constant] at all accurately' [Euler, 1755]. (The determination, not merely of an approximate value, but of the true value, is effected by using Wallis's product for $\pi/2$ in §158.)[6] It is, however, uncertain as to whether Euler appreciated that the series would diverge for *all* values of x, no matter how large[7].

Thus it may well be that Bayes was the first to note the asymptotic nature of the relationship between $\log z!$ and the right-hand side of (6.1), and one might well ask whether he was perhaps drawn to this conclusion by some knowledge of Euler's work. There is, however, no evidence in this paper of Bayes's having been aware of Euler's writings on this matter, and we show later that a similar negative conclusion must be drawn from the Notebook.

Although attention has usually been focused on the second paragraph of Bayes's *Letter*, something should be said about the third. Like Euler, Bayes notes the inappropriateness of taking $z = 1$ in the series, which would in fact amount to an attempt at evaluating 2π. Bromwich indeed writes

> The reader may be warned against attempting to deduce Wallis's product from Stirling's formula; this would be an illustration of the old fallacy *ignotum per ignotius*. [1931, p. 329]

Bayes also comments on the divergence of the appropriate series for $\log(2z - 1)!!$ (where $n!! = n(n-2)!!$ and $n!! = 1$ for $n = 0$ or 1). No details of this are provided; here is one proof. From

$$\log(2n)! = \log(2n-1)!! + \log(2^n n!)$$

we have, on using (6.1) and (6.2),

$$\log(2n-1)!! = \log(2n)! - n\log 2 - \log n!$$

$$= (n + \tfrac{1}{2})\log 2 + n\log n - n - (1/24n) + O(1/n^3).$$

(This series is in fact given in MacLaurin [1742, §840].)

The last matter on which Bayes comments concerns series typified by $\sum_n (k/n^r)$, one which is well known to converge for $r > 1$ and to diverge for $r \leq 1$. But his concern is not so much with the convergence or divergence

6.3. Commentary

of such series, but rather with their behaviour *when expressed in the form given before*. For example, consider the series[8]

$$\frac{1}{n^2} + \frac{1}{(n+1)^2} + \frac{1}{(n+2)^2} + \cdots = \frac{1}{n} + \frac{1}{2n^2} + \frac{B_1}{n^3} - \frac{B_2}{n^5} + \frac{B_3}{n^7} - \cdots \quad (6.3)$$

(see Bromwich [1931, p. 326]). Now

$$\frac{B_r}{B_{r-1}} = \frac{2r(2r-1)}{4\pi^2} \sum \frac{1}{n^{2r}} \bigg/ \sum \frac{1}{n^{2r-2}},$$

and, for $r > 3$,

$$\sum \frac{1}{n^{2r-2}} < 1 \bigg/ \left(1 - \frac{1}{2^4}\right) \quad \text{and} \quad \sum \frac{1}{n^{2r}} > 1.$$

Thus the ratio of successive terms in (6.3) gives

$$\frac{B_r}{n^{2r+1}} \bigg/ \frac{B_{r-1}}{n^{2r-1}} > \frac{15(2r-1)^2}{64\pi^2 n^2},$$

from which the increase of successive terms, after an appropriate value of r, is clear.

Although Stirling's series is suitable for numerical supputation, there are, of course, series for $\log z!$ that do converge. For instance, Dingle gives

$$\ln x! = -\gamma x + \sum_{\nu=1}^{\infty} \{(x/\nu) - \ln(1 + (x/\nu))\}$$

$$= -\gamma x + \sum_{s=2}^{\infty} s^{-1} \zeta(s)(-x)^s, \quad |x| < 1,$$

where γ is the Euler constant, $\gamma = \lim_{n\to\infty}(1 + \frac{1}{2} + \cdots + \frac{1}{n} - \log n)$, and $\zeta(\cdot)$ is the Riemann zeta function defined by $\zeta(s) = \sum_{n=1}^{\infty}(1/n^s)$. In 1917 William Burnside gave the rapidly convergent formula

$$\log N! = (N + \frac{1}{2})\log(N + \frac{1}{2}) - N + \log\sqrt{(2\pi/e)}$$

$$- \sum_{r=1}^{\infty} \sum_{n=N+1}^{\infty} \frac{1}{2r(2r+1)2^{2r}n^{2r}}.$$

7
The Essay on Chances

Rem viderunt, causam non viderunt.

Augustine.
Contra Pelagium.

7.1 Introduction

In the introduction to the first volume of their *Breakthroughs in Statistics* Kotz and Johnson listed eleven works, up to and including Galton's *Natural Inheritance*, that have had lasting and fundamental effects on the direction of statistical thought and practice. One of these is Thomas Bayes's *Essay towards solving a Problem in the Doctrine of Chances*.

Other opinions[1] of this tract were similarly laudatory: for example, Cajori[2] and von Wright[3], respectively, referred to the *Essay* as 'a meritorious article' and 'a masterpiece of mathematical elegance', and R.A. Fisher, who is well known for his criticism of views contrary to his own[4], described this work[5] as 'celebrated', and noted further that 'Bayes excelled in logical penetration', a comment with which we, after our examination of Bayes's work on fluxions, can agree wholeheartedly. Yet again, in his preface to Clero's translation of the *Essay*, Bru described the latter as 'l'un des textes les plus subtilement raffinés de toute la littérature statistique', and in his notes to the German translation of the *Essay* Timerding wrote of 'eine der merkwürdigsten mathematischen Entdeckungen'. And in his review of the facsimile edition of the *Essay* published by Molina and Deming in 1940, Lidstone wrote

> It is thought that readers of the essay will arise from its perusal with the definite impression that Bayes had a clear and penetrating mind with considerable power of algebraic manipulation, orderly arrangement and clear exposition. [p. 177]

Fisher indeed went even further than Kotz and Johnson in his assessment of the *Essay*: besides noting, in his *Statistical Methods and Scientific*

Inference[6], that '[Bayes's] mathematical contributions to the *Philosophical Transactions* show him to have been in the first rank of independent thinkers', he expressed the opinion in *The Design of Experiments*[7] that this *Essay* was sufficient to ensure Bayes a place in the history of science.

It has been suggested that Bayes's books and papers were left, on his death, to the Reverend William Johnson, his successor in the Presbyterian Meeting-house in Tunbridge Wells[8]. Whether this is indeed true is uncertain (there is no mention of such a bequest in Thomas's will, as we show in a later chapter), but it certainly appears that several of his papers found their way into the hands of Richard Price: a Notebook of Thomas's is still preserved in the muniment room of the Equitable Life Assurance Society in England (presumably through the interest shown in that society by Price and his nephew William Morgan), and Price was responsible for the communication of the *Essay* presented in this chapter to the Royal Society. D.O. Thomas [1977, p. 128] indeed suggests that it was his publication of Bayes's papers that set in train Price's 'increasing involvement in insurance, demography, and financial and political reform'.

Some authors have suggested that it was an overweening modesty (if we may be permitted such an oxymoron) that prevented Bayes from publishing his *Essay*. It would appear that this pudency was first ascribed to Thomas by William Morgan[9], who, in his biography of his uncle, wrote

> On the death of his friend Mr. Bayes of Tunbridge Wells in the year 1761 he [i.e., Price] was requested by the relatives of that truly ingenious man, to examine the papers which he had written on different subjects, and which his own modesty would never suffer him to make public. [1815, p. 24]

Whether such modesty, if indeed possessed by Bayes, should be considered a virtue is debatable — though we must of course recall Aristotle's saying, in the *Nicomachean Ethics*, Bk. iv, ch. 9, sec. 1, that 'Modesty cannot properly be described as a virtue, for it is a feeling rather than a disposition — a kind of fear of disrepute'. Whether it should even be regarded as meritorious is arguable; for although La Bruyère wrote 'La modestie est au mérite ce que les ombres sont aux figures dans un tableau: elle lui donne de la force et du relief' (*Les Caractères*, Sec. 2.), Sydney Smith could find the only connexion of modesty with merit in the observation that both began with an 'm'[10].

If Bayes's modesty was shown only by his reluctance to publish, then he was in good company, for Fermat and Newton shared that trait[11]. Raphael, in his preface to a re-issue of Price's *A Review of the Principal Questions in Morals*, wrote

As for acknowledgement of his debt to [Ralph] Cudworth (and others), Price is vastly more explicit than was customary in his time. His comparative scrupulousness in this regard is a mark of his modesty and candour alike, qualities that he possessed to a fault. [Price, 1787/1974, p. vi]

And Price himself wrote, in the introduction to this same work,

There are in truth none who are possessed of that cool and dispassionate temper, that freedom from all wrong byasses, that habit of attention and patience of thought, and, also, that penetration and sagacity of mind, which are the proper securities against error. How much then do modesty and diffidence become us? how open ought we to be to conviction, and how candid to those of different sentiments? [1787/1974, p. 10]

Commenting further on Price's modesty, Laboucheix wrote

Il fut modeste, cela est sûr; mais cette modestie n'était pas, en fait, incompatible avec la conscience parfaitement claire d'une force qui n'avait pas besoin d'être au grand jour pour s'affirmer, ni d'être au pouvoir pour s'exercer. [1970, p. 55]

It offends one's sense of the fitness of things to observe that in the Notes to both Laboucheix's original work and the 1982 English translation by Raphael & Raphael, and in the list of Price's Works, both the *Essay* and the *Supplement* are attributed to Price! Indeed, Note 39 on page 101 of Laboucheix's book[12] contains, in referring to the *Essay*, the words

Dans cet essai, Price résout, pour la première fois, le problème suivant: étant donné le nombre de fois qu'un événement inconnu est arrivé, et celui qu'il n'est pas arrivé, on demande quel sera la probabilité que cet événement arrive dès la première tentative, entre deux degrés quelconques donnés de probabilité.

A similar, though perhaps more guarded, statement was made by Jeremy, who wrote

In 1761 a mere accident led him [i.e., Price] to investigate a problem connected with the doctrine of chances. The result of his studies on this subject, extending over two years, was presented by his friend Mr. Canton to the Royal Society, and was published in their Transactions in 1763. [1885, p. 151]

Modern authors have suggested other reasons for the *Essay* not having been published by its author: thus Stigler [1986, p. 130] finds a possible ascription to the difficulty of the evaluation of the integral in the eighth proposition, and in his more recent *Statistics on the Table* he suggests that both Bayes and Jakob Bernoulli deferred publication because of 'the lack of an accepted standard of reference that could tell them how close to certainty is "good enough" ', both of them 'initiating work that would help set the standard of acceptable uncertainty' [Stigler, 1999, p. 375]. We show later in this chapter that Bayes did indeed attempt to derive bounds for the incomplete beta-integral used in the calculation of a posterior probability, but these bounds were rather wide. Good [1988] suggests several possible reasons for non-publication, viz. (i) the tacit assumption of a discrete uniform prior for the number of successes implies a continuous uniform prior for the (physical) probability of a success in each trial, (ii) the essential equivalence of these two priors when the number of trials is large, and (iii) the first ball is essentially a red herring.

Would we have 'Bayes's Theorem' today had Price not seen the need for the preservation of Bayes's paper and submitted it to John Canton? The answer is probably 'Yes', for in 1774 Laplace presented the fundamentals of inverse probability in his *Mémoire sur la probabilité des causes par les événements*. This work was apparently carried out in ignorance of Bayes's *Essay*, for it was only in subsequent papers that Laplace acknowledged the contribution to the topic made by his predecessor[13]. Von Mises in fact goes so far as to write

> It may be worth while mentioning that we owe to Bayes only the statement of the problem and the principle of its solution. The theorem itself was first formulated by Laplace[14].
> [1981, p. 126]

Timerding, however, attributed somewhat more than the formulation of the theorem to Laplace, writing

> Aber an *Bayes* liegt die Schuld nicht, daß diese Mißverständnisse aufgekommen sind, sie sind wesentlich in *Laplaces* Darstellung (Mémoires présentés par divers savants, tome VI, 1774) begründet.
> [1908, p. 44]

Although the importance of Laplace's work in this matter should not be underestimated, it is unfortunate that many have relied on it for information about Bayes's results, and have not consulted the original source. This was emphasized by Hogben, when he wrote

> In justice to Bayes, one may say that one form of the theorem attributed to him certainly does not occur in his works, and the other, which is a modern interpretation of the *ipsissima verba*, is dubiously consistent with the author's intentions. Such misunderstanding would be merely of lexicographical interest, were

it not also true that later generations have chosen to identify, fairly or falsely, the views of Laplace with those of Bayes himself. [1957, p. 115]

Forms of a result described as 'Bayes's Theorem' or 'Bayes's Rule' are almost as many as the writers who mention it. Perhaps most common is the form

$$P(B_i \mid A) = P(A \mid B_i)P(B_i)/\sum_i P(A \mid B_i)P(B_i),$$

though, as we show, this version does not appear in Bayes's work — it appears for the first time in Laplace's *Mémoire sur la probabilité des causes par les événements*. It is perhaps worthy of note that although Bayes's geometric approach yielded a result for continuous probabilities, Laplace began with an urn containing a finite number of balls and *then* passed to the case of an urn with an infinite number of balls.

We also find simple versions of the 'theorem' such as that given in de Finetti's *La Prevision: ses lois logiques, ses sources subjectives* as

On obtient comme corollaire immédiat de (1) [i.e. $P(E'.E'') = P(E').P(E''/E')$] le théorème de Bayes sous la forme

$$P(E''/E') = P(E'') \frac{P(E'/E'')}{P(E')},$$

qui se prête à l'énonce suivant, particulièrement significatif: *la probabilité de E', subordonée à E'', se modifie dans le même sens et dans la même mesure que la probabilité de E'' considérèe comme subordonée à E'.* [1937, p. 15]

This version perhaps supports Kyburg's view of the Rule as 'Basically ... a restatement of the multiplication axiom or the definition of conditional probability' [1970, p. 20]. Popper in fact goes so far as to write 'The various theorems which may be connected with the name of Bayes are all special cases of the division theorem' [1968, p. 288], and on the same page he gives the general form of these latter theorems as

$$_{\alpha.\beta}F''(\gamma) = {}_\alpha F''(\beta.\gamma)/{}_\alpha F''(\beta).$$

A form of the theorem, giving the posterior odds in terms of prior odds and likelihoods, may be found in Edwards [1972, p. 46].

More complicated versions, in which additional information is considered, have also been given — Good [1950, §3.3], for example[15], gives

$$\Pr[F \mid E.H] \propto \Pr[E \mid F.H]/\Pr[E \mid H].$$

A particularly sophisticated, and useful, version involving martingales is given by Bingham and Kiesel [1998, p. 167] as

> Assume $\widetilde{\mathbf{P}}$ is absolutely continuous with respect to $\mathbf{P}$ and Z its Radon-Nikodým derivative. If Y is bounded (or $\widetilde{\mathbf{P}}$-integrable) and $\mathcal{F}_t$ measurable, then
>
> $$\widetilde{\mathbf{E}}(Y \mid \mathcal{F}_s) \;=\; \frac{1}{Z(s)}\, \mathbf{E}(Y\, Z(t) \mid \mathcal{F}_s) \quad \text{a.s.} \quad s \leq t.$$

There are also far more esoteric versions, perhaps less accessible to the probabilist or mathematical statistician, as, for example, the forms given by von Wright [1960] in terms of sequences of properties, by Carnap [1962, pp. 330–333] in terms of c-functions (i.e., functions considered as explicata for probability), by Shafer [1976] in terms of Belief Functions, and by Chuaqui [1991, pp. 375–376] in the setting of logical languages.

We might also note that von Wright gives his form of the result, when the a priori probabilities are equal, as

> If, in the Problem of Bayes, the probabilities *a priori* are equal, then the probability *a posteriori* that a thing in which m of n given properties are present and the rest absent, will belong to a field of measurement, in which the probability of the individual properties is p_i, is the greater, the less the ratio $m:n$ differs from p_i.
> This we shall call the Inverse Principle of Maximum Probability. [1960, p. 209]

(Jeffreys finds the use of the term *a priori* unfortunate in this context, in view of its common connotation in logic. He suggests rather the word *prior*, since 'the prior probability is intended to express simply the probability at the start of an investigation and may have been influenced by many previous investigations' [1973, p. 31].)

Similar attention was drawn to maximization by Kyburg, who wrote 'Bayes' Rule (maximize expected utility) ...' [1970, pp. 187–188] and also Jeffrey, who wrote 'The Bayesian principle, then, is to *choose an act of maximum estimated desirability*' [1983, p. 1]. Bearing in mind the importance of Bayesian methods in statistical inference[16], one may be tempted on reading this statement (especially when it is taken in conjunction with the subject of Jeffrey's book) to consider the rôle of decision in scientific inference. Such consideration might, however, well be tempered in the light of the following statement from M.G. Kendall's obituary of R.A. Fisher,

> a man's attitude towards inference, like his attitude towards religion, is determined by his emotional make-up, not by reason or by mathematics.
>
> Fisher would have maintained (and in my opinion rightly) that inference in science is not a matter of decision, and that, in any case, criteria for choice in decision based on pay-offs of one kind or another are not available. [1963, p. 4]

Whether 'Bayes's Rule' or 'Bayes's Theorem' should be used to describe the result of interest here (and omitting the question of the correct form of the genitive or possessive) seems a matter of individual taste. Lucas goes so far as first to give the 'Rule' as

$$\text{Prob}[H(fg)] = \text{Prob}[F(hg)] \times \frac{\text{Prob}[H(g)]}{\text{Prob}[F(g)]},$$

(where, for instance, $H(g)$ represents the propositional function that a g — an individual in the universe of discourse of G-type things — has feature H), and then to give 'Bayes' two Inversion Theorems' [1970, p. 67]. The first of these, 'sometimes known as The Inverse Principle of Maximum Likelihood[17]', is framed as follows.

> If a propositional function has come true on m out of n occasions that are similar and independent of one another, then the most probable value for the probability of the propositional function is around m/n, provided that all hypotheses assigning a probability-value to the propositional function are each, apart from the information that the propositional function has come true on m out of n occasions, as probable as any other hypotheses. [1970, p. 151]

The second theorem, 'sometimes known as the Inverse Law of Great Numbers', is given as

> If the relative frequency with which a propositional function comes true is m/n, on a number, n, of occasions that are similar and independent of one another, then the probability of the hypothesis assigning to the propositional function the probability-value m/n approaches as a limit the maximum value 1, as n increases indefinitely, provided that the probability of that hypothesis, apart from the information that the relative frequency is m/n, does not have a probability-value 0. [1970, p. 156]

Bayes's Theorem is called *The Second Law of Large Numbers*[18] by von Mises, who frames it as follows.

7.1. Introduction

if, for a great number of different dice, each one has given n_1 results 6 in a game of n casts, where n is a sufficiently large number, *nearly all of these dice* must have almost the same limiting values of the relative frequencies of the result 6, namely, values only slightly different from the observed ratio n_1/n. [1981, p. 125]

The First Law of Large Numbers, or the Bernoulli–Poisson Theorem, is given by von Mises:

if the game of n casts is repeated again and again with the same die, and n is a sufficiently large number, *nearly all games* will yield nearly the same value of the ratio n_1/n. [1981, p. 125]

Howson and Urbach [1989, p. 26] also give two forms of Bayes's Theorem. The first of these, attributed by them to Bayes, is

$$P(a \mid b) = \frac{P(b \mid a) P(a)}{P(b)} \quad \text{where } P(a), P(b) > 0,$$

and the second, ascribed to Laplace, is given as follows.

If $P(b_1 \vee \ldots \vee b_n) = 1$ and $b_i \vdash\sim b_j$ for $i \neq j$, and $P(b_i), P(a) > 0$ then

$$P(b_k \mid a) = \frac{P(a \mid b_k) P(b_k)}{\sum_{i=1}^{n} P(a \mid b_i) P(b_i)}.$$

So much has been written on Bayesian statistics that I try to avoid saying more than is absolutely necessary here[19]. I also try to refrain from extended discussion of Inverse Probability *per se*[20], although it is a subject to which Bayes's *Essay* is fundamental; the reader may be referred to Dale [1999] for a study of this topic.

In his *A Treatise on Induction and Probability* von Wright notes that the relevance of Inverse Probability to induction displays itself in several manifestations: the first is the problem of the probability of causes[21]; the second is in connexion with the relevance of statistical samples to statistical laws[22] (via maximum probability and the law of large numbers); and the third problem concerns the probability of future events. Price's Appendix is certainly concerned with the third problem, and one can perhaps see something of the first in the *Essay* itself. There seems, however, to be nothing here that is relevant to the second, although that is the one that has become most important in modern statistics. Indeed, commenting on Bayes's problem in general, Savage remarks that[23]

> [it] is of the kind we now associate with Bayes's name, but it is confined from the outset to the special problem of drawing the Bayesian inference, not about an arbitrary sort of parameter, but about a "degree of probability" only. [1960]

Despite the restrictive nature of Bayes's original work, there have been those, in addition to Price, who saw its wider applicability. Thus Salmon wrote

> I shall claim, nevertheless, that Bayes' theorem provides the appropriate logical schema to characterize inferences designed to establish scientific hypotheses. The hypothetico-deductive method[24] is, I think, an oversimplification of Bayes' theorem. It is fallacious as it stands, but it can be rectified by supplementing it with the remaining elements required for application of Bayes' theorem [1967, p. 117]

whereas Fisher, perhaps, as in so many cases, seeing further than others, wrote in *The Design of Experiments*[25]:

> he [i.e., Bayes] seems to have been the first man in Europe to have seen the importance of developing an exact and quantitative theory of inductive reasoning, of arguing from observational facts to the theories which might explain them. [p. 6]

The relevance and successful application of Bayes's (and Price's) work on induction are debatable. For instance, de Finetti wrote

> The vital element in the inductive process, and the key to every constructive activity of the human mind, is then seen to be Bayes's theorem [1975, vol. II, p. 199]

and in commenting on Bayes's ninth proposition Hald says:

> [Bayes] has thus shown that a strictly logical probabilistic induction is possible for the physical experiment in question. He does not mention Hume, but this result may be considered as the first part of his refutation of Hume's postulate that induction can have no logical basis[26]. [1998, pp. 141–142]

Horwich finds at best a 'pseudo-solution to the problem of induction' in Bayesian methods [1982, pp. 63 et seqq.], noting the difficulty that arises in considering Goodman's horripilant *grue* paradox[27] (one that Boudot [1972, p. 84] finds to cast doubt on the correctness of Laplace's Rule of Succession).

7.1. Introduction

The simplest inductive argument, as least as far as we are concerned here, is perhaps typified by the drawing of balls from an urn. Suppose that the urn contains N balls, each either red or white, and each possible number of red balls having the same probability $1/(N+1)$. Each ball has the same chance of being taken on each draw, and each is replaced in the urn before another draw is made. If n draws have resulted in the appearance of r red balls, the probability of the next draw yielding another red is $(r+1)/(n+2)$. This is Laplace's *Rule of Succession*[28], or the Bayes–Laplace scheme, one in which the evaluation is the same as in the Pólya (perhaps more correctly Pólya–Eggenberger) urn model of contagion.

Hogben was somewhat dismissive of Laplace's derivation of the Rule of Succession, and he constructed a model that did in fact lead to Laplace's solution [1957, pp. 134–138, 149–153].

Similarly, the success of Price's application of probability in induction was viewed with some reserve by Raphael, who wrote

> ... when Price talks of probability, it is always mathematical probability, which seems to have had a special fascination for him.
> In the First Edition [of Price's *A Review of the Principal Questions in Morals*] the [concluding] chapter struck an even more ludicrous note, with elaborate fractions of the mathematical chances involved. This preoccupation with mathematical probability prevents Price from seeing the force of Hume's observations on probable reasoning. He was able to put his finger on Hume's mistakes, but Hume's greatness escaped him.
> All the same, Price is not very clear about probability in relation to induction (he knows a good deal about mathematical probability, as may be seen from some of his notes), and has failed to appreciate the force of Hume's arguments on this question. Thus his positive exposition of epistemology is faulty. [Price, 1787/1974, pp. xvi, xvii, xxiv]

Reichenbach rather more assertively writes[29]

> Similarly, the general inductive inference from observational data to the validity of a given scientific theory must be regarded as an inference in terms of Bayes's rule [1971, p. 95]

whereas Fetzer, on the other hand, seriously queries whether inductive rules of inference *are* probabilistic [1981, pp. 21–22].

What may the reader expect to find in this *Essay*? As regards probability, he will expect, of course, some or other version of what has become known as 'Bayes's Theorem'; and such an expectation will indeed be met.

In addition he will find a clear discussion of the binomial distribution[30], and if he should probe even deeper he will find, as Hailperin did, '(implicitly) the first occurrence of a probability logic result involving conditional probability' [1996, p. 14]. The *Essay* should also be of interest to mathematicians for the evaluation of the incomplete beta-function. We note too the use of approximations to various integrals made here and in the *Supplement* by both Bayes and Price, and the attention paid to the question of the error incurred in the making of such approximations. The *Essay*, then, mainly, and perhaps justly, remembered for the solution of the problem posed by Bayes, should also be remembered for its contribution to pure mathematics.

I believe, and I hope that such a belief will be justified in our Commentary in §7.3, that a strict interpretation and careful study of the *Essay* shows that Bayes's work is correct and reasonable as it stands, and that many of the criticisms that have been levelled at it over the years are due either to a misinterpretation of what Bayes himself wrote or to an extension of his results to cases and situations he did not consider.

In a paper delivered at the sesquicentennial celebrations of the Royal Statistical Society in 1984, D.J. Newell drew attention to the fact that the achievements of Florence Nightingale in hospital design rendered nugatory further developments in this field for several decades, and Isaac Newton's success in the development of the differential calculus effectively put a damper on developments in British mathematics for a hundred years. We have mentioned before the fact that Bayes's *Essay* attracted little attention until the twentieth century; perhaps, in the light of the preceding remarks, this was not a bad thing for statistics as a whole.

LII. *An Essay towards solving a Problem in the Doctrine of Chances. By the late Rev. Mr.* Bayes, *F.R.S. communicated by Mr.* Price, *in a Letter to* John Canton, *A.M. F.R.S.*

Dear Sir,

Read Dec. 23, 1763. I Now send you an essay which I have found among the papers of our deceased friend Mr. Bayes, and which, in my opinion, has great merit, and well deserves to be preserved. Experimental philosophy, you will find, is nearly interested in the subject of it; and on this account there seems to be particular reason for thinking that a communication of it to the Royal Society cannot be improper.

He had, you know, the honour of being a member of that illustrious Society, and was much esteemed by many in it as a very able mathematician. In an introduction which he has writ to this Essay, he says, that his design at first in thinking on the subject of it was, to find out a method by which we might judge concerning the probability that an event has to happen, in given circumstances, upon supposition that we know nothing concerning it but that, under the same circumstances, it has happened a certain number of times, and failed a certain other number of times. He adds, that he soon perceived that it would not be very difficult to do this, provided some rule could be found according to which we ought to estimate the chance that the probability for the happening of an event perfectly unknown, should lie between any two named degrees of probability, antecedently to any experiments made about it; and that it appeared to him that the rule must be to suppose the chance the same that it should lie between any two equidifferent degrees; which, if it were allowed, all the rest might be easily calculated in the common method of proceeding in the doctrine of chances. Accordingly, I find among his papers a very ingenious solution of this problem in this way. But he afterwards considered, that the *postulate* on which he had argued might not perhaps be looked upon by all as reasonable; and therefore he chose to lay down in another form the proposition in which he thought the solution of the problem is contained, and in a *scholium* to subjoin the reasons why he thought so, rather than to take into his mathematical reasoning any thing that might admit dispute. This, you will observe, is the method which he has pursued in this essay.

Every judicious person will be sensible that the problem now mentioned is by no means merely a curious speculation in the doctrine of chances, but necessary to be solved in order to a sure foundation for all our reasonings concerning past facts, and what is likely to be hereafter. Common sense is indeed sufficient to shew us that, from the observation of what has in former instances been the consequence of a certain cause or action, one may make a judgment what is likely to be the consequence of it another time, and that the larger number of experiments we have to support a conclusion, so much the more reason we have to take it for granted. But it is certain that we cannot determine, at least not to any nicety, in what degree repeated experiments confirm a conclusion, without the particular discussion of the beforementioned problem; which, therefore, is necessary to be considered by any one who would give a clear account of the strength of *analogical* or *inductive reasoning*; concerning, which at present, we seem to know little more than that it does sometimes in fact convince us, and at other times not; and that, as it is the means of [a]cquainting us with many truths, of which otherwise we must have been ignorant; so it is, in all probability, the source of many errors, which perhaps might in some measure be avoided, if the force that this sort of reasoning ought to have with us were more distinctly and clearly understood.

These observations prove that the problem enquired after in this essay is no less important than it is curious. It may be safely added, I fancy, that it is also a problem that has never before been solved. Mr. De Moivre, indeed, the great improver of this part of mathematics, has in his *Laws of chance**, after Bernoulli, and to a greater degree of exactness, given rules to find the probability there is, that if a very great number of trials be made concerning any event, **[372]** the proportion of the number of times it will happen, to the number of times it will fail in those trials, should differ less than by small assigned limits from the proportion of the probability of its happening to the probability of its failing in one single trial. But I know of no person who has shewn how to deduce the solution of the converse problem to this; namely, "the number of times an unknown event has "happened and failed being given, to find the chance that the probability "of its happening should lie somewhere between any two named degrees of "probability." What Mr. De Moivre has done therefore cannot be thought sufficient to make the consideration of this point unnecessary: especially, as the rules he has given are not pretended to be rigorously exact, except on supposition that the number of trials made are infinite; from whence it is not obvious how large the number of trials must be in order to make them exact enough to be depended on in practice.

Mr. De Moivre calls the problem he has thus solved, the hardest that can be proposed on the subject of chance. His solution he has applied to a very important purpose, and thereby shewn that those are much mistaken who have insinuated that the Doctrine of Chances in mathematics is of trivial consequence, and cannot have a place in any serious enquiry[†]. The

---

*See Mr. De Moivre's *Doctrine of Chances*, p. 243, &c. He has omitted the demonstrations of his rules, but these have been since supplied by Mr. Simpson at the conclusion of his treatise on *The Nature and Laws of Chance*.

[†]See his Doctrine of Chances, p. 252, &c.

purpose I mean is, to shew what reason we have for believing that there are in the constitution of things fixt laws according to which events happen, and that, therefore, the frame of the world must be **[373]** the effect of the wisdom and power of an intelligent cause; and thus to confirm the argument taken from final causes for the existence of the Deity. It will be easy to see that the converse problem solved in this essay is more directly applicable to this purpose; for it shews us, with distinctness and precision, in every case of any particular order or recurrency of events, what reason there is to think that such recurrency or order is derived from stable causes or regulations in nature, and not from any of the irregularities of chance.

The two last rules in this essay are given without the deductions of them. I have chosen to do this because these deductions, taking up a good deal of room, would swell the essay too much; and also because these rules, though of considerable use, do not answer the purpose for which they are given as perfectly as could be wished. They are however ready to be produced, if a communication of them should be thought proper. I have in some places writ short notes, and to the whole I have added an application of the rules in the essay to some particular cases, in order to convey a clearer idea of the nature of the problem, and to shew how far the solution of it has been carried.

I am sensible that your time is so much taken up that I cannot reasonably expect that you should minutely examine every part of what I now send you. Some of the calculations, particularly in the Appendix, no one can make without a good deal of labour. I have taken so much care about them, that I believe there can be no material error in any of them; but should there be any such errors, I am the only person who ought to be considered as answerable for them. **[374]**

Mr. Bayes has thought fit to begin his work with a brief demonstration of the general laws of chance. His reason for doing this, as he says in his introduction, was not merely that his reader might not have the trouble of searching elsewhere for the principles on which he has argued, but because he did not know whither to refer him for a clear demonstration of them. He has also made an apology for the peculiar definition he has given of the word *chance* or *probability*. His design herein was to cut off all dispute about the meaning of the word, which in common language is used in different senses by persons of different opinions, and according as it is applied to *past* or *future* facts. But whatever different senses it may have, all (he observes) will allow that an expectation depending on the truth of any *past* fact, or the happening of any *future* event, ought to be estimated so much the more valuable as the fact is more likely to be true, or the event more likely to happen. Instead therefore, of the proper sense of the word *probability*, he has given that which all will allow to be its proper measure in every case where the word is used. But it is time to conclude this letter. Experimental philosophy is indebted to you for several discoveries and improvements; and, therefore, I cannot help thinking that there is a peculiar propriety in directing to you the following essay and appendix. That your enquiries may be rewarded with many further successes, and that you may enjoy every every valuable blessing, is the sincere wish of, Sir,

your very humble servant,

Newington-Green,
Nov. 10, 1763.

Richard Price.

[375]

PROBLEM.

Given the number of times in which an unknown event has happened and failed: *Required* the chance that the probability of its happening in a single trial lies somewhere between any two degrees of probability that can be named.

SECTION I.

DEFINITION 1. Several events are *inconsistent*, when if one of them happens, none of the rest can.

2. Two events are *contrary* when one, or other of them must; and both together cannot happen.

3. An event is said to *fail*, when it cannot happen; or, which comes to the same thing, when its contrary has happened.

4. An event is said to be determined when it has either happened or failed.

5. The *probability of any event* is the ratio between the value at which an expectation depending on the happening of the event ought to be computed, and the value of the thing expected upon it's happening.

6. By *chance* I mean the same as probability.

7. Events are independent when the happening of any one of them does neither increase nor abate the probability of the rest.

PROP. 1.

When several events are inconsistent the probability of the happening of one or other of them is the sum of the probabilities of each of them. **[376]**

Suppose there be three such events, and which ever of them happens I am to receive N, and that the probability of the 1st, 2d, and 3d are respectively $\frac{a}{N}$, $\frac{b}{N}$, $\frac{c}{N}$. Then (by the definition of probability) the value of my expectation from the 1st will be a, from the 2d b, and from the 3d c. Wherefore the value of my expectations from all three will be $a+b+c$. But the sum of my expectations from all three is in this case an expectation of receiving N upon the happening of one or other of them. Wherefore (by definition 5) the probability of one or other of them is $\frac{a+b+c}{N}$ or $\frac{a}{N}+\frac{b}{N}+\frac{c}{N}$. The sum of the probabilities of each of them.

Corollary. If it be certain that one or other of the three events must happen, then $a+b+c = N$. For in this case all the expectations together amounting to a certain expectation of receiving N, their values together must be equal to N. And from hence it is plain that the probability of an event added to the probability of its failure (or of its contrary) is the ratio of equality. For these are two inconsistent events, one of which necessarily happens. Wherefore if the probability of an event is $\frac{P}{N}$ that of it's failure will be $\frac{N-P}{N}$.

PROP. 2.

If a person has an expectation depending on the happening of an event, the probability of the event is to the probability of its failure as his loss if it fails to his gain if it happens.

Suppose a person has an expectation of receiving N, depending on an event the probability of which **[377]** is $\frac{P}{N}$. Then (by definition 5) the value of his expectation is P, and therefore if the event fail, he loses that which in value is P; and if it happens he receives N, but his expectation ceases. His gain therefore is N − P. Likewise since the probability of the event is $\frac{P}{N}$, that of its failure (by corollary prop. 1) is $\frac{N-P}{N}$. But $\frac{P}{N}$ is to $\frac{N-P}{N}$ as P is to N − P, i.e. the probability of the event is to the probability of it's failure, as his loss if it fails to his gain if it happens.

PROP. 3.

The probability that two subsequent events will both happen is a ratio compounded of the probability of the 1st, and the probability of the 2d on supposition the 1st happens.

Suppose that, if both events happen, I am to receive N, that the probability both will happen is $\frac{P}{N}$, that the 1st will is $\frac{a}{N}$ (and consequently that the 1st will not is $\frac{N-a}{N}$) and that the 2d will happen upon supposition the 1st does is $\frac{b}{N}$. Then (by definition 5) P will be the value of my expectation, which will become b if the 1st happens. Consequently if the 1st happens, my gain by it is $b - P$, and if it fails my loss is P. Wherefore, by the foregoing proposition, $\frac{a}{N}$ is to $\frac{N-a}{N}$, i.e. a is to $N - a$ as P is to $b - P$. Wherefore (componendo inversè) a is to N as P is to b. But the ratio of P to N is compounded of the ratio of P to b, and that of b to N. Wherefore the **[378]** same ratio of P to N is compounded of the ratio of a to N and that of b to N, i.e. the probability that the two subsequent events will both happen is compounded of the probability of the 1st and the probability of the 2d on supposition the 1st happens.

Corollary. Hence if of two subsequent events the probability of the 1st be $\frac{a}{N}$, and the probability of both together be $\frac{P}{N}$, then the probability of the 2d on supposition the 1st happens is $\frac{P}{a}$.

PROP. 4.

If there be two subsequent events to be determined every day, and each day the probability of the 2d is $\frac{b}{N}$ and the probability of both $\frac{P}{N}$, and I am to receive N if both the events happen the 1st day on which the 2d does; I say, according to these conditions, the probability of my obtaining N is $\frac{P}{b}$. For if not, let the probability of my obtaining N be $\frac{x}{N}$ and let y be to x as N − b to N. Then since $\frac{x}{N}$ is the probability of my obtaining N (by definition 1) x is the value of my expectation. And again, because according to the foregoing conditions the 1st day I have an expectation of obtaining N depending on the happening of both the events together, the probability of which is $\frac{P}{N}$, the value of this expectation is P. Likewise, if this coincident should not happen I have an expectation of being reinstated in my former circumstances, i.e. of receiving that which in value is x depending **[379]** on the failure of the 2d event the probability of which (by cor. prop. 1) is $\frac{N-b}{N}$ or $\frac{y}{x}$, because y is to x as N − b to N. Wherefore since x is the thing expected and $\frac{y}{x}$ the probability of obtaining it, the value of this expectation is y. But these two last expectations together are evidently the same with my original expectation, the value of which is x, and therefore P + y = x. But y is to x as N − b is to N. Wherefore x is to P as N is to b, and $\frac{x}{N}$ (the probability of my obtaining N) is $\frac{P}{b}$.

Cor. Suppose after the expectation given me in the foregoing proposition, and before it is at all known whether the 1st event has happened or not, I should find that the 2d event has happened; from hence I can only infer that the event is determined on which my expectation depended, and have no reason to esteem the value of my expectation either greater or less than it was before. For if I have reason to think it less, it would be reasonable for me to give something to be reinstated in my former circumstances, and this over and over again as often as I should be informed that the 2d event had happened, which is evidently absurd. And the like absurdity plainly follows if you say I ought to set a greater value on my expectation than before, for then it would be reasonable for me to refuse something if offered me upon condition I would relinquish it, and be reinstated in my former circumstances; and this likewise over and over again as often as (nothing being known concerning the 1st event) it should appear that the 2d had happened. Notwithstanding therefore this discovery that the 2d **[380]** event has happened, my expectation ought to be esteemed the same in value as before, i.e. x, and consequently the probability of my obtaining N is (by definition 5) still $\frac{x}{N}$ or $\frac{P}{b}$*. But after this discovery the probability of my obtaining N is the probability that the 1st of two subsequent events has happened upon the supposition that the 2d has, whose probabilities were as before specified. But the probability that an event has happened is the same as the probability I have to guess right if I guess it has happened. Wherefore the following proposition is evident.

PROP. 5.

If there be two subsequent events, the probability of the 2d $\frac{b}{N}$ and the probability of both together $\frac{P}{N}$, and it being 1st discovered that the 2d event has happened, from hence I guess that the 1st event has also happened, the probability I am in the right is $\frac{P}{b}$†. [381]

PROP. 6.

The probability that several independent events shall all happen is a ratio compounded of the probabilities of each.

For from the nature of independent events, the probability that any one happens is not altered by the happening or failing of any of the rest, and consequently the probability that the 2d event happens on supposition the 1st does is the same with its original probability; but the probability that any two events happen is a ratio compounded of the probability of the 1st event, and the probability of the 2d on supposition the 1st happens by prop. 3. Wherefore the probability that any two independent events both happen is a ratio compounded of the probability of the 1st and the probability of the 2d. And in like manner considering the 1st and 2d event together as

---

*What is here said may perhaps be a little illustrated by considering that all that can be lost by the happening of the 2d event is the chance I should have had of being reinstated in my former circumstances, if the event on which my expectation depended had been determined in the manner expressed in the proposition. But this chance is always as much *against* me as it is *for* me. If the 1st event happens, it is *against* me, and equal to the chance for the 2d event's failing. If the 1st event does not happen, it is *for* me, and equal also to the chance for the 2d event's failing. The loss of it, therefore, can be no disadvantage.

†What is proved by Mr. Bayes in this and the preceding proposition is the same with the answer to the following question. What is the probability that a certain event, when it happens, will be accompanied with another to be determined at the same time? In this case, as one of the events is given, nothing can be due for the expectation of it; and, consequently, the value of an expectation depending on the happening of both events must be the same with the value of an expectation depending on the happening of one of them. In other words; the probability that, when one of two events happens, the other will, is the same with the probability of this other. Call x then the probability of this other, and if $\frac{b}{N}$ be the probability of the given event, and $\frac{p}{N}$ the probability of both, because $\frac{p}{N} = \frac{b}{N} \times x$, $x = \frac{p}{b}$ = the probability mentioned in these propositions.

one event; the probability that three independent events all happen is a ratio compounded of the probability that the two 1st both happen and the probability of the 3d. And thus you **[382]** may proceed if there be ever so many such events; from whence the proposition is manifest.

Cor. 1. If there be several independent events, the probability that the 1st happens the 2d fails, the 3d fails and the 4th happens, &c. is a ratio compounded of the probability of the 1st, and the probability of the failure of the 2d, and the probability of the failure of the 3d, and the probability of the 4th, &c. For the failure of an event may always be considered as the happening of its contrary.

Cor. 2. If there be several independent events, and the probability of each one be a, and that of its failing be b, the probability that the 1st happens and the 2d fails, and the 3d fails and the 4th happens, &c. will be $abba$, &c. For, according to the algebraic way of notation, if a denote any ratio and b another, $abba$ denotes the ratio compounded of the ratios a, b, b, a. This corollary therefore is only a particular case of the foregoing.

Definition. If in consequence of certain data there arises a probability that a certain event should happen, its happening or failing, in consequence of these data, I call it's happening or failing in the 1st trial. And if the same data be again repeated, the happening or failing of the event in consequence of them I call its happening or failing in the 2d trial; and so on as often as the same data are repeated. And hence it is manifest that the happening or failing of the same event in so many diffe-[rent ?] trials, is in reality the happening or failing of so many distinct independent events exactly similar to each other. **[383]**

P R O P. 7.

If the probability of an event be a, and that of its failure be b in each single trial, the probability of its happening p times, and failing q times in $p+q$ trials is $E\,a^p b^q$ if E be the coefficient of the term in which occurs $a^p b^q$ when the binomial $\overline{a+b}\,|^{p+q}$ is expanded.

For the happening or failing of an event in different trials are so many independent events. Wherefore (by cor. 2. prop. 6.) the probability that the event happens the 1st trial, fails the 2d and 3d, and happens the 4th, fails the 5th, &c. (thus happening and failing till the number of times it happens be p and the number it fails be q) is $abbab$ &c. till the number of a's be p and the number of b's be q, that is; 'tis $a^p b^q$. In like manner if you consider the event as happening p times and failing q times in any other particular order, the probability for it is $a^p b^q$; but the number of different orders according to which an event may happen or fail, so as in all to happen p times and fail q, in $p+q$ trials is equal to the number of permutations that $aaaa\,bbb$ admit of when the number of a's is p, and

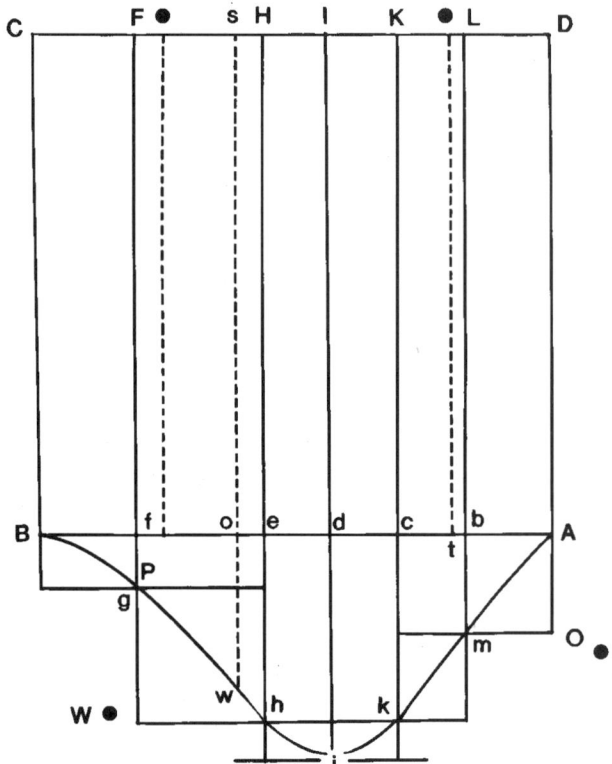

the number of b's is q. And this number is equal to E, the coefficient of the term in which occurs $a^p b^q$ when $\overline{a+b}\,|^{p+q}$ is expanded. The event therefore may happen p times and fail q in $p+q$ trials E different ways and no more, and its happening and failing these several different ways are so many inconsistent events, the probability for each of which is $a^p b^q$, and therefore by **[384]** prop. 1. the probability that some way or other it happens p times and fails q times in $p+q$ trials is $\mathrm{E}\,a^p b^q$.

SECTION II.

Postulate. 1. I suppose the square table or plane ABCD to be so made and levelled, that if either of the balls o or W be thrown upon it, there shall be the same probability that it rests upon any one equal part of the plane as another, and that it must necessarily rest somewhere upon it.

2. I suppose that the ball W shall be 1st thrown, and through the point

where it rests a line os shall be drawn parallel to AD, and meeting CD and AB in s and o; and that afterwards the ball O shall be thrown $p+q$ or n times, and that its resting between AD and os after a single throw be called the happening of the event M in a single trial. These things supposed,

Lem. 1. The probability that the point o will fall between any two points in the line AB is the ratio of the distance between the two points to the whole line AB.

Let any two points be named, as f and b in the line AB, and through them parallel to AD draw fF, bL meeting CD in F and L. Then if the rectangles Cf, Fb, LA are **[385]** commensurable to each other, they may each be divided into the same equal parts, which being done, and the ball W thrown, the probability it will rest somewhere upon any number of these equal parts will be the sum of the probabilities it has to rest upon each one of them, because its resting upon any different parts of the plane AC are so many inconsistent events; and this sum, because the probability it should rest upon any one equal part as another is the same, is the probability it should rest upon any one equal part multiplied by the number of parts. Consequently, the probability there is that the ball W should rest somewhere upon Fb is the probability it has to rest upon one equal part multiplied by the number of equal parts in Fb; and the probability it rests somewhere upon Cf or LA, i.e. that it dont rest upon Fb (because it must rest somewhere upon AC) is the probability it rests upon one equal part multiplied by the number of equal parts in Cf, LA taken together. Wherefore, the probability it rests upon Fb is to the probability it dont as the number of equal parts in Fb is to the number of equal parts in Cf, LA together, or as Fb to Cf, LA together, or as fb to Bf, Ab together. Wherefore the probability it rest upon Fb is to the probability it dont as fb to Bf, Ab together. And (*componendo inverse*) the probability it rests upon Fb is to the probability it rests upon Fb added to the probability it dont, as fb to AB, or as the ratio of fb to AB to the ratio of AB to AB. But the probability of any event added to the probability of its failure is the ratio of equality; wherefore, the probability it rest upon Fb is to the ratio of equality as the ratio of fb to AB to the ratio of AB to AB, or the ratio of equality; and therefore the probability it rest upon **[386]** Fb is the ratio of fb to AB. But *ex hypothesi* according as the ball W falls upon Fb or not the point o will lie between f and b or not, and therefore the probability the point o will lie between f and b is the ratio of fb to AB.

Again; if the rectangles Cf, Fb, LA are not commensurable, yet the last mentioned probability can be neither greater nor less than the ratio of fb to AB; for, if it be less, let it be the ratio of fc to AB, and upon the line fb take the points p and t, so that pt shall be greater than fc, and the three lines Bp, pt, tA commensurable (which it is evident may be always done by dividing AB into equal parts less than half cb, and taking p and t the nearest points of division to f and c that lie upon fb). Then because Bp, pt, tA are commensurable, so are the rectangles Cp, Dt, and that upon pt compleating the square AB. Wherefore, by what has been said, the probability that the point o will lie between p and t is the ratio of pt to AB. But if it lies between p and t it must lie between f and b. Wherefore, the probability it should lie between f and b cannot be less than the ratio of pt to AB, and therefore must be greater than the ratio of fc to AB (since pt is greater than fc). And after the same manner you may prove that the forementioned probability cannot be greater than the ratio of fb to AB, it must therefore be the same.

Lem. 2. The ball W having being thrown, and the line os drawn, the probability of the event M in a single trial is the ratio of Ao to AB.

For, in the same manner as in the foregoing lemma, the probability that the ball o being thrown shall **[387]** rest somewhere upon Do or between AD and so is is the ratio of Ao to AB. But the resting of the ball o between AD and so after a single throw is the happening of the event M in a single trial. Wherefore the lemma is manifest.

P R O P. 8.

If upon BA you erect the figure B$ghikm$A whose property is this, that (the base BA being divided into any two parts, as Ab, and Bb and at the point of division b a perpendicular being erected and terminated by the figure in m; and y, x, r representing respectively the ratio of bm, Ab, and Bb to AB, and E being the coefficient of the term in which occurs $a^p b^q$ when the binomial $\overline{a+b}|^{p+q}$ is expanded) $y =$E$\,x^p r^q$. I say that before the ball W is thrown, the probability the point o should fall between f and b, any two points named in the line AB, and withall that the event M should happen p times and fail q in $p+q$ trials, is the ratio of $fghikmb$, the part of the figure B$ghikm$A intercepted between the perpendiculars fg, bm raised upon the line AB, to CA the square upon AB.

D E M O N S T R A T I O N.

For if not; 1st let it be the ratio of D a figure greater than $fghikmb$ to CA, and through the points e, d, c draw perpendiculars to fb meeting the curve AmigB in h, i, k; the point d being so placed that di shall be the longest of the perpendiculars **[388]** terminated by the line fb, and the curve AmigB; and the points e, d, c being so many and so placed that the rectangles, bk, ci, ei, fh taken together shall differ less from $fghikmb$ than D does; all which may be easily done by the help of the equation of the curve, and the difference between D and the figure $fghikmb$ given. Then since di is the longest of the perpendicular ordinates that insist upon fb, the rest will gradually decrease as they are farther and farther from it on each side, as appears from the construction of the figure, and consequently eh is greater than gf or any other ordinate that insists upon ef.

Now if Ao were equal to Ae, then by lem. 2. the probability of the event M in a single trial would be the ratio of Ae to AB, and consequently by cor. Prop. 1. the probability of it's failure would be the ratio of Be to AB. Wherefore, if x and r be the two forementioned ratios respectively, by Prop. 7. the probability of the event M happening p times and failing q in $p+q$ trials would be $\mathrm{E}\,x^p r^q$. But x and r being respectively the ratios of Ae to AB and Be to AB, if y is the ratio of eh to AB, then, by construction of the figure AiB, $y = \mathrm{E}\,x^p r^q$. Wherefore, if Ao were equal to Ae the probability of the event M happening p times and failing q in $p+q$ trials would be y, or the ratio of eh to AB. And if Ao were equal to Af, or were any mean between Ae and Af, the last mentioned probability for the same reasons would be the ratio of fg or some other of the ordinates insisting upon ef, to AB. But eh is the greatest of all the ordinates that insist upon ef. Wherefore, upon supposition the point should lie **[389]** any where between f and e, the probability that the event M happens p times and fails q in $p+q$ trials can't be greater than the ratio of eh to AB. There then being these two subsequent events, the 1st that the point o will lie between e and f, the 2d that the event M will happen p times and fail q in $p+q$ trials, and the probability of the 1st (by lemma 1st) is the ratio of ef to AB, and upon supposition the 1st happens, by what has been now proved, the probability of the 2d cannot be greater than the ratio of eh to AB, it evidently follows (from Prop. 3.) that the probability both together will happen cannot be greater than the ratio compounded of that of ef to AB and that of eh to AB, which compound ratio is the ratio of fh to CA. Wherefore, the probability that the point o will lie between f and e, and the event M happen p times and fail q, is not greater than the ratio of fh to CA. And in like, manner the probability the point o will lie between e and d, and the event M happen and fail as before, cannot be greater than the ratio of ei to CA. And again, the probability the point o will lie between d and c, and the event M happen and fail as before, cannot be greater than the ratio of ci to CA. And lastly, the probability that the point o will lie between c and b, and the event M happen and fail as before, cannot be greater than the ratio of bk to CA. Add now all these several probabilities together, and their sum (by Prop. 1.) will be the probability that the point will lie somewhere between f and b, and the event M happen p times and fail q in $p+q$ trials. Add likewise the correspondent ratios together, and their sum will be the ratio of the sum of the antecedents **[390]** to their common consequent, i.e. the ratio of fh, ei, ci, bk together to CA; which ratio is less than that of D to CA, because D is greater than fh, ei, ci, bk together. And therefore, the probability that the point o will lie between f and b, and withal that the event M will happen p times and fail q in $p+q$ trials, is *less* than the ratio of D to CA; but it was supposed the same which is absurd. And in like manner, by inscribing rectangles within the figure, as eg, dh, dk, cm, you may prove that the last mentioned probability is *greater* than the ratio of any figure less than $fghikmb$ to CA.

Wherefore, that probability must be the ratio of $fghikmb$ to CA.

Cor. Before the ball W is thrown the probability that the point o will lie somewhere between A and B, or somewhere upon the line AB, and withal that the event M will happen p times, and fail q in $p+q$ trials is the ratio of the whole figure AiB to CA. But it is certain that the point o will lie somewhere upon AB. Wherefore, before the ball W is thrown the probability the event M will happen p times and fail q in $p+q$ trials is the ratio of AiB to CA.

P R O P. 9.

If before any thing is discovered concerning the place of the point o, it should appear that the event M had happened p times and failed q in $p+q$ trials, and from hence I guess that the point o lies between any two points in the line AB, as f and b, and consequently that the probability of the event M in a single trial was somewhere between the ratio of Ab to AB and that of Af to AB: the probability I am in **[391]** the right is the ratio of that part of the figure AiB described as before which is intercepted between perpendiculars erected upon AB at the points f and b, to the whole figure AiB.

For, there being these two subsequent events, the first that the point o will lie between f and b; the second that the event M should happen p times and fail q in $p+q$ trials; and (by cor. prop. 8.) the original probability of the second is the ratio of AiB to CA, and (by prop. 8.) the probability of both is the ratio of $fghimb$ to CA; wherefore (by prop. 5) it being first discovered that the second has happened, and from hence I guess that the first has happened also, the probability I am in the right is the ratio of $fghimb$ to AiB, the point which was to be proved.

Cor. The same things supposed, if I guess that the probability of the event M lies somewhere between o and the ratio of Ab to AB, my chance to be in the right is the ratio of Abm to AiB.

SCHOLIUM.

From the preceding proposition it is plain, that in the case of such an event as I there call M, from the number of times it happens and fails in a certain number of trials, without knowing any thing more concerning it, one may give a guess whereabouts it's probability is, and, by the usual methods computing the magnitudes of the areas there mentioned, see the chance that the guess is right. And that the same rule is the proper one to be used in the case of an event concerning the probability of which **[392]** we absolutely know nothing antecedently to any trials made concerning it, seems to appear from the following consideration; viz. that concerning such an event I have no reason to think that, in a certain number of trials, it should rather happen any one possible number of times than another. For, on this account, I may justly reason concerning it as if its probability had been at first unfixed, and then determined in such a manner as to give me no reason to think that, in a certain number of trials, it should rather happen any one possible number of times than another. But this is exactly the case of the event M. For before the ball W is thrown, which determines it's probability in a single trial, (by cor. prop. 8.) the probability it has to happen p times and fail q in $p+q$ or n trials is the ratio of AiB to CA, which ratio is the same when $p+q$ or n is given, whatever number p is; as will appear by computing the magnitude of AiB by the method* of fluxions. And consequently before the place of the point o is discovered or the number of times the event M has happened in n trials, I can have no reason to think it should rather happen one possible number of times than another.

In what follows therefore I shall take for granted that the rule given concerning the event M in prop. 9. is also the rule to be used in relation to any event concerning the probability of which nothing **[393]** at all is known antecedently to any trials made or observed concerning it. And such an event I shall call an unknown event.

Cor. Hence, by supposing the ordinates in the figure AiB to be contracted in the ratio of E to one, which makes no alteration in the proportion of the parts of the figure intercepted between them, and applying what is said of the event M to an unknown event, we have the following proposition, which gives the rules for finding the probability of an event from the number of times it actually happens and fails.

P R O P. 10.

If a figure be described upon any base AH (Vid. Fig.) having for it's equation $y = x^p r^q$; where y, x, r are respectively the ratios of an ordinate of the figure insisting on the base at right angles, of the segment of the base intercepted between the ordinate and A the beginning of the base, and of the other segment of the base lying between the ordinate and the point H,

---

*It will be proved presently in art. 4. by computing in the method here mentioned that AiB contracted in the ratio of E to 1 is to CA as 1 to $\overline{n+1} \times$E: from whence it plainly follows that, antecedently to this contraction, AiB must be to CA in the ratio of 1 to $n+1$, which is a constant ratio when n is given, whatever p is.

7. The Essay on Chances

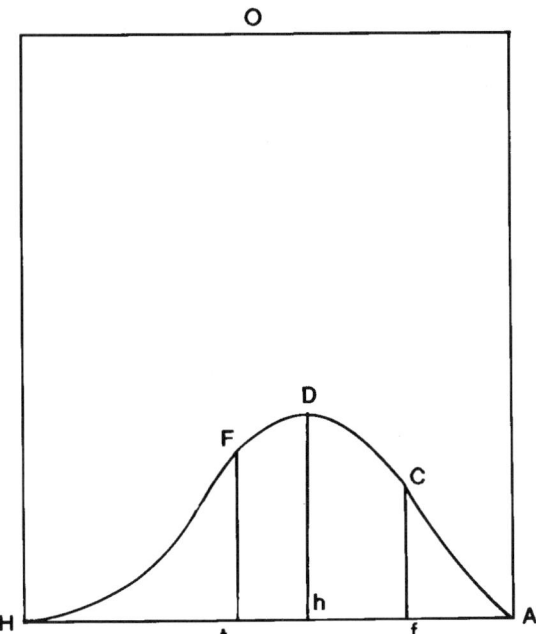

to the base as their common consequent. I say then that if an unknown event has happened p times and failed q in $p+q$ trials, and in the base AH taking any two points as f and t you erect the ordinates fc, tF at right angles with it, the chance that the probability of the event lies somewhere between the ratio of Af to AH and that of At to AH, is the ratio of tFCf, that part of the before-described figure which is intercepted between the two ordinates, to ACFH the whole figure insisting on the base AH.

This is evident from prop. 9. and the remarks made in the foregoing scholium and corollary. **[394]**

Now, in order to reduce the foregoing rule to practice, we must find the value of the area of the figure described and the several parts of it separated, by ordinates perpendicular to its base. For which purpose, suppose AH $= 1$ and HO the square upon AH likewise $= 1$, and CF will be $= y$, and A$f = x$, and H$f = r$, because y, x and r denote the ratios of Cf, Af, and Hf respectively to AH. And by the equation of the curve $y = x^p r^q$ and (because Af + fH = AH) $r + x = 1$. Wherefore $y = x^p \times \overline{1-x}\,]^q = x^p - qx^{p+1} + q \times \frac{q-1}{2} \times x^{p+2} - q \times \frac{q-1}{2} \times \frac{q-2}{3} \times x^{p+3} + \&c.$
Now the abscisse being x and the ordinate x^p the correspondent area is

$\frac{x^{p+1}}{p+1}$ (by prop. 10. cas. 1. Quadrat. Newt.)* and the ordinate being qx^{p+1} the area is $\frac{qx^{p+2}}{p+2}$; and in like manner **[395]** of the rest. Wherefore, the abscisse being x and the ordinate y or $x^p - qx^{p+1}+$ &c. the correspondent area is $\frac{x^{p+1}}{p+1} - \frac{q}{p+2} \times x^{p+2} + q \times \frac{q-1}{2} \times \frac{x^{p+3}}{p+3} - q \times \frac{q-1}{2} \times \frac{q-2}{3} \times \frac{x^{p+4}}{p+4} +$ &c. Wherefore, if $x = \mathrm{A}f = \frac{\mathrm{A}f}{\mathrm{AH}}$, and $y = \mathrm{C}f = \frac{\mathrm{C}f}{\mathrm{AH}}$, then $\mathrm{AC}f = \frac{\mathrm{AC}f}{\mathrm{HO}} = \frac{x^{p+1}}{p+1} - \frac{q}{p+2} \times x^{p+2} + q \times \frac{q-1}{2} \times \frac{x^{p+3}}{p+3} -$ &c.

From which equation, if q be a small number, it is easy to find the value of the ratio of $\mathrm{AC}f$ to HO. and in like manner as that was found out, it will appear that the ratio of $\mathrm{HC}f$ to HO is $\frac{r^{q+1}}{q+1} - p \times \frac{r^{q+2}}{q+2} + p \times \frac{p-1}{2} \times \frac{r^{q+3}}{q+3} - p \times \frac{p-1}{2} \times \frac{p-2}{3} \times \frac{r^{q+4}}{q+4}$ &c. which series will consist of few terms and therefore is to be used when p is small.

2. The same things supposed as before, the ratio of $\mathrm{AC}f$ to HO is $\frac{x^{p+1}r^q}{p+1} + \frac{q}{p+1} \times \frac{x^{p+2}r^{q-1}}{p+2} + \frac{q}{p+1} \times \frac{q-1}{p+2} \times \frac{x^{p+3}r^{q-2}}{p+3} + \frac{q}{p+1} \times \frac{q-1}{p+2} \times \frac{q-2}{p+3} \times \frac{x^{p+4}r^{q-3}}{p+4} +$ **[396]** &c. $+ \frac{x^{n+1}}{n+1} \times \frac{q}{p+1} \times \frac{q-1}{p+2} \times$ &c. $\times \frac{1}{n}$ where $n = p + q$. For this series is the same with $\frac{x^{p+1}}{p+1} - q \times \frac{x^{p+2}}{p+2}$ &c. set down in Art. 1st. as the value of the ratio of $\mathrm{AC}f$ to HO; as will easily be seen by putting in the former instead of r its value $1 - x$, and expanding the terms and ordering them according to the powers of x. Or, more readily, by comparing the fluxions of the two series, and in the former instead of $\dot{r}$ substituting $-\dot{x}$ †. **[397]**

3. In like manner, the ratio of $\mathrm{HC}f$ to HO is $\frac{r^{q+1}x^p}{q+1} + \frac{p}{q+1} \times \frac{r^{q+2}x^{p-1}}{q+2} + \frac{p}{q+1} \times \frac{p-1}{q+2} \times \frac{r^{q+3}x^{p-2}}{q+3} +$ &c.

4. If E be the coefficient of that term of the binomial $\overline{a+b}\,|^{p+q}$ expanded in which occurs $a^p b^q$, the ratio of the whole figure ACFH to HO is $\frac{1}{n+1} \times \frac{1}{\mathrm{E}}$, n being $= p+q$. For, when $\mathrm{A}f = \mathrm{AH}\ x = 1$, $r = 0$. Wherefore, all the terms

---

*Tis very evident here, without having recourse to Sir Isaac Newton, that the fluxion of the area $\mathrm{AC}f$ being $y\dot{x} = x^p \dot{x} - qx^{p+1}\dot{x} + q \times \frac{q-1}{2} x^{p+2}\dot{x}$ &c. the fluent or area itself is $\frac{x^{p+1}}{p+1} - q \times \frac{x^{p+2}}{p+2} + q \times \frac{q-1}{2} \times \frac{x^{p+3}}{p+3}$ &c.

†The fluxion of the first series is $x^p r^q \dot{x} + \frac{qx^{p+1}r^{p-1}\dot{r}}{p+1} + \frac{qx^{p+1}r^{q-1}\dot{x}}{p+1} + q \times \frac{q-1}{p+1} \times \frac{x^{p+2}r^{q-2}\dot{r}}{p+2} + \frac{q}{p+1} \times \frac{q-1}{p+2} \times x^{p+2}r^{q-2}\dot{x} + \frac{q}{p+1} \times \frac{q-1}{p+2} \times \frac{q-3}{p+3} \times x^{p+3}r^{q-3}\dot{r}$ &c. or, substituting $-\dot{x}$ for $\dot{r}$, $x^p r^q \dot{x} - \frac{qx^{p+1}r^{q-1}\dot{x}}{p+1} + \frac{qx^{p+1}r^{q-1}\dot{x}}{p+1} - q \times \frac{q-1}{p+1} \times \frac{x^{p+2}r^{q-2}\dot{x}}{p+2} + q \times \frac{q-1}{p+1} \times \frac{x^{p+2}r^{q-2}\dot{x}}{p+2}$ &c. which, as all the terms after the first destroy one another, is equal to $x^p r^q \dot{x} = x^p \times \overline{1-x}\,|^q \dot{x} = x^p \dot{x} \times \overline{1 - qx + q \times \frac{q-1}{2}x^2}$ &c. $= x^p \dot{x} - qx^{p+1}\dot{x} + q \times \frac{q-1}{2} x^{p+2}\dot{x}$ &c. = the fluxion of the latter series or of $\frac{x^{p\,|\,1}}{p+1} - q \times \frac{x^{p\,|\,2}}{p+2}$ &c. The two series therefore are the same.

of the series set down in Art. 2. as expressing the ratio of ACf to HO will vanish except the last, and that becomes $\frac{1}{n+1} \times \frac{q}{p+1} \times \frac{q-1}{p+2} \times$ &c. $\times \frac{1}{n}$. But E being the coefficient of that term in the binomial $\overline{a+b}\,|^n$ expanded in which occurs $a^p b^q$ is equal to $\frac{p+1}{q} \times \frac{p+2}{q-1} \times$ &c. $\times \frac{n}{1}$. And, because Af is supposed to become $=$ AH, AC$f =$ ACH. From whence this article is plain.

5. The ratio of ACf to the whole figure ACFH is (by Art. 1. and 4.) $\overline{n+1} \times$ E $\times \frac{x^{p+1}}{p+1} - q \times \frac{x^{p+2}}{p+2} + q \times \frac{q-1}{2} \times \frac{x^{p+3}}{p+3}$ &c. and if, as x expresses the ratio of Af to AH, X should express the ratio of At to AH; the ratio of AFt to ACFH would be $\overline{n+1} \times$ E $\times \frac{\mathrm{X}^{p+1}}{p+1} - q\frac{\mathrm{X}^{p+2}}{p+2} + q \times \frac{q-1}{2} \times \frac{\mathrm{X}^{p+3}}{p+3} - $ &c. and consequently the ratio of tFCf to ACFH is $\overline{n+1} \times$ E $\times^d$ into the difference [398] between the two series. Compare this with prop. 10. and we shall have the following practical rule.

RULE 1.

If nothing is known concerning an event but that it has happened p times and failed q in $p+q$ or n trials, and from hence I guess that the probability of its happening in a single trial lies somewhere between any two degrees of probability as X and x, the chance I am in the right in my guess is $\overline{n+1} \times$ E $\times^d$ into the difference between the series $\frac{\mathrm{X}^{p+1}}{p+1} - q\frac{\mathrm{X}^{p+2}}{p+2} + q \times \frac{q-1}{2} \times \frac{\mathrm{X}^{p+3}}{p+3} - $ &c. and the series $\frac{x^{p+1}}{p+1} - q\frac{x^{p+2}}{p+2} + q \times \frac{q-1}{2} \times \frac{x^{p+3}}{p+3} - $ &c. E being the coefficient of $a^p b^q$ when $\overline{a+b}\,|^n$ is expanded.

This is the proper rule to be used when q is a small number; but if q is large and p small, change every where in the series here set down p into q and q into p and x into r or $1-x$, and X into R $= 1-$ X; which will not make any alteration in the difference between the two serieses.

Thus far Mr. Bayes's essay.

With respect to the rule here given, it is further to be observed, that when both p and q are very large numbers, it will not be possible to apply it to practice on account of the multitude of terms which the serieses in it will contain. Mr. Bayes, therefore, by [399] an investigation which it would be too tedious to give here, has deduced from this rule another, which is as follows.

RULE 2.

If nothing is known concerning an event but that it has happened p times and failed q in $p+q$ or n trials, and from hence I guess that the probability of its happening in a single trial lies between $\frac{p}{n} + z$ and $\frac{p}{n} - z$; if $m^2 = \frac{n^3}{pq}$ $a = \frac{p}{n}$, $b = \frac{q}{n}$, E the coefficient of the term in which occurs $a^p b^q$ when $\overline{a+b}\,|^n$ is expanded, and $\Sigma = \frac{n+1}{n} \times \frac{\sqrt{2pq}}{\sqrt{n}} \times \mathrm{E}\, a^p b^q \times d$ by the series

$$mz - \frac{m^3 z^3}{3} + \frac{n-2}{2n} \times \frac{m^5 z^5}{5} - \frac{\overline{n-2} \times \overline{n-4}}{2n \times 3n} \times \frac{m^7 z^7}{7} + \frac{n-2}{2n} \times \frac{n-4}{3n} \times \frac{n-6}{4n} \times \frac{m^9 z^9}{9}$$

&c. my chance to be in the right is greater than $\dfrac{2\Sigma}{1 + 2\mathrm{E}\, a^p b^q + \frac{2\mathrm{E}\, a^p b^q}{n}}$ * and less than $\dfrac{2\Sigma}{1 - 2\mathrm{E}\, a^p b^q - \frac{2\mathrm{E}\, a^p b^q}{n}}$. And if $p=q$ my chance is 2Σ exactly. **[400]**

In order to render this rule fit for use in all cases it is only necessary to know how to find within sufficient nearness the value of $\mathrm{E}\, a^p b^q$ and also of the series $mz - \frac{m^3 z^3}{3}$ &c.[†]. With respect to the former Mr. Bayes has proved that, supposing K to signify the ratio of the quadrantal arc to it's radius, $\mathrm{E}\, a^p b^q$ will be equal to $\dfrac{\sqrt{n}}{2\sqrt{Kpq}} \times$ by the *ratio* whose *hyperbolic* logarithm is $\frac{1}{12} \times \overline{\frac{1}{n} - \frac{1}{p} - \frac{1}{q}} - \frac{1}{360} \times \overline{\frac{1}{n^3} - \frac{1}{p^3}[-]\frac{1}{q^3}} + \frac{1}{1260} \times \overline{\frac{1}{n^5} - \frac{1}{p^5} - \frac{1}{q^5}} - \frac{1}{1680} \times \overline{\frac{1}{n^7} - \frac{1}{p^7} - \frac{1}{q^7}} + \frac{1}{1188} \times \overline{\frac{1}{n^9} - \frac{1}{p^9} - \frac{1}{q^9}}$ &c. where the numeral coefficients may be found in the following manner. Call them A, B, C, D, E, &c. Then $\mathrm{A} = \frac{1}{2.2.3} = \frac{1}{12}$. $\mathrm{B} = \frac{1}{3.4} - \frac{\mathrm{A}}{3}$. $\mathrm{C} = \frac{1}{2.4.5} - \frac{1}{2.6.7} - \frac{10\mathrm{B}+\mathrm{A}}{5}$. $\mathrm{D} = \frac{1}{2.8.9} - \frac{35\mathrm{C}+21\mathrm{B}+\mathrm{A}}{7}$. $\mathrm{E} = \frac{1}{2.10.11} - \frac{126\mathrm{C}+84\mathrm{D}+36\mathrm{B}+\mathrm{A}}{9}$. $\mathrm{F} = \frac{1}{2.12.13} -$ **[401]** $\frac{462\mathrm{D}+330\mathrm{C}+165\mathrm{E}+55\mathrm{B}+\mathrm{A}}{11}$ &c. where the coefficients of B, C, D, E, F, &c. in the values of D, E, F, &c. are the 2, 3, 4, &c. highest coefficients in $\overline{a+b}\,|^7$, $\overline{a+b}\,|^9$, $\overline{a+b}\,|^{11}$, &c.

---

*In Mr. Bayes's manuscript this chance is made to be greater than $\dfrac{2\Sigma}{1+2\mathrm{E}\,a^p b^q}$ and less than $\dfrac{2\Sigma}{1-2\mathrm{E}\,a^p b^q}$. The third term in the two divisors, as I have given them, being omitted. But this being evidently owing to a small oversight in the deduction of this rule, which I have reason to think Mr. Bayes had himself discovered, I have ventured to correct his copy, and to give the rule as I am satisfied it ought to be given.

†A very few terms of this series will generally give the hyperbolic logarithm to a sufficient degree of exactness. A similar series has been given by Mr. De Moivre, Mr. Simpson and other eminent mathematicians in an expression for the sum of the logarithms of the numbers 1, 2, 3, 4, 5 to x, which sum they have asserted to be equal to $\frac{1}{2}\log.c + \overline{x + \frac{1}{2}} \times \log.x - x + \frac{1}{12x} - \frac{1}{360x^3} + \frac{1}{1260x^5}$ &c. c denoting the circumference of a circle whose radius is unity. But Mr. Bayes, in a preceding paper in this volume, has demonstrated that, though this expression will very nearly approach to the value of this sum when only a proper number of the first terms is taken, the whole series cannot express any quantity at all, because, let x be what it will, there will be always a part of the series where it will begin to diverge. This observation, though it does not much affect the use of this series, seems well worth the notice of mathematicians.

expanded; affixing in every particular value the least of these coefficients to B, the next in magnitude to the furthest letter from B, the next to C, the next to the furthest but one, the next to D, the next to the furthest but two, and so on*.

With respect to the value of the series $mz - \frac{m^3 z^3}{3} + \frac{n-2}{2n} \times \frac{m^5 z^5}{5}$ &c. he has observed that it may be calculated directly when mz is less than 1, or even not greater than $\sqrt{3}$: but when mz is much larger it becomes impracticable to do this; in which case he shews a way of easily finding two values of it very nearly equal between which it's true value must lie.

The theorem he gives for this purpose is as follows.

Let K, as before, stand for the ratio of the quadrantal arc to its radius, and H for the ratio whose hyperbolic logarithm is $\frac{2^2-1}{2n} - \frac{2^4-1}{360\, n^3} + \frac{2^6-1}{1260\, n^5} - \frac{2^8-1}{1680\, n^7}$ &c. Then the series $mz - \frac{m^3 z^3}{3}$ &c. will be greater or less than the series $\frac{Hn}{n+1} \times \frac{\sqrt{K}}{\sqrt{2}} - \frac{n}{n+2} \times \frac{\left[1 - \frac{2m^2 z^2}{n}\right]^{\frac{n}{2}+1}}{2mz} + \frac{n^2}{n+2} \times \frac{\left[1 - \frac{2m^2 z^2}{n}\right]^{\frac{n}{2}+2}}{n+4 \times 4 m^3 z^3} +$

[402] $\frac{3n^3}{n+2} \times \frac{\left[1 - \frac{2m^2 z^2}{n}\right]^{\frac{n}{2}+3}}{n+4 \times n+6 \times 8 m^5 z^5} + \frac{3 \times 5 \times n^4}{n+2} \times \frac{\left[1 - \frac{2m^2 z^2}{n}\right]^{\frac{n}{2}+4}}{n+4 \times n+6 \times n+8 \times 16 z^7 m^7} -$ &c. continued to any number of terms, according as the last term has a positive or a negative sign before it.

From substituting these values of $E\, a^p b^q$ and $mz - \frac{m^3 z^3}{3} + \frac{n-2}{2n} \times \frac{m^5 z^5}{5}$ &c. in the 2d rule arises a 3d rule, which is the rule to be used when mz is of some considerable magnitude.

RULE 3.

If nothing is known of an event but that it has happened p times and failed q in $p+q$ or n trials, and from hence I judge that the probability of it's happening in a single trial lies between $\frac{p}{n} + z$ and $\frac{p}{n} - z$ my chance to be right is *greater* than

$$\frac{\sqrt{Kpq} \times h}{2\sqrt{Kpq} + hn^{\frac{1}{2}} + hn^{-\frac{1}{2}}} \times \overline{2H - \frac{\sqrt{2}}{\sqrt{K}} \times \frac{n+1}{n+2} \times \frac{1}{mz} \times \left[1 - \frac{2m^2 z^2}{n}\right]^{\frac{n}{2}+1}}$$

and *less* than $\frac{\sqrt{Kpq} \times h}{2\sqrt{Kpq} - hn^{\frac{1}{2}} - hn^{-\frac{1}{2}}}$ multiplied by the 3 terms $2H - \frac{\sqrt{2}}{\sqrt{K}} \times \frac{n+1}{n+2} \times \frac{1}{mz} \times \overline{\left[1 - \frac{2m^2 z^2}{n}\right]^{\frac{n}{2}+1}} + \frac{\sqrt{2}}{\sqrt{K}} \times \frac{n}{n+2} \times \frac{n+1}{n+4} \times \frac{1}{2m^3 z^3} \times \overline{\left[1 - \frac{2m^2 z^2}{n}\right]^{\frac{n}{2}+2}}$

where m^2, K, h and H stand for the quantities already explained. [403]

---

*This method of finding these coefficients I have deduced from the demonstration of the third lemma at the end of Mr. Simpson's Treatise on the Nature and Laws of Chance.

An APPENDIX.

CONTAINING

An Application of the foregoing Rules to some particular Cases.

THE first rule gives a direct and perfect solution in all cases; and the two following rules are only particular methods of approximating to the solution given in the first rule, when the labour of applying it becomes too great.

The first rule may be used in all cases where either p or q are nothing or not large. The second rule may be used in all cases where mz is less than $\sqrt{3}$; and the 3d in all cases where $m^2 z^2$ is greater than 1 and less than $\frac{n}{2}$, if n is an even number and very large. If n is not large this last rule cannot be much wanted, because, m decreasing continually as n is diminished, the value of z may in this case be taken large, (and therefore a considerable interval had between $\frac{p}{n} - z$ and $\frac{p}{n} + z$,) and yet the operation be carried on by the 2d rule; or mz may not exceed $\sqrt{3}$.

But in order to shew distinctly and fully the nature of the present problem, and how far Mr. Bayes has carried the solution of it; I shall give the result of this solution in a few cases, beginning with the lowest and most simple. **[404]**

Let us then first suppose, of such an event as that called M in the essay, or an event about the probability of which, antecedently to trials, we know nothing, that it has happened *once*, and that it is enquired what conclusion we may draw from hence with respect to the probability of it's happening on a *second* trial.

The answer is that there would be an odds of three to one for somewhat more than an even chance that it would happen on a second trial.

For in this case, and in all others where q is nothing, the expression $\overline{n+1} \times \overline{\frac{X^{p+1}}{p+1} - \frac{x^{p+1}}{p+1}}$ or $X^{p+1} - x^{p+1}$ gives the solution, as will appear from considering the first rule. Put therefore in this expression $\overline{p+1} = 2$, $X = 1$ and $x = \frac{1}{2}$ and it will be $1 - \overline{\frac{1}{2}}|^2$ or $\frac{3}{4}$; which shews the chance there is that the probability of an event that has happened once lies somewhere between 1 and $\frac{1}{2}$; or (which is the same) the odds that it is somewhat more than an even chance that it will happen on a second trial*.

In the same manner it will appear that if the event has happened twice, the odds now mentioned will be seven to one; if thrice, fifteen to one; and in general, if the event has happened p times, there will be an odds of $2^{p+1} - 1$

---

*There can, I suppose, be no reason for observing that on this subject unity is always made to stand for certainty, and $\frac{1}{2}$ for an even chance.

to one, for *more* than an equal chance that it will happen on further trials.

Again, suppose all I know of an event to be that it has happened ten times without failing, and the **[405]** enquiry to be what reason we shall have to think we are right if we guess that the probability of it's happening in a single trial lies somewhere between $\frac{16}{17}$ and $\frac{2}{3}$, or that the ratio of the causes of it's happening to those of it's failure is some ratio between that of sixteen to one and two to one.

Here $p+1 = 11$, $X = \frac{16}{17}$ and $x = \frac{2}{3}$ and $X^{p+1} - x^{p+1} = \overline{\frac{16}{17}}\Big|^{11} - \overline{\frac{2}{3}}\Big|^{11}$
$= .5013$ &c. The answer therefore is, that we shall have very nearly an equal chance for being right.

In this manner we may determine in any case what conclusion we ought to draw from a given number of experiments which are unopposed by contrary experiments. Every one sees in general that there is reason to expect an event with more or less confidence according to the greater or less number of times in which, under given circumstances, it has happened without failing; but we here see exactly what this reason is, on what principles it is founded, and how we ought to regulate our expectations.

But it will be proper to dwell longer on this head.

Suppose a solid or die of whose number of sides and constitution we know nothing; and that we are to judge of these from experiments made in throwing it.

In this case, it should be observed, that it would be in the highest degree improbable that the solid should, in the first trial, turn any one side which could be assigned before hand; because it would be known that some side it must turn, and that there was an infinity of other sides, or sides otherwise marked, which it was equally likely that it should turn. The first **[406]** throw only shews that *it has* the side then thrown, without giving any reason to think that it has it any one number of times rather than any other. It will appear, therefore, that *after* the first throw and not before, we should be in the circumstances required by the conditions of the present problem, and that the whole effect of this throw would be to bring us into these circumstances. That is: the turning the side first thrown in any subsequent single trial would be an event about the probability or improbability of which we could form no judgment, and of which we should know no more than that it lay somewhere between nothing and certainty. With the second trial then our calculations must begin; and if in that trial the supposed solid turns again the same side, there will arise the probability of three to one that it has more of that sort of sides than of *all* others; or (which comes to the same) that there is somewhat in its constitution disposing it to turn that side oftenest: And this probability will increase, in the manner already explained, with the number of times in which that side has been thrown without failing. It should not, however, be imagined that any number of such experiments can give sufficient reason for thinking that it would *never* turn any other side. For, suppose it has turned the same side in every trial a million of times. In these circumstances there would be an improbability that it had *less* than 1.400,000 more of these sides than all others; but there would also be an improbability that it had *above* 1.600,000 times more. The chance for the latter is expressed by $\frac{1600000}{1600001}$ raised to the millionth power subtracted from unity, which is equal to .4647 &c. and [407] the chance for the former is equal to $\frac{1400000}{1400001}$ raised to the same power, or to .4895; which, being both less than an equal chance, proves what I have said. But though it would be thus improbable that it had *above* 1.600,000 times more or *less* than 1.400,000 times *more* of these sides than of all others, it by no means follows that we have any reason for judging that the true proportion in this case lies somewhere between that of 1.600,000 to one and 1.400,000 to one. For he that will take the pains to make the calculation will find that there is nearly the probability expressed by .527, or but little more than an equal chance, that it lies somewhere between that of 600,000 to one and three millions to one. It may deserve to be added, that it is more probable that this proportion lies somewhere between that of 900,000 to 1 and 1.900,000 to 1 than between any other two proportions whose antecedents are to one another as 900,000 to 1.900,000, and consequents unity.

I have made these observations chiefly because they are all strictly applicable to the events and appearances of nature. Antecedently to all experience, it would be improbable as infinite to one, that any particular event, before-hand imagined, should follow the application of any one natural object to another; because there would be an equal chance for any one of an infinity of other events. But if we had once seen any particular effects, as the burning of wood on putting it into fire, or the falling of a stone on detaching it from all contiguous objects, then the conclusions to be drawn from any number of subsequent events of the same kind would be to be determined in the same manner with the conclusions just mentioned relating to the constitution of the solid I have [408] supposed. —— In other words. The first experiment supposed to be ever made on any natural object would only inform us of one event that may follow a particular change in the circumstances of those objects; but it would not suggest to us any ideas of uniformity in nature, or give us the least reason to apprehend that it was, in that instance or in any other, regular rather than irregular in its operations. But if the same event has followed without interruption in any one or more subsequent experiments, then some degree of uniformity will be observed; reason will be given to expect the same success in further experiments, and the calculations directed by the solution of this problem may be made.

One example here it will not be amiss to give.

Let us imagine to ourselves the case of a person just brought forth into this, world and left to collect from his observation of the order and course of events what powers and causes take place in it. The Sun would, probably, be the first object that would engage his attention; but after losing it the first night he would be entirely ignorant whether he should ever see it again. He would therefore be in the cond[i]tion of a person making a first experiment about an event entirely unknown to him. But let him see a second appearance or one *return* of the Sun, and an expectation would be raised in him of a second return, and he might know that there was an odds of 3 to 1 for *some* probability of this. This odds would increase, as before represented, with the number of returns to which he was witness. But no finite number of returns would be sufficient to produce absolute or physical certainty. For let it be supposed that he has seen it return at regular and stated intervals a million of times. The conclusions **[409]** this would warrant would be such as follow —— There would be the odds of the millionth power of 2, to one, that it was likely that it would return again at the end of the usual interval. There would be the probability expressed by .5352, that the odds for this was not *greater* than 1.600,000 to 1; And the probability expressed by .5105, that it was not *less* than 1.400,000 to 1.

It should be carefully remembered that these deductions suppose a previous total ignorance of nature. After having observed for some time the course of events it would be found that the operations of nature are in general regular, and that the powers and laws which prevail in it are stable and parmanent. The consideration of this will cause one or a few experiments often to produce a much stronger expectation of success in further experiments than would otherwise have been reasonable; just as the frequent observation that things of a sort are disposed together in any place would lead us to conclude, upon discovering there any object of a particular sort, that there are laid up with it many others of the same sort. It is obvious that this, so far from contradicting the foregoing deductions, is only one particular case to which they are to be applied.

What has been said seems sufficient to shew us what conclusions to draw from *uniform* experience. It demonstrates, particularly, that instead of proving that events will *always* happen agreeably to it, there will be always reason against this conclusion. In other words, where the course of nature has been the most constant, we can have only reason to reckon upon a recurrency of events proportioned to the degree of **[410]** this constancy; but we can have no reason for thinking that there are no causes in nature which will *ever* interfere with the operations of the causes from which this constancy is derived, or no circumstances of the world in which it will fail. And if this is true, supposing our only *data* derived from experience, we shall find additional reason for thinking thus if we apply other principles, or have recourse to such considerations as reason, independently of experience, can suggest.

But I have gone further than I intended here; and it is time to turn our thoughts to another branch of this subject: I mean, to cases where an experiment has sometimes succeeded and sometimes failed.

Here, again, in order to be as plain and explicit as possible, it will be proper to put the following case, which is the easiest and simplest I can think of.

Let us then imagine a person present at the drawing of a lottery, who knows nothing of its scheme or of the proportion of *Blanks* to *Prizes* in it. Let it further be supposed, that he is obliged to infer this from the number of *blanks* he hears drawn compared with the number of *prizes*; and that it is enquired what conclusions in these circumstances he may reasonably make.

Let him first hear *ten* blanks drawn and *one* prize, and let it be enquired what chance he will have for being right if he guesses that the proportion of *blanks* to *prizes* in the lottery lies somewhere between the proportions of 9 to 1 and 11 to 1.

Here taking $X = \frac{11}{12}$, $x = \frac{9}{10}$, $p = 10$, $q = 1$, $n = 11$, $E = 11$, the required chance, according to the first [411] rule, is $\overline{n+1} \times E$ into the difference between $\overline{\frac{X^{p+1}}{p+1} - \frac{qX^{p+2}}{p+2}}$ and $\overline{\frac{x^{p+1}}{p+1} - \frac{qx^{p+2}}{p+2}} = 12 \times 11 \times \overline{\frac{\frac{11}{12}\rceil^{11}}{11} - \frac{\frac{11}{12}\rceil^{12}}{12}} - \frac{\frac{9}{10}\rceil^{11}}{11} - \frac{\frac{9}{10}\rceil^{12}}{12} = .07699$ &c. There would therefore be an odds of about 923 to 76, or nearly 12 to 1 *against* his being right. Had he guessed only in general that there were less than 9 blanks to a prize, there would have been a probability of his being right equal to .6589, or the odds of 65 to 34.

Again, suppose that he has heard 20 *blanks* drawn and 2 *prizes*; what chance will he have for being right if he makes the same guess?

Here X and x being the same, we have $n = 22$, $p = 20$, $q = 2$, $E = 231$, and the required chance equal to $\overline{n+1} \times E \times \overline{\frac{X^{p+1}}{p+1} - q\frac{X^{p+2}}{p+2} + q \times \frac{q-1}{2} \times \frac{X^{p+3}}{p+3}} - \overline{\frac{x^{p+1}}{p+1} - q\frac{x^{p+2}}{p+2} + q \times \frac{q-1}{2} \times \frac{x^{p+3}}{p+3}} = .10843$ &c.

He will, therefore, have a better chance for being right than in the former instance, the odds against him now being 892 to 108 or about 9 to 1. But should he only guess in general, as before, that there were less than 9 blanks to a prize, his chance for being right will be worse; for instead of .6589 or an odds of near two to one, it will be .584, or an odds of 584 to 415. [412]

Suppose, further, that he has heard 40 *blanks* drawn and 4 *prizes*; what will the before-mentioned chances be?

The answer here is .1525, for the former of these chances; and .527, for the latter. There will, therefore, now be an odds of only $5\frac{1}{2}$ to 1 against the proportion of blanks to prizes lying between 9 to 1 and 11 to 1; and but little more than an equal chance that it is less than 9 to 1.

Once more. Suppose he has heard 100 *blanks* drawn and 10 *prizes*.

The answer here may still be found by the first rule; and the chance for a proportion of blanks to prizes *less* than 9 to 1 will be .44109, and for a proportion *greater* than 11 to 1.3082. It would therefore be likely that there were not *fewer* than 9 or *more* than 11 blanks to a prize. But at the same time it will remain unlikely* that the true proportion should lie between 9 to 1 and 11 to 1, the chance for this being .2506 &c. There will therefore be still an odds of near 3 to 1 against this.

From these calculations it appears that, in the circumstances I have supposed, the chance for being right in guessing the proportion of *blanks* to *prizes* to be nearly the same with that of the number of *blanks* **[413]** drawn in a given time to the number of prizes drawn, is continually increasing as these numbers increase; and that therefore, when they are considerably large, this conclusion may be looked upon as morally certain. By parity of reason, it follows universally, with respect to every event about which a great number of experiments has been made, that the causes of its happening bear the same proportion to the causes of its failing, with the number of happenings to the number of failures; and that, if an event whose causes are supposed to be known, happens oftener or seldomer than is agreeable to this conclusion, there will be reason to believe that there are some unknown causes which disturb the operations of the known ones. With respect, therefore, particularly to the course of events in nature, it appears, that there is demonstrative evidence to prove that they are derived from permanent causes, or laws originally established in the constitution of nature in order to produce that order of events which we observe, and not from any of the powers of chance†. This is just as evident as it would be, in the case I have insisted on, that the reason of drawing 10 times more *blanks* than *prizes* in millions of trials, was, that there were in the wheel about so many more *blanks* than *prizes*.

But to proceed a little further in the demonstration of this point.

We have seen that supposing a person, ignorant of the whole scheme of a lottery, should be led to conjecture, from hearing 100 *blanks* and 10 prizes drawn, **[414]** that the proportion of *blanks* to *prizes* in the lottery was somewhere between 9 to 1 and 11 to 1, the chance for his being right would be .2506 &c. Let now enquire what this chance would be in some higher cases.

---

*I suppose no attentive person will find any difficulty in this. It is only saying that, supposing the interval between nothing and certainty divided into a hundred equal chances, there will be 44 of them for a less proportion of blanks to prizes than 9 to 1, 31 for a greater than 11 to 1, and 25 for some proportion between 9 to 1 and 11 to 1; in which it is obvious that, though one of these suppositions must be true, yet, having each of them more chances against them than for them, they are all separately unlikely.

†See Mr. De Moivre's Doctrine of Chances, pag. 250.

7.2. The Tract   295

Let it be supposed that *blanks* have been drawn 1000 times, and prizes 100 times in 1100 trials.

In this case the powers of X and x rise so high, and the number of terms in the two serieses $\frac{X^{p+1}}{p+1} - \frac{qX^{p+1}}{p+2}$ &c. and $\frac{x^{p+1}}{p+1} - \frac{qx^{p+2}}{p+2}$ &c. become so numerous that it would require immense labour to obtain the answer by the first rule. 'Tis necessary, therefore, to have recourse to the second rule. But in order to make use of it, the interval between X and x must be a little altered. $\frac{10}{11} - \frac{9}{10}$ is $\frac{1}{110}$, and therefore the interval between $\frac{10}{11} - \frac{1}{110}$ and $\frac{10}{11} + \frac{1}{110}$ will be nearly the same with the interval between $\frac{9}{10}$ and $\frac{11}{12}$, only somewhat larger. If then we make the question to be; what chance there would be (supposing no more known than that blanks have been drawn 1000 times and prizes 100 times in 1100 trials) that the probability of drawing a blank in a single trial would lie somewhere between $\frac{10}{11} - \frac{1}{110}$ and $\frac{10}{11} + \frac{1}{110}$ we shall have a question of the same kind with the preceding questions, and deviate but little from the limits assigned in them.

The answer, according to the second rule, is that this chance is greater than $\frac{2\Sigma}{1-2\mathrm{E}\,a^p b^q + \frac{2\mathrm{E}\,a^p b^q}{n}}$ **[415]** and less than $\frac{2\Sigma}{1-2\mathrm{E}\,a^p b^q - \frac{2\mathrm{E}\,a^p b^q}{n}}$, Σ being $\frac{n+1}{n} \times \frac{\sqrt{2pq}}{\sqrt{n}} \times \mathrm{E}\,a^p b^q \times \overline{mz - \frac{m^3 z^3}{3} + \frac{n-2}{2n} \times \frac{m^5 z^5}{5}}$ &c.

By making here $1000 = p$ $100 = q$ $1100 = n$ $\frac{1}{110} = z$, $m = \frac{\sqrt{n^3}}{\sqrt{pq}} = 1.048808$, $\mathrm{E}\,a^p b^q = \frac{h}{2} \times \frac{\sqrt{n}}{\sqrt{Kpq}}$, h being the ratio whose hyperbolic logarithm is $\frac{1}{12} \times \overline{\frac{1}{n} - \frac{1}{p} - \frac{1}{q}} - \frac{1}{360} \times \overline{\frac{1}{n^3} - \frac{1}{p^3} - \frac{1}{q^3}} + \frac{1}{1260} \times \overline{\frac{1}{n^5} - \frac{1}{p^5} - \frac{1}{q^5}}$ &c. and K the ratio of the quadrantal arc to radius; the former of these expressions will be found to be .7953, and the latter .9405 &c. The chance enquired after, therefore, is greater than .7953, and less than .9405. That is; there will be an odds for being right in guessing that the proportion of blanks to prizes lies *nearly* between 9 to 1 and 11 to 1, (or *exactly* between 9 to 1 and 1111 to 99) which is greater than 4 to 1, and less than 16 to 1.

Suppose, again, that no more is known than that *blanks* have been drawn 10,000 times and *prizes* 1000 times in 11000 trials; what will the chance now mentioned be?

Here the second as well as the first rule becomes useless, the value of mz being so great as to render it scarcely possible to calculate directly the series $\overline{mz} - \frac{m^3 z^3}{3} + \frac{n-2}{2n} \times \frac{m^5 z^5 [5]}{5}$ — &c. The third rule, therefore, must be used; and the information it gives us is, that the required chance is greater than .97421, or more than an odds of 40 to 1. **[416]**

By calculations similar to these may be determined universally, what expectations are warranted by any experiments, according to the different number of times in which they have succeeded and failed; or what should be thought of the probability that any particular cause in nature, with which we have any acquaintance, will or will not, in any single trial, produce an effect that has been conjoined with it.

Most persons, probably, might expect that the chances in the specimen I have given would have been greater than I have found them. But this only shews how liable we are to error when we judge on this subject independently of calculation. One thing, however, should be remembered here; and that is, the narrowness of the interval between $\frac{9}{10}$ and $\frac{11}{12}$, or between $\frac{10}{11} + \frac{1}{110}$ and $\frac{10}{11} - \frac{1}{110}$. Had this interval been taken a little larger, there would have been a considerable difference in the results of the calculations. Thus had it been taken double, or $z = \frac{1}{55}$, it would have been found in the fourth instance that instead of odds against there were odds for being right in judging that the probability of drawing a blank in a single trial lies between $\frac{10}{11} + \frac{1}{55}$ and $\frac{10}{11} - \frac{1}{55}$.

The foregoing calculations further shew us the uses and defects of the rules laid down in the essay. 'Tis evident that the two last rules do not give us the required chances within such narrow limits as could be wished. But here again it should be considered, that these limits become narrower and narrower as q is taken larger in respect of p; and when p and q are equal, the exact solution is given in all cases by the second rule. These two rules therefore afford **[417]** a direction to our judgment that may be of considerable use till some person shall discover a better approximation to the value of the two series's in the first rule[†].

But what most of all recommends the solution in this *Essay* is, that it is compleat in those cases where information is most wanted, and where Mr. De Moivre's solution of the inverse problem can give little or no direction; I mean, in all cases where either p or q are of no considerable magnitude. In other cases, or when both p and q are very considerable, it is not difficult to perceive the truth of what has been here demonstrated, or that there is reason to believe in general that the chances for the happening of an event are to the chances for its failure in the same *ratio* with that of p to q. But we shall be greatly deceived if we judge in this manner when either p or q are small. And tho' in such cases the *Data* are not sufficient to discover the exact probability of an event, yet it is very agreeable to be able to find the limits between which it is reasonable to think it must lie, and also to be able to determine the precise degree of assent which is due to any conclusions or assertions relating to them.

[†]Since this was written I have found out a method of considerably improving the approximation in the 2d and 3d rules by demonstrating that the expression $\frac{2\Sigma}{1+2\mathrm{E}\,a^p b^q + \frac{2\mathrm{E}\,a^p b^q}{n}}$ comes almost as near to the true value wanted as there is reason to desire, only always somewhat less. It seems necessary to hint this here; though the proof of it cannot be given.

The following list of errata was published at the end of this volume (i.e., 53) of the *Philosophical Transactions*.

p. **375**   2 from the bottom, *dele* every

p. **397**   first line in the note, for r^{p-1} read r^{q-1}

p. **401**   erase the asterisk in the 4th line from the top, and place it in the 10th line

p. **405**   6 and 10, *for* on *read* in

p. **415**   4 from the bottom, *for* 1−2E *read* 1+2E

p. **418**   3d and 4th lines in the note, *for* comes almost as near *read* comes, in most cases, almost as near.

Note that the correction to be made to **[401]** in fact refers to our sign (†) on that page, which, in terms of this correction, should be moved to follow the first '&c.' on that page. Further, in the expression for the ratio H preceding Rule 3, the denominator in the first term of the hyperbolic logarithm should be $12n$ rather than $2n$.

7.3 Commentary

In view of the fairly detailed treatment that parts of the *Essay* have recently received in Dale [1999] and Hald [1998] (to which the reader may be referred for more information), we pass somewhat briefly over certain aspects that a full discussion would require: indeed, we limit ourselves, at least in our discussion of the first section of the *Essay*, to a consideration of Bayes's definitions of probability and independence.

Although the *Essay* is often viewed as a whole, it is necessary, I feel, to consider Price's Introduction and his Appendix apart from Bayes's text — and perhaps even separately. The matters raised by Price are qualitatively different to those considered in the *Essay*, and there might well be some doubt as to whether Price's concerns are in fact met by Bayes's work.

Let us begin, then, with Price's introductory letter to the *Essay*, a letter that, useful though it is, is in some sense to be regretted inasmuch as it replaced the proem written by Bayes himself. We must, however, believe that Price recorded here what he considered to be useful and pertinent from Bayes's original introduction.

From consideration of Price's statement of Bayes's main result, viz.

> to find out a method by which we might judge concerning the probability that an event has to happen, in given circumstances, upon supposition that we know nothing concerning it but that, under the same circumstances, it has happened a certain number of times, and failed a certain other number of times

[370-371]

certain points emerge: firstly, the event of current concern is supposed to take place *under the same circumstances* as it has in the past, and secondly, what meaning is to be attached to the phrase 'judge concerning the probability'?

In connexion with the first of these matters it should be noted that the italicised phrase occurs neither in Bayes's own statement of the problem nor in his Scholium, though Price does indicate that it was used in the suppressed introduction. Consideration of the use of the word *trial* in the postulate at the beginning of Bayes's Section II and in the propositions occurring there suggests that the phrase is at least implicit in the *Essay*.

In considering the second matter, it is useful to take it in conjunction with the later phrase 'to estimate the chance that the probability...'. Edwards [1974] has suggested (correctly, I feel) that Bayes was in fact only interested in a (possibly vague) inference about the probability; and to effect this inference he considered the estimation of the chance of the probability being between any two named degrees of probability[31].

Price remarks in his introductory letter that Bayes

thought fit to begin his work with a brief demonstration of the
general laws of chance. His reason for doing this, as he says in
his introduction, was not merely that his reader might not have
the trouble of searching elsewhere for the principles on which
he has argued, but because he did not know whither to refer
him for a clear demonstration of them. **[375]**

Although one may see this as a rather curious statement, bearing in mind that three editions of de Moivre's *The Doctrine of Chances* were published during Bayes's lifetime — the last of these editions containing an Introduction expounding and illustrating the main rules of the subject — we show later that Bayes's definition of probability differs considerably from de Moivre's, and this could well be the reason for the detailed first section of the *Essay*[32].

The *Essay* itself opens with a lucid statement of the problem posed by Bayes, which, for ease of reference, we repeat here[33]:

Given the number of times in which an unknown event has
happened and failed: *Required* the chance that the probability
of its happening in a single trial lies somewhere between any
two degrees of probability that can be named. **[376]**

In modern notation, the solution — given in Proposition 10 — to this problem can be expressed as follows.

$\Pr[x_1 \leq x \leq x_2 \mid p$ happenings and q failures of the unknown event$]$

$$= \int_{x_1}^{x_2} x^p (1-x)^q \, dx \bigg/ \int_0^1 x^p (1-x)^q \, dx.$$

We comment on this solution in due course.

The first section of the *Essay* opens with seven definitions[34]. Writing of this section Hogben says

The first [part] sets out as Euclidean propositions, ... a few
school certificate level tautologies of the classical theory. It is
notable only because one of them, one which prompted the editor
to comments seemingly inconsistent with the author's intention and one which prompts the first historian of probability to
bewildered reflection on its ostensible novelty and possible self-evidence, plays an important role in the section which follows.
It there provides a platform for a highly sophisticated pun; and
its examination will help us to see the pitfalls of a symbolism
which does service to those who locate probability *in* the mind.
[1957, pp. 116–117]

Whether he is completely justified in these remarks emerges in our following discussions.

Bayes's definition of the probability of an event, given in terms of expectation, runs as follows.

> The *probability of any event* is the ratio between the value at which an expectation depending on the happening of the event ought to be computed, and the value of the thing expected upon it's happening. **[76]**

On this point we might note the following passage by Hogben.

> Bayes did recognise a nicety which is at the very core of what is most controversial at the present time, viz. the distinction between statements about the probability of events and statements about the probability of making correct assertions about events. ... [Bayes] never explicitly locates probability in the nebulous domain of the *mind*. [1957, p. 119]

The approach to probability via expectation, although perhaps unusual, was not unknown before Bayes's time. Christiaan Huygens was among the first to approach the subject in this way, writing in his *Van Reeckening in Speelen van Geluck*[35] of 1660

> Ick neeme tot beyder fondament, dat in het speelen de kansse, die yemant ergens toe heeft, even soo veel weerdt is als het geen, het welck hebbende hy weder tot deselfde kansse kan geraecken met rechtmatigh spel, dat is, daer in niemandt verlies geboden werdt.[36]

In the Latin version[37], *De ratiociniis in ludo aleæ*, as reprinted in the first part of Jakob Bernoulli's *Ars conjectandi* of 1713, this passage appeared as

> Hoc autem utrobique utar fundamento: nimirum, in aleæ ludo tanti æstimandam esse cujusque sortem seu expectationem ad aliquid obtinendum, quantum si habeat, possit denuò ad similem sortem sive expectationem pervenire, æquâ conditione certans

which John Arbuthnot, in his translation of 1692, gave as

> In both Cases I shall make use of this Principle, *One's Hazard or Expectation to gain any Thing is worth so much, as, if he had it, he could purchase the like Hazard or Expectation again in a just and equal Game.*

Freudenthal [1980] provides a translation of parts of the *Van Rekeningh in Spelen van Geluck* (to give the title in more modern Dutch), in which 'chance' is used with three meanings[38], viz. the pay-off matrix (*kansse; sors seu expectatio; expectatio*), in the context of 'equal chance' (*kans; aeque facile; aequa sors; similis expectatio*), and in the context of 'number of chances' (*numerus casuum*; [number] *expectationes*).

A careful translation of parts of Huygens's original Dutch text is given by van der Wærden in his *Historische Einleitung* to Bernoulli [1975]. We give here, for comparison, the Dutch, Latin, English (Arbuthnot), and German (van der Wærden — the first from the Dutch and the second from the Latin) versions of Huygens's first proposition[39]:

> Als ick gelijcke kans hebbe om a of b te hebben, dit is my so veel weerdt als $(a+b)/2$.
>
> Si a vel b expectem, quorum utrumvis æquè facilè mihi obtingere possit, expectatio mea dicenda est valere $(a+b)/2$.
>
> If I expect a or b, either of which, with equal Probability, may fall to me, then my Expectation is worth $(a+b)/2$ that is, the half Sum of a and b.
>
> Wenn ich gleiche Chancen habe, um a order b zu haben, ist mir das so viel wert wie $(a+b)/2$.
>
> Wenn ich a oder b erwarte, von denen ich jedes ebenso leicht erhalten kann, so ist meine Erwartung $(a+b)/2$ wert.

Arbuthnot is fairly consistent in his translation. Thus *sors*[40] is translated by 'hazard', *aeque facile* by 'with equal probability', *similis expectatio* by 'equal expectation', and *casuum* by 'chances'. However, Arbuthnot, perhaps with a prevision of Humpty Dumpty's question to Alice as to who should be master, does not sacrifice freedom of speech on the altar of accuracy; thus we also find

Huygens	Arbuthnot
quanti exspectatio	Value of [his] Expectation
aequam expectationem	as good a Chance
expectatio	Expectation
pro valore meae expectationis	for the Value of my Expectation
aequâ expectatione	with equal Hopes.

Thus although he in general preserves the distinction noted by Freudenthal, Arbuthnot is prepared to use synonyms (or things that are more or less synonyms) from time to time. However, it is not altogether clear whether Arbuthnot distinguished between *an expectation* and *the value of an expectation* — unlike the White Knight, who took pains to explain to Alice the difference between the name of the song, what the name of the song is called, what the name is, what the song is called, and what the song is.

Hacking [1975, p. 94] notes that certain questions present themselves when we come to consider the idea of expectation in Huygens's work:

1. Is mathematical expectation the fair price for a gamble in the "one-off" case?
2. How does one justify the answer to Question 1?

Although Huygens clearly had no problem in answering the first question in the affirmative, Hacking notes that, when considering the second,

> Huygens was not trying to justify expectation; he was trying to justify a method for pricing gambles which happens to be the same as what we call mathematical expectation. [1975, p. 95]

Taking as basic the fair lottery (in which the fair price for the game is supposed known), Huygens supposes that the lots are symmetric, so that any lot can be drawn as easily as any other. (This symmetry is regarded as a primitive notion.) Then he invents *equivalent gambles*, in which the price of a ticket T is decided by the finding of a fair lottery such that one is indifferent between a ticket in that lottery and T. The comparison between this idea and F.P. Ramsey's utility and probability theory is plain; and indeed Huygens tacitly assumes a number of principles of utility theory[41], viz.

(a) fair prices are additive;

(b) there is no discount for quantity as regards the purchase of tickets;

(c) lotteries may be compounded (i.e., the prize in one lottery may be a ticket in another);

(d) fair prices are unaffected by the introduction of consolation prizes.

There is indeed an intimate connexion between probability and utility. Modern decision theory postulates the existence of a set of *rewards* $\mathcal{R}$ and a set of probability distributions $\mathcal{P}$, the reward $r \in \mathcal{R}$ being identified with the probability distribution in $\mathcal{P}$ that assigns probability 1 to r. A utility function $U(\cdot)$ is a function whose expectation $E^P[U(r)]$ gives the 'value' of $P \in \mathcal{P}$. It is perhaps not easy to see which concept, if indeed either, is to be taken as primitive, though Edgeworth had no hesitation in plumping for utility[42].

As we have seen, Bayes had apparently rejected de Moivre's definition of probability as given in *The Doctrine of Chances*, a definition framed as follows.

> ¶1. The Probability of an Event is greater or less, according to the number of Chances by which it may happen, compared with the whole number of Chances by which it may either happen or fail.

> ¶2. Wherefore, if we constitute a Fraction whereof the Numerator be the number of Chances whereby an Event may happen, and the Denominator the number of all the Chances whereby it may either happen or fail, that Fraction will be a proper designation of the Probability of happening. [1756, pp. 1–2]

Similarly, de Moivre had earlier rejected the approach adopted by Huygens, writing in his Preface

> As for the French Book [*L'Analyse des Jeux de Hazard*], I had run it over but cursorily, by reason I had observed that the Author chiefly insisted on the Method of *Huygens*, which I was absolutely resolved to reject, as not seeming to me to be the genuine and natural way of coming at the Solution of Problems of this kind

though it seems that by the 'Method of Huygens' de Moivre meant the method Huygens had used in his consideration of the problem of points rather than the expectation approach to probability[43].

A definition that is not dissimilar to that in de Moivre's ¶1 above was given by Arbuthnot in the introduction to his translation of Huygens's *De ratiociniis in ludo aleæ*. (English versions of this work by Arbuthnot and by Browne appeared in 1692 and 1714, respectively, and we can only suppose that Bayes, if he knew of the original or of these translations, found the theoretical level of the work unsuitable for the solution of his problem[44].) Here we read

> I call that Chance, which is nothing but want of Art; that only which is left to me, is to wager where there are the greatest Number of Chances, and consequently the greatest Probability to gain. [1770, p. 259]

Notice the introduction of the *wager*.

Although a definition like de Moivre's of probability as a (proper) fraction was not uncommon at that time or in later years, Hald has noted that the approach taken by Bayes provided something of a watershed:

> Before Bayes, nearly all probability models assume that the parameter space is discrete; exceptions are some models in insurance mathematics by Nicholas Bernoulli (1709), de Moivre (1725), and Simpson (1742), and the description of an experiment by Buffon (1735)... Bayes invented a new physical model with a continuously varying probability of success.... He thus gives a geometrical definition of probability as the ratio of two areas. [1998, p. 138]

For further comparison let us say something about expectation and its connexion with probability, in the works of other writers of this time and of the twentieth century. Thus, in *The Doctrine of Chances* de Moivre wrote

> ¶4. If upon the happening of an Event, I be intitled to a Sum of Money, my Expectation of obtaining that Sum has a determinate value before the happening of the Event ...
> ¶5. In all cases, the Expectation of obtaining any Sum is estimated by multiplying the value of the Sum expected by the Fraction which represents the Probability of obtaining it.
> [1756, pp. 2–3]

Here we see that the expectation is determined *before* the event occurs, that probability is prior to expectation, and — something that is reinforced by later work — that *expectation* and *value of an expectation* (and similarly *sum* and *value of a sum*) are not, as in Huygens, distinguished. In support of this last point, notice that we find, in ¶5,

$$\mathcal{E}(\text{obtaining } S) = \text{Val}(\text{sum expected}) \times \Pr[S], \qquad (7.1)$$

where $\mathcal{E}$ denotes expectation (my '=' sign here is given by de Moivre as 'is estimated by'). An example following gives

$$\text{Present value of expectation} = 100^L \times (3/5)$$

$$= S \times \Pr[\text{winning } S],$$

and later on we find the words

> the value of the Expectation of any Sum, is determined by multiplying the Sum expected by the Probability of obtaining it.
> [de Moivre, 1756, p. 3] $\qquad (7.2)$

Note that the probability is no longer 'represented' by a fraction.

We seem, in (7.1) and (7.2), to have slightly different expressions, viz.

$$\mathcal{E}(S) = \text{Val}(S) \times \Pr[S] \quad \text{and} \quad \text{Val}(\mathcal{E}(S)) = S \times \Pr[S].$$

I believe, however, that de Moivre does not make any distinction between the two, and this is borne out by the numerical examples he adduces: indeed, further on [de Moivre, 1756 p. 4] we find, in essence,

$$\text{Val}(\mathcal{E}(S))/\text{Val}(S) = \Pr[S].$$

The connexion of *value* with *chance* had been made earlier by Richard Cumberland, who wrote[45]

This Difference between the Values of the Chances can and may, with great Propriety, be estimated and rated as the Chance, the Gain, i.e. The natural Reward of the wiser, of a more prudent choice. [1750, p. 467]

Stigler notes that Daniel Bernoulli confused two very different properties in his tract of 1777:

(a) that larger errors of observation could reasonably be supposed to be less frequent than smaller ... And (b) observations that are less accurate (that is, further from the true point) are hence less valuable for the purposes of estimation than those closer to the truth. [1999, p. 310]

Worth, or value, was thus identified with probability, or frequency: 'And yet the two properties ... bear no necessary relation to one another' (loc. cit.). In an unpublished paper of 1769 Bernoulli referred to ' "the value or probability" of an observation' (Stigler [1999, p. 310]), regarding these terms as the same. It is clear from what we have said above that de Moivre did not fall into this trap — nor did Bayes.

We also note, from de Moivre [1756, ¶4], that the expectation of obtaining a sum is conditional upon the occurrence of an event, and indeed the first few examples in the third edition of *The Doctrine of Chances* are devoted to the finding of the expectation of sums (cf. ¶5) (de Moivre also shows that $\mathcal{E}(\sum S_i) = \sum \mathcal{E}(S_i)$). Gradually, however, the idea of expectation as ascribable to *events* intrudes[46], and indeed on page 8 we read 'my Expectation of the first Event' (though this *might* be an elision for 'my Expectation of the Sum to be expected on the occurrence of the first Event'). If we consider the example following this quotation, in which $\Pr[A_1] = 3/5, \Pr[A_2] = 4/9$, and where 'I am to have $90^{L\cdot}$ once for all for the happening of one or the other of the two afore-mentioned Events', then since 'my Expectation of the second [Event] will cease upon the happening of the first', what we essentially have is

$$\begin{aligned}\mathcal{E}(A_1 \cup A_2) &= \mathcal{E}(A_1) + \mathcal{E}(A_2) \\ &= \mathcal{E}(A_1) + \mathcal{E}(A_2 \mid A_1)\mathcal{E}(A_1) + \mathcal{E}(A_2 \mid \overline{A}_1)\mathcal{E}(\overline{A}_1) \\ &= \left(\tfrac{3}{5} \times 90\right) + 0 + \tfrac{4}{9} \times \left(\tfrac{2}{5} \times 90\right) \\ &= 70^{L\cdot}.\end{aligned} \quad (7.3)$$

This use of the expectation operator $\mathcal{E}$ looks a little odd at first sight. However, Edwards, in his 1983 paper on the 'Gambler's Ruin' problem,

notes that $E(a,b)$, the expectation (or expected winnings) of player A when A and B lack a and b points, respectively, in a fair game, satisfies the relation
$$E(a,b) = \tfrac{1}{2}E(a-1,b) + \tfrac{1}{2}E(a,b-1), \tag{7.4}$$
an assertion made (though not in this form) by Pascal. If the probabilities that A and B win a point are p and q, respectively, then (7.4) becomes
$$E(a,b) = pE(a-1,b) + qE(a,b-1),$$
a result bearing obvious comparison with the use of $\mathcal{E}$ we have made above. Moreover, if one recalls the use of the symbol **P** in de Finetti [1974] for both Probability and Expectation (and called 'Prevision' by de Finetti's translators), one sees that the use of $\mathcal{E}$ in (7.3) is perhaps not so strange after all. One might also note Kamlah's [1987, p. 112] assertion that the older native German speaker actually used 'Wahrscheinlichkeit' for a degree of expectation.

Bayes in fact takes as fundamental in his *Essay* the idea of 'expectation of benefit'. Following Jeffreys [1973, §2.4], let us take a to denote the possible benefit that will be received if and only if a certain event q happens, let p denote the evidence available before any action is taken, and let $E(a,q \mid p)$ be the expectation of benefit. Bayes assumes essentially that $E(a,q \mid p)$ is proportional to a, and hence we may define $\Pr[q \mid p]$ by
$$E(a,q \mid p) = a \Pr[q \mid p].$$
Jeffreys (loc. cit.) notes that the fundamental rules of probability theory then follow immediately. He also notes, however, that although this approach avoids some problems that occur in considering equally possible and mutually exclusive alternatives, other problems arise in connexion with the additivity of expectations[47].

After Bayes the taking of expectation rather than probability as basic fell into disuse. Daston in fact writes 'Bayes was the last great representative of the expectation tradition in mathematical probability until the twentieth century' [1988, p. 33]. Once we reach the twentieth century, however, the defining of probability in terms of expectation once again takes place. Bartlett [1962], for instance, cites Ramsey's work as an early example, and further impetus to this approach was given by de Finetti (see, for instance, his [1972], [1974], and [1975]), who, however, preferred the term 'prevision' to 'expectation'[48]. Walley has in fact shown that, for the mathematical probabilist, whether probability or expectation is taken as fundamental signifies little[49]:

> Because of the one-to-one correspondence between linear previsions and additive probabilities, Bayesian theories can take probability as the fundamental concept, and define prevision or expectation as an integral with respect to a probability measure. [1991, p. 91]

Walley also points out (loc. cit.) the analogy between these two approaches to probability theory and the Lebesgue and Daniell approaches to the construction of a theory of integration.

Before leaving this point we might note that, in the comparison of Bayes's and Laplace's definitions of probability in his *Statistical Methods and Scientific Inference*, Fisher wrote

> I have, for myself, no doubt that Bayes' definition is the more satisfactory, being not only in accordance with the ideas upon which the *Doctrine of Chances* of his own time was built, but in connecting the comparatively modern notion of *probability*, which seems to have been unknown to the Islamic and to the Greek mathematicians[50], with the much more ancient notion of an *expectation*, capable of being bought, sold and evaluated.
> [p. 16]

The second point to be noted in connexion with Bayes's definitions concerns the definition of independence. The seventh definition reads as follows.

> Events are independent when the happening of any one of them does neither increase nor abate the probability of the rest. **[376]**

This might be interpreted as implying that no distinction was seen between 'independence' and 'pairwise independence' [Savage, 1960]. However, it has been suggested by Lindley that Savage was possibly wrong on this point; for details of his argument see Dale [1999, §4.3]. Notice that Bayes does not define *dependence* — in line with the modern trend — though he shows essentially that

$$\Pr[E_1 \cap E_2] = \Pr[E_1]\Pr[E_2 \mid E_1].$$

The idea of independence is first mentioned in ¶8 of *The Doctrine of Chances*, where we read

> If the obtaining of any Sum requires the happening of several Events that are independent on each other, then the Value of the Expectation of that Sum is found by multiplying together the several Probabilities of happening, and again multiplying the product by the Value of the Sum expected. [1756, p. 5]

It should be noted here that the 'expectation' at one stage can serve as the 'sum' at another: compare Huygens's conditional lotteries.

Specialization to two events then follows, together with consideration of things like $A_1 \overline{A}_2$. It is only now, 'for fear that ... the terms independent or dependent may occasion some obscurity' [1756, p. 6], that de Moivre provides the following definitions.

> Two Events are independent, when they have no connexion one with the other, and that the happening of one neither forwards nor obstructs the happening of the other.
> Two Events are dependent, when they are so connected together as that the Probability of either's happening is altered by the happening of the other. [1756, p. 6]

In a résumé, introduced because

> it frequently happening that some truths, when represented to the mind under a particular Idea, may be more easily apprehended than when represented under another [1756, p. 18]

de Moivre extends his earlier definition, noting that

> the Probability of the happening of several Events independent, is the product of all the particular Probabilities whereby each particular Event may be produced [1756, p. 21]

that is, $\Pr[\cap A_i] = \prod \Pr[A_i]$: this adds a lot to his earlier definition.

Notice that the definition of dependence does not specify *how* the probability is altered; however, if we use de Moivre's definition of *independence*, as given in this résumé, and negate it, we get

$$A_1, A_2 \text{ not independent} \quad \Rightarrow \quad \Pr[A_1 \cap A_2] \neq \Pr[A_1]\Pr[A_2]$$

$$\Rightarrow \quad \Pr[A_2] \neq \Pr[A_1 \cap A_2]/\Pr[A_1]$$

$$\Rightarrow \quad \Pr[A_2] \neq \Pr[A_2 \mid A_1],$$

that is, the probability of A_2 is altered by the occurrence of A_1, or A_1 and A_2 are dependent.

However, de Moivre *infers* that

$$\Pr[A_1 \cap A_2] = \Pr[A_1]\Pr[A_2 \mid A_1] \tag{7.5}$$

for dependent events, and indeed more generally [1756, pp. 7–8] that

$$\Pr[A_1 \cap A_2 \cap \ldots \cap A_n] = \Pr[A_1] \prod_{j=2}^{n} \Pr[A_j \mid A_1 \cap A_2 \ldots \cap A_{j-1}].$$

The notion of *comparative*, as opposed to *stochastic*, independence, axiomatized in Fine [1973], is not determined by Fine's axioms for quantitative comparative conditional probability (see his ¶II F): these axioms 'merely assert compatibility requirements between independence and comparative probability' [Fine, 1973, p. 34][51]. Bearing in mind the weakness of de Moivre's system in comparison to Fine's, one is not surprised to find that (7.5), although reasonable in de Moivre's theory, does not follow from his definitions.

The fundamental importance of the notion of 'independence' in probability theory was stressed in 1933 by Kolmogoroff, who wrote[52]

> Geschichtlich ist die Unabhängigkeit von Versuchen und züfalligen Größen derjenige mathematische Begriff, welcher der Wahrscheinlichkeitsrechnung ihr eigenartiges Gepräge gibt. ... Es ist dementsprechend eine der wichtigsten Aufgaben der Philosophie der Naturwissenschaften, ... die Voraussetzungen zu präzisieren, bei denen man irgendwelche gegebene reele Erscheinungen für gegenseitig unabhängig halten kann. [pp. 8–9]

It might be noted here that although we must be aware of the importance and power of independence (and also of its restrictiveness), we must be prepared to relax its assumption from time to time. Thus in the second volume of his *Theory of Probability* de Finetti notes that

> In order to retain the right of being influenced by experience, it will therefore be necessary to express an initial opinion differing from that of independence. [1975, §11.3.1]

This leads him to exchangeability and partial exchangeability.

Modern authors tend to be fairly consistent in choosing to define two events A and B as *independent* if $\Pr[A \cap B] = \Pr[A]\Pr[B]$ — or sometimes $\Pr[A \mid B] = \Pr[A]$, though this latter formulation is perhaps best avoided because of the necessity for the additional requirement that $\Pr[B]$ be positive. The idea of *dependence* is usually not explored, though when it does occur it is often treated as merely the negation of independence.

Like de Moivre, William Emerson, in 1776, defined *dependence* in terms of probability and *independence* without using probability:

> Def. V.
> Events are *independent* when they have no manner of connection with one another; or when the happening of one neither forwards nor obstructs the happening of any other of them.
> Def. VI.
> An event is *dependent* when the probability of its happening is altered by the happening of some other.

These differences are summarized in the following table.

	dependent	independent
de Moivre	probability	no probability
Emerson	probability	no probability
Bayes	—	probability

Perhaps, though, the starting with *expectation* rather than *probability* makes the definition of independence difficult. Whittle [1976, p. 93], for example, defines two observables X and Y as *statistically independent* if [53]

$$E[H(X)K(Y)] = E[H(X)]E[K(Y)]$$

for all H and K for which the expectations on the right-hand side exist. From this the (more usual) definition in terms of probability is easily derived by choosing H and K to be indicator functions.

Fine, on the other hand, does not favour a purely probability-based definition. He suggests that

> the usual definition of SI [stochastic independence] is necessary but not sufficient for a characterization of our informal or intuitive idea of independence. [1973, p. 142]

Fine in fact takes independence as a binary relation between events (recall de Moivre's approach). Intuitively reasonable axioms, together with axioms for comparative probability, show that

$$A, B \text{ independent} \Rightarrow \Pr[A \cap B] = G(\Pr[A], \Pr[B]),$$

where G is symmetric, associative, and increasing in each variable. However, the existence of an additive $\Pr$ and $G(x, y) = xy$ involves 'intuitively unacceptable constructions' [Fine, 1973, p. 81].

So much for the definitions in the first section of the *Essay*. Before considering some of the propositions and corollaries that follow — however routine they may appear today — let us make a few general remarks about the contents of this section.

First of all it should be noted that one finds here a clear definition of the binomial distribution, a distribution that is first encountered in Pascal's *Traité du triangle arithmétique* of 1665. Secondly, it would appear that Bayes seems to have regarded the failure of an event as the same thing as the happening of its contrary **[376, 383, 386]**, a fact that has bearing on the question of additivity of degrees of belief. And thirdly, Bayes remarks that the happening or failing of the same event, in different trials (i.e., as a result of certain repeated data), is the same thing as the happening or failing of as many distinct independent events, all similar **[383]**.

7.3. Commentary 311

Bayes's first proposition gives in essence the additive property of a probability function: that is, if $\{E_i\}$ is a sequence of mutually exclusive events, then $\Pr[\cup E_i] = \sum \Pr[E_i]$. Bayes was apparently the first to mention this property in print[54].

Also requiring comment are Propositions 3 (and its corollary) and 5. Once again we cite them in this section for easy referral.

> Proposition 3. The probability that two subsequent events will both happen is a ratio compounded of the probability of the 1st, and the probability of the 2d on supposition the 1st happens. **[378]**
>
> Corollary. Hence if of two subsequent events the probability of the 1st be a/N, and the probability of both together be P/N, then the probability of the 2d on supposition the 1st happens is P/a. **[379]**
>
> Proposition 5. If there be two subsequent events, the probability of the 2d b/N and the probability of both together P/N, and it being 1st discovered that the 2d event has happened, from hence I guess that the 1st event has also happened, the probability I am in the right is P/b. **[381]**

A casual reading of these results suggests that the Corollary to Proposition 3 and Proposition 5 are merely different ways of saying the same thing. Thus if E_1 and E_2 are two events (with E_1 preceding E_2 in time), one might be tempted to express these two results as

$$\Pr[E_2 \mid E_1] = \Pr[E_1 \cap E_2]/\Pr[E_1]$$

$$\Pr[E_1 \mid E_2] = \Pr[E_1 \cap E_2]/\Pr[E_2].$$

In a careful study of this part of the *Essay* Shafer [1982] has suggested, in contrast to generally received opinion, that the timing of events indeed affects the concept of conditional probability, and that although the validity of the corollary to Proposition 3 can be established by an argument using rooted trees, the truth of Proposition 5 cannot be shown in this way. And since this latter proposition plays a crucial rôle in the proof of the ninth proposition, Shafer's criticism is important.

In Bayes's fifth proposition the order in which one *learns* about the happening of the events is introduced as a new factor. Shafer concludes that this result

> seems unconvincing unless we assume foreknowledge of the conditions under which the discovery of B's [the second event's] having happened will be made. [1982, p. 1080]

If, however, the fifth proposition is read as referring to expectations that are *subjectively* determined, Shafer concludes that 'the fifth proposition would then become merely a subjective version of the third' [1982, p. 1086]. Such knowledge, I suggest, in fact obtains in cases in which Bayes uses this result, and I therefore believe this proposition to be correct[55].

Hailperin [1996, p. 77] too draws attention to the fact that Bayes does not explain the meaning he attaches to the words 'subsequent event', but differs from Shafer in supposing that what is meant is that a 'subsequent' event is one that is not yet determined, 'i.e. that it is a contingent or chance event'[56].

In the Postscript to his French edition of the *Essay* Clero devotes several pages to a philosophical study of time and causality as treated by Bayes. He notes that

> les formules de Bayes — et surtout leurs demonstrations — mettent en relief une certaine dissymétrie de valeur de nos informations selon qu'elles nous permettent de nous tourner vers l'avenir ou vers la passé [1988, p. 135]

and he comments in a note to this passage that

> La probabilité de l'événement passé A si l'événement B a eu lieu n'est pas égale à celle de l'événement futur B si l'événement A a eu lieu (sauf dans le cas que l'on n'a aucune raison de supposer fréquent où $P(A) = P(B)$. [1988, p. 152]

Like Shafer, Clero pays particular attention to Bayes's third and fourth propositions, and remarks that

> lorsque l'on enquête vers le futur, que l'événement A s'est produit et que l'on attend l'événement B, l'espérance de B s'est accrue; mais lorsque l'on enquête vers le passé, que l'événement B s'est produit et que l'on se demande si l'événement A s'est produit, la réalisation de l'événement B n'a, dans cette situation, pas fait accroître notre espérance. [1988, p. 153]

In the light of these remarks it seems that we may proceed with Bayes's theory, though such proceeding should be carefully executed.

Since the other propositions of this section do not seem to require special attention, we pass on to Section II, in which the real meat of the *Essay* lies.

An examination of the proofs of the results of this section shows that the postulate given at the beginning may, without loss of generality, be expressed in the form

(i) a single value x is drawn from a uniform distribution concentrated on [0,1], and

(ii) a sequence of Bernoulli trials, with success probability x, is generated.

Written in modern notation, with no assumption being made as to the table $ABCD$ being of unit area, the results of this section run as follows[57].

<u>Lemma 1.</u> $\Pr[b < o < f] = (f - b)/AB$.

Hald [1998, p. 140] suggests that Bayes thought that this lemma needed a proof 'presumably because of the ongoing discussion on the infinite divisibilty of a finite line', and he also notes that Bayes gave an indirect proof for the case in which the bounds on the parameter were irrational.

<u>Lemma 2.</u> $\Pr[M \text{ in a single trial} \mid W] = \Pr[1 \text{ success} \mid W] = Ao/AB$.

<u>Proposition 8.</u> Let $y = Ex^p r^q$, where $E = \binom{p+q}{p}$. Then

$$\Pr[b < o < f \ \& \ p, q] = \int_b^f Ex^p r^q \, dx \Big/ \text{area } ABCD.$$

A detailed proof of this result may be found in Dale [1999, §4.4], where it is shown that there is no loss of generality in supposing the table $ABCD$ to be of unit area (note that Timerding [1908, p. 47] has pointed out that the two-dimensionality of the table plays no part in Bayes's construction). Commenting on this proposition Hald cogently notes that

> ... Bayes speaks about probability as an idealized relative frequency in a repeatable physical experiment, just as in classical probability theory on games of chance; he does not allude to his own definition of probability. [1998, p. 139]

A similar sentiment was expressed by Fisher in his *Statistical Methods and Scientific Inference*:

> Subject to the latent stipulation of fair use, or of homogeneity in the series of tests, his [i.e., Bayes's] definition is therefore equivalent to the limiting value of the relative frequency of success. [p. 14]

Jeffreys, on the other hand, was less certain of the rôle of frequency in Bayes's definitions, writing

> It is often said that some frequency definition is implicit in the work of Bernoulli, and even of Bayes and Laplace. This seems

out of the question. Bayes constructed the elaborate argument in terms of expectation of benefit to derive the product rule, which he could have written down in one line by elementary algebra if he was using the De Moivre definition. [1983, p. 403]

From the *Essay* [**388**] we have $y = bm/AB$, $x = Ab/AB$, and $r = Bb/AB$.

<u>Corollary.</u> $\Pr[A < o < B \ \& \ p, q] = \int_A^B Ex^p r^q \, dx / \text{area } ABCD$.

Bayes notes on page [**393**] (in a reference to 'art. 4', which in turn can be found on [**398**]) that were the table to be of unit area this corollary would yield the value $1/(n+1)$, independent of x. Here the indirectness of Bayes's proof appears: he supposes that the probability is not that given on the right-hand side in the statement of the corollary and deduces a contradiction (for further details see Hald [1998, pp. 140–141]). Hald provides the following interpretation of the corollary.

> Let us imagine the two-stage experiment[58] carried out a large number of times giving the results $\theta_1, \theta_2, \ldots$ at the first stage and $a_1, a_2, \ldots$ at the second stage. According to Bernoulli's theorem $a_1/n, a_2/n, \ldots$ converge in probability to $\theta_1, \theta_2, \ldots$, wherefore the a's like the θ's will be uniformly distributed. This result is also to be expected from symmetry considerations, since θ may be considered as just one among $n+1$ uniformly distributed positions on the table. [1998, p. 141]

<u>Proposition</u>[59] 9. Let $P(M)$ denote the probability of M. Then

$$\Pr[b < o < f \mid p, q] \equiv \Pr[Ab/AB < P(M) < Af/AB \mid p, q]$$

$$= \int_b^f Ex^p r^q \, dx \Big/ \int_A^B Ex^p r^q \, dx.$$

<u>Corollary.</u>

$$\Pr[Ab/AB < P(M) < o \mid p, q] = \int_b^o Ex^p r^q \, dx \Big/ \int_A^B Ex^p r^q \, dx.$$

<u>Proposition 10.</u> Let N be an 'unknown event' with probability $P(N)$. Then

$$\Pr[Af/AH < P(N) < At/AH \mid p, q] = \int_f^t Ex^p r^q \, dx \Big/ \int_A^H Ex^p r^q \, dx.$$

For the sake of a quick insight into the minor differences in formulation that arise when it is assumed that the table $ABCD$ is of unit area, we give the 'normalized' forms of Propositions 9 and 10.

<u>Proposition 9.</u> For any x_1, x_2 such that $0 \leq x_1 < x_2 \leq 1$,

$$\Pr\left[x_1 < x < x_2 \mid p \text{ successes and } q \text{ failures in } p+q \text{ trials}\right]$$

$$= \int_{x_1}^{x_2} \binom{p+q}{p} x^p (1-x)^q \, dx \bigg/ \int_0^1 \binom{p+q}{p} x^p (1-x)^q \, dx$$

$$\left(= \frac{(p+q+1)!}{p!\,q!} \int_{x_1}^{x_2} x^p (1-x)^q \, dx\right).$$

<u>Proposition 10.</u> Let x be the (prior) probability of an unknown event A. Then

$\Pr\left[x_1 < x < x_2 \mid A \text{ has happened } p \text{ times and failed } q \text{ times in } p+q \text{ trials}\right]$

$$= \int_{x_1}^{x_2} \binom{p+q}{p} x^p (1-x)^q \, dx \bigg/ \int_0^1 \binom{p+q}{p} x^p (1-x)^q \, dx.$$

In passing, we might note that Hogben, in a characteristically enthusiastic and flamboyant passage, though one that perhaps has a hint of acerbity, writes as follows of Bayes's experiment.

> In defiance of the, not as yet established, first law of thermodynamics, perfectly spherical and perfectly smooth billiard balls roll at the behest of the author's pen on frictionless planes as the theme of Euclidean demonstrations in the grand manner of the *Principia*. [1957, p. 116]

Two conditions are needed for the application of Proposition 10: the observations should have independent binomial distributions with the same parameter, and either there should be no reason to suppose that the number of times the unknown event occurs in n trials is not uniformly distributed, or that the $n+1$ possible outcomes may a priori be taken to be equally probable. Hald [1998, p. 143] refers to this latter condition as 'Bayes's criterion of ignorance.'

This criterion may be seen as related to the *principle of insufficient* (or *non-sufficient*) *reason* or the *principle of indifference* as Keynes later called it, devoting an entire chapter of his *A Treatise on Probability* to its discussion. Simple though this principle might appear, its practical application has long been a matter of controversy: Laplace took pains to point out that, if events were not 'également possibles', on déterminera d'abord leurs possibilités respectives dont la juste appréciation est un des points les plus délicats de la théorie des hasards. [1814, p. 7]

The passage from 'equally possible' to 'equally probable' is not one that should be lightly taken by the casual day-tripper. For example, in commenting on Sir Edward Eizat's *Apollo Mathematicus* of 1695 Stigler writes

to say that cases are equally possible does not imply without further argument that they must be equally probable.
Eizat put his finger on what is still often a weak point in arguments that apply probability in nonexperimental settings: what *are* the equally likely cases? [1999, pp. 227, 238]

Keynes emphasized the care that had to be taken in applying the principle of indifference, writing

the Principle of Indifference is not applicable to a pair of alternatives, if we know that either of them is capable of being further split up into a pair of possible but incompatible alternatives of the same form as the original pair.
[1921, Chap. IV, §21]

The ignoring of the truth expressed in this statement had earlier caused considerable problems in the work of the Danish actuary F. Bing in the second half of the nineteenth century[60]. In his reply to criticism of his earlier work by L. Lorenz, Bing noted that re-parameterization of the model could well result in a different posterior distribution, the indifference principle[61] thus leading to conflicting results. Hald notes the following reactions to this observation,

(1) rejection of the theory of inverse probability, (2) replacement of the indifference principle by an invariance principle, and (3) introduction of a subjective prior for the preferred parameters. [1998, p. 271]

All three of these reactions have been espoused by various writers at various times, but the principle has generally been cavalierly dismissed. Hogben for instance writes

The absurdities into which the principle of insufficient reason leads us arise less from the circumstance that it is gratuitous than because it is wholly irrelevant to most situations in which even its opponents would readily concede its convenience if susceptible of proof. [1957, p. 143]

Edwards [1972, §4.4] notes that in order that the principle of indifference be correctly applied it is necessary to ensure that an appropriate statistical model be entertained for the generation of hypotheses or events — 'the real issue is whether *any* probability statement about the parameter is valid.' In an example involving the drawing of cards from a pack from which one card has been removed and a duplicate of one of the remaining cards substituted, Edwards states that he, for one, refuses to equate the *ignorance* of how this procedure was carried out to the *knowledge* that cards were chosen at random.

Although we must note Edwards's remarks, it is also worth recording that in his *Statistical Methods and Scientific Inference* Fisher asks whether Laplace did not pass unawares from proposition (a) to proposition (b) below:

> (a) A possible outcome must be assigned equal probabilities in different future throws, because we can draw no relevant distinction between these in advance.
>
> (b) Hypotheses must be judged equally probable *a priori* if no relevant distinction can be drawn between them. [p. 36]

Bayes's Theorem is an important component in learning from experience[62], without which statistical inference would be nigh impossible[63]: Savage in fact goes so far as to say, 'Inference means for us the changes in opinion induced by evidence on the application of Bayes' theorem' [1961, pp. 577–578], and here the principle of indifference also plays a rôle. Kyburg, for instance, notes that a relatively natural way to use the principle of indifference to derive prior probabilities is to apply it either to *state descriptions* (for example, the statement that each ball in an urn has a specific one of n colours), or to *structure descriptions* (the statement that the proportion of red balls, say, is some specified value). He then concludes that, when the principle of indifference is applied to state descriptions, 'it is easy to show, as Carnap does, that Bayes' Theorem does not enable us to learn from experience in the way we expect it to' [1970, p. 35]. This he demonstrates by supposing that the four initial possible compositions of an urn containing two balls, each of which may be either white or black, are equally probable. Bayes's Theorem then shows, for instance, that $\Pr[B_2 \mid B_1] = \Pr[B_2]$, the probability that the second ball is black being unaffected by the result of the first draw.

In considering the passing from the probability of an event $\mathrm{P}(E)$ to that of $\mathrm{P}(E \mid H)$, de Finetti [1974, vol. I, §4.5.3] notes that

> The acquisition of a further piece of information, H — in other words, *experience*, since experience is nothing more than the acquisition of further information — acts always and only in the

way we have just described: *suppressing the alternatives that turn out to be no longer possible* ... As a result of this, the probabilities are the P($E \mid H$) instead of the P(E), but *not because experience has forced us to modify or correct them, or has taught us to evaluate them in a better way* ... the probabilities are the same as before ... *except for the disappearance of those which dropped out and the consequent normalization of those which remained.* [p. 141]

In his Scholium Bayes supposes one knows how often a success has occurred (and how often it has not occurred) in n trials. One may then 'give a guess whereabouts it's probability is', and hence (by Proposition 9) find 'the chance that the guess is right' **[392]**. In what is perhaps a controversial part of his argument, Bayes now asserts that the same rule should be used in the consideration of an event whose probability, antecedent to any trial, is unknown. In support of this assertion he argues as follows. Let us suppose that knowing nothing of the (antecedent) probability is equivalent to being indifferent between the possible number of successes in n trials — i.e., each possible number of successes is as probable as any other[64]. Writing then of 'an event concerning the probability of which we absolutely know nothing antecedently to any trials made concerning it' **[392 − 393]**, he says

> that concerning such an event I have no reason to think that, in a certain number of trials, it should rather happen any one possible number of times than another. **[393]**

The Corollary to Proposition 8 then implies that this situation obtains in the proposed model, and Bayes then adds

> In what follows therefore I shall take for granted that the rule given concerning the event M [i.e. success] in prop. 9. is also the rule to be used in relation to any event concerning the probability of which nothing at all is known antecedently to any trials made or observed concerning it. And such an event I shall call an unknown event. **[393 − 394]**

A careful examination of the propositions of this section shows that Bayes states his results sometimes in terms of 'the probability that the point o should fall [in a certain interval]' (Proposition 8, its corollary, and Proposition 9) and sometimes in terms of 'the probability of the event M [is in a certain interval]' (Proposition 9 and its corollary). Bayes's assumptions, however, result in these formulations being identical (see Edwards [1978] and Hald [1998, pp. 143–144]). Should (the position of) the first ball *not* be uniformly distributed, however, the *probability* will still have a uniform distribution, an assertion that may be verified as follows. Suppose that the first ball has the distribution $dF(\cdot)$, and let z denote the position on the table (more correctly, the ordinate of the position) at which this ball comes to rest. Then

$$\theta \equiv F(z) = \int_0^z f(x)\,dx.$$

From Bayes's postulate, the probability that the first ball lies in the interval $(z, z + dz)$ and that the next n throws result in p successes and q failures is

$$\binom{p+q}{p}[F(z)]^p[1-F(z)]^q\,dF(z),$$

the resulting posterior distribution then being

$$f(z \mid p, q) = \frac{[F(z)]^p[1-F(z)]^q\,dF(z)}{\int_0^1 [F(z)]^p[1-F(z)]^q\,dF(z)}.$$

Since $d\theta = dF(z)$, we thus have

$$f(\theta \mid p, q) = \frac{\theta^p(1-\theta)^q\,d\theta}{\int_0^1 \theta^p(1-\theta)^q\,d\theta}.$$

Note that the posterior distribution of z depends on $F(z)$ and hence on the density $f(z)$, whereas the posterior density of θ is independent of $f(z)$. Thus the posterior distribution of θ is a beta distribution whether or not the first ball has a uniform distribution[65].

We may give a paraphrase of the Scholium as follows. It follows from the ninth proposition that, given the number of times the event M happens and fails in a certain number of trials, 'one may give a guess whereabouts it's probability is, and, by the usual methods computing the magnitudes of the areas there mentioned, see the chance that the guess is right' **[392]**. If our concern is with an event of whose probability, prior to any trials, we are ignorant, we should follow the same rule; for, writes Bayes, 'concerning such an event I have no reason to think that, in a certain number of trials, it should rather happen any one possible number of times than another' **[393]**. The representation of complete absence of knowledge about p is thus achieved by our assuming that p has a uniform prior distribution. Bayes completes this discussion with the following words.

> In what follows therefore I shall take for granted that the rule given concerning the event M in prop. 9. is also the rule to be used in relation to any event concerning the probability of which nothing at all is known antecedently to any trials made

or observed concerning it. And such an event I shall call an unknown event. **[393 − 394]**

One must also, however, bear in mind that the meaning of the word 'guess' has changed over the years. The *Oxford English Dictionary* notes that in the fourteenth century the word was the usual rendering of the Latin word *æstimare*, and later it was used to indicate the forming of an approximate judgement (of, for example, a number or amount) without any actual calculation.

To complete Bayes's argument in full there is one aspect that still needs attention: it is necessary to show that only the uniform distribution for p has the property stated in the corollary to Propostion 8. This problem was examined by Murray, who posed it as follows,

> the assumption "all values of p are equally likely" is *equivalent* to the assumption "any number x of successes in n trials is just as likely as any other number y, $x \leq n$, $y \leq n$". [1930, p. 129]

That part of Murray's proof (a proof carried out using moments) that was not given in the *Essay* may be more briefly put by noting that the uniform distribution does indeed provide the required sequence of moments, and then invoking the uniqueness theorem for moment generating functions (for further details see Dale [1999, §4.5]).

Bayes's Postulate and Scholium have been the subject of considerable and heated discussion over the years. Among the main critics is Hacking[66]. A more sympathetic view has been taken by Stigler, who claimed that 'Bayes's actual argument is free from the principal defect it has been charged with' [1982, p. 250] (see also Stigler [1986, pp. 126–129]), and Geisser [1988] presented three different versions — the 'received', the 'revised', and the 'stringent' — of Bayes's main result. These arguments, and others, are discussed in Dale [1999, §4.5].

It is worth noting here that Fisher believed that Bayes did not use his Axiom and Scholium in his actual mathematics (possibly regarding these as disputatious), but chose rather to rely on an auxiliary experiment (see, for example, Fisher [1962] and his *Statistical Methods and Scientific Inference*, p. 132]). Probability statements a posteriori are then based solely on observation, although the two types of observations play different rôles in Bayes's Theorem. This is, of course, the basis of Fisher's fiducial approach. The reverse use of Bayes's Theorem was discussed by Good [1950].

In addition to its importance in probability, Bayes's tenth proposition, one that leads to the beta distribution

$$B(p) \equiv \Pr[dp \mid s, n] = (n+1)\binom{n}{s} p^s (1-p)^{n-s} dp,$$

is noteworthy for its consideration of the (incomplete) beta-function[67]. Once he has effected the integration, over $[0, x]$, of $x^p(1-x)^q$ by expanding the binomial and integrating term-by-term (essentially repeated integration by parts), Bayes introduces a transformation that yields

$$\int_0^x x^p(1-x)^q \, dx = \frac{x^{p+1}r^q}{p+1} + \frac{q}{p+1} \times \frac{x^{p+2}r^{q-1}}{p+2} + \&c.$$

with $r = 1 - x$. Multiplication of the left- and right-hand sides of this expression by the left- and right-hand sides of the expression

$$\left[\int_0^1 x^p(1-x)^q \, dx\right]^{-1} = \frac{(p+q+1)!}{p!\,q!},$$

essentially yields an expression for the beta probability integral

$$I_x(p+1, q+1) \equiv \int_0^x u^p(1-u)^q \, du \Big/ \int_0^1 u^p(1-u)^q \, du$$

$$= \sum_{i=0}^{q} \binom{p+q+1}{q-i} x^{p+1+i}(1-x)^{q-i}$$

$$= \sum_{i=p+1}^{p+q+1} \binom{p+q+1}{i} x^i (1-x)^{p+q+1-i},$$

which not only shows that the ratio of the two integrals is equal to the sum of the last $(q+1)$ terms of the binomial $(x + (1-x))^{p+q+1}$, but also gives a formula for finding the lower tail of the beta density. Bayes is thus seen to be using his result to evaluate the integral as a sum of terms in the binomial expansion, whereas Laplace, on the other hand, summed the binomial terms by using the ratio of the integrals. It is thus apparent that the problem of evaluating the incomplete beta-function

$$B_x(a, b) = \int_0^x x^{a-1}(1-x)^{b-1} \, dx, \quad a > 0, \ b > 0,$$

is essentially equivalent to the evaluation of cumulative binomial sums[68].

Dutka [1981, p. 14 et seqq.] has noted that Bayes carried out his evaluation essentially by using Newton's methods: he also notes that Laplace gave an asymptotic series for the expansion of $\int_0^x x^p(1-x)^q \, dx$ — as did Bayes and Price of course — for large values of p and q. A 'long and detailed investigation' by Uspensky in 1937 gave an improved version of Laplace's result for an approximation to a sum of binomial probabilities, and tables, monumental in extent and of immense usefulness, of the incomplete beta-function were published in 1934 by Karl Pearson and his collaborators.

Bayes's main result is of course derived under the assumption that the prior distribution is uniform; that it indeed holds for (almost) any prior has been noted by many authors: we instance only Cournot [1843, §95], Edgeworth [1884, p. 228], Mill [1843, Book III, Chapter XVIII. §6], and Bowley[69]. As an example, let us mention the version given by the last-mentioned:

> except in the neighbourhood of the central value, it is indifferent what distribution of *a priori* probabilities of p we suppose. Over the small, important central region the assumption that the *a priori* probability of p over a region is proportional to that region is likely to be a good first approximation, whatever the actual law. [1926, p. 414]

We can also note the observation of the 'swamping' of the effect of the prior distribution by the influence of empirical observations in Kyburg [1970, p. 72]. Although Salmon [1967, p. 129] notes that large inaccuracies in prior probabilities may have negligible effects on the posteriors 'if a reasonable amount of confirmatory evidence is accumulated', Maxwell [1975, p. 148] sounds a cautionary note, remarking, among other things, that should the prior probability not be near 1, the new evidence obtained should be different *in kind* to the old. See also Weatherford [1982, p. 226], and Blair [1975] for a discussion of the change effected under Bayes's Theorem in the probability of a scientific hypothesis as the number of confirming tests increases.

In his approach to the problem caused by the unknown prior distribution function $\Pr[\theta]$, von Mises [1942] assumed that it lay between certain limits and then derived bounds for a ratio of posteriors under a uniform prior and under the prior initially chosen. Bounds were then derived, in a specific case, for the posterior given by Bayes's Theorem.

In his *A Treatise on Differential Equations* George Boole considered the series

$$p^a \left\{ 1 + aq + \frac{a(a+1)}{1.2} q^2 \cdots + \frac{a(a+1)\ldots(a+b-1)}{1.2\ldots b} q^b \right\},$$

> [which] occurs as the expression of the probability that an event whose probability of occurrence in a single trial is p, and of failure q, will occur at least a times in $a + b$ trials.
> [Chap. XVII, Art. 11]

Denoting the series in brackets by $u = \sum u_m \epsilon^{m\theta}$, Boole established that

$$mu_m - (m + a - 1)u_{m-1} = 0,$$

and solving this by the symbolical D method he showed that the probability P sought is[70]

$$P = 1 - \frac{a(a+1)\ldots(a+b)}{1.2\ldots b} \int_0^q q^b(1-q)^{a-1}\,dq$$

$$= \int_q^1 q^b(1-q)^{a-1}\,dq \Big/ \int_0^1 q^b(1-q)^{a-1}\,dq.$$

Starting with the cumulative series for the incomplete negative binomial distribution, using contour integrals, Wise has derived a rapidly converging expansion for the percentage points of the incomplete beta-function.

Three Rules are given in the *Essay* for the evaluation of the incomplete beta-integral, the first of which may be given as follows.

Rule 1. $\Pr[x_1 < x < x_2 \mid p \text{ successes and } q \text{ failures}]$

$$= (n+1)\binom{p+q}{p}\left[\frac{x_2^{p+1}}{p+1} - \binom{q}{1}\frac{x_2^{p+2}}{p+2} + \binom{q}{2}\frac{x_2^{p+3}}{p+3} - \&c.\right.$$

$$\left. - \left\{\frac{x_1^{p+1}}{p+1} - \binom{q}{1}\frac{x_1^{p+2}}{p+2} + \binom{q}{2}\frac{x_1^{p+3}}{p+3} - \&c.\right\}\right].$$

We may write this alternatively as

$$\Pr[x_1 < x < x_2 \mid p, q] = (n+1)\binom{n}{p}\sum_{i=0}^{q}\binom{q}{i}\frac{1}{p+1+i}(x_2^{p+1+i} - x_1^{p+1+i}).$$

Bayes notes that although this form is to be used for p large and q small, something similar holds for q large and p small, the required expression being that given above with p and q interchanged and x_1 and x_2 replaced by $1 - x_2$ and $1 - x_1$, respectively (this resulting from a simple change of variable in the beta-integral).

The proof of the first Rule runs essentially as follows. Expansion of the binomial and term-by-term integration results in

$$\int_0^x u^p(1-u)^q\,du = \sum_{i=0}^q (-1)^i \binom{q}{i}\frac{x^{p+i+1}}{p+i+1} \qquad (7.6)$$

$$= \sum_{i=0}^q \frac{(q)_i}{(p+i+1)_{i+1}} x^{p+i+1}(1-x)^{q-i}, \qquad (7.7)$$

where, for r a positive integer, $(x)_r = x(x-1)\ldots(x-r+1)$ and $(x)_0 = 1$. The equivalence of the two series may be seen either by expanding $(1-x)^{q-i}$ and collecting terms in the same powers of x, or by differentiation. The setting of $x = 1$ in (7.7) results in

$$\int_0^1 u^p(1-u)^q\, du = 1 \bigg/ (n+1)\binom{n}{p}, \qquad (7.8)$$

and (7.6) and (7.8) lead to Bayes's first rule as given above.

Bayes's contribution to the work essentially finishes with this Rule, the remainder of the paper (up to the Appendix) being 'Bayes as edited by Price'. Two further rules follow here, to be used when p and q are large, in which case the first rule becomes impracticable. The proof of the second rule, briefly given at this point in the *Essay*, is postponed to the *Supplement*, and the proof of the third, deduced from the second, is merely indicated here. Although we consider these last rules in our next chapter, we might note here that in his first rule Bayes gives a result in which the desired probability lies between X and x, limits that become $\frac{p}{n} + z$ and $\frac{p}{n} - z$ in the second and third rules.

As a final remark on Bayes's proofs let us note that in his results (be they his own proofs or those by Price) relations between integrals are proved by examining the corresponding relations between the integrands, these latter relations being in turn investigated by the comparing of the derivatives of the integrands, due regard being taken of monotonicity (see Hald [1990b, p. 144]). We have in fact seen in an earlier chapter that Bayes adopted the same technique in his *An Introduction to the Doctrine of Fluxions*.

Part of the difficulty of the *Essay* and the *Supplement* lies in the Rules and their proofs. Both Bayes and Price, having given various approximations to the incomplete beta-integral, were concerned with the accuracy of their approximations (as we have already mentioned, this concern has sometimes been given as a reason for Bayes's hesitation in publishing his work); and Hald [1990b, p. 140] has noted that especial attention is paid in the *Essay* to the maximum error that may be incurred in the making of these approximations.

In his Introduction Price noted that de Moivre's result was not 'rigorously exact' for finite n. Now the integral limit theorem, as given by de Moivre, may be written in the form

$$\Pr\left[-z \leq \frac{\mu/n - p}{\sqrt{(pq)/n}} \leq z\right] \to \frac{2}{\sqrt{2\pi}} \int_0^z e^{-x^2/2}\, dx,$$

the term μ/n representing the relative frequency and pq/n its variance. If we ignore Price's objection and, anticipating somewhat, consider the limit as $n \to \infty$ in the Bayesian formula, we obtain

$$\Pr\left[-z \le \frac{\overline{p} - a}{\sqrt{(pq/n^{3/2})}} \le z\right] \to \frac{2}{\sqrt{2\pi}} \int_0^z e^{-x^2/2}\, dx,$$

where $a = p/n$ and $\overline{p}$ is the estimate of the binomial probability p. (Sheynin [1971a, p. 236] notes that this is given in a slightly different form in Timerding, but that the first approximation $a = E(\overline{p})$ and $qp/n^{3/2} = Var(\overline{p})$ is missing there.)

Hald [1990b, p. 147] notes in fact that the combination of the right-hand side of

$$I_x(p+1, q+1) = \sum_{i=p+1}^{p+q+1} \binom{p+q+1}{i} x^i (1-x)^{p+q+1-i}$$

with de Moivre's Normal approximation, and the replacement of $n+1$ by n, yields a Normal approximation to the beta distribution.

One must also note that de Moivre had proved that the *symmetric* binomial tends, as $n \to \infty$, to the Normal distribution, and that the latter could therefore be used as an approximation to a cumulative binomial. Although he later showed that the same limit obtained for the *skew* binomial, he did not substantiate his assertion that

> the Problems relating to the Sum of the Terms of the Binomial[71] $(a+b)^n$ will be solved with the same facility as those in which the Probabilities of happening and failing are in a Ratio of Equality. [de Moivre, 1756, p. 250]

Thus although Bayes and Price took peculiar pains — like London's Noble Fire-brigade in Belloc's *Matilda* — over the accuracy of their approximations, de Moivre (and indeed also Laplace) merely(!) derived the Normal distribution as the limiting form of the binomial and beta distributions, respectively.

Finally let us turn our attention to the Appendix, in which Price considers a number of illustrative examples. Almost immediately the results of the *Essay* are applied to the occurrence of future events, as, for example, Price's first illustration shows:

> Let us first suppose, of such an event as that called M in the essay, or an event about the probability of which, antecedently to trials, we know nothing, that it has happened *once*, and that

326   7. The Essay on Chances

> it is enquired what conclusion we may draw from hence with respect to the probability of it's happening on a *second* trial. The answer is that there would be an odds of three to one for somewhat more than an even chance that it would happen on a second trial. **[405]**

Direct application of Bayes's first rule yields the desired solution, and Price then writes

> which shews the chance there is that the probability of an event that has happened once lies somewhere between 1 and $\frac{1}{2}$; or (which is the same) the odds that it is somewhat more than an even chance that it will happen on a second trial. **[405]**

Although a solution to Price's problem might be obtained by using Laplace's Rule of Succession[72], in terms of which the probability of a second occurrence of the event M would be

$$\int_0^1 x^2\,dx \bigg/ \int_0^1 x\,dx = \frac{2}{3},$$

I do not think that this would be a correct interpretation of the question. For no cognisance would be taken of the requirement that there should be 'more than an even chance that it will happen on a second trial'. The incorporation of such a requirement, I have suggested elsewhere (see Dale [1999, §4.6]), can be accomplished by consideration of Condorcet's *Essai sur l'application de l'analyse à la probabilité des décisions rendues à la pluralité des voix*, Problem IV, pp. 180–183.

It is, however, possible to obtain Price's result from Bayes's theory by an appropriate interpretation. The rôle of W, the first ball thrown onto Bayes's table, being filled by the event described by Price as having happened once, the latter is then to be regarded as the initial event with reference to which successes and failures are to be defined. Price's problem is then seen to be the finding of the probability that the next throw results in a 'success' (say), in falling in the interval $[\frac{1}{2}, 1]$, where the lower limit of this interval is determined by the position of the first ball. The solution is then immediately given for *one* success as

$$\int_{1/2}^1 x\,dx \bigg/ \int_0^1 x\,dx = \frac{3}{4},$$

as Price showed.

Price next turns his attention to the problem of finding the odds when the event of interest happens once again after having happened twice, thrice, ..., p times, these odds, $2^{p+1} - 1$ to 1, being given in the general case by

$$\int_{1/2}^{1} x^p \, dx \Big/ \int_{0}^{1} x^p \, dx = 1 - 1/2^{p+1}.$$

This is indeed the solution given by Bayes's tenth proposition. However, it might perhaps be wrong to follow Price and interpret it as the odds'for *more* than an equal chance that it will happen on further trials' **[405]**.

In the next example considered in the Appendix Price interprets the results of the *Essay* even more broadly. If it is only known that an event has happened ten times without failing, he supposes the

> enquiry to be what reason we shall have to think we are right if we guess that the probability of it's happening in a single trial lies somewhere between 16/17 and 2/3, or that the ratio of the causes of it's happening to those of it's failure is some ratio between that of sixteen to one and two to one **[406]**

a quaesitum we may write as

$$\Pr[\tfrac{2}{3} < x < \tfrac{16}{17} \mid 10 \text{ successes and 0 failures}]$$

or

$$\Pr[\tfrac{2}{3} < C(E)/C(\overline{E}) < \tfrac{16}{17} \text{ given 10 successes and 0 failures}],$$

where $C(E)$ denotes the 'causes of E'. Although the first formulation is in agreement with Bayes's Proposition 9, the second, in terms of causes, has no parallel in the *Essay*.

In his next example, important because of the attention that is placed there upon the initial event, Price supposes there to be a die of unknown number of faces and unknown constitution (we may suppose the faces to be numbered $n_1, n_2, ..., n_k$ — not necessarily distinct). Should the first throw of the die result in the appearing of the face n_i (say), then all that is known is that the die has this face. It is only at this stage, i.e., *after* the first throw, that the situation described in the *Essay* obtains, and the occurrence of n_i in any subsequent trial is then an event of whose probability we are completely ignorant. Should the face labelled n_i again appear on the second trial, then by an earlier example in the Appendix, the odds will be three to one on that n_i is favoured (either through being more numerous, or (equivalently) because of the constitution of the die).

Price next turns to the problem of the probability of the rising of the sun. Using an argument entirely analogous to that advanced in his die-tossing example, he explains that the first sinking of the sun a sentient person who has newly arrived in this world would see (i.e., a person with no prior

328   7. The Essay on Chances

knowledge of the situation), would leave him 'entirely ignorant whether he should ever see it again' **[409]**. Thus, according to Price,

> let him see a second appearance or one *return* of the Sun, and an expectation would be raised in him of a second return, and he might know that there was an odds of 3 to 1 for *some* probability of this **[409]**

a result we may express symbolically as

$$\Pr[(1/2) < x < 1 \mid \text{one return}] = \int_{1/2}^{1} x\,dx \bigg/ \int_{0}^{1} x\,dx = 3/4.$$

We might note here that Price adopts in his appendix the common interpretation of 'probable' ('improbable') as a probability greater than (less than) $\frac{1}{2}$. This, Hald has suggested [1998, p. 146], is the reason for his examining things such as '$x > \frac{1}{2}$' and 'a probability of nearly $\frac{1}{2}$'.

Although this problem of the rising of the sun had been addressed by Hume, Sir Edward Eizat had written earlier about it in his *A Discourse of Certainty. Wherein you have a Further Proof Of the Power of the Mathematicks, and of The Profound Knowledge of A.P. M.D.*, a work bound separately at the end of his *Apollo Mathematicus*. Here Eizat wrote

> For the Certainty that I have, that the Sun rose Yesterday, and that I have that it will rise to Morrow, tho vastly different, are both stiled by them Indubitable; whereas only the former is such: For tho the latter be so certain that I have no such reason to doubt of it, as to be in continual fear of the contrary; yet it comes not up to the certainty of the former, of which I have as great assurance, and every whit as well grounded, as of any Proposition in Euclid. [1695, pp. 8–9]

Price next extends his example to the case of several occurrences[73]:

> let it be supposed that he has seen it return at regular and stated intervals a million of times. The conclusions this would warrant would be such as follow — There would be the odds of the millionth power of 2, to one, that it was likely that it would return again at the end of the usual interval. **[409 – 410]**

There is in fact a slight error here: since the number of occurrences of the event n is 1,000,000, the odds should be $2^{1,000,001}$ to 1 on a reappearance of the sun. The correct exponent of 2 is $(n+1)$ — i.e., the number of *risings* of the sun — and not n, which is the number of *returns* of the sun[74].

7.3. Commentary

One might note here that, although Price is perhaps correctly applying Bayes's result, he is applying it to future events. There is no explicit mention in the *Essay* of the applicability of Bayes's result to the case of a 'single throw' *after* experience, and I have suggested elsewhere (see Dale [1999, Chapter 7]) that Bayes's result is in accord with *not* interpreting this 'single trial' in a predictive sense — though it is not obvious from the quotation from **[393 – 394]** given above ('In what follows') that Bayes meant his result to be used only in a retrodictive sense: Price in fact writes quite explicitly in his introduction that Bayes's original intent was to find the probability of an event given a number of occurrences and failures.

In connexion with the question of Bayes's Theorem and its applicability to future events, one might note the following result by Watanabe, in which the distribution of the prior is seen to be unimportant.

> If an event whose probability is unknown à priori, has been observed to have occurred r times in n independent trials made on the event, n being a very large number, the probability of the event happening s times out of next m trials (m being small compared with n) can be approximated by
>
> $$ {}_mC_s \left(\frac{r}{n}\right)^s \left(1 - \frac{r}{n}\right)^{m-s}. $$

[1933, p. 390]

It should be noted that, for the objectivist, the prior distribution may be viewed as a posterior (posterior to the acquisition of relevant prior information), Bayes's Theorem then being used in reverse to derive the prior from the posterior (see O'Hagan [1994, §5.33]).

One might also note that Price passes easily from the application of probability in games of chance (his die-tossing example) to its use in connexion with physical phenomena (the problem of the rising of the sun). Whether the notion that is applicable in the case of the tossing of a die is also applicable in the case of natural phenomena could be debated: the analogy would be rejected by some, and others would perhaps accept it in connexion with matters such as birth ratios but not accept it — or at least query its fitness — in matters such as the example discussed by Price here.

As his next example Price considers 'cases where an experiment has sometimes succeeded and sometimes failed' **[411]**. Illustrating the general ideas involved, he considers the drawing of blanks and prizes from a lottery, fixing his attention on what we might write as

$$ \Pr\left[x_1 < x < x_2 \mid p \text{ blanks and } q \text{ prizes drawn}\right], $$

where x is the (presumably true) proportion of blanks to prizes in the lottery. This again is a direct (and correct) application of Bayes's results[75].

After remarking on the probability of causes, Price notes that 'The foregoing calculations further shew us the uses and defects of the rules laid down in the essay' **[417]**. The defects, as noted by Price, seem to be that the second and third rules 'do not give us the required chances within such narrow limits as could be wished' **[417]**. These limits, however, contract as q increases with respect to p, the exact solution being given by the second rule when $p = q$.

In a footnote at this point Price states that he had found an improvement of the approximation given in the second and third rules, by showing that

$$2\Sigma / (1 + 2Ea^p b^q + 2Ea^p b^q / n)$$

'comes almost as near to the true value wanted as there is reason to desire, only always somewhat less' **[418]**. We take this up in our next chapter.

Let us end this discussion by recalling that in his introduction to the *Essay* Price had commented on de Moivre's rules,

> to find the probability there is, that if a very great number of trials be made concerning any event, the proportion of the number of times it will happen, to the number of times it will fail in those trials, should differ less than by small assigned limits from the proportion of the probability of its failing in one single trial.

He also noted that, to the best of his knowledge, no one had to that time shown how to solve the converse problem, viz.

> the number of times an unknown event has happened and failed being given, to find the chance that the probability of its happening should lie between any two named degrees of probability.
> **[373]**

De Moivre's work was thus insufficient to make consideration of this point unnecessary, and Price concludes the Appendix by noting that

> what most of all recommends the solution in this *Essay* is, that it is compleat in those cases where information is most wanted, and where Mr. De Moivre's solution of the inverse problem[76] can give little or no direction; I mean, in all cases where either p or q are of no considerable magnitude. **[418]**

Let us, in conclusion, say something briefly about the relationship between Bayes's Theorem and similar results.

Although I have dwelt on the matter at some length before, it might be opportune to recall here that Bayes's Theorem is sometimes seen as an inverse to Bernoulli's (my reasons for dissenting from such a view are given in my [1999, §1.3]). In his *Ars conjectandi* Bernoulli gave a precise proof of the direct result and perhaps a hint of the inverse theorem, though I believe that the latter was first precisely stated by David Hartley in 1749. Having mentioned de Moivre's generalization of Bernoulli's result, Hartley went on to say

> An ingenious Friend has communicated to me a Solution of the inverse Problem, in which he has shewn what the Expectation is, when an Event has happened p times, and failed q times, that the original Ratio of the Causes for the Happening or Failing of an Event should deviate in any given Degree from that of p to q. And it appears from this Solution, that where the Number of Trials is very great, the Deviation must be inconsiderable.
> [pp. 338–339]

The identity of this 'ingenious Friend' is uncertain (see Dale [1988], Edwards [1986], and Stigler [1983] for a discussion): however, if we note Hartley's use of *expectation* and *cause* here, and bear in mind that Bayes's work in the *Essay* was based on expectations and that Price certainly considered the probability of causes, we might assign a non-zero probability to the statement that the friend was Bayes. Further support, albeit very slight, for this opinion is perhaps provided by noting that Hartley was acquainted with the novelist Samuel Richardson (see Barbauld [1804, vol. II, pp. 25, 46]) and, as we have seen in an earlier chapter, the latter is known to have visited Tunbridge Wells while Bayes was there.

I have explored the connexion between Bernoulli's Theorem, its inverse and Bayes's Theorem in my [1999, §1.3]; nevertheless, it might be useful to have a brief summary here of part of the appropriate passage[77]. To that end, let us consider a binary experiment with constant probability p of success and suppose that n independent trials of this experiment have resulted in S_n successes. Then

$$\Pr[S_n = s \mid n, p] = \binom{n}{s} p^s (1-p)^{n-s}, \quad s \in \{0, 1, 2, \ldots, n\}.$$

According to Bernoulli's Theorem, as $n \to \infty$ the observed frequency of successes, $f = s/n$, tends to p in the sense that

$$(\forall \varepsilon > 0) \quad \Pr[|f - p| < \varepsilon] \to 1.$$

No mention is made in this result, however, of how large n must be for any specified accuracy to be reached. The answer to this problem is given by the de Moivre–Laplace limit theorem[78], from which one finds that

$$\Pr[df \mid p, n] \sim \left[\frac{n}{2\pi p(1-p)}\right]^{1/2} \exp\left[-\frac{n(f-p)^2}{2p(1-p)}\right] df. \qquad (7.9)$$

However, under the conditions mentioned above, Bayes's Theorem leads to the beta distribution

$$B(p) \equiv \Pr[dp \mid s, n] = (n+1)\binom{n}{s} p^s (1-p)^{n-s} dp.$$

Since $\ln B(p)$ is maximized by $\hat{p} = s/n = f$, we find, on expanding $\ln B(p)$ in a Taylor series about $\hat{p}$,

$$\ln B(p) \sim \{\ln[(n+1)!/s!(n-s)!] + s \ln f + (n-s)\ln(1-f)\}$$

$$+ \left[-\frac{n}{f(1-f)}\right]\frac{(p-f)^2}{2},$$

whence

$$B(p) \sim (n+1)\binom{n}{s} f^s (1-f)^{n-s} \exp\left[-\frac{n(p-f)^2}{2f(1-f)}\right].$$

On using the Stirling–de Moivre formula to approximate the factorials, and noting that, for large n,

$$(n+1)^{n+1+1/2} \sim n^{n+1+1/2}$$

$$e^{-(n+1)} \sim e^{-n},$$

we find that

$$\Pr[dp \mid s, n] = B(p) \sim \left[\frac{n}{2\pi f(1-f)}\right]^{1/2} \exp\left[-\frac{n(p-f)^2}{2f(1-f)}\right] dp. \qquad (7.10)$$

The symmetry between the probability of p given f and of f given p is seen on comparing (7.10) with (7.9), the comparison also showing the solution of Bernoulli's inversion problem in the Normal case (Jaynes [1979, p. 19]).

We have already mentioned that the approaches taken by Bayes and Laplace, in their consideration of what we might loosely call a limiting inverse result to Bernoulli's Theorem, differed from each other. To summarize the French investigations, let us suppose that an event having unknown probability x of occurring on a single trial has happened p times and failed to occur q times in $n = p + q$ (independent) trials. As Todhunter [1865, Art. 997] notes, Laplace then *states* that

$$\Pr\left[\frac{p}{n} - \frac{z\sqrt{(2pq)}}{n\sqrt{n}} < x < \frac{p}{n} + \frac{z\sqrt{(2pq)}}{n\sqrt{n}}\right]$$
$$= \frac{2}{\sqrt{\pi}} \int_0^z \exp(-t^2)\, dt + \frac{\sqrt{n}}{\sqrt{(2\pi pq)}} \exp(-z^2). \qquad (7.11)$$

However, in deriving an approximation to the incomplete beta-function

$$\int_{x_1}^{x_2} x^p (1-x)^q\, dx \bigg/ \int_0^1 x^p (1-x)^q\, dx, \qquad (7.12)$$

Laplace lets

$$x_1 = \frac{p}{n} - \frac{z\sqrt{(2pq)}}{n\sqrt{n}}, \qquad x_2 = \frac{p}{n} + \frac{z\sqrt{(2pq)}}{n\sqrt{n}},$$

and the ratio (7.12) (i.e., the probability on the left-hand side in (7.11), by Bayes's Theorem) is then found to be given approximately by

$$\frac{2}{\sqrt{\pi}} \int_0^z \exp(-t^2)\, dt. \qquad (7.13)$$

That is, expression (7.11) is derived by what we might term an 'inversion' of Bernoulli's Theorem, whereas (7.13) follows from Bayes's Theorem.

Adopting the technique of Laplace's direct method, Monro shows that

> upon the hypothesis of equally probable values within equal ranges, the inversion [of Bernoulli's Theorem] is so far legitimate, that either theorem [i.e., (7.11) or (7.13)] may be inferred from the other with little calculation, and in particular without the approximate evaluation of a general integral, and accordingly that the two solutions are identical in principle.
> [1874, p. 75]

See also Dempster [1966, pp. 356–357].

Todhunter (loc. cit.) also notes that Poisson gave another result in addition to deriving (7.11): he showed that the probability

$$\Pr\left[\frac{p}{n} - \frac{(z+dz)\sqrt{(2pq)}}{n\sqrt{n}} < x < \frac{p}{n} - \frac{z\sqrt{(2pq)}}{n\sqrt{n}}\right]$$

is given approximately by $Z\,dz$, where Z is defined by

$$Z = \frac{1}{\sqrt{\pi}} \exp(-z^2) - \frac{2(p-q)z^3}{3\sqrt{(2\pi npq)}} \exp(-z^2).$$

Moreover, by carrying the approximation further than Laplace did, he reached this same result in 1830 as an approximation to (7.12). (One must also note[79] that de Morgan [1838b] queried Poisson's results, though Todhunter noted that their differences were occasioned by de Morgan's not having carried the needed approximations far enough.)

7.4 Editions of the *Essay*

There have been a number of translations and editions of the *Essay* since it was first published. Those I have traced are listed here.

An early copy. 1764. The 1918 catalogue of the Printed Books in the Edinburgh University Library lists, as number 0*22.14/1, an item entitled 'A Method of Calculating the Exact Probability of All Conclusions founded on Induction. By the late Rev. Mr. Thomas Bayes, F.R.S.' [This is a reprint of the *Essay*. With it is bound, as 0*22.14/2 in the Library, the *Supplement*[80].]

Timerding, H.E. (Ed.) 1908. *Versuch zur Lösung eines Problems der Wahrscheinlichkeit von Thomas Bayes.* Ostwald's Klassiker der Exakten Wissenschaften, No. 169. Leipzig: Wilhelm Engelmann.

Molina, E.C. & Deming, W.E. 1940. *Facsimiles of two papers by Bayes.* Washington, D.C.: The Graduate School, The Department of Agriculture.

Barnard, G.A. 1958. Thomas Bayes — a biographical note. *Biometrika* 45: 293–295. Followed by a copy of the *Essay*, pp. 296–315.

Clero, J.P. 1988. *Thomas Bayes. Essai en vue de resoudre un probleme de la doctrine des chances.* With a preface by B. Bru. Cahiers d'Histoire et de Philosophie des Sciences, N. 18. Paris: Société Française d'Histoire des Sciences et des Techniques. [This edition contains a photocopy of the original.]

Press, S.J. 1989. *Bayesian Statistics: Principles, Models, and Applications.* New York: John Wiley & Sons. [The *Essay* is reprinted as an Appendix, the version given being that from *Biometrika* mentioned above.]

Timerding's edition presents a translation of the *Essay* in several parts: the first covers pages 376 to 399, the second pages 298 to 310 of the *Supplement*, and the third pages 401 to 403 of the *Essay*. Price's introductory letters to the *Essay* and to the *Supplement* are omitted, as is the Appendix to the *Essay* and the footnotes. The notation is slightly changed — for example, some expressions are given in terms of dx rather than $\dot{x}$ — though Timerding's amplifications of the formulae etc. are useful. The third part is concerned in particular with the evaluation of Bayes's coefficient E, the discussion giving considerable more detail than the original.

Press's book carries a photograph of a portrait reputed to be of Thomas Bayes. The publication of the portrait has been traced back to O'Donnell [1936], where it appears on page 335 above the legend 'Rev. T. Bayes Improver of the Columnar Method developed by Barrett." I have given elsewhere[81] reasons for my disbelief that this is indeed a portrait of Bayes.

Of the above texts, the photographic reprint issued by the Department of Agriculture is perhaps the most useful to the English reader. That published in *Biometrika* underwent some editing — more, indeed, than in the present book: certain printing errors and archaisms were removed, and certain other small changes were made. Thus, for instance, Bayes's 'which ever' **[377]** becomes 'whichever'; '2d' and '3d' on the same page become '2nd' and '3rd' (and elsewhere 'second' and 'third'); '&c.' is often given as 'etc.'; 'any where' and 'can't' on **[390]** become 'anywhere' and 'cannot'; 'any thing' on **[392]** becomes 'anything'; '$\times^d$', 'every where' and 'serieses' on **[399]** are given as 'multiplied' (and elsewhere), 'everywhere' and 'series'; 'it's' on **[401]** (and elsewhere) is given as 'its', and on the same page Barnard inserts '[h]' after the word '*ratio*'; 'parmanent' is changed to 'permanent' on **[410]**; 'Let now enquire' on **[415]** becomes 'Let [us] now enquire'; the expression $m = \frac{\sqrt{n^3}}{\sqrt{pq}}$ on **[416]** is given as $mz = z\frac{\sqrt{n^3}}{\sqrt{pq}}$ (a change needed to get the correct numerical value). On three occasions on **[386]** Bayes's subjunctive is changed: for example, 'Wherefore the probability it rest upon Fb is to the probability it dont as fb to Bf' appears as 'Wherefore the probability it rests upon Fb is to the probability it does not as fb to Bf'.

Barnard does, however, seem to have incorporated most of the changes indicated in the set of errata given here at the conclusion of the tract, apart from those to be made to **[405]** and **[418]**. He has also effected the corrections noted by Price in the *Supplement to the Essay*, though he has not changed the definition of m^2 on **[400]** and **[416]** from n^3/pq to $n^3/2pq$ as indicated in the *Supplement*. In his statement of the third Rule Barnard gives the term $\sqrt{Kpq} \times h/[2\sqrt{Kpq} + hn^{\frac{1}{2}} + hn^{-\frac{1}{2}}]$ appearing in the first formula as $\frac{1}{2}\sqrt{Kpq} \times h/[\sqrt{Kpq} + hn^{\frac{1}{2}} + hn^{-\frac{1}{2}}]$, and similarly in the following formula.

8

The Supplement to the Essay

Every man has his price.

Edward Bulwer Lytton.
Walpole.

8.1 Introduction

On the 26th November 1764 Richard Price sent a Supplement to Bayes's *Essay towards solving a Problem in the Doctrine of Chances* to John Canton. Devoted in the main to proofs and some elaboration of the Rules set out in the *Essay*, this Supplement may be seen as a mathematical appendix to the earlier paper. The first part of the *Supplement* is apparently due to Bayes himself, for before Section 13 one finds the words 'Thus far I have transcribed Mr. Bayes.': the rest of the results are due to Price, and consist mainly of the obtaining of tighter bounds for certain probabilities than those given in Bayes's second rule.

As we noted in the previous chapter, Price, in his letter accompanying the submission of the *Essay* to Canton, expressed some slight dissatisfaction with some of the results given by de Moivre. The latter had first proved that the *symmetric* binomial distribution tends to the Normal probability distribution function, and had then shown that the *skew* binomial distribution behaves in the same way. A similar limiting result for the beta distribution was later derived by Laplace[1]. Bayes and Price, on the other hand, in the *Essay* and more particularly in the *Supplement*, obtained an approximation to the two-sided beta probability integral that is substantially better than the Normal approximation. It was shown that the skew beta probability density function may be approximated by a symmetric beta probability density function multiplied by a factor that tends to 1 as $n \to \infty$, the skew and symmetric beta densities having the same maximum and points of inflexion.

This approximation of the (skew) beta distribution by a non-Normal distribution is an instance of a matter that is still of concern to statisticians[2].

It is frequently necessary for the Bayesian to evaluate various characteristics of posterior distributions, for example, their densities, means, or variances. Although such an evaluation may be possible in certain cases, the problem becomes complicated when a conjugate prior-likelihood pair is not involved. Tierney & Kadane [1986] have given a method for appropriate approximation in such a case, and Morris [1988] has discussed the fitting of a two-parameter Pearson family (a family that contains the beta distribution) when the sample size is small, this method requiring the calculation only of the first two derivatives of the logarithm of the posterior density. Such methods may well result in approximations that are better than those obtained by using the Normal distribution (as may be inferred from — or at least suggested by — the tract under discussion here), and the approach is more general than that using the Laplace method, which provides only the approximation of integrals. Again then, we find that there are aspects of Bayes's work that are the seeds of later harvests.

Although neither de Moivre nor Laplace discussed the accuracy of his approximations, Bayes and Price took pains to find the maximum error that might be incurred, this being accomplished by the finding of an appropriate measure of skewness. These investigations of course result in formulae that are considerably more complicated than those given by de Moivre and Laplace, whose concern was essentially more with the obtaining of a limiting distribution to the binomial and beta distributions, respectively.

Letting $X \sim b(n, p)$, we may write de Moivre's limiting result as

$$\Pr[-z < (X/n) - p < z \mid n, p] \sim \frac{2}{\sqrt{\pi}} \int_0^t \exp(-u^2)\, du, \qquad (8.1)$$

where $t = (z\sqrt{n})/\sqrt{2pq}$ and $q = 1 - p$. On expanding the integrand in (8.1) and integrating term-by-term, de Moivre found the expansion

$$\frac{2}{\sqrt{\pi}} \int_0^t \exp(-u^2)\, du = \frac{2}{\sqrt{\pi}} \sum_{k=0}^{\infty} \frac{(-1)^k t^{2k+1}}{k!\,(2k+1)}.$$

We show that in the *Supplement* Price derived the same series as an expansion of the Normal approximation to the posterior distribution, the latter of course being that arrived at in the *Essay*. This posterior being a beta distribution, Price's results anticipated those published in Laplace's *Mémoire sur les probabilités* in 1781.

LII. *A Demonstration of the Second Rule in the Essay towards the Solution of a Problem in the Doctrine of Chances, published in the Philosophical Transactions,* Vol. LIII. *Communicated by the Rev. Mr.* Richard Price, *in a Letter to Mr.* John Canton, *M. A. F. R. S.*

Dear Sir, Nov. 26, 1764.

Read Dec. 6, 1764. I Send you the following *Supplement to the Essay on a Problem in the Doctrine of Chances*, hoping that you may not think it improper to be communicated to the Royal Society. I should not have troubled you again in this way had I not found that some additions to my former papers were necessary in order to explain some passages in them, and particularly what is hinted in the note at the end of the Appendix. "I have first given the deduction of Mr. Bayes's second rule chiefly in his own words; and then added, as briefly as possible, the demonstrations of several propositions, which seem to improve considerably the solution of the problem, and to throw light on the nature of the curve by the quadrature of which this solution is obtained." Perhaps, there is no reason for being very anxious about proceeding to further improvements. It would, however, be very agreeable to me to see a yet easier and nearer approximation to the value of the two series's in the first rule: but this I must leave abler persons to seek, chusing now entirely to drop this subject. **[296]** The solution of the problem enquired after in the papers I have sent you has, I think, been hitherto a *desideratum* in philosophy of some consequence. To this we are now in a great measure helped by the abilities and skill of our late worthy friend; and thus are furnished with a necessary guide in determining the nature and proportions of unknown causes from their effects, and an effectual guard against one great danger to which philosophers are subject; I mean, the danger of founding conclusions on an insufficient induction, and of receiving just conclusions with more assurance than the number of experiments will warrant. I am, under a sense of the value of your friendship, heartily yours,

Richard Price.
[297]

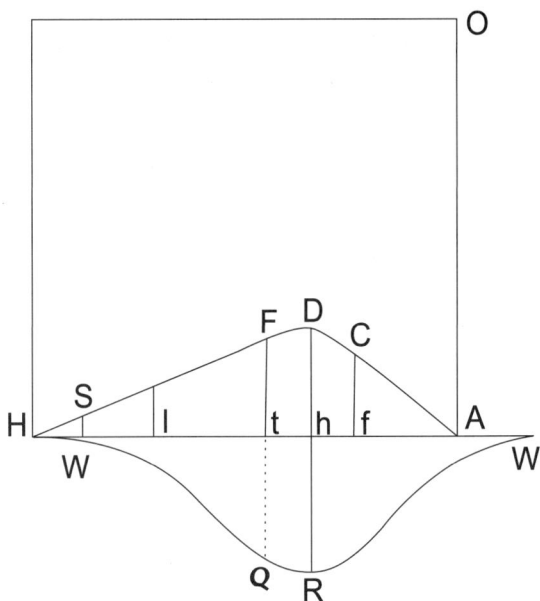

ART. 1. If the curve ADH be divided into two parts by the ordinate Dh making Ah to Hh as p is to q; then taking $a = \frac{p}{n}$ and $b = \frac{q}{n}$ the ratio of the Area ADh to HO will be $\frac{a^{p+1} \times b^q}{p+1} \times \overline{1 + \frac{q}{p+2} \times \frac{p}{q} + \frac{q \times \overline{q-1} \times p^2}{p+2 \times \overline{p+3} \times q^2} + \frac{q \times \overline{q-1} \times \overline{q-2} \times p^3}{p+2 \times \overline{p+3} \times \overline{p+4} \times q^3}}$ +&c. For the series $\frac{x^{p+1} r^q}{p+1} + \frac{q}{p+1} \times \frac{x^{p+2} \times r^{q-1}}{p+2} +$ &c. in Prop. 10. Art. 2. of the Essay, which expresses the ratio of ACf to HO, becomes the series when $x = a = \frac{p}{n}$, $b = r = \frac{q}{n}$; that is when **[298]** Cf has moved till it coincides with Dh and ACf becomes ADh. In like manner, from Art. 3. in the Essay, it appears that the ratio of HDh to HO is $\frac{a^p b^{q+1}}{q+1} \times \overline{1 + \frac{p}{q+2} \times \frac{q}{p} + \frac{p}{q+2} \times \frac{p-1}{q+3} \times \frac{q^2}{p^2}}$+ &c.

From hence it follows that the ratio of the difference between ADh and HDh to HO is $\frac{a^p b^q}{n}$ multiplied by the difference between the series $\frac{p}{p+1} + \frac{q}{p+1} \times \frac{p^2}{pq+2q} + \frac{q \times \overline{q-1} \times p^3}{p+1 \times \overline{p+2} \times pq^2 + 3q^2} +$ &c. and the series $\frac{q}{q+1} + \frac{p \times q^2}{q+1 \times pq+2p} + \frac{p \times \overline{p-1} \times q^3}{q+1 \times \overline{q+2} \times p^2 q + 3p^2} +$ &c. the former series being to be subtracted from the latter, if HDh is greater than ADh, and *vice versa*.

2. The ratio of any term in the former of the two foregoing series to that which next but one follows the correspondent term in the latter is

340   8. The Supplement to the Essay

$\frac{pq+p}{p\times q} \times \frac{pq+2p}{p\times q} \times \frac{p\times q}{qp+q} \times \frac{pq+3p}{pq-q} \times \frac{pq}{pq+2q} \times \frac{pq+4p}{pq-2q} \times \frac{pq-p}{pq+3q} \times \frac{pq+5p}{pq-3q} \times \frac{pq-2p}{pq+4q} \times \frac{pq+6p}{pq-4q}$
&c. taking twice as many terms and four over as there are units in the number which expresses the distance of the term in the former series from its first term; which **[299]** ratio if q be greater than p, it is evident must be greater than the ratio of equality. Wherefore, if from the second series you subtract the two first terms which together are less than two, the remainder is less than the former series; and of consequence, the former series subtracted from the latter cannot leave a remainder so great as two. And therefore in this case, that is, when q is greater than p, by the preceding article, the ratio of HDh−ADh to HO cannot be so great as $\frac{2a^p b^q}{n}$.

3. The curve ADh being as before divided into two parts ADh and HDh, let the ordinates Cf and Ft be placed on each side of Dh and at the same distance from it, and let z be the ratio of hf or ht to AH. Then if y, x and r be respectively the ratios of Cf, Af and Hf to AH, by the nature of the curve $y = x^p r^q$. But because the ratio of Ah to AH is a, and that of hf to AH is z, the ratio of A$h - hf$ (=Af) to AH is $a - z$. Wherefore $a - z = x$. And in like manner $b + z = r$. But $y = x^p r^q$, and y is the ratio of Cf to AH. Wherefore the ratio of Cf to AH is $\overline{a-z}|^p \times \overline{b+z}|^q$. And in like manner the ratio of Ft to AH is $\overline{a+z}|^p \times \overline{b-z}|^q$. And consequently C$f$ is to Ft as $\overline{a-z}|^p \times \overline{b+z}|^q$ is to $\overline{a+z}|^p \times \overline{b-z}|^q$.

4. If q is greater than p, $\overline{a+z}|^p \times \overline{b-z}|^q$ is greater than $\overline{a-z}|^p \times \overline{b+z}|^q$, and the ratio between them increases as z increases. For the hyperbolic logarithm **[300]** of that ratio taken as usual, and then instead of p and q putting na and nb because $a = \frac{p}{n}$ and $b = \frac{q}{n}$ (Vid. Art. 1.) you will find to be $2n$ multiplied by the series $\frac{b^2-a^2}{3b^2 a^2} \times z^3 + \frac{b^4-a^4}{5b^4 a^4} \times z^5 + \frac{b^6-a^6}{7b^6 a^6} \times z^7$ &c. which logarithm when q is greater than p, and therefore b greater than a has all its terms positive, and so much the greater as z is greater; and therefore it is the logarithm of a ratio greater than that of equality, and which increases as z increases.

5. By Art. 3. Ft is to Cf as $\overline{a+z}|^p \times \overline{b-z}|^q$ is to $\overline{a-z}|^p \times \overline{b+z}|^q$. And by Art. 4. $\overline{a+z}|^p \times \overline{b-z}|^q$ is greater than $\overline{a-z}|^p \times \overline{b+z}|^q$, and the ratio between them increases as z increases, if q is greater than p. Wherefore, upon this supposition, also Ft is greater than Cf, and the ratio between them increases as z or ht and hf increases, and consequently this will be true also concerning the areas described by them as their equal abscisses ht and hf increase. Wherefore, when q is greater than p, DhtF is greater than DhfC, and the ratio between them increases as $hf = ht$ increases.

8.2. The Tract   341

6. Because Ah is as to Hh as p is to q, when q is greater than p, Hh is greater than Ah. In Hh therefore taking hl equal to Ah, by the preceding Art. the part of the figure HDh which insists upon hl will be greater than ADh, and the ratio of that part of HDh to ADh will be greater than the ratio of DhtF to DhfC. Consequently, much more (q being greater than p) the whole figure HDh is [301] greater than ADh, and the ratio of HDh to ADh is greater than that of DhtF to DhfC.

7. When q is greater than p, $\overline{1 - \frac{n^2 z^2}{pq}}\Big|^{\frac{n}{2}}$ is greater that $\overline{1 - \frac{nz}{p}}\Big|^p \times \overline{1 + \frac{nz}{q}}\Big|^q$ and less than $\overline{1 - \frac{nz}{q}}\Big|^q \times \overline{1 + \frac{nz}{p}}\Big|^p$. For the fluxion of $\overline{1 - \frac{n^2 z^2}{pq}}\Big|^{\frac{n}{2}}$ is $-\frac{n^3 z \dot{z}}{pq} \times \overline{1 - \frac{n^2 z^2}{pq}}\Big|^{\frac{n}{2} - 1}$ and the fluxion of $\overline{1 - \frac{nz}{p}}\Big|^p \times \overline{1 + \frac{nz}{q}}\Big|^q$ (because $p + q = n$) is $-\frac{n^3 z \dot{z}}{pq} \times \overline{1 - \frac{nz}{p}}\Big|^{p-1} \times \overline{1 + \frac{nz}{q}}\Big|^{q-1}$. Wherefore $\overline{1 - \frac{n^2 z^2}{pq}}\Big|^{\frac{n}{2}}$ is to $\overline{1 - \frac{nz}{p}}\Big|^p \times \overline{1 + \frac{nz}{q}}\Big|^q$ as the fluxion of the former multiplied by $\overline{1 - \frac{n^2 z^2}{pq}}$ to the fluxion of the latter multiplied by ($\overline{1 - \frac{nz}{p}} \times \overline{1 + \frac{nz}{q}}$ or) $1 - \frac{nz}{p} + \frac{nz}{q} - \frac{n^2 z^2}{pq}$. From which analogy, because q is greater than p, it is plain that $\overline{1 - \frac{nz}{p}}\Big|^p \times \overline{1 + \frac{nz}{q}}\Big|^q$ varies at a greater rate in respect of its own magnitude than $\overline{1 - \frac{n^2 z^2}{pq}}\Big|^{\frac{n}{2}}$ does. And, because their fluxions as found out before have a negative sign before them, they both decrease as z increases; [302] consequently, if they are ever equal, as z increases the latter must be the largest. But when $z = 0$ they are each equal to 1. In like manner the other part of this article appears. And hence, since $a = \frac{p}{n}$ and $b = \frac{q}{n}$, it is manifest that $a^p b^q \times \overline{1 - \frac{n^2 z^2}{pq}}\Big|^{\frac{n}{2}}$ is greater than $\overline{a - z}\big|^p \times \overline{b + z}\big|^q$ and less than $\overline{a + z}\big|^p \times \overline{b - z}\big|^q$, when q is greater than p.

8. Suppose now further that the curve RQW be described meeting the lines Dh, Ft, ht produced in R, Q, W, in such manner that Ft, which is to Cf as $\overline{a + z}\big|^p \times \overline{b - z}\big|^q$ to $\overline{a - z}\big|^p \times \overline{b + z}\big|^q$ (Vid. Art. 3.) shall be to Qt as $\overline{a + z}\big|^p \times \overline{b - z}\big|^q$ to $a^p b^q \times \overline{1 - \frac{n^2 z^2}{pq}}\Big|^{\frac{n}{2}}$ wherever the points t and f fall at equal distances from h. And it is manifest by the foregoing Art. that Qt must always be greater than Cf, and less than Ft. And of consequence the same must be true concerning the areas described by their motion while

their equal abscisses increase. Wherefore RhtQ is greater than DhfC, and less than DhtF.

9. Since Ft is to Qt as $\overline{a+z}|^p \times \overline{b-z}|^q$ to $a^p b^q \times \overline{1 - \frac{n^2 z^2}{pq}}\Big|^{\frac{n}{2}}$; and $\overline{a+z}|^p \times \overline{b-z}|^q$ (by [303] Art. 3.) expresses the ratio of Ft to AH; the ratio of Qt to AH must be $a^p b^q \times \overline{1 - \frac{n^2 z^2}{pq}}\Big|^{\frac{n}{2}}$, and as has been all along supposed z is the ratio of ht to AH. Wherefore, by squaring the curve RhtQ, it will appear that the ratio of RhtQ to HO is $a^p b^q \times \overline{z - \frac{n^3 z^3}{2.3pq} + \frac{n-2}{4} \times \frac{n^5 z^5}{2.5 p^2 q^2}}$ $-\frac{n-2}{4} \times \frac{n-4}{6} \times \frac{n^7 z^7}{2.7 p^3 q^3} +$ &c. which (if $m^2 = \frac{n^3}{2pq}$) is $a^p b^q \times \frac{\sqrt{2pq}}{n\sqrt{n}} \times \overline{mz - \frac{m^3 z^3}{3}}$ $+\frac{n-2}{2n} \times \frac{m^5 z^5}{5} - \frac{n-2}{2n} \times \frac{n-4}{3n} \times \frac{m^7 z^7}{7} + \frac{n-2}{2n} \times \frac{n-4}{3n} \times \frac{n-6}{4n} \times \frac{m^9 z^9}{9} -$ &c. Which last series when $\frac{n^2 z^2}{pq} = 1$, and consequently the ordinate Qt vanishes, becomes $B - \frac{B^3}{3} + \frac{B^2 - 1}{2B^2} \times \frac{B^5}{5} - \frac{B^2 - 1}{2B^2} \times \frac{B^2 - 2}{3B^2} \times \frac{B^7}{7} +$ &c. taking $B^2 = \frac{n}{2}$.

10. If $B^2 = \frac{n}{2}$ the ratio of the whole figure RQWh to HO is $\frac{\sqrt{2pq}}{n\sqrt{n}} \times a^p b^p \times \overline{B - \frac{B^3}{3} + \frac{B^2 - 1}{2B^2} \times \frac{B^5}{5}} -$ &c. Now, (by Prop. 10. Art. 4. of the Essay) the ratio of ACFH to HO is $\frac{1}{n+1} \times \frac{1}{E}$, [304] E being the coefficient of that term of the binomial $\overline{a+b}|^n$ expanded in which occurs $a^p b^q$. Wherefore, the ratio of RQWh to ACFH is $\frac{n+1}{n} \times \frac{\sqrt{2pq}}{\sqrt{n}} \times E a^p b^q \times \overline{B - \frac{B^3}{3} + \frac{B^2 - 1}{2B^2} \times \frac{B^5}{5}}$ &c. Put G now for the coefficient of the middle term of the same binomial, and if $p = q = \frac{n}{2}$, E = G, $a = \frac{1}{2} = b$ the area RQWh is equal to half ACFH; for then Qt, Ft, Cf are all equal, and consequently the areas RQWh, HDh and ADh. Wherefore, the series $B - \frac{B^3}{3} +$ &c. is equal to $\frac{\sqrt{2n}}{n+1} \times \frac{2^{n-1}}{G}$. But the series $B - \frac{B^3}{3} +$ &c. (because $B^2 = \frac{n}{2}$) does not alter whatever p and q are, whilst their sum n remains the same. Wherefore, in all cases, the ratio of RQWh to ACFH is $\frac{\sqrt{pq}}{n} \times \frac{E a^p b^q}{G} \times 2^n$.

11. By Prop. 10. Art. 4. of the Essay, the ratio of ACFH to HO* is $\frac{1}{n+1} \times \frac{1}{E}$; and by Art. 9. the ratio of RhtQ to HO is $a^p b^q \times \frac{\sqrt{2pq}}{n\sqrt{n}} \times \overline{mz - \frac{m^3 z^3}{3} + \frac{n-2}{2n} \times \frac{m^5 z^5}{5}}$ &c. Wherefore, the ratio of [305] RhtQ to ACFH

---

*It is hoped that the imperfection of the figure all along referred to will be excused. The lines Rh and Dh should appear equal; and it will be found presently, that the curve line ACDFH should have been drawn from F and C convex towards AH.

is $\frac{n+1}{n} \times \frac{\sqrt{2pq}}{\sqrt{n}} \times \mathrm{E}a^p b^q \times \overline{mz - \frac{m^3 z^3}{3} + \frac{n-2}{2n} \times \frac{m^5 z^5}{5} - \frac{n-2}{2n} \times \frac{n-4}{3n} \times \frac{m^7 z^7}{7} +}$
&c. Likewise, by Art. 10. the ratio of RQWh to ACFH is $\frac{\sqrt{pq}}{n} \times \frac{\mathrm{E}a^p b^q}{\mathrm{G}} \times 2^n$. Wherefore the ratio of RhtQ to RQWh is $\frac{n+1}{\sqrt{n}} \times \frac{\sqrt{2}}{2^n} \times \mathrm{G} \times \overline{mz - \frac{m^3 z^3}{3}} +$ &c.

12. By Art. 2.6. When q is greater than p, the ratio of HDh − ADh to HO is less than $\frac{2a^p b^q}{n}$. And by Prop. 10. Art. 4. of the Essay, the ratio of ACFH or HDh + ADh to HO is $\frac{1}{n+1} \times \frac{1}{\mathrm{E}}$. Wherefore, the sum of these two ratios, or the ratio of 2HDh to HO, is less than $\frac{1}{n+1} \times \frac{1}{\mathrm{E}} + \frac{2a^p b^q}{n}$; and the difference between them, or the ratio of 2ADh to HO is greater than $\frac{1}{n+1} \times \frac{1}{\mathrm{E}} - \frac{2a^p b^q}{n}$. Wherefore, the ratio of 2HDh to 2ADh, or that of HDh to ADh, is less than that of $\frac{1}{n+1} \times \frac{1}{\mathrm{E}} + \frac{2a^p b^q}{n}$ to $\frac{1}{n+1} \times \frac{1}{\mathrm{E}} - \frac{2a^p b^q}{n}$, which is equal to the ratio of $1 \times 2\mathrm{E}a^p b^q + \frac{2\mathrm{E}a^p b^q}{n}$ to $1 - 2\mathrm{E}a^p b^q - \frac{2\mathrm{E}a^p b^q}{n}$. [306] But the ratio of HDh to ADh, by Art. 6. is greater than the ratio of DhtF to DhfC, when q is greater than p. Wherefore, much more when q is greater than p, the ratio of DhtF to DhfC will be less than that of $1 + 2\mathrm{E}a^p b^q + \frac{2\mathrm{E}a^p b^q}{n}$ to $1 - 2\mathrm{E}a^p b^q - \frac{2\mathrm{E}a^p b^q}{n}$. And because, by Art. 8. RhtQ is a mean between DhtF and DhfC, the ratio of DhtF to RhtQ will be less than that of $1 + 2\mathrm{E}a^p b^q + \frac{2\mathrm{E}a^p b^q}{n}$ to $1 - 2\mathrm{E}a^p b^q - \frac{2\mathrm{E}a^p b^q}{n}$. And the ratio of D$hf$C to R$ht$Q will be greater than that of $1 - 2\mathrm{E}a^p b^q - \frac{2\mathrm{E}a^p b^q}{n}$ to $1 + 2\mathrm{E}a^p b^q + \frac{2\mathrm{E}a^p b^q}{n}$.

RULE II.

If nothing is known of an event but that it has happened p times and failed q in $p+q$ or n trials, and q be greater than p; and from hence I judge that the probability of its happening in a single trial lies between $\frac{p}{n}$ and $\frac{p}{n} + z$, (if $m^2 = \frac{n^3}{2pq}$, $a = \frac{p}{n}$, $b = \frac{q}{n}$, E the coefficient of the term in which occurs $a^p b^q$ when $\overline{a+b}|^n$ is expanded, and $\Sigma = \frac{n+1}{n} \times \frac{\sqrt{2pq}}{\sqrt{n}} \times \mathrm{E}a^p b^q \times \overline{mz - \frac{m^3 z^3}{3} + \frac{n-2}{2n} \times \frac{m^5 z^5}{5} - \frac{n-2}{2n} \times \frac{n-4}{3n} \times \frac{m^7 z^7}{7}} +$ &c.) [307] my chance to be in the right is greater than Σ, and less than $\Sigma \times \dfrac{1 + 2\mathrm{E}a^p b^q + \frac{2\mathrm{E}a^p b^q}{n}}{1 - 2\mathrm{E}a^p b^q - \frac{2\mathrm{E}a^p b^q}{n}}$.

For by Art. 11. compared with the value of Σ here set down, the ratio of RhtQ to ACFH is Σ. But by Art. 8. DhtF is greater than RhtQ, and by Art. 12. the ratio of DhtF to RhtQ is less than that of $1 + 2\mathrm{E}a^p b^q + \frac{2\mathrm{E}a^p b^q}{n}$ to $1 - 2\mathrm{E}a^p b^q - \frac{2\mathrm{E}a^p b^q}{n}$. From whence it is plain that the ratio of DhtF to ACFH is greater than Σ, and less than $\Sigma \times \dfrac{1 + 2\mathrm{E}a^p b^q + \frac{2\mathrm{E}a^p b^q}{n}}{1 - 2\mathrm{E}a^p b^q - \frac{2\mathrm{E}a^p b^q}{n}}$. But, as appears from the 10th Proposition in the Essay, the chance that the probability of the event lies between $\frac{p}{n}$ and $\frac{p}{n} + z$ (that is, between the ratio of Ah to AH, and that of At to AH) is the ratio of DhtF to ACFH. Wherefore, the chance I am right in my guess is greater than Σ and less than $\Sigma \times \dfrac{1 + 2\mathrm{E}a^p b^q + \frac{2\mathrm{E}a^p b^q}{n}}{1 - 2\mathrm{E}a^p b^q - \frac{2\mathrm{E}a^p b^q}{n}}$. **[308]**

In like manner, 2dly, the same things supposed, if I judge that the probability of the event lies between $\frac{p}{n}$ and $\frac{p}{n} - z$, my chance to be right is less than Σ, and greater than $\Sigma \times \dfrac{1 - 2\mathrm{E}a^p b^q - \frac{2\mathrm{E}a^p b^q}{n}}{1 + 2\mathrm{E}a^p b^q + \frac{2\mathrm{E}a^p b^q}{n}}$. This is manifest as the other case, because DhfC is less than RhtQ, but the ratio between them is greater than that of $1 - 2\mathrm{E}a^p b^q - \frac{2\mathrm{E}a^p b^q}{n}$ to $1 + 2\mathrm{E}a^p b^q + \frac{2\mathrm{E}a^p b^q}{n}$.

3dly, If, other things supposed as before, p is greater than q, and I judge the probability of the event lies between $\frac{p}{n}$ and $\frac{p}{n} + z$, my chance to be right is less than Σ, and greater than $\Sigma \times \dfrac{1 - 2\mathrm{E}a^p b^q - \frac{2\mathrm{E}a^p b^q}{n}}{1 + 2\mathrm{E}a^p b^q + \frac{2\mathrm{E}a^p b^q}{n}}$. But if I judge it lies between $\frac{p}{n}$ and $\frac{p}{n} - z$, my chance is greater than Σ, and less than $\Sigma \times \dfrac{1 + 2\mathrm{E}a^p b^q + \frac{2\mathrm{E}a^p b^q}{n}}{1 - 2\mathrm{E}a^p b^q - \frac{2\mathrm{E}a^p b^q}{n}}$. And if $p = q$, **[309]** which ever of these ways I guess, my chance is Σ exactly. This may be proved in the same manner with the foregoing cases, where q is greater than p, or may be proved from them by considering the happening and failing of an event, as the same with the failing and happening of its contrary.

4thly, Other things supposed the same, whether q be greater or less than p, and I judge that the probability of the event lies between $\frac{p}{n} + z$ and $\frac{p}{n} - z$, my chance is greater than $\dfrac{2\Sigma}{1 + 2\mathrm{E}a^p b^q + \frac{2\mathrm{E}a^p b^q}{n}}$, and less than $\dfrac{2\Sigma}{1 - 2\mathrm{E}a^p b^q - \frac{2\mathrm{E}a^p b^q}{n}}$. This is an evident corollary from the cases already determined. And here, if $p = q$, my chance is 2Σ exactly.

Thus far I have transcribed Mr. Bayes.

It appears, from the Appendix to the Essay, that the rule here demonstrated, though of great use, does not give the required chance within limits sufficiently narrow. It is therefore necessary to look out for a contraction of these limits; and this, I think, we shall discover by the help of the following deductions; which, for the sake of greater distinctness, I shall give as a continuation of the foregoing Articles. **[310]**

13. The ratio of the fluxion of $\overline{1-\frac{n^2z^2}{pq}}\Big|^{\frac{n}{2}}$ to the fluxion of $\overline{1+\frac{nz}{p}}\Big|^{p} \times \overline{1-\frac{nz}{q}}\Big|^{q}$ is $\dfrac{\overline{1-\frac{n^2z^2}{pq}}\Big|^{\frac{n}{2}-1}}{\overline{1+\frac{nz}{p}}\big|^{p-1} \times \overline{1-\frac{nz}{q}}\big|^{q-1}}$; and the ratio of the fluxion of $\overline{1-\frac{nz}{p}}\Big|^{p} \times \overline{1+\frac{nz}{q}}\Big|^{q}$ to the fluxion of $\overline{1-\frac{n^2z^2}{pq}}\Big|^{\frac{n}{2}}$ is $\dfrac{\overline{1-\frac{nz}{p}}\big|^{p-1} \times \overline{1+\frac{nz}{q}}\big|^{q-1}}{\overline{1-\frac{n^2z^2}{pq}}\Big|^{\frac{n}{2}-1}}$. This will immediately appear from Art. 7.

14. While z is increasing from nothing till $\frac{n^2z^2}{pq}$ becomes equal to unity, these two ratios are at first greater than the ratio of equality, and increase as z increases, till they come to a *maximum*. Afterwards they decrease untill they become first equal to the ratio of equality, and then less. This is proved by finding the hyperbolic logarithms of these ratios. The hyperbolic logarithm of the first is the series $\overline{\frac{q-1}{q} - \frac{p-1}{p}} \times nz + \overline{\frac{q-1}{q^2} + \frac{p-1}{p^2} - \frac{n-2}{pq}} \times \frac{n^2z^2}{2} + \overline{\frac{q-1}{q^3} - \frac{p-1}{p^3}} \times \frac{n^3z^3}{3} + \overline{\frac{q-1}{q^4} + \frac{p-1}{p^4} - \frac{n-2}{p^2q^2}} \times \frac{n^4z^4}{4} + \overline{\frac{q-1}{q^5} - \frac{p-1}{p^5}} \times \frac{n^5z^5}{5} + \overline{\frac{q-1}{q^6} + \frac{p-1}{p^6} - \frac{n-2}{p^3q^3}} \times \frac{n^6z^6}{6} + $ &c. **[311]** The hyperbolic logarithm of the second ratio is the series $\overline{\frac{q-1}{q} - \frac{p-1}{p}} \times nz - \overline{\frac{q-1}{q^2} + \frac{p-1}{p^2} - \frac{n-2}{pq}} \times \frac{n^2z^2}{2} + \overline{\frac{q-1}{q^3} - \frac{p-1}{p^3}} \times \frac{n^3z^3}{3} - \overline{\frac{q-1}{q^4} + \frac{p-1}{p^4} - \frac{n-2}{p^2q^2}} \times \frac{n^4z^4}{4} + $ &c. It will appear from examining these two serieses (q all along supposed greater than p) that while z is small the value of each of them is positive, and increases as z increases till it becomes a *maximum*, after which it decreases till it becomes nothing, and after that negative; which demonstrates this article.

15. The former of the two ratios in Art. 13. (q being greater than p) is at first, while z is increasing from nothing, less than the second ratio; and does not become equal to it, till some time after both ratios have been the greatest possible.

Upon considering the two serieses in the last Art. it will appear that the first term of the first series is always positive, the second negative, the third also negative, after which the terms become alternately positive and negative. On the other hand, it will appear that in the second series the two first terms are always positive, and all that follow negative. But as the serieses converge very fast when z is small, the second term being negative in the first series and positive in the second, has a greater effect in making the first series less than the second, than can be compensated for by the terms being afterwards alternately negative and **[312]** positive in the one, and all negative in the other. It will further appear from considering these serieses, that the first must continue less than the second 'till z becomes so large as to make the fourth term equal to the second, in which circumstances the two serieses are nearly equal. Afterwards, as z goes on to increase, the value of both lessens continually; but the second now decreasing fastest, as before it increased fastest, becomes first nothing. After which, the other series becomes nothing; and after that both remain negative. From hence it is easy to infer this Article.

16. What has been now shewn of the ratio of the fluxion of $\overline{1 - \frac{n^2 z^2}{pq}}\Big|^{\frac{n}{2}}$ to the fluxion of $\overline{1 + \frac{nz}{p}}\Big|^p \times \overline{1 - \frac{nz}{q}}\Big|^q$ compared with the ratio of the fluxion of $\overline{1 - \frac{nz}{p}}\Big|^p \times \overline{1 + \frac{nz}{q}}\Big|^q$ to the fluxion of $\overline{1 - \frac{n^2 z^2}{pq}}\Big|^{\frac{n}{2}}$ is also true of the ratio of the fluxion of $a^p b^q \times \overline{1 - \frac{n^2 z^2}{pq}}\Big|^{\frac{n}{2}}$ (or Qt in the figure) to the fluxion of $\overline{a + z}|^p \times \overline{b - z}|^q$ (or Ft) compared with the ratio of the fluxion of $\overline{a - z}|^p \times \overline{b + z}|^q$ (or Cf) to the fluxion of $a^p b^q \times \overline{1 - \frac{n^2 z^2}{pq}}\Big|^{\frac{n}{2}}$ or Qt; the latter quantities being only the former multiplied by the common and permanent quantity $a^p b^q$. It appears, therefore, that if we conceive Ft, Qt, Cf (Vid. **[313]** Fig.) to move with equal and uniform velocities, from Dh and Rh along AH, in order to generate the areas HDh, RWh, ADh; Cf will at first not only decrease faster than Qt, and Qt than Ft; but the ratio of the rate at which Cf decreases to the rate at which Qt decreases, will be greater than the ratio of the rate at which Qt decreases to the rate at which Ft decreases. It appears also that after some time, first Cf and Qt, and then Qt and Ft will come to decrease equally; after which, Qt will decrease faster than Cf, and Ft faster than Qt.

17. The curves DFH, RQW, DCA, have each of them a point of contrary flexure; and the value of z, or of the equal abscisses at that point, is in all three $\frac{\sqrt{pq}}{\sqrt{n^3 - n^2}}$. This may be found in the common manner, by putting the second fluxions of the ordinates equal to nothing. In the single case, when either p or q is equal to unity, one of these points vanishes, or coincides with A or H.

18. At the points of contrary flexure (q being greater than p) the ratio of the fluxion of Qt to the fluxion of Ft is a *maximum*; and the same is true of the ratio of the fluxion of Cf to the fluxion of Qt. This is found by making

the fluxions of the logarithms of these ratios, or of $\dfrac{\overline{1-\dfrac{n^2z^2}{pq}}\Big|^{\frac{n}{2}-1}}{\overline{1+\dfrac{nz}{p}}\Big|^{p-1}\times \overline{1-\dfrac{nz}{q}}\Big|^{q-1}}$,

and $\dfrac{\overline{1-\dfrac{nz}{p}}\Big|^{p-1}\times \overline{1+\dfrac{nz}{q}}\Big|^{q-1}}{\overline{1-\dfrac{n^2z^2}{pq}}\Big|^{\frac{n}{2}-1}}$ [314] equal to nothing: which will give the

value of z equal to $\dfrac{\sqrt{pq}}{\sqrt{n^3-n^2}}$, or the same with the value of z at the points of contrary flexure.

19. At the points of contrary flexure, the ratio of the fluxion of Cf to the fluxion of Qt, is greatest in comparison of the ratio of the fluxion of Qt to the fluxion of Ft. This is proved by finding the value of z when the fluxion of the former ratio divided by the latter, or of $\dfrac{\overline{1-\dfrac{n^2z^2}{p^2}}\Big|^{p-1}\times \overline{1-\dfrac{n^2z^2}{q^2}}\Big|^{q-1}}{\overline{1-\dfrac{n^2z^2}{pq}}\Big|^{n-2}}$

is nothing, which will still give $z = \dfrac{\sqrt{pq}}{\sqrt{n^3-n^2}}$. The reason, therefore, in the nature of the curve, which, as the ordinates flow, keeps at first the excess of Ft above Qt less than the excess of Qt above Cf, operates with its greatest force at the points of contrary flexure.

20. The greatest part of the area $RQWh$ lies between Rh, and the ordinate at the point of contrary flexure. By Art. 11 the ratio of $RhtQ$ to $RQWh$ is $\dfrac{n+1}{\sqrt{n}} \times \dfrac{\sqrt{2}}{2^n} \times G \times \overline{mz - \dfrac{m^3z^3}{3} + \dfrac{n-2}{2n} \times \dfrac{m^5z^5}{5} -}$ &c. Substitute here $\sqrt{\dfrac{pq}{n^3-n^2}}$ for z, and $\dfrac{2^n}{\sqrt{nK \times H}}$* for G (K being the ratio of the quadrantal arc to [315] radius, and H the ratio whose hyperbolic logarithm is $\dfrac{3}{12n} - \dfrac{15}{360n^3} + \dfrac{63}{1260n^5}$ †

&c.) and the ratio of $RhtQ$ to $RQWh$ at the point of contrary flexure, will be

$\dfrac{n+1}{\sqrt{n}\times\sqrt{n-1}} \times \dfrac{.797884}{H} \times \overline{1 - \dfrac{n}{2.3.\overline{n-1}} + \dfrac{n\times\overline{n-2}}{2.5.4.\overline{n-1}|^2} - \dfrac{n.\overline{n-2}.\overline{n-4}}{2.3.7.8.\overline{n-1}|^3} + \dfrac{n.n-2.\overline{n-4}.\overline{n-6}}{2.3.4.9.16.\overline{n-1}|^4} -}$

&c. Now when n is little, the value of this expression will be considerably greater than .6822. It approaches to this continually as n increases; and when n is large, it may be taken for this exactly. Thus when $n = 6$, this expression is equal to .804. When $n = 110$, it is equal to .6903. If we would know the ratio of $RhtQ$ to $RQWh$, when Cf comes to decrease no faster in respect of Qt, than Qt decreases in respect of Ft; that is, when the excess of Qt above Cf, is greatest in comparison of the excess of Ft above Qt, it may

---

*This is always the true value of G; but it would be too tedious to give the demonstration of this here.

†Vid. the Second Rule in the Essay, Phil. Trans. Vol. LIII.

348   8. The Supplement to the Essay

be found (by putting the fourth term of the series in the 14th Art. equal to the second term, and then finding the value of z) to be about .8426, when n, p and q are considerable; and in other cases greater.

Coroll. 'Tis easy to gather from hence that in like manner the greatest part of the area ADH lies between the two ordinates at the points of contrary flexure.*[316]

21. RhtQ is greater than the arithmetical mean between DhtF and DhfC. This appears from the latter part of Art. 19. for what is there proved of the ordinates must hold true of the contemporary areas generated by them. And though beyond the points at which the ratio of the decrease of Qt to the decrease of Ft comes to an equality with the ratio of the decrease of Qt to the decrease of Cf, the excess of Ft above Qt begins to grow larger than before in respect of the excess of Qt above Cf; yet as it appears from the last article, that above five sixths of the areas RQWh and ACFH are generated before the ordinates come to these points, and as also beyond these points the said ratios, 'till they become [317] negative and for some time afterwards, are but small; the effect produced before towards rendering the excess of DhtF above RhtQ always less than the excess of RhtQ above DhfC, will be such as cannot be compensated for afterwards.

A further proof of this will appear from considering that even when RhtQ is increased to RQWh, it is but little short of the arithmetical mean between ADh and HDh. For from Art. 11. and 20. it may be inferred that the ratio of the whole area RQWh to this mean, or to $\frac{ACFH}{2}$, is $h \times$ H, which is never far from the ratio of equality, but when both p and q are of any considerable magnitude, it is very nearly the ratio of equality. For example; when $n = 110$, $q = 100$, $p = 10$, it is .9938.

---

*From this Article may be inferred a method of finding at once, without any labour, whereabouts it is reasonable to judge the probability of an unknown event lies, about which a given number of experiments have been made. For when neither p nor q are very small, or even not less than 10, it will be nearly an equal chance, that the probability of the event lies between $\frac{p}{n} + \frac{\sqrt{pq}}{\sqrt{2n^3 - 2n^2}}$ and $\frac{p}{n} - \frac{\sqrt{pq}}{\sqrt{2n^3 - 2n^2}}$. It will be the odds of two to one that it lies between $\frac{p}{n} + \frac{\sqrt{pq}}{\sqrt{n^3 - n^2}}$ and $\frac{p}{n} - \frac{\sqrt{pq}}{\sqrt{n^3 - n^2}}$; and the odds of five to one that it lies between $\frac{p}{n} + \frac{\sqrt{2pq}}{\sqrt{n^3 - n^2}}$ and $\frac{p}{n} - \frac{\sqrt{2pq}}{\sqrt{n^3 - n^2}}$. For instance; when $p = 1000$, $q = 100$, there will be nearly an equal chance, that the probability of the event lies between $\frac{10}{11} + \frac{1}{163}$ and $\frac{10}{11} - \frac{1}{163}$; two to one that it lies between $\frac{10}{11} + \frac{1}{115}$ and $\frac{10}{11} - \frac{1}{115}$; and five to one that it lies between $\frac{10}{11} + \frac{1}{81}$ and $\frac{10}{11} - \frac{1}{81}$.

22. The ratio of DhtF to RhtQ is less than that of $1 + 2\mathrm{E}a^p b^q + \frac{2\mathrm{E}a^p b^q}{n}$ to one. For by Art. 12. the ratio of DhtF to DhfC is less than that of $1 + 2\mathrm{E}a^p b^q + \frac{2\mathrm{E}a^p b^q}{n}$ to $1 - 2\mathrm{E}a^p b^q - \frac{2\mathrm{E}a^p b^q}{n}$. But by the last Art. R$h$tQ is greater than the arithmetical mean between DhtF and DhfC, and 1 is exactly the arithmetical mean between $1 + 2\mathrm{E}a^p b^q + \frac{2\mathrm{E}a^p b^q}{n}$ and $1 - 2\mathrm{E}a^p b^q - \frac{2\mathrm{E}a^p b^q}{n}$. From whence this Article is plain. **[318]**

23. The ratio of DhtF to ACFH is greater than Σ, and less than $\Sigma \times \overline{1 + 2\mathrm{E}a^p b^q + \frac{2\mathrm{E}a^p b^q}{n}}$. For D$h$tF being greater than R$h$tQ, the ratio of it to ACFH must be greater than the ratio of RhtQ to ACFH, or greater than Σ. Also; since the ratio of RhtQ to ACFH is equal to Σ; and the ratio of DhtF to RhtQ is less than the ratio of $1 + 2\mathrm{E}a^p b^q + \frac{2\mathrm{E}a^p b^q}{n}$ to 1; it follows that the ratio compounded of the ratio of RhtQ to ACFH, and of DhtF to RhtQ, that is, the ratio of DhtF to ACFH must be less than $\Sigma \times \overline{1 + 2\mathrm{E}a^p b^q + \frac{2\mathrm{E}a^p b^q}{n}}$.

24. The ratio of DhtF + DhfC to ACFH (that is, the chance for being right in judging that the probability of an event perfectly unknown, which has happened p and failed q times in $p + q$ or n trials, lies somewhere between $\frac{p}{n} + z$ and $\frac{p}{n} - z$) is greater than $\dfrac{2\Sigma}{1 + 2\mathrm{E}a^p b^q + \frac{2\mathrm{E}a^p b^q}{n}}$, and less than 2Σ. The former part of this Art. has been already proved, Art. 12. The latter part is evident from Art. 21. For RhtQ being greater than the arithmetical mean between DhtF and DhfC, 2RhtQ must be greater than DhtF + DhfC; and consequently **[319]** the ratio of 2RhtQ to ACFH, greater than the ratio of DhtF + DhfC to ACFH*.

It will be easily seen that this Article improves considerably the rule given in Art. 12. But we may determine within still narrower limits whereabouts the required chance must lie, as will appear from the following Articles.

25. If c and d stand for any two fractions, whenever the fluxion of $c \times \mathrm{F}t$ is greater than the fluxion of $d \times \mathrm{C}f$ (Vid. fig.) $c \times \mathrm{F}t + d \times \mathrm{C}f$ will be greater than Qt. For in the same manner with Art. 6. it will appear that $c \times \mathrm{F}t + d \times \mathrm{C}f$ is to Qt, as the fluxion of $c \times \mathrm{F}t \times \overline{1 + \frac{nz}{p}} \times \overline{1 - \frac{nz}{q}}$ together with the fluxion of $d \times \mathrm{C}f \times \overline{1 - \frac{nz}{p}} \times \overline{1 + \frac{nz}{q}}$ to the fluxion of Q$t \times \overline{1 - \frac{n^2 z^2}{pq}}$. Now since $1 - \frac{n^2 z^2}{pq}$ is the arithmetical mean between $\overline{1 + \frac{nz}{p}} \times \overline{1 - \frac{nz}{q}}$

---

*This Art. is true, whether p be greater or less than q.

and $\overline{1-\frac{nz}{p}} \times \overline{1+\frac{nz}{q}}$, it is plain, that were the fluxion of $c \times Ft$ equal to the fluxion of $d \times Cf$, $c \times Ft + d \times Cf$ would decrease in respect of its own magnitude at the same rate with Qt; and, therefore, since at first equal, they would always continue equal. But the fluxion of $c \times Ft$ being greater than the fluxion of $d \times Cf$ by supposition, and (since q greater than p) $\overline{1+\frac{nz}{p}} \times \overline{1-\frac{nz}{q}}$, [320] also greater than $\overline{1-\frac{nz}{p}} \times \overline{1+\frac{nz}{q}}$, it follows that the fluxion of $c \times Ft \times \overline{1+\frac{nz}{p}} \times \overline{1-\frac{nz}{q}}$ added to the fluxion of $d \times Cf \times \overline{1-\frac{nz}{p}} \times \overline{1+\frac{nz}{q}}$ is greater than these two fluxions multiplied by $\overline{1-\frac{n^2z^2}{pq}}$; and, therefore, greater than the fluxion of $Qt \times \overline{1-\frac{n^2z^2}{pq}}$; and, therefore, $c \times Ft + d \times Cf$ greater than Qt.

26. If we suppose three continued arithmetical means between Cf and Ft $\left(\frac{3Cf+Ft}{4}, \frac{Cf+Ft}{2}, \frac{3Ft+Cf}{4}\right)$ Qt will be greater than the second, and less than the third, if p is greater than 1. That Qt will be greater than the second has been already proved; and that it will be less than the third, will be an immediate consequence from the last Article, if it can be shewn that the fluxion of $\frac{3Ft}{4}$ is greater than the fluxion of $\frac{Cf}{4}$. This will appear in the following manner. The ratio of the fluxion of Cf to the fluxion of Ft is by Art. 7. and 14. $\dfrac{\overline{1-\frac{nz}{p}}^{p-1} \times \overline{1+\frac{nz}{q}}^{q-1}}{\overline{1+\frac{nz}{p}}^{p-1} \times \overline{1-\frac{nz}{q}}^{q-1}}$. The hyperbolic logarithm [321] of this ratio is $\frac{1}{p} - \frac{1}{q} \times 2nz - \overline{\frac{1}{p^2} - \frac{1}{p^3} - \frac{1}{q^2} + \frac{1}{q^3}} \times \frac{2n^3z^3}{3} - \overline{\frac{1}{p^4} - \frac{1}{p^5} - \frac{1}{q^4} + \frac{1}{q^5}} \times \frac{2n^5z^5}{5}$, &c. This ratio by Art. 18. is greatest at the point of contrary flexure, or when $z = \frac{\sqrt{pq}}{\sqrt{n^3-n^2}}$. Substitute this for z in the series, and it will become $\frac{1}{p} - \frac{1}{q} \times \frac{2\sqrt{pq}}{\sqrt{n-1}} - \overline{\frac{1}{p^2} - \frac{1}{p^3} - \frac{1}{q^2} + \frac{1}{q^3}} \times \frac{2p^{3/2} \times q^{3/2}}{3 \times \overline{n-1}^{3/2}}$, &c. which, therefore, expresses the logarithm of the ratio when greatest, and will easily discover it in every case. 'Tis apparent that the value of this series is greatest when p is least in respect of q. Suppose then $p = 2$, and q infinite. In this case, the value of the series will be 1.072, and the number answering to this logarithm is not greater than 2.92. The fluxion, therefore, of Cf, when greatest, cannot be three times the contemporary fluxion of Ft; from whence it follows that the fluxion of $\frac{3Ft}{4}$ must be greater than the fluxion of $\frac{Cf}{4}$.

It is easy to see how these demonstrations are to be varied when q is less than p, and how in this case similar conclusions may be drawn. Or, the same conclusions will in this case immediately appear, by changing p into q and q into p, which will not make any difference in the demonstrations.

In the manner specified in this Article we may always find within certain limits how near the value of Qt comes to the arithmetical mean between

Ft and Cf, which limits grow narrower and narrower, as **[322]** p and q are taken larger, or their ratio comes nearer to that of equality, 'till at last, when p and q are either very great or equal, Qt coincides with this mean. Thus, if either p or q is not less than 10; that is, in all cases, where it is not practicable without great difficulty to find the required chance exactly by the first rule, Qt will be greater than the fourth, and less than the fifth of seven arithmetical means between Cf and Ft.

27. The arithmetical means mentioned in the last Article may be conceived as ordinates describing areas at the same time with Qt; and what has been proved concerning them is true also of the areas described by them compared with $RhtQ$.

28. If either p or q is greater than 1, the true chance that the probability of an unknown event which has happened p times and failed q in $\overline{p+q}$ or n trials, should lie somewhere between $\frac{p}{n} + z$ and $\frac{p}{n} - z$ is less than 2Σ, and greaterer than $\Sigma + \dfrac{\Sigma \times \overline{1 - 2Ea^pb^q - \frac{2Ea^pb^q}{n}}}{1 + Ea^pb^q + \frac{Ea^pb^q}{n}}$. If either p or q is greater than 10, this chance is less than 2Σ, and greater than $\Sigma + \dfrac{\Sigma \times \overline{1 - 2Ea^pb^q - \frac{2Ea^pb^q}{n}}}{1 + \frac{1}{2}Ea^pb^q + \frac{Ea^pb^q}{2n}}$. **[323]** This is easily proved in the same manner with Art. 12, 23, 24.

That it may appear how far what has been now demonstrated improves the solution of the present problem, let us take the fifth case mentioned in the Appendix to the Essay, and enquire what reason there is for judging that the probability of an event concerning which nothing is known, but that it has happened 100 times and failed 1000 times in 1100 trials, lies between $\frac{10}{11} + \frac{1}{110}$ and $\frac{10}{11} - \frac{1}{110}$. The second rule as given in Art. 12. informs us, that the chance* for this must lie between .6512, (or the odds of 186 to 100) and .7700, (or the odds of 334 to 100). But from the last Art. it will appear that the required chance in this case must lie between 2Σ, and $\Sigma + \Sigma \times \dfrac{1 - Ea^pb^q - \frac{2Ea^pb^q}{n}}{1 + \frac{1}{10}Ea^pb^q + \frac{Ea^pb^q}{10n}}$; or, between .6748 and .7057; that is, between the odds of 239 to 100, and 207 to 100.

In all cases when z is small, and also whenever the disparity between

---

*In the Appendix, this chance, as discovered by Mr. Bayes's second rule, is given wrong, in consequence of making m^2 equal to $\frac{n^3}{pq}$, whereas it should have been taken equal to $\frac{n^3}{2pq}$ as appears from Article 8.

p and q is not great 2Σ is almost exactly the true chance required. And I have reason to think, that even in all other cases, 2Σ gives the **[324]** true chance nearer than within the limits now determined. But not to pursue this subject any further; I shall only add that the value of 2Σ may be always calculated very nearly, and without great difficulty; for the approximations to the value of $Ea^p b^q$, and of the series $m - z\frac{m^3 z^3}{3} + \frac{n-2}{2n} \times \frac{m^5 z^5}{5}$, &c.* given in the Essay, are sufficiently accurate in all cases where it is necessary to use them.

---

*In the expression for this last approximation there is an error of the press which should be corrected; for the sign before the fourth term should be $-$ and not $+$.

8.3 Commentary

Although the *Essay towards solving a Problem in the Doctrine of Chances* has frequently been commented upon, little has been said on the paper given in this chapter. Some attention was paid to it in Timerding's 1908 translation of the *Essay*, and more detail was provided by Sheynin in 1969, and Hald [1990b], [1998, §8.6] gave a detailed study of the work by Bayes and Price, in both the *Essay* and the *Supplement*, on the approximation of the beta distribution. It seems meet, however, to pay rather more attention to it here, and we accordingly develop some of the perhaps terse statements made by Bayes and Price.

We have already gleaned from the Appendix to the *Essay* that Price probably believed Bayes's results to be applicable in a causal setting. In his introductory letter to the present paper Price stated this belief more baldly, writing

> The solution of the problem enquired after in the papers I have sent you has, I think, been hitherto a *desideratum* in philosophy of some consequence. To this we are now in a great measure helped by the abilities and skill of our late worthy friend; and thus are furnished with a necessary guide in determining the nature and proportions of unknown causes from their effects, and an effectual guard against one great danger to which philosophers are subject; I mean, the danger of founding conclusions on an insufficient induction, and of receiving just conclusions with more assurance than the number of experiments will warrant.

As we have seen, however, there is little (if anything) in Bayes's *Essay* to warrant such an extension of its results. As it is, at all events, no causal application is called for in this paper, and the matter is accordingly of no importance in the present context.

It should be noted, in considering the sketch at the beginning of this paper, that the 'origin' is taken at the right-hand end of the horizontal HW axis, and that the 'positive' direction is to the left. Further, the 'upper' and 'lower' curves are (essentially) skew and symmetric beta-functions, respectively, with the same maximum and points of inflexion.

The first Article seems clear — provided, that is, that the appropriate series in the *Essay* are recalled. It might help to note that, although the maximum of $f(x) = x^p(1-x)^q, x \in (0,1)$ (essentially the well-known probability density function of a random variable having a beta distribution) occurs at $x = p/(p+q)$, the transformation $y = (p+q)x$ yields a function on $(0, p+q)$ whose maximum is at $y = p$, which is the case for Bayes's function. We might also mention that the reference 'from Art. 3 in the Essay' refers to the third Article in Proposition 10.

In connexion with the second Article, let us denote the first series given in the last paragraph of the first Article by S_1; that is,

$$S_1 = \frac{p}{p+1} + \frac{q}{p+1} \times \frac{p^2}{pq+2q} + \frac{q \times \overline{q-1} \times p^3}{p+1 \times p+2 \times pq^2+3q^2} + \&c.$$

$$= \sum_{k=0} \frac{p^{k+1} (q)_k}{q^k (p+k+1)_{k+1}},$$

where $(n)_k$ is the descending factorial, defined for all real n by

$$(n)_0 = 1 \ ; \quad (n)_k = n(n-1)\ldots(n-k+1), \quad k \geq 1.$$

The second series, S_2, say, in the last paragraph of Article 1 may be similarly expressed, *mutatis mutandis*. Denoting by r_k the ratio of the kth term in S_1 to the $(k+2)$nd term in S_2 we have

$$r_k = \frac{p^{k+1}(q)_k}{q^k(p+k+1)_{k+1}} \bigg/ \frac{q^{k+3}(p)_{k+2}}{p^{k+2}(q+k+3)_{k+3}}$$

$$= \frac{p^{2k+3}(q+k+3)_{2k+3}}{q^{2k+3}(p+k+1)_{2k+3}}.$$

This may be checked by setting $k = 2$ (i.e., considering the second term in the first series), our ratio then yielding the first eight terms in Bayes's product given in Article 2. Note that the method of construction of the series ensures that each r_k contains an even number of terms. Further, $(p+k+1)_{2k+3}$ and $(q+k+3)_{2k+3}$ have the same number of terms.

The fact that, for q greater than p (a global assumption that holds *presque partout*), this ratio is 'greater than the ratio of equality' may be seen as follows. Write the first few terms in Bayes's expression for the ratio, perhaps most suggestively, as

$$\left[\frac{p(q+1)}{q(p+0)} \times \frac{p(q+2)}{q(p-0)}\right] \times \left[\frac{p(q+0)}{q(p+1)} \times \frac{p(q+3)}{q(p-1)}\right] \times \left[\frac{p(q-0)}{q(p+2)} \times \frac{p(q+4)}{q(p-2)}\right]$$

$$\times \left[\frac{p(q-1)}{q(p+3)} \times \frac{p(q+5)}{q(p-3)}\right] \times \left[\frac{p(q-2)}{q(p+4)} \times \frac{p(q+6)}{q(p-4)}\right]. \quad (8.2)$$

To show that each of the n paired terms, $n \in \{1,\ldots,5\}$, is greater than 1, write them in the form

$$\frac{p[(q+3/2)-(n-1/2)]}{q[p+(n-1)]} \times \frac{p[(q+3/2)+(n-1/2)]}{q[p-(n-1)]}, \quad n \in \{1,2\}$$

$$\frac{p[(q+2)-(n-1)]}{q[p+(n-1)]} \times \frac{p[(q+2)+(n-1)]}{q[p-(n-1)]}, \quad n \in \{3,4,5\}$$

The first two bracketed terms in (8.2) seem to require a special formulation (although they can of course be given in other ways, that adopted here is perhaps more suggestive, in its symmetry, of the pattern followed by the remaining terms); all the rest are expressible as in the latter of the preceding two expressions. It is then easy to show that each of the paired terms is greater than 1, and hence so is their product.

That is, assuming that all the 'factors' are positive, we have

$$r_k = \frac{p^{2k+3}}{(p+k+1)_{2k+3}} \bigg/ \frac{q^{2k+3}}{(q+k+3)_{2k+3}} > 1,$$

as asserted.

Recalling that $q > p$, it is easy to see that the sum of the first two terms in S_2, viz. $q/(q+1) + pq^2/(q+1)(pq+2p)$, is indeed less than 2.

Let us for the moment write S_1 and S_2 as $S_1 = \sum a_i$ and $S_2 = \sum b_i$. In view of the ordering of the terms shown above, we have

$$\begin{aligned} S_2 - (b_1 + b_2) &= b_3 + b_4 + \cdots \\ &< a_1 + a_2 + \cdots = \\ &< S_1. \end{aligned}$$

Thus $S_2 - S_1 < b_1 + b_2 < 2$, and from the fact that

$$\frac{|ADh - HDh|}{HO} = \frac{a^p b^q}{n} |S_1 - S_2|,$$

where $a = p/n$ and $b = q/n$, it thus follows that

$$\frac{|ADh - HDh|}{HO} < \frac{2a^p b^q}{n},$$

as asserted at the end of this Article. When $q > p$ the absolute value is unnecessary, and the result may be written in modern terms as[3]

$$\int_a^1 u^p (1-u)^q \, du - \int_0^a u^p (1-u)^q \, du < \frac{2a^p b^q}{n}.$$

Having shown in the third Article that $Cf : AH = (a-z)^p \times (b+z)^q$ and $Ft : AH = (a+z)^p \times (b-z)^q$, and hence that

$$Cf : Ft :: (a-z)^p \times (b+z)^q : (a+z)^p \times (b-z)^q,$$

Bayes turns his attention in the fourth to proving that $(a+z)^p \times (b-z)^q$ is greater than $(a-z)^p \times (b+z)^q$. This is effected by consideration of the (hyperbolic) logarithm of the ratio of the first of these products to the second. Expansion of this logarithm, and the replacement in the expansion of p and q by na and nb, respectively, gives

$$\log \frac{(a+z)^p (b-z)^q}{(a-z)^p (b+z)^q} = 2n \left(\frac{b^2-a^2}{3a^2b^2} z^3 + \frac{b^4-a^4}{5a^4b^4} z^5 + \frac{b^6-a^6}{7a^6b^6} z^7 + \cdots \right)$$

as asserted. Now $q > p$ implies that $b > a$, and hence each term within the parentheses on the right-hand side is positive. Thus the logarithm of the ratio is positive, and hence the ratio itself is greater than 1. The asserted conclusion is then evident.

Articles 5 and 6 seem fairly clear; let us turn our attention then to Number 7. Consider the three functions

$$f_1(z) \equiv \left(1 - \frac{nz}{p}\right)^p \left(1 + \frac{nz}{q}\right)^q$$

$$f_2(z) \equiv \left(1 - \frac{n^2 z^2}{pq}\right)^{\frac{n}{2}}$$

$$f_3(z) \equiv \left(1 - \frac{nz}{q}\right)^q \left(1 + \frac{nz}{p}\right)^p.$$

(Although $f_3(z) = f_1(-z)$ we find it convenient to use both f_1 and f_3.) Here Bayes is essentially showing that the ordering $f_1(z) < f_2(z) < f_3(z)$ can be established by considering the ordering of the fluxions $\dot{f}_1, \dot{f}_2,$ and $\dot{f}_3$. These being as given in the text, we have

$$\frac{\dot{f}_2}{\dot{f}_1} = \frac{\dot{f}_2}{\dot{f}_1} \times \frac{\left(1 - \frac{n^2 z^2}{pq}\right)}{\left(1 - \frac{n^2 z^2}{pq}\right) + \left(\frac{nz}{q} - \frac{nz}{p}\right)}.$$

Since $q > p$, the second term in the denominator is negative, and hence

$$\frac{\dot{f}_1}{f_1} > \frac{\dot{f}_2}{f_2},$$

or '...$[f_1]$ varies at a greater rate in respect of its own magnitude than $[f_2]$ does'.

Since both fluxions are negative, they decrease with increasing z, and hence, if they ever coincide, $\dot{f}_2$ must be greater than $\dot{f}_1$. But they coincide when $z = 0$, and hence we must have $f_2 > f_1$. The other part of the inequality, viz. $f_2 < f_3$, may be similarly established. Recollection of the definitions of a and b in terms of p and q results in the last statement of this Article.

In the eighth Article Bayes uses the inequalities developed in the seventh to conclude that the area DhtF is greater than RhtQ which in turn is greater than DhfC.

In the ninth Article Bayes recalls that the ratio Ft : Qt is as

$$(a+z)^p \times (b-z)^q : a^p b^q \times (1 - n^2 z^2 / pq)^{n/2}$$

and further that
$$(a+z)^p \times (b-z)^q :: \text{F}t : \text{AH}.$$

He notes that one then has

$$\text{Q}t : \text{AH} :: a^p b^q \times (1 - n^2 z^2/pq)^{n/2},$$

as z is the ratio of ht to AH. The series derived from 'squaring the curve RhtQ' is found by expanding $(1 - n^2 z^2/pq)^{n/2}$ in a series and then taking the fluxion, the result being the ratio of RhtQ to HO. (I found it easiest to do this by writing the given binomial in the form

$$(1 - \alpha^2 z^2)^\beta \equiv (1 - \alpha z)^\beta (1 + \alpha z)^\beta,$$

expanding each term on the right-hand side in a MacLaurin series, and then integrating.)

One might note here that the phrase 'the ordinate Qt vanishes' does not mean that Qt becomes nought, but rather that Qt moves to Rh. It should also be noted that there are a few misprints in this Article: the first occurrence of z^5 is printed as 2^5, and terms such as $2B^2$ appear as $^2 B^2$.

There is one small typographical error in the tenth Article: the first vinculum should also cover 'B−'. One must also note that since p and q are required to be integers, the setting of $p = q = \frac{n}{2}$ will ensure that there is indeed a middle term (Bayes's 'G') in the expansion of the binomial $(a+b)^n$.

After considerable manipulation Bayes arrives in the twelfth Article at the following inequalities[4], which summarize his work on the skewness of the beta probability integral.

$$\frac{\text{D}ht\text{F}}{\text{D}hf\text{C}} < \frac{1 + 2Ea^p b^q + 2Ea^p b^q/n}{1 - 2Ea^p b^q - 2Ea^p b^q/n}$$

$$\frac{\text{D}ht\text{F}}{\text{R}ht\text{Q}} < \frac{1 + 2Ea^p b^q + 2Ea^p b^q/n}{1 - 2Ea^p b^q - 2Ea^p b^q/n}$$

$$\frac{\text{D}hf\text{C}}{\text{R}ht\text{Q}} > \frac{1 - 2Ea^p b^q - 2Ea^p b^q/n}{1 + 2Ea^p b^q + 2Ea^p b^q/n}.$$

The first of these inequalities may be written in integral notation (with similar formulations for the others) as

$$\int_{(p/n)}^{(p/n)+z} u^p(1-u)^q \, du \Big/ \int_{(p/n)-z}^{(p/n)} u^p(1-u)^q \, du < \frac{1+\alpha}{1-\alpha},$$

where $\alpha = 2(n+1)Ea^p b^q/n$. Using the discussion of the second rule in the *Essay*, Hald [1990b, §3] has shown that α may be written in the form

$$\alpha = 2\frac{n+1}{n}\frac{1}{\sqrt{2\pi npq}} e^\gamma, \tag{8.3}$$

where

$$\gamma = \frac{1}{12n}(1 - p^{-1} - q^{-1}) - \frac{1}{360n^3}(1 - p^{-3} - q^{-3}) + \cdots. \tag{8.4}$$

Does the occurrence of this series here provide a reason for Bayes's paper on the semi-convergent series?

The only other point that perhaps requires noting before Rule II is reached is that the formula in Article 12 given as $1 \times 2Ea^p b^q + \frac{2Ea^p b^q}{n}$, should be $1 + 2Ea^p b^q + \frac{2Ea^p b^q}{n}$.

In the notation of the *Essay* the second rule, as given here, runs as follows.

Rule II. Let $z > 0$.

(a) For $q > p$,

$$\Sigma < \Pr[p/n < x < (p/n) + z \mid p, q] < \frac{1+\alpha}{1-\alpha}\Sigma$$

and

$$\frac{1-\alpha}{1+\alpha}\Sigma < \Pr[(p/n) - z < x < p/n \mid p, q] < \Sigma.$$

(b) For $p \neq q$,

$$\frac{2\Sigma}{1+\alpha} < \Pr[(p/n) - z < x < (p/n) + z \mid p, q] < \frac{2\Sigma}{1-\alpha}.$$

(c) For $p = q$,

$$\Pr[(p/n) - z < x < (p/n) + z \mid p, q] = 2\Sigma.$$

8.3. Commentary   359

It should be noted that the discussion of this Rule given here encompasses more cases than it does in the *Essay*: it is the last two cases that are given in the latter.

This completes Bayes's work. The remainder of the tract contains Price's refinement of the limits derived by Bayes.

Articles 13 to 15 are perhaps best considered as one. Let us write the ratios mentioned in the first of these as

$$R_1(z;p,q) \equiv \frac{\left(1 - \frac{n^2 z^2}{pq}\right)^{(n/2)-1}}{\left(1 + \frac{nz}{p}\right)^{p-1} \times \left(1 - \frac{nz}{q}\right)^{q-1}}$$

$$R_2(z;p,q) \equiv \frac{\left(1 - \frac{nz}{p}\right)^{p-1} \times \left(1 + \frac{nz}{q}\right)^{q-1}}{\left(1 - \frac{n^2 z^2}{pq}\right)^{(n/2)-1}}.$$

The hyperbolic logarithm of R_1 is the series

$$g_1(z;p,q) \equiv \overline{\frac{q-1}{q} - \frac{p-1}{p}} \times nz + \overline{\frac{q-1}{q^2} + \frac{p-1}{p^2} - \frac{n-2}{pq}} \times \frac{n^2 z^2}{2}$$

$$+ \overline{\frac{q-1}{q^3} - \frac{p-1}{p^3}} \times \frac{n^3 z^3}{3} + \overline{\frac{q-1}{q^4} + \frac{p-1}{p^4} - \frac{n-2}{p^2 q^2}} \times \frac{n^4 z^4}{4}$$

$$+ \overline{\frac{q-1}{q^5} - \frac{p-1}{p^5}} \times \frac{n^5 z^5}{5} + \overline{\frac{q-1}{q^6} + \frac{p-1}{p^6} - \frac{n-2}{p^3 q^3}} \times \frac{n^6 z^6}{6} + \text{etc.},$$

and the hyperbolic logarithm of R_2 is

$$g_2(z;p,q) \equiv \overline{\frac{q-1}{q} - \frac{p-1}{p}} \times nz - \overline{\frac{q-1}{q^2} + \frac{p-1}{p^2} - \frac{n-2}{pq}} \times \frac{n^2 z^2}{2}$$

$$+ \overline{\frac{q-1}{q^3} - \frac{p-1}{p^3}} \times \frac{n^3 z^3}{3} - \overline{\frac{q-1}{q^4} + \frac{p-1}{p^4} - \frac{n-2}{p^2 q^2}} \times \frac{n^4 z^4}{4} + \text{etc.}$$

On noting that

$$R_2(z;p,q) = [R_1(z;q,p)]^{-1} = [R_1(-z;p,q)]^{-1},$$

we obtain, on differentiating,

$$\frac{d}{dz} g_1(z;p,q) = -\frac{d}{dz} g_2(z;q,p).$$

On setting $\frac{d}{dz} g_1(z;p,q)$ equal to zero one finds that $z = \pm\sqrt{pq/(n^3 - n^2)}$, only the positive root (z_m, say) being relevant since z is required to be non-negative. It is similarly easy to verify the following results.

(a) $R_1(0; p, q) = 1$.

(b) $R_1(z_m; p, q) > 0$.

(c) $R_1(\sqrt{pq/n^2}; p, q) = 0$ and $R_2(p/n; p, q) = 0$.

(d) $p/n < \sqrt{pq/n^2}$, and so $R_2(z; p, q)$ reaches the horizontal (or z) axis before $R_1(z; p, q)$ does.

(e) $R_1(z_m; p, q) < R_2(z_m; p, q)$ (the inequality may in fact be reduced to $(n-2)^{n-2} < (n-1)^{n-2}$, which is clearly true).

(f) $\sqrt{pq/(n^3 - n^2)} < \sqrt{pq/n^2}$.

Thus the graph of $R_1(z; p, q)$ starts off at 1 when $z = 0$, rises to a maximum of

$$(n-2)^{(n/2)-1} \Big/ \left(\sqrt{n-1} + \sqrt{q/p}\right)^{p-1} \left(\sqrt{n-1} - \sqrt{p/q}\right)^{q-1}$$

at $z = \sqrt{pq/(n^3 - n^2)}$, is zero when $z = \sqrt{pq/n^2}$, and then continues below the horizontal axis, since there are no other turning points. The behaviour of $R_2(z; p, q)$ may be similarly discussed.

On the signs of the terms in the two series, notice firstly that it is easy to show that the first two terms in the first series have coefficients that are positive and negative, respectively (recall that $q > p$ and $n = p + q$). Further, since

$$\frac{q-1}{q^2} < \frac{p-1}{p^2} \iff q(p-1) > p,$$

and since the right-hand inequality is true, it is easy to show by induction that $(q-1)/q^n < (p-1)/p^n$ for $n \in \{3, 5, 7, \ldots\}$. As regards the other terms, note that

$$\frac{q-1}{q^{2k}} + \frac{p-1}{p^{2k}} - \frac{n-2}{p^k q^k} > 0 \iff q^k(p-1) > p^k(q-1).$$

For $k = 2$ the last inequality reduces to $pq > (p + q)$, which is clearly true for $p > 2$, and induction then completes the proof. Similar remarks apply to the second series.

As regards the equality of the two ratios, notice that on setting $R_1(z; p, q)$ equal to $R_2(z; p, q)$ we get

$$\left(1 - \frac{n^2 z^2}{pq}\right)^{n-2} = \left(1 - \frac{n^2 z^2}{p^2}\right)^{p-1} \left(1 - \frac{n^2 z^2}{q^2}\right)^{q-1},$$

which, on expansion and rearrangement, may be written in the form

$$\sum_{k=1}^{n-2}(-1)^k\left[\binom{n-2}{k}\left(\frac{1}{pq}\right)^k - \gamma_k\right](n^2z^2)^k = 0,$$

where, as usual, $\binom{x}{r} = 0$ for $r > x$ and

$$\gamma_k = \sum_{j=0}^{k}\binom{p-1}{j}\binom{q-1}{k-j}\left(\frac{1}{p^2}\right)^j\left(\frac{1}{q^2}\right)^{k-j}.$$

It may be of interest, though perhaps not of much use, to note that if we consider the convolution $\{\alpha_k\} * \{\beta_k\}$, where

$$\alpha_k = \binom{p-1}{k}\left(\frac{1}{p^2}\right)^k, \quad \beta_k = \binom{q-1}{k}\left(\frac{1}{q^2}\right)^k,$$

with associated generating functions

$$A(s) = \sum\alpha_k s^k \text{ and } B(s) = \sum\beta_k s^k,$$

then $\{\gamma_k\} = \{\alpha_k\} * \{\beta_k\}$. Further, if $\nu_k = \binom{n-2}{k}(1/pq)^k$, with generating function $N(s) = \sum\nu_k s^k$, then $N(s) = A(sp/q)B(sq/p)$.

In Article 15 Price writes of the taking of z 'so large as to make the fourth term equal to the second'; this equality holds for $z = \sqrt{2p^2q^2/(n^3(pq-n))}$ (we say more on this later).

As an example, consider the case in which $p = 2$ and $q = 3$ (this is merely illustrative: the behaviour may well be very different for other values of p and q). Then $R_1(z;p,q) = R_2(z;p,q)$ yields the roots

$$z_1 = 0; \quad z_2 = -\sqrt{5-\sqrt{7}}/5; \quad z_3 = -\sqrt{5+\sqrt{7}}/5$$

$$z_4 = \sqrt{5-\sqrt{7}}/5; \quad z_5 = \sqrt{5+\sqrt{7}}/5.$$

The behaviour of $R_1(z;p,q)$ and $R_2(z;p,q)$ at $z = 0$ has already been considered. The roots z_2 and z_3 may be ignored, since z is required to be non-negative, and z_5 may be ignored since $R_1(z_5;2,3)$ is imaginary (on account of the exponent $(n/2) - 1 = (3/2)$). The only significant root is then z_4, for which $R_1(z_4;2,3) = 0.8905$. Note too that the maximum $z_m = \sqrt{pq/(n^3-n^2)} = 0.2449$, with $R_1(z_m;2,3) = 1.150396$ and

$R_2(z_m; 2, 3) = 1.183535$. Since $p/n = 0.4$, $\sqrt{pq/n^2} = 0.48989$, and $z_4 = 0.30687$, we have

$$z_m < z_4 < p/n < \sqrt{pq/n^2},$$

as expected. Thus, in summary, R_2 and R_1 reach the z-axis at 0.4 and 0.48989, respectively, and $R_1(z; 2, 3) = R_2(z; 2, 3)$ for $z = \sqrt{5 - \sqrt{7}}/5$.

In his twentieth Article Price shows that the ratio RhtQ : RQWh is equal to

$$\frac{n+1}{\sqrt{n}} \times \frac{\sqrt{2}}{2n} \times G \times \overline{mz - \frac{m^3 z^3}{3} + \frac{n-2}{2n} \times \frac{m^5 z^5}{5} - } \&c.$$

On taking G, K, and H as indicated in this Article, replacing K by its modern expression $\pi/2$ and letting n be large, we obtain (see Hald [1990b, p. 147])

$$\Pr\left[\frac{p}{n} - z\sqrt{\frac{2pq}{n^3}} < x < \frac{p}{n} + z\sqrt{\frac{2pq}{n^3}} \,\middle|\, p, q\right] \approx \frac{2}{\sqrt{\pi}} \sum_{k=0}^{\infty} (-1)^k \frac{z^{2k+1}}{k!\,(2k+1)}.$$

This therefore derives the series expansion for the Normal distribution as the limit of the beta posterior distribution.

At 'the point of contrary flexure', viz. $z = \sqrt{pq/(n^3 - n^2)}$, this probability is evaluated by Price as 0.6822, a modern evaluation by Hald [1990b, p. 148] yielding 0.6827.

Further in this same Article Price writes '... by putting the fourth term of the series in the 14th Art. equal to the second term, and then finding the value of z'. Now some care is needed in carrying out this procedure: from the discussion in Article 15 of the signs of the terms in the series in Article 14, it may be seen that what we have to do is solve the following for z.

$$-\left[\frac{q-1}{q^2} + \frac{p-1}{p^2} - \frac{n-2}{pq}\right] \cdot \frac{n^2 z^2}{2} = \left[\frac{q-1}{q^4} + \frac{p-1}{p^4} - \frac{n-2}{p^2 q^2}\right] \cdot \frac{n^4 z^4}{4}.$$

Since z is to be taken as positive we obtain

$$z = \sqrt{\frac{2(p/n)(q/n)}{n}} \times \sqrt{\frac{pq}{pq - n}},$$

the second factor on the right-hand side being approximately 1 'when n, p, and q are considerable'. In this case the desired probability has as lower limit the value 0.8426, which Hald [1990b, p. 148] notes is the value of the 'Normal' series for $z = 1$.

In the footnote to his corollary to Article 20 Price notes that it follows from that Article that in the case in which neither p nor q is small (or even not less than 10) the probability x of the event satisfies

(i) $\quad \Pr\left[\dfrac{p}{n} - \dfrac{1}{\sqrt{2}}\gamma < x < \dfrac{p}{n} + \dfrac{1}{\sqrt{2}}\gamma\right] \approx \dfrac{1}{2},$

(ii) $\quad \Pr\left[\dfrac{p}{n} - \gamma < x < \dfrac{p}{n} + \gamma\right] \approx \dfrac{2}{3},$

(iii) $\quad \Pr\left[\dfrac{p}{n} - \sqrt{2}\gamma < x < \dfrac{p}{n} + \sqrt{2}\gamma\right] \approx \dfrac{5}{6},$

where $\gamma = \sqrt{pq/(n^3 - n^2)}$ is 'the point of contrary flexure' [Art. 26]. This point is presumably found by 'differentiating' $x^p(1-x)^q$ twice, though this actually gives $(p/n) \pm \sqrt{pq/(n^3 - n^2)}$. A numerical example, with $p = 1000$ and $q = 100$, follows.

Perhaps somewhat more generally Price shows in Article 28 that

$$\Sigma + \Sigma \times \dfrac{1-\alpha}{1+(\alpha/2)} < \Pr[(p/n) - z < x < (p/n) + z \mid p, q] < 2\Sigma$$

for either p or q greater than 1, whereas if either p or q is greater than 10,

$$\Sigma + \Sigma \times \dfrac{1-\alpha}{1+(\alpha/4)} < \Pr[(p/n) - z < x < (p/n) + z \mid p, q] < 2\Sigma.$$

In the final paragraphs, **[324 − 325]**, Price re-examines an example he had considered in his Appendix to the *Essay* concerned with the finding of $\Pr[(10/11) - (1/110) < x < (10/11) + (1/110)]$. In addition to remarking in a footnote that an error had occurred in the calculation of this chance in the Appendix consequent on the setting of m^2 equal to $n^3/(pq)$ instead of $n^3/(2pq)$, he notes that the bounds given by Bayes's Rule II as given in this *Supplement* are 0.6512 and 0.7700, whereas the (improved) Rule of his Article 28 gives the limits 0.6748 and 0.7957.

We have seen in the preceding chapter that

$$\Pr\left[\dfrac{p}{n} - \dfrac{\tau\sqrt{2pq}}{n\sqrt{n}} < x < \dfrac{p}{n} + \dfrac{\tau\sqrt{2pq}}{n\sqrt{n}}\right] \approx \dfrac{2}{\sqrt{\pi}}\int_0^\tau e^{-t^2}\,dt. \qquad (8.5)$$

With $\tau = \frac{1}{2}\sqrt{n/(n-1)}$, the left-hand side of (8.5) becomes the left-hand side of (i), and hence the latter is approximately equal to

$$\dfrac{2}{\sqrt{\pi}}\int_0^{\sqrt{n/(n-1)}/2} e^{-t^2}\,dt = 2\Phi(\sqrt{n/2(n-1)}) - 1, \qquad (8.6)$$

where $\Phi(\cdot)$ is the cumulative distribution function of a random variable having the standard Normal distribution. If n is large, then $n/(n-1) \approx 1$, and (8.6) becomes[5]

$$2\Phi(1/\sqrt{2}) - 1 = 0.52,$$

which accords reasonably well with (i).

In his twenty-fourth Article Price finally presents his improvement of Bayes's Rule II: for $p \neq q$,

$$\frac{2\Sigma}{1+\alpha} < \Pr[(p/n) - z < x < (p/n) + z \mid p, q] < 2\Sigma.$$

In his discussion of the *Supplement* Hald presents a derivation of this Rule that, although essentially following Price's proof, (a) improves on the lower limit, (b) shows what values of z are permissible, and (c) amends an error made by Price. Hald's derivation [1990b, §7] runs as follows. Note firstly that, for $p > 1$ and $0 < z < z_0 \approx \sqrt{2pq/n^3}$,

$$f_3(z) - f_2(z) < f_2(-z) - f_3(-z), \tag{8.7}$$

where $f_2(\cdot)$ and $f_3(\cdot)$ are as defined before. Since $f_2(z) = f_2(-z)$, we may write this as

$$\frac{1}{2}f_3(-z) + \frac{1}{2}f_3(z) < f_2(z). \tag{8.8}$$

Further, for suitable values of β,

$$f_2(z) < (1-\beta)f_3(-z) + \beta f_3(z). \tag{8.9}$$

On combining (8.8) and (8.9) we get

$$\tfrac{1}{2}f_3(-z) + \tfrac{1}{2}f_3(z) < f_2(z) < (1-\beta)f_3(-z) + \beta f_3(z). \tag{8.10}$$

Next

$$\int_0^d f_3(z)\,dz = \left(\frac{n}{p}\right)^p \left(\frac{n}{q}\right)^q \int_{p/n}^{(p/n)+d} t^p (1-t)^q \, dt,$$

and similarly for $\int_{-d}^0 f_3(z)\,dz$. Thus, as we have already shown,

$$\int_0^d f_3(z)\,dz \Big/ \int_{-d}^0 f_3(z)\,dz < \frac{1+\alpha}{1-\alpha}. \tag{8.11}$$

On integrating (8.10) over $[0, d]$ (and abridging the integral notation somewhat), we get

$$\frac{1}{2}\int_{-d}^{d} f_3 < \int_0^d f_2 < (1-\beta)\int_{-d}^0 f_3 + \beta \int_0^d f_3. \quad (8.12)$$

Now

$$\int_{-d}^{d} f_3 = \int_{-d}^0 f_3 + \int_0^d f_3$$

$$= \left[(1-\beta)\int_{-d}^0 f_3 + \beta \int_0^d f_3\right] + \left[\beta \int_{-d}^0 f_3 + (1-\beta)\int_0^d f_3\right]. \quad (8.13)$$

On using the right-hand inequality in (8.12), inverting and multiplying by $\int_{-d}^d f_3$, we get

$$\int_{-d}^{d} f_3 \Big/ \int_0^d f_2 > \int_{-d}^d f_3 \Big/ \left[(1-\beta)\int_{-d}^0 f_3 + \beta \int_0^d f_3\right].$$

Now replace the numerator on the right-hand side of this last inequality by the expansion (8.13) and divide by 2. This gives

$$\frac{\int_{-d}^d f_3}{2\int_0^d f_2} > \frac{1}{2}\frac{\left[(1-\beta)\int_{-d}^0 f_3 + \beta \int_0^d f_3\right] + \left[\beta \int_{-d}^0 f_3 + (1-\beta)\int_0^d f_3\right]}{(1-\beta)\int_{-d}^0 f_3 + \beta \int_0^d f_3}$$

$$> \frac{1}{2} + \frac{1}{2}\frac{\beta \int_{-d}^0 f_3 + (1-\beta)\int_0^d f_3}{(1-\beta)\int_{-d}^0 f_3 + \beta \int_0^d f_3} =$$

$$> \frac{1}{2} + \frac{1}{2}\frac{\beta + (1-\beta)\int_0^d f_3 \Big/ \int_{-d}^0 f_3}{(1-\beta) + \beta \int_0^d f_3 \Big/ \int_{-d}^0 f_3}.$$

Using (8.11) in the last denominator, and the fact that

$$\int_0^d f_3 > \int_{-d}^0 f_3,$$

one may show, with a little manipulation, that

$$\int_{-d}^{d} f_3 \Big/ 2\int_{0}^{d} f_2 > \frac{1}{2} + \frac{1}{2}\frac{\beta + (1-\beta)}{(1-\beta) + \beta(1+\alpha)/(1-\alpha)} =$$

$$> \frac{1}{2}\left[1 + \frac{1-\alpha}{1+\alpha(2\beta-1)}\right].$$

The combination of this last expression with (8.12) yields the desired result.

In his proof Price supposes that $\int_{-z}^{z} f_2 > \int_{-z}^{z} f_3$ for all values of z, arguing that this follows from the fact that (a) $f_2(z) > 0.8247$ for $z = \sqrt{2pq/n^3}$ (which follows from his limit theorem) and (b) that (8.7) holds also for $z \in [z_0, z_1]$, and hence that $\int_{-z}^{z} f_2 > \int_{-z}^{z} f_3$ for $z > z_1$. Hald (loc. cit.) provides a sketch showing the error in this argument, though he notes that it works reasonably well in practice. Hald notes too that the right-hand side of his form of Price's Rule II holds for $0 < z < z_0$, where $z_0 \approx 3.2\sqrt{pq/n^3}$, and the left-hand side holds for $\beta > \hat{\beta}$, where $\hat{\beta}$ satisfies

$$\hat{\beta}/(1-\hat{\beta}) = e^{2\sqrt{p}}[(\sqrt{p}-1)/(\sqrt{p}+1)]^{p-1}.$$

In the *Essay* Bayes showed that Σ can be given in the form

$$\Sigma = \frac{n+1}{n} \times \frac{\sqrt{2pq}}{\sqrt{n}} \times \mathrm{E}\, a^p b^q \times \Sigma_n(t),$$

where

$$\Sigma_n(t) = \sum_{k=0}^{\infty} (-1)^k \binom{n/2}{k} \left(\frac{2}{n}\right)^k \frac{t^{2k+1}}{2k+1},$$

the series being found essentially by integrating $(1-(2u^2)/n)^{n/2}$ over $[0,t]$. The use of the expression for α given in (8.3) then permits the writing of Σ in the form

$$\Sigma = \frac{n+1}{n\sqrt{\pi}} \Sigma_n(mz)\, e^{\gamma}.$$

(See also Timerding [1908, p. 38].) Hald [1990b, p. 147] has shown that Bayes's Σ in fact tends to the Normal probability integral.

The proof of Bayes's third rule, a proof merely indicated in the *Essay*, may be completed as follows, as Hald [1990b, §5] shows. Write the series $\Sigma_n(t)$ as

$$\Sigma_n(t) = \Sigma_n(\sqrt{n/2}) - \frac{n}{(n+2)2mz}\left(1 - \frac{2t^2}{n}\right)^{(n/2)+1} + \mathfrak{S}_n(t), \quad (8.14)$$

where $\Sigma_n(\sqrt{n/2})$ and $\mathfrak{S}_n(t)$ are the terms defined by

$$\Sigma_n(\sqrt{n/2}) = \frac{\sqrt{\pi}}{2} \frac{n}{n+1} e^{-\zeta}$$

$$\mathfrak{S}_n(t) = \sum_{k=2}^{\infty} (-1)^k \frac{(2k-3)!!}{(2k+n)!!} \frac{n^k}{2^k t^{2k-1}} \left(1 - \frac{2t^2}{n}\right)^{(n/2)+1}$$

and ζ is our γ given in (8.4) with $p = q = n/2$. (Here $n!! = n(n-2)!!$ with $0!! = 1!! = 1$.) The lower and upper bounds in the third rule are then given by taking the first two and the first three terms, respectively, in $\Sigma_n(mz)$.

Timerding [1908, pp. 37–42] derived (8.14) by comparing the derivatives of the integral representation of $\Sigma_n(mz)$ and (8.14). We have already seen that this is a technique often used by Bayes, and it might well be the case that this was how his unpublished proof was indeed effected.

Wishart [1927, p. 10] notes that in the third rule the term h in the expression $\sqrt{2\pi pq} + h(n^{1/2} + n^{-(1/2)})$ should in fact be $2h$, where $h = e^{\gamma}$.

Note that the example given by Wishart illustrating the third rule provides an upper bound to the probability sought that is greater than 1; he goes on to say

> Now the required chance cannot be greater than unity, but we see that values greater than this are not excluded from Bayes' results. We suspect that this is why Price, in one of the examples he worked, only gave the *lower* limit to the answer.
> [1927, p. 11]

Wishart also gives here Price's improvement of the third rule. He notes too (loc. cit.) that whereas Bayes 'tried to minimise his error by determining the area between two ordinates equidistant from the mode' (a good result in this case being obtained by replacing the skew curve by a symmetric one), 'In actual practice we require the areas, or the frequencies, on either side of the mode separately and not together.' He also notes [p. 17] that 'The problem is not easy, for complexities are introduced as soon as we pass from the symmetrical to the asymmetrical curve.'

Hald [1990b, p. 149] notes that, as $n \to \infty$, the asymptotic expansion for $\Sigma_n(t)$ tends to[6]

$$\int_0^t \exp(-u^2)\, du = \frac{\sqrt{\pi}}{2} + (mz)\exp(-t^2) \sum_{k=1}^{\infty} (-1)^k \frac{(2k-3)!!}{(2t^2)^k}.$$

He also notes here that from the expression for $\Sigma_n(\sqrt{n/2})$ given above it follows that

$$\int_0^\infty \exp(-t^2)\,dt = \sqrt{\pi}/2.$$

In the *Supplement* Price in fact derived (speaking somewhat crudely) the Normal approximation to Bayes's probability (see also Timerding [1908, pp. 58–59]). More precisely, he showed that

$$\Pr\left[\frac{p}{n} - z\sqrt{\frac{2pq}{n^3}} < x < \frac{p}{n} + z\sqrt{\frac{2pq}{n^3}} \,\bigg|\, p,q\right] \to \frac{2}{\sqrt{\pi}} \sum_{k=0}^\infty (-1)^k \frac{z^{2k+1}}{k!\,(2k+1)}.$$

As we mentioned in §8.1, the series on the right-hand side is that obtained earlier by de Moivre as a representation of $(2/\sqrt{\pi}) \int_0^\infty \exp(-t^2)\,dt$. Thus the same series (or integral) is obtained as a limiting result in both the direct and the inverse problems. Timerding [1908, pp. 57–59] has noted that $\Sigma_n(t)$ (as given in Bayes's Rule II), tends to the standard Normal probability integral[7].

In his introductory letter Price wrote

> Perhaps, there is no reason for being very anxious about proceeding to further improvements. It would, however, be very agreeable to me to see a yet easier and nearer approximation to the value of the two series's in the first rule: but this I must leave abler persons to seek, chusing now entirely to drop this subject.

And indeed the importance of the incomplete beta-function, particularly in statistics, is such that it has been the subject of no inconsiderable study over the years. Karl Pearson and his collaborators published in 1934 extensive tables of the function, and several mathematical tracts were devoted to the quadrature. A number of other series expansions for the incomplete beta-function are known today; particularly relevant to the work by Bayes, Price, and Laplace, and in addition to that by Wishart already cited, is Wise [1950]. Using contour integration and earlier results in Molina [1932], Wise presented a rapidly converging series for the percentage points of the incomplete beta-function that permits the finding of either p or x from

$$\int_0^x t^{p-1}(1-t)^{q-1}\,dt \,\bigg/\, \int_0^1 t^{p-1}(1-t)^{q-1}\,dt = P,$$

the solution being given in terms of percentage points of the χ^2 distribution. For a deep study of the history of the incomplete beta-function see Dutka [1981].

Price also gave tighter bounds on the desired probability as an improvement on the second rule. Hald [1990b, §7] gives a detailed discussion of this improvement, including a correction of an error made by Price and a generalization of one of his subsidiary results. Hald also notes (p. 155) that Bayes's 2Σ can be written as the product

$$\Pr\left[\frac{1}{2} - \frac{z}{2\sqrt{ab}} < x < \frac{1}{2} + \frac{z}{2\sqrt{ab}} \,\bigg|\, \frac{n}{2}, \frac{n}{2}\right] \times \left[1 - \frac{1-4ab}{12nab} + \cdots\right],$$

where, as before, $a = p/n$ and $b = q/n$. Here the first term tends to the standard Normal probability integral from above, and the second tends to 1 from below as $n \to \infty$ for fixed a.

Hald [1990b, p. 155] compares the exact tail probability, the approximate tail probability, and the tail probability given by the Normal approximation. It may be concluded from his investigations that the 2Σ approximation given in the *Supplement* is considerably better than the Normal approximation.

9
Letters from John Ward

> *'That's rather a sudden pull up, ain't it, Sammy?'* inquired Mr. Weller.
> *'Not a bit on it,'* said Sam; *'she'll vish there wos more, and that's the great art o' letter writin'.'*
>
> Charles Dickens.
> Pickwick Papers.

9.1 Introduction

John Ward (1679?–1758), son of the dissenting minister John Ward, was a clerk in the navy office until leaving it for a schoolmaster's post in Tenter Alley, London, in 1710. A number of notebooks containing fair copies of letters written by Ward are preserved in the British Library: one of these letters is addressed to Thomas Bayes, and it is presented here as the only letter to Bayes of which I am aware. Written on the same day as this letter was a letter to Skinner Smith: this second letter, being relevant to the first, is also given here.

Ward was elected Professor of Rhetoric in Gresham College on the 1st of September 1720. It is possible, therefore, that at the time of the writing of these letters, he still occupied his academical position; and, as we have already suggested, it is also possible that the young Thomas Bayes had attended the academy with which Ward was associated, a possibility that perhaps gains some slight strength from the first of these letters.

9.2 Letters to Bayes and Skinner Smith

J.W. THOMAE BAYSIO. Qui sit rerum tuarum status, studiorumque quibus versaris ratio, es amicissima tua epistola libenter intellexi; nec stylo sane et orationis figura inter legendum parum delectabar. Eam autem Latinae linguae puritatem et elegantiam in tanta, ut dicis, scribendi desuetudine retinuisse miror aeque ac laudo. Sed si me audies, omnes styli exercendi occasiones captabis. Nam quid ad literarum jucunditatem accommodatius, vel ad laudem splendidius, quam pura et ornata scribendi ratio? *Stylus* enim, ut ait Cicero noster, est *optimus dicendi effector et magister.*[*] Novi temporis angustias, novi negotiorum molem, novi etiam turpem multorum in ea re incuriam et negligentiam. Veruntamen ut persuasum habeo te non ex eorum numero esse, qui malis aliorum exemplis ad incommodum tuum pertraharis; ita eo majori alacritate in hanc commendationem incumbere debes, quo cum paucioribus tibi communis sit futura. Nec operae profecto aut temporis multum requirit, modo constans et assiduus usus accedat; frequentia enim potius, quam longo et continenti labore, stylus excolitur. Materia autem, ni fallor, quotidie se offeret, dum aut ea quae legis excerpas, aut professorum dictata describas, aut tuas ipsius commentationes Latinis literis mandes. Crede mihi, vix concipi potest, quantam sermonis facilitatem ac ubertatem quotidiana istiusmodi exercitatio brevi secum afferat; dum ex contrario vetus illud dictum, *non progredi est regredi,* nulla alia in re, quam in hac, verius esse plurimorum experientia docet. Quod Graecos autem scriptores cum Latinis legendo conjungis, et inter se confers, meo sane judicio optime tibi consulis; nec dubito, quin ex ea legendi ratione fructus uberes sis percepturus. In reliquis studiis ordinem, quem sequeris, nequeo quin magnopere probem. Dum mathesi enim et logicae simul et conjuncta opera incumbis, quid et quantum utraque illa egregia instrumenta ad cogitationes sensusque animi dirigendos adjumenti adferant, clarius ac facilius percipies. Patri tuo librum de rerum antiquarum capitibus quibusdam partim mea, partim amici opera et studio conscriptum, nuper tradidi; quem is brevi, ut opinor, ad te mittendum curabit. Quae contineat, et cui commodo inserviat, ex epistola, quam ad assiduum tuum comitem de isto argumento hodie scripsi, ediscas. Tu vero ut amoris nostri in te singularis munusculum accipies. Cognato tuo C. meis verbis plurimum salvere jubeas, certioremque facias me literis ejus humanissimis brevi responsurum, quod et jam facerem, ni temporis augustiae impedirent. Vale. Dat. 10. kal. Maii 1720.

J.W. SKINNERO FABRO SUO. Quanquam epistola tua multis nominibus mihi jucundissima advenit, unum tamen, ut verum fatear, defuit, quod non potui quin maximopere desiderarem. Etenim cum tot amoris tui in

---

[*]De orat. 150.

nos singularis indicia, studiorumque rationem tam utilem et laudabilem percepi; quid in Latino stylo expoliendo profeceris, ex ea quoque intelligere valde cupiebam. Cujus rei quae sit laus atque utilitas, in epistola, quam individuo tuo comiti idem tradet tabellarius, significavi; quod ne hic iterum declarare in me suscipiam monet proverbium, quo docemur φίλων πάντα κοινά. Inter vos emin, tot necessitudinis vinculis conjunctos, veritatem ejus vel maxime comprobari nequaquam dubitare licet. Illud igitur potius: Cum in libro de capitibus quibusdam diversarum gentium antiquitatum una cum amico componendo, quicquid temporis a quotidiano munere mihi concessum fuit, aliquamdiu nuper impenderim, exemplar illius ad te mittere volui; quocum quicquid a me proficisci potest, quod tuae in literis humanioribus utilitati aliquid omnino tribuat, nequeo quin libenter communicem. Multa in eo reperies, quae ad veteres scriptores illustrandos haud parum conferant, et alibi forsan frustra quaerantur; mythologiam quoque, seu gentilium theologiam, melius ac certius, opinor, quam ex scriptis ejusmodi, quae signorum fide haud nituntur, illinc edisces. Proptianorum etiam hominum, et qui primis christianae ecclesiae seculis Christi fidem impuris gentium dogmatibus miscuerunt, nefandos ritus ac mysteria, non ex adversariorum scriptis, sed suis ipsorum notis ac tesselis cognosces; unde innumeris propemodum locis apud veteres ecclesiae scriptores, obscuris alioquin et difficillimis, mirum quantum lucis adfertur. Item plurima alia, tam ad mores publicos quam ad usum privatum atque domesticum spectantia, invenies, quae antiqui scriptores Romani, utpote ubique tunc obvia, ac bene nota, leviter tantum ac in transitu fere attingunt. Quae de vasis enim, lucernis, fibulis, annulis, sepulcris, urnis, numinis, aliisque non paucis scripta reliquerunt, ita fere sunt obscura et tenebris obducta, ut sine rerum ipsarum conspectu et tractione difficile plerumque sit certi aliquid de illis concipere, nedum exacte et accurate descibere. Sed nolo te diutius morari ea scribendo, quae tu melius evolvendo librum ipse percipies. Nequeo tamen finem hisce literis imponere, quin valetudinem recuperatam ex animo tibi gratuler; quam ut divini numinis clementia tueri pergat, obnixe precor. Dat. 10 kal. Maii 1720.

9.3 Translation of Letters

9.3.1 The letter to Bayes

I learned with pleasure from your most friendly letter the condition of your circumstances, and the nature of the studies in which you are engaged; and I was not a little delighted on reading it by the sober and the eloquent figures of speech. Moreover I look on it with admiration, and I likewise commend you for holding fast to the purity and elegance of the Latin language, your writing of which, you say, is rather rusty. But if you will take my advice, you will catch at every opportunity to practise style. For what contributes more to the pleasure of literature, or is more praiseworthy, than a plain and unornamented way of writing? For *style*, as our Cicero says, *is the best producer and teacher of eloquence*. I know the shortness of time, I know the demands made on one by business, I even know the shameful indifference and negligence of many in that respect. But nevertheless, inasmuch as I am convinced that you are not one of those who are led astray, to your disadvantage, by the bad example of others; consequently you ought to pay attention to this recommendation with the greater alacrity, so that in the future you may be one of a select few. And really it requires neither much work nor much time, provided that constant and assiduous practice is engaged in; for style is polished rather by frequent practice than by long and painstaking labour. Moreover, if I am not mistaken, material will present itself daily, provided that you either select from that which you read, or transcribe the lessons of the professors, or write down your own thoughts in Latin. Believe me, one can scarcely comprehend how great a facility and fruitfulness of expression daily practice of such a kind produces within a short space of time; while on the contrary the experience of many teaches us that in no matter other than this is that old saw, *not to progress is to regress*, more true. In my considered opinion you are wise in combining Greek with Latin authors in your reading, and in comparing them with one another; nor do I doubt, but that you will benefit abundantly from this way of reading. I cannot but greatly approve of the order in which you pursue the rest of your studies. For while you apply yourself simultaneously to mathematics and logic and to allied matters, you will comprehend more clearly and more easily what and how much each of these admirable disciplines contributes to the disposition of reasoning and one's state of mind. I have lately given to your father a book, partly by me and partly by a friend, about some of the most important aspects of antiquity, compiled by hard work and assiduity together; I believe he will shortly send this to you. What it contains, and what purpose it may serve, you may learn from the letter that I have written today about this matter to your constant companion. Please accept this as a small token of our affection toward you. My kindest regards to your kinsman C., and assure him that I shall soon answer his most affable letter, which I would now do, were I not prevented by shortness of time. Farewell. Dated 22nd April 1720.

9.3.2 The letter to Skinner Smith

Although your letter was in many respects most pleasing to me, one thing however, to tell the truth, was lacking, a thing that I could not but long for very greatly. And indeed, while I perceived so much evidence of your singular affection toward us, and the extremely useful and laudable account of your studies; I was very keen also to learn from it what progress you had made in polishing your Latin style. I have indicated in a letter that the same postman will deliver to your inseparable associate, what the merit and usefulness of such a polishing may be; the proverb that teaches us that

All things are held in common among friends admonishes me not to take it upon myself to discuss these things again here. For there can be no doubt of its established truth, especially between you two, joined as you are by so many close ties. Therefore let me rather write as follows: since for a fairly long time recently I have expended whatever available time I had each day in composing along with a friend a book about some aspects of the antiquities of different peoples, I wished to send you a copy of it; I cannot but willingly share whatever may originate with me, that may be of any use at all in your Humanity studies. You will discover many things in it, that may contribute not a little to the elucidation of the ancient authors, and that may be sought elsewhere perhaps in vain; also I am of the opinion that you will learn about mythology, or heathen systems of teaching about the gods [theology], better and more surely from this book than from writings of this kind that are not based upon the authenticity [true meaning] of symbols [visible signs of past actions]. You will also become acquainted with the impious rites and mysteries of profane [secular] men, and with those who in the early Christian church mixed faith in Christ with the impure doctrines of the heathen; these things you will learn, not from the writings of their adversaries, but from their own notes and tokens; whence an extraordinary amount of light is shed on almost innumerable passages in the works of old church writers. Likewise you will find many other things with regard both to public customs and to private and familial [domestic] use, which the ancient Roman writers, inasmuch as it was then everywhere commonplace and well known, mentioned only slightly and more or less in passing. For the writings that they have left on vessels, lamps, brooches, rings, graves, urns, coins and many other things, are almost so obscure and covered in darkness, that without seeing and touching the things themselves it may generally be difficult to take in any definite idea of them, to say nothing of describing them exactly and accurately. But I do not wish to hold your attention any longer by writing of that which you will see better for yourself by studying the book. I am unable however to end this letter without congratulating you on regaining good health; which I pray with all my might that the mercy of the Divine Will may continue to preserve. Dated 22nd April 1720.

9.4 Commentary

It seems from Ward's letter to Bayes that the young man had written to him in Latin, a tongue in which, as far as writing went, Bayes admitted to being 'rather rusty'. It is perhaps interesting to note that, as its first source of the use of 'rusty' in the sense of 'of knowledge, accomplishments, etc.: impaired by neglect', the *Oxford English Dictionary* gives 'For the benefit of those whose Greek is rather rusty with disuse, I have added a Latin version' (Porson, 1796) — though how those whose knowledge of these tongues was like Shakespeare's (at least if Ben Jonson is to be believed) were supposed to manage, is unclear.

There is some difficulty about the exact meaning to be attached to the Latin word *stylus*: Lewis [1937] gives '*stilus, not stylus*', defining it as meaning 'writing' or 'composition', and giving this quotation from Cicero as an example of its use. White [1880] gives '*stilus (sty–)*', meaning 'a setting down in writing; composition; manner of writing; mode of composition; style in writing', once again with this quotation as an illustration. The *Oxford Latin Dictionary*, however, instances this quotation as an example of the word *stilus* as 'the (use of the) stylus (i.e. the action or practice of writing, especially as a branch of education)'. The question thus reduces to whether *stilus* should be taken in our context as meaning *writing* in the sense of *style*, or in the sense of *penmanship*, the original 'Roman hand'[1].

The determination of the appropriate translation is forwarded by the following passage from the thirty-fifth lecture, 'Of Stile, and its different Characters' from Ward's *A System of Oratory*[2] of 1759.

> The word *stile* properly signifies the instrument, which the antients used in writing ... But tho this be the first sense of the word, yet afterwards it came to denote the manner of expression. [pp. 110–111]

The importance of style in the curriculum of the College of Edinburgh in the late sixteenth and early seventeenth centuries was noted by Grant in his *The Story of the University of Edinburgh* [1884, vol. I, p. 150] in a comparison of the course laid down there with its antecedents:

> It differed from the mediæval degree system in Scotland — (1) By making Greek an indispensable part of University study; ... (2) By the spirit of humanism which it exhibited, great attention being paid to purity of style both in Greek and Latin. (3) By its modernising tendency, in the admission of Ramus, and Talæus, and Hunter's Cosmography, and descriptive Anatomy.

Whether the emphasis insisted upon in (2) was still in effect in Bayes's time at the College is uncertain: it does however suggest that our translation of *stilus* as 'style', rather than 'penmanship', is correct.

Although Ward advocates the use of a 'plain and unornamented way of writing' (I feel strongly that the *ornata*, 'beautiful' or 'embellished', in the fair copy, should be *inornata*, and I have accordingly translated it as such), he suggests that style can be honed by the writing of one's own thoughts or the copying of lectures in Latin. We show in Chapter 11 that Bayes felt comfortable enough in writing Latin to use it extensively in a Notebook kept in his later years.

The matter of style is one that was of considerable importance to writers of earlier centuries. If one looks through the twenty-one entries under *Style* in Russell's *The Book of Authors* (and supplying some additional unlisted references) one finds that authors' styles were commended for being striking, pleasingly innovative, and with happy temerities (Samuel Johnson on Sir Thomas Browne (1605–1682)); for their splendid imagery (Hughes on Jeremy Taylor (1613–1667)); for a very gentlemanly acquaintance with the subject-matter (Thackeray on Sir William Temple (1628–1700)); for plainness and strength — along with a certain coarseness and rusticity, in this case (Southey on John Bunyan (1628–1688)); for being pure, transparently clear, and free from both levity and stiffness (Macaulay on Bishop Tillotson (1630–1694)); for a happy application of Scriptural quotation and language (Sydney Smith on Hugh Blair (1718–1800)); for being pure and clear, with a fine tone of chiaro-oscuro (Hazlitt on Charles Lamb (1775–1834))[3]. On the other hand, Gibbon's 'loaded and luxuriant style' was found by Bishop Hurd to be disgusting, and Hazlitt said that Jeremy Bentham's style was

> unpopular, not to say unintelligible ... His works have been translated into French — they ought to be translated into English. [Russell, 187-?, p. 327]

Hazlitt, in his early nineteenth-century essay *On familiar style*, writes

> To write a genuine familiar or truly English style, is to write as any one would speak in common conversation who had a thorough command and choice of words, or who could discourse with ease, force, and perspicuity, setting aside all pedantic and oratorical flourishes. [1822, II, p. 85]

As Hazlitt notes, such a style is neither quaint, vulgar (that is, 'coarse' rather than 'common')[4], nor florid. Of course, what a 'thorough command and choice of words' means is open to interpretation, and what passes as such in a specific era may be scornfully, if not derisively, viewed in another.

John Ward himself lectured on style, these lectures being posthumously published in two volumes as *A System of Oratory* in 1759. As illustration of the sort of thing then considered important, we give the following (rather long) quotation from Lecture XXXII, 'Of Figures of Sentences. Particularly those suited for Proof', from the second volume of the *System*[5].

> In my last discourse I treated on *verbal Figures*, and shall now, as I then promised, procede to *Figures of Sentences*. Among the

antients, as Isocrates is thought chiefly to have excelled in the beauties and delicacies of the former, so Demosthenes is most celebrated for expressing the force and energy of the latter. And Cicero lais so great stress upon these *Figures*, that he represents them as the brightest parts of oratory; [p. 65]

Prolepsis[6], or *Anticipation*, is so called, when the orator first starts an objection, which he forsees may be made either against his conduct or cause, and then answers it. Its use is to forestall an adversary, and prevent his exceptions, which cannot afterwards be introduced with so good a grace. Tho it has likewise a farther advantage, as it serves to conciliate the audience, while the speaker appears desirous to represent matters fairly, and not to conceal any objection, which may be made against him. [p. 66]

The figure *Hypobole* or *Subjection* is not much unlike the former. And is so called, when several things are mentioned, that seem to make for the contrary side, and each of them refuted in order. It consists of three parts when complete; a proposition, an enumeration of particulars with their answers, and a conclusion. [p. 70]

The next *Figure* in order is *Anacoinosis*, or *Communication*, by which the speaker deliberates either with the judges, the hearers, or the adversary himself. ... This *Figure* carries in it an air of modesty and condescension, when the speaker seems unwilling to determine in his own cause, but refers it to the opinion of others. It likewise shows a persuasion of the equity of his cause, that he can leave it to their arbitration; as serves very much to conciliate their minds, while he joins them, as it were, with himself, and makes them of his party. [pp. 72–73]

Another *Figure*, that comes under this head, is *Epitrope* or *Concession* which grants one thing to obtain another more advantageous. It is either real or feigned. And either the whole of a thing, or a part only is granted. [p. 73]

The next *Figure* of this kind is *Parabole* or *Similitude*, which illustrates a thing by comparing it with some other, to which it bears a resemblance. Similitudes are indeed generally but weak arguments, tho often beautiful and fine ornaments. [p. 76]

The last Figure above mentioned was *Antithesis* or Opposition, by which things contrary or different are compared to render them more evident. [p. 77]

To this *Figure* may also be refered *Oxymoron*, or *seeming Contradiction*, that is, when the parts of a sentence disagree in sound, but are consistent in sense. [p. 78]

Elsewhere in his *System* Ward set down other thoughts which we find expressed in the letters given here:

> To be master of a good style therefore it seems necessary, that a person should be endued with a vigorous mind and lively fancy, a strong memory, and a good judgment. [p. 113]
>
> And therefore it is not without reason, that Cicero recommends to all such, who are candidates for eloquence, and desirous to become masters of a good stile, to write much. [A marginal note here says 'De Orat. Lib. 1. c. 35] [p. 117]

Buffon has an entire *Discourse sur le style*, one described by Nisard as the only academic discourse having 'l'autorité d'un ouvrage d'enseignement' [18–, vol. 4, p. 409], and one which, Nisard claims, could have for its own epigraph Buffon's well-known maxim[7]

Le style est l'homme même,

though a similar thought had been expressed by Robert Burton much earlier in the introductory *Democritus Junior to the Reader* to his *Anatomy of Melancholy* as

> It is most true, *stilus virum arguit*, our style bewrays us, and as hunters find their game by the trace, so is a man's *genius* descried by his works.

As to how one goes about writing, Buffon says[8]

> Quand vous avez un sujet à traiter, n'ouvrez aucun livre, tirez tout de votre tête. ... Si l'on y joint encore de la défiance pour son premier mouvement, du mépris pour tout ce qui n'est que brillant, et une répugnance constante pour l'équivoque et la plaisanterie, le style aura de la gravité, il aura même de la majesté.

Taking slight exception to this opinion of Buffon's, Nisard notes that

> cette théorie du style pèche par un point. Elle ne tient pas compte de la diversité des voies de l'esprit. ... Buffon semble nier les bonheurs du premier jet, suspecter la verve, exclure la peinture à fresque, aussi charmante dans les passages d'un livre que sur les murs d'une coupole. [op. cit., p. 412]

Hugh Blair, the excellence of whose style (at least in Sydney Smith's opinion) has already been noted, was so persuaded of the importance of style that he published an entire treatise, composed of lectures given over twenty-four years at Edinburgh University, on it. Concerned mainly with 'the Eloquence suited to the Pulpit', Blair [1784, pp. 167, 168, 176] distinguished various kinds of style[9]:

> A Plain Style ... A writer of this character, employs very little ornament of any kind, and rests, almost, entirely upon his sense.
> A Neat Style ... here we are got into the region of ornament; but that ornament not of the highest or most sparkling kind.
> An Elegant Style is a character, expressing a higher degree of ornament than a neat one; and, indeed, is the term usually applied to Style, when possessing all the virtues of ornament, without any of its excesses or defects. ... In a word, an elegant writer is one who pleases the fancy and the ear, while he informs the understanding; and who gives us his ideas clothed with all the beauty of expression, but not overcharged with any of its misplaced finery.
> When the ornaments, applied to Style, are too rich and gaudy in proportion to the subject; when they return upon us too fast, and strike us either with a dazzling lustre, or a false brilliancy, this forms what is called a Florid Style; a term commonly used to signify the excess of ornament.
> An ostentatious, a feeble, a harsh, or an obscure Style, for instance, are always faults; and Perspicuity, Strength, Neatness, and Simplicity, are beauties to be always aimed at.

Blair did not restrict himself to the mere description of various styles, but also gave directions for the attaining of a good style:

> The first direction which I give for this purpose, is, to study clear ideas on the subject concerning which we are to write or speak.
> In the second place, in order to form a good Style, the frequent practice of composing is indispensably necessary.
> In the third place, with respect to the assistance that is to be gained from the writings of others, it is obvious, that we ought to render ourselves well acquainted with the Style of the best authors.
> In the fourth place, I must caution, at the same time, against a servile imitation of any one author whatever.
> In the fifth place, it is an obvious, but material rule, with respect to Style, that we always study to adapt it to the subject, and also to the capacity of our hearers, if we are to speak in public.
> In the last place, ... attention to Style must not engross us so much, as to distract from a higher degree of attention to the thoughts: "Curam verborum," says the great Roman Critic, "rerum volo esse solicitudinem." [Footnote: "To your expression be attentive; but about your matter be solicitous."]
> [1784, pp. 176, 177, 178]

Similar instructions were given by Buffon:

> Pour bien écrire, il faut donc posséder pleinement son sujet; il faut y réfléchir assez pour voir clairement l'ordre de ses pensees, et en former une suite, une chaîne continue, dont chaque point représente une idée; et lorsqu'on aura pris la plume, il faudra la conduire successivement sur ce premier trait, sans lui permettre de s'en écarter, sans l'appuyer trop inégalement, sans lui donner d'autre mouvement que celui qui sera détermineé par l'espace qu'elle doit parcourir. C'est en cela que consiste la sévérité du style, c'est aussi ce qui en sera l'unité & ce qui en réglera la rapidité; et cela seul aussi suffira pour le rendre précis & simple, égal et clair, vif et suivi.
>
> Bien écrire, c'est tout à la fois bien penser, bien sentir et bien rendre; c'est avoir en même temps de l'esprit, de l'âme et du goût. Le style suppose la réunion et l'exercice de toutes les facultés intellectuelles: les idées seules forment le fond du style; l'harmonie des paroles n'en est que l'accessoire, & ne dépend que de la sensibilité des organes. [Cahour, 1854, pp. 417–418]

Many of these directions are to be seen in Ward's letters.

Perhaps the whole matter was best summarized by Alexander Pope in his *Essay on Criticism*:

> True ease in writing comes from art, not chance,
> As those move easiest who have learn'd to dance.
> [Part ii, l. 162]

Important though style might have been in earlier times, the founders of the Royal Society were not slaves to its pursuit. One of the original statutes of 1663 read in part

> In all Reports of Experiments to be brought into the Society, the matter of fact shall be barely stated, without any prefaces, apologies, or rhetorical flourishes; and entered so in the Register-book, by order of the Society.

Four years later Thomas Sprat, in his history of the Society, noted that the latter had resolved on

> bringing all Things as near the mathematical Plainness as they can; and preferring the Language of Artizans, Countrymen, and Merchants before that of Wits, or Scholars ... to reject all the Amplifications, Digressions, and Swellings of Style.
> [1722, p. 113]

Jones [1963, p. 15] suggests that the attitude of the Royal Society in this respect was in fact hostile to the anti-Ciceronian style.

In earlier chapters we have mentioned both Bacon and Boyle; and although we would not wish to boil our bacon or to carp at a sprat, it is perhaps worth noting that the styles of the Lord Chancellor and the eminent scientist were fundamentally opposed. The experimental philosophers, followers of Bacon, downplayed the rôle of imagination, insisting on the importance of accurate observation: their stylistic attitudes were those that prevailed in the Royal Society as Sprat has noted. The atomists or mechanical philosophers, on the other hand, wrote as gentlemen for gentlemen, their aristocratic scientific spirit contrasting with the perhaps more democratic one of the experimental philosophers[10].

But enough of this stylistic excursion: before the reader can exclaim *Toujours perdrix!* let us return to Ward's letters[11].

The quotation from Cicero, perhaps given by Ward from memory in his letter to Bayes, is given by Leeman as *stilus optimus et praestantissimus dicendi effector ac magister* [1986, p. 118], and translated on p. 416, Note 27 of this work, as *The pen is the best and most eminent author and teacher of eloquence*[12].

One may also glean from Ward's letter to Bayes some information about the latter's studies. It seems that he was reading both Greek and Latin, a habit he clearly kept up in later years, for, as we have already noted in Chapter 3, Richard Onely, a Speldhurst clergyman, is reputed to have said that Bayes was the best Greek scholar he had ever met. We also find here confirmation of the fact that Bayes studied logic and (what is more important to us) mathematics. We might also note that one of the rules pertaining to academical studies at Doddridge's Academy, framed in 1743, was the following.

> Four classics, viz.: one Greek and one Latin poet and one Greek and one Latin prose writer as appointed by the Tutor are to be read by each student in his study and observations are to be written upon them to be kept in a distinct book, and communicated to the Tutor whenever he shall think fit.
> [Parker, 1914, p. 148]

It is possible that a similar rule had operated in the academy in which Ward had taught, and in which Bayes might have been a student, in which case (as indicated by Ward's letter) Bayes would merely have been continuing in the way in which he had been brought up.

Although Ward stresses the importance of writing Latin, he does not explicitly mention the importance of the translation of Greek into Latin, something that was stressed by Erasmus:

> In his "Ratio Studiorum," he strongly recommends translation from Greek into Latin, as giving insight into the comparative powers and idioms of each language, and showing what *we* have

in common with the Greek. ... The corresponding exercise in the present day would be careful translation from the classics into the mother tongue. [Parker, 1867, pp. 22–23]

Perhaps by this time the need for such translation had become attenuated.

The book given by Ward to Joshua Bayes has been identified by Bellhouse [1992] as *Monumenta Vetustatis Kempiana, ex vetustis scriptoribus illustrata, eosque vicissim illustrantia* ..., an account and catalogue by Robert Ainsworth of the classical antiquities collected by John Kemp, published in 1720. Two of the essays in this compilation have been attributed to Ward, viz. *De Asse et Partibus ejus commentarius* (published anonymously in 1719) and *De Vasis et Lucernis, de Amuletis, de Annulis et Fibulis*[13].

Bellhouse [1992, p. 225] has suggested that 'C.' could be Bayes's cousin, Nathaniel Carpenter. The son of Anne Bayes's brother (also Nathaniel), Nathaniel entered Edinburgh University in 1719, being recommended by Dr Calamy. He delivered homilies in April and December 1720, and was licensed but not ordained.

The Greek proverb φίλων πάντα κοινά cited by Ward is from Diogenes Laertius's *Vitae Philosophorum*[14]. Writing of this author, Sir Thomas Browne, in his *Hydriotaphia, Urne-buriall, or, A Discourse of the Sepulchrall Urnes lately found in Norfolk*, writes 'There is scarce any philosopher but dies twice or thrice in Laërtius' [chap. 3], and the same opinion on repetition can be held on quotations from this author. The Greek quotation given by Ward appears in the *Vitae Philosophorum* in various forms, e.g.[15],

Κοινὰ τὰ φίλων. [IV. 53]
[Friends have all things in common.]

τῶν θεῶν ἐστι πάντα· φίλοι δὲ οἱ σοφοὶ τοῖς θεοῖς· κοινὰ δὲ τὰ τῶν φίλων. πάντ' ἄρα ἐστὶ τῶν σοφῶν. [VI. 37]
[Everything comes from the gods; and the gods are friends to the wise; and all things are held in common among friends; therefore everything is for the wise.]

πάντα τῶν θεῶν ἐστι· φίλοι δὲ τοῖς σοφοῖς οἱ θεοί· κοινὰ δὲ τὰ τῶν φίλων. πάντα ἄρα τῶν σοφῶν. [VI. 72]
[Everything comes from the gods; and the gods are friends to the wise; and all things are held in common among friends; therefore everything is for the wise.]

In his letter to Skinner Smith, Ward uses the phrase *in literis humanioribus*, one that I have translated as 'Humanity studies'. On this matter the following is pertinent. In 1597 the Town Council and the College of Justice of Edinburgh appointed to the College of King James its first 'private Professor of Humanity' [Grant, 1884, vol. I, p. 190]. The adjective indicates that the occupant of this post was essentially a tutor, or 'an infra-Academical teacher of the subject' [Grant, loc. cit.]. This *Regens humaniorum literarum*, in addition to teaching Latin, had to instruct his pupils in Greek grammar. His task was thus to prepare students for the other four Regents to whom he was thus a subsidiary. (A Regent, unlike a Professor, instructed all his pupils in all the subjects in a prescribed currriculum.) The Regent of Humanity was turned into a Professor of Humanity (i.e., Latin) in 1708[16]. It thus seems reasonable to translate Ward's phrase *literis humanioribus* as I have done here.[17]

The translation of the word *tesselis* may be justified as follows. In Georges Lafaye's article on *tessera* and its diminutive *tessela* in Saglio's *Dictionnaire des antiquités grecques et romaines d'après les textes et les monuments* of 1919, we read that the former word was often applied by the ancients to counters [tallies, tokens, or vouchers for admission to theatres, etc.], tablets, and small plates of metal of different kinds. It is, moreover, clear that the distinction between a small *tabella* [tablet, note, voting-ticket, votive tablet] and a large *tessera* [die, tablet, token, mosaic piece] was very fine. Although a systematic classification cannot be given, one finds among the number amulets, oracles, religious prescriptions, funerary tablets, commemorative plaques, medallions of eminent persons, etc. Even at the time of the writing of this dictionary the use of some of these objects was uncertain.

10
Miscellaneous Items

> *Boswell: 'But of what use will it be, Sir?'*
> *Johnson: 'Never mind the use; do it.'*
>
> *James Boswell.*
> *The Life of Samuel Johnson, LL.D.*

10.1 Introduction

There are several manuscript items by Thomas Bayes in the Library of the Royal Society, viz. (i) a letter to John Canton on infinite series that was published in the *Philosophical Transactions*, (ii) another letter, also to Canton, commenting on some remarks by Thomas Simpson on errors in observations, and (iii) some notes on electricity, commenting on Hoadly and Wilson's *Observations on a Series of Electrical Experiments*. The first of these items is in the Royal Society's Miscellaneous Manuscripts collection [MM.1.17], and is printed elsewhere in the present treatise; the remaining two are in the Canton papers [Ca.2.32].

Recently D.R. Bellhouse discovered a set of manuscripts, some in Bayes's hand and others mentioning his name, among the Stanhope of Chevening papers [UC1590/C21] in the Centre for Kentish Studies in Maidstone, Kent, England. Bellhouse has given details of this important find elsewhere; nevertheless, for the sake of completeness of the present study, we give a brief discussion of these manuscripts here.

As regards the presentation of the Royal Society documents here, the reader is asked to note the following conventions, the first of which holds throughout the first few sections of this chapter and the others are particularly pertinent to §10.4. Words in crotchets preceded by the sign of equality, [= text], are expansions of abbreviations used by Bayes, and words in crotchets, [text], are words in the original that are missing from Bayes's version. Things such as 'w[h]ere' indicate that the encrotcheted matter is missing in the original manuscript. Words that are underlined appear in Bayes's manuscript but not in Hoadly and Wilson's book, whereas words in braces, {text}, are as in Hoadly and Wilson's text, Bayes's word (or words) preceding this formulation. Note that Bayes uses the symbol ∵ as a general punctuation mark for a comma, a colon, a semi-colon or a full stop. Bayesian abbreviations, such as 'rem.' for 'remember' and 'appear^{nces}' for 'appearances', have not been written out in full where the words abbreviated are thought to be patent and, in §10.4, where they occur in the expanded form in Hoadly and Wilson's book.

10.2 A Letter from Bayes to Canton

S^r.

Y^{ou} may rem. a few days ago we were speaking of M^r. Simpson[1] attempt to show y^e great advantage of taking y^e mean between several Astron. observations rather than trusting to a single observation carefully made in order to diminish y^e errors arising fro⁻ [= from] y^e imp[er]fection of instrum^{ts}. & y^e organs of sense. We both agreed that y^e former method was undoubtedly the best upon y^e whole & prtlrly [= particularly[2]] adapted to prev^t any considerable error, wch might possibly be committed in a single observation. But I really think ~~he~~ M^r. Simpson has not justly represent^{ted} its advantage: neither is it by far so great as he seems to make it.

According to him by multiplying our observations & taking the mean we always diminish ye probability of any given error, & that very fast. e.g, if a single observatiō may be relied on to 5″, & you take the mean of six observations it is above 5000 to 1 that your conclusion do's not differ 3″ from the truth, & by sufficiently increasing the number of observations you may make it as probable as you please that the result do's not differ from the truth above a single second or any small quantity whatsoever. Now that the Errors arising from the imperfection of instrum^{ts} & ye organs of sense shou'd be thus reduced to nothing or next to nothing only by multiplying the number of observations seems to me extremely incredible. On the contrary the more observations you make with an imperfect instrum^t the more certain it seems to be that the error in your conclusion will be proportional to the imperfection of the instrum^t made use of. for were it otherwise there wou'd be little or no advantage in making your observations with a very accurate instrum^t rather than with a more ordinary one, in those cases w[h]ere the observation cou'd be very often repeated: & yet this I think is what no one will pretend to say. Hence therefore as I see no mistakes in Mr Simpsons calculations I will venture to say that there is one in ye hypothesis upon which he proceeds. And I think it is manifestly this, when we observe with imperfect instruments or organs; he supposes that the chances for the same errors in excess, or defect are exactly the same, & upon this hypothesis only has he shown the incredible advantage, which he wou'd prove arises from taking the mean of a great many observations. Indeed Mr Simpson says that if instead of that series of numbers which he uses to express the respective chances for the different errors to which any single observation is subject any other series whatever shou'd be assumed the result will turn out greatly in favor of the method now practised by taking a mean value. But this I apprehend is only true where the chances for the errors of the same magnitude in excess or defect are upon an average nearly equal. for if the chances for the errors in excess are much greater than for those in defect, by taking the mean of many observations I shall only more surely commit a certain error in excess. & vice versâ. This I think is manifest without any particular calculation. & conseq^{tly} the errors which arise from the imperfection of the instrument with which you make a careful observation cannot in many cases be much diminished by repeating the observation ever so often & taking the mean.

10.3 Commentary on the letter to Canton

Bayes's letter to Canton is undated, but the contents suggest that the conversation referred to, dealing with Simpson's attempt to show the advantage of taking the mean of several observations over the observing of a single value, could have been prompted either by Simpson's *Letter to the Right Honourable* George *Earl of* Macclesfield, *President of the* Royal Society, *on the* Advantage *of taking the Mean of a Number of Observations, in practical Astronomy*, published in the *Philosophical Transactions* in 1756 (in the volume for 1755), or to its republication, with minor changes, as the first part of Simpson's *An Attempt to shew the Advantage arising by Taking the Mean of a Number of Observations, in practical Astronomy*, published in his *Miscellaneous Tracts* of 1757. Bayes's letter does not make the reference clear: if written as a response to the 1755 paper, the lack of any appropriate comment or correction by Simpson in his tract would suggest that he either chose to ignore the letter or was not informed of its contents; if, on the other hand, it was written as a response to the tract, then one cannot expect to find any action being taken.

There are substantial additions to the 1755 paper in the 1757 version, and Clarke is somewhat inaccurate in writing, in her biography of Simpson,

> The fourth paper [in Simpson's *Miscellaneous Tracts*] is an attempt to show mathematically that the mean of a number of astronomical observations is more exact than a single observation. ... In this *Tract* Simpson has somewhat changed the

original paper, but he gives the same two hypotheses that the distribution of errors might be a rectangle, each magnitude of error being equally probable, or that it might be an equilateral triangle, the probability of an error being inversely proportional to its magnitude. [1929, p. 187]

The 1755 paper consists essentially of two propositions and some examples. We cite the propositions here, with an indication of their proofs, that the reader may have easy access to Simpson's exact results and hence see the point of Bayes's remarks.

PROPOSITION I.

Supposing that the several chances for the different errors that any single observation can admit of, are expressed by the terms of the progression $r^{-v} \ldots r^{-3}, r^{-2}, r^{-1}, r^0, r^1, r^2, r^3 \ldots r^v$ (where the exponents denote the quantities and qualities of the particular errors, and the terms themselves the respective chances for their happening): 'tis proposed to determine the probability, or odds, that the error, by taking the Mean of a given number (n) of observations, exceeds not a given quantity (m/n).
[1755, p. 84]

It is perhaps not obvious from the phrase 'to determine the probability, or odds' whether Simpson intends the nouns to be synonymous or whether one is required to determine (a) the probability or (b) the odds (i.e., one of two distinct things). If, however, one turns to Simpson's *Nature and Laws of Chance* of 1740, one finds that *probability* is defined as the ratio of the number of favourable chances to the total number of chances, whereas the term *odds* is introduced in the solution of the following example.

Imagine a Heap of 16 Counters, whereof 6 are red, and the rest black; and a Person to draw out 2 of them blindfold: To find the Odds that one or both of those shall be red ones.
[1740, p. 5]

Simpson's solution, viz. odds of 5 to 3 (obtained by his using expectations, and perhaps more expeditiously found by consideration of the complementary event), is also given in terms of the probability 5/8. Thus it seems that *odds* and *probability* are different, rather than synonymous, terms that are connected in the usual way. However in the Note to the solution of Problem XI Simpson writes of 'the Probability, or Odds of winning the Game' [1740, p. 29], and to complicate the matter still further, the solution to this problem begins 'Since the Odds, or Chances, that any assigned Bowl ...'.

One may recall that in his *Essay towards solving a Problem in the Doctrine of Chances* Bayes carefully stated 'By *chance* I mean the same as probability.' Simpson, however, was less precise: he writes, on page 91 (and elsewhere) of his paper of 'the probability, or chance' of various errors, reflecting the title of his 1740 book. And although one might think that *chance* and *chances* have different connotations, one finds in this last work, in a discussion of ordered arrangements of letters, the words 'there is but one Chance, or Way, for all the Letters, or Things, to come out in that Order' [Simpson, 1740, p. 9], and in Problem VII we have

> Supposing a great, but given Number of each of two Sorts of Things to be put promiscuously together; To find how many must be taken out of the Whole, to make it an equal Chance that they shall all come out of one given Sort. [1740, p. 22]

So *chance* is either a *probability* or a *way in which things turn out*.

The ambiguity persists in the use of the plural. Thus in his first definition Simpson has

> The *Probability* of the Happening of an Event is to be understood as the Ratio of the Chances by which that Event may happen, to all the Chances by which it may either happen or fail [1740, p. 1]

whereas in Problem XIII he writes of '*A* and *B*, whose Proportion of Skill, or Chances for winning any assigned Game' (see also Problem XXV), *chances* now being related to ability.

This ambiguity is of course still found today, where one might well say one has three *chances* out of six of getting an even number when a die is tossed, and also assert that the *chance* of an even number is one half.

It would appear then that one should neither seek nor expect to find either precision or uniformity in the use of the terms *chance, chances, odds*, and *probability* ('variety amidst uniformity', one might be tempted to say in an inversion of Francis Hutcheson's aphorism cited in an earlier chapter); one should perhaps rather be guided by common usage and the presentation of the solution of the problem.

Revenons à ces moutons. To determine the probability sought in Proposition I, Simpson expands $(\sum_{k=-v}^{v} r^k)^n$ by writing it in the form

$$[r^{-v}(1 + r + \cdots + r^{2v})]^n = r^{-nv}(1 - r^{2v+1})^n(1 - r)^{-n}.$$

Then,

> to find from hence the sum of all the chances whereby the excess of the positive errors above the negative ones can amount, precisely, to a given number *m* [1755, p. 85]

he considers only those terms in the expansion that contain r^m. The first few terms are explicitly given: a modern expansion shows that the term in r^m may be written

$$\sum_{k=0}^{n} \binom{-n+k-1}{k}\binom{n+q-kw-1}{q-kw} r^m,$$

where, following Simpson, we have set $w = 2v+1$ and $q = m+nv$.

> From which general expression, by expounding m by 0, +1, −1, +2, −2, &c. successively, the sum of all the chances, whereby the difference of the positive and negative errors can fall within the proposed limits, will be found; which, divided by $r^{-vn} \times (1-r^w)^n \times (1-r)^{-n}$, will give the true measure of the probability required: from whence the advantage of taking the Mean of several observations might be shewn. [1755, p. 86]

Note that Simpson writes here of 'the true *measure of the probability*', whereas in his *Nature and Laws of Chance*, as we have already noted, he defines *probability* as the ratio of the number of favourable chances to the total number of chances. Similarly, in the introductory *Essai philosophique sur les probabilités* to his *Théorie analytique des probabilités* of 1820 Laplace defines such a ratio as 'la mesure de cette probabilité' in the section *De la probabilité*, and just as 'la probabilité' in the following section[3] (the first of these formulations also being adopted by Poisson in his *Recherches sur la probabilité des jugements en matière criminelle et en matière civile, précédés des règles générales du calcul des probabilités* [1837, p. 31]).

The second proposition differs from the first in having a different progression.

PROPOSITION II.

> Supposing the respective chances, for the different errors which any single observation can admit of, to be expressed by the terms of the series $r^{-v} + 2r^{1-v} + 3r^{2-v} \cdots + \overline{v+1}.r^0 \cdots + 3r^{v-2} + 2r^{v-1} + r^v$ (whereof the coefficients, from the middle one $(v+1)$, decrease, both ways, according to the terms of an arithmetical progression): 'tis proposed to determine the probability, or odds, that the error, by taking the Mean of a given number (t) of observations, exceeds not a given quantity (m/t).
> [1755, p. 87]

Simpson cleverly notes that the given series may be written as the square of the geometric progression

$$r^{-v/2}(1 + r + r^2 + \cdots + r^v),$$

the sum of this progression being easily seen to be given by

$$r^{-v}(1 - r^{v+1})^2/(1-r)^2.$$

The desired answer is thus obtained as before, with $n = 2t$, $w = v+1$, and $q = tv + m$.

After particular mention of the case in which $r = 1$ (the problem then being analogous to that of obtaining the points with n dice, each having $2v+1$ faces — a question solved by Simpson as Problem XXII of his *Nature and Laws of Chance*)[4], Simpson notes that, with respect to the series giving the answer,

> The difference between which and half (w^n), the sum of all the chances, (which difference I shall denote by D), will consequently be the number of the chances whereby the errors in excess (or in defect) can fall within the given limit m: so that $D/\frac{1}{2}w^n$ will be the true measure of the required probability, that the error, by taking the Mean of t observations, exceeds not the quantity m/t, proposed. [1755, p. 90]

(Note that by the 'difference' here is meant $|\text{Series} - (1/2)w^n|$.)

As a specific example, and having noted that the limits expressed by v 'depend on the goodness of the instrument, and the skill of the observer' [1755, p. 91], Simpson supposes that the observations may be relied on to five seconds, and that the chances for the errors $-5''$, $-4''$, ..., $+4''$, $+5''$ are 'respectively proportional to the terms of the series' 1, 2, ..., 2, 1,

> which series seems much better adapted than if all the terms were to be equal, since it is highly reasonable to suppose, that the chances for the different errors decrease, as the errors themselves increase. [1755, p. 91]

Supposing that six observations have been taken, Simpson now finds 'the probability, or chance' of the various errors. He shows, for instance, that the odds are (roughly) $2\frac{2}{3}$ to 1 ($2\frac{1}{2}$ to 1 in the 1757 tract) that the error incurred by taking the mean of six observations exceeds not a single second, these odds being 16 to 20 when only one observation is taken. Proceeding in this way, Simpson concludes the paper by saying

> ... it appears, that the taking of the Mean of a number of observations, greatly diminishes the chances for all the smaller errors, and cuts off almost all possibility of any great ones: which last consideration, alone, seems sufficient to recommend the use of the method, not only to astronomers, but to all others

concerned in making of experiments of any kind (to which the above reasoning is equally applicable). And the more observations or experiments there are made, the less will the conclusion be liable to err, provided they admit of being repeated under the same circumstances. [1755, pp. 92–93]

In connexion with Simpson's remarks that his advocated method should be used by all experimenters, one might recall Poincaré's writing

On voit que la méthode des moindres carrés n'est pas légitime dans tous les cas; en général, les physiciens s'en défient plus que les astronomes. Cela tient sans doute à ce que ces derniers, outre les erreurs systématiques qu'ils rencontrent comme les physiciens, ont à lutter avec une cause d'erreur extrêmement importante et qui est tout à fait accidentalle; je veux parler des ondulations atmosphériques. Aussi il est très curieux d'entendre un physicien discuter avec un astronome au sujet d'une méthode d'observation: le physicien, persuadé qu'une bonne mesure vaut mieux que beaucoup de mauvaises, se préoccupe avant tout d'éliminer à force de précautions le dernières erreurs systématiques et l'astronome lui répond: <<Mais vous ne pourrez observer ainsi qu'un petit nombre d'etoiles; les erreurs accidentelles ne disparaitront pas>>. [1903, pp. 241–242]

As we have already said, this paper was essentially incorporated into the 1757 tract, the new material there present being prefaced by the following remarks.

In the preceding calculations, the different errors to which any observation is supposed subject, are restrained to whole quantities, or a certain, precise, number of seconds; it being impossible, from the most exact instruments, to take off the quantity of an angle to a *geometrical exactness*. But I shall now shew how the chances may be computed, when the error admits of any value whatever, whole or broken, within the proposed limits, or when the result of each observation is supposed to be *accurately* known. [1757a, p. 71]

In the introduction to his 1755 paper, Simpson says that in order to prosecute his design he had

been obliged to make use of an hypothesis, or to assume a series of numbers, to express the respective chances for the different errors to which any single observation is subject; which series, to me, seems not ill-adapted [1755, p. 83]

and further,

> Should not the assumption, which I have made use of, appear to your Lordship so well chosen as some others might be, it will, however, be sufficient to answer the intended purpose: and your Lordship will find, on calculation, that, whatever series is assumed for the chances of the happening of the different errors, the result will turn out greatly in favour of the method now practised, by taking a mean value. [1755, p. 83]

This assumption was replaced in [1757a, p. 64] by the following suppositions.

> 1. That there is nothing in the construction, or position of the instrument whereby the errors are constantly made to tend the same way, but that the respective chances for their happening in excess, and in defect, are either accurately, or nearly, the same.
> 2. That there are certain assignable limits between which all these errors may be supposed to fall; which limits depend on the goodness of the instrument and the skill of the observer.

This certainly seems a more satisfactory ground on which to proceed than the hypothesis of the paper; it does not, however, address Bayes's concern with the possible inaccuracy of the measuring instrument (accidental *vs.* systematic errors?)[5], nor does it avoid his other point about the symmetric distribution of errors. (Incidentally Bayes's ratio of 'above' 5000 : 1 is more nearly 6681.49577 : 1.) One might recall Whitehead's citation [1967, p. 243] of Poincaré's (perhaps slightly paradoxical) assertion to the effect that 'instruments of precision, used unseasonably, may hinder the advance of science'.

It might perhaps be of interest to note that in one of the appendices to his *Mathematical Psychics* Edgeworth says something similar, viz.

> greater uncertainty of hedonimetry in the case of others' pleasures may be compensated by the greater number of measurements, a wider average; just as, according to the theory of probabilities, greater accuracy may be attained by more numerous observations with a less perfect instrument. [1881, p. 102]

The reader may recall that in our Preface we grouped Bayes, Boscovich, and Simpson together, remarking that each is remembered by statisticians for a single work. Bayes's letter to Canton weakly connects Bayes with Simpson, and it is perhaps worth noting that there was a stronger connexion between Simpson and Boscovich. In 1973 Sheynin published details of an undated manuscript by Boscovich on observational errors, noting that if this manuscript had been written before 1756 (the year of publication of

10.3. Commentary on the letter to Canton

Simpson's first writings on this topic), then Boscovich could be regarded as a precursor of Simpson (and of Lagrange) on this matter[6]. Sheynin [1973b, p. 318] framed Boscovich's problem as follows.

> In terms of probability (not chances) BOSCOVICH's problem could be stated thus: equiprobable values of each of n random quantities $\xi_1, \xi_2, \ldots, \xi_n$ are 0, 1, and -1. To find, the law of distribution of their sum (ξ). The generally known formula, the equivalent of which BOSCOVICH actually knew, is
>
> $$P\{\xi = a\} = \sum [P\{\xi_1 = x\} P\{\xi_2 = y\} \ldots P\{\xi_n = w\}],$$
>
> the summation extending over all the values of $x, y, \ldots, w$ complying with the conditions
>
> $$x + y + \cdots + w = a, \quad x, y, \ldots, w = 1, \text{or } 0, \text{or } -1.$$

In 1984 Stigler discussed a manuscript fragment by Simpson headed 'A problem proposed to me by M. Boscowitze', that problem being one in least absolute deviations regression. This fragment Stigler dated as 1760, the year in which Boscovich visited London, and in 1990 Farebrother presented evidence to show that Boscovich had personally received Simpson's solution in June 1760.

Although it would be inappropriate to spend too much time on Simpson's paper and tract themselves, a few general comments might not come amiss[7]. First of all, one should note that Simpson essentially used de Moivre's method of generating functions[8] in what was tantamount to the derivation of the distribution of a sum of independent errors, each having a discrete probability distribution, this latter being uniform in Proposition I and triangular in Proposition II[9]. In the latter portion of his tract Simpson introduced a continuous triangular distribution as an approximation to the earlier discrete one, and it is worth noting that he is sometimes described as the first to consider not only the concept of error distributions, but also that of continuous distributions[10]; Shoesmith [1985], however, notes that, before Simpson, J. Bernoulli, N. Bernoulli, and de Moivre had all derived continuous approximations to the binomial distribution[11]. It is important to note that estimation theory arose from the consideration of problems in which most of, if not all, the statistical variability was occasioned by errors of measurement, and not from problems in which the data themselves exhibit a large measure of internal variability (Huber [1972, p. 1042]). Simpson's work was continued later by Lagrange and still later by Laplace[12].

There are those who have perhaps read somewhat more into Simpson's work than was actually achieved. Sheynin [1968] notes that Simpson could have derived the Normal distribution (later than de Moivre) and could have been the first to sketch this distribution (the sketch of his limiting distribution is not in fact that of a Normal density). Lancaster [1994] in fact goes further, stating that Simpson was the first to consider the Normal distribution as a possible error density function.

In his discussion of Boscovich's determinism, Sheynin [1973b, p. 321] suggests that in his *Philosophiæ naturalis theoria* of 1758 Boscovich might have been considering a discrete uniform distribution of velocities

$$v, v \pm \Delta v, v \pm 2\Delta v, \ldots, v \pm n\Delta v.$$

If this indeed be so, then this consideration may be viewed as a generalization of the distribution considered in the undated manuscript by Boscovich mentioned before, and hence the latter might well pre-date the 1758 book.

In discussing Boscovich's 1757 criteria for the determination of the best-fitting straight line to a set of data, Eisenhart [1961, p. 200] instances the following conditions imposed by Boscovich.

> (I) The sums of the positive and negative corrections ... shall be equal.
>
> (II) The sum of (the absolute values of) all of the corrections, positive and negative, shall be as small as possible.

Although different to Simpson's (as is not surprising, in view of the difference in the problems addressed), these criteria are of interest in showing the direction being taken by those investigating error theory at that time. One might also note Eisenhart's statement that Boscovich believed his first criterion to be required by

> the *traditional*[13] assumption that positive and negative errors are equally probable. [1961, p. 209]

The fact that the arithmetic mean of a set of observations is not always the best summarizing statistic was discussed by Daniel Bernoulli in 1777. Here Bernoulli took it as axiomatic that errors in excess or in defect of the true value were equally possible, whereas values nearer the true value were more probable than those removed from it. Errors greater than some maximum value were regarded as impossible. And although Simpson assumed a triangular distribution, Bernoulli took the probability curve to be a semi-circle of given radius.[14]

An interesting woodcut from a 1535 book by Jacob Köbel on surveying is reproduced in Chapter 20 of Stigler [1999], an illustration showing the feet of sixteen church-goers being used to determine the length of a 'right and lawful rood'. Stigler notes this woodcut demonstrates (a) the act of taking a sample where chance enters into the selection, (b) the recognition of the benefit to be derived from compensatory errors, and (c) a rudimentary acquaintance with the well-known concepts of statistical sufficiency and exchangeability. For our purposes it is (b) that is of greatest relevance.

10.4 Item on Electricity[15]

Pag. 1. Takes it for granted that according to Sr Isaac N. there is an Æther which is dispersed thro' all space & which is the cause of gravity.

Obs. [= Observation] S^r Is. in his Opt. 3^d Edit^n in order to shew that he did not take gravity for an Essential property of Bodies added a question concerning its cause. And what S^r Is. makes a mere query, in support of which he owns he had no experim^{ts} is here represented as a point decided by him. Whereas it wou'd be very ~~injurious~~ unfair on the account of what he has said, to assert it was Sr Isaac's opinion that an Ætherial fluid was the cause of gravity; for according to him it is only a possible way of accounting for the Gravity of bodys but as yet supported by no experim^{ts}

Pag. 3. It being taken for granted that the Phænomena of Elasticity arise from the force & action of a certain fluid, it is proposed to be inquired into whether this Electrical fluid & the Æther be not one & the same fluid.

Obs. From hence one wou'd fancy that it was the principal design of the Treatise to shew this. But I can scarcely see any thing like an attempt to prove this.

P. 17. W^{never} a body is electrifyed either plus or minus, & remains so after ye experim^t is over there are similar Atmospheres of ye electrical fluid surrounding them that are ready to expand themselves into any body that approaches that resist less than the Air. & this is the reason why bodys give very nearly the same signs, when they are Electrifyed either plus or minus.

Observ. On the contrary it seems rather to follow that because bodys Electrifyed plus or minus give the same signs, therefore they are not surrounded with similar Atmospheres ready to expand themselves into any approaching body which resists less than the air.

P. 19. When two balls ~~are~~ Electrifyed plus & suspended by two silk strings {lines} are brought near to each other; they repel each other & stand for some time at a distance from each other; because ye two Atmospheres, each of them exerting their indeavors to expand into the Air, want more room to do it in; when y^e weight of y^e balls is not sufficient to prevent it must naturally drive them asunder; till these Atmospheres are dissipated. &c.

Obs. From this paragraph I am really at a loss to know what it is which according to our A. [= Authors] must naturally drive the balls asunder. Will it be said y^t y^e two Atmospheres drive y^e balls asunder that they may have room to exert their indeavors to expand into y^e air? Or do they do this that they may actually expand into y^e Air? One or other of these must be the meaning of the parag. I think if it has any meaning at all. But neither of these inform me in what manner or by what force the two atmospheres drive the balls asunder. And it is to observed that according to our author whether a ball be Electrifyed plus or minus it is surrounded with a similar Atmosphere exerting its indeavor to expand into the air & consequently reading this paragr. without the word plus the reasoning in it will appear full as good & plain as it do's now & will prove as well that any two Electrifyed balls must repel each other as w^n. both are Electrifyed plus.

P. 19. Par. 2. W^n. two bodys are both Electrifyed minus, suspended by silk strings & brought near one another, they repel each & stand f. [= for] some time at a distance {from each other}; because the condensed Electrical fluid in y^e air in order to force it self in at y^e surfaces of the balls between their two centers, crouds in, & forces them asunder.: till their Atmospheres get all into the balls.

[Here an observation is heavily scored out — illegible.]

P. 19. Par. 3. When two balls are electrifyed one plus and ye other minus are brought near one another; they will gradually come together & unelectrify each other. Because the atmosphere of ye ball electrifyed plus is indeavoring to dissipate it self from the center of the ball outwards; & ye atmosphere of ye ball Electrifyed minus is indeavoring to dilate it self from the air inwards to the center of ye ball. The common atmospheres therefore of the two balls thus brought near together must exert their forces in one & ye same direction between them; the flow of the Electrical fluid into ye ball Electrified minus is facilitated by ye indeavor of ye Electrical fluid to get out of ye ball electrified plus; & vice versâ; & so the two balls & ye air between them very readily return to their natural states.

[Another illegible and scored-out observation.]

P. 20. Thus then it appears that when we only know [that] a body is Electrified ∴ without knowing in what manner it was electrified there is no criterion to form our judgment upon whether it was Elect {electrified} plus or minus because the common appear^{nces} are the same in both cases when we unelectrify it ∴ But by pursuing the same chain {train} of reasoning we shall easily {readily} obtain a certain method to know whether a body is electrified plus or minus [even] without unelectrifying it ∴ Fasten 2 {two} cork balls one at each end to a peice {piece} of thread about eight inches long and doubling the thread over the bar before it is electrified make the balls to hang as near together {to one another} under the bar as they can and now the bar the thread {threads} and the balls should be considered as one body equally ready to be electrified either plus or minus ∴ 1mo. {1°}

We will suppose this bar &c. to be electrified minus and in consequence of their being electrified at all the balls to repel one another and hang at a greater distance from each other than {and} they did naturally ∴ Now let an excited Tube be brought to a certain distance under y^{se} balls in y^{se} crcmstncs and they will at first repel each other more ∴ because the force of the excited Tube will condense the atmospheres about {around} the balls still more till the resistnce at their surfaces is over come ∴ which will take up some little time during which their atmospheres being encreasing they will repel each other more than before the Tube was brought near them ∴ But so soon as ever this resistance is [once] over come ∴ the excited Tube drives the atmospheres into the balls &c and consequently begins to unelectrify y^m i.e. to render them less forcibly electrified minus and on withdrawing the Tube the balls will not repel each other near so much and so hang [very] visibly nearer together than they did before the Tube was first brought near them ∴

2do {2°} We will suppose the bar &c [to be] electrified plus and in consequence [of their being electrified at all] that the balls [to] repel each other ∴ Now when the {an} excited Tube is brought to a certain distance under the balls in y^{se} circumstances they will repel each other less forcibly and come nearer together ∴ because the resist[a]nce of the Air alone to the Electrical {electrified} fluids {fluid's} escaping out of the balls electrified plus has been seen not to be able entirely to prevent it ∴ as an atmosphere is made and supported round the balls till they are reduced to their natural state ∴ But now the excited Tube acts in concert with the Air, this Atmosphere wch had prevailed against the Air alone must on this additional f[o]rce acting against it retire again into the ball and continue to do so for some small time ∴ And after it is all retired into the balls again so long as the bar &c can be electrified more and more plus this appearance will continue whilst the Tube remains in action ∴ and when the Tube is withdrawn the balls will repel each [other] more forcibly than at first because they will remain more forcibly electrified plus. Thus then it appears that though the balls repel each other when the bar is electrified either plus or minus ∴ yet when an excited Tube is brought near them in [this] their repulsive state they will in one case have their repulsive force encreased at first but very soon after gradually diminished ∴ and in the other case be diminished at first but very soon after increased. And when we see the appear^{nces} {appearance} in the first case we may safely conclude the bar has {had} been electrified minus and when we see the appearance in the other case we may conclude the bar has {had} been electrified plus. N.B. A string with balls by the assistance of a glass Tube may be easily hung on an electrified bar without unelectrifying it {For the string, with the balls affixed to its ends, may, with the assistance of a glass tube, be easily hung on the bar without unelectrifying it, supposing it to have been electrified at first without the string and balls}.

P. 23. I cannot help observing here that in the first case after the excited Tube has been held so long near the balls that on its being with drawn they no longer repel each other if it be again presented to them it will electrify them and the bar plus and they will be put into a repulsive state again ∴ Whence it was reasonable to conclude that if the excited Tube in our first experim^t after it had been presented at a proper distance to the extended surface of the bar and so had electrified it minus had been brought nearer and nearer to the bar so as at last to touch it ∴ it will {would} have electrified it plus. And accordingly w^n the experim^t was tried it succeeded and the bar was electrified plus. W^{nce} there must be some middle distance between the situation where it is electrified plus & where minus {Now the consequence of this is, that there must be some middle distance of the excited tube from the bar, between its situation, where it electrified it minus, and its situation, where it electrified it plus;} at which middle distance the bar will be reduced to its natural state and not be electrified any more than if the Tube was not there ∴ It will {This would} appear <u>as</u> a most amazing paradox &c that {... paradox in electricity, if it was told in general terms to any one, who did not know that bodies were capable of being electrified plus and minus; viz.} the same excited Tube brought near a body electrifies [it] and after that (without ever with drawing it) brought nearer ceases to electrify it. and after that brought still nearer electrifies it again Now as this was deduced fm {from} our way of reasoning the ev^t answering on making the experim^t greatly confirms the truth of the reasoning.

P. 69. It is likewise improper to call this fluid (viz Electrical) Fire. Air may just as properly be called sound as this fluid can be called fire. W^n sound is produced ye particles of the Air are put into so regular a motion as to convey such sensations by means of the Ear as raise the Idea of sound. But Air is not therefore sound.

In the same manner when a body has all its component particles thrown into such agitat^{ns} in the Air by the force and action of this fluid within it & {and} without it, y^t it grows hot & {and} shines, & {and} glows & {and} consumes away in smoak & {and} flame, we say ye body is on fire or burns; but y^s fluid is not therefore fire: nor can it without confounding our Ideas have y^t name given to it ∴ Nor indeed can fire be called a Principle or Elem^t in ye Chemists sense of ye w[or]d any more than sound can.

P. 71. A Spark (by a flint and steel) {Now if such a spark be} caught on a sheet of paper and examined in a microscope {it} will be found to be a piece {either} of the flint or of the steel struck off so exactly spherical and polished that the windows of the room may be seen in it in the same manner as they are in a large polished sphere of metal or glass ... And they could not be so spherical and well polished as they are found to be if they had not been melted and kept in this form by the cohæsion of their component particles.

10.5 Commentary on the Item on Electricity

In the eighteenth century electricity was a matter of considerable interest to natural philosophers. The range of phenomena considered was so vast, however, and the number of experiments carried out so large, that in the one hundred and thirty-eighth of his letters to a German princess, the one dated 20th June 1761, on the principal phenomena of electricity, Leonhard Euler — one whose name is to writers on eighteenth-century mathematics what King Charles's head was to Mr Dick[16] — wrote

> La matiere, sur laquelle je voudrois à présent entretenir V. A. me fait presque peur. La variété en est surprenante, & le dénombrement des faits sert plûtot à nous éblouïr qu'à nous éclairer. C'est de l'Electricité dont je parle, & qui depuis quelque tems est deveniie un article si important dans la Physique, qu'il n'est presque plus permis à personne d'en ignorer les effets. [1770b, II, p. 227]

Euler had an answer to the problem, however, in Letter CXL of the 27th June, 1761:

> Cela posé, j'ose avancer, que tous les phénomènes de l'électricité sont une suite naturelle du défaut de l'équilibre dans l'éther, de sorte que partout, où l'équilbre de l'éther est troublé, les phénomènes de l'électricité en doivent résulter; ou bien je dis, que l'électricité n'est autre chose qu'un dérangement dans l'équilibre de l'éther. [1770b, II, p. 286]

There were, in fact, almost as many explanations of electrical phenomena as there were experimenters. Noting this diversity, George Adams, mathematical instrument maker to His Majesty King George III, wrote

> As electricity is in its infancy, when considered as a science, its definitions and axioms cannot be stated with geometric accuracy [1799, p. 33]

the truth of the first part of which is preserved by the noting that electricity is 'considered as a science', for Mottelay, in his *Bibliographical History of Electricity and Magnetism chronologically arranged*, instances the first investigations into the subject under the heading B.C. 600–580, writing

> Thales of Miletus ... is said to have been the first to observe the electricity developed by friction in amber[17].
>
> Thales, Theophrastus, Solinus, Priscian and Pliny, as well as other writers, Greek and Roman, mention the fact that when a vivifying heat is applied to amber it will attract straws, dried leaves, and other light bodies in the same way that a magnet attracts iron. [1922, p. 7]

Adams went on further to describe this esotericism as follows.

> a field of inquiry, wherein fancy has, indeed, sufficiently exhibited her luxuriance. No other science has had more admirers, nor been subject in so short a space of time, to so great a variety of hopotheses [sic]. [1799, p. 92]

As examples of the different theories proposed let us consider two expounded by Adams: those of Benjamin Franklin and Henry Eeles. Franklin's theory was reduced to the following principles.

> 1. That the atmosphere and all terrestrial substances are full of the electric fluid.
> 2. That the operations of electricity depend on the uncompounded action of a simple fluid of a peculiar nature, extremely subtile and elastic.
> 3. Glass and other electric substances, though they contain a great deal of electric matter, are nevertheless impermeable to this fluid.
> 4. That the electric matter violently repels itself, and attracts all other matter.
> 5. By the excitation of an electric[18] the equilibrium of the contained fluid is broken; and one part becomes overloaded with electricity, while the other contains too little.
> 6. Conducting substances are permeable to the electric matter through their whole substance.
> 7. Positive electricity is when a body has more than its natural state of the electric fluid; and negative electricity, when it has less than its natural share. [1799, pp. 100-101]

These principles Adams found inferior to those presented by Eeles in his *Philosophical Essays*, summarizing the latter as follows.

> 1. The two electric powers exist together in all bodies.
> 2. As they counteract each other when united, they can be rendered evident to the senses only by their separation.
> 3. The two powers are separated in non-electrics by the excitation of electrics, or by the application of excited electrics.
> 4. The two powers cannot be altogether separated in electrics.

> 5. The two electricities attract each other strongly through the substances of electrics.
> 6. Electric substances are impervious to the two electricities.
> 7. Either power, when applied to an unelectrified body, repels the power of the same sort, and attracts the contrary.
> [1799, p. 108]

Claiming that electricity was real matter, and not a mere property [p. 48], Adams continued

> Some have supposed the electric matter to be a kind of unctuous effluvia[19], arising by means of friction from substances termed electrics *per se*; others, the ether pointed out by Sir *Isaac Newton*, in the effects of which a certain subtile medium was concerned. Some called it elementary fire, and imagined it to be a modification of the fire they termed an element; while others conceived it to be a fluid distinct from chemical fire, but of a nature greatly resembling it. [1799, p. 92]

What the 'electric fluid' mentioned above was, was uncertain. Adams described it as follows. 'There is a natural agent or power, generally called the *electric fluid*, which by friction, or other means, is excited or brought into action' [1799, p. 1], though he then went on to say

> Though the electric fluid has been known and studied for many years, we are altogether ignorant of its real essence and nature; to me it seems probable, that it is fire or light connected with some terrestrial base. [1799, p. 49]

Some fifty years before this, however, in a paper published in the *Philosophical Transactions*, William Watson wrote quite categorically

> That what we call Electricity is the Effect of a very subtil and elastic Fluid, diffused throughout all Bodies in Contact with the terraqueous Globe (those Substances hitherto termed Electrics *per se* probably excepted), and every-where, in its natural State of some Degree of Density. [1748, p. 95]

By the turn of the eighteenth century we find Thomas Young commenting less dogmatically on this fluid as follows.

> It is supposed that a peculiar etherial fluid pervades the pores, if not the actual substance, of the earth and of all other material bodies, passing through them with more or less facility, according to their different powers of conducting it: that the particles of this fluid repel each other, and are attracted by the particles

of common matter: that the particles of common matter also repel each other: and that these attractions and repulsions are equal among themselves, and vary inversely as the squares of the distances of the particles.

The effects of this fluid are distinguished from those of all other substances by an attractive or repulsive quality, which it appears to communicate to different bodies, and which differs in general from other attractions and repulsions, by its immediate diminution or cessation, when the bodies, acting on each other, come into contact, or when they are touched by other bodies. ... In general a body is said to be electrified, when it contains, either as a whole, or in any of its parts, more or less of the electric fluid than is natural to it; and it is supposed that what is called positive electricity depends on a redundancy, and negative electricity on a deficiency of the fluid.
[1807, vol. I, p. 659]

So it seems that the 'electric fluid' was to be identified with the 'peculiar etherial fluid'.

Euler, however, was perhaps a little dismissive of the electric fluid. In his letter CL of the 1st August 1761 he wrote

La plupart des auteurs, qui en ont écrit, embrouillent tellement les expériences, qu'à la fin on n'y comprend absolument rien, & sur tout quand ils veulent en donner une explication. Tous ont recours à une certaine matiere subtile, qu'ils nomment *le fluide électrique*, auquel ils attribuent des qualités si bizarres, que nôtre esprit en est tout-à-fait revolté; & au bout du compte ils sont rien moins que suffisans pour nous procurer une connoissance solide de ces phénomènes importans de la nature.
[1770b, II, p. 332]

How bodies became electrified was unclear. Having noted that a body becomes electrical when the ether in its pores becomes more or less elastic than that in adjacent bodies, Euler, in his letter CXLII of the 4th July 1761, wrote

De là V. A. voit qu'un corps peut devenir électrique en deux manieres différentes, selon l'éther contenu dans ses pores devient plus ou moins élastique que celui de dehors; d'où une double électricité peut avoir lieu: L'une, où l'éther se trouve plus élastique ou plus comprimé, est nommée *l'électricité en plus* ou bien *électricité positive*; l'autre, où l'éther est moins élastique ou plus raréfié, est nommée *l'électricité en moins*, ou *électricité négative*. Les phénomenes de l'une & de l'autre sont à peu près les mêmes; on n'y remarque qu'une légere différence dont je parlerai dans la suite. [1770b, II, p. 295]

10.5. Commentary on the Item on Electricity

The ether itself was a subject of some discussion. We say more on the matter in our examination of Bayes's comments on Hoadly and Wilson's book, contenting ourselves for the moment with some quotations: the first, from Euler's Letter CXLI of the 30th June 1761, runs as follows.

> V. A. n'a qu'à s'en tenir à l'ideé de l'éther, que je viens d'etablir, & qui est cette matiere extrêmement subtile & élastique, repandue non seulement par tous les espaces vuides du monde, mais aussi dans les moindres pores de tous les corps, dans lesquels il est tantôt plus, tantôt moins engagé, selon que ces pores sont plus ou moins fermés. [1770b, II, pp. 289–290]

This, incidentally, is not a view with which Franklin would have agreed, for in his 'Further Remarks, by a Gentleman of New-York' of the 2nd April, 1754, he wrote

> Sir *Isaac Newton* says, there are many phænomena to prove the existence of such a fluid [i.e., a subtle elastic fluid]; and this opinion has my assent to it. I shall only observe that it is essentially different from that which I call æther; for æther, properly speaking, is neither a fluid nor elastic; its power consists in reacting any action communicated to it, with the same force it receives the action. [1774, p. 285]

The second quotation, from Whittaker's *History of the Theories of Aether and Electricity*, is rather more poetically phrased:

> The aether is the solitary tenant of the universe, save for that infinitesimal fraction of space which is occupied by ordinary matter. [1910, p. 1]

Perhaps the state of the matter is best summarized in Young's words

> It must be confessed that the whole science of electricity is yet in a very imperfect state. [1807, vol. I, pp. 683–684]

But now, On to Hecuba! It has been suggested (see Home [1974–1975, p. 82]) that Bayes's Note might have been passed on to Canton by Richard Price after Bayes's death. Price himself is known to have had some interest in matters electrical: in the introduction to his *History and Present State of Electricity*, Joseph Priestley wrote

> My grateful acknowledgments are also due to the Rev. Mr. Price F.R.S. and to the Rev. Mr. Holt, our professor of Natural Philosophy at Warrington, for the attention they have given to the work, and for the many important services they have rendered me with respect to it. [1767, p. viii][20]

Further, in Part IV, Section II, *Queries and Hints concerning Electrics and Conductors*, we find

> As thunder generally happens in a sultry state of the air, when it seems replenished with some sulphureous vapours; may not the electric matter then in the clouds be generated by the fermentation of sulphureous vapours with mineral or acid vapours in the air. *Mr. Price*[21]. [1767, p. 498]

This lends weight to Home's suggestion.

Hoadly and Wilson's book was first published in 1756, with a second edition in 1759, the Advertisement to the latter stating that it was by Wilson only, Hoadly having died. Wilson in fact published a number of works on electricity around this time[22], his aim being the establishment of a theory of electricity based on the all-pervading ether whose existence Newton had postulated in the second edition of his *Opticks*. Home asserts further that Wilson's development 'of electrical atmospheres as accumulations of electrical matter surrounding electrified bodies' [1974–1975, p. 85] followed that given in a series of letters[23] sent by Benjamin Franklin to the Royal Society, written from Philadelphia from 1747 to 1754. It was in these letters that Franklin also distinguished between the *plus* and *minus* modes of electrification, this being recorded by Mottelay as follows.

> To electrise *plus* or *minus* no more needs to be known than this, that the parts of the tube or sphere that are rubbed do, in the instant of the friction, attract the electrical fire, and therefore take it from the thing rubbing; the same parts, immediately as the friction upon them ceases, are disposed to give the fire they have received to any body that has less. [1922, pp. 196–197]

Priestley remarks

> In another paper, read at the Royal Society November the 13th. 1760, Mr. Wilson recites some curious experiments, which, he says, shew that a *plus* electricity may be produced by means of a *minus* electricity[24]. [1767, p. 226]

During the American Revolution the relationship between Franklin and the Royal Society became somewhat strained. The details, as reported by Mottelay, run as follows[25].

> "During the year 1777 a dispute arose among the members of the Royal Society relative to the form which should be given to electrical conductors so as to render them most efficacious in protecting buildings from the destructive effects of lightning. Franklin had previously recommended the use of points, and the

propriety of this recommendation had been acknowledged and sanctioned by the Society at large. But, after the breaking out of the American Revolution, Franklin was no longer regarded by many of the members in any other light than an enemy of England, and, as such, it appears to have been repugnant to their feelings to act otherwise than in disparagement of his scientific discoveries. Among this number was their patron George III, who, according to a story current at the time, and of the substantial truth of which there is no doubt, on its being proposed to substitute knobs instead of points, requested that Sir John Pringle would likewise advocate their introduction. The latter hinted that the laws and operations of nature could not be reversed at royal pleasure; whereupon it was intimated to him that a President of the Royal Society entertaining such an opinion ought to resign, and he resigned accordingly."
In Benjamin Franklin's letter to Dr. Ingen-housz, dated Passy, Oct. 14, 1777, occurs the following: 'The King's changing his *pointed* conductors for *blunt* ones is therefore a matter of small importance to me. If I had a wish about it, it would be that he had rejected them altogether as ineffectual.' It was shortly after the occurrence above alluded to that the following epigram was written by a friend of Dr. Franklin:

> "While you Great George, for knowledge hunt,
> And sharp conductors change for blunt,
>   The nation's out of joint:
> Franklin a wiser course pursues,
> And all your thunder useless views,
>   By keeping to the *point*."

[1922, p. 251]

In the case of positively charged bodies, the conclusions reached by Hoadly and Wilson were similar to Franklin's; in the case of negatively charged bodies, Hoadly and Wilson went beyond Franklin, and concluded that bodies charged either positively or negatively would display essentially the same effects. Franklin pointed out that any plausibility there might be in the account of the formation of an atmosphere about a body that was negatively charged, lay only in the cases in which an insulated body was charged by induction. Equally difficult to deal with was the reconciliation of the theory with the observation that positive and negative charges, when they came into contact, cancelled each other out. (This, as we have noticed in the preceding section, was a matter that also concerned Bayes.)

Home finds that Bayes's objections were to more than just the 'negative atmospheres' proposed by Hoadly and Wilson, and he writes 'the position he [i.e., Bayes] adopted with respect to the whole question of atmospheres was a quite radical one' [1974–1975, p. 86]. For although Franklin was prepared to concede that electrical attractions and repulsions[26] could be satisfactorily explained in terms of atmospheres,

> Bayes would not even concede this much: simply to talk about atmospheres was no substitute, he insisted, for providing a detailed explanation of the forces that had been found to act in these cases, and such an explanation had not been given.
> [Home, 1974–1975, p. 86]

Nor was Bayes prepared 'to accept without argument the identification of the electrical fluid with Newton's æther' [Home, loc. cit., p. 87].

Bayes was certainly not alone in his refusal to accept the identification of the electrical fluid with the ether. In his Letter CXXXIX of the 23rd June 1761 to a German princess, Euler wrote

> Ils [la plûpart des Physiciens] y reconnoissent bien une matiere subtile, qui en est le prémier agent, & qu'ils nomment la matiere électrique, mais ils sont si embarrassés d'en déterminer la nature & les propriétés, que cette grande partie de la Physique en devient plutôt embrouillée qu'éclaircie. ... Cette même matiere subtile qu'on nomme l'Ether, & dont j'ai déjà eu l'honneur de prouver la réalité à V. A.* est suffisante pour expliquer très naturellement tous les effets étranges, que nous observons dans l'électricité.
> *Voyez Lettre XIX. [1770b, II, pp. 281–282]

Tiberius Cavallo similarly refused to accept the identity of the electrical and the etherial fluids, writing in the second edition of his *Complete Treatise on Electricity*,

> As to the identity of the Electric, and the ethereal fluid, it seems to me quite an improbable, or rather a futile, and insignificant hypothesis; for this ether is not a real, existing, but merely an *hypothetical fluid*, supposed by different Philosophers to be endued with different properties, and to be an element of several principles. Some suppose it to be the element of fire itself, others make it the cause of attraction, others again derive animal spirits from it, &c.; but the truth is, that not only the essence, or properties, of this fluid, but even the reality of its existence is absolutely unknown.
> According to SIR ISAAC NEWTON'S supposition, this ether is an exceedingly subtle, and elastic fluid, dispersed throughout all the universe, and whose particles repel the particles of other matter. But on this supposition the electric fluid is different

from ether; for, although the former is subtle, and elastic, like the latter, yet (as DR. PRIESTLEY observes) it is not repulsive like the ether, but attractive of all other matter.
[1782, pp. 121–122]

The position adopted towards the ether in the first edition of the *Encyclopædia Britannica*, however, was rather different. Here we read

> ÆTHER, the name of an imaginary fluid, supposed by several authors, both ancient and modern, to be the cause of gravity, heat, light, muscular motion, sensation, and, in a word, of every phænomenon in nature. Anaxagoras maintained that æther was of a similar nature with fire; Perrault represents it as 7200 times more rare than air; and Hook makes it more dense than gold itself. ... It must indeed be acknowledged, that there is a propensity in the human mind, which, unless it be properly restrained, has a direct tendency both to corrupt science, and to retard our progress in it. Not contented with the examination of objects which readily fall within the sphere of our observation, we feel a strong desire to account for things which, from their very nature, must, and ever will, elude our researches. Even Sir Isaac Newton himself was not proof against this temptation. ...
> He had recourse to a subtile elastic æther, not much different from that of the ancients, and by it accounted for every thing he did not know, such as the cause of gravitation, muscular motion, sensation, &c.
> Notwithstanding the reputation of Sir Isaac, philosophers have generally looked upon this attempt as the foible of a great man, or, at least, as the most useless part of his works; and accordingly peruse it rather as a dream or a romance, than as having any connection with science. [1771, vol. I, pp. 31–32]

The article on electricity in this same work begins

> The word ELECTRICITY signifies, in general, the effects of a very subtile fluid matter, different in its properties from every other fluid we are acquainted with. This fluid is capable of uniting with almost every body, but unites more readily with some particular bodies than with others: its motion is amazingly quick, is regulated by peculiar laws, and produces a vast variety of singular phenomena, the principal of which shall be enumerated in this article.
> As we are entirely ignorant of the nature of the electrical fluid, it is impossible to define it but by its principal properties: that of repelling and attracting light bodies, is one of the most remarkable. [1771, vol. II, p. 471]

From Bayes's observation on the quotation from Page 1 of Hoadly and Wilson's book, we can deduce that he used the first edition for his notes and commentary. The actual passage in the first edition runs

> There is a very fine Fluid, of the same nature with Air, but extremely more subtle and elastic, according to Sir *Isaac Newton*, every where dispersed through all space, which in his Optics he calls *Æther*.

In the second edition, issued by Wilson alone in 1759 after Hoadly's death, this sentence was changed to the following.

> *Sir *Isaac Newton* has supposed that there is a fluid of the same nature with Air, but extremely more subtle and elastic, every where dispersed through all space. This fluid he calls Æther ...
>
> *This alteration in the beginning is chiefly to do justice to Sir *Isaac Newton*; as in the former edition he was made to assert, rather than suppose, the existence of a fluid, but which in the following sheets we shall endeavour to make probable.

Could Bayes's observations perhaps have influenced this alteration?

It is perhaps curious that the reference to Newton's work should have been framed in the 1756 edition as it was, for in the second edition of Wilson's *Treatise on Electricity* published four years before, we find

> SIR *Isaac Newton* supposes that there is an exceedingly subtle and elastic fluid, which readily pervades all bodies, and is by its elastic force expanded throughout the universe ... This fluid, which he calls *æther* .... [1752, pp. 95, 96]

One might also note that in the *Advertisement II* to the second edition of his *Opticks* Newton wrote

> And to shew that I do not take Gravity for an essential Property of Bodies, I have added one Question concerning its Cause, chusing to propose it by way of a Question, because I am not yet satisfied about it for want of Experiments.

Further, in Question 18 Newton writes of 'the Vibrations of a much subtiler Medium than Air ... And is not this Medium exceedingly more rare and subtile than the Air, and exceedingly more elastic and active?' In Question 21 he writes 'I do not know what this *Æther* is', and in the following Question he identifies the Æthereal Medium with *Æther*[27].

The connexion between gravitation[28] and electricity was also noted by Young, who wrote

> the electrical forces differ from the common repulsion which operates between the particles of elastic fluids, and resemble more

10.5. Commentary on the Item on Electricity

nearly that of gravitation. .... unless we choose to consider gravitation itself as arising from a comparatively slight inequality between the electrical attractions and repulsions.
[1807, vol. I, pp. 659, 660]

The passage from which Bayes's second extract is given runs in full

> Now, as it is universally agreed among those, who are most conversant with electrical experiments, that the appearances, which occur in those experiments, arise from the force and action of a fluid of the same elastic nature, communicating, and freely passing in and out at the surface of the earth, and pervading likewise the the pores of bodies: and as the clearest definition of what we mean, when we say a body is electrified, is this, that either the body has by the force of the experiment made in order to electrify it, been forced to part with a share of this electrical fluid, that naturally belonged to it during the experiment, and to remain without it sometime after the experiment is over: or to admit more than it naturally had within it, during the experiment, and to remain so overloaded, some time after the experiment is over: *it will be worth our while to enquire whether this electrical fluid, and the æther, be not one and the same fluid.* [pp. 2–3]

(The passage italicised above is italicised in the second but not the first edition.) Home [1974–1975, p. 87] agrees with Bayes's observation here, finding not only no demonstration by Hoadly and Wilson of the identity of the electrical fluid with the ether, but also no attempt at such a demonstration.

One might note that this matter also concerned Wilson in *A Treatise on Electricity*, for in Part II, Section VIII he wrote

> we may look upon it [i.e., the electric matter discussed in the first part] as the *æther* joined with grosser particles of matter propelled from bodies by the force and vigour of its action. [1752, p. 97]

The next quotation given by Bayes is taken almost exactly from Hoadly and Wilson, there being but the omission of 'therefore' after 'W^{never}' and the giving of 'resists' and 'bodies' as 'resist' and 'bodys', respectively. In his comment on Bayes's observation on this passage, Home writes

> Bayes may even have been prepared to deny the existence of atmospheres altogether; at least, this is a possible interpretation of his otherwise obscure argument that 'because bodys Electrifyed plus or minus give the same signs, therefore they are not

410   10. Miscellaneous Items

> surrounded with similar Atmospheres ready to expand themselves into any approaching body which resists less than the air'. [1974–1975, p. 87]

The next citation, from page 19, is given as follows in the original.

> When therefore two balls are both of them electrified *plus*, suspended by two silk lines, and brought near one another; they repel each other, and stand for some time at a distance from each other; because the two atmospheres, each of them exerting their endeavours to expand into the air, want more room to do it in; and when the weight of the balls is not sufficient to prevent it, must naturally drive them asunder, till these atmospheres are dissipated, and the weight of the balls takes place again.

Although I have suggested that Bayes's 'A.' stands for 'Authors' (having identified the text used as the first edition of Hoadly and Wilson's book), one must note that later in the observation on this passage Bayes uses the singular 'author'.

The passage quoted from page 19, paragraph 2, runs in full as follows.

> And when two bodies are both of them electrified *minus*, suspended by silk strings, and brought near one another, they likewise repel each other, and for some time at a distance from each other; because the condensed electrical fluid in the air, in order to force itself in at the surfaces of the balls between their two centers, crouds in, and forces them asunder, till the atmospheres get all into the balls, and their weight then takes place again.

The extract from page 19, paragraph 3, runs as follows in the original.

> But two balls are in the same circumstances, one electrified *plus*, and the other *minus*, and brought near one another; they will gradually come together and unelectrify each other. Because the atmosphere of the ball electrified *plus*, is endeavouring to dissipate itself from the center of the ball outwards; and the atmosphere of the ball electrified *minus*, is endeavouring to dilate itself from the air inwards to the center of the ball. The common atmospheres therefore of the two balls, thus brought near together, exert their forces in one and the same direction between them; the flow of the electrical fluid into the ball electrified *minus*, is facilitated by the endeavour of the electrical fluid to get out of the ball, electrified *plus*; and *vice versa*: and so the two balls and the air between them very readily return to their natural states.

10.5. Commentary on the Item on Electricity   411

In the second edition this was amplified in the following footnote.

> The reader is desired to take notice, that though two electrical atmospheres seem necessary in this experiment to restore the equilibrium in the balls, yet he is not to conclude therefore that they both enter the balls, because the electrical signs of both disappear at the same time. But this seems to follow from the principles we have already deduced, that a quantity, equal to what was in the ball electrified *plus*, is at the end of the experiment diffused in the balls electrified *minus*; and the remaining quantity which form'd the atmospheres is diffused in the air, or the bodies from whence it at first flow'd, in order to crowd into the body electrified *minus*. For if this was not the case such bodies, or the air, would remain in an unnatural state, because the equilibrium of the electric fluid in such circumstances would not be restored. [1759, p. 20]

The vast majority of the text following the last illegible observation in Bayes's note was given in shorthand.

In the quotation from page 20 one should note that in the shorthand used by Bayes it is impossible to distinguish between 'judgment' and 'judgement', unless great care is taken in the insertion of the vowels. Further, in the second edition of 1759 the phrase 'because the force of the excited Tube will condense the atmospheres about {around} the balls still more till the resistnce at their surfaces is over come' has the words 'on account of the resistance of the air' after 'still more'. Finally, the words 'not repel each other near so much and so' in the sentence preceding '2do.' are missing in the second edition.

The first sentence in the quotation from page 71 is preceded in the original by the following words, whose insertion here will make the reference to 'such a spark' clear.

> When a flint and steel are struck together with sufficient force and velocity, a spark of fire, as we call it, is produced, which readily fires gunpowder, or lights tinder: but soon cools, if left to itself. [1756, p. 71]

A fitting summary of Bayes's note is provided by Home in commenting on the difficulty experienced by many of Franklin's contemporaries in trying to follow his explanation of electrical atmospheres:

> The attempt by Hoadly and Wilson was but one of many that were made in the 1750s and '60s to improve upon what Franklin had done in this regard. In the end, however, their account was no more successful than Franklin's had been, and the criticisms levelled at it in the document we have been considering [i.e.,

10.6 Papers in the Stanhope Collection

We have had occasion in an earlier chapter to mention Philip Stanhope, a visitor to Tunbridge Wells during Bayes's ministry and one of the sponsors of Bayes's election to the Royal Society, and we also find his name occuring in the next chapter[29]. That the manuscripts to be noted in this section were carefully preserved in Stanhope's papers shows that there was more than a nodding acquaintance between the Earl and the clergyman; and Bellhouse has indeed suggested that Bayes might quite possibly have assisted Stanhope as a commentator on, and critic of, mathematical papers (a view that is supported by Bayes's letter to Canton on Thomas Simpson's work on the arithmetic mean).

Two preliminary versions of Bayes's posthumous paper on the semi-convergent series for log z! (see Chapter 6) are to be found in this collection. The first of these we look at bears on one side the words 'Mathematical paper of Mr Bayes's communicated Septr 1st 1747'. It begins

> It is said that y^e integral of Log, $\overline{\tfrac{z+1}{z}}\bigg|^{z+\tfrac{1}{2}}$ is
>
> $$= z - \frac{1}{12z} + \frac{1}{360z^3} - \frac{1}{1260z^5} + \frac{1}{1680z^7} - \&c$$
>
> But in y^e following manner it will evid^{tly} appear that this series do's not converge.

The proof of this proceeds essentially as in the published version, and finishes with the words, 'From whence it is manifest that the subsequent terms of the series increase in infinitum'.

Although the superscription on the above-mentioned paper indicates that Bayes and Stanhope were acquainted (and Mrs Elizabeth Montague records in her letters that the latter visited Tunbridge Wells as early as 1736), it might appear that Stanhope did not regard his estate at Chevening, some twenty miles north-west of Tunbridge Wells, as sufficiently near the latter for him to call on Bayes when the latter's mathematical advice was needed. For in a letter begun on the 27th March 1752 and finished on the 31st, Stanhope wrote to Patrick Murdoch 'And my perplexity is the greater, as this neighbourhood will afford me no mathematical adviser'. We have already noted (see Chapter 3) that Bayes resigned his ministry in 1752, and although a letter written by him to Stanhope from Tunbridge Wells in 1755 is mentioned later in this section, it is perhaps possible that ill-health had forced his resignation and that he was absent from Tunbridge Wells for some time to recuperate.

The second version of the paper, like the published one, begins 'It has been asserted by several eminent Mathematicians' and, unlike the previously mentioned draft, considers the series mentioned above with the additional term $\log \sqrt{2\pi}$ (as we should write it today) that is required for the expansion of $\log z!$. The proof proceeds essentially as in the published paper, and concludes with a paragraph indicating the inappropriateness of setting $z = 1$ in the series in an attempt to evaluate $\log \sqrt{2\pi}$.

Also in this collection of papers is a manuscript in which Bayes provides a proof of Stirling's approximation to $z!$ that does not use the semi-convergent series. He begins by supposing that I is the ratio 'whose hyperbolic Logarithm is $= 1$'. Letting k be a constant ratio and setting $v = z!$ (Bayes of course uses products rather than factorials), he supposes further that one always has

$$I^p = (k \times z^z \times \sqrt{z})/v.$$

Supposing further that $\dot{z} = 1 = \ddot{z}$, we find 'By the method of increments' that $\dot{v} = \overline{z+1} \times v$ (note that it would be more accurate to interpret Bayes's v as $v(z)$, so that $\dot{v}$ is $v(z+1)$) and hence

$$I^{\dot{p}} = (k \times (z+1)^z \times \sqrt{z+1})/\dot{v}.$$

From this last expression and the expression given for I^p it follows that

$$I^{\dot{p}} = ((z+1)/z)^{z+(1/2)}$$

and hence, on taking the fluxion,

$$\dot{p} = \text{Log},((z+1)/z)^{z+(1/2)}.$$

In the second article Bayes supposes that

$$v = \dot{x} = \ddot{x} - (1/2)\ddot{x} + (1/3)\dddot{x} - (1/4)\ddddot{x} + \&\text{c};$$

and then, 'vice versa' (and omitting part of the series as given by Bayes)

$$\ddot{x} = \dot{v} + (1/2)\ddot{v} - (1/12)\dddot{v} + (1/24)\ddddot{v} - (19/720)\dddddot{v} + \&\text{c};$$

the coefficients in the second series being found either by division of 1 by the series $1 - \frac{x}{2} + \frac{x^2}{3} - \frac{x^3}{4} + \&c$ or by a method that we find in the Notebook.

In Article 3 Bayes sets $x = \text{Log},z$ '& other things as before' and derives a series for $\text{Log},((z+1)/z)^{z+(1/2)}$. 'Wherefore the Integral of the' latter function is

$$z - \frac{1}{12z} + \frac{1}{360} \times \frac{1}{z \times \overline{z+1} \times \overline{z+2}} + \frac{1}{120} \times \frac{1}{z \times \overline{z+1}\,(4)}$$

$$+ \frac{5}{168} \times \frac{1}{z \times \overline{z+1}\,(5)} + \frac{11}{84} \times \frac{1}{z \times \overline{z+1}\,(6)} + \&c = p,$$

since $p = \text{Log},((z+1)/z)^{z+(1/2)}$.

As a corollary Bayes notes that

p is always less than z & greater than $z - 1/(12z)$ & therefore when z is infinite $p = z$ & $v = (k \times z^z \times \sqrt{z})/\mathbf{I}^z$.

Attention is next turned to the determination of the constant k. Supposing that G is the middle term (more accurately, the *coefficient* of that term) in the expansion of $(a+b)^n$ (and hence n must be even) Bayes notes that $\sqrt{n}\,G/2^n$ is equal to

$$\sqrt{\frac{1}{2} \times \frac{9}{8} \times \frac{25}{24} \times \frac{49}{48} \times \&c}$$

'taking as many terms as there are units in $n/2$'. We do not give Bayes's proof here; but note only that he shows that the last term under the surd displayed above is $(n^2 - 2n + 1)/(n^2 - 2n)$, the number of terms being $n/2$. The following corollary is then stated.

If $q = \frac{2}{1} \times \frac{8}{9} \times \frac{24}{25} \times \frac{48}{49} \times \&c$ in infinitū which series is known to express the ratio of the quadrantal arch of a circle to its radius, then when n is infinite $\sqrt{n}G/(2^n) = 1/\sqrt{q}$ & $G = 2^n/\sqrt{nq}$.

In the last article Bayes supposes that n is infinite and equal to $2z$. Using this identification and the Corollary to Article 3 he deduces that

$$n!/(z!)^2 = G = 2^n\sqrt{2}/(k\sqrt{z}),$$

and since $G = 2^n/\sqrt{nq}$, (by the Corollary to Article 4), it follows that

$$2^n/\sqrt{nq} = 2^n\sqrt{2}/(k\sqrt{z}),$$

whence $k = \sqrt{4q}$. Consequently 'k^2 expresses the ratio of the whole circumference of a circle to the radius' (i.e., $k^2 = 2\pi$). On combining this result with the value for p given in Article 3, one obtains

$$z! = (kz^z\sqrt{z})/\mathbf{P}^p,$$

the writing of which in the form $z! = \sqrt{2\pi}z^{z+(1/2)}e^{-z}$ clearly exhibits Stirling's well-known approximation.

Bayes's suggested rôle as a mathematical commentator gains further credence in the light of a letter from him to Stanhope of the 25th April 1755. Here Bayes, ascribing his delay in replying to Stanhope's communication to his having been 'not well' when the the latter was received, thanks the Earl for the copy of some observations by Patrick Murdoch. Bayes did not see himself in a position to comment further on the paper without putting Murdoch to too much trouble, and concluded his letter with the tantalising mention of 'the adjoining paper', one that does not seem to have survived the passage of time. Bayes's comments and the cited paper were apparently forwarded to Murdoch, for the latter replied to Stanhope on the 11th of May writing

> I do not think any answer [to the paper] necessary, seeing we seem to be agreed on that point. That Mr Bayes misunderstood me was certainly my own fault . . . .

In another letter, of the 18th March 17_5 (the year is unclear, but since Murdoch writes here 'The Edition which Mr de Moivre desired me to make of his Chances is now almost printed' 1755 seems to be correct) Murdoch wrote

> I am ashamed not to have sooner acknowledged your Lordship's goodness in Communicating to me Mr Bayes' paper which I received from Dr Pringle[30]... I have now returned it inclosed, with my answer on the blank page: which I wish your Lordship and Mr Bayes may find satisfactory.

It was presumably in response to this letter that the letter from Bayes of the 25th April was written. Incidentally, Bayes's commenting on his illness perhaps supports our earlier suggestion (in Chapter 3) that ill-health had contributed to his resigning his clerical duties in Tunbridge Wells.

Another item bears the heading 'Mr Bayes's Demonstration of a Theorem which I found lately & told him of. Septr 1747'. The result considered (again in Bayes's handwriting) is the following.

To find $(1/x)$ when $\dot{z} = 1$ and

$$x = 1 + \frac{z}{2} + \frac{z^2}{2.3} + \frac{z^3}{2.3.4} + \frac{z^5}{2.3.4.5} + \&c.$$

Bayes's proof consists in the multiplication of both sides of this equality by z and the taking of fluxions, resulting in $\dot{x}z + x = 1 + xz$. Hence he deduces essentially that
$$\frac{1}{x} = \frac{z}{e^z - 1},$$
and we show in Chapter 11 that the analogous expression
$$\frac{ze^z}{e^z - 1} - \frac{z}{2}$$
is useful in finding a series for $\log z!$.

Another manuscript is concerned with the recursive determination of infinite series for the expansion of $(\arcsin z)^n$. The first Article runs as follows.

If x be the arch & z the sine the radius being unity. &

$$x^n = z^n + \frac{Az^{n+2}}{\overline{n+1} \times \overline{n+2}} + \frac{Bz^{n+4}}{\overline{n+1} \times \overline{n+2} \times \&c \times \overline{n+4}}$$

$$+ \frac{Cz^{n+6}}{\overline{n+1} \times \overline{n+2} \times \&c \times \overline{n+6}} + \&c$$

&

$$x^{n-2} = z^{n-2} + \frac{az^n}{\overline{n-1} \times n} + \frac{bz^{n+2}}{\overline{n-1} \times n \times \&c \times \overline{n+2}}$$

$$+ \frac{cz^{n+4}}{\overline{n-1} \times n \times \&c \times \overline{n+4}} + \&c$$

Then $A = a + n^2 \quad B = b + (n+2)^2 A \quad C = c + (n+4)^2 B$
$D = d + (n+6)^2 C$ & so on.
Also
$$\frac{x^{n-1}}{\sqrt{1-z^2}} = z^{n-1} + \frac{Az^{n+1}}{n \times \overline{n+1}} + \frac{Bz^{n+3}}{n \times \overline{n+1} \times \&c \times \overline{n+3}}$$

$$+ \frac{Cz^{n+5}}{n \times \overline{n+1} \times \&c \times \overline{n+5}} + \&c.$$

(Bayes's expression for $x^{n-1}/\sqrt{1-z^2}$ presumably comes from finding the fluxion of x^n.) I found it easiest to derive the relationship between the coefficients in the two series by (a) differentiating the series for x^n (which results in a series for $nx^{n-1}/\sqrt{1-z^2}$, and (b) cross-multiplying by $\sqrt{1-z^2}$, differentiating a second time, and 'tidying up' to obtain the series for x^{n-2}.

In the second article Bayes uses the first to find a series, in powers of z, for x^2, with $a = b = c = \cdots = 0$ (or, rather crudely, integrate $x/\sqrt{1-z^2}$ with respect to z and multiply by 2), whereas in Articles 3 and 4 series for x^3 and x^4 are given. In these last two cases Bayes notes that the coefficients are such that the rule for the continuation of the series (beyond the terms in z^{11} and z^{10}, respectively) is not 'immediately apparent' (the numerators of the terms given by Bayes contain prime numbers).

We find, in our later discussion of Bayes's Notebook, a page bearing the legend 'According to Ld Stanhopes notation'. More details of the matter discussed there are to be found in one of the manuscripts in the collection at present under discussion: it begins as follows.

> Def. $\boxed{abcd\,\&c}^{\,n}$ signifys all y^{e} factors of y^{e} dimension n, which can be formed of of [sic] y^{e} quantitys a, b, c, d &c, added together.

In the first of the three Articles in which this manuscript is written, one finds expressions of the form given in our §11.6. For example,

$$\boxed{abc}^{\,n} = \boxed{ab}^{\,n} + \boxed{ab}^{\,n-1} c + \boxed{ab}^{\,n-2} c^2 + \&c + \left(\boxed{ab}^{\,0} c^n = \right) c^n.$$

In his second article Bayes proves results of the form

$$\boxed{ab}^{\,n} = (a^{n+1} - b^{n+1})/(a-b)$$

and

$$\left(\boxed{ab}^{\,n+1} - \boxed{cb}^{\,n+1}\right)/(a-c) = \boxed{abc}^{\,n}.$$

In Article 3 Bayes considers the equation $x^4 - Ax^3 + Bx^2 - Cx + D = 0$, with roots a, b, c, and d. Substitution of a for x and multiplication by a^{n-4} results in

$$a^n - Aa^{n-1} + Ba^{n-2} - Ca^{n-3} + Da^{n-4} = 0,$$

with a similar expression with b substituted for x. Subtraction of the second of these from the first and division by $a - b$ yields, by Article 2,

$$\boxed{ab}^{\,n-1} - A\boxed{ab}^{\,n-2} + B\boxed{ab}^{\,n-3} - C\boxed{ab}^{\,n-4} + D\boxed{ab}^{\,n-5} = 0.$$

A similar expression is determined in terms of $\boxed{bc}^{\,n-1}$, subtraction of which from $\boxed{ab}^{\,n-1}$ and division by $a-c$ yields

$$\boxed{abc}^{\,n-2} - A\boxed{abc}^{\,n-3} + B\boxed{abc}^{\,n-4} - C\boxed{abc}^{\,n-5} + D\boxed{abc}^{\,n-6} = 0.$$

418   10. Miscellaneous Items

Carrying this procedure one step further Bayes arrives eventually at

$$\boxed{abcd}^{n-3} - A\boxed{abcd}^{n-4} + B\boxed{abcd}^{n-5} - C\boxed{abcd}^{n-6} + D\boxed{abcd}^{n-7} = 0.$$

I have difficulty in understanding the reason for this investigation: could any subsequent pages have been lost?

A very short note is headed 'Theorem mentioned to me at Tunbridge Wells by Mr Bayes Aug. 12. 1747', the pertinent result being the following.

$$\dot{y} = y - \frac{1}{2}\underset{\cdot\cdot}{y} + \frac{1}{3}\underset{\cdot\cdot\cdot}{y} - \frac{1}{4}\underset{\cdot\cdot\cdot\cdot}{y} + \frac{1}{5}\underset{\cdot\cdot\cdot\cdot\cdot}{y} - \frac{1}{6}\underset{\cdot\cdot\cdot\cdot\cdot\cdot}{y} + \&\text{c},$$

the subscripted and superscripted dots indicating differences and fluxions, respectively. This, as we show in a later chapter, is a result that is also given in Bayes's Notebook. No proof is provided. Bellhouse finds the earliest published proof to have been given by Lagrange in 1772 (see Goldstine [1977, pp. 164–165]).

Although the manuscripts we have discussed so far are the only ones in this bundle of the Stanhope papers that are in Bayes's handwriting, there are several others here of at least peripheral interest. Let us, for completeness, say a few words on these.

One manuscript is a letter, with seal, addressed to Stanhope, that begins

$$\dot{y}/\dot{x} = \left(y - \frac{1}{2}\underset{\cdot\cdot}{y} + \frac{1}{3}\underset{\cdot\cdot\cdot}{y} - \frac{1}{4}\underset{\cdot\cdot\cdot\cdot}{y}\right)/x + \&\text{c};$$

the proof (in Stanhope's hand?) being concluded with the words (written with a different quill to the proof) 'This is a Theorem shewn to me by the late Mr Bayes.'

Another manuscript bears the legend 'The Reverend Mr Bayes's Paper concerning Trinomial divisors'. The paper is in Stanhope's hand, and Bellhouse has suggested that it might be a transcription of a paper originally by Bayes (there are also two pages of mathematical calculations that seem to be attempts at verifying certain statements in the manuscript). Using geometrical arguments MacLaurin [1742, Articles 765–768] presented an expansion of $x^{2n} - 2\cos(\theta)\,x^n + 1$, an earlier inductive generalization having been given by de Moivre in 1730.

Bayes's proof is in essence a combination of these two approaches: like MacLaurin, he considers a circle of unit radius whose circumference is divided into arcs of equal length, and like de Moivre he proceeds inductively. He shows first that if a, b, c, d, &c. represent, respectively 'the Chords of an arch, its double, its triple, its quadruple &c.', and if $A = a^2 - 2$, $B = b^2 - 2$, $C = c^2 - 2$ &c., then $A + C + A \times B = 0$, $B + D + A \times C = 0$, &c. He then proves that if the trinomial $s^2 + As + 1$ divide the first two of the trinomials $s^{2n} + Ms^n + 1$, $s^{2n+2} + Ns^{n+1} + 1$ and $s^{2n+4} + Ps^{n+2} + 1$ then it will divide the third, provided that $M + P + A \times N = 0$. With A, B, C &c. as before, he shows that $s^2 + As + 1$ will exactly divide any of the trinomials $s^4 + Bs^2 + 1$, $s^6 + Cs^3 + 1$, $s^8 + Ds^4 + 1$ &c., from which the factorization of $s^{2n} + Qs^n + 1$ into its simple trinomials follows.

11

The Notebook

> *This Booke (of what worth I say not; but more men, I feare, will commend it, then will know how to make use of it:) after it had for so many ages undeservedly beene buryed in darknesse, is now first, if I may not say brought unto light, yet at least made common and intelligible.*
>
> Meric Casaubon.
> The Meditations of Marcus Aurelius
> Antoninus.

11.1 Introduction

In the muniment room of the Equitable Life Assurance Society in London, England, lies a notebook bearing on its first page the following words.

> This book appears to be a mathematical notebook by Rev. Thomas Bayes, F.R.S. The handwriting agrees very well with papers by him in the Canton papers of the Royal Society, Vol. 2. p. 32.

This note, dated 21-1-1947, is signed by M.E. Ogborn, formerly general manager and actuary of the Equitable.

It is this notebook, written in both longhand and shorthand, and in English, French, and Latin, that is under examination here[1]. In the next two sections we say something about the attribution of the manuscript and the shorthand used therein, and then pass on to a discussion of the contents.

11.2 Attribution of the notebook

The writer of the notebook cannot be conclusively identified. However, there is sufficient evidence, both within the notebook itself and in other sources, for us to be comfortably confident in ascribing this book to Bayes.

Firstly, as indicated in Ogborn's inscription, the handwriting in the manuscript accords with that in certain papers in the Royal Society collection. Secondly, the shorthand used both in Bayes's notes on Hoadly & Wilson's *Observations on a Series of Electrical Experiments* (see Chapter 10) and in the notebook is the same, and is apparently not one of those commonly used in the first half of the eighteenth century. In the third place the notebook contains a proof of one of the rules published in Bayes's posthumous *Essay towards solving a Problem in the Doctrine of Chances.* Further evidence is, I believe, provided by the presence in the notebook of some work on the asymptotic expansion of log $z!$, the substance of Bayes's published letter on infinite series.

In Thomas Bayes's will[2], signed 12th December 1760, an amount of £200 was left to be shared equally between John Hoyle and Richard Price, or the whole to the survivor. It seems not impossible that Price should have been entrusted with an examination of Bayes's papers on the death of his friend, and the notebook could then easily have found its way to the Equitable, a society (then named *The Society for Equitable Assurances on Lives and Survivorships*) to which Price gave actuarial advice.

We thus henceforth assume that the ascription of the notebook to Thomas Bayes is correct.

11.3 The shorthand

According to Holland [1962], [1965], and [1968], the shorthand used by Bayes was identified by J.I. Mason as being that of Thomas Shelton[3] as modified by Elisha Coles. Shelton published two shorthand books: *Tachygraphy* and *Zeiglographia*; it is the latter of these that is most closely related to Coles's work of 1674.

The transcription of the alphabet used by Coles runs as follows.

a	b	c	d	e	f	g	h	i
<	∩		\	e	L	∧	o	.

j	k	l	m	n	o	p	q	r
Γ	⊂	—	⊃	/	∪	ρ	q	r

s	t	u	v	w	x	y	z	ch
σ	|	.	V	c	ℓ	γ	z	δ

Note that the symbol used for 'y' is actually more 'open' in the lower half than shown here, and that used for 'x' is more like that for 'y' rotated about a horizontal axis through its middle. The symbols given here are from Coles's book itself: in Rockwell's *Shorthand Instruction and Practice* of 1893 Coles's symbol for 'q' is given as ч, and his symbol for 'ch' resembles ȣ, or that given above reflected about a vertical axis.

Bayes's system differs from Coles's in the presence of the following symbols.

c: ⌐ or c ;   w: C ;   i: ⟨ ;   z: ⌐

u: like Coles's symbol for 'r', but with the 'arms' more widely spread;
v: like the symbol for 'u' given above, but with the 'arms' still more widely spread.

Further, the symbol used by Coles for 'x' is also used by Bayes for 'ex'.

Although symbols for the vowels are given here, these symbols are customarily omitted (or replaced by a dot) in a word, the vowel being indicated by the position of the following letter: thus

bit : ∩ı ;   bat : ∩̇ ;   but : ∩̣

Specifically, the positions of the vowels are as follows.

$$\begin{array}{ccc} & a & e \\ \times & & i/y \\ & u & o \end{array}$$

Special symbols were used for commonly occurring words, and even for groups of letters. Thus

the: | ;   and: / ;   -ing: ⋀ ;   according: ⌐⌐

not (as used by Bayes): ⌐ ;   not (as used by Coles): ⌐

As examples of some words we give the following from the notebook.

foot: L.ı ;   head: o˙ˋ ;   profound: ρ ;   palms: ·ρ

The dot at the start of a word indicates a plural (regular or irregular), the superlative ending of an adjective, any final accidental(s) whatsoever, or the third person singular of a verb. Special rules were also given by Coles for the handling of double vowels and for the omission of certain letters, and special symbols were introduced for prepositions, terminations, and the more common words.

11.4 General plan of the notebook

The notebook begins with an unheaded table of contents, in which the majority of the topics discussed are noted. Although the (double) pages of the notebook are numbered, the numbering is at times erratic: pages 24 and 25 are blank, some pages are numbered (differently) on left-hand and right-hand sides, and others are given the additional lettering (a), (b), (c), etc. The listing in the table of contents is not always exact: articles may start a page before or after that listed. In all the pages are numbered from 1 to 125.

It is possible to identify the sources of many of the matters discussed. The earliest work cited is by Roger Cotes of 1722; the latest is by Thomas Allen of 1760. Of course, many of the books cited appeared in several editions, and it is usually not possible to say which of these editions were used by Bayes.

The topics discussed in the notebook can be divided — roughly and arbitrarily — into several groups, viz. (i) mathematics, (ii) natural philosophy, (iii) celestial mechanics, and (iv) miscellaneous matters. The boundaries between, or within, the groups are by no means clearly defined: for example, should work on logarithms appear under 'series' or 'fluxions' — if either? Here a reasonable classification is attempted.

Throughout this chapter, references of the form 'Of the Pyramid measured by Greaves [C: 26L–28R]' indicate that the matter discussed is mentioned by Bayes in his table of contents (the C), and is to be found on the left-hand side of page 26 to the right-hand side of page 28.

11.5 Mathematics

A large portion of the notebook is devoted to mathematical topics, grouped here into eight classes.

11.5.1 Probability

On pages [81R–83R] a Latin passage may be found that close examination shows to be a proof of one of the rules in Bayes's *Essay* of 1763. Although this passage has been printed fairly recently (see Dale [1986] and [1999]), its importance in the determination of the correct attribution to Bayes of the result generally ascribed to him warrants, I feel, its being reprinted here. Both the original Latin version and a translation by the present author (see Appendix 11.1) are provided, certain obvious slips in the former being corrected in the latter.

The Extract

Si $\frac{\dot{S}}{S} > \frac{\dot{V}}{V}$ semper atq$_3^4$ S & V ambo crescunt Ratio $\frac{S}{V} = x$ semper crescit. Si vero ambo decrescunt tum ratio $\frac{V}{S}$ semper crescit.

Art. 2. Sit $\overline{1 - \frac{nz}{p}}\Big|^p \times \overline{1 + \frac{nz}{q}}\Big|^q = A$ semper $\overline{1 + \frac{nz}{p}}\Big|^p \times \overline{1 - \frac{nz}{q}}\Big|^q = B$ semper $n = p + q$ & $\boxed{A\dot{z}} + \boxed{B\dot{z}} = \frac{n^n}{n+1 \times Ep^p q^q}$ si E sit coefficiens termini in quo occurrit $x^p r^q$ q^{do} $\overline{x + r}\Big|^n$ expanditur si sumatur $\boxed{A\dot{z}}$ q^{do} $A = 0$ & $\boxed{B\dot{z}}$ q^{do} $B = 0$.

3. Iisdem positis B est $> \overline{1 - \frac{n^2 z^2}{q^2}}\Big|^{\frac{nq}{2p}} = D$ & A est $< \overline{1 - \frac{n^2 z^2}{p^2}}\Big|^{\frac{np}{2q}} = \Delta$ quando $q > p$. Est enim

$$\frac{\dot{A}}{A} = -\frac{n^3 z \dot{z}}{pq} \times \overline{1 - \frac{nz}{p} + \frac{nz}{q} - \frac{n^2 z^2}{pq}}\Big|^{-1}$$

$$\frac{\dot{B}}{B} = -\frac{n^3 z \dot{z}}{pq} \times \overline{1 - \frac{nz}{q} + \frac{nz}{p} - \frac{n^2 z^2}{qp}}\Big|^{-1}$$

$$\frac{\dot{D}}{D} = -\frac{n^3 z \dot{z}}{pq} \times \overline{1 - \frac{n^2 z^2}{q^2}}\Big|^{-1} \quad \&$$

$$\frac{\dot{\Delta}}{\Delta} = -\frac{n^3 z \dot{z}}{pq} \times \overline{1 - \frac{n^2 z^2}{p^2}}\Big|^{-1}.$$

Est igitur

$$\frac{\dot{B}}{B} : \frac{\dot{D}}{D} \;::\; \overline{1 - \frac{n^2 z^2}{q^2}} : \overline{1 - \frac{nz}{q}}\Big|^q \times \overline{1 + \frac{nz}{p}}\Big|^p$$

$$::\; 1 + \frac{nz}{q} : 1 + \frac{nz}{p}.$$

Sed $q > p$ Quare $\frac{\dot{D}}{D} > \frac{\dot{B}}{B}$. Sed quū $z = 0$ $D = B$ adeoq$_3$ z crescente magis decrescit D quā B adeoq$_3$ semper postea $B > D$ & ita de cæteris

$$\Delta^2 = \overline{1 - \frac{n^2 z^2}{p^2}}\Big|^{\frac{np}{q}} \qquad AB = \overline{1 - \frac{n^2 z^2}{p^2}}\Big|^p \times \overline{1 - \frac{n^2 z^2}{q^2}}\Big|^q$$

adeoq$_3$

$$\Delta^2 : AB \;::\; \overline{1 - \frac{n^2 z^2}{p^2}}\Big|^{\frac{np - qp}{q} = \frac{p^2}{q}} : \overline{1 - \frac{n^2 z^2}{q^2}}\Big|^q \quad \&$$

$$\Delta^{2q} : \overline{AB}\,]^q \;::\; \overline{1 - \frac{n^2z^2}{p}}\,]^{p^2} \;:\; \overline{1 - \frac{n^2z^2}{q^2}}\,]^{q^2}$$

Quare $\Delta^2 < AB$ & multo magis $2\Delta < A + B$.

Art. 4. Hinc si eventus aliter prorsus incognitus accidit tempora p, neq$_3$ plura in tentaminibu $p + q = n$ & hinc conjicio quod ejus probabilitas accidendi in uno tentamine cadit inter $\frac{p}{n}$ & $\frac{p}{n} - z$. probabilitas q$^\text{d}$ hæc conjectura est recta erit æqualis fluente $\widehat{\tau o \nu}$

$$\overline{n+1} \times \frac{E p^p q^q}{n^n} \times \overline{1 - \frac{nz}{p}}\,]^p \times \overline{1 + \frac{nz}{q}}\,]^q \dot{z}$$

ita sumptæ ut sit $= 0$ quando $z = 0$.

Fluens prædicta est minor fluente $\widehat{\tau o \nu}$

$$\overline{n+1} \times \frac{E p^p q^q}{n^n} \times \overline{1 - \frac{n^2 z^2}{p^2}}\,]^{\frac{np}{2q}} \dot{z}$$

quum q est major quā p.

2$^\text{do}$. Iisdem positis si conjicio q$^\text{d}$ probabilitas accidendi in uno tentamine cadit inter $\frac{p}{n}$ & $\frac{p}{n} + z$. probabilitas quod hæc conjectura est justa erit fluens $\widehat{\tau o \nu}$

$$\overline{n+1} \times \frac{E p^p q^q}{n^n} \times \overline{1 - \frac{nz}{q}}\,]^q \times \overline{1 + \frac{nz}{p}}\,]^p \dot{z}$$

ita sumpta ut sit $= 0$ quū $z = 0$. & hæc fluens est major fluente $\widehat{\tau o \nu}$

$$\overline{n+1} \times \frac{E p^p q^q}{n^n} \times \overline{1 - \frac{n^2 z^2}{q^2}}\,]^{\frac{nq}{2p}} \dot{z}$$

eodem modo sumptâ.

3°. Iisdem positis si conjicio quod prædicta probabilitas cadit inter $\frac{p}{n} - z$ & $\frac{p}{n} + z$ probabilitas q$^\text{d}$ hæc conjectura sit justa erit fluens

$$\overline{n+1} \times \frac{E p^p q^q}{n^n} \times \overline{1 - \frac{nz}{p}}\,]^p \times \overline{1 + \frac{nz}{q}}\,]^q \dot{z} + \overline{1 - \frac{nz}{q}}\,]^q \times \overline{1 - \frac{nz}{p}}\,]^p \dot{z}$$

& hæc fluens est major fluente

$$\overline{n+1} \times \frac{E p^p q^q}{n^n} \times \overline{1 - \frac{n^2 z^2}{p^2}}\,]^{\frac{np}{2q}} \times 2\dot{z}$$

NB. *E* hic est ubiq$_3$ coefficiens termini in quo occurrit $x^p r^q$ quū $\overline{x+r}\,\Big|^n$ expanditur, & fluentes omnes ita sumi debent ut sint $= 0$ quū $z = 0$.

5. Sit *c* ratio peripheriæ circuli ad radiū eritq$_3$ $\frac{E p^p q^q}{n^n}$ minor q̄ $\frac{\sqrt{n}}{\sqrt{pqc}}$ major vero quā $N \times \frac{\sqrt{n}}{\sqrt{pqc}}$ si N sit ratio cui Log,$= \frac{pq-n^2}{12npq}$.

6. Fluens τοῦ $\overline{1 - \frac{n^2 z^2}{p^2}}\,\Big|^{\frac{np}{2q}} \dot{z}$ quando $\frac{nz^2}{p^2} = 1$ est paulo major quam $\frac{\sqrt{pqc}}{2\sqrt{n}} \times \frac{p}{np+q}$ & fluens τοῦ $\overline{1 - \frac{n^2 z^2}{q^2}}\,\Big|^{\frac{nq}{2p}}$ est paulum major quam $\frac{\sqrt{pqc}}{2\sqrt{n}} \times \frac{q}{nq+p}$ quando $\frac{n^2 z^2}{q^2} = 1$.

7. Hinc & per 3. Art. 4. Si conjicio q^{d} probabilitas Eventus cadit inter 0 & $\frac{2p}{n}$ probabilitas quod hæc conjectura est justa erit major quā

$$N \times \frac{\sqrt{n}}{\sqrt{pqc}} \times \frac{\sqrt{pqc}}{2\sqrt{n}} \times \frac{p}{np+q} \times 2 \times \overline{n+1}$$

$$= N \times \frac{np+p}{np+q}.$$

ubi si & *p* & *q* sint magnæ $N \times \frac{np+p}{np+q}$ vix differt ab unitate unde in hoc casu h.e. q^{do} *q* & *p* sint ambæ magnæ & *q* major quā *p* si conjicio q^{d} prob. eventus cadit inter $\frac{p}{n} - z$ & $\frac{p}{n} + z$ prob. q^{d} hæc conjectura sit justa erit sine errore sensibili

$$\overline{2n+2} \times \frac{\sqrt{n}}{\sqrt{pqc}} \times FL, \overline{1 - \frac{n^2 z^2}{p^2}}\,\Big|^{\frac{np}{2q}} \dot{z}.$$

8. Si $x : r :: p : q$ et binomiū $\overline{x+r}\,\Big|^n$ expandatur in

$$x^n + n x^{n-1} r + n \times \frac{n-1}{2} x^{n-2} r^2 \ \&c$$

terminus in quo occurrit $x^p r^q$ est ad proximū præcedentem & reliquos 1 ad

$$\frac{q}{p+1} \times \frac{p}{q} \ \& \ \frac{q}{p+1} \times \frac{q-1}{p+2} \times \frac{p^2}{q^2} \ \&c$$

item est ad proximū subsequentem & reliquos ut 1 ad

$$\frac{p}{q+1} \times \frac{q}{p} \ \& \ \frac{p}{q+1} \times \frac{p-1}{q+2} \times \frac{q^2}{p^2} \ \&c.$$

11.5. Mathematics 427

Quare si $x : r :: p : q$ & $\overline{x+r}\,|^n$ expandatur ut antea terminus inquo occurrit $x^p r^q$ unâ cum omnibus præcedentibus erit ad omnes subsequentes sicut

$$1 + \frac{q}{p+1} \times \frac{q}{p} + \frac{q}{p+1} \times \frac{q-1}{p+2} \times \frac{p^2}{q^2} + \frac{q}{p+1} \times \frac{q-1}{p+2} \times \frac{q-2}{p+3} \times \frac{p^3}{q^3} + \&c$$

ad

$$\frac{p}{q+1} \times \frac{q}{p} + \frac{p}{q+1} \times \frac{p-1}{q+2} \times \frac{q^2}{p^2} + \frac{p}{q+1} \times \frac{p-1}{q+2} \times \frac{p-2}{q+3} \times \frac{q^3}{p^3} + \&c$$

Iam si q est major quam p,

$$1 > \frac{p}{q+1} \times \frac{q}{p}, \quad \frac{q}{p+1} \times \frac{p}{q} > \frac{p}{q+1} \times \frac{p-1}{q+2} \times \frac{q^2}{p^2}$$

& ita de reliquis item ob p minorē posterior series prius abrumpit quare series prior est major secundâ. Si vero a priore serie dematur 1 erit 2da series illâ major.

9. Invenire fluentē $\widehat{\tau o \nu}$

$$\overline{n+1} \times \frac{E p^p q^q}{n^n} \times \overline{1 - \frac{nz}{p}}\,\Big|^p \times \overline{1 + \frac{nz}{q}}\,\Big|^q \dot{z} = \dot{V}$$

fiat $x = \frac{p}{n} - z$ & $r = \frac{q}{n} + z$. Eritq$_3$ $V = \overline{n+1} \times E x^p r^q \times \dot{r}$ fiat

$$V = r^{n+1} + \overline{n+1}\, r^n x + \overline{n+1} \times \frac{n}{2} r^{n-1} x^2 + \&c + F r^{q+1} x^p.$$

Eritq$_3$

$$\dot{V} = \overline{n+1}\, r^n \dot{r} - \overline{n+1}\, r^n \dot{x} + \overline{n+1} \times n r^{n-1} x \dot{r} - \overline{n+1} \times n \times r^{n-1} x \dot{r}$$

$$+ \&c + \overline{q+1} \times F r^q x^p \dot{r}$$

$$= \overline{q+1} \times F r^q x^p \dot{r}.$$

Sed si E sit terminus binomii in quo occurrit $x^p r^q$ erit

$$\overline{q+1} \times F = \overline{n+1} \times E.$$

Quare

$$V = r^{n+1} + \overline{n+1}\, r^n x + \&c + \frac{\overline{n+1}}{q+1} \times E x^p r^{q+1}.$$

Quando $z = 0$ sit $V = B$ eritq₃

$$B = F\frac{p^p q^{q+1}}{n^{n+1}} \times 1 + \frac{p}{q+2} \times \frac{q}{p} + \frac{p}{q+2} \times \frac{p-1}{q+3} \times \frac{q^2}{p^2}$$

$$+ \frac{p}{q+2} \times \frac{p-1}{q+3} \times \frac{p-2}{q+4} \times \frac{q^3}{p^3} + \&c$$

sive

$$B = \frac{n+1}{n} \times \frac{q}{q+1} \times \frac{Ep^p q^q}{n^n} \times \text{series } 1 + \frac{p}{q+2} \times \frac{q}{p}$$

$$+ \frac{p}{q+2} \times \frac{p-1}{q+3} \times \frac{q^2}{p^2} + \&c.$$

Adeoq₃ fluens prædicta (ita sumpta ut sit $= 0$ q^{do} $z = 0$) $= V - B$. Item fluens prædicta ita sumpta ut sit $= 0$ q^{do} $x = 0$ quoniā tum $V = 1$ fit

$$1 - V = x^{n+1} + \overline{n+1}\, x^n r + \&c + \frac{n+1}{p+1} \times Ex^{p+1} r^q.$$

Quod (q^{do} $z = 0$) fit

$$\frac{n+1}{n} \times \frac{p}{q} \times \frac{Ep^p q^q}{n^n} \times \frac{1+q}{p+2} \times \frac{p}{q} + \frac{q}{p+2} \times \frac{q-1}{p+3} \times \frac{p^2}{q} + \&c = A$$

adeoq₃ Fluens prædicta ita sumpta ut sit $= 0$ q^{do}: $z = 0$ fit $A + V - 1$.

Commentary

In 1749 David Hartley published his *Observations on Man, His Frame, His Duty, and His Expectations* in which the following passage occurs.

> Mr. *de Moivre* has shewn, that where the Causes of the Happening of an Event bear a fixed Ratio to those of its Failure, the Happenings must bear nearly the same Ratio to the Failures, if the Number of Trials be sufficient; and that the last Ratio approaches to the first indefinitely, as the Number of Trials increases. This may be considered as an elegant Method of accounting for that Order and Proportion, which we every where see in the Phænomena of Nature. ...
>
> An ingenious Friend has communicated to me a Solution of the inverse Problem, in which he has shewn what the Expectation is, when an Event has happened p times, and failed q times, that the original Ratio of the Causes for the Happening

or Failing of an Event should deviate in any given Degree from that of p to q. And it appears from this Solution, that where the Number of Trials is very great, the Deviation must be inconsiderable; Which shews that we may hope to determine the Proportions, and, by degrees, the whole Nature, of unknown Causes, by a sufficient Observation of their Effects.

The Inferences here drawn from these two Problems are evident to attentive Persons, in a gross general way, from common Methods of Reasoning. [1749, Part I, Chap. III, §II]

Now it has been suggested that Bayes's Theorem (the tenth proposition in his *Essay*) might have been discovered by someone else (see Stigler [1983], a paper in which the attention of the statistical community was drawn to Hartley's work). The notebook extract presented here is unfortunately undated. However, the entry on p. 51(a) is headed 'Paris July 4, N.S. 1746', and that on p. 86 is headed 'Estimate of the National debt upon 31 December 1749'. It thus seems reasonable to date this extract as being written somewhere in 1746–1749, and although this does not *prove* that Bayes is the correct eponym, it certainly indicates that the result was known to him.

11.5.2 *Trigonometry*

The first passage [C: 22L–23L] is labelled in the table of contents 'Dato sinu invenire arcū', the fuller heading at the top of p. 22L being 'Dato sinû recto arcus invenire ipsum arcus'. However, other related matters are also discussed.

Bayes begins by finding what is essentially a series expansion for arcsin z (cf. §11.5.5). Letting x denote the sine of an angle in a circle of radius r, he sets

$$\log(x/r) + 3.5362738828 = \log A$$

$$\log(x^2/r^2) + \log(A/10) + 0.2218487497 = \log B$$

and so on, the logarithms being taken to base 10. This results in a series 'in minutis primis' (i.e., in minutes) of the form $A + B + \cdots$.

The following argument explains the constants in the above expressions. Consider the series expansion, for $|z| < 1$,

$$\arcsin z = z + \frac{1}{2.3} z^3 + \frac{1.3}{2.4.5} z^5 + \frac{1.3.5}{2.4.6.7} z^7 + \cdots .$$

Bearing in mind Bayes's giving of the series as $A + B + C + \cdots$, we set $A = z$, whence $\log A = \log z$. Conversion of radians to minutes requires the replacement of z by $(x/r) \times (180 \times 60)/\pi$, whence we have

$$\log z = \log(x/r) + \log((180 \times 60)/\pi)$$
$$= \log(x/r) + 3.536273882,$$

as stated by Bayes. Similarly,

$$B = \frac{1}{2.3} z^3 = z^2 A \frac{1}{2.3},$$

whence

$$\log B = \log z^2 + \log \tfrac{A}{10} + \log 10 + \log \tfrac{1}{2.3}$$
$$= \log z^2 + \log \tfrac{A}{10} + 0.221848749,$$

which is Bayes's expression for $\log B$ when z is written in the 'radius-dependent' form x/r.

A separate (finite) series is given for angles less than $3°$. The two terms A and B whose sum yields the desired approximation are given by

$$\log(x/r) + 5.3144251 = \log A$$

$$-12 + 0.5929986 + 3 \log A = \log B,$$

the constant in the expression for $\log A$ being $\log((180 \times 60 \times 60)/\pi)$. This yields an answer 'in minutis 2^{dis}' (i.e., in seconds), and tables are given for several values.

Solutions are also given for the problems:

(a) the finding of $\sin A$ from given A and vice versa, and

(b) the finding of $\tan A$ from given A and vice versa, with special attention again being paid to the case of angles less than $3°$.

The next passage [C: 48R] is described in the table of contents as 'Datis A & y invenire sinū, $A + y$'. Here, without proof, we find the infinite series that we can write as

$$\sin(A + y) = \sin A + \sum_{n=1}^{n} 2^n \sin(n\pi/2 + A) \prod_{k=1}^{n} \sin(y/2^k).$$

Bayes gives the first five terms of the series, followed by '&c'. This expansion can be verified by noting that

$$\sin(A + y) - \sin A = 2 \cos(A + y/2) \sin(y/2),$$

and then writing the cosine term in the form
$$\cos(A + y/2) = [\cos(A + y/2) - \cos A] + \cos A$$
and applying an addition formula to the bracketed difference. No proof of, or motivation for, this formula is given, but we may conjecture that it was to be used in finding $\sin(A + y)$ for known $\sin A$ and small y.

The next passage on trigonometry [C: 54L–54R] is listed as 'Datis sin,A & cos,A invenire sin,$2A$, s,$3A$ &c '. (Bayes frequently uses 's,x' for 'sin x'.[5]) The passage opens with a statement of addition formulae that seem curious from today's viewpoint. As an example we mention the first such formula, viz.
$$\text{s}, M \times \text{s}, N = (R/2) \times \overline{\cos, M - N} - \overline{\cos, M + N}.$$
However, when one recalls that it was common practice at the time Bayes was writing to define trigonometric ratios in a 'radius-dependent' sense (see Appendix 11.2), the presence of the factor 'R' above is not surprising. This factor can be removed by replacing cos and sin in the above formula by R cos and R sin, respectively, and the formula then takes on a more familiar appearance.

Despite the description of the passage in the table of contents, Bayes in fact does not provide explicit expressions for $\sin 2A$, $\sin 3A$, etc. Rather, we have expressions analogous to that given above, for the product of three, of four, and of five sines, followed by series for $(\sin A)^n$ (i) in terms of $\sin(nA)$ for $n \in \{3, 5, 7, 9\}$, and (ii) in terms of $\cos(nA)$ for $n \in \{2, 4, 6, 8, 10\}$. Thus, for example, we have for the product s, $A \times$ s, $B \times$ s, $C \times$ s, D the expressions
$$(R^2/4) \times \text{s}, \overline{A - B + C} \times \text{s}, D - \text{s}, \overline{A - B - C} \times \text{s}, D$$
$$- \text{s}, \overline{A + B + C} \times \text{s}, D + \text{s}, \overline{A + B - C} \times \text{s}, D$$
and
$$(R^3/8) \times \cos, \overline{A - B + C - D} - \cos, \overline{A - B + C + D} - \cos, \overline{A - B - C - D}$$
$$+ \cos, \overline{A - B - C + D} - \cos, \overline{A + B + C - D} + \cos, \overline{A + B + C + D}$$
$$+ \cos, \overline{A + B - C - D} - \cos, \overline{A + B - C + D},$$
while
$$(32/R^5) \times \text{s},^6 A = 10R - 15 \cos, 2A + 6 \cos, 4A - \cos, 6A$$
and
$$(64/R^6).\text{s},^7 A = 35.\text{s}, A - 21.\text{s}, 3A + 7.\text{s}, 5A - \text{s}, 7A.$$
See Bromwich [1931, chapter 9] for a detailed discussion of such series.

The next trigonometric article [66L] presents some integration formulae for sines and cosines when the radius is 1. Given formulae range from the simple[6]

$$\boxed{\dot{a} \times \cos, a} = \sin, a$$

to the more complicated

$$\boxed{32\dot{a} \times \cos, 6a} = 10a + \frac{15 \text{ s}, 2a}{2} + \frac{6 \text{ s}, 4a}{4} + \frac{\text{s}, 6a}{6}.$$

There next follows [C: 96R, 97R] a passage labelled 'De Æstimatione Errorū', though examination shows it merely to deal with the solution of triangles. The passage begins with a statement of two lemmata:

Lem. 1. $\dot{\text{s}}, A = \dot{A} \times \cos, A$

Lem. 2. $\dot{\text{T}}, A \times \cos^2 A = \dot{A}$.

Formulae are next given for some relationships between the sides and angles of triangles, both rectilinear and spherical, when one is given (a) two sides and the included angle, (b) two sides and the angles opposite each of them, and (c) one side and all angles. Some formulae are presented, in a corollary, for the case in which all sides and angles are known.

This passage is preceded, on p. 96L, by three lines giving formulae we would today write as

$$\frac{\sin A}{a} = \frac{\sin B}{b},$$

and identified as being for rectilinear, spherical, and polar triangles.

The final passage on trigonometry [121L–121R] is headed 'Ex Machin. Phil. Trans. N° 447.' The reference is to John Machin's paper 'The solution of Kepler's problem', published in volume 40 [1738] of the *Philosophical Transactions*, pp. 205–230, the problem in question being the following.

> To divide the Area of a Semicircle into given Parts, by a Line from a given Point of the Diameter, in order to find an universal Rule for the Motion of a Body in an Elliptic Orbit. [p. 205]

The method attributed by Machin to Kepler for the solution of this problem runs as follows.

> To compute a Table for some Part of the Orbit, and therein examine if the Time to which the Place is required, will fall out any-where in that Part. [p. 206]

The first part of Machin's paper is devoted to the mathematical tools needed for the solution, and Bayes quotes the result of the first lemma (though in a slightly different form); viz. if r is the radius of a circular arc having tangent t, sine y, and cosine z, then for given integral n,

$$\text{Tang}, nA = \frac{\overline{r + \sqrt{-t^2}}^n - \overline{r - \sqrt{-t^2}}^n}{\overline{r + \sqrt{-t^2}}^n + \overline{r - \sqrt{-t^2}}^n} \times \sqrt{-r^2}$$

(the exponent n in the second term in the numerator is missing in the notebook). As a corollary Bayes gives

$$\sin, nA = \frac{\overline{z + \sqrt{-y^2}}^n - \overline{z - \sqrt{-y^2}}^n}{2r^n} \times \sqrt{-r^2}$$

$$\cos, nA = \frac{\overline{z + \sqrt{-y^2}}^n + \overline{z - \sqrt{-y^2}}^n}{2r^{n-1}}.$$

Bayes expands the binomials and, by similar expansions applied to Machin's expressions for $\sin nA$ and $\cos nA$, he finds series for these functions. Special mention is made of the case in which $r = 1$ (not given by Machin), results of the form

$$s, 2A = 2yz, \quad \cos, 2A = 1 - 2y^2$$

obtaining.

One must take care, in checking the formulae given by Bayes and Machin, not to take (for a positive) $\sqrt{-a^2}$ as ia. Such an interpretation would give, for example, $-\tan A$ for the right-hand side of the expression for Tang, nA when $n = 1$. The surds are to be interpreted rather as satisfying

$$\sqrt{-a^2} \times \sqrt{-b^2} = \sqrt{(-a^2)(-b^2)} = \sqrt{a^2 b^2} = ab.$$

In proceeding in this way Bayes and Machin are in good company, for even Euler, in his *Vollständige Anleitung zur Algebra* of 1770 remarked that $\sqrt{-1} \times \sqrt{-4} = \sqrt{4} = 2$ since $\sqrt{a} \times \sqrt{b} = \sqrt{ab}$. In the same vein we find Simpson writing in 1762,

> ... both $-a$ and $-b$, as they stand here independently, are as much impossible in one sense, as the imaginary surd quantitites $\sqrt{-b}$ and $\sqrt{-c}$; since the sign $-$, according to the established Rules of notation, shews that the quantity to which it is prefix'd, is to be subtracted; but, to subtract something from nothing is impossible, and the notion, or supposition of a quantity less than nothing, absurd and shocking to the imagination ...

434   11. The Notebook

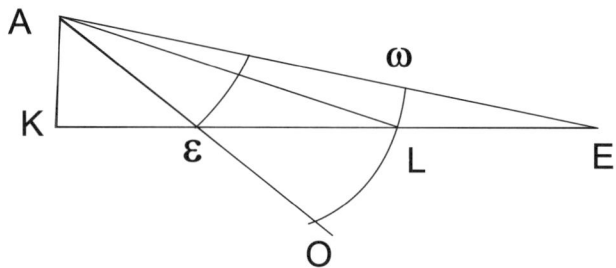

FIGURE 11.1 Sketch for the first geometrical problem.

> ... First Principles, and the more established Rules of notation; according to which the signs + and − are relative only to the magnitudes of quantities, as composed of different terms or members, and not to any future operations to be performed by them ...
> There can, therefore, be no such things as negative numbers, or quantities absolutely negative in pure Algebra, whose Object is Number .... [1762, pp. 24, 25, 25]

For further discussion of this matter see Kline [1990, vol. 2, p. 594].

11.5.3 Geometry

There are two passages in the notebook that are devoted to this topic. In the first of these [93L–95R] several problems are discussed, each being accompanied by a sketch. We instance only the second problem here (the others would require more detailed discussion and complicated diagrams), the attendant sketch being shown in Figure 11.1.

After some work Bayes writes

> Hinc si AK sit $\perp^{\text{is}}$ in KL & puncta A, K, L, dentur E vero punctū alicubi in rectâ KL sitq$_3$ $AE \times AL - KE \times KL = \mu$ semper erit μ minimū coincidente punctis E & L quoniā LE semper est major quam $E\omega$.

In the second passage [118(a)L–119R] two problems are considered. The first of these (see Figure 11.2) runs as follows.

> Si a 4 punctis lineæ rectæ A, B, C, D inflectantur ad (?) datū punctū H, quatuor (?) rect. (?) AH, BH, CH, DH & hisce lineis

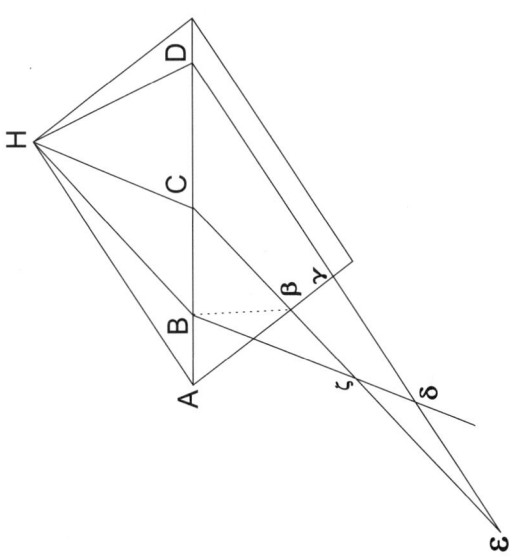

FIGURE 11.2 Sketch for the second geometrical problem.

respectivē parallelæ ducantur $D\gamma\delta\varepsilon$, $C\beta\zeta\varepsilon$, $B\alpha\zeta\delta$, $A\alpha\beta\gamma$. singulæ harū linearum in eadem ratione secabuntur a tribus reliquis, ac secta est abs (?) iisdem linea $ABCD$ erit nempe $\alpha\beta : \beta\gamma ::$ $BC : CD$, $\alpha\zeta : \zeta\delta :: AC : DC$, $\varepsilon\zeta : \beta\zeta :: BD : AB$, $\varepsilon\delta : \delta\gamma ::$ $BC : AB$.

(The writing is obscure in places.) A corollary, in the same vein, follows.

The second problem, on p. 119, is headed 'Calculus trigonometricus'. It is also concerned with the relative proportions between several line segments.

11.5.4 Solution of equations

The first passage [C: 48R–49L] is labelled 'Invenire minorē ex radicibus æquationis $x^2 - ax + r = 0$'. The actual rule, given on p. 49L, runs

Regula ad inveniendum minorē ex radicibus æquationis $x^2 - ax + r = 0$. sume $r/a = A$, $(a^2 - 2r)/r = b$, $A/b = B$, $c = b^2 - 2$, $B/c = C$, $d = c^2 - 2$, $C/d = D$ & ita deinceps eritq$_3$ $x = A + B + C + D +$ &c

(notation slightly altered). Bayes's proof of this rule runs as follows. If

$$f(x) \equiv x^2 - ax + r = 0, \qquad (11.1)$$

then

$$\begin{aligned} 0 &= (x^2 - ax + r)(x^2 + ax + r) \\ &= x^4 - bx^2 + r^2, \end{aligned}$$

where $b = a^2 - 2r$. Similarly,

$$\begin{aligned} 0 &= (x^4 - bx^2 + r^2)(x^4 + bx^2 + r^2) \\ &= x^8 - cx^4 + r^4, \end{aligned}$$

where $c = b^2 - 2r^2$. Proceeding in this way Bayes obtains

$$x^{16} - dx^8 + r^8 = 0 \qquad (11.2)$$

$$x^{32} - ex^{16} + r^{16} = 0 \qquad (11.3)$$

$$x^{64} - fx^{32} + r^{32} = 0. \qquad (11.4)$$

Substitution of x^{32} from (11.3) in (11.4), and similar substitution for x^{16}, from (11.2), in the resulting expression, etc., results finally in

$$x - \frac{x^{64}}{abcdef} = \frac{r}{a} + \frac{r^2}{ab} + \frac{r^4}{abc} + \frac{r^8}{abcd} + \frac{r^{16}}{abcde} + \frac{r^{32}}{abcdef}. \quad (11.5)$$

Now if m and n are the roots of (11.1) then $(x - m)(x - n) = 0$. Thus $a = m + n$ and $r = mn$, and hence

$$b \equiv a^2 - 2r = (m+n)^2 - 2(mn) = m^2 + n^2.$$

Similarly it follows that $c = m^4 + n^4$, $d = m^8 + n^8$, $e = m^{16} + n^{16}$, and $f = m^{32} + n^{32}$.

Now

$$n - \frac{n^{64}}{abcdef} \equiv n - \frac{n^2}{a} \times \frac{n^2}{b} \times \frac{n^4}{c} \times \frac{n^8}{d} \times \frac{n^{16}}{e} \times \frac{n^{32}}{f},$$

and since n is a root of $f(x) = 0$ one finds, on combining this latter result with (11.5), that, as Bayes puts it,

$$n = \frac{r}{a} + \frac{A}{b}r + \frac{B}{c}r^2 + \frac{C}{d}r^4 + \frac{D}{e}r^8 + \&c \text{ in infinitū}, \quad (11.6)$$

where $A = r/a$, $B = r^2/(ab)$, $C = r^4/(abc)$, and $D = r^8/(abcd)$ and where, as we have seen, a, b, c, d, and e may be expressed as functions of m and n. Presumably the unexpressed part of (11.6) consists of terms in r^{16} and r^{32} plus the 'remainder'

$$\frac{n^2}{a} \times \frac{n^2}{b} \times \frac{n^4}{c} \times \frac{n^8}{d} \times \frac{n^{16}}{e} \times \frac{n^{32}}{f}.$$

On writing this 'remainder' in the form

$$\frac{n^2}{m+n} \times \frac{n^2}{m^2+n^2} \times \frac{n^4}{m^4+n^4} \times \frac{n^8}{m^8+n^8} \times \frac{n^{16}}{m^{16}+n^{16}} \times \frac{n^{32}}{m^{32}+n^{32}}$$

one sees that this product will be small if the larger root m is very much bigger than n — or, as Bayes has it,

> Et ex hâc fluet sequens regula quæ tanto celerius convergit quanto major est proportio inter radices æquationis.

With a small change in notation (11.6) is the expression given in Bayes's Rule.

That Bayes's Rule is merely Newton's Method in disguise may be seen from the following considerations. Let $x_0 = r/a$ and, for $n \in \{1, 2, \ldots\}$, let x_n be defined recursively by

$$x_n = x_{n-1} - f(x_{n-1})/f'(x_{n-1}).$$

Then $x_0 = A$, $x_1 = A + B$, $x_2 = A + B + C$, etc.[7]. Moreover, since the maximum of $f(x)$ occurs at $x = a/2$, and since the discriminant $a^2 - 4r$ is positive (recall that Bayes writes of the 'smaller' of the roots, and hence a repeated root cannot occur), it follows that $(a/2) > (r/a)$. Hence the approximation approaches this smaller root from below.

The second item [58L–63L] consists of several results on the solution of specific equations. Several terms are undefined, and this makes it difficult to understand what is going on. We therefore content ourselves with giving only a sketch here.

Bayes begins by considering the equation

$$z^n - Az^{n-1} + Bz^{n-2} - Cz^{n-3} + Dz^{n-4} - \&c - G = 0, \tag{11.7}$$

where the coefficients vary in such a manner that $\dot{A} = n\dot{z}$, $\dot{B} = \overline{n-1}A\dot{z}$, $\dot{C} = \overline{n-2}B\dot{z}, \ldots, \dot{G} = F$. He then concludes, by 'taking y^e fluxions', that (11.7) is in fact an identity, and finds, by the same method, relationships between the fluxions of sums of various integral powers (both positive and negative) of the roots[8]. (We might note that Bayes uses 'φ,' in his proofs here for 'fluxion'.)

As a further example from this section we cite the consideration of the quintic equation

$$z^5 - az^4 + bz^3 - cz^2 + dz - e = 0.$$

An equation (of tenth degree) expressing the sum of any two roots of this equation is derived. The exercise is also carried out for an equation of sixth degree, the sum of any two roots now being found to satisfy an equation of fifteenth degree[9].

The next passage on this topic, [91L], is headed 'Pro Æquationibus ubi omnes radices sunt positivi'. Three articles are considered here, the first running as follows.

> Maxima coefficiens cui signū negativū unâ cum unitate (?) major est maximâ positivarū radicū.

The result is illustrated by an example involving the quartic equation

$$x^4 - 7x^3 + 13x^2 - 14x + 8 = 0,$$

one that clearly has no negative roots in view of the alternating signs of the coefficients (see Cajori [1927, p. 9]).

To prove the general result, consider

$$f(x) \equiv x^n + a_1 x^{n-1} + a_2 x^{n-2} + \cdots + a_n = 0.$$

Let $\mathcal{N}$ denote the set of negative coefficients appearing in $f(x)$, and let $m = \max\{|\nu|, \nu \in \mathcal{N}\}$. Then $f(x) > 0$ for any positive value x_0 of x for which

$$x_0^n - m(x_0^{n-1} + x_0^{n-2} + \cdots + 1) > 0, \qquad (11.8)$$

since

$$f(x_0) > (a_1 + m)x_0^{n-1} + (a_2 + m)x_0^{n-2} + \cdots + (a_n + m) > 0.$$

Re-writing (11.8) in the form

$$x_0^n - m \frac{x_0^n - 1}{x_0 - 1} > 0,$$

one sees a fortiori that $f(x_0) > 0$ if

$$(x_0^n - 1) - m \frac{x_0^n - 1}{x_0 - 1} > 0,$$

or

$$(x_0^n - 1)\left(1 - \frac{m}{x_0 - 1}\right) > 0.$$

Now this last expression is always positive if $m < x_0 - 1$, that is, if $x_0 > m + 1$. Thus $f(x)$ is positive for any real value of x greater than $m+1$, and hence any (real) value of x for which $f(x) = 0$ cannot exceed $m + 1$; that is, $m + 1$ is an upper bound[10] on the real positive roots of the equation $f(x) = 0$.

In the second article it is stated (correctly) that if $Ax^2 + B$ and $ax^4 + bx^2 + c$ have a common factor of the form $x^2 + p$, then $aB^2 - ABb + A^2c = 0$. The third article exhibits a similar relationship between the coefficients of two quadratics in x^2.

The next item on this matter, [C: 96L], is described as 'De limitibus Æquationū'. Here an equation of fourth degree, with four (positive) roots, is considered, it being ascertained whether the graph is above or below the horizontal axis on intervals whose end-points are zero and the consecutive roots.

There is one further item, [C: 18L–18R], which, although perhaps not strictly speaking concerned with the solution of equations, fits here as well as anywhere else. It is described in the table of contents as follows.

> Data æquatione in solū y & x si habeas ÿ q^do ẍ = 0 invenire ẍ si ÿ = 0.

In his solution of this problem Bayes proves the following result[11].

> Si quantitates duæ fluentes ambæ crescentes aut ambæ decrescentes dato momento temporis sint æquales, habeantq₃ fluxiones primas æquales, illa erit major & ante & post datū momentū cui est major fluxio 2da positiva aut minor negativa.

11.5.5 Series

A fair part of the notebook is taken up with work on series. We consider the minor matters first, leaving the more interesting work on $\log z!$ until the end.

The first passage, [20L], is devoted to a proof, with text in Latin, that[12]

$$x \equiv \frac{p}{p+1} \times \frac{p-1}{p+2} \times \cdots \times \frac{p-a+1}{p+a} > \left(1 - a^2/p^2\right)^p$$

with $p \geq a$ and $\dot{p} = 1$. Having stated that this inequality clearly holds for $p = a$, Bayes takes the logarithm of the left-hand side and finds the fluxion, obtaining

$$\dot{x}/x = \frac{1}{p^2 + p} + \frac{3}{p^2 + p - 2} + \cdots + \frac{2a-1}{p^2 + p - a^2 + a}.$$

Since $1 + 3 + 5 + \cdots + (2a-1) = a^2$, it follows that

$$\dot{x}/x < a^2/(p^2 - a^2).$$

Then, letting

$$y = (1 - a^2/p^2)^p,$$

he finds, on proceeding as before, that

$$\dot{y}/y > a^2/(p^2 - a^2).$$

It is then concluded that $\dot{x}/x < \dot{y}/y$. Further, since $x = y$ when p is infinite, it follows from the increasing nature of the relevant functions that $x > y$.

Another Latin note, [20R], is concerned with the series

$$x = Az^n + Bz^m + \&c., \quad \dot{z} = 1, \quad \ddot{z} = 1,$$

where $\dot{z}$ denotes the fluxion and $\underset{\prime}{z}$ the first finite difference of z. Here Bayes expands $\dot{x}$, $\underset{\prime}{x}$, etc. as series in $\dot{x}$, $\ddot{x}$, ... and claims that if v and $\underset{\prime}{v}$ denote the numerators of any two consecutive coefficients in the series for $\dot{x}$, then $\underset{\prime}{v} = 2v + 1$. Similarly, the numerators of any three consecutive terms in the series for $\underset{\prime\prime}{x}$ satisfy $\underset{\prime\prime}{v} = 5\underset{\prime}{v} - 6v + 1$.

The third item, [C: 49L–49R], is labelled 'Item de reversione Serierū'. Here Bayes considers the equations

$$\frac{x^{n-1}\dot{x}}{z^{n-1}\dot{z}} = a + bz + cz^2/2 + dz^3/2.3 + ez^4/2.3.4 + \&c$$

$$\frac{z^{n-1}\dot{z}}{x^{n-1}\dot{x}} = A + Bx + Cx^2/2 + Dx^3/2.3 + Ex^4/2.3.4 + \&c,$$

and discusses the finding of the coefficients in the second of these equations in terms of those in the first. Special mention is made of the case $n = 2$ and of the system

$$x = az + bz^2/2 + cz^3/2.3 + \cdots$$

$$z = Ax + Bx^2/2 + Cx^3/2.3 + \cdots.$$

He also provides expressions for $A, B, C, \ldots$ in terms of $a, b, c, \ldots$ when these coefficients satisfy

$$x = a + bz + cz^2/2 + dz^3/2.3 + \&c$$

$$1/x = A + Bz + Cz^2/2 + Dz^3/2.3 + \&c.$$

The next passage, [C: 57R], is described as 'De valore quadrati logarithmi $1 + z$.' Here Bayes presents two methods for finding the coefficients in the series for $[\log(z+1)]^2$ in terms of those in the series for $\log(z+1)$.

In [108R–109L] Bayes considers an infinite series for $y_n = \sin(nA)$ (for integral n) as

$$y_n = z^n \sum_{k=0} (-1)^k \frac{n(n-1)\ldots(n-2k)}{(2k+1)!} t^{2k+1},$$

where $t = \tan A$ and $z = \cos A$. The series is obtained by noting that, if x, $\underset{\prime}{x}$, and $\underset{\prime\prime}{x}$ denote the sines of nA, $(n+1)A$, and $(n+2)A$, respectively, then

$$\underset{\prime\prime}{x} = 2xz - xz^2(1+t^2).$$

(It should be noted that the radius is taken as 1 throughout this passage.)

The desired series is then obtained 'by proceeding in the common way by the method of increments' (p. 108R). Similar expansions are derived for $\cos(nA)$ (in terms of t) and for $\sin(nA)$ and $\cos(nA)$ in terms of $\cot A$. In each instance some special cases (e.g., $nA = 90°$ or $180°$) are considered.

The next passage, [109R–112(a)L], follows hard on the heels of the preceding one. Here Bayes considers the series that we may write as

$$x = p^n + n \sum_{k=1}^{k-1} (-1)^k p^{n-2k} \prod_{j=1}^{} (n - 2k + j)/k!, \qquad (11.9)$$

where p is 'permanent' and the increment of n is always 1. (Here, for any sequence $\{\alpha_k\}$, $\prod_{j=1}^{0} \alpha_j = 1$ by definition, and the series terminates when a term becomes zero.) In the first article it is shown that $\underset{\sim\sim}{x} = p\underset{\sim}{x} - x$, where $\underset{\sim}{x}$ and $\underset{\sim\sim}{x}$ are the above series for x with n replaced by $(n+1)$ and $(n+2)$, respectively. A special proof is given for $n = 1$ and 2, in which cases we find that $\underset{\sim}{x} = p^3 - 3p$ and $\underset{\sim\sim}{x} = p^4 - 4p^2 + 2$.

In the second article in this passage Bayes supposes that the 'Radius or modulus sinuum' is 2. Recalling his definitions of the trigonometric ratios (see §11.5.2), we note that this leads to the correct formulation of the identity used in this article, viz.

$$\cos(n+1)A \times \cos A = \cos(n+2)A + \cos nA.$$

Then our equation (11.9), arising from Bayes's first article, is seen to be satisfied on our setting $p = \cos A$ and $x = \cos nA$.

In the third article Bayes chooses two numbers r and b such that $rb = 1$ and $r + b = p$. Then $x = r^n + b^n$ is seen to satisfy (11.9). In an addendum similar series are established for $r^n - b^n$ and for $(r^{2n+2} - b^{2n+2})/(r^2 - b^2)$ (in the latter case the conditions are $rb = 1$ and $r^2 + b^2 = p$).

A similar result is established in the fourth article for $rb = 1$ and $r - b = p$, the cases 'n odd' and 'n even' being considered separately.

In the final (unnumbered) article Bayes returns to the notation of the third article, replacing r, b, and p by r^2, b^2, and $z^2 + p$, respectively. This leads to a series for $r^{2n} + b^{2n}$ in powers of $z^2 + p$, and it is noted that, if this series is expanded and ordered in powers of z^2, the last term in which z^2 does not appear is precisely the original series (11.9). With p and x as in the second article it follows that 'the product made by multiplying all the values of z^2 together will be $= $ to $r^{2n} + b^{2n} - \cos, nA$'. Further results in a similar vein follow.

One might see a connexion between the work detailed in the preceding paragraphs and that on the solution of equations discussed in our §11.5.4.

On page 111 the equation $\underset{\sim}{x} = a\underset{\sim}{x} - x$ is considered, under the conditions that when $n = 1$, $x = a$, and when $n = 2$, $x = a^2 - 1$. In this case x is given by the series

$$x = a^n + \sum_{k=1}(-1)^k a^{n-2k} \prod_{j=1}^{k}(n - 2k + j)/k!.$$

A similar expression for other pairs of values for x and n is given, and results analogous to those obtained in earlier articles are derived.

The next relevant passage, [114L–114R], again is concerned with a series for a trigonometric ratio. Under the assumption that x is an angle ('y$^\text{e}$ arch'), z its sine, and $\dot{z} = 1$ always, expressions are derived for fluxions of x of various orders. Evaluating these when $z = 0$, Bayes deduces that

$$x = z + \frac{z^3}{2.3} + \frac{9\,z^5}{2.3.4.5} + \frac{9.25\,z^7}{2.3.4.5.6.7} + \&c$$

(i.e., the usual series for arcsin z — cf. §11.5.2). Similar series are also derived for x^2 and x^3.

Note that if one takes x to be $\pi/2$ one obtains the series

$$\frac{\pi}{2} = 1 + \frac{1}{2.3} + \frac{3^2}{2.3.4.5} + \frac{3^2.5^2}{2.3.4.5.6.7} + \&c.$$

Now in Proposition XXXIII of his *Methodus Differentialis: sive Tractatus de Summatione et Interpolatione Serierum Infinitarum* of 1730, James Stirling, in considering the problem of interpolation in the sequence $\{2^{2n}/\binom{2n}{n}\}$, expanded

$$T^2 = \left\{2^n \Big/ \binom{n}{n/2}\right\}^2,$$

for even n, as

$$An\left[1 + \frac{(1)^2}{2(n+2)} + \frac{(1.3)^2}{2.4(n+2)(n+4)} + \frac{(1.3.5)^2}{2.4.6(n+2)(n+4)(n+6)} + \&c.\right];$$

and, by noting that the encrotcheted series is asymptotic to An, showed, using Wallis's product,

$$\frac{4}{\pi} = \frac{3}{2} \cdot \frac{3}{4} \cdot \frac{5}{4} \cdot \frac{5}{6} \cdot \frac{7}{6} \cdot \frac{7}{8} \cdots,$$

that $A = \pi/2$. Were we allowed to set $n = 1$, T^2 would become $(\pi/2)^2$ and Stirling's series would agree with the above special case of Bayes's.

The most important work on series, however, is that relating to $\log z!$. Although I have discussed this in detail elsewhere (see Dale [1991]), it is, I feel, sufficiently important to be given attention here.

The first passage, [C: 1R–10L], is described in the table of contents as follows.

> Pag. primæ novem — pertinent ad problemata
> Data area invenire summā æquidistantium ordinatarū & vice versa sive quod eodem redit.
> Datâ fluxione invenire incrementū aut vice versa.
> 10. Invenire Log, $\overline{\dfrac{z+1}{z}}\Big|^{z+\frac{1}{2}}$ ejusq₃ integrale.

In these first ten pages we find the following;

(a) a derivation of what is essentially the Euler–MacLaurin sum formula;

(b) an extract, partly in shorthand, from MacLaurin's *A Treatise of Fluxions*[13] of 1742, in which series are given for

   (i) $\log(m + z) - \log m$, where m is given;

   (ii) various formulae derived from (i);

   (iii) $\sum_{1}^{n-1} \log k$;

   (iv) $\varepsilon N/(N-1)$ and $\varepsilon N/(N^2-1)$, where N is the number whose hyperbolic logarithm is ε.

(c) various series for $\log\left(\dfrac{z+1}{z}\right)^{z+(1/2)}$ and the sum of such things;

(d) a series for $\sum \log(z-1)$.

Some of these results have a direct bearing on Bayes's published letter on $\log z!$, and we accordingly give a fairly detailed discussion of them.

Fundamental to this work is the following result of Bayes's first section.

Let t be a uniformly flowing quantity $\dot t = \ddot t = 1$

$$x = N + at + \frac{bt^2}{2} + \frac{ct^3}{2.3} + \frac{dt^4}{2.3.4} + \&c. + \frac{kt^n}{2.3.4.5.\&c.n}$$

&

$$x = \dot x + \frac{\ddot x}{2} + \frac{\dddot x}{2.3} + \frac{\ddddot x}{2.3.4} + \&c.$$

as will be evident by find[ing] $\underset{\cdot}{x}$ by y^e method of increments & $\dot{x}, \ddot{x}$ &c by y^e method of fluxions. Also in y^e same manner

$$\dot{x} = \underset{\cdot}{x} - \underset{\cdot\cdot}{x}/2 + \underset{\cdot\cdot\cdot}{x}/3 - \underset{\cdot\cdot\cdot\cdot}{x}/4 + \underset{\cdot\cdot\cdot\cdot\cdot}{x}/5 - \&c$$

And thus also y^e relation between $\ddot{x}$ and $\underset{\cdot\cdot}{x}$ & so on may be found.

(Here letters that are 'pricked' above and below denote fluxions[14] and finite differences, respectively.) The expression for $\ddot{x}$ may be found as follows. From the expression given for $\underset{\cdot}{x}$ we have, by transposition,

$$\dot{x} = \underset{\cdot}{x} - \frac{\underset{\cdot\cdot}{x}}{2} - \frac{\dot{\underset{\cdot\cdot}{x}}}{2.3} - \frac{\ddot{\underset{\cdot\cdot}{x}}}{2.3.4} - \&c. \tag{11.10}$$

It also follows from the expression for $\underset{\cdot}{x}$ that

$$\underset{\cdot\cdot}{x} = (\underset{\cdot}{x})^{\cdot} + \frac{1}{2}(\underset{\cdot}{x})^{\cdot\cdot} + \frac{1}{2.3}(\underset{\cdot}{x})^{\dot{\cdot\cdot}} + \frac{1}{2.3.4}(\underset{\cdot}{x})^{\ddot{\cdot\cdot}} + \&c.,$$

where $(\underset{\cdot}{x})^{\cdot}$ denotes the fluxion of $\underset{\cdot}{x}$, etc. Substitution of the given expression for $\underset{\cdot}{x}$ in this last series yields

$$\underset{\cdot\cdot}{x} = \left(\ddot{x} + \frac{\dot{\underset{\cdot\cdot}{x}}}{2} + \cdots\right) + \frac{1}{2}\left(\dot{\ddot{x}} + \frac{\ddot{\underset{\cdot\cdot}{x}}}{2} + \cdots\right) + \cdots.$$

Substitution in (11.10) of $\ddot{x}$ obtained from this expression gives

$$\dot{x} = \underset{\cdot}{x} - \underset{\cdot\cdot}{x}/2 + f(\dot{\ddot{x}}, \ddot{\ddot{x}}, \ldots).$$

This process may then be repeated.

In the first two pages Bayes derived, using this result, what is essentially the Euler–MacLaurin sum formula. This he stated as follows.

> if upon y^e base at equal distances each $= z = 1$ you erect any number of ordinates, call y^e 1st x & y^e last y y^e area between y^e 1st and y^e last A & y^e sum of all y^e ordinates except y^e last S. y^t
>
> $$S = A - \frac{\dot{A}}{2} + \frac{\ddot{A}}{12} - \frac{\dddot{A}}{720} + \frac{\ddddot{A}}{30240} - \frac{\dddddot{A}}{1209600} + \&c.$$
>
> & $\dot{A} = y - z$ [should be $y - x$]. But note that y^{re} are some exceptions to this rule.

In fact, on writing the Euler–MacLaurin sum formula[15] in the form

$$\sum_{1}^{n} f(m) \sim \int_{a}^{n} f(x)dx + C + \frac{1}{2}f(n) + \sum_{r=1}^{\infty}(-1)^{r-1}\frac{B_r}{(2r)!}f^{(2r-1)}(n),$$

where the B_r are the Bernoulli numbers[16], we find that the first sum and the integral given here correspond, respectively, to Bayes's S and A, with $C = -f(n)$.

This result is then used in §3 to deduce that

$$\log\left(\frac{z+1}{z}\right) = \frac{2}{2z+1}\left[1 - \frac{1}{12}\left(\frac{1}{z+1} - \frac{1}{z}\right) + \frac{2!}{720}\left(\frac{1}{(z+1)^3} - \frac{1}{z^3}\right)\right.$$
$$\left. + \frac{4!}{30240}\left(\frac{1}{(z+1)^5} - \frac{1}{z^5}\right) + \cdots\right].$$

(The '$+\cdots$' is in fact not given by Bayes, but its presence seems indicated by his '&c.' and the Euler–MacLaurin sum formula.) This result we write for later reference in the form[17]

$$\left.\begin{aligned}\log\left(\tfrac{z+1}{z}\right)^{z+(1/2)} &= 1 + \sum_{r=1}^{\infty}(-1)^r \tfrac{B_r}{(2r-1)(2r)}\Delta(1/z^{2r-1}) \\ &= 1 + \sum_{r=1}^{\infty}(-1)^r \tfrac{B_r}{(2r)!}\Delta\left((1/z)^{(2r-2)}\right) \\ &= 1 + \sum_{r=1}^{\infty}(-1)^r \tfrac{B_r}{(2r)!}[\Delta(1/z)]^{(2r-2)}\end{aligned}\right\}, \quad (11.11)$$

where $\Delta f(z) = f(z+1) - f(z)$ and $f^{(n)}(z) \equiv (f(z))^{(n)}$ denotes the nth derivative of $f(z)$.

The next pages of the notebook contain (partly in shorthand) extracts from §§837, 839, 842, and 847 (concluding with a very brief extract from §827) of MacLaurin's *A Treatise of Fluxions* of 1742, in which series as detailed in (b)(i) to (iv) above are given; we do not give any details here.

On page [7L] of the notebook the following problem is considered.

Sit $z =$ Log,x & ex data equatione $x = 1 + \frac{2z}{2v-z}$ invenire v ex data z.

Now on the face of it this seems a rather curious problem. However, on solving for v we find that

$$v = \frac{z(x+1)}{2(x-1)} = \frac{x+1}{2(x-1)}\log x.$$

Setting $x = (t+1)/t$ and writing $v \equiv v(x)$ as a function of t, we find that $v(t) = \log\left(\frac{t+1}{t}\right)^{t+(1/2)}$. But from $v = z(x+1)/2(x-1)$, and on recalling that $z = \log x$, we see that we can write $v \equiv v(z)$ as

$$v(z) = \frac{z(e^z+1)}{2(e^z-1)} = \frac{ze^z}{e^z-1} - \frac{z}{2},$$

the first term on the right-hand side here being the first of the series mentioned in (b)(iv) above; and in fact this is a series that generates the Bernoulli numbers.

Using the above expression for $v(t)$ we find that, for $n \in \mathbb{N}$,

$$\sum_1^n v(t) = \sum_1^n \log\left(\tfrac{t+1}{t}\right)^{t+(1/2)} = (n+\tfrac{1}{2})\log(n+1) - \sum_1^n \log k.$$

Thus

$$\begin{aligned}
\log n! &= \sum_1^n \log k \\
&= (n+\tfrac{1}{2})\log(n+1) - \sum_1^n v(t) \\
&= (n+\tfrac{1}{2})\log(n+1) - \sum_1^{n-1} v(t) - v(n) \\
&= (n+\tfrac{1}{2})\log(n+1) - \sum_1^{n-1} v(t) - \log\left(\tfrac{n+1}{n}\right)^{n+(1/2)} \\
&= (n+\tfrac{1}{2})\log n - \sum_1^{n-1} v(t).
\end{aligned}$$

Thus, if one is interested in series for $\log n!$, $v(\cdot)$ is an eminently reasonable function to consider. Indeed, we show in what follows that Bayes gave expansions not only for $v(\cdot)$, but also for $\log n$ and $\sum_1^n \log k$.

Bayes also used the equation $x = 1 + 2z/(2v-z)$ to deduce the coefficients in the series for $(z+\tfrac{1}{2})\log\left(\tfrac{z+1}{z}\right)$. Setting $\dot{z} = 1$ and using $z = \log x$, he found that $1 = \dot{x}/x$. The finding of the fluxion[18] of the initial expression for x and the equating of the result to x then yields

$$v^2 + \dot{v}z = v + z^2/4. \qquad (11.12)$$

Bayes now assumed that

$$v = 1 + az^2 - bz^4 + cz^6 - dz^8 + ez^{10} - fz^{12} + \&c.,$$

and substitution of this expression, together with $\dot{v}$, in (11.12) resulted in an identity from which the coefficients $a, b, c, \ldots$ can be found. Bayes, in fact, gave the coefficients, up to that of the term in z^{14}, as

$$a = 1/12, \qquad b = 1/720, \qquad c = 1/30240,$$

$$d = 1/1209600, \qquad e = 1/47900160,$$

$$f = 691/2^{11}.3^6.5^3.7^2.11.13$$

$$g = 1147/2^9.3^8.5^4.7^2.11.13.$$

These are also given by Bayes in the alternative forms

$$a = 1/12, \quad 2!b = 1/360, \quad 4!c = 1/1260,$$

$$6!d = 1/680, \quad 8!e = 1/1188,$$

$$10!f = 691/360360, \quad 12!g = 2294/61425,$$

where we have replaced Bayes's products by factorials.

That all these coefficients were carefully evaluated suggests that Bayes was aware of the behaviour of the coefficients with increasing powers of z, a behaviour to which, as we have already seen in Chapter 6, he drew attention in his 'Letter to Canton.'

Only one expansion is given for $\log(z-1)$, one that we can write in an admittedly anachronistic notation as

$$\log(z-1) = \Delta f(z-1) - \frac{1}{2}\Delta^2 f(z) + 2\left(\frac{1}{3}S - \frac{1}{4}\Delta S + \frac{1}{5}\Delta^2 S - \cdots\right), \quad (11.13)$$

where $f(z) = \log z^z$ and $S = \frac{1}{2}\Delta(1/z) + \frac{1}{12}\Delta(1/z^3) + \frac{1}{30}\Delta(1/z^5) + \cdots$.

This Bayes derived by writing (in our notation)

$$1/(z-1)^2 = -\Delta(1/(z-1)) + \frac{1}{2}\Delta^2(1/(z-1)) - \frac{1}{3}\Delta^3(1/(z-1)) + \cdots \quad (11.14)$$

and then by noting that, since

$$\Delta^2(1/(z-1)) = 2/z^3 + 2/z^5 + 2/z^7 + \cdots,$$

it follows that

$$\begin{aligned}\Delta^3(2/(z-1)) &= \Delta(\Delta^2(1/(z-1))) \\ &= \Delta(2/z^3 + 2/z^5 + 2/z^7 + \cdots) \\ &= 2/z^3 - 2/(z+1)^3 + \cdots,\end{aligned}$$

and so on. Then, on taking (11.14) and 'sumendo fluentes bis' (or, as one might say, by integrating twice) Bayes arrived at (11.13). Notice, incidentally, that no arbitrary constants enter into this integration.

Immediately after this derivation Bayes found a series for $\Sigma \log(z-1)$, preceding the actual series with the words 'Quare integrale logarithmi $\widehat{\tau o\nu}$ $z-1$.' (That is, 'Whereby the sum of the logarithm of $z-1$ is.') This is really equivalent to saying that $\Delta^{-1} \log z = \log \Gamma(z) + c$. It is clear from what follows that the operation concerned here is in fact one that is inverse to that of finite difference. Once again, though, no arbitrary constants appear. We may write the given series as

$$\log(z-1)^{z-1} - \frac{1}{2}\Delta \log(z-1)^{z-1} + \frac{1}{3}\left[\frac{1}{z} + \frac{1}{6z^3} + \frac{1}{15z^5} + \frac{1}{28z^7} + \cdots\right]$$

$$-\frac{1}{4}\left[\Delta^2(1/(z-1)) + \frac{1}{6}\Delta^2(1/(z-1)^3) + \frac{1}{15}\Delta^2(1/(z-1)^5) + \cdots\right]$$

$$+\frac{1}{5}\left[\Delta^3(1/(z-1)) + \frac{1}{6}\Delta^3(1/(z-1)^3) + \frac{1}{15}\Delta^3(1/(z-1)^5) + \cdots\right] + \cdots .$$

This series, after one further 'integration' (and my correction of some trifling errors), is rewritten at the top of page 7L in the form

$$K + z + \log(z-1)^{(3z-3)/2} - (1/2)\log z^z + x/3 - x/4 + x/5 - x/6 + \&c.,$$

where $x = 1/z + 1/6z^3 + 1/15z^5 + \cdots$.

The presence of $K+z$ here requires some comment. It would appear that, in accordance with the custom of his time, Bayes adopted a somewhat cavalier attitude to constants of integration[19]. Thus although no constant appears on passing from $\dot{y} = \dot{x}/x$ to $y = \log x$, one further 'integration' yields a constant that in turn, on being operated upon by Δ^{-1}, yields $K+z$ or, more strictly, $K + cz$.

Five series are given in the notebook for $\log\left(\frac{z+1}{z}\right)^{z+(1/2)}$. Written anachronistically, they are the following.

S1. $-\frac{1}{12}\Delta(1/z) + \frac{1}{360}\Delta(1/z^3) - \frac{1}{1260}\Delta(1/z^5) + \cdots$

S2. $1 - \frac{1}{3.4}\Delta^3(\log z^z) + \frac{2}{4.6}\Delta^4(\log z^z) + \frac{3}{5.8}\Delta^5(\log z^z) + \cdots$

S3. $1 - \frac{1}{12}\Delta(1/z) + \frac{1}{120.3!}\Delta^3(1/z) - \frac{1}{30.5!}\Delta^5(1/z) + \cdots$

S4. $1 - \frac{1}{12}\Delta(1/z) - \frac{1}{10.12}[\Delta(1/z)]^2 + \frac{1}{7.10.12}[\Delta(1/z)]^3 + \cdots$

S5. $1 - \sum_{r=1}(-1)^r B_r z^{2r}/(2r)!.$

(S5 is, in fact, given as a series for $v(z)$, a series that Bayes obtained by the assumption that $v(z) = 1 + az^2 - bz^4 + cz^6 - \cdots$ and by appropriate 'differentiation' of the original expression (14) in v, z, and x. Notice that this is $\varepsilon/2$ plus the series for MacLaurin's $\varepsilon N/(N-1)$ mentioned in (b)(iv) above.) The series S2 does not seem particularly useful. The series S1, S3, and S4, the series in powers of $1/z$, can be reconciled by using known properties of the difference operator (compare the series for $\log\left(\frac{z+1}{z}\right)^{z+(1/2)}$ given earlier in this section).

Finally, on the tenth page of his notebook Bayes gave a series for

$$\sum \log\left(\frac{z+1}{z}\right)^{z+(1/2)},$$

prefacing it again by the word 'integrale' and deducing it by applying this operation to S3.

In his *Methodus Differentialis* of 1730 Stirling gave series similar to those discussed above. More exactly he wrote

$$x(x+1)\ldots(x+n-1) = C_n^0 x^n + C_n^1 x^{n-1} + \cdots + C_n^{n-1} x^n$$

$$1/x(x+1)\ldots(x+n-1) = \sum_{k=0}^{\infty} (-1)^k \Gamma_n^k / x^{n+k}$$

(the notation here following Tweedie [1922, p. 31]), where the C_r^n and Γ_n^k are the Stirling numbers of the first and second kind[20]. From this he deduced that

$$x^n = \Gamma_2^{n-1} x + \Gamma_3^{n-1} x(x-1) + \cdots + \Gamma_{n+1}^0 x(x-1)\ldots(x-n+1)$$

$$1/x^n = \sum_{r=n-1}^{\infty} C_r^{r-n+1}/x(x+1)\ldots(x+r)$$

(the connexion between the first and the fourth of these expressions, and between the second and the third, is discussed in Jordan [1965], pp. 142 and 214 and pp. 175 and 168, respectively).

As we have already mentioned, some six years before Bayes died Euler, in §§157 to 159 of the second, and last, part of his *Institutiones calculi differentialis*, had noted the divergence of the factorial series for $z = 1$. It is not clear, however, whether he was aware that the series would diverge[21] for *all* values of z.

11.5.6 The differential method

In several passages in the notebook — in particular, in those on series — Bayes makes use of fluxions and increments. However there are a few comments on 'differentials', and it is to this topic that we now turn.

The first passage, [72L–74R], is headed 'De method. Differentiali'. Starting with the equation 'ad curvā sit $A+x$ abscissa v ordinata' (is the origin of the co-ordinate system thus taken at A?) Bayes considers the equation

$$v = A + Bx + Cx^2 + Dx^3 + Ex^4,$$

setting $x = a, b, c, d, e$ in succession and denoting the resultant values of v by $\alpha, \beta, \gamma, \delta, \varepsilon$. He then defines $\overset{\backprime}{\alpha}, \overset{\backprime\backprime}{\alpha}, \ldots$ (with similar expressions for the backprimed values of $\beta, \gamma, \ldots$) by

$$\overset{\backprime}{\alpha} = (\alpha - \beta)/(a - b)\,,\ \overset{\backprime\backprime}{\alpha} = (\overset{\backprime}{\alpha} - \overset{\backprime}{\beta})/(a - c)\,,\ \overset{\backprime\backprime\backprime}{\alpha} = (\overset{\backprime\backprime}{\alpha} - \overset{\backprime\backprime}{\beta})/(a - d)\,,\ldots.$$

This allows the writing of A, B, C, D, E in terms of $\overset{\backprime}{\alpha}, \overset{\backprime\backprime}{\alpha}, \ldots$, and it then follows that the general equation of the curve is

$$v = \alpha + \overset{\backprime}{\alpha}(x-a) + \overset{\backprime\backprime}{\alpha}(x-a)(x-b) + \overset{\backprime\backprime\backprime}{\alpha}(x-a)(x-b)(x-c)$$
$$+ \overset{\backprime\backprime\backprime\backprime}{\alpha}(x-a)(x-b)(x-c)(x-d).$$

It might be mentioned in passing that, in deriving the expressions for B, C, D, E, Bayes introduces a notation that he later ascribes to Lord Stanhope[22] (see §§10.6 and 11.5.8), viz.

$$\boxed{ab}^{\,n-1} = (a^n - b^n)/(a - b)\,.$$

A similar expression is given for more than two 'arguments', and on page 73L it is shown that

$$\left(\boxed{abcd}^{\,n} - \boxed{bcde}^{\,n}\right)\Big/(a - e) = \boxed{abcde}^{\,n-1}.$$

Most of page 72R is crossed out: all that remains is a corollary at the foot of the page, in which specific numerical values are given for $a, b, \ldots$, and the relationships between $\overset{\prime}{\alpha}$ and $\underset{\cdot}{\alpha}$, etc. are stated: for example, $\overset{\prime}{\alpha} = \underset{\cdot}{\alpha}$, $\overset{\prime\prime}{\alpha} = \underset{\cdot}{\alpha}/2$, and so on (the backprimes now becoming primes).

On page 73R several relations are established between $\alpha, \beta, \gamma, \ldots$ and their various primes. This is continued on the next page, and Bayes concludes from these expressions that

Et hinc patet si dantur a, b, c, d, e & $\alpha, \beta, \gamma, \delta, \varepsilon$ invenire possunt $\overset{\prime}{\alpha}, \overset{\prime\prime}{\alpha}, \overset{\prime\prime\prime}{\alpha}, \overset{\prime\prime\prime\prime}{\alpha}$.

Some special cases are considered on the final page.

The next passage, [C: 77L–78R], has the same label in the table of contents as that just discussed. Bayes begins by supposing that a, b, c, d, e are 'series cujus $2^{\text{dæ}}$ differ^{tiæ} sunt æquales'. It is further supposed that

$$a = x, \quad b = x + \dot{x}, \quad c = x + 2\dot{x} + \ddot{x}$$
$$d = x + 3\dot{x} + 3\ddot{x}, \quad e = x + 4\dot{x} + 6\ddot{x}.$$

Expressions are then given for b and d for known a, c, and e; viz.

$$b = (6c + 3a - e)/8 \quad \text{and} \quad 2d = c + e - \ddot{x}.$$

A similar study is made in the case of a number of series whose third or fourth differences are equal.

Although in the same vein, the first passage on page 78L differs in being concerned with the interpolation of series. Thus it is supposed that $AaBbCcDdE$ is a series whose fourth differences are equal. Bayes sets

$$B - A = \dot{A}, \quad C - B = \dot{B}, \quad D - C = \dot{C}, \quad E - D = \dot{D}$$
$$\dot{B} - \dot{A} = \ddot{A}, \quad \dot{C} - \dot{B} = \ddot{B}, \quad \dot{D} - \dot{C} = \ddot{C}$$
$$\ddot{B} - \ddot{A} = \dddot{A}, \quad \ddot{C} - \ddot{B} = \dddot{B}$$
$$\dddot{B} - \dddot{A} = \ddddot{A}.$$

Then, 'si z sit dist. Termini ab A et terminus ipse N est si $\dot{z} = 1$', he shows that

$$N = a, \quad N + \dot{N} = b, \quad N + 2\dot{N} + \ddot{N} = c$$
$$N + 3\dot{N} + 3\ddot{N} + \dddot{N} = d.$$

A reference here says 'Vid. P. 91' : we take this up in due course.

Another passage on page 78L, ruled off from the one just discussed, but on the same topic, is headed 'Idem secundū Methodum Cotesii de constr. Tab. Prop. 1.' The reference seems to be to Roger Cotes's *Canonotechnia, sive constructio tabularum per differentias*, a paper that occupies pp. 35 to 71 of his *Harmonia Mensurarum* of 1722. The extract is concerned with the interpolation at appropriate distances in a particular series. Bayes gives his answers both in terms of Cotes's *rotundus* and *quadratus* numbers (see Appendix 11.4), and in terms of the more usual 'pricked below' finite differences.

This subject is continued on p. 91R (which bears at the foot the reference 'Vid. Pag. 77,78'). Here Bayes considers the insertion of terms in the series A, A^3, A^5, etc. when the fourth differences (*differentiæ*) are equal.

In reading the above passages from the notebook one must bear in mind the meaning that was attached to the term 'differential' in the eighteenth and early nineteenth centuries. Thus Cotes published a paper entitled 'De Methodo Differentiali Newtoniana' (see Cotes [1722]), Stirling's *Methodus Differentialis* of 1730 deals with the calculus of finite differences, and MacLaurin wrote of 'Sir *Isaac Newton's differential method*' [1801, §827]. Further, in the preface to his translation of 1730 of L'Hospital's *Analyse des infiniment petits, pour l'intelligence des lignes courbes* of 1696, Edmund Stone used 'differential' and 'increment' synonymously (either term being able to be used for a fluxion), and also declared his intention to replace the foreign d by Newton's $\dot{x}$ — this despite the admonition fifteen years before against identifying the finite velocity $\dot{x}$ with the infinitesimal dx (see Anon. [$171\frac{4}{5}$]). Taking more care than Stone, Hodgson pointed out the difference between the differential and the fluxionary methods in 1736 as follows[23].

> The *Differential* Method teaches us to consider Magnitudes as made up of an infinite Number of very small constituent Parts put together; whereas the *Fluxionary* Method teaches us to consider Magnitudes as generated by Motion. [pp. v–vi]

Writing in 1852 de Morgan drew attention to the early equating of differentials and fluxions and the confusion this engendered in fluxionary writings[24]. That this persisted to late in the nineteenth century is shown by the following quotation from Buckingham.

> The function which Leibnitz terms 'differential' and which Newton designates as a 'fluxion' is the concrete symbol which represents the rate of change in the variable. [1880, p. 42]

In his *Observations on Man, His Frame, His Duty, and His Expectations*, Hartley shows more than a slight acquaintance with Newton's differential method, comparing it 'with that of arguing from experiments and observations, by induction and analogy'. He compares the finding of a general curve fitting specified ordinates and abscissae to the drawing of general conclusions from particular effects, noting that

> the mathematical conclusion drawn by the differential method, though formed in a way that is strictly just, and so as to have

the greatest possible probability in its favour, is, however, liable to the same uncertainties, both in kind and degree, as the general maxims of natural philosophy drawn from natural history, experiments, &c. [1749, Part I, Chap. III, Sect. II]

11.5.7 Numbers

Several pages of the notebook are filled with numbers. In no case, however, is any verbal explanation provided, so that it is only by trying to identify the various patterns that we can arrive at the matters under investigation.

The first set of numbers, [C: 75(a)L–76L], is labelled 'Logarithmi' in the table of contents. Here, to 23 decimal places, are logarithms to the base 10 of all integers in the set $\{10^k - n, 10^k + n\}$ with $k \in \{3, 4, \ldots, 11\}$ and $n \in \{1, 2, \ldots, 9\}$. At the foot of p. 75(a)L values for $2^n M/n$ are given for $n \in \{1, 2, \ldots, 6\}$ ranging from 20 decimal figures (for $n = 1$ or 2) to 6 (for $n = 6$). The value $2M = 0.86858\ldots$ shows that $M = \log_{10} e$ — i.e., M is the reciprocal of the modulus of Briggs's system of logarithms to base 10. Although there is a reference, on [22R], to 'Tabulas Sherwini', the figures given in these tables do not appear to be from Henry Sherwin's *Mathematical Tables* of 1706.

The second item, [87R], consists of numbers set out in the form $kN = k$, where $k \in \{1, 2, \ldots, 9\}$ and $N \in \{199, 198, \ldots, 176\}$. I can proffer no explanation for the consideration of these figures. There is a possible connexion with the matter, pertaining to harmony, discussed on [81L] (see §11.6.4), a footnote there saying 'vid. P. 87': but the precise connexion is not clear to me.

The third numerical item, [90(a)L–90(b)R], is concerned with discovering whether every number of the form $4n + 2$, for $n \in \{0, 1, \ldots, 250\}$, can be written, not necessarily uniquely, as a sum of three squares. (Note that '.' rather than '+' is used here to denote addition.) In each case Bayes successfully manages the decomposition. The result, of course, is merely a special case of the theorem[25] stating that every positive integer n can be written as a sum of three squares if and only if n is not of the form $4^a(8b + 7)$, where $a \geq 0$ and $b \geq 0$; clearly $4n + 2$ is not of this form.

On [100R] three number-theoretic problems are considered, viz.

(i) If 4 Numbers are proportional y^e sum of their Squares are = to y^e sum of 2 Sq^{re} N^{rs} ;

(ii) 2 Sq^{re} N^o ×^{d} into 2 Sq^{re} N^o make a product = to 2 Square numbers;

(iii) If 2 Sq^{re} N^{os} can be divided by y^e sum of 2 Sq N^s y^e quotient will be composed of 2 Sq N^s.

Supposing that $a : b :: c : d$, Bayes notes that in the first case

$$(a-d)^2 + (b+c)^2 = (a+d)^2 + (b-c)^2 = a^2 + b^2 + c^2 + d^2,$$

and in the second

$$\begin{aligned}(a^2+b^2) \times (c^2+d^2) &= a^2c^2 + b^2c^2 + a^2d^2 + b^2d^2 \\ &= (ac-bd)^2 + (ad+bc)^2.\end{aligned}$$

This we recognize as a special case of the Lagrange identity

$$\left(\sum_1^n x_i^2\right)\left(\sum_1^n y_i^2\right) - \left(\sum_1^n x_i y_i\right)^2 = \sum_{i<j}(x_i y_j - x_j y_i)^2$$

(see Beckenbach & Bellman [1971, p. 3]).

However, the proof of the third problem is defective. Bayes first notes (correctly) that if $(M^2+N^2)/(a^2+b^2) = G$, then $(M^2 a^2 - N^2 b^2)/(a^2+b^2) = (M^2 - b^2 G)$. He then states

> W^{ref} if a^2+b^2 divides M^2+N^2 it will divide also $M^2 a^2 - N^2 b^2$ & conseq^{tly} it will also divide either $Ma - Nb$ or $Ma + Nb$.

Supposing $Ma - Nb$ to be divisible by $a^2 + b^2$, Bayes sets $Ma - Nb = h(a^2+b^2)$ and $Mb + Na = (p/q)(a^2+b^2)$, where p and q are presumably integers. Squaring then results in $G = (M^2+N^2)/(a^2+b^2) = h^2 + p^2/q^2$, and it is argued that, since G and h^2 are integers, p^2/q^2 is also an integer. The second possibility, in which $Ma + Nb$ rather than $Ma - Nb$ is assumed to be divisible by $a^2 + b^2$, follows similarly.

This argument, however, is faulty. Firstly, note that the statement

$$(a^2+b^2) \text{ divides } (M^2 a^2 - N^2 b^2)$$

does not necessarily imply that

$$(a^2+b^2) \text{ divides } (Ma - Nb) \text{ or } (a^2+b^2) \text{ divides } (Ma+Nb);$$

for example, let $a = 6$, $b = 7$, $M = 77$, $N = 36$. Furthermore, the example $4(1^2+1^2) = 2^2 + 2^2$ shows that G ($= 4$ here) need not be the sum of the squares of two (non-zero) integers. Of course, one can always argue as follows; from Bayes's problem (ii),

$$(a^2+b^2)(c^2+d^2) = (ac-bd)^2 + (ad+bc)^2.$$

Now let $M = |ad+bc|$ and $N = |ac-bd|$, and find c and d. This will provide an answer to problem (iii), though not necessarily, as we have seen, one in which c and d are positive integers.

The next item in this category, [101L–102L], is headed 'odd N^{os} = 2 Sq^{rs} not end^{ng} in 5'. A footnote on [102L] says 'Vid. Pag. 112.', and the sequence under examination is in fact continued on [112(a)R to 112(d)L]. The numbers whose decompositions are sought range from 1 to 3,013, and all seem to be of the form $4n+1$. Some perfect squares (e.g., $49 = 0^2 + 7^2$) are missing from the table, whereas others (e.g., $169 = 0^2 + 13^2$) are given. In other instances the desired decomposition into a sum of two squares is not achieved: for example, we find $1797 = 3 \times 599$ (it should be mentioned that Bayes uses both '.' and '×' to denote multiplication here) and $1953 = 7.9.31$. Other possibilities are missing — for example, $145 = 1^2 + 12^2$ — but on the whole a fairly complete coverage of integers of the form $4n+1$, in the range 1 to 3,013, that are expressible as the sum of two squares, is given.

The failure of some of the attempted decompositions is not surprising. It is known (see Landau [1966, p. 129]) that a positive integer n can be written as the sum of two squares if and only if n has no prime factor of the form $4m+3$ of odd multiplicity.

On what theory Bayes constructed his table is unknown. He might have noted that, for $4n+1$ to be representable as x^2+y^2, exactly one of x and y must be even. This leads to $4n+1 = 4(u^2+v^2+u)+1$, or $n = u^2+v^2+u$. It would then only be a matter of letting u and v 'run' to obtain the lists. But this is mere conjecture.

The final item is to be found on [113(a)L to 113(i)R; 124R to 125R]. The first page is headed 'Tabula 1^{morū} & compositorū sine 2.3.5.7.' (where '1^{morū}' probably stands for 'primorum') whereas the heading on [124R] is 'vid. antea pag. 113.' The table lists the odd prime numbers up to 10,001: occasionally an odd number that is not prime is given with one or all of its factors, for example, $(851 = 23n), (917 = 7n)$, and $(8143 = 17\ )$.

On [79R] Bayes considers the Diophantine equation $Mz - Ny = 1$. The passage seems to be related to what follows, and we accordingly postpone its consideration to the next subsection.

11.5.8 Miscellaneous mathematics

There are a number of mathematical topics discussed in the notebook that receive but scant attention. We consider all of these briefly in this section.

The first passage, [C: 55R], is described as 'Item datis Log,M & Log,N invenire Log,$M + N$ & Log,$M - N$ per Tab. Sinuū &c.' Three different methods are given here, the whole argument running as follows.

$$M : N :: R^2 : s^2, a \ \& \ M - N : N :: \cos^2, a : s^2, a :: \cot^2, a : R^2.$$
But if $M : N :: R^2 : T^2, a \ \ M+N : N :: R^2 : s^2, a.$ Or thus make
$$M : N :: R : s, A \ \& \ M + N : N :: \cot, \frac{90-A}{2} : T, A \ \& \ M - N : N :: T, \frac{90-A}{2} : T, A.$$

We note next a passage, [79R–80(a)L], on number theory, a passage that begins with the following statement.

> Suppose $Mz - Ny = 1$, M & N be whole numbers the least in y^e same ratio, & z & y whole numbers also & it is required to find z & y from M & N given.

Bayes considers two examples: in the first $M = 277$, $N = 67$, and in the second $M = 1000$, $N = 99$. No general method for the solution of this linear Diophantine equation is suggested, but it seems that Euclid's algorithm is used (see Burton [1980]). This is followed by three corollaries, which, together with their proofs, run as follows.

> Cor. To find a number which M perfectly divides & divided by N leave 1 for a remainder.

Sketch of proof: if Mz is the desired number, then $(Mz - 1)/N = y$, and integral values of y and z can be found by the main result 'if M & N are the least in y^e same ratio, & otherwise y^e problem is impossible.'

<u>Remark</u>: the conclusion quoted would be expressed today by saying that the linear Diophantine equation $Mz + Ny = c$ has a solution if and only if the greatest common divisor of M and N divides c.

> Cor. 2. To find a number which divided by M leaves 3 for a remainder & divided by N leaves 2.

Proof: let $Mz - Ny = 1$ and $Nv - Ms = 1$. Then $2Mz + 3Nv$ is a number of the required form.

> Cor. 3. In like manner you may find a number which divided by M leaves m & divided by N leaves n wch will be $nMz + mNy$.

The next passage, [85L–85R], in this miscellaneous section is devoted to the solution of several 'fluxion-difference' equations; viz.

(i) $\dot{x}/x = -\dot{z}/zz,$ (ii) $\dot{x}/x = -z/zz,$

(iii) $\dot{x}/x = -2\dot{z}/zz\,z,$ (iv) $\dot{x}/x = -nz/zz.$

In each case the solution x is given in the form of a series one may write as $\sum \alpha_n f(1/z, 1/z\!\!\smile, \ldots, 1/\underbrace{z\!\!\smile \cdots z}_{n})$ (the series for (i) also involves $\underset{\smile\smile}{z}$), under the condition that z is a uniformly flowing quantity. Elsewhere in the notebook it is clear that $\underset{\smile}{z}$ stands for $(z+1)$, $\underset{\smile\smile}{z}$ for $(z+2)$, etc. Although this may be meant here, it is also possible that the under-written grave accents merely indicate successive terms — e.g., $\underset{\smile}{z} = z + h$, $\underset{\smile\smile}{z} = z + 2h$, etc.

Two passages in the notebook are devoted to continued fractions. The first, [C: 105L–105R], is labelled 'Of continued Fractions'. The functions expanded in such fractions are y/x and $\sqrt{1+q/p}$, together with a terminating expansion for x from $x^2 = a^2 + a^2 q/p$.

There are some results here that seem to have little to do with the topic. They run as follows,

(i) let $a^2 - pb^2 = m$ and $p\varepsilon^2 - \delta^2 = g$. Further, let

$$y = (2pb\varepsilon - 2a\delta)/m, \quad x = (2a\delta - 2pb\varepsilon)/g.$$

If $E = \varepsilon + yb$ and $D = \delta + ya$, then $pE^2 - D^2 = g$. Similarly, if $A = a + x\delta$ and $B = b + x\varepsilon$, then $A^2 - pB^2 = m$.

(ii) If $A + z : B :: C : D + z$ & $D - nC = E$ & $E + z : C :: F : G + z$, then $G = A + D - E = A + nC$ & $G = B + n(E - A)$.

The second passage on continued fractions, [122L–122R], is concerned with the identity

$$C = x = (n+1) \times \frac{(n+1)(n+5)}{(n+3)^2} \times \frac{(n+5)(n+9)}{(n+7)^2} \times \frac{(n+9)(n+13)}{(n+11)^2} \ \&c.,$$

where C is the continued fraction we might write as

$$C = n + \frac{1}{2n\ +}\ \frac{9}{2n\ +}\ \frac{25}{2n\ +}\ \frac{49}{2n\ +}\ \&c.$$

The values denoted by x^a, x^b, x^c, and x^d are obtained by setting $n = 0, 1, 3$, and 2, respectively, in the infinite product, x^c turning out to be twice Wallis's product for $\pi/2$. More generally, by choosing n to be $2p - 1$ Bayes obtains an expression for $\sqrt{x}$ as the product

$$\sqrt{x} = \sqrt{2} \times \frac{p}{p+1} \times \frac{p+2}{p+3} \times \frac{p+4}{p+5} \times \frac{p+6}{p+7} \times \sqrt{p+2n-4}.$$

A similar expression is given for $\sqrt{x}$ when $n = 2p$, and the ratio of the two (up to the fourth term) is given as n tends to infinity.

The connexion between continued fractions and infinite series was made by Euler in 1748 (Jones & Thron [1980, §2.3.1] and Kjeldsen [1993, p. 22]). It is also possible to express an infinite series as an infinite product (see Knopp [1928, §30]), and so Bayes's equivalence seems acceptable.

The next miscellaneous passage, [115L], is headed 'According to Ld Stanhopes notation'. A recent discovery by David Bellhouse in the Centre for Kentish Studies in Maidstone, England, of manuscripts considered elsewhere in the present book indicates that the 'Stanhope' referred to is Philip Dormer, fourth Earl of Chesterfield and second Lord Stanhope (1694–1773), who was noted for his mathematical ability[26]. Samples of the notation are:

$$\begin{aligned}
{}^1\boxed{abcd}^m &= a \times a^m + b \times b^m + c \times c^m + d \times d^m \\
&= a^{m+1} + b^{m+1} + c^{m+1} + d^{m+1} \\
{}^2\boxed{abcd}^m &= ab(a^m + b^m) + ac(a^m + c^m) + \cdots + cd(c^m + d^m) \\
{}^1\boxed{abcd}^0 &= a + b + c + d \\
{}^3\boxed{abcd}^0 &= 3abc + 3abd + 3acd + 3bcd.
\end{aligned}$$

The final item in this miscellaneous section, [120L–120R], is headed 'De methodo Falsæ Positiones'. The first equation considered is $Ax^2 + Bx = D$, and a second example is devoted to the system

$$z = ax + by, \qquad w = \beta x - \gamma y$$

with x and y being sought when $z = M + d$ and $v = N + \varepsilon$ (is $v = w$?). Two corollaries conclude this passage.

11.6 Natural philosophy

11.6.1 Electricity

This item, [C: 51(a)R–51L], listed in the table of contents merely as 'Item of Electricity', is headed 'Paris. July 4. N.S. 1746'. It turns out to be an extract from a letter from Turbervill Needham[27] to Martin Folkes, which was published in the *Philosophical Transactions* in 1746. The letter deals with some experiments made with an electrifying spheroid by Le Monnier[28]. Eleven experiments were reported in the original paper, though not all of these are recorded by Bayes. Those that are noted, however, are given in some detail. Several passages are in shorthand.

460   11. The Notebook

In the following paragraphs we quote, as much to draw attention to Bayes's interest in this matter as to honour Le Monnier's pioneering work, some of the passages copied into the notebook. For the rest of this subsection, passages in crotchets, [...], are words or letters either omitted by Bayes or given by him in an alternative form (the distinction should in every case be clear), and passages in braces, {...}, are given in shorthand by Bayes. Words in italics are Bayes's additions, whereas the footnotes on these pages (indicated by parenthesized common footnote symbols) are my clarifications or comments on the text.

'The Electrifying glass used by Monsieur le Monnier is an oblong Sphæroid, whose diameter from pole to pole is 4 or 5 inches longer y^n. that at ye Equator which is about twelve inches. Each of these poles is terminated in a stem, or portion of a hollow Cylinder about 3 inches in length, & one in diameter, spirally embossed on ye outside into a large male screw: to each of these male screws is adapted a female screw of wood closed at one extremity, with a piece of steel excavated in ye center, to receive ye steel pivots upon wch. ye Electrifying Glass turns. These female screws of wood are so formed at their open extremity, y^t. they grasp & cover as much at ye Poles as nearly renders w^t. appears of ye glass sphæroid a perfect Sphere. This wth a design y^t. ye wood may fix ye more effectually & embrace ye Electrifying glass. From ye exterior surface of one of these wooden female screws a circular ledge rises & projects to y^e height of about 2 inches, y^e ambitus of wch. ledge is excavated to recieve y^e cord that turns y^e. Electrifying glass. ...

'The Electrifying Sphæroid is turned by means of a Wheel near five inches [four feet]$^{(*)}$ in diameter with the same motion, & exactly in y^e same manner as y^e spindle is turned round by y^e spinning wheel, allowing a due proportion to y^e frame upon wch. y^e glass sphæroid is mounted, that it may answer to y^e wheel y^t turns it. Y^e sides of y^s. frame wch. stand perpendicular to y^e horizon are near as strong {and} as large every way as y^e posts of ordinary closet door, [and] wth y^e ledges y^t join them at top & bottom form a rectangular Pgr^m. $^{(\dagger)}$

'The front of y^s. frame is provided with silken loops conveniently disposed in several places to bring to, & fix at a contact with y^e. Electrifying glass, wires, thread[s] packthread or w^{tever} else is to be Electrifyed. Into one side of this frame at about $\frac{1}{2}$ its height, the pivot y^t. receives one of y^e

---

$^{(*)}$ The slip here is forgiveable: in the original paper the words 'four feet' appear on p. 248, and directly across from them, on p. 249, are the words 'five inches'.

$^{(\dagger)}$ parallelogram

Poles of y^e glass Sphæroid is fixed. y^e other pivot on y^e opposite side is a round long bar of Iron screwed into & passing thro' the post in order to fix or give liberty of removing y^e Electrifying glass.

'This bar of Iron for the convenience [conveniency] of turning it has another in ye nature of a lever, wch passes thro' its extremity at right angles {with it}. The whole Machine is mounted upon a floor of boards, wheel, frame, glass &c. & employs two men, one to turn ye wheel & another to sit behind ye glass Sphæroid, & apply ye concave of each hand to its lower convex surface; for it is by y^s friction y^t ye Electricity is excited. W^n ye Electrifying glass has been sometime in motion, ye person who desires to be Electrifyed, applys ye extremities of ye nails of one hand, & stands not upon cakes of wax as in England but within ye area of a Square drawer or box about 5 inches deep & filled with 5 parts pitch, 4 of resin & one of bees wax. ... ye pitch is placed next to ye sides of ye box & rises almost to a level with them, the resin in ye middle is on a level with ye pitch & ye wax forms a thin surface covering both to a level with ye box it self. {However I suppose this to be in it self very indifferent, and that any one body of the} Electrics per se {would answer equally}.

'1. Exp. The person Electrifyed by y^s Machine, not only emits fire from all parts of its [his] body upon y^e touch of another wth more vigor, & in a much more sensible manner, thā w^n electrifyed by a common tube; but fires [also] sp$^{t(*)}$ of wine wth such ease, that w^n y^e Sp^{ts} have been but once simply set on fire by a match or lighted paper, & y^e flame has been instantly blowed [blown] out, they will wth that small degree of heat y^{ey} have acquired take fire by his touch 10 or 20 times successively without failing once. {I am told here, that they have frequently attempted in vain to fire} Sp^{ts} {with a common} tube of Glass; {so that I believe the use of the tube has been more improved in} England {than in any other place; but it is a down-right slavery, and much inferior in its effects [in its effects many degrees inferior] to this machine. I should have thought, as this [so] much exceeds [in strength] the common tube, that many glass} Sphæroids {acting at once upon the same body, would have considerably increased the effect; but} monsieur de Buffon {tells me, that} [Monsieur] le Monnier {had found, upon trial, that they answered not his expectations; so that it might seem *that* there is a} ne plus ultra {in the} intensity of Electricity {as well as in the heat, which is communicated to boiling water}.

$^{(*)}$ spirits

'2.$^{(*)}$ If y$^\text{e}$ pson$^{(\dagger)}$ Electrifyed holds a sword in one hand, y$^\text{e}$ chamber being darkened, a continual flame issues out at y$^\text{e}$ point in smell & color resembling y$^\text{e}$ fumes of Phosphorous & near as strong as y$^\text{t}$ of an enamellar's lamp; with y$^\text{s}$ differ$^\text{nce}$ y$^\text{t}$ w$^\text{n}$ any other of y$^\text{e}$ company applys a hand even to y$^\text{e}$ very point w$^\text{re}$ y$^\text{e}$ concentred rays begin to diverge it burns not nor is any otherwise sensible to y$^\text{e}$ feeling, y$^\text{n}$ as a continual blast of wind.

'4. Exp. [The most surprising of all,] is that of [Mr] Muschenbroeck improved by [Monsieur] le Mo͠nier a muschet [musquet] barrel is suspended ∥$^{(\ddagger)}$ to y$^\text{e}$ horizon by silken threads within reach. At y$^\text{e}$ breech end about 3 inches fm$^{(\S)}$ y$^\text{e}$ extremity is hung by a ring of iron worked into y$^\text{e}$ barrel it self a small iron chain ab$^\text{t}$ $\frac{1}{2}$ a foot in length. A glass Phial resembling in size & shape a common vinegar crewet is y$^\text{n}$ prepared full of water & well corked with an iron wire running thro' y$^\text{e}$ cork almost to y$^\text{e}$ bottom, & emerging some 2 or 3 inches above it out of y$^\text{e}$ top of y$^\text{e}$ Phial. The head of this wire is bent, to catch in y$^\text{e}$ lowest link of y$^\text{e}$ chain, & is to be y$^\text{re}$ suspended when it has been Electrifyed. From y$^\text{e}$ mouth of y$^\text{e}$ barrel w$^\text{ch}$ is pointed in a line parallel to y$^\text{e}$ Equatorial plane of y$^\text{e}$ revolving Sphæroid comes a long iron wire inserted into y$^\text{e}$ barrel it self as far as $\frac{1}{3}$ of its length & thence proceeding till it touches y$^\text{e}$ glass Sphæroid, to a contact w$^\text{th}$ w$^\text{ch}$ it is determined by one of y$^\text{e}$ silken loops [I] mentioned above in y$^\text{e}$ description of y$^\text{e}$ apparatus. Every thing [being] thus disposed the gun barrel is to be Electrifyed by repeated revolutions of y$^\text{e}$ glass Sphæroid, w$^\text{ch}$ is to be in [a] continual contact w$^\text{th}$ y$^\text{e}$ [long] wire y$^\text{t}$ proceeds from it. The Phial is at y$^\text{e}$ same time to be Electrifyed by y$^\text{e}$ Operator, who takes hold of y$^\text{e}$ body of y$^\text{e}$ barrel & applies to y$^\text{e}$ [electrifying] Sphæroid y$^\text{e}$ bent extremity of y$^\text{t}$ wire w$^\text{ch}$ passes from near y$^\text{e}$ bottom of y$^\text{e}$ vessel [phial] thro' y$^\text{e}$ cork. ... The operator must take care not to touch y$^\text{e}$ wire it self, {whilst [while] he endeavours to electrify the} Phial, {otherwise he will [would] be in the case of one, who should endeavour [aim] to electrify himself, without standing upon some [one of the] body [bodies] electrified [that are electrics]} per se. W$^\text{n}$ y$^\text{e}$ Phial is sufficiently electrified w$^\text{ch}$ will be done in ab$^\text{t}$ 8 or 10 revolutions of y$^\text{e}$ Sphæroid, for I wou'd not have any one [be] too free in bestowing such an efficacy upon it by too long an application as might perhaps occasion his receiving a more violent shock than he wou'd be willing

---

$^{(*)}$ i.e., Le Monnier's second experiment.
$^{(\dagger)}$ person
$^{(\ddagger)}$ parallel
$^{(\S)}$ from

to feel prtly$^{(*)}$ if y^e glass sphæroid has been any time in action & has been heated thereby: y^e Phial is y^n to be suspended by y^e iron chain, the glass [spheroid] still continuing to revol[ve] & to electrify y^e gun barrel. The pson then who has courage eno' to suffer y^e experim^t grasps y^e bottom of y^e [electrified] Phial with one hand & with ye other touches ye gun barrel. At that instant a great part of ye nervous system receives such a shock [a shock so violent] y^t it wou'd force ye strongest man to quit his hold & turn him half round. ... A boy ~~thought~~ said he imagined ye instant he touched ye gun barrel his arms had been broke[n] short of[f] at ye elbows & y^t he had been cut in two parts just below ye breast. {another of the company, with a sort of pun, termed it being broken upon the wheel.} In effect so far the boy seemed [was] in ye right, y^t ye shock in ye arms seems to extend no farther y^n ye elbows, & y^t of ye body no lower than ye breast without affecting however [in the least] ye head or seeming to reach beyond ye outwd expansion of ye nerves: yet it is not to be termed a pain {for there is not the least sense of that sort in it,} but a more sudden convulsionary motion, or rather a shock, w^{ch} [surprises much, and] is indeed an uneasy but not a painful sensation. {In this experiment the improvement of} le Monnier {consists in the invention and application of the electrified} Phial {which considerably augments the force of the communicated electricity.} [In this Experiment, it is very remarkable how greatly the Force of the communicated Electricity is augmented, by the Application of the electrified Phial:] {But the most surprising property of the} Electrified Phial is y^t it loses not its Electricity for several minutes, {and I am told} in a frost will retain it six & thirty hours. [But the most surprising circumstance attending the use thereof, and which, I believe, is, among all the bodies that are susceptible of Electricity, peculiar to this alone, is, that it loses not entirely its Efficacy under several Minutes; and I am told, that in a Frost it will retain it for six-and-thirty Hours together.] le Monnier {has electrified it at home, and brought it in his hand through many streets from the College of} Harcourt {to the} King's Garden {without any sensible diminution of its efficacy}. [Monsieur de Buffon, who informed me that Monsieur le Monnier was the first who discovered this Particular, has also assured me, that this same Gentleman, had frequently electrified the Phial at home, and brought it in his Hand through many Streets from the College of Harcourt in the King's Garden[29], without any very sensible Diminution of of its Efficacy.]

$^{(*)}$ particularly

(*) 'At y^e grand convent of y^e Carthusians at Paris y^e whole community formed [a] line of 900 toises by means of iron wires between every [two ...] y^e effect was y^t w^n y^e 2 extremitys of y^s long line met in contact with y^e electrified Phial y^e whole company at y^e same instant gave a spring & felt y^e shock. ...

'Le Monnier [The other Phænomenon was the result of a late Experiment of Abbè Nollet's[30]. He] fixed at y^e 2 extremitys of a brass ruler a sparrow {and} [a] chaffinch. y^s ruler had a handle or pedestal fixed to it in y^e middle f$^{(\dagger)}$ y^e conveniency [convenience] of holding it. W^n y^e gun barrel & Phial had been sufficiently electrifyed he applied y^e head of y^e sparrow to y^e suspended Phial & y^e head of y^e chaffinch to y^e barrel. Y^e consequence was upon y^e 1st trial they were both struck lifeless & motionless but recovered a few minutes afterwds. upon ye 2^d trial ye sparrow was struck dead & upon examination found livid without as if killed wth a flash of lightning most of ye blood vessels within being burst with ye shock ye Chaffinch revived as before.'

11.6.2 *On the weight of a body*

The next pertinent passage, [C: 52R–53L], is described as follows: 'De pondere aquæ pluvialis. Item datis pondere globi in ære & in aquâ invenire diametrū globi & pondus ejus in vacuo.' In fact the passage itself covers somewhat more than is suggested by this description.

In the first problem, assuming the density of water (in Troy ounces per cubic inch) and the weight (in grains) of a sphere of unit diameter and filled with water in air and in a vacuum to be known, Bayes finds expressions for the weight *in vacuo* and the diameter of a sphere whose weight in water and in air is known. Notice that Bayes begins his discussion by saying

> Aquæ pluvialis digitus solidus (pedis Londinensis) continet 19/36 uncias libræ Romanæ seu grana $253\frac{1}{3}$ & Globus aqueus cujus diameter est digitus unus continet grana 132,645 in medio aeris vel grana 132,8 in vacuo.

These figures allow the identification of 'uncias' as Troy ounces, since the equating of (19/36) uncias to $253\frac{1}{3}$ grains shows that 1 grain is equal to (1/480) uncias, and it is known that 1 Troy ounce = 480 grains (note that on page 29L Bayes states that 'The ancient & modern Roman Ounce containeth 438 English grains'). As regards the conversion of the density

---

(*) The eleventh experiment is succeeded by the following remarks.
($\dagger$) for

to the modern value of 62.4 lbs/foot^3, note that several sets of weights and measures are given in the notebook (see §11.8.2), a number of different ratios for the conversion of Troy weights to Avoirdupois being recorded — for instance, on page 29R we have

(a) the English Avoirdupois Pound weighs 7004 Troy grains (rather than the more usual 7000 grains);

(b) the Avoirdupois ounce (16 to the pound) is 437.5 Troy grains (should be 437.75);

(c) the Troy pound to the Avoirdupois pound as 88 to 107;

(d) the Troy ounce to the Avoirdupois ounce as 80 to 73;

and on page 88R we find

(e) the Troy pound to the Avoirdupois pound as 14 to 17;

(f) the Troy ounce to the Avoirdupois ounce as 56 to 51.

It is thus not easy to try to convert Bayes's Troy weights to Avoirdupois, and we content ourselves here merely with noting, in addition to what we have already said, that 62.4lbs/foot^3 is equal to 252.9$\dot{2}$ grains/inch^3 (compare Bayes's figure of $253\frac{1}{3}$ given above)[31]. Moreover, Bayes's figure of 132.645 grains comes from

$$4/3 \times \pi \, (d/2)^3 \times 253\tfrac{1}{3},$$

with diameter $d = 1$.

In the second problem, which follows close on the first, the falling of a body under gravity is considered. An expression is derived for the time it takes for a body of known mass to fall, with known terminal velocity, through a given distance.[32]

11.6.3 Optics

There are three passages in the notebook dealing with this topic. In the first of these, [C: 50L–50R], entitled 'Of ye image seen thro several glasses' in the table of contents, the following problem is discussed: a linear object is placed parallel to a sequence of concave glasses of known focal distances. Where is the final image, and what is its size? Bayes approaches the problem via series, noting that the solution obtained will hold equally well for convex glasses, provided that focal distances are regarded as negative in this case.

There is an annotation 'See Smith's Optics' at the top of page 50R.

The reference is to Robert Smith's *A Compleat System of Opticks* [1738], and although the extract in the notebook is not taken from this work, the matter is discussed there in Book 2: *A Mathematical Treatise*, chap. 5, entitled 'To determine the apparent distance, magnitude, situation, degree of distinctness and brightness, the greatest angle of vision and visible area, of an object seen by rays successively reflected from any number of plane or spherical surfaces; or successively refracted through any number of lenses of any sort, or through any number of different mediums whose surfaces are plane or spherical. With an application to Telescopes and Microscopes.'

The second item on optics, [C: 56L–56L], is to be found on pages that are curiously numbered: a left-hand page numbered '56' is followed in turn by a right-hand '56(a)', '56' (L), '56(b)' (R), and '56' (L). The table of contents lists '56. Data Sphærica superficie dirimente duo media invenire focos'. This in fact occupies the final p. 56 (with the addition of the word 'præcipuos' after 'focos'), the preceding pages being devoted to a problem given on the first p. 56 as follows. 'Prob. generale. Dato foco radiorum incidentiū fere normaliter in superficiem sphæricā invenire focum post refractionem.' Two situations are considered here, viz.

> Quando radii incidant ex medio rariore in densius seu magis refringens casus sunt sequentes,

and

> Quando radii exeunt ex medio densiore in rarius seu minus refringens casus sunt sequentes.

Under each of these conditions several cases are considered: for the first condition the cases are:

> Cas. 1. Quum radii incidant paralleloos ~~in superficiem convexam~~;
> Cas. 2. Quum radii divergentes incidant in convexam, aut convergentes in concavam;
> Cas. 3. Quando radii convergentes incidant in superf. convexam, aut divergentes in concavam;

and the cases for the second condition are

> Cas. 1. Quum radii incidentes sunt paralleli;
> Cas. 2. Quum radii divergentes incidunt in convexam aut convergentes in concavam;
> Cas. 3. Quū radii convergentes incidunt in convexam aut divergentes in concavam.

In each case the (very short) proof that is given makes use of the ratio of the sine of the refracted ray to that of the incident ray. Some attention is also given, on the last page 56, to the question of conjugate foci.

In the third, and final, entry on optics, [102R], Bayes considers some relationships between (i) the sines of the angles of incidence and refraction, and (ii) the velocities of two differently coloured rays before and after refraction, special attention being paid to the passage of the rays (i) from air to glass, and (ii) from air to water. As an example we mention the following. Consider two differently coloured rays, say red and violet, passing the same refracting surface. Let I and y denote the sines of the (angle of) incidence of the red and violet rays, respectively, let R and r denote the sines of the respective angles of refraction, and let W and V denote the respective velocities before incidence. Then

$$W^2 : V^2 :: \frac{R^2}{I^2 - R^2} : \frac{r^2}{y^2 - r^2}.$$

11.6.4 Harmony

We have already mentioned (see §11.5.8) the passage, [79R–80(a)L], in which Bayes examines the solution, in integers, of $Mz - Ny = 1$, where M and N are relatively prime. This work seems to be undertaken only for its use in a study of some aspects of harmony[33], this investigation being found on [80(a)L–81R].

The first of the propositions, given without proof and following hard on the heels of the third corollary dealing with the above Diophantine equation, runs as follows.

> Prop. 1. Let V be y^e base, & v y^e treble of a tempered unison. & $V : v :: 1 + x : 1 :: p + px : p$. Then $px = 1$ (?). pV will be y^e period of y^e pulses of this tempered unison or y^e time between 2 successive beat.

In the proof of the second proposition some use of the preceding mathematical investigations seems to be made. The pertinent passage is as follows.

> Prop. 2. V & v being as before $3V$ will be y^e Base & $2v$ y^e treble of tempered fifth, & y^e period of the least imperfection of this fifth will be y^e same with that of y^e preceding unison very nearly. For $3V : 2v :: 3 + 3x : 2 :: 3P + 3Px : 2P$. Then if P be a number divisible by 6. The proof proceeds thus. Whilst $3V$ makes $\frac{2P+6}{6}$ vibrations $2v$ makes $\frac{3P+12}{6}$ vibrations, which are whole numbers because P is divisible by 6. W^{refore} $P = 6q$. W^{ref} after $3V$ has made $2q + 1$ vibrations, The vibrat^{ns} of y^e base & treble of y^e fifth coincide ag^{n}. But whilst $3V$ makes $2q + 1$

vibrations V makes $6q + 3 = P + 3$, & $\text{y}^{\text{t}}_{\cdot}$ is nearly y^{e} time of the period of V, $\text{y}^{\text{s}}_{\cdot}$ period being PV & that time $\overline{P+3} \times V$.

Further propositions, in similar vein and dealing with a tempered fifth, a perfect consonance, an octave divided into 5 mean tones and 2 limmas, etc., follow.

On [80(f)L] we find the following result.

> Harm. Pag. 111. Si T & $\frac{n}{m}T$, V & $\frac{s}{r}V$ sint vibrationes duarū perfect. consonantiū, $\text{q}^{\text{d}}_{\cdot}$ attemperantur, numeri pulsuū in temporibus $=^{\text{libus}}$ erunt in ratione compositâ ex inversâ ratione T ad V, directâ ratione m ad r & directâ ratione temperamentorū.

This, as we show, is almost a direct translation from a passage in Robert Smith's *Harmonics, or the philosophy of musical sounds*, first published in 1749. In that work it appears as Corollary 1, (p. 111), following Scholium 1, which in turn succeeds Proposition XI. Smith's seventh corollary to this Scholium, though this time unreferenced and in English, is given, slightly altered, on [80(f)R].

Robert Smith, educated at Leicester Grammar School and Trinity College, Cambridge, succeeded his cousin Roger Cotes as professor in 1716 and became Master of Trinity College in 1742. In addition to works on optics and on harmony (mentioned in the present work) he wrote on gunnery, gunpowder, the resistance of the air, and presented a proposal to increase the strength of the British Navy. His person was described[34] by one of his pupils, Richard Cumberland, as follows.

> [Smith was] of a thin spare habit, a nose prominently aquiline, and an eye penetrating as that of the bird, the semblance of whose beak marked the character of his face: the tone of his voice was shrill and nasal, and his manner of speaking such as denoted forethought and deliberation.

Eschewing comment on Smith's physical composition, Kassler records of his musical ability,

> Smith was not unmusical, for he was proficient as a performer on the violincello and harpsichord ... Smith thought that the greatest improvement made by modern musicians was in tempering the ancient scales. [1979, II, p. 949]

To see the relation between Smith's statements and Bayes's work we give some passages from the *Harmonics* here (proofs omitted).

The first relevant results are Proposition IX and its first corollary, which run as follows.

PROPOSITION IX.

If either of the vibrations of imperfect unisons, and any aliquot part or parts of the other, be the vibrations of an imperfect consonance, the length of the period of its least imperfections, will be the same part of the period of the imperfect unisons. [p. 99]

Coroll. 1. Fig. 37. If T and $\frac{n}{m}T$, V and $\frac{s}{r}V$ be the vibrations of any two perfect consonances, the periods of their least imperfections when tempered, will be in the ratio composed of the direct ratio of T to V, the inverse ratio of the denominators of the fractions $\frac{n}{m}$, $\frac{s}{r}$, and the inverse ratio of the temperaments. [pp. 100–101]

Several other corollaries, which need not concern us here, follow. The next relevant result is

PROPOSITION X.

When either of the sounds of a perfect consonance is tempered flat, the length of the period of its least imperfections is the same as it would have been, if the other sound had been equally tempered flat; and when either of the given sounds is equally tempered sharp, the period will be of another given length; and if the flat and sharp temperaments be equal, the former equal periods will be longer than the latter, in the ratio of the vibrations which terminate the given temperament. [p. 103]

Coroll. 1. The given vibrations of any perfect consonance being t and $\frac{n}{m}t$, the period of its least imperfections, when tempered flat or sharp by any part or parts of a comma, denoted by $\frac{q}{p}$, will be respectively $\frac{161p+q}{2q} \times \frac{t}{m}$ or $\frac{161p-q}{2q} \times \frac{t}{m}$ very nearly. [pp. 104–105]

The relationship between periods and beats is explained by Smith as follows.

Scholium 1.

The ratio of the number of beats made by any two consonances in any equal times, being the inverse ratio of the times between their successive beats, is also the inverse ratio of the periods of their least imperfections; and is therefore determined in all cases by the corollaries to prop. IX, which however are translated hither with that alteration, to prevent the frequent repetition of this observation. [p. 111]

This is immediately followed by

> Coroll. 1. If T and $\frac{n}{m}T$, V and $\frac{s}{r}V$ be the vibrations of any two perfect consonances, when they are tempered the numbers of their beats, made in equal times, will be in the ratio composed of the inverse ratio of T to V, the direct ratio of the denominators of the fractions $\frac{n}{m}$, $\frac{s}{r}$, and the direct ratio of the temperaments [p. 111]

which should be compared with Bayes's statement given above. Note too the relationship between Corollary 1 to Proposition IX and that just cited, whereas the analogue to Corollary 1 to Proposition X is given as the following corollary to the Scholium.

> Coroll. 7. Let the single vibrations of any perfect consonances be t and $\frac{n}{m}t$, and the numbers of them made in one second of time, be N and M respectively; then the number of the beats of that consonance, when tempered flat by any part or parts of a comma denoted by $\frac{q}{p}$, will be $\frac{2q}{161p+q} \times mN$, or $\frac{2q}{161p+q} \times Mn$ in one second of time very nearly; and when tempered sharp by the same quantity, the number will be $\frac{2q}{161p-q} \times mN$, or $\frac{2q}{161p-q} \times Mn$ in one second very nearly. [p. 113]

This last result is stated by Bayes [80(f)R] as follows.

> If T and $\frac{n}{M}T$; be y^e vibrations of any perfect consonance & y^e number of them made in one second of time be N and M respectively then y^e number of y^e beats of that consequence when either of y^e sounds are tempered sharp in y^e ratio of 1 to $1+x$ or flat in y^e ratio of 1 to $1-x$ in one 2^d of time will be xmN or xnM.

This is extended to a result not explicitly given by Smith in which the vibrations of two perfect consonances are compared, viz.

> If T & $\frac{n}{m}T$; V & $\frac{s}{r}V$ be y^e vibrations of two perfect consonances, then if one of the sounds of y^e former be either tempered sharp in y^e ratio of 1 to $1+x$ or flat in y^e ratio of 1 to $1-x$; & one of y^e sounds of y^e latter be either tempered sharp in y^e ratio of 1 to $1+z$ or flat in y^e ratio of 1 to $1-z$ y^e number of their beats made in equal times will be as xmV to zrT.

On page 87L there appears a further discussion on this topic, this time involving imperfect and perfect consonances. The page bears the shorthand heading 'part of a letter to Dr. Smith' (presumably a letter from Bayes).

Noting that opinions about Smith's *Harmonics* were divided in the period 1749–1830, and that this is still the case today, Kassler writes:

> As for Smith's theory of tuning, this likewise had its champions and detractors. Sir John Hawkins thought that it was 'far from being clear that any benefit can result to music from the division of the octave which Dr. Smith recommends; but this is certain, that his book is so obscurely written, that few who have read it can be found who will venture to say that they understand it.' Nor was John Worgan taken with the work, for having been a practical tuner early in life and having made this acquisition a partial means of his subsistence, he did not proceed far in reading the HARMONICS, soon finding 'it was not what I wanted.' [1979, p. 952]

11.7 Celestial mechanics

A fair part of the notebook is taken up with matters that, broadly speaking, fall under the above heading. Most of these passages contain rather complicated sketches, and to avoid filling the present work with figures essential to the discussions, we content ourselves with a somewhat cursory description of the relevant matters.

The first passage, [C: 10R–17R], in fact contains several problems, described in the table of contents as follows.

> a 10 usq$_3$ ad 14. De motu corporū in plano viribus agitat.
>
> a 14 ad 17. Data vi centrali invenire vim tangentialē ut corpus moveatur in curva data; & speciatim in Ellipsi.

As part of the first of these passages we find, on pages 11L and 11R,

> Defin. Velocitas ST, S_____T notet velocitatem qua corpus tempore dato e.g. unâ horâ percurrere potest lineam ST
>
> 2. Vis ST notē vim quæ uniformiter agens unâ horâ generabit velocitatem ST.
>
> 3. Si corporis utcunq$_3$ moventis centrum gravitatis describit lineam quamvis erit fluxio lineæ istius æqualis corporis velocitati, & 2da fluxio lineæ istius æqualis est vi qua corpus in via sua acceleratur aut retardatur Et eodem modo acceleratio aut retardatio areæ descripta est ejus fluxio 2$^{da}_{.}$
>
> 4. Corpore in plano movente fluxio 2$^{da}_{.}$ distantiæ ejus a rectâ quâvis æqualis est vi qua corpus accelaratur $^{versus}_{aut}$ retardatur a rectâ ista.

The next passage on this matter, [19L], consists of a single axiom in two parts, viz.

> Si corpus in plano movens describat lineā quamcunq$_3$ curvam, & vis totalis qua semper urgetur, resolvatur in vim versus rectā positione datam & vim cujus directio sit alteri normalis, erit vis versus rectā positione datā semper ut fluxio 2da distantiæ corporis a rectâ; accelaratur autem corpus versus rectā, nisi motus ejus sit a rectâ, & in illo casu retardatur.
> 2. Quantitas fluens non potest evanescere ad datū temporis momentū, nisi pro aliquo tempore ante momentum istud decrescit; neque potest fluens ex nihilo oriri nisi pro aliquo tempore a momento ortus crescit. Hæc patent quoniam ex naturâ Fluentis est, ut nullum accipiat instantaneū incrementum aut decrementū.

The next passage, [C: 21L–21R], described as 'De paralaxibus', presents expressions for the latitudinal and longitudinal parallax of a planet. Two expressions for the former are given, 'quando latitudo est Borealis & ... quando est Australis'.

Then follows a long passage, [C: 35L–48R], on celestial motion. The topics discussed are listed as follows.

> 35-40 De motu Nodorū Lunæ.
> 40-46 De motu nodorū Annuli.
> 47.48. De Elliptica Æquatione planetarū. Vide 51,

the entry for p. 51 being described as 'De. Ellipt Æquation. Planetarū vide etiam 47'. However, the problems discussed in the text do not fall neatly into the above classification, and we therefore describe them *seriatim*.

The first problem studied runs as follows.

> Ponantur lunā semper a Terrâ quiescente eandem distantiā servare & in orbitâ suâ uniformiter moveri; determinare motus ejus in longitudine & latitudine, ex data longitudine lunæ a nodo & orbitæ inclinatione ad Eccliptica.

The next two problems are related to this one: we do not give them here as fairly complicated sketches are needed for their understanding. Problems involving annuli start on page 39, and the work on the elliptic equation of the planets is continued, as stated, on p. 51L, the problems here being given as

> Methodus inveniendi M anomaliā mediā & V verā ex data distantia δ

and

> Invenire æquationem maximā.

11.7. Celestial mechanics

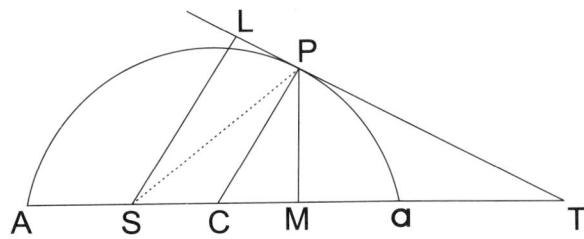

FIGURE 11.3

The next problem, [53R], with its accompanying sketch (here shown as Figure 11.3), runs as follows.

> Si corpus P movens in Orbita aPA vi tendente ad punctū S in suâ Orbita retineatur, tum punctis A, a positione datis ducatur recta $ASCMaT$ &c. Vocēturq$_3$ PM, y ; AM, x ; SM, z ; $AP = u$ eritq$_3$ ut notū $\ddot{y} : \ddot{x} :: PM : SM :: y : z$ & dico etiā erit $-\ddot{u} : \ddot{y} :: PL : PM$.

The next passage, [C: 55L–55R], is described as 'De motu Nodorū'. It is possible that this is a continuation of earlier work (perhaps from p. 48), though one should note that the beginning of the present passage appears, crossed out, at the foot of page 54R, and prefixed by 'Mem.'

The next passage, [64L–65R], opens with a sketch (see Figure 11.4) and the following words.

> Let y^e circle EBC represt y^e earth's Disk, i.e. its section by a plane thro' its center, & $\perp^r$ to y^e line joining y^e Center of y^e Earth & Sun; & upon y^s disk imagine all y^e enlightened Hemisphere of the Earth together with y^e centers of y^e Sun & Moon to be Orthographically projected.

Supposing further that

> M be y^e place of y^e Moon's Center on y^e Disk & Z y^e situatiō of an[y] particular place of y^e Earth upon it at y^e same instant of time (e.g. London),

Bayes proves several results, of which we mention the following.

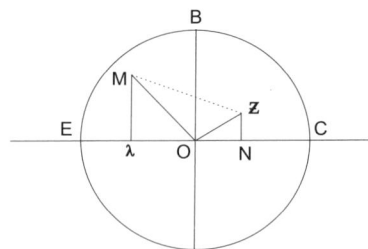

FIGURE 11.4

Cor. Hence the time of yᵉ true conjunction being given at any time near it yᵉ place M may be found upon yᵉ Disk if the Longitude of yᵉ Sun & Long. & Lat. of yᵉ moon be given.
Cor. 2. Hence yᵉ Sun's place, the Nonagessime degree, & its Altitude being given, & yᵉ time when yᵉ Sun is in Nonagᵐᵉ you have yᵉ Place of Z upon the disk.
Cor. 3. The place of M & Z upon the Disk being known to find yᵉ distance of yᵉ Sun's & Moon's centers view'd from yᵉ place in yᵉ Earth's surface of which Z is the projection.

When stating, [65R], that the Earth's diameter is to the Moon's as 1716 to 469, Bayes adds a reference 'vid. Smith's Optics Article 1168'. The reference is to Robert Smith's *A Compleat System of Opticks* of 1738.

We have already noted comment on Smith's *Harmonics*: Benjamin Robins, whose work on fluxions we have considered in an earlier chapter, was not enamoured of the *Compleat System of Opticks*: his *Remarks on Dr. Smith's Compleat System of Opticks* of 1739, republished in 1761 on pp. 222–227 of his *Mathematical Tracts*, criticized Smith severely. We mention only the following remarks, among many such, here.

> This treatise sets out with a very unskilful representation of a capital proposition in Sir Isaac Newton's Opticks, by omitting an essential part of it, which Sir Isaac Newton had expressed under the form of an exception. [§2, p. 223]
>
> There soon follows, at art. 17, another instance of this author's imperfect knowledge of the true theory of the action between light and bodies .... [§3, p. 223]
>
> However besides this gentleman's inconsistency, the method he has taken to determine the place of images made by reflection or refraction simply from the magnitude of the angle, under which they are seen, is certainly erroneous. [§49, p. 251]

11.7. Celestial mechanics

But in the second book we find perpetual proofs of this unskilfulness in regard to geometrical demonstrations. [§55, p. 256]
Indeed the demonstrations in the first chapter of the foci, besides the imperfection of the principle, upon which they are grounded, are on another account also inconclusive.
[§58, p. 258]

There then follows, [C: 65R, 67L–68R], a passage described as 'Data ratione inter tangentes arcuū invenire sinus eorum differentiæ', the matter being discussed within the context of celestial mechanics. (It is not clear whether all the matters discussed here should be regarded as forming one or several passages.) There are references on p. 68L to pages 48 and 54, the latter calling for an expression for $\cos^n f$. Mention is also made, on p. 65R, to *Ward's hypothesis*, the reference being most probably to the amendment of Kepler's second law proposed by Seth Ward, successor in the Savilian professorship to John Greaves.

I have found references to Ward's hypothesis hard to come by[35]; the following passage, from the first part, chapter 1, 'Astronomia Elliptica Principia quedam Generalia, & propositio operis designati', of the first book, *Astronomia Solaris*, of Ward's *Astronomiæ Geometricæ. Liber secundus de Astronomia Coelesti seu Reliquorum in Planetarum primariorum* of 1656 seems pertinent:

> Hujus Ellipseos, cum focus alter sit sol, (Motuum planetariorum verum atque physicum Instrumentum) super alterum interim focum, ita temperatur planetae cujusq;[36] motus, ut temporibus aequalibus, aequales illic angulos absolvat.
>
> Quare cum super focum *unum* Ellipseos sit motus *aequalis*, necesse est ut sit super *alterum*, atque etiam in ipsâ Ellipsi, *inaequalis*: neque in Motu Elliptico (quantum ad orbitam suam) alia quaerenda est inaequalitas praeter illam quae a motu medio (seu aequali) regulatur.

Next we find, [69L–71R], a fairly long investigation of the relation between the true anomaly and the mean, the following question being typical.

> Cor. 2. Hinc data anom. vera = V invenire mediam = M cum excentricitas = ε. semiaxis major = 1 & minor = λ.

The next relevant section, [92L–93L], is concerned with the resolution of the forces acting on various heavenly bodies. The first article, typical of the matter discussed, runs as follows.

> 1. Si corpus P movens utcunq$_3$ in orbitâ $CPAD$ in ea retineatur vi quæ componitur ex vi αP versus datum punctū T & vi δP in

> directione Tangentis sitq$_3$ A = areæ CPT bis sumptæ, $p = \perp^\circ$
> TQ $q = QP$ parti Tangentis per TQ abscissæ & r = radio curvaturæ ad punctū P, atq$_3$ $TP = z$ erit vis $\alpha P = \dot{A}^2 z / r p^3$ & vis $\delta P = \ddot{A}/P$.

In the fourth article of this passage, where forces in operation on the moon are considered, we find a parenthesized reference 'modo ab ill. Newtono posito. prop. 26. lib 3^i.' The reference is to 'De Mundi Systemati liber tertius' in Newton's *Philosophiæ Naturalis Principia Mathematica*, where we have[37]

> Propositio XXVI. Problema VI. Invenire incrementum areæ quam Luna radio ad Terram ducto describit. [1687]

The next passage, [97L], involves the finding of relationships between various trigonometric ratios of the latitude and the longitude (presumably of some heavenly body). This is followed by some pages, [C: 98L–98R], entitled 'Of the 4th Equation of the Moon'. The passage starts off by referring to 'the construction given by S^r Is. p. 423.424'. The edition of the *Principia* used by Bayes in this instance appears not to have been the third, for the figure given in the notebook (one that does not appear in the first edition of the *Principia*) is to be found in Book III in the Scholium to Proposition XXXV, Problem XVI (pp. 462–463 of that edition).

We now find a long passage, [103(a)R–104R], in eight articles, dealing with gravitational variation on the earth's surface. No reference is given to the source from which the passage, written in both longhand and shorthand, is taken, but a reference to 'Mr Stirling' in the introductory article has enabled us to identify the notes as a very full extract (though not a copy) with some minor changes, of a paper published by James Stirling in the *Philosophical Transactions* in 1735. In addition to the variation of gravity on the surface of the earth, Stirling discussed here also the shape of the earth, and recorded the results of some pertinent experiments.

On [103(b)R] we find a parenthesized reference to 'cor. 2 Prop. 4 lib. 1. Princ.' The work referenced has been identified as Newton's *Principia*, the second corollary to the fourth proposition in Book 1 running as follows[38].

> Et reciproce ut quadrata temporum periodicorum applicata ad radios ita sunt hæ vires inter se. Id est (ut cum Geometris loquar) hæ vires sunt in ratione composita ex duplicata ratione velocitatum directe & ratione simplici radiorum inverse: necnon in ratione composita ex ratione simplici radiorum directe & ratione duplicat temporum periodicorum inverse.

11.7. Celestial mechanics

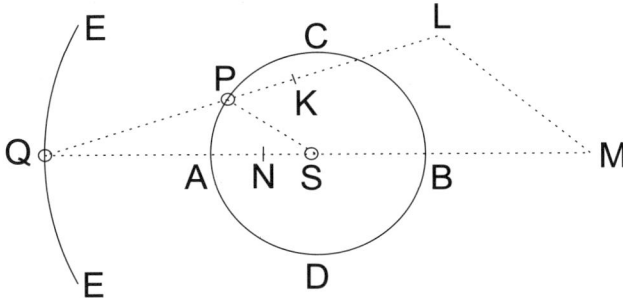

FIGURE 11.5

There is also a reference here to 'cor. 17. Prop. 66. lib. 1'. This reference from the *Principia* runs as follows (see Figure 11.5).

> Cum autem linea LM nunc major fit nunc minor quam radius PS, Exponantur vis mediocris LM per radium illum PS, & erit hæc vim mediocrem QK vel QN (quam exponere licet per QS) ut longitudo PS ad longitudinem QS. Est autem vis mediocris QN vel QS, qua corpus retinetur in orbe suo circum Q, ad vim qua corpus P retinetur in Orbe suo circum S, in ratione composita ex ratione radii QS ad radium PS, & ratione duplicata temporis periodici corporis P circum S ad tempus periodicum corporis S circum Q. Et ex æquo, vis mediocris LM, ad vim qua corpus P retinetur in Orbe suo circum S (quave corpus idem P eodem tempore periodico circum punctum quodvis immobile S ad distantiam PS revolvi posset) est in ratione illa duplicata periodicorum temporum. Datis igitur temporibus periodicis una cum distantia PS, datur vis mediocris LM; & ea data datur etiam vis MN quamproxime per analogiam linearum PS, MN.

In order to give some indication of the content of Stirling's paper, we mention the following passages from the notebook (see Figure 11.6).

'The centrifugal force arising from ye diurnal motion of ye Earth brings it into an oblate Sphæroidical figure. Yet the kind of y^t figure has not hitherto been discovered; & y^{refore} Mr Stirling supposes it to be the common Sphæroid generated by y^e rotation of an Ellipsis round its lesser axis, tho'

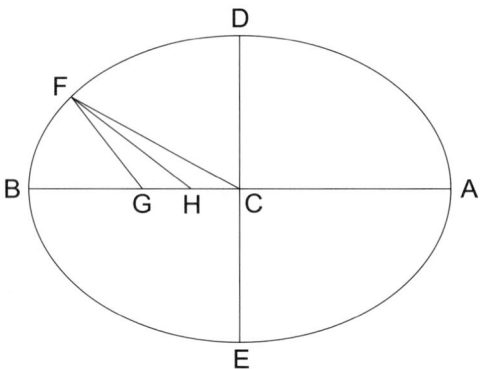

FIGURE 11.6

he finds by computation[s] it is not accurately but only nearly so. He likewise supposes the density to be every where y^e same fm the Centre to the Surface & the mutual gravitation of the particles towards one another to decrease in y^e duplicate ratio of their distances & hence from the nature of the Sphæroid derives the following rules.

'1. Let $ADBE$ be the Meridian of an Oblate Sphæroid, DE the axis, AB the diameter of the Equator & C y^e centre. Take any point on the surface, as F, from which draw FC to the centre FG perpendicular to the surface at F [, meeting CB in G,] & FH cutting the line CGB [CG] so that $CH : GH :: 3 : 2$. A body y^n at F will gravitate in the direction FH & the mean force of gravity on the surface will be to y^e excess of the gravity at the Pole above that at F, as the mean diameter $\times^d$ into y^e Square of y^e Radius is to $\frac{1}{5}$ of y^e difference of y^e longest & shortest diameter $\times^d$ into [the square of] y^e cosine of y^e Latitude at F.

(*) i.e. Mean gravity : gravity at $D-$ gravity at $F :: \frac{BC+DC}{2(?)} \times R^2 : \frac{BC-CD}{5} \times \cos, BCF$

'2. The decrem^t of gravity fm y^e Pole to y^e Equator is proportional to y^e Square of y^e cosine of latitude. ...

'Hitherto Mr Stirling considers y^e variation of gravity arising frō y^e Sphæroidical figure whilst it do's not turn round its axis. But if it doth y^e direction of gravity will be in y^e line $FG \perp^r$ to the surface & its variation now arising both frō y^e figure & centrifugal force will be 5 times greater than

---

(*) This passage is not given by Stirling.

11.7. Celestial mechanics

what arises frō y^e figure alone; as will appear frō y^e proportion of y^e lines FH & FG y^e former being to the latter as the whole force of gravity at F, while y^e Sphæroid is at rest, to the force with which the body descends at F while it turns round its axis.

'3. Hence $\frac{1}{5}$ of the variatiō of gravity is occasioned by the figure of y^e sphæroid & y^e remaining $\frac{4}{5}$ by the centrifugal force. ...

'4. The mean force of gravity on y^e Surface is to the centrifugal force at any point F as a rectangle under y^e Radius & mean diameter to a rectangle under y^e cosine of Latitude, & $\frac{4}{5}$ of the difference of y^e longest & shortest diameters ... This article is found from y^e proportion of y^e lines FH & GH; y^e former being to the latter as y^e force of gravity to y^e centrifugal force.'

Another reference, on [103(c)R], is 'vid. Phil. Trans. N°. 432. P. 302'. This refers to a paper published in the volume for 1733 and 1734 of that journal, entitled 'An account of some observations made in London ...' and concerned with the differences between the lengths of two isochronal pendulums in different areas.

An extract on [106L–106R] deals with centripetal and centrifugal forces.

The entry on [116(a)R–117R] bears the heading 'vid. prob. 56. Pag. 223. Newt. Algebra'. The reference is to Newton's *Arithmetica Universalis; sive de compositione et resolutione arithmetica liber* of 1707, Bayes's reference to 'Algebra' being explained by the fact that the pages in this edition, from page 18 onwards, bear the headings 'Algebrae' (on the left-hand side) and 'Elementa' (on the right-hand side). Problem 56 of this work is stated by Bayes as follows,

> Ex Cometæ motu uniformi rectilineo per Cælū trajicientis locis
> 4 observatis distantiā a ⊙ & motûs determinatiōe colligere &c.

the only significant difference from the original being the omission of the words 'in Hypothesi Copernicaea' before 'colligere'. Only a few sentences in this passage are from Newton's book.

On [121R] the following problem is considered,

> Given the periodical time of two planets T & L to find the time
> & place of their conjunction,

and the final entry on celestial mechanics is to be found on [123L]. Here, under the heading 'Thomas Allen. Lond. Magaz. Nov. 1760. Pag. 613', is to be found a slightly abbreviated copy of some astronomical characteristics of Venus (e.g., its mean anomaly and heliocentric place in the ecliptic) as well as details of its predicted position and first contact in 1761.

11.8 Miscellaneous matters

11.8.1 Of the pyramid measured by Greaves

This is fairly long extract, [C: 26L–28R], from a tract entitled 'Pyramidographia: or, a description of the pyramids in Aegypt', by John Greaves and published in 1736. The passage copied is chiefly concerned with the measurements of the passages and inner chambers of the pyramid of Cheops. The copying is of but part of the original, and is written partly in shorthand. In each case the descriptions of the various parts of the pyramid are enhanced (or at least attended) by detailed notes concerning the pertinent lengths, breadths, heights, etc. A reference to the source is given on p. 28R as 'Greav. Pyram. graph.'

Greaves's 'Pyramidographia' was in fact first printed in 1646, being followed a year later by his 'A Discourse of the Roman Foot and Denarius'[39]. Ward [1740, p. 149] describes Birch's publication of Greaves's *Miscellaneous Works* as 'a curious collection of treatises'. Because of the rarity of this miscellany, or at least because of the difficulty of finding a copy for consultation, we give a fairly full account of the matter recorded in the notebook.

Bayes's extract from this work opens with a sketch (here eschewed) of the Pyramid of Cheops, showing passages, galleries, anticlosets, and chambers. The description of the pyramid starts as follows. (*)

'On ye North of ye fairest Egyptian Pyramid ascending thirty [thirty eight] feet upon an artificial bank of Earth {there is a square and narrow passage leading into the} Pyramid {through the mouth of which being equidistant from the two sides of the} Pyramid {we enter as it were down the} steep of a hill declining {with} an angle of twenty six degrees. The breadth of y^s entrance is exactly three feet & $\frac{463}{1000}$, y^e length of it [beginning] from y^e 1st declivity wch is some ten palms without to y^e utmost extremity of y^e neck or streight within, where it contracts it self almost nine feet continued with scarce half y^e depth it had at its first entrance tho' it keep y^e same breadth is ninety two feet & a half.'

Following Greaves[40], Bayes then leads us through this memorial through the first gallery to 'an arched vault or little chamber', then to the second gallery, whence, via a square hole, we move on to two antichambers or anticlosets 'lined wth {a rich and speckled kind of} Thebaic marble'. After a full description of these chambers we move on, again through a square

---

(*) Crotchets and braces are used here as in §11.6.1.

hole, to a 'very sumptuous {and} well proportioned room' as it were in the heart of the pyramid.

'Within y^s glorious room ... stands ye monument of Cheops or Chemmis of one piece of Marble hollow within & uncovered at top & sounding like a bell ...
'This monument ... stands exactly in y^e Meridian North & south & is as it were equidistant from all y^e sides of y^e chamber except ye East from w^{nce} it is doubly remoter than from y^e West. ... There are two inlets or spaces in y^e south & north sides of this Chamber just opposite to one another. that on y^e north was in breadth $\frac{700}{10000}$ of a [the English] foot & in depth $\frac{400}{10000}$ of a foot ... & running in a strait line six feet, & farther, into y^e wall. That on y^e South is larger & somewhat round not so long as the former & by y^e blackness within [it] seems to have been a recepticle for y^e burning of lamps. ...
'The pyramid whose inside has been described stands upon a Square base each of whose sides are 693 feet; wherefore ye area of ye base is 480249 Square feet i.e. eleven english Acres & 1089 of 43560 parts of an acre.
'The height is 499 feet, from ye angle of ye base to angle of ye Square at ye top, is 693 feet.
'The ascent to ye top of ye Pyramid is thus contrived [in this manner]. From all [the] sides without we ascend by degrees; ye lower-most degree is near 4 foot in height, & 3 in breadth. Y^s runs about ye Pyramid upon [in] a level; & at ye 1st w^n ye stones were intire made on every side of it a long but narrow walk. The 2^d degree is like ye 1st each stone amounting to almost 4 feet in height & 3 in breadth; it retires inward from y^e 1st near 3 feet & this runs about y^e Pyramid in a level, as y^e former. In y^e same manner is y^e 3^d row placed upon y^e 2^d & so in order y^e rest like so many stairs rise one above [another] to y^e top, wch ends not in a point ... but a little flat or square ... whose side is 13 feet & $\frac{280}{10000}$.
'The degrees ... are not all equal, but seem so to lessen as you go to y^e top, that a right line extended from any part of y^e basis without to y^e top will equally touch y^e outward angle of every degree.'

11.8.2 Of weights and measures

Here is an extract, [C: 28R–29R], from another work by Greaves — this time from his 'A Discourse of the Roman Foot and Denarius: From whence, as from two Principles, The Measures and Weights used by the Ancients may be deduced' of 1736. The first passage contains a comparison of ancient and modern (i.e., eighteenth century) European and Middle Eastern units of distance with the English foot; the second contains a comparison of similar gold and silver weights with the English grain. Two references are given: 'Greaves of ye Roman foot' [29L] and 'Greaves of ye Denarius' [29R].

At the foot of p. 29R is a listing of comparative Parisian and English measures of weight and distance, the reference here being merely 'Phil. Trans.' This section is repeated on p. 89R, and we have identified the material as coming from an anonymous article entitled 'An account of the proportions of the English and French measures and weights, from the standards of the same, kept at the Royal Society', published in volume 42 of the *Philosophical Transactions*.

A further listing, [C: 88R], headed 'English Monys, weights Measures &c' in the text, contains some old English measures of distance, area, and weight, and also some measures of money (for instance, one Mark = £1. 10s., and one Noble = £3). Finally, on [89R] is a list of wine, beer, and ale measures. No reference for any of these listings has been found, apart from that mentioned earlier, though some similar matters are discussed in Hooper [1721].

11.8.3 Extract from a dissertation upon cubits by Sir Isaac Newton

This is a very full (though often paraphrastic) extract, [C: 30L–34R], from a Latin work by Newton first translated and published in Greaves [1737], volume 2. It is concerned with various measures of the cubit: the Roman, the Greek, the Persian, and (in great detail) that obtained from measurements made by Greaves of the pyramid at Memphis. There is also much discussion of the sacred and vulgar cubits of the Jews, and the connexion with the Sabbath-day's journey. A major part of the extract is in shorthand. Once again we give a fairly comprehensive copy.

'The Roman & Greek cubits were a foot & a half & like ye sacred cubit {of} ye Jews consisted of six Palms or [and] 24 digits. For ye Roman & Greek feet contained 4 Palms or [and] 16 digits. The Roman foot was likewise divided into 12 inches, unciæ, or pollices & was equal to 0.967 of ye English foot. ... The Roman cubit therefore is 1.4505 of ye English foot.'

There follow details and comparison of the Greek stadium, the Roman milliare or mile, the Attic foot, the cubits of Memphis, Babylon, and Persia, and the Persian parasang. The main theme of Newton's tract, viz. the Jewish cubit, is recorded by Bayes as follows.

'Thus among ye Jews ye Kibrath Terræ or pasture land sufficient I think for ye flock of one shepherd was determined by ye space of a 1000 Cubits & a Sabbath days journey by y^t of 2000 Cubits.'

11.8. Miscellaneous matters

Noting that measurements change over the centuries, Bayes records

'And it is no wonder y^t a Measure shd somew^t increase in ye Space of 3000 years. Y^e measures of feet & cubits now far exceed ye Proportion of human member[s] & yet Mr Greaves shows from ye Egyptian monuments y^t ye human stature was ye same ab^t [above] 3000 years ago as now. Y^e measures {therefore} are increased, ye reasons of wch may {be} assigned. Y^e instrum^{ts} wch used to be preserved as standards, by contracting rust are increased. Iron beaten by ye hammer may insensibly relax in a long Space of time. Artificers likewise in making instrum^{ts} choose to err in ye excess of ye materials & w^n by filing y^{ey} attain any measure wch y^{ey} think sufficient y^{ey} stop knowing y^t any little excess may soon be corrected by filing, but y^{ey} cannot remedy a defect. . . .

'{The} Cubits {of the} Eastern {nations with which the} Jews {were surrounded being thus determined [determined in this manner] we may from hence form a conjecture concerning the magnitude of the} Jewish cubit. {The} vulgar Jewish Cubit {ought not to be greater than them all nor the sacred cubit less than them all.} . . .

'A Sabbath's days journey was 2000 cubits {by the [unanimous] consent of the Talmudists and all the} Jews {and in describing this journey the} Jews {instead of cubits frequently substitute} paces {by which we are not to understand} Rom̄ or Greek paces {but such as a man takes when travelling [upon a sabbath] not with speed as in} Rom. paces {nor too slowly but moderately [in the manner of those who travel on the Sabbath-day]. Now men of a middling stature in walking thus go every step more than} 2 Rom. feet {and less than} $2\frac{1}{3}$ {and within these limits was the sacred cubit circumscribed.} . . .

'{It is agreeable to reason to suppose that the} Jews {when they passed out of} Chaldea {carried with them into} Syria {the cubit which they had received from their ancestors. This is confirmed both by the dimensions of} Noahs {ark preserved by tradition in this cubit and by the agreement of this cubit with the two cubits which the} Talmudists {say were engraved on the sides of the city} Susan {during the empire of the} Persians {and that one of them exceeded the sacred cubit half a digit and the other a whole digit.} Susan {was a city of} Babylon {and consequently these cubits were} Chaldean [Chaldaic]. {We may conceive one of them to be the cubit of the royal city} Susan {the other that of the city of} Babylon. {The sacred cubit therefore agreed with the cubits of divers provinces of} Babylon {as far as they agreed with one another [each other] and the difference was so small that they might all be derived [in different countries] from the same primitive cubit, the} Jewish {cubit being less enlarged *than the rest* after sacred things began to be detemined by it. This therefore was the proper and principal cubit of the} Jews. {But now in} Egypt {they must needs learn the} Egyptian {cubit.} [But that people afterwards going down into Ægypt, and living for above two hundred years under the dominion of the Ægyptians, and enduring an hard service under them, especially in building, where the measures came daily under consideration; they must necessarily learn the Ægyptian cubit.] {Hence came the two [double] cubit of the} Jews: {that of their own country and this [the] adventitious one which [from its] being used upon ordinary occasions only was esteemed vulgar and profane. This} hypothesis {is confirmed by the proportion of the cubits to each other. For the} Babylonian cubit = 2 Eng. f. is to y^e cubit of Memphis = 1.719 Eng. f. as 6 to 5.157 i.e. {as the sacred cubit of the} Jews {to the vulgar cubit very near. The small fraction} [of] $\frac{157}{1000}$ {might arise either from [from either] the difference of the} Babylonian {cubits or the greater antiquity of the} Babylonian {building above [than of] the} Pyramid {or the dimension of the brick, expressed not in [the] exact but the nearest round numbers. Suppose the thickness of the brick to be} $6\frac{3}{16}$ Eng. inch., {the breadth} $8\frac{1}{4}$ [inches and] {the length} $12\frac{3}{8}$ [inches] {and a cubit double that length will be to [the] cubit of} Memphis {as six to five.} I {am inclined therefore to think that the cubit of} Memphis {at the time when the} Jews {went down to [into]} Egypt {was equal to five palms of the} Chaldæo-Hebraic {cubit and that the} Jews {thus determining the magnitude of that cubit the palms of} Memphis {at last began to be neglected [... magnitude of that cubit by five Palms of the proper Cubit, the Palms of Memphis became at last neglected] and the double cubit with only a simple palm remained among the} Jews. {Besides as it is reasonable to suppose that the profane and adventitious cubit agreed with the cubits of the nations round about.} viz. {that [those] of} Memphis, Samos {and} Persia {so it appears from the following argument that this [cubit] was the same with that of} Memphis. {The different measure[s] of the cubit of} Memphis {taken from different parts of the} pyramid {were} 1.717 [$1\frac{727}{1000}$], 1.719 & 1.732 Eng. f. {To this [these] measure[s] in the proportion of the sacred cubit to the vulgar} Jewish {cubit are the measures} 2.0624——2.0628——2.0784 Eng. f. {which in [the]} unc. [unciae] of {the} Rom. f. are 25.57——25.60——25.79 {and consequently fall in the middle of these limits with which we have before circumscribed the sacred cubit and which were 24 and 27} unciæ

{of the} Rom. f. {Supposing therefore that the vulgar cubit of the} Jews {was this cubit of} Memphis {the sacred cubit was not less than 25.57 nor greater than} 25.79 Rom. unc. And fm hence I {will infer taking the measure of the cubit of} Memphis {from the length of the chamber in the middle of the} pyramid {where the king's monument stood to the} Sacred cubit of Moses was 25 unc. Rom. {and} $\frac{6}{10}$ of an uncia or w^t is equivalent y^t it had y^e same proportion to 2 Rom. f. as 16 to 15. $^{(*)}$ [Supposing therefore that the Jews learned the Cubit of Memphis in Ægypt, and that it was their vulgar Cubit, and consequently that in the time of Moses, and soon after, when, as Mr. Greaves contends, the Pyramids were built, the vulgar Cubit was of the same magnitude with that of Memphis; the sacred Cubit in those times was not less than $25\frac{57}{100}$, nor greater than $25\frac{79}{100}$ Unciae of the Roman Foot. Those, who shall hereafter examine the Pyramid, by measuring and comparing together with great accuracy more dimensions of the stones in it, will be able to determine with greater exactness the true measure of the Cubit of Memphis, and from thence likewise of the sacred Cubit. In the mean time for the precise determination of the Cubit of Memphis, I should choose to pitch upon the length of the chamber in the middle of the Pyramid, where the king's monument stood, being very large, and built with admirable skill; which length was the twentieth part of the length of the whole Pyramid, and contained 20 Cubits, and which was very carefully measured Mr. Greaves, as he informs us himself. And from hence I would infer, that the sacred Cubit of Moses was equal to 25 Unciae of the Roman Foot, and $\frac{6}{10}$ of an Unciae; or, what is equivalent, that it had the same proportion to two Roman Feet as 16 to 15.]
'Mersennus {writes thus} [in his treatise] de mensuris Prop. 1. cor. 4. 'I {find that the cubit} (upon {which a learned} Jewish {writer [, which I received by the favour of the illustrious Hugenius, Knight of the order of St. Michael,] supposes the dimensions of the temple were formed) answers to} $23\frac{1}{4}$ Paris inches [of our inches].' {The Paris foot} [, with which Mersennus compares this Cubit,] is = 1.068 [of the] Eng. f. {according to} [Mr.] Greaves & [consequently] {to the} Rom. f. {as} 1068 {to} 967. {In the same proportion} reciprocally {are} $23\frac{1}{4}$ & 25.68. {That cubit therefore [is equal to $25\frac{68}{100}$ Unciae of the Roman Foot, and consequently] falls within the middle of the limits 25.57 & 25.79 with which we have just circumscribed the sacred cubit; so that} I {suspect this cubit was taken from some authentic model preserved in a secret manner from the knowledge of the Christians.

$^{(*)}$ The following encrotcheted passage succeeds the sentence given above ending '24 and 27 unciæ of the Roman foot.'

Lest any one [person] should be surprized that the cubit which we have concluded to have been in the time of} Moses 25.6 inches {should not have more increased [increased more] in} 3000 {years he may observe that the palms used in building at} Rome {which were [was] anciently [authentically]} 9 Rom. unc. {are [is] now} 0.732 Eng. f. i.e. $9\frac{1}{12}$ Rom. unc. {and consequently it has not increased above} $\frac{1}{12}$ unc. {in} 1500 years [that in fifteen hundred years it has increased but $\frac{1}{12}$ of an Unciae] {though [it was] not preserved in a religious manner.}

'Y^e Roman Cubit is to y^e sacred cubit in round numbers as 2 to 3. And y^s is y^e Proportion generally used by Josephus in writing to y^e Romans. W^{refore} by this proportion y^e numbers of cubits in him are to be reduced to sacred cubits.'(*)

11.8.4   Aulay Macaulay's shorthand

This entry, [C: 84R], contains the complete key to another eighteenth-century shorthand (first published in 1747, according to Rockwell [1893]). The symbols are considerably different from those given by Coles, and a study of the shorthand used in the notebook yields no sign of Bayes's having used this system.

11.8.5   Estimate of the National debt upon 31 Dec. 1749

This is a fairly complete extract, [86L–86R], from pp. 150–151 of the *London Magazine*, vol. 19 [1750], of the British National debt in that year: the total is recorded with astonishing precision as £78,497,791. 3*s*. $8\frac{3}{4}d$.

11.8.6   A shorthand verse

This is a most curious entry, [88L]. The shorthand in which the doggerel (for such it seems to be) is written is often exceedingly difficult to decipher, and what follows cannot be seen as anything more than a preliminary attempt at decoding.

> From foot to head 'tis all profound,
> Not sworn to rise, it creeps on ground.
> On dirty ground as can be found
> 'Twixt howler's hole and gillie's pond.
> Ah me, my giddy head turns round,

(*) This last paragraph seems to be due to Bayes himself.

>   There leaden hammers quite confound.
>   And what tho' types will knock me down,
>   Of men of county and of town.
>   O Londoners of great renown,
>   This Poet Laureate must be crowned.

The following alternative readings are possible.

> line 2. 'Not sworn' could perhaps be 'It scorns', but there is no '.' at the start of the second word to indicate the final s in 'scorns'.

> line 4. The second word might well start with an m, and the second last word with an l ('lily' ?). The last word might be 'pound'.

> line 6. The first word is obscure: could it be 'No'?

> line 7. 'And what tho' ' could be 'And where the'.

11.8.7 Warburton's syllogism

Most of this passage, [99L], together with part of the title, is written in shorthand. It proves to be an extract from a book first published by William Warburton in 1738 under the title *The Divine Legation of Moses*. The complete extract, taken from Book 1 (of nine), Section 1 (and paraphrased by Bayes), is preceded by the words 'That therefore the law of Moses is of divine original. Which, one or both of the two following syllogisms will evince', and runs as follows[41].

> I. Whatsoever religion and society have no future state for their support, must be supported by an extraordinary providence. The Jewish religion, and society had no future state for their support; Therefore, the Jewish religion and society were supported by an extraordinary providence.
> And again II. The ancient lawgivers universally believed that such a religion could be supported only by an extraordinary providence. Moses, who instituted such a religion, was an ancient lawgiver. Therefore, Moses believed his religion was supported by an extraordinary providence.

11.8.8 Lettres Provinciales

This is a fairly complete extract, [107L–108L], all in longhand, of the last five pages of Note 3, by Guillaume Wendrock, to the 16th letter of Pascal's *Lettres Provinciales* (new edition of 1735). Identification is easily made from the words 'Lett. provinciales not. 3 sur la XVI Lettre' on [108L], Bayes's extract being from p. 276 et seqq. from this edition.

Since this note is missing from many modern editions of Pascal's works, I quote Bayes's transcription (leaving acute and grave accents *presque partout* as Bayes had them) in full; an English summary may be found in Dale [1995, p. 184].

'Le Monastère de Port Royal ... Elles avoient parmi leur[s] pensionere [pensionaires] une jeune Demoiselle [nommée] Marguerite Perier, niece de M^r. Pascal, qui depuis trois ans & demi ètoit dangereusement malade d'une ægylops ou fistule lacrymale. Les plus fameux Chirurgiens de Paris avoient inutilem^t. employè [emploié] tout leur art pour la guèrir. La Malignitè de mal l'emportoit sur l'habilitè [l'habileté] des Medecins. La matiere sanieuse avoit deja cariè l'os du nez, & le pus qui sortoit de son œil s'ètoit percè une passage au travers du palais: en sorte qu'une partie decouloit sur le visage, & l'autre partie se dechargeoit dans la gorge. Cette fille etoit devenue par la si affreuse qu'elle faisoit horreur a tout la monde ... Les Chirurgiens etoient donc pres d'y appliquer les derniers remides [remedes] & on etoit resolu d'y mettre le feu. On avoit deja mandé son pere pour etre present a ce triste spectacle, lorsq$_3$ Dieu par une prodige suprenant délivra tout d'un coup cette jeune fille de cette maladie. ...

'Il y a à Paris un excellent Prêtre nommè Mr de la Poterie, egalement illustre par sa naissance & par sa piété. La veneration singulier[e] qu'il a pour les reliques des Saints, l'a porté à en amasser un si grand nombre des plus approuvées dans sa chapelle, qu'il n'y a point de particulier dans toute l'Europe qui en ait autant que lui. Il avoit eu depuis peu une Épine de la Couronne de notre Sgr. [Seigneur.] Pleusieurs [Plusieurs] Monasteres de filles de Paris avoient obtenu de lui qu'il la leur envoyàt [envoiât] pour [l']honorer. ...

'Les Religieuses de Port Royal l'ayant appris, & étant touchées des memes sentiments de Piété, le prierent de leur faire la memes [même] grace.: ce qu'il leur accorda.

'Elles reçurent cette precieuse relique le vendredi 24 Mars [de l'année] 1656. Elles [l']exposerent aussi tôt a la veneration de toute leur maison & le religieuses allerit [allerent] toutes la baiser chacune en son rang. Mademoiselle Perier s'etant approchée a son tour, la religieuse qui en avoit soin jetta par hazard les yeux sur elle & l'ayant trouvée plus horrible & plus defigurée q'a [qu'à] l'ordinaire, elle se sentoit [sentit] touchée de compassion & lui dit de faire toucher son œil a la Ste. [Sainte] Epine.

'Cette fille obeit sans songer a autre chose qu'a faire ce qu'on lui disoit. Mais ce qui paroit incroyable, dans ce mom^t. [moment] même elle fut entierem^t. guérie. Le trou que cette [cet] ulcére avoit fait a son palais, fut aussitot

11.8. Miscellaneous matters   489

[aussi tôt] refermé: l'os qui etoit carié, fut retabli en son premier état: enfin il ne resta pas la moindre marque d'un mal qui etoit si affreux. On fit venir peu de temps apres les Medecins & Chirurgiens qui l'avoit [l'avoient] vue pendant sa maladie. A peine croyoient-ils ce qu'ils voyoient, à peine pouvoient-ils reconnoître la malade d'avec les autres Pensionnaires: tant la guerison etoit parfaite & entiere!

'Les Medecins & les Chirurgiens touchez d'une si grande merveille, que les religieuses tenoient secrete, se crurent obligez de la divulger [divulguer]. Le bruit s'en repandit aussitot [aussi tôt] dans tout Paris, & on vit tout le monde accourir en foule à ce Monastere pour y honorer cette Ste [Sainte] Epine. J'etois pour lors a Paris ... j'avois fait une grande liaison avec Mr Pascal ...

'Il etoit oncle de cette Demoiselle & temoin irreprochable de ce miracle. J'allai comme les autre[s] à Port-Royal, & je demandai à voir cette fille; etant bien aise, si je m'en etois rapportè pour sa maladie au temoignage de Mr Pascal, qui etoit une homme digne de toute créance, & a celui des Medecins & des Chirurgiens, de me m'en rapporter qu'a moimême [moi même pour sa [la] guerison. Enfin ... l'autoritè de l'eglise acheva de confirmer ce Miracle. Il fut examiné avec toute l'exactitude possible par le[s] Grands Vicaires de Mr Archevêque de Paris assistez de plusieurs Docteurs de Sorbonne. Ils declarerent par leur sentence de [du] 22 Oct. 1656 de l'avis de ces Docteurs, que cette guerison étoit très-certainement surnaturelle, & un miracle de la toute-puissance de Dieu. ....

'Un apres un soigneux examen fut declarè autentic [authentique le] 14 Dec. 1756 [1656] par l'eglise de Sens, & l'autre par celle de Paris 29. Août 1657.

'Il faut remarquer [Mais qu'il faut encore plus remarquer c'est] que ces derniers miracles sont arrivez depuis la dispute qui s'etoit elevée au sujet des miracles precedens, depuis la constitution d'Alexandre VII. & enfin depuis [que] plusieurs auteurs avoient publiè hautem^t [hautement] par toute la France que la foi de religieuses de Port Royal etoit justifié[e] par ces miracles & les calomnies des Jesuites détruites par l'autorite de Dieu meme.'

Why this passage, with its emphasis on the value of relics, should have been of interest to the (English) Presbyterian Bayes is uncertain; perhaps it is merely a token of Bayes's catholic interests, though one might note that 'Presbyterian' has as anagrams 'nearby priest' and 'best in prayer'.

It should be noted that Marguerite Périer (born on the 5th of April, 1646 according to Shiokawa [1977, p. 79]), died only in 1733. Full details of the claimed miracle were given in a letter from Mother Angélique to Louise Marie de Gonzaga, Queen of Poland, a copy of which letter, with considerable amplification, may be found in the second volume of 1954 of Sainte-Beuve's *Port-Royal*. A whole chapter of Shiokawa's *Pascal et les miracles* is devoted to this occurrence.

Sainte-Beuve, incidentally, did not share the Jansenists' delight in this phenomenon, for he wrote

> Les Jansénistes y voyaient le triomphe de leur cause: j'y vois surtout l'humiliation de l'esprit humain. [1954, p. 179]

In his *History of Statistics*, [1978, p. 681], Karl Pearson noted that consideration of the supernatural requires cognisance of certain factors, viz. (i) the events that are reported as having happened, and (ii) the interpretation of such events. In the case of Mlle. Périer, he conceded that one may well accept that she was cured, this cure being effected neither by remedies nor by natural processes. This gives (i). But it is (ii) that gives rise to the miracle. Pearson was, however, anticipated, and indeed perhaps influenced in his observation, by Thomas Huxley, who, in his essay 'The value of witness to the miraculous' of 1889, wrote:

> When a man testifies to a miracle, he not only states a fact, but he adds an interpretation of the fact. [1909, p. 187]

11.8.9 *A prescription*

This entry [111L] runs as follows.

> ℞ Pul. Sem. cymin. chamomel. ʒij
> Sol. c.c. ʒss
> camph. (solut. in Sp. Terebinth. ʒij) Ɔi.
> ung. Sambuc. ʒiii
> sapon. nigr. Commun. ʒi
> M. ft. liniment.

That is, take two drachms of the powdered seed of cumin and chamomile (both carminative); 1/2 drachm of a solution of calcium carbonate; 1 scruple of camphor, in a solution of 2 drachms of spirits of turpentine (turpentine oil is a rubefacient used in liniments for rheumatic pain and stiffness); mix together, to one fluid ounce, an unguent of 3 fluid ounces of sambucus (the elder flower — has an astringent action on the skin) with 1 fluid ounce of saponis nigra (liquid drawn from lye soap — an emulsifying agent). Mix and use as a liniment. This is probably a prescription for stiffness of the joints or rheumatism.

11.9 Appendix 11.1

Translation of the passage on probability

Firstly, let $S/V = x$. Then $\dot{x} = (S/V)(\dot{S}/S - \dot{V}/V)$. Thus if $\dot{S}/S > \dot{V}/V$ and S and V are both increasing (and of the same sign), $\dot{x} > 0$, and so x is increasing. Similarly, if $y = V/S$ and S and V are both decreasing (and of opposite sign) then V/S is increasing.

<u>Art. 2.</u> Let
$$A = (1 - nz/p)^p (1 + nz/q)^q$$
$$B = (1 + nz/p)^p (1 - nz/q)^q,$$

where $n = p + q$. Then

$$\int_{-q/n}^{p/n} A\, dz = n^n \left[(n+1)\binom{n}{p} p^p q^q\right]^{-1}.$$

The integral of B from $z = -p/n$ to q/n reduces to the same expression.

<u>Art. 3.</u> With A and B as given above, and D and Δ defined by

$$D = (1 - n^2 z^2/q^2)^{nq/2p}, \qquad \Delta = (1 - n^2 z^2/p^2)^{np/2q},$$

we find that

$$\dot{B}/B \; : \; \dot{D}/D \; :: \; (1 - n^2 z^2/q^2) : (1 - nz/q)(1 + nz/p)$$
$$:: \; (1 + nz/q) : (1 + nz/p).$$

Since $q > p$ and $\dot{D}/D$ is negative, it follows that $\dot{B}/B > \dot{D}/D$. (Examination of the subsequent Articles shows that attention is restricted to $z \geq 0$.) Hence, since B and D are both decreasing functions of z, we find from the first Article that B/D is increasing; and since $B = D = 1$ when $z = 0$, we may conclude that $B > D$ — and similarly that $A < \Delta$. Moreover, since

$$\Delta^2 = (1 - n^2 z^2/p^2)^{np/q} \text{ and } AB = (1 - n^2 z^2/p^2)^p (1 - n^2 z^2/q^2)^q,$$

it follows that

$$\Delta^2 \; : \; AB \; :: \; (1 - n^2 z^2/p^2)^{p^2/q} \; : \; (1 - n^2 z^2/q^2)^q.$$

Thus
$$\Delta^{2q} : (AB)^q :: (1-n^2z^2/p^2)^{p^2} : (1-n^2z^2/q^2)^{q^2}$$
and hence $\Delta^2 < AB$ and $2\Delta < A+B$.

<u>Art. 4.</u> Let A be an unknown event with prior probability x, and let $A^n_{p,q}$ denote the event that A has happened exactly p times in $n=p+q$ trials. By Proposition 10 of the *Essay* it follows that, for $z \geq 0$,

$$\begin{aligned} P_1 &\equiv \Pr\left[p/n - z \leq x \leq p/n \mid A^n_{p,q}\right] \\ &= \int_{p/n-z}^{p/n} \binom{n}{p} x^p(1-x)^q \, dx \bigg/ \int_0^1 \binom{n}{p} x^p(1-x)^q \, dx \\ &= (n+1)\binom{n}{p} \int_0^z (p/n - u)^p (q/n + u)^q \, du \\ &= (n+1)\binom{n}{p} (p^p q^q / n^n) \int_0^z (1 - nu/p)^p (1 + nu/q)^q \, du. \quad (11.15) \end{aligned}$$

Having noticed in Article 3 that
$$(1-nu/p)^p(1+nu/q)^q < (1-n^2u^2/p^2)^{np/2q},$$
we see that
$$P_1 \leq (n+1)\binom{n}{p}(p^p q^q/n^n) \int_0^z (1-n^2u^2/p^2)^{np/2q} \, du.$$

2°. Under the same hypotheses it follows that
$$\begin{aligned} P_2 &\equiv \Pr\left[p/n \leq x \leq p/n + z \mid A^n_{p,q}\right] \\ &= (n+1)\binom{n}{p}(p^p q^q/n^n) \int_0^z (1-nu/q)^q(1+nu/p)^p \, du \quad (11.16) \\ &\geq (n+1)\binom{n}{p}(p^p q^q/n^n) \int_0^z (1-n^2u^2/p^2)^{nq/2p} \, du. \end{aligned}$$

3°. Finally,
$$P_3 \equiv \Pr\left[p/n - z \leq x \leq p/n + z \mid A^n_{p,q}\right]$$
$$= (n+1)\binom{n}{p}(p^p q^q/n^n) \int_0^z \left[\frac{(1-nu/p)^p}{(1+nu/q)^q} + \frac{(1-nu/q)^q}{(1+nu/p)^p}\right] du$$

$$\geq 2(n+1)\binom{n}{p}(p^p q^q/n^n)\int_0^z (1-n^2 u^2/p^2)^{np/2q}\,du \quad (\text{since } A+B > 2\Delta).$$

In subsequent Articles certain approximations to these integrals are given.

<u>Art. 5.</u> From the expression

$$n! = \sqrt{2\pi}\, n^{n+(1/2)} e^{-n} e^{\theta/12n}, \quad 0 < \theta < 1$$

one finds that

$$\binom{n}{p} p^p q^q / n^n = \sqrt{\frac{n}{2\pi pq}}\, e^{\theta(pq-np-nq)/12npq}.$$

Since $(pq - n^2) < 0$, we may conclude that

$$N\sqrt{\frac{n}{2\pi pq}} < \binom{n}{p} p^p q^q / n^n < \sqrt{\frac{n}{2\pi pq}},$$

where $\ln N = (pq - n^2)/12npq$.

(Bayes does not give details of his derivation of this last expression: I have tried to reconstruct his argument.)

<u>Art. 6.</u>

$$\int_{-p/n}^{p/n} (1-n^2 u^2/p^2)^{np/2q}\,du = \frac{2^{(np+q)/q}\, p\, [\Gamma((np/2q)+1)]^2}{n\,\Gamma((np/q)+2)}.$$

On our using the approximation

$$\Gamma(z+1) \sim \sqrt{2\pi}\, z^{z+(1/2)} e^{-z},$$

we obtain

$$\int_{-p/n}^{p/n} (1-n^2 u^2/p^2)^{np/2q}\,du \sim \frac{ep}{np+q}\sqrt{\frac{2\pi pq}{n}}\,(1+q/np)^{-(np/q+(1/2))}$$

$$> \frac{p}{2(np+q)}\sqrt{\frac{2\pi pq}{n}}.$$

Similarly

$$\int_{-q/n}^{q/n} (1-n^2 u^2/q^2)^{nq/2p}\,du > \frac{q}{2(nq+p)}\sqrt{\frac{2\pi pq}{n}}.$$

Art. 7. From Articles 6 and 4.3 it follows that

$$\Pr[0 \leq x \leq 2p/n \mid A^n_{p,q}]$$

$$> (n+1)\binom{n}{p}(p^p q^q/n^n)\int_0^{p/n}(1-n^2u^2/p^2)^{np/2q}\,du$$

$$> N(n+1)p/(np+q).$$

If p and q are both large, and $q > p$, then the right-hand side of this last expression is approximately 1. Furthermore, under such conditions on p and q,

$$\Pr[p/n - z \leq x \leq p/n + z \mid A^n_{p,q}]$$

$$\sim 2(n+1)\sqrt{n/(2\pi pq)}\int_{-p/n}^{p/n}(1-n^2u^2/p^2)^{np/2q}\,du$$

without appreciable error.

In the last two Articles of his work Bayes turns his attention to the evaluation of the integrals in (11.15) and (11.16).

Art. 8. If $x : r :: p : q$ and $(x+r)^n$ is expanded, then

$$\binom{n}{q}x^p r^q : \binom{n}{q-1}x^{p+1}r^{q-1} :: 1 : (q/(p+1))(p/q)$$

$$\binom{n}{q}x^p r^q : \binom{n}{q-2}x^{p+2}r^{q-2} :: 1 : \frac{q(q-1)}{(p+1)(p+2)}(p/q)^2 \quad \&c.$$

Similarly

$$\binom{n}{q}x^p r^q : \binom{n}{q+1}x^{p-1}r^{q+1} :: 1 : (p/(q+1))(q/p)$$

$$\binom{n}{q}x^p r^q : \binom{n}{q+2}x^{p-2}r^{q+2} :: 1 : \frac{p(p-1)}{(q+1)(q+2)}(q/p)^2 \quad \&c.$$

Thus

$$\sum_{k \leq q}\binom{n}{k}x^{n-k}r^k \Big/ \sum_{k > q}\binom{n}{k}x^{n-k}r^k$$

$$= \frac{\left[1 + \frac{q}{p+1}\left(\frac{p}{q}\right) + \frac{q(q-1)}{(p+1)(p+2)}\left(\frac{p}{q}\right)^2 + \frac{q(q-1)(q-2)}{(p+1)(p+2)(p+3)}\left(\frac{p}{q}\right)^3 + \cdots\right]}{Q},$$

where the denominator Q is given by

$$\frac{p}{q+1}\left(\frac{q}{p}\right) + \frac{p(p-1)}{(q+1)(q+2)}\left(\frac{q}{p}\right)^2 + \frac{p(p-1)(p-2)}{(q+1)(q+2)(q+3)}\left(\frac{q}{p}\right)^3 + \cdots.$$

Now if $q > p$ the kth term of the numerator on the right-hand side of this last expression is greater than the corresponding term of the denominator. For small p the series in the denominator terminates first, wherefore the series in the numerator is greater than that in the denominator. If in fact 1 is subtracted from the series in the numerator, the series in the denominator will be larger than that in the new numerator.

<u>Art. 9.</u> Attention is next turned to the evaluation of the integral

$$\begin{aligned} I &\equiv \int_{-q/n}^{p/n} (n+1)\binom{n}{p}(p^p q^q/n^n)(1-nz/p)^p(1+nz/q)^q\,dz \\ &= \int_{-q/n}^{p/n} \left[(n+1)\binom{n}{p}(p/n-z)^p(q/n+z)^q\right]dz \\ &= \int_{-q/n}^{0} (n+1)\binom{n}{p}x^p r^q\,dz + \int_{0}^{p/n}(n+1)\binom{n}{p}x^p r^q\,dz \\ &\equiv I_1 + I_2, \end{aligned}$$

say, where $x = p/n - z$ and $r = q/n + z$.
To evaluate I_1, consider the series

$$V = r^{n+1} + \binom{n+1}{1}r^n x + \cdots + Fr^{q+1}x^p.$$

Since $\dot{r} = \dot{z} = -\dot{x}$,

$$\dot{V} = (n+1)r^n\dot{r} + \binom{n+1}{1}\left[nr^{n-1}\dot{r}\,x + r^n\,\dot{x}\right] + \binom{n+1}{2}\left[(n-1)r^{n-2}\dot{r}\,x^2\right.$$

$$\left. + r^{n-1}2x\dot{x}\right] + \cdots + F\left[(q+1)r^q\dot{r}\,x^p + r^{q+1}px^{p-1}\dot{x}\right]$$

$$= (n+1)r^n\dot{r} + \binom{n+1}{1}\left[nr^{n-1}\dot{r}\,x - r^n\dot{r}\right] + \binom{n+1}{2}\left[(n-1)r^{n-2}\dot{r}\,x^2\right.$$

$$\left. - r^{n-1}2x\dot{r}\right] + \cdots + F\left[(q+1)r^q\dot{r}\,x^p - r^{q+1}px^{p-1}\dot{r}\right]$$

$$= (q+1)Fr^q x^p \dot{r}.$$

Noting that F is the coefficient of $r^{q+1}x^p$ in the expansion of $(x+r)^{n+1}$, we find that $(q+1)F = (n+1)\binom{n}{p}$, and hence V can be written in the form

$$V = r^{n+1} + \binom{n+1}{1}r^n x + \cdots + [(n+1)/(q+1)]\binom{n}{p}r^{q+1}x^p,$$

a series that reduces, when $z = 0$, to

$$B \equiv \frac{(n+1)}{n}\frac{q}{q+1}\binom{n}{p}\frac{p^p q^q}{n^n}\left[1 + \frac{pq}{(q+2)p} + \frac{p(p-1)q^2}{(q+2)(q+3)p^2} + \cdots\right].$$

Thus

$$I_1 = \int_{-q/n}^{0} (n+1)\binom{n}{p}x^p r^q \, dz$$

$$= \int_{0}^{q/n} (n+1)\binom{n}{p}(1-r)^p r^q \, dr \quad (= B)$$

$$= (n+1)\binom{n}{p}B_{q/n}(q+1, p+1),$$

$B_a(b,c)$ denoting the incomplete beta-function[42]. Similarly

$$I_2 = \int_{0}^{p/n} (n+1)\binom{n}{p}x^p r^q \, dz$$

$$= (n+1)\binom{n}{p}B_{p/n}(p+1, q+1).$$

11.10 Appendix 11.2

The development of trigonometric methods down to the end of the fifteenth century was discussed in detail by Bond in a series of papers published in *Isis* in the early 1920s. In the first part of this work we find the following remark.

> Abu'l-Wefa [Almagest — fl. 940-998, Bagdad] himself says: 'It is evident that if one takes the radius as unity the ratio of the sinus of an arc to the sinus of its complement is the tangent (first shadow), and the ratio of the sinus of the complement to the sinus of the arc is the cotangent (second shadow)' ... Braunmühl (p. 58) [Vorlesungen] remarks that the above quotation cannot be sufficiently emphasized, for in spite of its clear statement of the idea of setting $r = 1$, the radius was dragged along into the eighteenth century. [1921, pp. 308, 311]

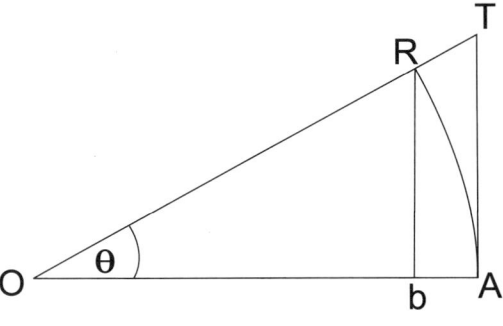

FIGURE 11.7

Trigonometric ratios were thus defined as indicated in Figure 11.7, where $AT = \tan\theta$, $bR = \sin\theta$, $OT = \sec\theta$, and $OA = OR = r$. As consequences we have
$$\cos\theta = r^2/\sec\theta, \qquad \sin^2\theta + \cos^2\theta = r^2.$$

The rebarbative custom mentioned above seems in fact to have continued well into the mid-nineteenth century, for we find Playfair writing

> The sine of an angle is defined above in the usual way, viz. the perpendicular drawn from one extremity of the arc, which measures the angle on that radius passing through the other; but in strictness the sine is not the perpendicular itself, but the ratio of that perpendicular to the radius, for it is this ratio which remains constant, while the angle continues the same though the radius vary. It might be convenient, therefore, to define the sine to be the quotient which arises from dividing the perpendicular just described by the radius of the circle. [1857, p. 316]

11.11 Appendix 11.3

It might be noted that both '∼' and '−' are used in this passage as signs of subtraction, though not with quite the same meaning, as the following argument from the notebook shows. From $\dot{u}^2 = \dot{x}^2 + \dot{y}^2$ it follows that $\dot{u}\ddot{u} = \dot{x}\ddot{x} + \dot{y}\ddot{y}$. Thus
$$(\dot{u}\ddot{u})^2 + (\dot{x}\ddot{y} \sim \dot{y}\ddot{x})^2 = (\dot{u})^2(\ddot{x}^2 + \ddot{y}^2),$$
and hence
$$\dot{x}\ddot{y} \sim \dot{y}\ddot{x} = \dot{u}\sqrt{\ddot{x}^2 + \ddot{y}^2 - \ddot{u}^2}.$$

Thus '$a \sim b$' seems to be used in the sense of the present-day modulus or absolute value $|a-b|$. Bayes is following Wallis [1657] here (see Cajori [1929], §487)[43].

11.12 Appendix 11.4

As mentioned in the text, Cotes's *Canonotechnia* is concerned with the insertion of terms, and in particular means, in a given series. The relevant definitions of his terms are as follows.

> Ordinis primarii differentiae divisae per numeros in ratione dupli crescentes dant alium differentiarum ordinem qui *Rotundus* appellari potest, & sic denotari ①, ②, ③, ④, ⑤, ⑥, &c. ...
> Ex ordine rotundo colligendus est differentiarum ordo tertius qui non incommodo *Quadratus* appellari potest & sic denotari ☐1, ☐2, ☐3, ☐4, ☐5, ☐6, &c.

12

Memento mori

Oh, come and see the skulls; come back and see the skulls!

Jerome K. Jerome.
Three Men in a Boat.

12.1 Introduction

Lying between the City Road and Bunhill Row in the Borough of Islington, London, and opposite the entrance to Wesley's Chapel in the City Road, is the Bunhill Fields Burial Ground (see Figure 12.1). Wheatley notes that the Field, 23 acres, 1 rod and 6 poles in extent, was to be found

> butting upon Chiswell Street on the south, and on the north upon the highway that leadeth from Wenlock's Barn to the well called Dame Agnes the Cleere. [1891, I, p. 302]

Hicks [1887] suggests that the name was originally Bon- or Bone-hill, but this is disputed by others, for the name seems to have been in use before the removal of bones from the Charnel House in St Paul's Churchyard to that site in 1549[1]. Meller [1981] states that the Corporation of London decided in 1665 to establish a churchyard at Bunhill for victims of the Plague, and as a result of negotiations with the tenant of the Finsbury Estate, one John Tyndall (Tindall or Tindale), a brick wall (at a cost of £140) was erected around a suitable portion of the estate in October of that year, a gate following probably in 1666.

There is apparently no direct evidence of the site actually having been used as a burial ground for those succumbing to the Plague, and Bell [1951, p. 211] in fact records that the ground keeper was expressly forbidden to dig Plague pits there. Citing Maitland [1756], Mrs Basil Holmes notes that

500   12. Memento Mori

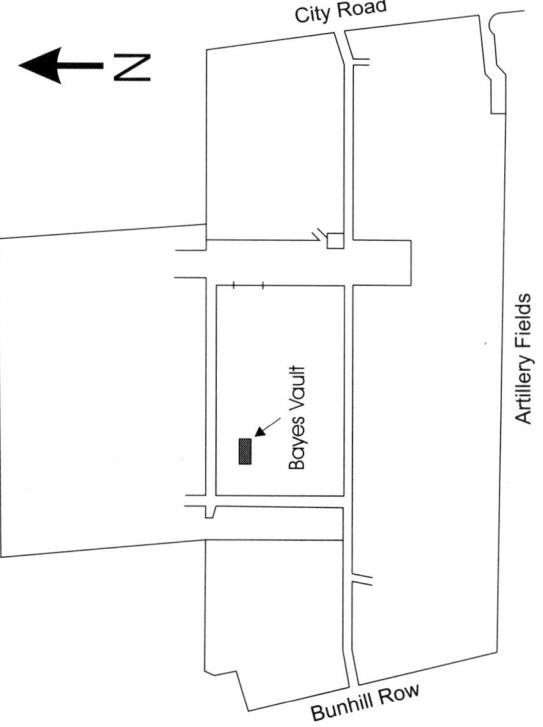

FIGURE 12.1 Bunhill Fields Burial Ground.

the Bunhill Fields Burial Ground was in fact set aside and consecrated in 1665 as a common cemetery for those corpses for which there was no room in the parochial burial grounds due to the pressure exerted by the Plague[2]. It was not, however, used on that occasion, and was therefore converted into a burial place for dissenters by Tyndall[3].

Whether Holmes is correct in asserting that the ground was consecrated is debatable. We show later that one reason for the use of Bunhill Fields as a burying ground for those of Nonconformist[4] persuasions was that the ground had *not* been consecrated. It is possible that only the southern part was consecrated, leaving the northern, free of Establishment taint, for the burial of dissenters.

Holmes summarizes the state of the Ground in the late nineteenth century as follows.

> *Bunhill Fields.*—5 or 6 acres. This was originally two grounds, the southern part having been intended for burials in the Great Plague, but not being used was let by the Corporation to a Mr. John Tyndall, who carried it on as a private cemetery. Subsequently the northern part was added, and the whole ground extensively used for the interment of Dissenters. The Corporation maintain it as a public garden, but the tombstones have not been moved, and only the gates at the eastern end are generally open. [1896, p. 291]

Meller [1981, p. 88] gives the size of the cemetery at the time of his writing as five acres: I myself, as the result of a very rough measurement in Roman paces, would put the size of the present public area (in 1999), in which the open places and remaining tombs are incorporated, as three to four acres, roughly 74% being 'under tombs'.

The change in Bunhill Fields from burying ground to cemetery, with some cognisance of the plague being taken on the way, was not without precedent. In the second edition of his *Survey of London* of 1603, John Stow relates

> A great pestilence entering this island, began first in Dorsetshire, then proceeded into Devonshire, Somersetshire, Gloucestershire, and Oxfordshire, and at length came to London, and overspread all England, so wasting the people, that scarce the tenth person of all sorts was left alive, and churchyards were not sufficient to receive the dead, but men were forced to choose out certain fields for burials; whereupon Ralph Stratford, bishop of London, in the year 1348, bought a piece of ground, called No Man's Land, which he inclosed with a wall of brick, and dedicated for burial of the dead, building thereupon a proper chapel, which is now enlarged and made a dwelling-house; and

this burying plot is become a fair garden retaining the old name of Pardon churchyard. [1912, p. 384]

The Great Plague of 1665, the last of such scourges to hit London, killed a greater number of Londoners, though perhaps not a greater percentage, than any of its predecessors[5]. The impact that it still has on us today is due in large measure to the writings of Daniel Defoe[6] and Samuel Pepys. In his article *Plague* in Latham & Matthews [1983, vol. X, pp. 328–337], Morris suggests that this Great Plague, probably bubonic rather than pneumonic, was one of the last efforts of the Black Death of 1348. It is also suggested that the Plague had assumed an endemic form between successive epidemics: there were outbreaks in London in 1563, 1593, 1603, and 1625, and in the sixty-four years preceding the 1665 outbreak there were only sixteen in which London's mortality records show fewer than ten deaths from the Plague, and twenty years in which they were 1,000 or more — though the numbers recorded in the years immediately preceding the outbreak were officially 13, 20, 12, 9, and 5.

In his contemporary account of the Great Plague — 'undoubtedly the best medical account of the great epidemic, which has been preserved' according to Payne [1894, p. vi] — William Boghurst notes that there was a succession of pestilences in the fifteenth, the sixteenth, and the seventeenth centuries, and he lists the numbers of deaths in the last of these as follows[7].

Deaths from Plague in London in the Seventeenth Century, according to the Bills of Mortality

1603 (epidemic)	33,347	1648	611
1609 (minor epidemic)	4,240	1650–1662 (average)	15
1625 (epidemic)	41,313	1663	12
1636 (minor epidemic)	10,400	1664	5
1637–1646 (average)	1,500	1665 (epidemic)	68,596
1647	3,597	1666	1,998

The break-down of the numbers of deaths in 1665 by month is given in the following table.

Monthly Mortality in 1665

January	0	July	6,137
February	1	August	17,036
March	0	September	26,230
April	2	October	14,373
May	43	November	3,449
June	590	December	734

Our forefathers, perhaps wiser than we, believed that tobacco was a prophylactic against this scourge[8]. Sextons smoked while engaged in their useful, if — at least to some — unsavoury task. Fragments of clay pipes found among skeletons often indicate the presence of a victim of the Plague (the less phlegmatic of us might today think, or at any rate wish to believe, that such fragments indicate rather that the deceased died of one of the ills that we like to ascribe to the (excessive) use of tobacco), and such things are apparently also to be seen in some tombstone symbols. Bell in fact records that the belief in the efficacy of tobacco as a personal disinfectant was such that one Tom Rogers, a scholar at Eton in 1665, told the antiquary Thomas Hearne that

> all the boys smoked in school by order, and that he was never whipped so much in his life as he was one morning for not smoking. [1951, p. 156].

In Defoe's *A Journal of the Plague Year* one finds the following description of the use made of tobacco and perhaps more exotic materials in combatting the Plague.

> He [i.e., John Hayward, an under-sexton] never used any Preservative against the Infection, other than holding *Garlick* and *Rue*[9] in his Mouth, and smoking Tobacco. ... and his Wife's Remedy was washing her Head in Vinegar, and sprinkling her Head-Cloths so with Vinegar, as to keep them always Moist; and if the smell of any of those she waitd on was more than ordinary Offensive, she snuft Vinegar up her Nose, and sprinkled Vinegar upon her Head-Cloths, and held a Handkerchief wetted with Vinegar to her Mouth.

Landa records in his Notes to the 1969 edition of Defoe's *Journal* that tobacco, either chewed ('chawed' by Pepys) or smoked, was recommended for all ages and sexes. It was indeed believed that no tobacconist died of the Plague in 1665.

Pepys carefully, if not ghoulishly, recorded the numbers of deaths from the Plague, but took pains to note that the figures reported might in fact have been understated[10]. In a detailed study of the Plague year Bell finds the return of 68,596 recorded deaths of the Plague in 1665 up to the 19th December to be considerably understated — perhaps by as much as 30,000 (the Bills of Mortality[11] admit 97,306 deaths in all to that date). Notification of death at that time was handled in a casual way: no official notification had to be made by doctor or relative, and it was left to the 'searchers of the dead' to ascertain the cause of death. These 'searchers' were usually illiterate, elderly female paupers, ignorant of disease, and appointed to the post to avoid their having to be supported by the parish.

A small fee, perhaps only a few pence, was payable to the searchers, who identified the need for their services by the hearing of the tolling of the passing bell or a message from the sexton appointed to dig the grave. (One is reminded of Thomas Hood's poem *Sally Brown*, which ends 'They went and told the sexton, and the sexton toll'd the bell'[12].) Many sectarians refused any communication with the established church: they buried their own dead, and consequently many Plague deaths, particularly in the early stages of the epidemic, went unreported. In the later stages the inhabitants of a house in which anyone had died of the Plague were forbidden to leave the house for forty days, the doors being padlocked and warders posted. Red crosses marked the infected houses, together with the subscription[13] *Lord have mercy upon us*. There would therefore certainly have been pressure, perhaps eased by monetary considerations, exerted on the searchers to identify Plague deaths as being of some other cause.

It had not of course been the official intention to appoint women of such a calibre to undertake this melancholy task: one of the orders published by the Lord Mayor and Aldermen of the City of London[14] at this time ran as follows.

> That there be a special care to appoint Women-Searchers in every Parish, such as are of honest Reputation, and of the best Sort as can be got in this kind; And these to be sworn to make due Search and true Report to the utmost of their Knowledge, whether the Persons whose Bodies they are appointed to Search, do die of the Infection, or of what other Diseases, as near as they can. And that the Physicians who shall be appointed for Cure and Prevention of the Infection, do call before them the said Searchers, who are, or shall be appointed for the several Parishes under their respective Cares; to the end they may consider, whether they are fitly qualified for that Employment; and charge them from time to time as they shall see Cause, if they appear defective in their Duties.

It was not only the searchers who were responsible for the misreporting of the cause of death. Bell [1951, p. 158] relates that

> A common device by nurses was to wrap the newly dead in wet cloths, the effect being to drive the spots in, and the death was given out to the searcher as not having been by Plague.

Not even parish clerks were immune from such suppression of facts: Pepys records on the 30th of August 1665 that he

> met with Hadley[15], our Clerke, who upon my asking how the plague goes, he told me it encreases much, and much in our parish: "For," says he, "there died nine this week, though I have

returned but six" — which is a very ill practice, and makes me think it is so in other places; and therefore the plague much greater than people take it to be.

The nurses were in fact responsible for the deaths of some whom the Plague might have spared. In his *Loimologia, or an Historical Account of the Plague in London in 1665* Nathaniel Hodge recorded

> But what greatly contributed to the loss of people thus shut up was the wicked practices of nurses, for they are not to be mentioned but in the most bitter terms. These wretches, out of greediness to plunder the dead, would strangle their patients and charge it to the distemper in their throats. Others would secretly convey the pestilential taint from sores of the infected to those who were well. And nothing indeed deterred these abandoned miscreants from prosecuting their avaricious purposes by all the methods their wickedness could invent, who, although they were without witnesses to accuse them, yet it is not doubted but divine vengeance will overtake such wicked barbarities with due punishment. [Browning, 1953, pp. 494–495]

Those who could afford it of course fled the City as the incidence of the pest increased: the Court, for instance, moved first to Hampton Court, then to Salisbury and finally to Oxford. Members of the Royal Society left London, and even William Sancroft, Dean of St Paul's, found Tunbridge Wells safer. (In mitigation it should be said that Sancroft had left London in July 1665 to take the waters, and after remaining for some time in Tunbridge Wells, he removed to Durham. Despite his absence from the City — he did not return until the worst was over — he sent money to London for the relief of the poor[16].)

In such absence Sancroft was not alone, and it was left largely to the Nonconformist clergy to minister both to the well and to the dying. Bell in fact regards these ministers as

> the real heroes of the Plague, the men whose golden example ennobles their great profession, and condemns the political Churchmen who made them outcasts [1951, p. 149]

and he notes further that one of the major outcomes of this period was that 'The Great Plague established English Nonconformity' [1951, p. 227]. Defoe writes in his *A Journal of the Plague Year* in 1722

> many of the best and most valuable Ministers and Preachers of the Dissenters, were suffer'd to go into the Churches, where the Incumbents were fled away, as many were, not being able to stand it; and the People flockt without Distinction to hear them

preach, not much inquiring who or what Opinion they were of.
But after the Sickness was over, that Spirit of Charity abated.

The Government indeed was not particularly impressed by the work of the Nonconformist clergy. In 1665 the Secretary of State, Lord Arlington, wrote to the Bishop of London[17]

> The King is informed that many ministers and lecturers having been absent from their posts during this time of contagion, nonconformists have thrust themselves into their pulpits, to preach sedition, and doctrines contrary to the Church; His Majesty wishes to prevent such mischiefs to Church and State.

It is indeed distressing to note that the selfless labours of the Nonconformist clergy, known opprobriously at this time as 'fanatics[18]', were rewarded by the passing of the Five Mile Act [Bell, 1951, p. 228].

Little help was obtained from the medical profession. Hospitals were few, and most suspected cases were pent up with their families with armed guards at the door. Suggested remedies were almost Gothic: one Dr George Thomson caught the disease in performing an autopsy on a Plague victim (showing, in so doing, a most commendable, if foolhardy, devotion to medical science), and miraculously survived the disease himself, even though his self-cure consisted in the placing of a dried toad on his chest. He recorded his treatment in 1666 as follows.

> I am sufficiently persuaded, That the adjunction of this Bufo nigh my Stomach, was of wonderful force to master and tame this Venom then domineering in me.

The use of amphibians was apparently common, as the following extract from a contemporary source (as noted in the Appendix to Defoe [1963]) shows.

> Here are many who weare amulets made of the poison of the toad, which, if there be no infection, workes nothing, but, upon any infection invadeing from time to time, raise a blister, wch a plaister heales, and so they are well.

On the 30th of August 1665 Pepys was overcome by a morbid curiosity to visit a Plague pit[19], writing in his diary

> I went forth and walked towards Moorfields to see (God forgive my presumption) whether I could see any dead Corps going to the grave; but as God would have it, did not.

Whether familiarity bred contempt or merely resignation in Pepys is uncertain: however, his fears and horror became less, or at least less expressed, and he was able to write on the 31st of December of that year 'I have never lived so merrily (besides that I never got so much) as I have done this plague-time.' Morris has pointed out that Pepys was perhaps better off than he knew in having elected to remain in the City: with many of the rich having left, houses were shut up or in the care of only a servant. Those engaged in trade were often discharged, and were forced to find accommodation in the badly infected outer parts of the City. The population in London was thus less dense (Defoe suggests in *A Journal of the Plague Year* that about 200,000 fled the City) than in suburbs such as Holborn, Cripplegate or Southwark, and deaths from the Plague were certainly greater in number, and persisted for longer, in the Out-parishes and Liberties.

Morris asserts too that it is unclear why the Plague died out. Most probably it was because either the rats or the humans developed an immunity to the bacillus. He suggests that the immunity may be attributable either to the extermination of the more susceptible stocks or, as seems to be the case, to the presence of a certain correlation between the possession of the 'A' blood group and susceptibility to the Plague. Some men seem to have blood that is less attractive to fleas than others, and if this be true it could well be that Pepys was among the first to note (or at least record) it:

> The Doctor [Timothy Clark] and I lay together at Wiards the Chyrurgeons in Portsmouth — his wife a very pretty woman. We lay very well and merrily. In the morning, concluding him to be of the eldest blood and house of the Clerkes, because that all the fleas came to him and not to me. [23rd April, 1662]

Whatever the cause might have been, the Plague did die out, and Morris states that 'The last recorded case in Britain occurred in 1679 in Rotherhithe' [Latham & Matthews, 1983, vol. X, p. 336].

Both during and after this visitation space was at a premium, then as always, in the City. Few Nonconformist meeting houses had the luxury of their own burial grounds (see, for example, the map in Chapter 2 showing the position of Joshua Bayes's meeting house in Leather Lane), and Bunhill Fields became the repository of dissenting remains. An added advantage to this site was that burials were allowed here without the services of an Anglican clergyman, the ground not having been consecrated. Indeed it became customary, in the days of the early private cemeteries in the nineteenth century, to have two chapels, usually Gothic in design with simple interiors, and linked by a single *porte cochère*. One of these would be for Anglican services and one for Nonconformist[20]. It is in fact noted in the *Bunhill Fields Burial Ground: proceedings in reference to its preservation* of 1867 that 'probably every denomination of Christians has here found a resting-place for its dead, including the Roman Catholic, and certainly the Established Church' [p. 26].

More land was added to the Bunhill Fields cemetery about 1700, and the Corporation of London itself took over the management in 1781. By the early part of the nineteenth century, however, cholera epidemics and the vast increase in population placed considerable pressure on London churchyards in general, the attendant unsanitary conditions causing Charles Dickens to write[21] that, in some churches, 'rot and mildew and dead citizens formed the uppermost scent'. In the same book he also gave the following description of a service.

> I then find, to my astonishment, that I have been, and still am, taking a strong kind of invisible snuff, up my nose, into my eyes, and down my throat. I wink, sneeze, and cough. The clerk sneezes; the clergyman winks; the unseen organist sneezes and coughs (and probably winks); all our little party wink, sneeze, and cough. The snuff seems to be made of the decay of matting, wood, cloth, stone, iron, earth, and something else. Is the something else the decay of dead citizens in the vaults below? As sure as Death it is! Not only in the cold damp February day do we cough and sneeze dead citizens, all through the service, but dead citizens have got into the very bellows of the organ, and half choked the same. We stamp our feet to warm them, and dead citizens arise in heavy clouds. Dead citizens stick upon the walls, and lie pulverised on the sounding-board over the clergyman's head, and, when a gust of air comes, tumble down upon him.

The unsavory nature of one private cemetery is described by Holmes as follows.

> One of the most notoriously offensive spots in London was Enon Chapel, Clement's Lane. The chapel was built, and the vaults under it were made, as a speculation by a dissenting minister named Howse. The burial-fees were small, and the place was resorted to by the poor, as many as nine or ten burials often taking place on a Sunday afternoon. The space available for coffins was, at the highest computation, 59 feet by 29 feet, with a depth of 6 feet, and no less than 20,000 coffins were deposited there. In order to accomplish this herculean task it was the common practice to burn the older coffins in the minister's house, under his copper and in his fireplaces. Between the coffins and the floor of the chapel there was nothing but the boards. In time the effluvium in the chapel became intolerable, and no one attended the services, but the vaults were still used for interments, so that "more money was made from the dead than from the living"—a state of affairs which existed in many of the private burial-places of the metropolis. [1896, pp. 193–194]

12.1. Introduction

Various Burial Acts empowered the Board of Health to establish new cemeteries, to close old churchyards, and, if necessary, to buy the private cemeteries[22] that had been started in the late eighteenth and early nineteenth centuries.

By 1852 more than 120,000 burials had taken place in Bunhill Fields[23], and an Order in Council led to the closure of the cemetery on the 29th of December 1853, the last person to be interred there being the fifteen-and-a-half year old Elizabeth Howell Oliver, who was buried[24] on the 5th of January 1854. Sir C. Read, chairman of the Bunhill Fields Preservation Committee, noted that

> From 1665 to 1832, when the ground was closed, 123,000 bodies were registered as buried here, and although only 5000 tombs are now discoverable, it is found that vaults are lying buried at depths varying from 6 feet to 12 feet beneath the surface. [Wheatley, 1891, I, p. 302]

As a result of various negotiations between the Ecclesiastical Commissioners for England and the London Corporation, the passing of the Bunhill Fields Burial Ground Act in 1867 ensured that the Ground would be maintained as a public open space. Various improvements were made during the nineteenth century, with less tasteful alterations being made by inimical pilots during the Second World War. As a result of this bombing the northern section[25] was cleared of the remains of monuments (a change from monuments of remains) and laid out as a garden: the Bayes & Cotton family vault is fortunately not in this section, and so it remains for our edification today (see Figure 12.3).

As we have already said, this burying ground was used extensively by Nonconformists and humanitarians. Interred here were Daniel Defoe, John Bunyan, Richard and Sarah Price[26] (Richard being referred to by Wheatley [1891, I, p. 303] as 'the great statistician'), John Eames, William Blake, Isaac Watts, and Susanna Wesley (the wife of Samuel, the mother of John and Charles, and the daughter of the Rev. Dr Samuel Annesley, who, as we have seen in Chapter 2, was one of the examiners of Joshua Bayes at his ordination)[27].

Where they exist, tombstones and vaults generally provide useful information. With the passage of time, however, such memorials become decayed, and their inscriptions, like those whom they commemorate, turn to dust. Restoration of such vaults is an expensive matter, and where there is little general interest in those entombed, one may expect the memorials to degenerate.

Although this is to some measure true of many of the tombs in the Bunhill Fields Burial Ground, the statistical community may rejoice in the fact that the Bayes family vault has been restored a number of times. Such rejoicing, however, can be no more than moderate[28], for with such restoration the original inscriptions have been changed, and indeed, as we show, some information has been lost.

It is fortunate that records of the burials in the 'Campo Santo', as Robert Southey called it, over the period of interest to us are preserved in the Public Record Office[29], London, and details of the inscriptions on the monuments are to be found in certain documents in the British Library and in some nineteenth-century books. It is less satisfying, however, to discover that even when these sources claim to be exact they differ from one to the other, and although consensus on some inscriptions can be reached we shall have to remain uncertain as to the original wording of others.

The co-ordinates of the actual position of the Bayes/Cotton vault also change. Thus we find the position given variously as $56\mathcal{E}W\ 38\mathcal{N}S$, $64\mathcal{E}W\ 50\mathcal{N}S$, $64\mathcal{E}W\ 51\mathcal{N}S$, and $66\mathcal{E}W\ 51\mathcal{N}S$. Some of these descriptions may have been occasioned by the re-labelling of the ground, however, others are probably erroneous.

12.2 A small old book

In the British Library, Add. MS. 28523, is a manuscript bearing the following description.

> *Bunhill Fields.*
> *A verbatim et literatim Copy of a small old Book found in a Cupboard at Bunhill Fields, being manifestly a Register of Burials there, in the part added thereto in or about the year 1748; & for many years there after call'd by the Name of "the New Ground" but not actually referred to by the Keeper of the Ground, on Search for a Register of Burials; but which it ought to be; it doubtless being a Register of the Burials in that part of the ground within the period it specifies viz. between Jany 25 1745 and June 8th 1753.*

Given the period of coverage, we cannot expect to find much of interest to us here. However, following this document, and in the same manuscript, we have the following.

> *An imperfect List of Names down to the letter N which occurr'd to John Rippon Son of $Rev^d\ D^r$ John Rippon on going thro' the Register of Burials.*

Here, under the heading *Names of Persons apparently worthy of note which have occurr'd to me in going thro' the Register*, we find the following.

Bays, Revd Joshua 1746   Bays, Revd Thomas 1761

12.3 Register of Burials in Bunhill Fields

Aprile y^e 15	The Revd Mr Bayes from Fonders Hall in a vault	$00 = 14 = 0$
Ap^r y^e 28	The Revd Mr Bayes from Chepside in a vault	$00 = 14 = 0$
Oct^{er} y^e 14	Mr Bays from Chep Side in a Vault	$00 = 14 = 0$

Ann West. 13th Jan 1789. Age 82. Brought from Stoke Newington.
    $\mathcal{E}\&\mathcal{W}$56 $\mathcal{N}\&\mathcal{S}$38 Vault opened $00 = 18 = 0$

Theodosia Bayes. Octr 2nd Aged 68. Brought from Clapham Common.
    $\mathcal{E}\&\mathcal{W}$64 $\mathcal{N}\&\mathcal{S}$51 Vault opened $00 = 18 = 0$

Samuel Bayes. Octr 20th Age 77. Brought from Clapham Common.
    $\mathcal{E}\&\mathcal{W}$64 $\mathcal{N}\&\mathcal{S}$51 Vault opened $00 = 18 = 0$

Thos Cotton. March 26 Age 87 brought from Hackney
    Vault opened. $\mathcal{E}$ and $\mathcal{W}$64 $\mathcal{N}$ and $\mathcal{S}$51 $00 = 18 = 0$

Rebecca Cotton. Feby. 14 Age 83 brought from Hackney
    vault $\mathcal{E}\&\mathcal{W}$64 $\mathcal{N}\&\mathcal{S}$51 $00 = 18 = 0$

Nathewell(?) Bays from Snow-hill in a vault $00 = 14 = 0$

February 16 Mr Bayes from Cripplegate parrich buried
    in a grave. $00 = 13 = 6$

Jan. 14 173$\frac{3}{4}$ Rec^d Seven Pounds Seven Shillings for the ground
    to build a Valt for Mrs Bayes from St Andrews[30]

February y^e 10 Mrs Mary Bayes from Stoke Newington in a
    Vault $00 = 14 = 0$

Aprile 8 1767 Mr Joshua Cottin from Hackney in a Vault
     $00 = 14 = 0$

Octr 7 1756. Mr Thos West from Fanchurch Street in a Vault
     $00 = 14 = 0$

This undated register[31] is preserved in the Public Record Office, London, microfilm RG4 3982. Writing of Bunhill Fields Strype records 'The Price of Burial in the Vaults, I am told is 15s.' [1720, Book 4, p. 54].

Although many of the years of burial are not given, inscriptions on the present vault allow the identification of those buried. Thus

(a) the first 'Rev$^\text{d}$ M$^\text{r}$ Bayes' is Thomas and the second Joshua,

(b) 'M$^\text{r}$ Bays from Chep Side' refers to John, the son of Joshua and Ann,

(c) 'Nathewell' is Nathaniel, and

(d) 'Mr Bayes from Cripplegate parrich' is John, Joshua's brother. This identification is strengthened by the fact that the burial of this person occurred on the 16th of February, the *Familiæ Minorum Gentium* giving John's death as 13 February 173$\frac{2}{3}$, whereas the vault was only erected in the following year.

12.4 Inscriptions on vault

12.4.1 Inscriptions from Rippon

In the British Library is a collection of manuscripts, Add. MSS. 28513–28523, under the general title *Collections relating to the Dissenters' burial-ground at Bunhill Fields, London. by John Rippon, D.D. F.S.A.* The collection is undated, but according to the *Dictionary of National Biography* Rippon began collecting materials relating to Bunhill Fields in 1800.

The third volume of this collection, [Add. MS. 28515], is *An Alphabetical List of all the Tombs and other Monuments in Bunhill-Fields Burial Ground in the City Road with their respective Situations and the last Lines of the Inscriptions thereon up to 18–*. Here we find the following.

Bayes and Cotton

Tomb E.W.64 N.S.51 S. Side of top } aged 59 years

S. Side N$^\text{r}$ Base } in the 83$^\text{d}$ year of her age

E. End or Edge of Top } aged 10 months

N. Side of Top   aged 76 years

N. Side N$^\text{r}$ Base } aged 77 years

W. End   aged 27 years

12.4. Inscriptions on vault

In the fourth volume of this collection, [Add. MS. 28516], the following inscriptions may be found; and although, as we have already said, one cannot say when the collection was compiled, it would appear, in view of the expected names that are missing, that this list was drawn up before 1789 when Theodosia and Samuel Bayes were buried (the list in MS. 28516 says, of the Bayes entry, 'This is now removed and a Tomb erected in its place'):

Bay^{e}s $\mathcal{H}.\mathcal{S}.$ 64$\mathcal{EW}$ 50$\mathcal{NS}$ (N° 448. Right S). In this Vault lies interr'd the Bodies of Mrs Ann Bays wife of the Revd Mr Joshua Bays who died 7 Jany 1733 aged 57

John Bays Esqr Son of the Revd Mr Jos: & Mrs Ann Bays died 11 Octr 1743 aged 38.

The Revd Mr Joshua Bayes died 24 April 1746 aged 75.

Mr Thos West died 30 Octr 1756 aged 53

The Rev^d Mr Thos Bays Son of Joshua & Ann died 7 of April 1761 aged 59

Mr Nathaniel Bays died 26 Nov. 1764 aged 42

Mrs Mary Bays died 2 Feb^{ry} 1780 aged 76

It might be noted that, in the entry given above for Thomas Bayes, the actual entry in the body of the text reads '... Bays (†) died ...' while an appropriately marked footnote runs 'Son of J., Ann'.

12.4.2 Further inscriptions from Rippon

The fifth volume, part 1, of Rippon's collection, [Add. MS. 28517A], is entitled 'Inscriptions A–BL'. Here we find the following details.

Bayes & Cotton Vault (Stone tomb, Cottage Top.) 64EW 51NS

South Side of Top

Mrs Ann Bayes, Wife of the Revd Joshua Bayes died Jany 7th 1733 aged 57 years

John Bayes Esqr Son of the said Joshua & Ann Bayes died Octr 11th 1743 aged 38 years

The Revd Joshua Bayes died April 24th 1746 aged 75 years

Mr Thomas West died Octr 3rd 1756 aged 52 years

The Revd Thomas Bayes Son of the said Joshua & Ann Bayes died April 7th 1761 aged 59 years

North Side of Top

Mr Nathaniel Bayes, Son of the said Joshua and Ann Bayes died Novr 26th 1764 aged 42 years

Joshua Cotton Esqr Grandson of the said Joshua & Ann Bayes died March 30th 1767 aged 28 years

Mrs Mary Bayes daughter of the said Joshua & Ann Bayes died Feby 2nd 1780 aged 76 years

North Side next the base

Mrs Ann West Widow of the said Mr Thomas West and Daughter of the said Joshua & Ann Bayes died Decr 31st 1788 aged 82 years

Theodosia Bayes, Wife of Samuel Bayes Esqr, died Septr 22nd 1789 aged 68 years

Samuel Bayes Esqr Son of the said Joshua & Ann Bayes died Octr 11th 1789 aged 77 years

South Side next to the Base

Thomas Cotton Esqr died March 23^d 1797, aged 87 years

Rebecca Cotton Widow of the said Thomas Cotton Esqr and daughter of the said Joshua & Ann Bayes died Feby 7th 1799 in the 83^d Year of her Age.

East Side

On Edge of Top

The Vault of the Families of Bayes & Cotton

Thomas Bayes Cotton Son of Bayes Cotton and Sarah his Wife & Great Grandson of the said Joshua & Ann Bayes died March 21st 1787 aged 10 Months

West End

Miss Decima Chance, late of Upton, upon Severn in the County of Worcester; Sister of the said Mrs Sarah Cotton, died April 12th 1795 aged 27 years.

Rippon ends this record with the words 'This Inscription taken before tomb was re-cut & repainted.' (Bayes Cotton was one of the mourners at Richard Price's burial in Bunhill Fields on the 26th of April, 1791.)

12.4.3 *Inscriptions on present vault*[32]

South Side of Top

> Mrs Ann Bayes wife of Rev Joshua Bayes (57) 7 Jan 1733
> Rev Joshua Bayes (75) 24 April 1746

Mr Nathaniel Bayes son of the said Joshua and Ann Bayes (42) 26 Nov 1764

Joshua Cotton great-grandson of the said Joshua and Ann Bayes (28) 30 March 1767

Mary Bayes daughter of the said Joshua and Ann Bayes (76) 2 Feb 1780

North Side of Top

Rev Thomas Bayes Son of the said Joshua and Ann Bayes (59)
7 April 1761
In recognition of Thomas Bayes's important work in probability
this vault was restored in 1969 with contributions received
from statisticians throughout the world

North Side

Mrs Ann West widow of the said Mr Thomas West
and daughter of the said Joshua and Ann Bayes
21 December 1758
Theodosia Bayes wife of Samuel Bayes Esq (68)
22 Sept 1769
Samuel Bayes Esq Son of the said Joshua and Ann Bayes (77)
11 Oct 1789

South Side

John Bayes Esq
Son of the said Joshua and Ann Bayes (38)
11 Oct 1743
Thomas West (52) 3 Oct 1756
Thomas Cotton

East Side

Vault of the families of
Bayes and Cotton
Thomas Bayes Cotton
Son of Bayes Cotton and Sarah
his wife and great-grandson of the said
Joshua and Ann Bayes (10)
21 March 1787

West Side

Miss Decima Cotton
Late of Upton upon Severn
In the County of Worcester
Sister of the said Sarah Cotton (27)
12 April 1795

Several comments on the above present inscriptions may be made. In the first place note that Joshua Cotton is mistakenly described as the *great-*grandson of Joshua and Ann. Next, Rebecca's name is missing from the vault, and the age at death of Thomas Bayes Cotton, great-grandson of Joshua and Ann, is given as 10 rather than 10 *months*. The year of Theodosia's death (1789) is mistakenly given here as 1769: had the latter been correct her death would doubtlessly have been given in Rippon's list quoted in §12.4.1. Finally, Decima Chance is incorrectly surnamed 'Cotton', and 'Worcester' is incorrectly given as 'co. Glouc.' in the *Familiæ Minorum Gentium*[33].

12.4.4 Inscriptions from Jones

In 1849 J.A. Jones[34] published a collection of inscriptions, taken from tombstones and vaults of 'three hundred ministers and other persons of note' in Bunhill Fields, prefacing his remarks with the following observations.

> In every *practicable* instance, the *inscriptions* on the tomb or gravestone are inserted *verbatim*, which this mark (†) will certify: and the *numbered intersections* of the ground, will conduct the footsteps direct to the desired object of search.
> [1849, pp. i, ii]

The following inscriptions, 'from authentic sources', are pertinent.

> †Rev. Joshua Bayes, many years Minister of the gospel in Leather Lane; at length, after a series of laborious and useful services, he was called home to his reward on the 24th April 1746, in the 76th year of his age, and the 53rd of his Ministry.
> [Tomb, $\mathcal{E}$. and $\mathcal{W}$. 66, — $\mathcal{N}$. and $\mathcal{S}$. 51]

> †The Rev. Thomas Bayes, son of the said Joshua, died April 7th, 1761, aged 59 years.

12.5 Bayesian wills

The wills of many members of the Bayes family may be found in the Public Record Office, London. We quote Joshua's and Thomas's in some detail, but restrict ourselves in the main only to some particulars of those of the latter's siblings.

In the Name of God Amen

I Joshua Bayes Minister of the Gospel now living with my Son at the Sign of the Black Lyon in Cheapside London Do make and constitute this my last Will and Testament in manner following that is to say I give and bequeath to my Eldest Son Thomas Bayes all my Books and Two Thousand pounds in Money. and to my Eldest Daughter Mary Bayes One Thousand Eight Hundred pounds and to my Daughter Ann West One Thousand Four Hundred pounds and to my Son Samuel Bayes One Thousand Four Hundred pounds and the Messuage or Tenement in Ingram Court in Fenchurch Street he paying the Rent and performing the Conuenants[35] in the Lease on the Tenants part. and to my Daughter Rebecca Cotton One Thousand Four Hundred pounds. and to my Son Nathaniel Bayes One Thousand Six Hundred pounds. And Whereas I have in my hands the Sum of Six Hundred pounds belonging to my Daughter Mary and the Sum of Five Hundred pounds belonging to my Son Nathaniel I will and appoint that the said Sums of Money be paid to them over and above their Legacys they giving proper discharges for the same. But without any interest I having been at the Charge of their Maintenance and Education and having given them the Legacys aforesaid and therefore I will that they Release my Executors from all demands of Interest and that my Executors release them from all demands for Maintenance and Education. And I further will that all my Rings Plate Linnen Household Goods Household Stuff and Furniture be equally divided among my Three Daughters Mary Ann and Rebecca. And I moreover give to my sister Elizabeth Delarose One Hundred pounds and to my sister in law Sophia Bayes Fifty pounds. And all the Rest and Residue of my Money Goods Chattles and Estate whatsoever I give and bequeath to my Son Thomas Bayes. And I make and constitute my said Son Thomas Bayes and my Son Samuel Bayes Executors of this my last Will revoking all former Wills by me made. In Witness whereof I the said Joshua Bayes have to this my last Will and Testament set my Hand the Fifteenth day of January in the year of our Lord 1744 *S. Joshua Bayes.*

The drawing up of this will was followed on the 25th February of the same year by a long codicil in which Rebecca's legacy was revoked, she being left now 'only the Sum of Forty pounds for Mourning'. Samuel and Thomas were left £1,400 in Trust 'so that the same may not be subject to his [i.e., Thomas Cotton's] Debts or Control(?)'. On Rebecca's death the trust money was to go to her son Joshua Cotton, or his children should he die before the age of twenty-one; and if these children should die before Rebecca, the money was to go to her other child or children. If Thomas Cotton were to die, Rebecca would inherit the whole amount, whereas if she predeceased her husband, and if their son Joshua was already dead, the money was to be shared between Thomas, Samuel, Mary, Ann, and Nathaniel. Finally the trustees were empowered to use some of the money to put out Joshua Cotton apprentice 'or otherwise for his advancement'.

One might infer from this codicil that something had gone wrong with Thomas Cotton's business affairs: according to the *Familiæ Minorum Gen-*

tium he was an attorney-at-law, and Joshua presumably wanted to ensure the financial security of his daughter and grandchild (or grandchildren).

𝕿𝖍𝖎𝖘 𝖎𝖘 𝖙𝖍𝖊 𝕷𝖆𝖘𝖙 𝖂𝖎𝖑𝖑 and Testament of me Thomas Bayes of Tonbridge Wells in the County of Kent. First I give and bequeath to my Sister Mary Bayes Five Hundred pounds. Also I give and bequeath to my Brother Nathaniel Bayes Four Hundred pounds. Also I give and bequeath to my Sister Rebekah Cotton Five Hundred pounds. Also I give and bequeath to my Nephew Bayes Cotton Five hundred pounds. Also I give and bequeath to my Aunt Wildman to my Cousin Elias Wordsworth and my Cousin Samuel Wildman twenty pounds apeice. Also I give and bequeath to John Hoyle late preacher at Newington and now I suppose at Norwich and Richard Price now I suppose Preacher at Newington Green two hundred pounds equally between them or the whole to the Survivor of them. Also I give and bequeath to Richard Jeffery and Sarah Jeffery Son and Daughter of Richard Jeffery of Mount Sion Tonbridge Wells One Hundred pounds equally between them or the whole to the Survivor of them. Also I give and bequeath to Sarah Jeffrey daughter of John Jeffrey living with her father at the corner of Jourdains Lane at or near Tonbridge Wells five Hundred pounds and my watch made by Ellicot and all my linnen and wearing apparell and Household Stuff. And as to and concerning all the rest and residue of my personal Estate not herein before given and disposed of after payments of my debts legacys and Funeral Expenses which I desire may be as Frugal as possible I give and bequeath the whole to my Nephew Joshua Cotton. And I do hereby constitute and appoint my Brother Nathaniel Bayes and my Nephew Joshua Cotton Executors of this my Will. In Witness whereof I hereunto set my hand this twelfth day of December in the year of our Lord one thousand seven hundred and Sixty. *Thomas Bayes.*

𝕬𝖕𝖕𝖊𝖆𝖗𝖊𝖉 𝕻𝖊𝖗𝖘𝖔𝖓𝖆𝖑𝖑𝖞 William Finney of the Parish of Saint Michael Le Quern London Linnen Draper and Samuel Bays of the Parish of Saint Mary Magdalen in Milk Street London Linnen Draper and made Oath that they knew and were well acquainted with Thomas Bayes late of Tonbridge Wells in the County of Kent deceased and have often seen him write and subscribe his Name whereby they became well acquainted with his manner and Character of handwriting and having now seen and carefully viewed and perused the paper writing hereunto annexed purporting to be the last Will and Testament of the said deceased beginning thus This is the last Will and Testament of me Thomas Bayes and ending thus In Witness whereof I hereunto set my hand this twelfth day of December in the year of our Lord One thousand seven Hundred and Sixty and thus Subscribed Thomas Bayes they do jointly and severally depose and say that they do verily and in their Consciences believe the whole Series and Contents of the said Will together with the said name Thomas Bayes thereto subscribed to be all of the proper manner and Character of handwriting and subscription of him the said Thomas Bayes deceased. W^m. *Finney.* Saml. *Bayes.* 30th. April 1761. The said William Finney and Samuel Bayes were sworn to the truth of this Affidavit Before me *James Marriott. Surr. pres^t Robt Longden. N. P.*[36]

This Will was proved at London on the Second day of May in the year of our Lord One Thousand Seven Hundred and Sixty One before the Worshipfull James Marriott Doctor of Laws and Surrogate of the Right Worshipfull Edward Simpson Doctor of Laws Master Keeper or Commissary of the Prerogative Court of Canterbury lawfully constituted by the oath of Nathaniel Bayes and Joshua Cotton the Executors named in the said Will To whom Administration was granted of all and singular the Goods Chattles and Credits of the deceased they having been first Sworn duly to Administer. Ex^d.

Although it may be deduced from his testamentary disposition that Thomas was comfortably off, one might infer that he was not as warm as his father. Possibly the life-style in Tunbridge Wells was more demanding than that in London; perhaps ill health had forced him to resign his cure, and resulted in the unwilling enrichment of the medical practitioner; or perhaps Thomas was merely frugal, did not enjoy a large stipend, and so managed to pass on to his surviving relatives most of what he had inherited from his father. We do not know: but the moneys specifically mentioned in his will were not insignificant (Roger Farthing suggested in 1990 that £2,000 might be as much as £500,000 at that time), and surely there would have been a reasonable sum left over for Joshua Cotton[37].

We have already reminded the reader of Sir Thomas Browne's concern with the song the sirens sang, and he who studies Thomas's will may be similarly perplexed by the identity of Sarah Jeffrey (Roger Farthing has suggested that this Sarah and the Richard and Sarah Jeffery also mentioned in the will were probably related — perhaps even cousins-german). In an earlier chapter we presented Farthing's theory that she might have been the daughter of the Jeffreys who kept a lodging house in which Thomas might have boarded, for as Britton wrote in one of his 'Hints to Visitors',

> Tunbridge Wells is calculated to afford domestic accommodation to almost every class of visitors, from the prince or princess,

with appropriate retinue and household, to the solitary batchelor in sulky singleness [1832, p. xii]

and it would not be inappropriate to pursue this notion a bit further here. In this pursuit we are in a large measure helped by Farthing's painstaking research in the history of Tunbridge Wells and in particular the myriad Jeffreys and Jefferys to be found there in the eighteenth century (see the genealogy compiled by Farthing and shown here as Figure 12.2).

One possible candidate is the Sarah Jeffrey listed in the *International Genealogical Index* — hereafter 'Sarah-IGI' for convenience — as having been baptized on the 29th January 1724. Farthing, in a private communication, notes more accurately that the original entry in the Speldhurst register records the year of birth of this 'daughter of John Jeffery and Sarah his wife', in the old style, as 1723. Farthing also records the marriage of John Jeffery of Tonbridge and Sarah Taylor in Bidborough Church on 1st January 1728 (o.s.), a date of which he is certain as there exists a marriage settlement, to which this last-named Sarah was party, in which the bridegroom's mother Bridget agrees to give her house at 69 London Road to her son and 'his intended wife Sarah Taylor'. Farthing also believes it unlikely that there would have been a birth several years before the marriage.

It would appear that the Sarah whom we seek ('Sarah-T') was unmarried on the 12th December 1760 when Thomas's will was drawn up; but in her father's will, dated 11th May 1769 and proved 5th July 1772, she is referred to as married. Sarah-IGI was married twice: the first such union, consecrated on the 3rd November 1762, was with Robert Jeffery, a wheelwright[38], and the second marriage was to the tallow-chandler Richard Jeffery on the 5th May 1775. It thus seems possible that this Sarah could have been the one remembered by Thomas, though documents in the Kent Archives indicate that she was still alive in 1812, which would make her a venerable age. (Farthing has found that a Sarah Jeffery was buried in Tonbridge on the 19th of December 1817 aged 94.)

The identification of Sarah-IGI with Sarah-T is, however, viewed with some suspicion by Farthing. He suggests that Sarah-T might have been thirty or forty years younger than her benefactor, which could have made her a teenager when Thomas died and in her early seventies in 1812. Another point against the seeing of Sarah-IGI as Sarah-T is the fact that the lady in question would have been almost thirty-nine years old when she married Robert the wheelwright, and since she is known to have had four children a marriage at that age seems unlikely.

The number of carriers of the surname 'Jeffrey', in one form or another, complicated by the apparent predilection for Christian names such as 'Richard' for the male line and 'Sarah' on the distaff side (in defiance of part of the line from the old music-hall song 'None could be fairer nor rarer

12.5. Bayesian wills 521

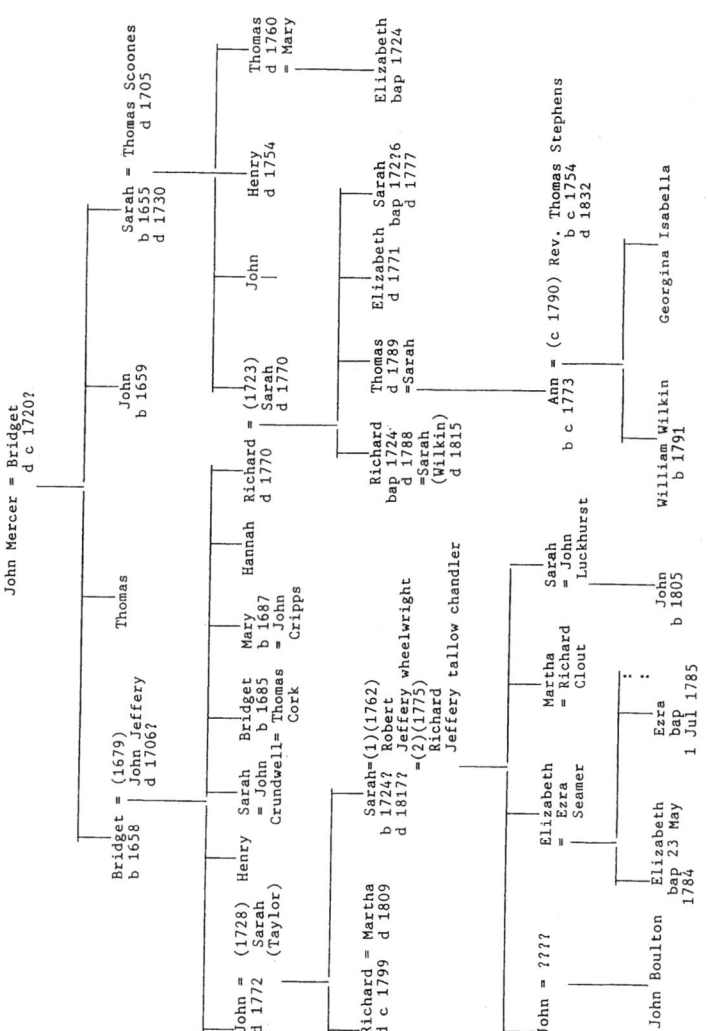

FIGURE 12.2 Jeffery & Jeffrey Family Tree.

than Sarah'[39]), makes it unlikely that we shall be able to identify Sarah-T with certainty. In the absence of any more likely candidate I would have to plump for Sarah-IGI, but the cogency of Farthing's argument against her cannot be ignored.

John Hoyle preceded Price as preacher at Newington Green, the latter spending twenty-five years there from 1758 [Laboucheix, 1982, p. 14].

We now pass on to the details of the wills of Thomas's siblings and, where possible, their spouses.

(a) John Bayes's will is a model of succinctness, and runs in full as follows.

> I John Bayes of Lincoln's Inn Esq. do hereby constitute and appoint my Brother Samuel Bayes my sole executor and give him all my Estate and Effects whatsoever. John Bayes June 5th 1742, of Clapham in the County Surrey, Esq.

(b) In Thomas West's will of the 11th(?) February 1744, Thomas Bayes was left £200.

(c) By the will of Nathaniel Bayes of Snow Hill, London, grocer, dated the 4th November 1764, £500 was left to each of Samuel, Mary, Ann, Rebecca, Joshua Cotton, Samuel Wildman (a cousin[40]), and Nathaniel's partner Mr Guy Warwick. (According to *Kent's Directory* the firm continued as 'Bayes & Warwick, Grocers, No. 48 Snow-hill' until 1788, when the name was changed to 'Warwick, Guy & Guy'.) To his nephew Bayes Cotton, Nathaniel left £1,000, with a further £100 to the Rev. Michael Pope (who was assistant to Joshua at the Leather Lane chapel from 1742 to 1746, and was then sole pastor there for forty-two years) and £20 to each of two journeymen. Further bequests were made to his two doctors and the son of one of these 'for cloaths and to put him out apprentice', and also to his servant, the servant of his sister Mary, and the servant and the footman of his sister Ann. Other bequests of £100 each went to a Mrs Mary Cromwell of Pater Noster Row and her two sons and one daughter, and £20 to each of two charity schools. The rest and residue of his personal estate and effects went to his brother Samuel. Finally, 'I desire to be interred in the same decent manner as my Brothers but not to have any Boys to sing at my funeral.'[41]

(d) Mary Bayes, spinster, of Stoke Newington in the County of Kent, by her will of the 8th July 1769, 'direct[ed] that my Funeral be plain and not expensive but I want and desire to be buried in the Burying Ground in Bunhill Fields'. Among other bequests she left £1,500 to each of Samuel and Rebecca, £1,000 to her sister Ann, £500 to her nephew Bayes Cotton, and £100 to the Revd. Michael Pope.

(e) Samuel Bayes left all his estate to his wife Theodosia. In the event of her predeceasing him, everything was to go to his nephew Bayes Cotton,

12.5. Bayesian wills 523

FIGURE 12.3 The Bayes Family Vault.

or to a Trust administered by Benjamin Cotton, Samuel's brother-in-law (and therefore presumably a brother of Thomas Cotton, the husband of Samuel's sister Rebecca). This will, dated the 1st September 1788, was followed by a codicil occasioned by the death of Samuel's sister Anne West, in which more was left to Theodosia and Bayes Cotton. It was dated the 3rd February 1789, and a further codicil of the 3rd October of that year established a trust for 'a youth called or known by the name of James Thomas ... son of a sister(?) of my late wife'.

It is quite clear from this last codicil that Theodosia predeceased Samuel. Further, I do not find a Benjamin Cotton in the Cotton stemma in the *Familiæ Minorum Gentium*: could Thomas Cotton be meant?

(f) Thomas Cotton, of Hackney, Middlesex, Gentleman, by his will of the 6th June 1792, left his wife Rebecca £500, with £300 and £200 going to Bayes Cotton and his wife Sarah respectively. And (a nice touch) 'to my dear Grandson Willm Cotton my pretty watch that I usually wear and to my dear Grandson Saml Cotton my gold watch.'

12.6 Speldhurst memorials

In Sprange [1797] records of the memorials to John Archer and William Johnson, both of whom were associated with Thomas Bayes, may be found. Archer's memorial is (or was, at the time Sprange was writing), 'On the South side of the Chancel, in the Church Yard ... on an Altar Tomb-stone'.

> Hic requiescit
> Quod mortale fuit
> D. JOHANNIS ARCHER,
> viri vere reverendi
> ob singularem vitæ modestiam,
> Illœsam morum integritatem
> Sinceram erga Deum religionem,
> liberalem in omnes benevolentiam,
> et veritatis Evangelicæ
> (apud aquas Tunbridgienses
> Inagro Cantiano)
> Fidelissimam prædicationem.
> Qui, dui oppressus viscerum doloribus
> Quos fortiter et patienter tulit
> Requiem futuri seculi ardenter expetivit,
> Quam tandem summo gaudio consecutus est;

> Virtutum rarissimarum
> Triste nobis desiderium relinquens,
> Nat. Martii 15, A.D. 1675,
> Den. Sep. 23, 1733.
>
> Hic jacet etiam
> ANNA ARCHER,
> Fidelissimo Conjux ejus
> Obiit July 14, 1750, ætatis suæ 78.

Immediately following his recording of this inscription Sprange notes that Johnson's memorial is 'underneath, Upon the before-mentioned Tomb'. The inscription runs as follows[42].

> Here are deposited the Remains of
> The Rev. WILLIAM JOHNSON, A.M.
> Twenty-four Years Pastor of the
> *Dissenting Congregation Meeting*,
> On Mount-Sion, Tunbridge-Wells.
>
> His Character,
> distinguished by Learning, Piety,
> Benevolence and Usefulness,
> will cause his Death to be
> long lamented by his Family,
> his Flock, and his numerous Friends.
>
> He departed this Life
> in the believing Hope of a better,
> on March 2, 1776;
> in the sixty-second Year of his Age.
>
> *Quis desideris sit pudor, aut modus tam chari*
> *capitis?*

Notes

Preface

1. See Kingsley Amis's *Lucky Jim*, [1954, chap. 1].
2. See Stigler [1984].
3. For various forms of Boscovich's Christian names see Farebrother [1999, §2.5].
4. See Stigler [1984] for further details.
5. For a detailed study of the calculus of observations see Farebrother [1999].
6. See Clarke [1929] and Whyte [1961].
7. See Lynd [1928, p. 2].
8. Charles Dickens, *Our Mutual Friend*, vol. I, Book II, chap. VIII.

Chapter 1

1. See F. Darwin [1887, vol. I, p. 83].
2. See F. Darwin [1887, vol. I, p. 52].
3. The quotation given here is from the 1819 edition (the fourth) of Laplace's *Essai philosophique sur les probabilités* as printed in the third edition of the *Théorie analytique des probabilités* of 1820: the wording changed over the years, but the sentiment was essentially the same (see Dale [1995, p. 114]).
4. See Russell [1961, p. 527].
5. A quotation from Sheridan's *The Rivals*, Act III, Sc. 3.
6. See the *Essai philosophique sur les probabilités* [1820, p. clvi].
7. Suetonius, in *The Twelve Caesars*, notes that Tiberius amused himself by asking these questions, together with things like 'Who was Hecuba's mother?', of professors of Greek literature. One wonders how often the latter thought, with Shakespeare, 'What's Hecuba to him, or he to Hecuba'.
8. See ΓΩΝΘΙ ΣÈΑΥΤΟΝ. *Know Yourself* in the second volume of Arbuthnot's *Miscellaneous Works* of 1770, p. 253.

9. Horace *Odes* iv. 8. 28. 'The Muse forbids the man worthy of praise to die' (would the Muse concerned be Clio?).
10. See Stigler [1984].

Chapter 2

1. See Passler [1971, p. xii].
2. See Stigler [1999, p. 190].
3. One should of course bear in mind that even the chaff may serve a useful purpose.
4. On the 12th December 1726 Philip Doddridge, then some twenty-four years old, wrote a letter to the Rev. Samuel Clark, in which he said

   > As I am a stranger at Thorp [near Peterborough], I could wish that you would send a few lines to Mr Goodrich, at Mrs King's, in Oundle, and to Mr Some at Mr Bayes's, in Harborough .... [1829, vol. II, p. 234]

   I do not know how this Bayes fits into the family under discussion here.

5. Before giving the Bayes family tree Clay mentions two people surnamed Bayes (or Base) who were apparently not related to this family. A.G. Matthews, in his 1934 revision of Calamy, gives the following details.

   > Robert Bayes. Vicar of Weybread, Suffolk. 1662.
   > d. 8 December 1702 aged 73.
   > Son of Thomas. Family from Norfolk.
   > Wife Rebecca.
   > Children: Bezalel, 28 September 1654.
   >   Rebecca, 5 September 1656.
   >   Sam. Bayes, perhaps son, V. of Weybread 1684.

   Whether these were members of the Bayes family of interest to us I cannot say.

6. Whitam notes that

   > The Company of Cutlers in Hallamshire, or to give it its correct title, Master, Wardens, Searchers, Assistants and Commonalty of the Company of Cutlers in Hallamshire in the County of York, was founded in the year 1624.
   > [1948, p. 3]

7. The figures given in this quotation, and a later one, as 0 12 0 stand for £0. 12*s*. 0*d*.

8. According to Venn & Venn [1922], there were three ranks in which students were matriculated in Cambridge University: the first was that of *Fellow-commoner*, the second that of *Pensioner*, and the third *Sizar*, the latter performing many menial tasks.
9. See Matthews [1934].
10. The restrictive and oppressive Acts passed between 1661 and 1670, aimed generally at curbing dissent, are known as the Clarendon Code. These statutes, the Corporation Act of 1661, the Act of Uniformity of 1662, the Five Mile Act of 1665, and the Conventicle Act of 1671, 'not merely defined the position of the Church but imposed heavy disabilities, both religious and political, on all who failed to conform to it' [Browning, 1953, p. 360].
11. The date of the Feast is variously given as August the 24th and 12th (the 14th edition of 1939 of the *Encyclopædia Britannica*), August the 22nd (W.B. Forbush's edition of *Fox's Book of Martyrs* of 1926), August the 23rd (Charles Dickens's *A Child's History of England*), and August the 24th (old style) and September the 3rd (*Haydn's Dictionary of Dates and Universal Information*, ed. B. Vincent, of 1904).
12. See Hastings [1967, vol. 9, p. 387].
13. I suspect that 'S' stands for 'South'.
14. Lewis described the living at this time as follows:

    > a discharged vicarage, valued in the king's books at £8; net income, £135; patrons and impropriators, the Master and Fellows of Trinity College, Cambridge.
    > [1849, vol. II, p. 338]

15. Presumably the 'a', 'r', and 'p' stand for the square measurements *acre*, *rood*, and (square) *pole*, respectively. (1 acre = 4 roods, 1 rood = 40 square poles, 1 square pole = $30\frac{1}{4}$ square yards.)
16. One who refused to take the oath was also forbidden to keep any school [Hastings, 1967, vol. 9, p. 389].
17. Dale [1907, p. 416] notes that a 'lecturer' was a clergyman appointed to preach but without charge of a parish.
18. It was only in 1689 that the Act of Toleration officially ended the persecution of dissenters. In 1688 the yearly income per family was £60 for 'eminent clergymen' and £45 for 'lesser clergymen'. Compare this with temporal lords (£2,800), gentlemen (£280), persons in sciences and liberal arts (£60), common seamen (£20), and common soldiers (£14) [Browning, 1953, p. 516]. Rogers [1978, pp. 51–52] records that the average family's annual income in the mid-eighteenth century was

around £40. The amount a clergyman of that time could earn is mentioned in Henry Fielding's *Joseph Andrews* of 1742, where the curate Abraham Adams, at the age of fifty, 'was provided with a handsome income of twenty-three pounds a year' [Book I, chap. III] (by the end of the book, however, Mr Adams has been presented 'with a living of one hundred and thirty pounds a year'). Somewhat later, Goldsmith describes the village preacher in *The Deserted Village* as follows.

> A man he was, to all the country dear,
> And passing rich with forty pounds a year.

These figures are not fictitious: Calamy records in his *Life* [1830, vol. I, p. 316], that in 1692 he was offered £100 a year to minister to a dissenting congregation at Bristol. For various reasons he rejected the call, taking up an assistantship to Matthew Sylvester at Blackfriars for £40 per annum (in 1685 men servants were paid £4. 10s. 0d. per annum, and maid servants got two pounds less. In 1683 a dozen flambeaux cost 16s., as did a peruke for the page in 1684: in 1696 a peruke 'for my master' cost £6 — see Browning [1953, pp. 469–472]). Even this amount proved a drain on the congregation, and Sylvester, though scrupulous in the quarterly payment of his assistant, frequently found himself in straitened circumstances.

19. See Latham & Matthews, vol. X [1983].
20. 'Pr.' may stand for 'Preacher', 'Private', or possibly even 'Presbyterian'.
21. Lewis provides the following description of Sankey.

> Sankey, Great, a chapelry, in the parish of Prescot, union of Warrington, hundred of West Derby, S. division of Lancashire, $2\frac{3}{4}$ miles (W.) from Warrington; containing 567 inhabitants. ... The township comprises 1909a. 25p., and is bounded by Sankey brook. ... The living is a perpetual curacy; net income, £103; patron, Lord Lilford; impropriators, the provost and fellows of King's College, Cambridge; the tithes belonging to the college have been commuted for £130, those of the rector of Warrington for £65, and those of the vicar of the parish for £75, per annum. [IV, p. 19]

Carlisle [1808] gives the population 'of this township' in 1801 as 431.

22. It would be interesting to know whether this Sarah Pearson was an ancestress of Karl Pearson, biometrician, statistician, philosopher of science, etc.

23. One John de la Rose was tutor at an academy that flourished at Attercliffe from 1691 to 1744 (these dates are given by Jeremy [1885, p. 25] as 1690–1696). Could they be the same person? (Creasy's *Index to the John Evans List* shows that Elizabeth's husband died in 1723.)
24. Copy in the Guildhall Library, London.
25. In addition to the sources cited in the text the reader may be referred to Goldstein [1996]: see also Parker [1914, p. 50].
26. The reader is referred to Appendix 2.1 of this chapter for a discussion of the differences between dissent and Nonconformity.
27. See Parker [1914, p. 65].
28. In connexion with the rivalry between the academies and the universities, Parker notes that, after the attempt to found a new university at Stamford in 1334, every graduate of Oxford or Cambridge was obliged to swear an oath in the following vein.

> You shall also swear that in that Faculty to which you are now admitted Graduate, you shall not solemnly perform your readings as in a University anywhere within this kingdom but here in Oxford or in Cambridge; nor shall you take degrees, as in a University, in any Faculty whatsoever, nor shall you consent that any person who hath taken his Degree elsewhere shall be admitted as a master here in the said Faculty, to which he shall be elsewhere admitted.
> You shall also swear that you will not read lectures, or hear them read, at Stamford, as in a University study, or college general. [1914, p. 66]

It was for the alleged breaking of this oath that Frankland, M.A. (Christ's College, 1655) was persecuted and indeed excommunicated, a decision that was later removed by command of William III.

Frankland was not unique in undergoing this experience: in 1733 an ecclesiastical prosecution was begun against Philip Doddridge 'for keeping an Academy in Northampton.' Once again the Crown intervened to stop it, George II saying 'During my reign there shall be no persecution for conscience' sake' [Dale, 1907, p. 518].

29. See Bogue & Bennett [1808, vol. I, p. 296].
30. See Dale [1907], Holland [1962, p. 452], and Parker [1914, p. 50].
31. From an editorial footnote to Calamy [1830, vol. I, p. 128].
32. Mallet [1924, vol. II] records that at the University of Oxford, students for the Bachelor of Arts degree were required by the Laudian Code (first copies printed in 1634, the revision of 1636 being first printed in 1768), to read Grammar (including Latin), Rhetoric, Logic, Metaphysics, Dialectic and Moral Philosophy, and Natural Philosophy.

33. As late as 1867 C.S. Parker could write 'the great English schools do not yet teach English' [p. 73].
34. Joseph Bennett, later of the Old Jewry Chapel, died on the 21st February 1726; Edmund Calamy, of Westminster, died on the 3rd June 1732; Thomas Reynolds died on the 25th August 1727; and William King died on the 4th of March 1769.
35. This meeting-house was described by White as 'of moderate size with three good galleries' [1901, part II, p. 35]. In a handwritten note to this Part of White's book in a copy in Dr Williams's Library the Rev. W.T. Whiteley writes

> Mr. White doubtless has excellent qualifications for the work that precedes this, on the old *churches* existing before the fire. But he is quite out of his element in dealing with the Dissenting Meeting Houses.

36. Colligan [1915, pp. 75–76].
37. This passage is also related, with minor differences, by Bogue and Bennett [1809, vol. II, pp. 121–122].
38. Presumably the 'usual questions' were those of either the Larger or the Shorter Westminster Catechism, drawn up by an assembly of divines at the instigation of the Long Parliament in 1646–1647. The Larger Catechism consisted of 196 questions and answers, beginning 'What is the chief and highest end of man? — Man's chief and highest end is to glorify God and fully to enjoy him forever.' Here are set out

> the doctrines of God, His decrees, Creation, the Fall, Sin and its Punishment, the Covenant of Grace, Christ the Mediator, His Offices, Humiliation, and Exaltation, and eternal Work, the Church, Membership in Christ, the Experience and Contents of Salvation, Future Judgment, the Commandments as Christian Duties, man's inability to keep them, the special aggravations of Sin, the Means of Grace — the Word, the Sacraments, and Prayer, with the proper meaning and use of each, the Lord's Prayer being expounded at the close. [W.A. Curtis, in Hastings, 1971, vol. 3, p. 877]

The Shorter Catechism, which appeared in 1648, was used rather than the Larger by the Church of Scotland as a 'Directory for catechizing such as are of a weaker capacity'. Here are 107 questions and answers, the first and last as in the Larger Catechism, but 'with even simpler and happier answers' [loc. cit.]: for instance, the answer given here to the first question is 'Man's chief end is to glorify God, and to enjoy him for ever'.

In an editorial footnote to this passage from Calamy's *Life* Rutt writes 'See "The Confession, &c. of Public Authority in the Church of Scotland," (1753) p. 536.'

39. The questions debated by the other candidates for ordination are given by Calamy [1830, vol. I, p. 350] as follows: Bennett: 'An Resurrectio Corporis sit Articulus Fidei fundamentalis?' *Aff.*; Reynolds: 'An Resipiscentia sit necessaria ad Peccatorum Remissionem?' *Aff.*; Calamy: 'An Christus Officio sacerdotali fungatur in Cœlis tantum?' *Neg.*; Hill: 'An omne Peccatum sit mortale?' *Aff.*; Bradshaw: 'An datur Notitia Dei Lumine Naturæ?' *Aff.* Calamy was unable to recover the thesis defended by King.

40. Carlisle [1808, vol. I] gives the population 'of this Hamlet' in 1801 as 779.

41. According to Calamy [1830, vol. II, p. 487], Sheffield died on the 24th January 1726.

42. According to Wilson,

> Mr. Batson was a very popular preacher, and always had a crowded auditory, insomuch that it was oftentimes difficult for a stranger to procure a seat. [1814, vol. IV, p. 312]

43. Wheatley described this area as follows.

> Baldwin's Garden's ... is said to have derived its name from Richard Baldwin, one of the royal gardeners, who built some houses here in 1589 ... Baldwin's Gardens acquired an evil reputation, but its character has greatly improved of late years. [1891, I, p. 92]

About a decade later, Besant & Mitton noted that

> [Baldwin's Buildings] consist largely of workmen's model dwellings, comfortable and convenient within, but with the peculiarly depressing exteriors of the utilitarian style. [1903, pp. 69–70]

Today (2000) anyone expecting to see more than a few trees and some struggling grass in Baldwin's Gardens is in for a disappointment.

44. According to Wheatley [1891, II, p. 347], the street took its name from Kirby Hall in Northamptonshire, one of the seats of the Earl of Winchelsea.

45. See Stow [1912, p. 159].

46. See Wilson [1967, p. 12].

47. See also Bell [1951, p. 99].

48. There is some discussion of Socinianism in Appendix 2.1; for further details see W.M. Clow's article on the topic in Hastings [1971, vol. 11, pp. 650–654].
49. See Harben [1918, pp. 516–517].
50. The word is used in a relative sense: this Calamy's father and grandfather were both Edmund, as were his son, grandson, and great-grandson.
51. In Thomas Carlyle's essay *Boswell's Life of Johnson* of 1832 there is an account of a court case in which the following exchange occurred.

> *Q.* What do you mean by 'respectable'? — *A.* He always kept a gig. [Carlyle, 1915a, p. 9]

(In his 1838 essay on Sir Walter Scott, Carlyle again writes 'where were all ambition, money-getting, respectability of gig or no gig' [1915b, p. 55].)

52. Dale in fact notes that after the reign of Queen Anne,

> in the towns the congregations of the second generation of Nonconformists consisted of merchants, manufacturers, and tradesmen; and in the country, of farmers, with here and there a country squire. [1907, p. 509]

53. See Appendix 2.1 for a discussion of these terms, and for yet more information see the appropriate entries in Hastings's *Encyclopædia of Religion and Ethics*.
54. Matthew Henry, aged 52, died on the 22nd June 1714, at Nantwich, of an apoplectic fit, while on a journey [Calamy, 1830, vol. II, p. 305].
55. Referring to a list of dissenting congregations drawn up by Josiah Thompson in 1772, Dale notes that

> Hereditary wealth and education were still with the Presbyterians; Independents, generally, belonged to an inferior social rank. [1907, p. 541]

Daniel Defoe had earlier expressed the same sentiment, noting in the first volume of *A Tour through England and Wales* [1724] that in Ipswich the Presbyterian chapel was a large and fine building, while that of the Independents was 'a handsome new-built building, but not so gay or so large as the other'.

56. The popularity of these establishments was perhaps to a large extent attributable to the 'virtues' seen in the beverage as described in the 1670 pamphlet *The Vertue of the Coffee Drink*:

> And about half a pint of it to be drunk fasting an hour before, and not eating an hour after, and to be taken as hot as possibly can be endured; the which will never fetch the skin off the mouth, or raise any blisters, by reason of the heat.
>
> This drink will very much quicken the spirits, and make the heart lightsom.
>
> It is very good against sore eyes, and the better, if you hold your head over it, and take in the steam that way.
>
> It is very good against a cough and cold.
>
> It is excellent to prevent and cure the Dropsy, Gout and the Scurvy.
>
> It is very good to prevent miscarryings in child-bearing Women.
>
> It will prevent drowziness.
>
> It is neither laxative nor restringent.

The reference to 'coffee-houses' is not new to the historian of statistics. Abraham de Moivre is associated with Slaughter's (later *Old Slaughter's*) Coffee House: originally established in 1692, it occupied St Martin's Lane, Number 77 (1799), 74–75 (1815–1838), and 75 (1839–1843). This establishment, together with Giles's, was a well-known meeting-place for Frenchmen. (After the King of France's death in 1712 Joseph Addison relates that he visited Giles's 'where I saw a Board of *French* Gentlemen sitting upon the Life and Death of their *Grand Monarque*' [Letter No. 403].) For details of coffee-houses of a later period see *The Spectator*.

57. See Appendix 2.1 for a description of Nonjurors.
58. Many details here are from Smith's Notes on Addison's first Letter in *The Spectator*.
59. The year is thus given by Blanchard [1963, p. 292].
60. See the notes to Blanchard [1955].
61. The author's name is given in the original as Tho. S^{t}Serfe.
62. The source cited is *Poems on Affairs of State, from the Reign of K. James I. to the year 1703*, vol. II, of 1716. The poem from which the quotation comes is *The Lover's Session. In Imitation of Sir John Suckling's 'Session of Poets'*, pp. 156–165.
63. In an advertisement for *Imperial Chocolate Made by a German Lately come into England* published in London in 1700, we read

> *Imperial Chocolate* excells all others; because it is made of Sugar. the spiritual part, and not the gross body of Spice; and the *Cacao Nut* first well cleaned from the Husk and Shells, then cured, and the earthy Substance taken from it.

64. Details from the *International Genealogical Index*.
65. Stow [1912, p. 74] relates that West Cheape had prior to his writing (in 1603) been the place where mercers and haberdashers kept their shops.
66. This work was published in two volumes, the list of subscribers appearing at the beginning of the second volume of 1752. This volume has the same title as the first, with the addition of the words 'With suitable Offices of Devotion'.
67. Pemberton's book was published in London and Dublin in 1728. The lists of subscribers are different, Thomas's name appearing only in the London edition.
68. It is possible that Price succeeded Foster, for Ogborn [1962, p. 86] and D.O. Thomas [1977, p. 15] note that, after leaving the Tenter Alley Academy, which he did at the age of twenty-one, Price spent the next twelve years as family chaplain to George Streatfield of Stoke Newington, and ministered from time to time to various congregations, one of these posts being as assistant to Samuel Chandler at the Presbyterian Chapel at Old Jewry.
69. The reader is referred to Bellhouse [1999] for further details.

Chapter 3

1. See Manuel [1968].
2. The Act substituting the Gregorian for the Julian Calendar (24 Geo. II, cap. 23.) was passed in 1751. In terms of this Act it was decreed

> That in and throughout all his Majesty's dominions and countries in *Europe, Asia, Africa* and *America*, belonging or subject to the crown of *Great Britain*, the said supputation, according to which the year of our Lord beginneth on the twenty-fifth day of *March*, shall not be made use of from and after the last day of *December* one thousand seven hundred and fifty-one; and that the first day of *January* next following the said last day of *December* shall be reckoned, taken, deemed and accounted to be the first day of the year of our Lord one thousand seven hundred and fifty-two ... and the feast of *Easter*, and other moveable feasts thereon depending, shall be ascertained according to the same method, as they now are, until the second day of *September* in the said year one thousand seven hundred and fifty-two inclusive; and that the natural day next immediately following the said second day of *September*, shall be called, reckoned and accounted to be the fourteenth day of

September, omitting for that time only the eleven intermediate nominal days of the common calendar; and that the several natural days, which shall follow and succeed next after the said fourteenth day of *September*, shall be respectively called, reckoned and numbered forwards in numerical order from the said fourteenth day of *September*, according to the order and succession of days now used in the present calendar. [Horn & Ransome, 1957, p. 239]

3. See Barnard [1964] and Hacking [1970].
4. Holland [1962, p. 452], Pearson [1978, p. 355] and Wilson [1814].
5. As I was writing this passage there arrived serendipitously in the post the December 1998 issue of the *Journal of the American Statistical Association*, in which there was an article by David Moore on Statistics as a Liberal Art. Moore claims that the Liberal Arts may be viewed from either the oratorical or the philosophical perspective, depending on whether they are seen as emphasizing a connexion with society (informing the enquirer about the virtues, rather than teaching how to search for them) or whether they are viewed as encouraging analytic and sceptical thinking. Because of its flexibility and generality as a mode of reasoning, statistics fits well into both these schools.
6. For pictures of sculptures showing the place held by mathematics in the liberal arts from Carolingean times to the eighteenth century see Artmann [1998].
7. Horne [1916, p. 250] describes Huxley's thoughts here as 'like apples of gold in pictures of silver'. The Latin Vulgate version of this quotation from *Proverbs* 25, v. 11, 'Mala aurea in lectis argenteis' (apples of gold in beds of silver), is perhaps happier. (In Charles Reade's *The Cloister and the Hearth* it is given yet more poetically as 'an apple of gold in a network of silver'.)
8. Parker [1867, p. 72] translates *enseignement secondaire professionnel* as 'Non-liberal secondary education'.
9. Whether *The Imperial Dictionary of Universal Biography* [Waller 1865], from which this reference is taken, is careful in making the distinction we have noted before between *academies* and *schools* is uncertain. Could Ward have been attached to one of the latter?
10. D.O. Thomas [1977, p. 11], citing M.E. Ogborn, writes 'for Bayes had been a student of Eames'. Ogborn himself was somewhat more careful: writing of Eames's being on the staff of Coward's Academy he says only, 'It seems a likely guess that Bayes as well as Price was educated at this academy and learned his mathematics from Eames' [1962, p. 89].

11. Several references are given here to papers in Welsh.
12. In another comment on Eames's valuable contribution to education, Doddridge wrote, in a letter to the Rev. John Mason of the 11th May 1724, 'I envy you the happiness of attending upon Mr Eames's lectures, which every body allows to be very valuable' [Doddridge, 1829, vol. I, p. 380].
13. Eames did in fact publish: for a comprehensive list of his works see Wallis & Wallis [1986].
14. See also Horn & Ransome [1957, pp. 565–566].
15. The passages of scripture mentioned here are the following.

> *Matthew 7.* 24. Therefore whosoever heareth these sayings of mine, and doeth them, I will liken him unto a wise man, which built his house upon a rock:
> 25. And the rain descended, and the floods came, and the winds blew, and beat upon that house; and it fell not: for it was founded upon a rock.
> 26. And everyone that heareth these sayings of mine, and doeth them not, shall be likened unto a foolish man, which built his house upon the sand:
> 27. And the rain descended, and the floods came, and the winds blew, and beat upon that house; and it fell: and great was the fall of it.
>
> *Matthew 11.* 29. Take my yoke upon you, and learn of me; for I am meek and lowly in heart: and ye shall find rest unto your souls.
> 30. For my yoke is easy, and my burden is light.

16. Despite Calamy's writing in his *Life* that his son Edmund went to Edinburgh in 1714, we find the latter's name given in the *List of Theologues in the College of Edinburgh* as entering the profession in 1716. No mention is made of the one who recommended him, though it was presumably his father. Nathaniel Carpenter entered both the college and the profession in 1719, at the recommendation of Calamy. Like Thomas Bayes, John Horsley and Isaac Maddox entered both the college and the profession in 1720, at the recommendation of Calamy and William Tong (a dissenting minister of Salter's Hall, who died on the 21st of March 1727), respectively. Skinner Smith is not given in the list.
17. The early name of Edinburgh University varied considerably. At the Disputation at Stirling in 1617 James VI said

> I am so well satisfied with this day's exercise, that I will be godfather to the Colledge of Edinburgh, and have it called the Colledge of King James [Grant, 1884, vol. I, p. 173]

while in a letter written from Paisley to the magistrates at Edinburgh shortly thereafter, he wrote

> these are to desyre you to order the said college to be callit in all times herafter by the name of KING JAMES'S COLLEGE. [J. Grant, 18–, vol. III, p. 10]

This gave the institution its official title of *Academia Jacobi Sexti* [Grant, 1884, vol. I, p. 131]. Grant (op. cit. p. 130) notes that the word *Academia* was used to mean both 'College' and 'University'.

Grant [1884] also lists in various places the names *College of Edinburgh* (used in 1695) [1884, vol. I, p. 238], *King James Universitie* (1688) [ibid. p. 254], and *University of Edinburgh* (1703) [ibid. p. 239].

18. See *The Life of Mr William Carstares* in M'Cormick [1774, p. 19].
19. See Arbuthnot [1770, vol. II, p. 186].
20. Grant [1884, vol. II, p. 262] says that the bounty was equally divided — but of course £250 is not divisible by seven.
21. Booth [1865, Notes, p. xxiii], citing *Notes and Queries*, records that the Scottish dramatic poet John Home, author of the tragedy *Douglas*, 'had the old Scottish prepossession in favour of claret and utterly detested port.' When high duties caused the expulsion of claret from the market Home wrote the following epigram.

> Firm and erect the Caledonian stood,
> Prime was his mutton, and his claret good;
> 'Let him drink port,' an English statesman cried;
> He drank the poison, and his spirit died.

22. I adopt this spelling rather than *Carstaires* unless the latter is called for in direct quotation.
23. See Calamy [1830, vol. II, p. 185].
24. The texts of the diplomas are given in an appendix in Calamy [1830, vol. II].
25. See also Bogue & Bennett [1810, vol. III, p. 454].
26. Grant says that Edinburgh University conferred the M.A. degree on Calamy on the 22nd of April, but there is no mention of this in Calamy's *Life*.
27. Grant notes that work on the *Memoirs* written by Thomas Craufurd (?–1662), Regent of Philosophy and Professor of Mathematics, was probably begun in 1626 when he first became a regent. Publication was first effected in 1808 under the title *History of the University of Edinburgh from 1580 to 1646*.

28. *The Concise Scots Dictionary* of 1985 gives Mes- or Mess-John as a term used from the seventeenth to the nineteenth centuries to describe a Presbyterian minister (in contrast, the *Oxford English Dictionary* says, to Anglican or Roman clergymen). The word comes from *maister*. Brewer's *Dictionary of Phrase and Fable* perhaps more suggestively gives the alternative *Mass-John*.

29. John Oldham (1653–1683) recorded such behaviour in his *A Satire. Addressed to a friend that is about to leave the university, and come abroad in the world*:

> Some think themselves exalted to the sky,
> If they light in some noble family;
> Diet, a horse, and thirty pounds a year,
> Besides the advantage of his lordship's ear,
> The credit of the business, and the state,
> Are things that in a youngster's sense sound great.
> Little the inexperienced wretch does know,
> What slavery he oft must undergo,
> Who, though in silken scarf and cassock dressed,
> Wears but a gayer livery at best;
> When dinner calls, the implement must wait,
> With holy words to consecrate the meat,
> But hold it for a favour seldom known,
> If he be deigned the honour to sit down.
> Soon as the tarts appear, Sir Crape, withdraw!
> Those dainties are not for a spiritual maw;
> Observe your distance, and be sure to stand
> Hard by the cistern with your cap in hand;
> There for diversion you may pick your teeth,
> Till the kind voider comes for your relief.
> For mere board wages such their freedom sell,
> Slaves to an hour, and vassals to a bell;
> And if the enjoyment of one day be stole,
> They are but prisoners out upon parole;
> Always the marks of slavery remain,
> And they, though loose, still drag about their chain.
>   And where's the mighty prospect after all,
> A chaplainship served up, and seven years' thrall?
> The menial thing, perhaps, for a reward,
> Is to some slender benefice preferred,
> With this proviso bound, that he must wed
> My lady's antiquated waiting maid,
> In dressing only skilled, and marmalade.

(Strictly speaking a *voider* was a container for carrying orts from the table, though the word was also used to describe one who cleared the table.) This was not a completely fictitious account: such behaviour was related of the chaplain to the Archbishop of Canterbury in the mid-eighteenth century (see Calamy [1830, vol. II, p. 217]).

30. See the footnotes in Oldham [1960, pp. 224–225] for further details on this score.
31. Calamy notes, however, that the ministers of the Established Church in Scotland were better off and less contemptuously treated.
32. Edmund Calamy ministered to a dissenting congregation in Westminster.
33. MacKinnon [1902] lists Onely as a Vicar of Speldhurst with the date 9th July 1768; note that this is some seven years after Bayes's death.
34. On this name Burr notes

    The parish of Speldhurst, or the learned wood (for such is the original meaning of this word, which was anciently wrote Spelherst) .... [1766, p. 155]

    Although MacKinnon [1930] believes the name to be Saxon in origin, giving the above derivation, a footnote by the reviser D. James notes that *speld* or *spelder* means 'splinter, or thin piece of wood'. Citing F. Edwards's *History of Names and Places* James also gives *speld* as 'Spald, a chief's name', and hence 'Spalda's possessions'.
35. I have no idea how competent a judge of Greek scholarship Onely was.
36. There seems to have been little pleasure in German schools at that time: Parker [1867, p. 42] in fact records 'The boys were forbidden to attend any public spectacle, unless it were to see heretics burnt' (*Neque ad publica spectacula, nec ad supplicia reorum, nisi forte hæreticorum, eant*, a quotation from the Jesuit work *Ratio et Institutio Studiorum*).
37. This work is referred to in a footnote in Calamy [1830, vol. II, p. 3], the surname of the author being given as 'Lesley' (and, elsewhere in Calamy, as 'Lesly'). (Perhaps, in view of its subject matter, *Shake in the Grass* might have been a better title.) The quotation *latet anguis in herba* (a snake lurks in the grass) is from Virgil's *Eclogue*, iii. 93.
38. After I had essentially finished writing this chapter I received, as a most welcome gift from the author, a copy of Stigler's *Statistics on the Table: The History of Statistical Concepts and Methods*. The eleventh chapter of this work is entitled *Apollo Mathematicus*, and contains not only a detailed study of Eizat's tract, but also an examination of the work by Pitcairne that occasioned its being written and various responses to it.

39. The Latin version runs as follows.

> Ego 𝒜 ℬ qui praesentibus subscribo, quandoquidem mihi potestas concessa est & privilegium fruendi beneficio & usu Bibliothecae Edinburgiae, ijs horis quibus constitutum est ut pateat; Ideo hâc meâ synographâ me obligo & Obstringo, conun(?) DEO OMNIPOTENTE, nec surrephirum(?) me, nec sponde contaminatorum, violaturum, aut quavis modô dammnô afferturum (affecturum?) quemvis memorata. Bibliothecae Librum; negs locco moturum aliquem absq₃ Librarij consensu. Spondeo etiam me pro conditione & facultate mea amicum fore & fautorum Academias Edinburgenae. Ita me DEUS amat(?).
> Iuro etiam, quod si contigent mihi temerè aliquam Librorum commaculare Labefintare(?); me id non celaturum Bibliothecarum, sed continuo ei significaturum, damnumq₃ prestaturum ad ad ejus arbitrium, vel novum pro eo librum reponendo, vel mult(?) damnum resarciendo. Assentior praeterea Bibliothecae legibus quae sequentur.

40. The full list, in the original Latin, translated in the text (with the suppression of the imperatives) by the present author, runs as follows.

   1. Bibliothecam nemo sine Librarij venia intrato vel Eo inscio exiti [page breaks here].

   2. Nemini nisi admisso & jurato Leg–(?) jus esto.
      [Up to the 12th October 1708: 'Juratus et Admissus est ...'; from 8th November 1708: 'Admissus est ....'.]

   3. Nemo Librum nisi Bibliothecario tradente attingito.

   4. Nemini Librum effere liceto.

   5. Nemo notam Libris vel atramentô vel reflexâ chartâ imprimito.

   6. Si quis per incuriam Librum maculaverit statim Bibliothecaris ostendito, & pro Bibliothecarij arbitrio damnum praestato, multaeq₃ hujus generis ad Bibliothecae usum à Bibliothecario conservantor.

   7. Nemo ad lychnum legito, vel Librum igni admoveto.

   8. Apertu collegio hieme singulis diebus, nisi Dominicis, ab hora decima in duodecimam, & à secunda pomeridiana in quartam, at aestate, etiam à septima matutina in nonam legendi jus esto.

   9. Signo dato ad nonam. duodecimam, & quartam horam pomeridianam, omnes sine mora exeunto.

10. Submissè secum quisque legito nec inter legendum alios interpellato, & si cui loquendum sit, in aurem proximi susurrato.

11. Qum quispiam post introitum nactus fuerit Librum si volet, in signum datum possideto, nec primo possidenti eripere fas esto.

41. See the Special Collections Department of Edinburgh University Library, item [Dc.1.36].
42. See item [Dc.1.4] in the Special Collections Department of Edinburgh University Library.
43. The quotation is from the entry for Samuel Horsley (John's son) in Volume IX. Note that the date given here as 24 Feb. is given as the 23rd of that month in the entry for Isaac Maddox in volume XII.
44. One is indebted for much of the following discussion to Roger Farthing's *Royal Tunbridge Wells*.
45. See Bryant [1960, p. 98]. The selection of Katherine de Braganza as Charles's queen has been attributed to the Earl of Clarendon, who, desiring to secure the succession for the Duke of York (later James II), chose the Infanta as she was reputed to be incapable of bearing children. As it happened she did once conceive, but miscarried (see H.G. Bohn's essay *The Personal History of Charles II*, pp. 418–453 in Hamilton [1859]).

    The Count Grammont, incidentally, was not very taken with the Queen and her train:

    > The new queen gave but little additional brilliancy to the court, either in her person, or in her retinue, which was then composed of the Countess de Panétra, who came over with her in quality of lady of the bedchamber; six frights, who called themselves maids of honour, and a duenna, another monster, who took the title of governess to these extraordinary beauties. [Hamilton, 1859, p. 109]

46. This work is described as follows in Letter No. 7 of the 26th April 1709 in *The Tatler*. 'The whole comedy is very just, and the low part of human life represented with much humour and wit.'
47. It is clear from the passage that the description is of Tunbridge Wells rather than Tunbridge itself.
48. Some of the terms in this passage are explained as follows by Walker.
    Phoebus: Appollo as the sun-god.
    Thetis: a sea-nymph, here taken to be the sea.
    trotted: Tunbridge Wells being relatively young at this time, visitors were usually accomodated at Rusthall, Tonbridge, or Southborough.

49. The psalm in question seems to be Psalm 48. The version given in *The Book of Common Prayer, and Administration of The Sacraments, and other Rites and Ceremonies of The Church, according to the use of The United Church of England and Ireland: together with the Psalter or Psalms of David, Pointed as they are to be sung or said in Churches* (Clarendon Press: Oxford, 1812), runs as follows.

> 1. The Lord, the only God, is great,
>   and greatly to be prais'd
> In Sion, on whose happy mount,
>   his sacred throne is rais'd.
> 2. Her tow'rs, the joy of all the earth,
>   with beauteous prospect rise:
> On her north side th' Almighty King's
>   imperial city lies.

Another metrical version was approved by the Church of Scotland in 1828: this runs

> 1. Great is the Lord, and greatly he
>   is to be praised still,
> Within the city of our God,
>   upon his holy hill.
> 2. Mount Sion stands most beautiful,
>   the joy of all the land;
> The city of the mighty King,
>   on her north side doth stand.

50. Private communication.
51. This Chapel became a church with a parish in 1889.
52. This Ditton was the fourteenth in line to bear the Christian name 'Humphrey'.
53. See Browning [1953, p. 359].
54. In his article on *Conventicle* in Hastings [1971, vol. 4], A.M. Hunter notes that the word originally signified merely an assembly. Later it came to refer specifically to gatherings of religious bodies for secret worship. Still later it gathered depreciative or reproachful overtones, being used to denote those who opposed the ruling ecclesiastical authorities. Finally, Hunter writes, it came to mean

> religious meetings of dissenters from an Established Church, held in places that were not recognized as specially intended for public worship or for the exercise of religious functions.
>
> [Hastings, 1971, vol. 4, p. 102]

55. See Blanchard [1955, p. 432] for details of an admonition to dissenters in an anonymous letter by Defoe.

56. According to Farthing [1990] it was in Mount Ephraim House that Charles II and Queen Catherine stayed on their visit to Tunbridge Wells in 1663 (the rest of the Court camped on the common).
57. There is some brief mention of Archer in Calamy's *Life*:

> I shall here take the pains to transcribe a paper, which I received from Mr. Archer, of Tunbridge, when in the summer of the year 1724, I spent some time there in drinking the water. [1830, vol. I, p. 195]

A footnote on page 387 of this work states that the visit occurred in August of that year.

58. Rough translations of the two Latin sentences in the following quotation run as follows: (1) *When the Dog-star cracks open the parched ploughed fields* and (2) *Don't drink the water when there's an 'R' in the month!*. I have been unable to identify the source of the first of these, though similar sentiments can be found in Horace's *Odes* I, 17, 17–18; in Virgil's *Georgics* II, 353; and Tibullus's *Poems* 1.7, 21–22.
59. See Pearce [—, p. 67].
60. Paul Amsinck became Master of Ceremonies in 1805 (see Farthing [1990, No. 71]).
61. See Appendix 2.1 for some remarks on anabaptism. More details may be found in the article on the subject by W.J. M'Glothlin in Hastings [1971, vol. 1, pp. 406–412].
62. This is probably Matthias Copper, who in 1755(?) wrote *Matthias Copper's Last Legacy and Advice to all his Surviving Brethren and Friends ... especially to the Church of Christ at Tunbridge Wells*.
63. According to Barton [1937, p. 304], the Mount Sion Chapel was abandoned 'in ruins' in 1814, and it remained in this state for some years. In 1826 Clifford wrote

> The Presbyterian chapel, a capacious structure, is still to be seen on Mount Sion — shut up, and in a dilapidated condition. [p. 37]

64. Bailey & Bailey.
65. The reference here is to *Augustus Carp Esq. By Himself. Being the Autobiography of a Really Good Man*, published thus in 1924, and now known to be by Dr Henry Howarth Bashford. In his capacity as an office-bearer in the Anti-Dramatic and Saltatory Union Carp handed out pamphlets like 'Did Wycliffe Waltz?' to 'the degenerate pleasure-seekers standing outside theatres'.
66. Elwig [1941, p. 54] notes that the garden was laid out by a Mr Eaglesham.

67. Note that Wilson writes here of *Tunbridge* rather than *Tunbridge Wells*: but in view of the association with Epsom, also a watering place, I assume that the latter is meant.
68. See Farthing [1990].
69. See Farthing [1990].
70. It had generally been thought that James Quin of Tunbridge Wells, a retired actor, wanted to succeed Richard Nash (who had died in 1761, the same year as Thomas Bayes) as Master of Ceremonies. However Nash was followed in this position by 'one Collett' [Melville 1912, p. 157], who resigned in 1763 (Melville gives 1673), Samuel Derrick being appointed in his room (1763–1769). A vigorous campaign was organised against Derrick by several visitors and inhabitants, including Quin, for having omitted to invite some distinguished visitor to a festivity at Bath, and when Quin was asked what action should be taken against Derrick, he said 'if you have a mind to put Derrick out, do it at once, and clap an extinguisher on it' [Melville, 1912, p. 161]. Derrick responded with the following lampoon.

> When Quinn, of all grace and dignity void,
> Murder'd Cato, the censor, and Brutus destroy'd;
> He strutted, he mouth'd, — you no passion cou'd trace
> In his action, deliv'ry, or plumb-pudding face;
> When he massacred Comus, the gay god of mirth,
> He was suffer'd, because we of actors had dearth,
> But when Foote, with strong judgement and genuine wit,
> Upon all his peculiar absurdities hit;
> When Garrick arose, with those talents and fire,
> Which nature and all the nine Muses inspire,
> Poor GUTS was neglected, or laugh'd off the stage;
> So, bursting with envy, and tortur'd with rage,
> He damn'd the whole town in a fury, and fled,
> *Little* BAYES *an extinguisher clapp'd on his head.*
> Yet we never shall Falstaff behold so well done,
> With such character, humour, such spirit and fun,
> So great that we knew not which most to admire,
> Glutton, parasite, pander, pimp, letcher, or liar; —
> He felt as he spoke; — nature's dictates are true;
> When he *acted the part*, his *own picture he drew.*
> [Melville, 1912, p. 162]

71. Farthing [1990] relates that Tunbridge Wells became a town in 1834.
72. The Philosopher of Botley, not known for not speaking his mind (he described Deal as 'a most villainous place' [1912, vol. I, p. 245]) had

little to say about Tunbridge Wells. Its nearness to London (the *Wen*, in Cobbett's view) prompted him to write 'I got clear of "*the Wells*," and out of the contagion of its Wen-engendered inhabitants' [1912, vol. I, p. 221].

Daniel Defoe was far more positive, noting in the first volume of *A Tour through England and Wales*, that 'company and diversion is in short the main business of the place' and commenting on the beauty of the ladies, the excellent air, healthful country, and reasonableness of provisions. He notes too the freedom of association there:

> Here you have all the liberty of conversation in the world, and any thing that looks like a gentleman, has an address agreeable, and behaves with decency and good manners, may single out whom he pleases, that does not appear engag'd, and may talk, rally, be merry, and say any decent thing to them; but all this makes no acquaintance, nor is it taken so, or understood to mean so. ... As for gaming, sharping, intrieguing; as also fops, fools, beaus, and the like, Tunbridge is as full of these, as can be desired. ... In a word, Tunbridge wants nothing that can add to the felicities of life, or that can make a man or woman compleatly happy, always provided they have money.

He also notes that the nobility and gentry frequented Tunbridge Wells, the merchants and rich citizens went to Epsome, whereas common people had to make do with Dullwich and Stretham.

73. See Linnell [1964, pp. 160–161] and Farthing [1990].
74. We shall find Smith's name recurring in Bayes's Notebook. Knipe notes

    > He [i.e., Smith] was a professor of astronomy, and a noted mathematician. Mrs. Montague could not understand why he was so fond of Tunbridge Wells — "for it seems not very agreeable to the nature of a Philosopher." [1916, p. 16]

75. Stewart describes Whiston as 'self-promoting' [1999, p. 135].
76. Reprinted in Volume III of Barbauld [1804].
77. See Farebrother [1999, §2.2] for some details of the longitude problem and the prize offered for its solution.
78. Ditton wrote a number of theological works in addition to mathematical tracts: for details of the latter see Wallis & Wallis [1986]. Although Ditton, in his capacity as a mathematical author, is generally looked upon with approbation by modern commentators, Stewart describes him as one of Newton's disciples 'who fancied himself a mathematician' [1999, p. 145].

79. Thomas Wilson relates in his diaries that on the occasion of his visit to Tunbridge Wells on the 13th of May 1750 he 'Preached for Mr Dowding and read prayers in the afternoon' [Linnell, 1964, p. 237].
80. See Besant & Mitton [1903, p. 67] and Force [1985, p. 87].
81. In ancient Rome, in both Republican and Imperial eras, games of chance involving astragali or dice were banned, except at the Saturnalia, from time to time by *leges aleariae*. The emperors themselves were not immune to Tyche's lures: Suetonius [Nero 30] relates that Nero would bet 4,000 gold pieces on each pip of a winning toss at a game of dice. For further details see David [1962, p. 6] and Ineichen [1996, p. 49].
82. Although the name of the editor is given in many editions as 'Edward Malone', his Christian name was actually 'Edmond'. One should perhaps note Wilson's remark [1967, p. 185] that Malone, perhaps best known as the editor of Dryden's work, was given to hasty statement.
83. Whether Samuel Johnson would have visited a spa for the bathing is doubtful, an opinion that is supported by the following anecdote related by a Mr Wickens (a respectable draper of Lichfield).

> We then came to a cold bath. I expatiated upon its salubrity. 'Sir,' said he, 'how do you do?' 'Very well, I thank you, Doctor.' 'Then, Sir, let well alone, and be content. I hate immersion ...'. [Page, 1987, p. 65]

84. The *Dictionary of National Biography*, vol. X, says that Johnson received an LL.D. from Dublin and Oxford. The source for the statement in our text is Malone [1807].
85. One trusts that the courtesy and kindness exhibited by Bayes were justly appreciated by the visitors, and one is reminded in this regard of the farewell taken by Lemuel Gulliver of the Houyhnhnms:

> my detractors are pleased to think it improbable, that so illustrious a person should descend to give so great a mark of distinction to a creature so inferior as I. Neither have I forgot, how apt some travellers are to boast of extraordinary favours they have received.

86. One must be somewhat careful in interpreting the word 'Indian': in Addison's letter of Friday the 27th April 1711 in *The Spectator* we read

> When the four *Indian* Kings were in this Country about a Twelvemonth ago, I often mixed with the Rabble, and followed them a whole Day together, being wonderfully struck with the Sight of every thing that is new or uncommon [Letter No. 50]

548   Notes: chapter 3

(see also his letter No. 56 and the footnote to Letter 171 in *The Tatler*). It is only in the Notes to the first of these letters, however, that one finds that these visitors were Iroquois chiefs. The same point is made in Note 64, p. 268 of Walker's edition of the Earl of Rochester's poems.

87. John Stuart Mill, in his discussion of grounds of disbelief, says

> It is with these uniformities [of mere coexistence, not proved to be dependent on causation] principally, that the marvellous stories related by travellers are apt to be at variance: as ... of ice, in the celebrated anecdote of the Dutch travellers and the King of Siam [1892, p. 367]

and on the necessity of confirmation before acceptance or rejection he writes

> Such ought to have been the conduct of the King of Siam when the Dutch travellers affirmed to him the existence of ice. [1892, p. 369]

Writing in the early part of the twentieth century, C.D. Broad remarked in similar vein, 'Negroes find it very hard to believe that water can become solid, because they have always lived in a warm climate' [1923, p. 18].

88. See Edgeworth [1881, p. 1].
89. The quotation, given in Addison's Letter No. 463 in *The Spectator*, occurs in Fergusson's *Scottish Proverbs* of 1641.
90. See *The Record of the Royal Society of London* [1940, p. 42].
91. By this time the 'admission-money' had risen from forty shillings to two guineas.
92. Jones [1849] records that Andrews was buried in Bunhill Fields.
93. Jean Mauldon, late of the Tunbridge Wells Library, has informed me that this Church Book has apparently been lost. Most of the quote given here is recorded in Strange [1949, p. 16].
    There seems to be some inconsistency in dates here: Bailey & Bailey give the date of the Church Book as 1743, yet the passage cited refers to events that took place some time after that.
94. *Our Mutual Friend*, Book I, Chapter 1.
95. Joseph Perkins also made an offer to the churchwardens of White-Chapel to preach there for nothing.
96. This bracketed comment is in the original.

97. The Fantoccini was an Italian dramatic performance by puppets. The device mentioned here was not unique to *The Rehearsal*: in Dryden's *Tyrranick Love; or, The Royal Martyr. A Tragedy*, first published in 1670, Valeria, daughter of Maximin, kills herself towards the end of the play. When Mrs Ellen (i.e., Nell Gwyn) acted this part, she said, as she was about to be borne off stage by the bearers,

> Hold; are you mad? you damn'd confounded Dog!
> I am to rise, and speak the Epilogue.

98. The Count Grammont was not impressed with Buckingham. In writing of the retinue that accompanied the Infanta of Portugal, Katherine de Braganza (later Queen Catherine), he said

> Among the men were ... one Taurauvédez, who called himself Don Pedro Francisco Correo de Silva, extremely handsome, but a greater fool than all the Portuguese put together: he was more vain of his names than of his person; but the Duke of Buckingham, a still greater fool than he, though more addicted to raillery, gave him the additional name of Peter of the Wood. He was so enraged at this, that, after many fruitless complaints and ineffectual menaces, poor Pedro de Silva was obliged to leave England, while the happy duke kept possession of a Portuguese nymph more hideous than the queen's maids of honour, whom he had taken from him, as well as two of his names.
> [Hamilton, 1859, p. 109]

Elsewhere he describes Buckingham as 'full of wit and vivacity, dissipated, without splendour' [op. cit. p. 106].

Calamy was similarly disenchanted with Dux Bucks (as Buckingham was referred to in the *House of Lords Journal*), writing, in his *Life*,

> The Duke of Buckingham was a man of no religion at all, and that gloried in his debaucheries. He was so addicted and abandoned to the most criminal pleasures, that he and his true associate, the Earl of Rochester, (whose Life was written by Burnet) seemed capable of corrupting any court in the world. He would, however, have been a great man, had he had any thing of steadiness or consistency in him; but he was of as mercurial a make as ever was known.
> [1830, vol. I, p. 98]

99. There were a number of satires and pamphlets published in the eighteenth century on *The New Tunbridge Wells*. The Wells referred to

here were at Islington, and their memory is preserved today in the name 'Sadler's Wells' and the survival of names such as 'Mount Pleasant' and 'Mount Ephraim' in the area. One of these pamphlets, *A Satyr on the New Tunbridge Wells*, bore on its title page the quotation from Rochester's *Tunbridge Wells* given here.

100. Walker explains some of Rochester's allusions as follows.

> *Cobb*: leading man.
>
> *Importance Comfortable*: Marvell quotes Parker, in his *Preface* to Bramhall's *Vindication*, as saying '*concerned ... in matters of a closer and more comfortable importance to himself and his own Affairs*'.
>
> *lazy dull distemper*: Marvell, in the Second Part of *The Rehearsal Transpros'd* wrote that Parker 'tells me: *I had heard from him sooner had he not, immediately after he undertook my Correction, been prevented by a dull and lazy distemper ...*'.
>
> *sweetnesse*: in the Second Part of *The Rehearsal Transpros'd* Marvell describes Parker as 'one of *so sweet a Nature.*'

101. For details see Marvell [1971].
102. *Symbolize* here means *agree*.

Chapter 4

1. Paley does not see any distinction between *ethics* and *morality*: the first chapter of his *The Principles of Moral and Political Philosophy* [1825, vol. IV, p. 1] begins:

   > MORAL PHILOSOPHY, Morality, Ethics, Casuistry, Natural Law, mean all the same thing; namely, *that science which teaches men their duty and the reasons of it.*

2. The connexion between theology and statistics is still of some interest today: see, for example, Bartholomew [1988].

3. For a selection from writings — ethical, moral, and theological — of eighteenth-century writers Selby-Bigge [1897] may be consulted. The annotated index to this two-volume work has some three pages devoted to *benevolence*, most of the entries being to this quality as it relates to man rather than to God.

4. See Thorburn [1918] for a discussion of Occam's razor and his unsuccessful search for it in the Singular and Invincible Doctor's works. Although Jeffreys remarked in his *Theory of Probability* that 'the simplest law is chosen because it is the most likely to give correct predictions' [1961, §1.0], Gillies noted in his discussion of Bartholomew

[1988] that Popper, in his *The Logic of Scientific Discovery*, argued that 'the simpler a hypothesis is the smaller its prior probability'.
5. The reader will find further reference to Tindal's work in §4.3.
6. Citing one 'Dr Chalmers' as his source, Bayne writes 'Butler is in theology what Bacon was in science' [Butler, 1906, p. xviii]. This is probably Thomas Chalmers (1780–1847), writer of various theological tracts and a book of lectures on Butler's *The Analogy of Religion*.
7. There can be no doubt that Bayes and Price were acquainted: in addition to the evidence of this quotation, we note that Bayes left a sum of money to Price in his will; and William Morgan (Price's nephew), in his biography of his uncle, states that, on Bayes's death, Price was asked by the relatives of the late lamented to examine his scientific papers.
8. Thomas would have been in his late twenties at the time of the writing of the tract.
9. Hartley gives the chief attributes of the divine nature 'that are deducible from, and explanatory of, the moral one, *viz.* of the divine benevolence' as 'holiness, justice, veracity, and mercy' [1834, Part II, chap. I].
10. The passage also occurs in *Isaiah* xxxviii. When Hezekiah was told by the prophet that the Lord would in fact heal him, he asked for a sign. Isaiah in turn asked him 'shall the shadow go forward ten degrees, or go back ten degrees?' Hezekiah chose the latter. Davidson notes in his commentary that 'there is no suggestion that the day was lengthened. The miracle was the miracle of a shadow, not of a long day' [1954, p. 329]. Velikovsky suggests that Hezekiah's illness was a bubonic affection [1957, p. 226]. He does not agree with Davidson's interpretation of the 'miracle'.
11. Paley has a chapter on 'Human Happiness' in his (still eminently readable) *The Principles of Moral and Political Philosophy*.
12. Was Paley acquainted with Bayes's tract? The answer, at least to me, is uncertain. Paley was remarkably reserved in citing his authorities, and the only reference I have managed to find is the following.

> But there is a class of properties, which may be said to be superadded from an intention expressly directed to happiness; an intention to give a happy existence distinct from the general intention of providing the means of existence; and that is, of capacities for pleasure, in cases wherein, so far as the conservation of the individual or of the species is concerned, they were not wanted, or wherein the purpose might have been secured by the operation of pain. The provision which is made of a variety of objects, not necessary

to life, and ministering only to our pleasures; and the properties given to the necessaries of life themselves, by which they contribute to pleasure as well as preservation; show a farther design, than that of giving existence*.

*See this topic considered in Dr. Balguy's Treatise upon the Divine Benevolence. This excellent author first, I think, proposed it; and nearly in the terms in which it is here stated. ...
[1825, vol. V, pp. 335–336]

The difficulty here is twofold: Paley's penchant for paraphrase rather than direct quotation makes it difficult to find the thoughts expressed here in Bayes's tract, and the reference is to 'Dr. Balguy's Treatise upon the Divine Benevolence' [sic].

13. On the place of Paley's works, in particular his *Natural Theology*, in Cambridge University in the nineteenth century see Fyfe [1997].
14. In a Note on this passage in The Religious Tract Society's edition of Butler's *Analogy*, Joseph Angus writes

> Some hold that the character of God is that of *simple benevolence*. But this is not probable; for a. This supposes that moral conduct is not regarded for its own sake, and so gives no place for *veracity* and justice, except as forms of benevolence; ... [1840(?), p. 49]

15. Edgeworth also provides the following definition.

> *Pleasure* is used for 'preferable feeling' in general (in deference to high authority, though the general term does not appear to call up with equal facility all the particulars which are meant to be included under it, but rather the grosser feelings than for instance the 'joy and felicity' of devotion). The term includes absence of pain.
> [1881, pp. 56–57]

16. The exact distinction between 'affections' and 'passions' is given by Price as follows.

> The former [i.e., affections], which we apply indiscriminately to all reasonable beings, may most properly signify the desires founded in the reasonable nature itself, and essential to it; such as self-love, benevolence, and the love of truth.—These, when strengthened by instinctive determinations, take the latter denomination; or are, properly, *passions*. —Those tendencies within us that are merely instinctive, such as hunger, thirst, &c, we commonly call *appetites* or *passions* indifferently, but seldom or never *affections*. [1787/1974, p. 74]

17. See the *Encyclopædia Britannica*, 14th edition.
18. This is presumably Simon Patrick (1626–1707), bishop of Chichester and later of Ely.
19. The eighteenth-century philosopher Moses Mendelssohn (whose connexion with inverse probability is discusssed in Dale [1999]), viewed Clarke, Locke, and perhaps Shaftesbury as the only true philosophers (see Altmann [1973, p. 31]).
20. The sentiment expressed here perhaps gained wider acceptance in *The Mikado*.
21. In a Note to The Religious Tract Society's edition of Butler's *Analogy* Angus writes

   > *Moral* government implies the rewarding of *virtue* and the punishment of vice; perfect moral government, an exact distribution of each. [1840(?), p. 49]

22. In his editorial introduction to Price's *A Review of the Principal Questions in Morals* Raphael finds Price to be weak on introspective psychology but strong on observational psychology, writing

   > Price no doubt knew from experience that the rational persuasion of the moralist and preacher can have some effect on the average, decent but weak, individual; for the hardened criminal it is, by itself, useless. [1787/1974, p. xxxii]

23. Of this parable Davidson [1954] notes that the words 'And whatsoever is right, that shall ye receive' are omitted by texts that are considered more reliable.
24. The Note on this passage in The Religious Tract Society's edition of the *Analogy* runs as follows.

   > Virtue is not wholly benevolence, or vice wholly the opposite; for if they were a. Benevolence would be approved equally to whomever it was shown, and falsehood would be condemned only as injurious. [1840(?), p. 322]

25. Incidentally, Price and Butler have an equestrian connexion: the former is remembered for riding a (blind) white horse (see Holland [1968, p. 49]), and the latter, according to Bayne, had the tradition of riding a black pony, and that very fast (Butler [1736/1906, p. xii]).
26. It is interesting that two of the three references to *benevolence* in the Index to Dent's edition of *The Spectator* are to letters from Henry Grove. Indeed, even in his 1734 tract Grove was not entirely dismissive of benevolence, as the following quotes show.

> Nothing whatsoever is of *value* but *happiness*, either the *agent*'s own, or the happiness of other beings; or that which hath some *relation* to happiness; either more *immediate* or *remote*, *necessary* or *voluntary*, to happiness *itself*, or to the *capacities* of it.
> ... the *end* or design of God in the creation, must be *happiness* ....
> We have had attempts to prove the *goodness of God*, as it signifies (in some mens idea of it) *a kind inclination or principle of benevolence in the Deity, without any reason for it*: but, I think, short of the mark. [1734, pp. 3, 6, 19]

(Here Grove cites Thomas Johnson's *An Essay on Moral Obligation, with a view towards settling the controversy concerning moral or positive duties* of 1734.)

27. Doddridge defines *rectitude* as follows.

    > <u>Lect. LIV.</u> Of God's moral Rectitude.
    > §1. Def. Moral rectitude is generally called HOLINESS, when applied to God, VIRTUE, when applied to the creature

    and Hartley writes 'Holiness may be defined by moral purity and rectitude' [1834, Part II, chap. I].

28. Grove's actual words were

    > When I examine my notion of the self-existent, all-perfect Being, I find, among others, these *two ideas*, of *infinite knowledge*, and a correspondent *energy*, or *active force*, included in it. The knowledge which this Being hath, of what is *fit* or *unfit*, to be chosen or done in every supposable circumstance, is what we mean by the *Wisdom* of God. These fitnesses and unfitnesses, are distinguished into *natural* and *moral. Natural fitness* is (*chiefly* at least) the fitness or subserviency of things and actions in their *own nature*, to some *good end. Moral fitness*, is the *fitness*, or congruity, that *intelligent and free agents*, should *make choice* of certain *ends* preferably to others, and of certain *actions*, as *means*, suited and adapted to these ends. [1734, p. 1]

29. Grove quite firmly writes 'God hath no such inclinations or affections as oppose the dictates of his infinite wisdom' [1734, p. iv].

30. Elsewhere Hartley in fact states:

    > And as justice in God is, by the same language, exalted into benevolence, he may inflict punishment, *i.e.* another

species of natural evil, justly, provided it be consistent with benevolence, *i.e.* with a balance of happiness.
[1834, Part II, chap. I]

31. Hartley distinguishes three kinds of self-interest, namely, gross self-interest, refined self-interest, and rational self-interest, and describes them briefly as follows.

   ... Gross self-interest sometimes excites persons to external acts of benevolence, and even of piety; and though there is much hypocrisy always in these cases, yet an imperfect benevolence or piety is sometimes generated in this way. ... As refined self-interest arises from benevolence, piety, and the moral sense; so, conversely, it promotes them in various ways. ... Rational self-interest puts us upon all the proper methods of ... begetting in ourselves the virtuous dispositions of benevolence, piety, and the moral sense.
   [1834, Part II, chap. III, Sect. V]

32. Thomas's reference here is to Hutcheson [1725].
33. Thomas cites as a footnote Bayes, Rutherforth [1744, p. 6] and Wollaston [1726, p. 52].
34. Pearson [1978, p. 359] also found that Bayes's views were less persuasive than those of Balguy and Grove, though one might wonder whether his opinion was influenced by Doddridge's.

Chapter 5

1. This reluctance to publish was not unique to Newton. There are those who have attributed Bayes's failure to publish his *Essay towards solving a Problem in the Doctrine of Chances* either to pudency or severe doubts as to the validity of his assumptions, and in his edition of the manuscripts of Henry Cavendish, James Clerk Maxwell wrote

   Cavendish cared more for investigation than for publication. He would undertake the most laborious researches in order to clear up a difficulty which no one but himself could appreciate, or was even aware of, and we cannot doubt that the result of his enquiries, when successful, gave him a certain degree of satisfaction. But it did not excite in him that desire to communicate the discovery to others which, in the case of ordinary men of science, generally ensures the publication of their results. How completely these researches of Cavendish remained unknown to other men of science is shown by the external history of electricity. [1879, p. xlv]

Perhaps, though, the hesitancy was but caused by too slavish a following of Horace's *Nonumque prematur in annum* (which may be freely rendered 'Put away your compositions for at least nine years before making them public') [*Ars Poetica*, 388]. The passage in which this occurs is translated by Francis Howes as follows.

> But, if at some chance hour you aught compose,
> See 'tis correct ere to the world it goes;
> Submit it first to Tarpa's critic ears,
> Your sire's, and mine; and keep your piece nine years.
> What is not published you can blot or burn;
> But words, once uttered, never can return.
> [Godolphin, 1949, p. 324]

2. In the *Account of the Commercium Epistolicum*, a work regarded by Cajori [1917, p. 145] as the most mathematically rigorous of all Newton's writings, we find the following remark.

> Mr. *Newton* published his *Treatise of Quadratures* in the Year 1704. This Treatise was written long before, many things being cited out of it in his Letters of Octob. 24 and Novemb. 8. 1676. [Newton, $171\frac{4}{5}$, p. 202]

3. The only version I have seen is the pirated edition of 1737, printed in London for T. Woodward and J. Millan. Whiteside in fact says of this edition, '... probably pirated from Colson's work but in our view more adequate' [1964, p. xviii].
4. This was intended for Newton's memorial in Westminster-Abbey.
5. The following lines are offered, not as an Alexandrine, but a popeling — for as Cervantes said, 'y debajo de ser hombre puedo venir a ser papa' (and being a man, I may come to be Pope) [*Don Quixote*, I, chap. XLVII].
6. Henry Pemberton, in an article published in *The History of the Works of the Learned* in January 1741, traced the origin of Newton's term *ultimate ratio* to Gregory of St Vincent. The pertinent passage runs:

> In particular, *Gregory* of St. *Vincent*, who avoids the use of indivisibles* first defines a limit thus† *Terminus* [the limit] *progressionis est serei finis, ad quem nulla progressio* [by the addition of its terms] *pertinget, licet in infinitum continuetur; sed quovis intervallo dato propius ad eum accedere poterit*; and then in his exposition of this defiinition has these Words; *Terminus igitur progressionis talis est, quemad modum explicuimus, cùm scilicet aggregatum, sive summa terminorum progressionis, quantumvis*

> *continuatæ, nunquam excedit quandam magnitudinem; excedit verò omne minus illa magnitudine, atque ita posset etiam dici productum sive quantitas totius, datæ progressionis, et magnitudo illa æqualis dicetur toti progressioni datæ; hoc est omnibus terminis proportionalibus simul sumptis.*
>
> ... and the ratio [i.e., of the quantitates nascentes], which is the limit according to the conditions specified in this Lemma [i.e., Newton's], is called *ultima ratio quantitatum evanescentium*, and *prima ratio nascentium*, though the quantities never actually bear that ratio.
> *See Prop. 45. Lib. de Duct. Plani in Planum.
> †Definit. 3. Libri de progression. Geometricis.
> [Pemberton, 1741, pp. 77–78]

(The words in crotchets in the first paragraph are thus given in the original.) See also Gibson [1898, p. 29].

7. The anonymous author of this paper was first identified as Newton by James Wilson in his *Appendix by the Publisher* to Benjamin Robins's *Mathematical Tracts II* [1761, p. 368]. See also de Morgan [1852a] and [1852b].

   On the account of the *Commercium Epistolicum* (and describing the latter work as one that 'relates to the invention of the methods of infinite series, moments and fluxions' [p. 368]) published in the *Philosophical Transactions*, Robins writes

   > This account is also published in Latin before the second edition of the Commercium Epistolicum. It was wrote by Sir Isaac Newton himself, and the arguments there used are unanswerable, which M. Bernoulli, in complaisance to M. Leibnitz, without having seen them, is pleased, in a letter to that gentleman, to call after him, *argutationes*, Tom. ii. p. 364, 367. [1761, vol. II, p. 368]

   See also Hall and Tilling [1976, p. 242].

   One should perhaps also mention Newton's consideration of physical intuition as one of the bases of the fluxionary theory (see Cajori [1991], p. 490]), though this does not concern us here.

8. See Guicciardini [1989, pp. 3–5].
9. See Guicciardini [1989, p. 17].
10. Such clarity at times seemed to emulate the opinion of Beauty expressed by Margaret Hungerford (or perhaps, as the second edition

of *The Oxford University Press Dictionary of Quotations* somewhat ambivalently suggests, by Lew Wallace).

11. *Essay on Criticism*, v. 9–10.
12. *Essay on Criticism*, v. 408–409.
13. Cajori [1991, p. 219] describes the publication of *The Analyst* as 'the most spectacular mathematical event of the eighteenth century in England'; see also Jesseph [1993, pp. 300] and Gibson [1898, p. 31].
14. Guicciardini [1989, p. 173, note 7] explicitly mentions some dozen papers written in response to *The Analyst*, and says that 'The list of "answers" to Berkeley is endless'. Jesseph [1993, p. 231] writes that, by his own count, 'more than a dozen publications in the period 1734–50 responded to Berkeley's criticisms of the calculus in one way or another', and de Morgan [1852b] adds to the twenty-six articles headed *Analyst Controversy* in the 1850 catalogue of the library of the Astronomical Society, eight further papers. Fraser writes

   > In the seven years that followed its [i.e., *The Analyst*'s] appearance nearly thirty pamphlets and articles were issued in attack or defence, some of the chief mathematicians of the time taking part in the fray. [1901, vol. III, p. 8]

15. Berkeley's first printed works, both anonymous, were in fact mathematical in nature: *Arithmetica* and *Miscellanea Mathematica*.
16. In his editorial introduction to Berkeley's *Philosophical Commentaries* (vol. I of Luce & Jessop [1948]) Luce points out that this work was misleadingly named the *Commonplace Book* by Fraser (the present title is Luce's). Further, the pair of notebooks making up this work were published by Fraser in the wrong order, they having been incorrectly bound together.
17. This short esssay, read before the Dublin (Philosophical) Society on the 19th November 1707, was discovered by S.P. Johnston in the library of Trinity College, Dublin, and published by him in 1901 in *Hermathena* XI, 180–185 (see Luce & Jessop [1957, vol. IX, pp. 143, 150]). (Fraser [1901, vol. I, p. 409] dated it to 1705 or 1706, and said that it was published in *Hermathena* vol. XXVI of 1900.) Jesseph [1993, p. 53] mistakenly gives the title repeatedly as 'Of Infinities'.
18. Luce, in the first volume of Luce & Jessop [1948], writes of these and similar remarks by Berkeley, 'Rather irresponsible remarks about mathematicians' [p. 123].
19. Johnston [1930], followed by Luce [1948, vol. I], identifies this 'M.' as 'moment'; Jesseph [1993, pp. 160–161], however, finds that such

a reading makes a nonsense of Berkeley's passage, and suggests that 'M.' stands for 'minimum'. I too believe that the latter reading is the more felicitous.

20. See Fraser [1901, vol. I, p. lxxiv & vol. III, p. 7].
21. I have been unable to identify this person: perhaps Benjamin Robins is meant.
22. Karl Pearson, writing of the opinions of a man who has won himself a name as a natural scientist, states that this person's judgements in other fields 'will be sound or not according as he has carried his scientific method into these fields' [1900, §2].
23. For a more detailed summary see Jesseph [1993, p. 183].
24. De Morgan in fact writes 'The Analyst is a tract which could not have been written except by a person who knew how to answer it' [1852b, p. 329]. Berkeley was no stranger to this rhetorical technique: commenting on its use in *A Treatise concerning the Principles of Human Knowledge*, Walmsley notes

> To the reader who recognizes it [i.e., *prolepsis*], it gives the impression not of emotional manipulation, but of the utmost candour. In raising and forcefully presenting all our objections to his thesis, Berkeley seems at his most disinterested, reflective, and philosophically impartial.
> [1990, p. 34]

John Ward, Professor of Rhetoric at Gresham College from the 1st September 1720 to his death on the 17th October 1758, in fact recommends *prolepsis* as a rhetorical tool since

> it serves to conciliate the audience, while the speaker appears desirous to represent matters fairly, and not to conceal any objection, which may be made against him.
> [1759, II, p. 66]

25. Jesseph [1993, p. 183] suggests that these queries were written in imitation of the thirty-one queries at the end of Newton's *Opticks*.
26. This work was published under the name Philalethes Cantabrigiensis. It was attributed by Cantor in the third part of the third volume of his *Vorlesungen über Geschichte der Mathematik*, to Conyers Middleton and Robert Smith, but is now known to be by Jurin (see Gibson [1898, pp. 16, 30]). Cantor's words were

> Kaum war *The Analyst* erscheinen, so traten Kämpfer für die Fluxionsrechnung auf, zunächst zwei Professoren der Universität Cambridge, Conyers Middleton und Robert Smith,

welche sich aber nicht nannten, sondern ihrer gemeinsam verfertigten Schrift den Titel beilegten: *Geometry no friend to infidelity by Philalethes Cantabrigiensis*, dann ein Dubliner Professor Walton, der Verfasser einer *Vindication of Sir Isaac Newton's Principles of Fluxions*. [1898, III, p. 718]

27. The version given here is in fact taken from Berkeley [1784, pp. xix–xx]; it is the same as that given by Stock in 1776 (?1735 edition), and we may thus assume that the anonymous biographical essay in the 1784 edition of Berkeley's *Works* is Stock's *Account*.

28. Fraser in fact boldly writes

> The *Analyst* is addressed to Edmund Halley (1646–1742), the famous astronomer, in the character of 'an infidel mathematician.' [1901, vol. III, p. 4]

The same statement is made by Gibson [1898, p. 11].

29. Cajori [1991, p. 220] in fact views this work, together with MacLaurin's *Fluxions*, as 'the topnotch of mathematical rigor, reached during the eighteenth century in the exposition of the calculus.' Note also Boyer's very similar remark '... the treatise of Maclaurin, which represents the high point in the rigorous interpretation of the calculus in terms of geometrical and mechanical notions' [1949, p. 235].

30. See Appendix 5.1.

31. The question of whether Newton's writings were to be interpreted as implying (if not positively affirming) that variables could reach their limits was one that caused considerable controversy in the eighteenth century. Jurin saw such an interpretation, whereas (according to Cajori [1991, p. 220]), Robins and Henry Pemberton did not. Gibson, however, states that

> Newton's method is the thoroughly sound one of limits, and Berkeley's criticism was really based on a misinterpretation of Newton's terminology. [1898, p. 21]

In his 1898 review of the third part of Cantor's *Vorlesungen über Geschichte der Mathematik* Gibson clearly finds Robins's contribution to the *Analyst* controversy more useful and accurate than that of Jurin. Indeed he writes

> There can be little doubt that Philalethes was in blank ignorance of the characteristic features of Newton's doctrine, as expounded, for example, in the introduction to the *Quadratura*. [1898, p. 18]

Gibson also says that had Robins's 'much more valuable' contributions been known to Cantor,

> these would have shown that the state of mathematical learning in England in the years immediately subsequent to the death of Newton was by no means so low as the incompetence of Philalethes would seem to indicate.
> [1898, p. 11]

More important to our present purpose, Gibson writes

> Robins gave a complete and masterly defence of fluxions, and, in particular, laid down in clear and unambiguous form the doctrine of limits as the basis of the Infinitesimal Calculus. ... To Robins, more than to Maclaurin, I think, is due the credit of expounding in systematic and consistent form the fundamental conception of a limit, and of freeing Newton's statements from the ambiguities which gave plausibility to Berkeley's attack. [1898, p. 11]

(See also Gibson, op. cit. pp. 30–31.)

32. See Weinstock [1982] and Westfall [1973] for further details of inadequacies in some of Newton's work. Westfall finds difficulties with Newton's treatment of the acceleration of gravity, the velocity of sound, and the precession of the equinoxes. In addition, he quotes a letter from Roger Cotes to Newton (written while the former was preparing the second edition of the *Principia*) in which, while editing Proposition XXXVII on tides, Cotes wrote

> This alteration will very much disturb Your Scholium of y^e 4th Proposition [the correlation of g with the moon] as it stands; neither will it agree with Proposition 9th [the precession]. [7th February $171\frac{1}{2}$]

In reply Newton stated that he had lost his rough notes and could not recall how to make the further correction, but

> If you can mend the numbers so as to make y^e precession of the Equinox about $50''$ or $51''$, it is sufficient.
> [12th February $171\frac{1}{2}$]

33. Holgate, in Bingham [1997, p. 162], notes that it was only by the 1920s that the validity of probability theory was being assessed 'in terms of its logical foundation rather than its practical applications', and Warnock [1953, p. 221] remarks that it was in fact in response to

the *logical* incoherencies and inadequacies of Newton's fluxional work that *The Analyst* was written.

34. The attentive reader will wonder why page [*v*] is followed by page [**7**]: page [**6**] merely contains a list of books printed for J. Noon.
35. For details of works on the fluxionary calculus appearing in the first part of the eighteenth century see Guicciardini [1989] and Jesseph [1993].
36. See Cajori [1919, p. 44]. The study of Newton's work referred to here was the *Institution of Fluxions* of 1706.
37. Berkeley was appointed Bishop of Cloyne in January 1734 and was consecrated in St Paul's Church, Dublin, on the 19th of May 1734: *The Analyst* was published in March of that year in London and in June in Dublin.
38. Bayes was correct in the restriction of his tract, for Berkeley's fundamental interest in *The Analyst* was mathematical in nature.
39. It is perhaps hardly necessary to recall that the word *analysis* had a slightly different meaning in the eighteenth century to that in common (mathematical) use today. The *General Dictionary of Arts and Sciences* (see Scott et al. [1765]) provides the following definition.

> ANALYSIS, among mathematicians, is the art of discovering the truth or falsehood of a proposition, or its possibility or impossibility, by supposing the hypothesis or propositions to be true; and, by examining what follows from thence, be enabled to come at some known truth or manifest impossibility, of which the first proposition is a necessary consequence whereby we establish the truth or impossibility of our first proposition. ... That of infinite quantities, called the new analysis, is particularly used for the method of fluxions, or the differential calculus.

The meaning changed considerably over the next century, however, and by the twentieth century the term had become a part of the everyday mathematical vocabulary. Thus Whittaker & Watson's classic *A Course of Modern Analysis* of 1902 had the sub-title *An introduction to the general theory of infinite processes and of analytic functions; with an account of the principal transcendental functions*. The index to Körner [1960] has 'analysis, mathematical *see* function, real number, theory of sets', which is not particularly useful, whereas Black says

> the infinitesimal calculus and the modern theory of functions which are usually grouped together as analysis, a term which excludes both arithmetic and geometry. ... the

method of analysis essentially consists of applying arithmetical methods to the manipulation of certain geometrical intuitions of continuity. [1933, pp. 85–86]

In his study of Berkeley's writings on the philosophy of mathematics Jesseph provides the following definitions.

> I use the term *analysis* to refer to a body of mathematical techniques and results concerned with the problems of finding tangents to curves, finding the area enclosed by a curve (known as "quadrature" in the parlance of the period), and finding the arc-length of a curve (known as "rectification"). I use the term *calculus* for the algorithmic procedures developed by Newton and Leibniz for constructing tangents and finding quadratures. [1993, p. 123]

40. See Jesseph [1993, p. 274].
41. Smith [1980, p. 381] cites §31 of *The Analyst* as pertinent; I think the following two of Berkeley's Queries are also relevant.

    > *Qu.* 22 Whether it be necessary to consider velocities of nascent or evanescent quantities, or moments, or infinitesimals? And whether the introducing of things so inconceivable be not a reproach to mathematics?
    > *Qu.* 30 Whether motion can be conceived in a point of space? And if motion cannot, whether velocity can? And if not, whether a first or last velocity can be conceived in a mere limit, either initial or final, of the described space?

42. Following John Wallis [1657] (at least at a distance), Bayes uses $\sim$ to denote the absolute difference (we find further use of this in our later examination of Bayes's Notebook). The symbol $\sim$, in a horizontally rotated form, was used in this sense by William Oughtred in 1648 (see Cajori [1993, §§487–488] for further details).
43. Although modern symbolism is often used in our discussion, the reader is cautioned against viewing our 'interpretation' as anything more than an explanation of how Bayes's work (and other work of that period) might be seen today.
44. This case is described by Jesseph [1993, p. 275] as 'degenerate'.
45. The fundamental definition in Robins's discussion of the calculus is:

    > we shall in the first place define an ultimate magnitude to be the limit, to which a varying magnitude can approach within any degree of nearness whatever, though it can never be made absolutely equal to it.
    > [Robins, 1735, p. 53; 1761, §110]

Jesseph [1993, p. 265] has drawn attention to the fact that in Robins's definition, unlike that in the modern theory of limits, a variable *never* attains its limit. Moreover,

> This definition of ultimate magnitudes leads to the fundamental proposition that when varying magnitudes remain in a constant ratio to one another their ultimate magnitudes stand in the same ratio. [Jesseph, 1993, pp. 265–266]

Robins then goes on to define the ultimate proportion or ultimate ratio of two quantities, the definition including the case in which the variable quantities do not bear the same ratio to each other as they converge to their ultimate magnitudes, though once again the ultimate ratio is not actually attained by the varying quantities. Moreover the $a(\cdot)$ and $b(\cdot)$ (in Bayes's notation) tend monotonically to their ultimate ratios (see Jesseph [loc. cit.] for further details).

46. John Walton, another respondent to Berkeley, was also one who viewed the 'ultimate ratio' as a limit of vanishing quantities, or as the ratio of two infinitely diminished quantities. Walton's explanation of ultimate ratios in fact runs as follows.

> The ultimate Ratios with which synchronal Increments of Quantities vanish, are not the Ratios of finite Increments, but Limits which the Ratios of those Increments attain, by having their Magnitudes infinitely diminish'd: The Proportions of Quantities which grow less and less by Motion, and at last cease to be, will, in most Cases, continually change, and become different in every successive Diminution of the Quantities themselves: And there are certain determinate Limits to which all such Proportions perpetually tend, and approach nearer than any assignable Difference, but never attain before the Quantities themselves are infinitely diminish'd; or 'till the Instant they evanesce and become nothing. These Limits are the last Ratios with which such Quantities or their Increments vanish or cease to exist; and they are the first Ratios with which Quantities or the Increments of Quantities, begin to arise or come into being.
> [1735, pp. 8–9]

And further

> The Dispute between the Followers of Sir *Isaac Newton*, and the Author of the *Analyst*, is not about the Principles of the *differential Calculus*, but about those of Fluxions; and it is whether these Principles in themselves are clear or obscure, and whether the inferences from them are just

> or unjust, true or false, scientific or otherwise: We are not concerned about Infinitesimals or minute Differences, but about the Ratios with which mathematical Quantities begin or cease to exist by Motion; and to consider the first or last Proportions of Quantities, does not imply that such Quantities have any finite Magnitudes. They are not the Proportions of first or last Quantities, but Limits of Ratios; which Limits the Ratios of Quantities attain *only* by an infinite Diminution of their Magnitudes, by which infinite Diminution of their Magnitudes they become evanescent and cease to exist. [1735, pp. 29–30]

Gibson finds Walton 'even less qualified than Jurin to demolish the arguments of *The Analyst*' [1898, p. 18], and writing somewhat later Cajori says

> Walton seemed to have a good intuitive grasp of fluxions, but lacked deep philosophic insight. He showed himself inexperienced in the conduct of controversies, and did not know how to protect himself against attack from a skilful adversary. [1919, p. 93]

47. Jesseph in fact writes that Bayes's approach is 'quite far removed from the modern theory of limits' [1993, p. 276], although Smith [1980, pp. 387–388] finds the prime and ultimate ratios to be 'closely analogous' to the modern concepts of right- and left-hand limits.
48. The translation is from the 'Quadrature of Curves, by Sir Is. Newton' in the second volume of John Harris's *Lexicon Technicum* of 1710.
49. The translation is from Cajori's revision of Motte's 1729 translation.
50. Note that the 1735 edition of this work of Robins (i.e., not the version that appears in his *Mathematical Tracts*) does *not* have section numbers, but only page numbers: §§95, 98, and 99 occur in the 1735 version on pp. 48, 48, and 49, respectively.
51. The translation is Motte's (Cajori's revision). The first edition differs from the third (a) in having *attigit* for the second occurrence of *attingit* in the third edition, (b) in a few changes in word order, and (c) in having the words *pergentis velocitatem ultimam.* where the third has *ubi motus finiatur, pervenientis velocitatem ultimam:*.
52. It is seldom clear whether Bayes intends a time-interval to be open or closed, and we are not too precise about this in our discussions.
53. We suppose that the fluents (functions?) are sufficiently 'nice' for us to be able to do what we want to with them.

54. Smith [1980, p. 386] suggests that Bayes's fourth proposition was introduced to enable him (i.e., Bayes) to improve on Newton's inadequate finding of the fluxion of a product; and although this may indeed be so, Bayes's proof relies only on his sixth and eighth propositions (and recall that although two proofs were given of the latter, only the second of these used Proposition IV).
55. Referring to this lemma as one involving the 'shifting of the hypothesis', Cajori [1991, p. 219] notes that it received approval of the English (British?) mathematicians only with the work of Robert Woodhouse in 1803. This hypothesis is mentioned in the preface to this work in the following words.

> These methods I now examine, and shall endeavour to shew that in all there is the same kind of difficulty, when the passage is made from finite to infinite, or from discrete to continued quantity: and that all are equally liable to the objection of Berkeley, concerning the fallacia suppositionis, or the *shifting of the hypothesis*. Of the infinitesimal calculus of Leibniz I shall take no notice, because its principles as formally laid down by him, are acknowledged to be inadmissible; and it is not necessary to controvert, what no man undertakes to defend. [p. ix]

56. The translation is from Harris [1710].
57. The proposition quoted here is given, with part of its proof, as follows in Newton's *Opticks* (the translation is from Harris [1710]).

> PROP. I. PROB. I. Data æquatione quotcunq; fluentes quantitates involvente, invenire fluxiones. [Having given an Equation involving any number of fluent or flowing Quantities, to find their Fluxions.]
> Demonstratio. ... Minuatur quantitas o in infinitum, & neglectis terminis evanescentibus restabit $3\dot{x}x^2 - \dot{x}yy - 2x\dot{y}y + aa\dot{z} = 0$. Q.E.D. (Note: Newton is considering here the equation $x^3 - xyy + aaz - b^3 = 0$.) [Demonstration. ... Let the Quantity o be lessened infinitely, and neglecting the Evanescent Terms there will remain $3\dot{x}x^2 - \dot{x}yy - 2x\dot{y}y + aa\dot{z} = 0$. Q.E.D.]

58. Gibson [1898, p. 13] has pointed out that, although Berkeley quotes from the introduction to Newton's *Quadratura Curvarum*, the description of moments that he gives is in fact from the *Principia*, Book II, Lemma II. This Gibson (loc. cit.) finds significant, since

one object of the introduction to the *Quadratura* is to show that the doctrine of fluxions is independent of infinitesimals, and the use of moments *there* would have endangered the contention.

Newton described his notation for fluxions and moments in a letter to John Keill on the 15th of May 1714 (reprinted in Hall & Tilling [VI, 1976]), as follows.

> Fluxions and moments are quantities of a different kind. Fluxions are finite motions, moments are infinitely little parts. I put letters with pricks for fluxions, & multiply fluxions by the letter o to make them become infinitely little & the rectangles I put for moments. And whenever prickt letters represent moments & are without the letter o this letter is always understood. ... The rectangles under fluxions & the moments o being my marks for moments are to be compared with the marks dx & dy of Mr Leibnitz & are much the older being used by me in the Analysis communicated by Dr Barrow to Mr Collins in the year 1669.

59. The English translation is from Cajori's notes to Motte's translation of the *Principia*.
60. The translation here is Motte's (Cajori's revision).
61. The definitions given in the *Oxford English Dictionary* run:
    Bigot. A hypocritical professor of religion, a hypocrite. A superstitious adherent of religion. (1598, 1664).
    A person obstinately and unreasonably wedded to a particular religious creed, opinion, or ritual. (1661, 1741, 1844).
    Enthusiast. One who is (really or seemingly) possessed by a god; one who is under the influence of prophetic frenzy. (1641, 1677).
    One who erroneously believes himself to be the recipient of special divine communications; in wider sense, one who holds extravagant and visionary religious opinions, or is characterized by ill-regulated fervour of religious emotion. (1609, 1856).
62. A number of versions of the saying 'Among the blind, the one-ey'd blinkard reigns', attributed to authors from Michael Apostolius to Frederick the Great, may be found in Stevenson [1958?]. (*Blinkard*: a reproachful name for one who habitually blinks or winks; one who has imperfect sight. *fig.* One who lacks intellectual perception.)
63. The word is apparently new. The best the *Oxford English Dictionary* can come up with is *cecutient*: partially blind, dim-sighted. My word is formed in analogy with *felicitous* (from 'felicity'), *necessitous* (from 'necessity') and *ubiquitous* (from 'ubiquity').

64. See Jesseph [1993, pp. 197–199].
65. For further discussion see Jesseph [1993, p. 277].
66. The passage is reprinted in Walmsley [1990, p. 1].
67. The word 'Siris' is from σειρά, meaning *a chain*. De Quincy said the name ought to have been 'Seiris'. (The title changed slightly from one edition to another: see Luce & Jessop, vol. IX [1957, p. 1148].)

Chapter 6

1. I have omitted here papers on the evaluation of logarithms by series, unless they include *specific* mention of infinite series.
2. Throughout this chapter 'log' denotes the natural logarithm.
3. I adopt this terminology from Hardy [1949, §13.7]. In the 1951 translation of the fourth edition of Knopp's *Theorie und Anwendung der unendlichen Reihen* of 1947 we read (p. 520) that Legendre introduced the term *semi-convergent*, one which Knopp claims to have been superseded by *asymptotic*. Discussing Poincaré's theory of asymptotic series, Bromwich writes 'Such series were often called *semiconvergent* by older writers' [1931, p. 342].
4. Tweedie [1922, p. 44] states that Legendre, writing the series as $\sum (-1)^{n+1} u_n$, gave the bound

$$u_{n+1}/u_n < (2n-1)2n/4\pi^2 z^2$$

$$< (n/3z)^2.$$

5. Tweedie [1922, pp. 203–205] provides a well-reasoned argument for ascribing the discovery of this theorem to Stirling alone, though he does say that one may well speak of de Moivre's *form* of Stirling's Theorem. For new derivations of Stirling's formula Patin [1989] and Marsaglia & Marsaglia [1990] may be consulted.
6. The relevant passage in the original runs as follows.

   Haec autem constans ponendo $x = 1$, quia fit $s = l1 = 0$, ita definietur, ut sit

   $$C\,[\text{onst.}] = 1 - \frac{A}{1.2} + \frac{B}{3.4} - \frac{C}{5.6} + \frac{D}{7.8} - \text{etc.,}$$

   quae series ob nimiam divergentiam est inepta ad valorem ipsius C[onst.] saltem proxime eruendum. §158. Non solum proximum, sed etiam ipsum verum valorem ipsius C[onst.]

inveniemus, si consideremus expressionem Wallisianam pro valore ipsius π inventam atque in *Introductione* demonstratam, quae erat

$$\frac{\pi}{2} = \frac{2.2.4.4.6.6.8.8.10.10.12.\text{etc}}{1.3.3.5.5.7.7.9.9.11.11.\text{etc}}.$$

The reference is to Euler [1748, vol. I, chap. XI].

7. Euler defined convergence and divergence of series in 1760.
8. The B_i are Bernoulli numbers.

Chapter 7

1. Still other views are cited in Seal [1978].
2. Cajori [1919, p. 157].
3. Von Wright [1960, p. 292].
4. According to Yates [1962, p. 281], Fisher 'early recognised the fallacies of the Bayesian approach'.
5. See Fisher's *Statistical Methods for Research Workers*, pp. 20, 21, as reprinted in Fisher [1995].
6. For an opinion on Fisher's *Statistical Methods and Scientific Inference* see Kendall [1963, p. 6].
7. As reprinted in Fisher [1995, p. 6].
8. See Dale [1999, p. 31] for a discussion.
9. See Dale [1999, p. 31] for further details.
10. Davies [1875, p. 347]. In *The Adventure of the Greek Interpreter* Sherlock Holmes said 'My dear Watson ... I cannot agree with those who rank modesty among the virtues'.
11. Hald [1990a, p. 67].
12. Page 174 of the translation.
13. Bayes was first cited, by d'Alembert, in 1780; see Rashed [1988, pp. 206, 226 et seqq.].
14. Todhunter in fact writes 'Bayes must have had a notion of the principle [for estimating the probabilities of the causes by which an observed event may have been produced]' [1865, Art. 868].
15. See also Jeffreys [1983, §1.2] and Nagel [1939, p. 29].

16. Nagel in fact writes 'Huygens, the Bernoullis, Montmort, De Moivre, and Bayes are the most prominent figures in the early history of the subject [statistical inference]' [1939, p. 8].

    Swijtink [1987, p. 284] notes that the expression 'Bayesian inference' does not convey the idea that Bayesians provide inferential rules, although they do provide criteria for the monitoring of one's own opinions. For a discussion of a correct usage of Bayes's Theorem in hypothesis testing see von Mises [1981, pp. 155–157].

    Fisher, in his *Statistical Methods and Scientific Inference*, described Bayes's work as follows,

    > the first serious attempt known to us to give a rational account of the process of scientific inference as a means of understanding the real world, in the sense in which this term is understood by experimental investigators. [p. 8]

    As regards Bayesian methods, notice O'Hagan's writing

    > it is quite possible to follow the Bayesian method in almost every respect without holding either a frequentist or a subjective view of probability. [1994, §5.32]

17. Compare the displayed quotation from von Wright [1960, p. 209] given earlier in this section.
18. For details see von Mises [1981, pp. 132–133].
19. See Kyburg & Smokler [1980, p. 207] for de Finetti's views on this topic.
20. Yates regarded the discussion of inverse probability in the first chapter of R.A. Fisher's *The Design of Experiments* as the best discussion of that subject he had ever read — see p. xxi of the Foreword to Fisher [1995].
21. Von Wright [1960, pp. 294–295] notes that the range of application of Inverse Probability should be restricted to cases analogous to some that arise in games of chance. Although von Wright and Fisher find Inverse Probability useful mainly (if not only) in games of chance, Savage [1961, p. 576] notes that the frequentist too is cut off from most applications of Bayes's Theorem.

    Von Wright finds that the Problem of Causes is best illustrated by questions concerning games of chance, writing 'Applications of the problem outside the realm of games of chance seem, therefore, to be practically out of the question' [1960, p. 280].

    There are, however, those who see a wider range of applicability for Bayes's Theorem, e.g., Reichenbach, who wrote

> The range of application for Bayes's rule is extremely wide, because nearly all inquiries into the causes of observed facts are performed in terms of this rule. The *method of indirect evidence*, as this form of inquiry is called, consists of inferences that on closer analysis can be shown to follow the structure of the rule of Bayes. The physician's inferences, leading from the observed symptoms to the diagnosis of a specified disease, are of this type; so are the inferences of the historian determining the historical events that must be assumed for the explanation of recorded observations; and, likewise, the inferences of the detective concluding criminal actions from inconspicuous observable data. In many instances the use of probability relations is not manifest because the probabilities occurring have either very high or very low values. [1971, p. 94]

22. See Jeffreys [1983, §3.1] for a discussion of problems caused by the assumption of a uniform prior distribution in sampling. See also Walley [1991, §5.5, & p. 377].
23. This 1960 note by L.J. Savage was first published in the first edition of my *A History of Inverse Probability from Thomas Bayes to Karl Pearson*.
24. See also Salmon [1967, pp. 120–121]. For a discussion of the hypothetico-deductive and other scientific methods see Medawar [1969].
25. In the Introduction to this same work Fisher noted that Bayes deserved 'honourable remembrance' for one fact in addition to those mentioned by de Morgan: his concern over the validity of his 'axiom' and consequent hesitation in publishing his work. Fisher was not completely dismissive of Bayes's method, finding it applicable when exact prior knowledge was available [op. cit. pp. 194, 198].
26. See Daston [1988, pp. 264–267] for a comparison of Hume's criticism of induction with Price's application of Bayes's work to inductive reasoning.
27. The Scots word *grue* (*v.* our more common *gruesome*) means a creeping of the flesh, a shudder, or shiver.
28. The term 'rule of succession' was introduced by John Venn in his *The Logic of Chance*. For a combinatorial consideration see de Finetti [1975, vol. II, §§11.4.3–4], and for a discussion of its use in a logical setting see von Wright [1960, pp. 213–215]. Walley [1991, §5.3.4] may also be consulted.
29. In a recent study Freedman [1999] has examined some problems of causal inference that arise from non-experimental data.
30. The binomial theorem was first given in print in a tract by Archibald Pitcairne in 1688 (see Stigler [1999, p. 211]).

31. On the 'probability of a probability' see von Wright [1960, p. 211] and Good [1983, p. 327].
32. Price remarks further that de Moivre 'has omitted the demonstrations of his rules'; Hald [1998, p. 145] has pointed out that this statement is incorrect. See Hald's §§2.2 and 2.3 for details of the history of the consideration of the direct problem.
33. See Walley [1991, p. 521] for comments on 'solutions' to this problem by various authors. On the use made by Bayes and Laplace of the words 'probability' and 'chance' see Jeffreys [1983, p. 51] and (more particularly Laplace) Hald [1998, pp. 161–162], and on the need for distinguishing between the two terms see Poisson [1837, p. 31].
34. Todhunter writes of the first section:

> this part of his essay is excessively obscure, and contrasts most unfavourably with the treatment of the same subject by De Moivre. [1865, Art. 544]

See also Shafer [1982, p. 1076].
35. This short tract was the Dutch version of the better known *Tractatus de ratiociniis in ludo aleæ* of 1657, the former being written before, but printed after, the latter. One reason for the delay in the printing of the Latin version was that the Dutch had to be translated and appropriate Latin terms coined, and although Huygens himself provided a translation, that which was published was by Francis van Schooten. It is therefore difficult to know which of these versions should be taken as the original, and even the few quotations we give in our text show that considerable care is needed in citing 'what Huygens wrote'.
36. This passage is described in Hald [1990a, p. 69] as a 'somewhat obscure' statement.
37. Stigler [1999, p. 239] notes that the title of this tract was given on the title page of van Schooten's book as *De Ratiociniis in Aleæ Ludo* and as *De Ratiociniis in Ludo Aleæ* on its own first page. One might be tempted to see something similar in Jakob Bernoulli's *Ars conjectandi*, the words *Artis conjectandi* appearing on its first page: a more careful examination shows, though, that the full pertinent latter phrase is *Artis conjectandi, pars prima*.
38. Van der Wærden, unlike Freudenthal, finds 'chance' (*kans*) used by Huygens in two senses, viz. (a) possibility (*Möglichkeit*), and (b) probability (*Wahrscheinlichkeit*) (see the introduction to Bernoulli [1975]). It is in this second sense that van der Wærden finds *kans* used in Huygens's first propostion.

39. See David [1962, p. 116] for a translation from the Latin. Hald [1990a, p. 69], giving an English translation of the propositions from the Dutch, says that 'there is no difference in meaning, merely slight differences in wording'. In such a statement he is perhaps a little optimistic. Todhunter's translation, on the other hand, follows the Latin in mentioning my 'expectation' [1865, Art. 31].
40. Notice that de Moivre also uses *sors* in his *De Mensura Sortis*, the word being translated by McClintock in Hald [1984] as 'expectation'.
41. An even earlier use of the importance of basing decisions in life on expected utility is found by Stigler [1999, p. 242] in the 1750 English translation of Cumberland's *De legibus naturae disquisitio philosophica*.
42. See, for example, the last paragraph of Edgeworth's *Metretike* of 1887.
43. See Hald [1990a, p. 73] and [1998, p. 138] for a discussion of 'Huygens's analytical method'.
44. Stigler [1999, p. 243] cites a 1678 work by Thomas Strode as possibly the first mathematical work on probability to be published in English.
45. See Stigler [1999, p. 242] for comments on Cumberland.
46. Hailperin [2000, p. 217] notes similarly that it is unclear in Borel [1909] whether probabilities have, of necessity, events associated with them.
47. See also Jeffreys [1983, §1.3].
48. Walley gives the following reasons for adopting this terminology:

> The main reason for regarding previsions as more fundamental than probabilities is that coherent lower probabilities may have many different extensions to coherent lower previsions.
> We prefer de Finetti's term 'prevision' rather than 'expectation', so that the latter can be used in the standard sense of a functional constructed from probability.
> ... previsions may be easier to assess than probabilities.
> [1991, pp. 53, 493, 497]

49. See also Walley [1991, p. 81]. An excellent text based on the expectation approach is Whittle [1976].
50. See David [1962] and Ineichen [1996] for a discussion of probability in antiquity.
51. Writing of the negative factors of the 'space-measure' formulation of probability, de Finetti says

Even more dangerous is the fact that *stochastic independence* — $P(AB) = P(A)P(B)$ — is considered as being a property of the events ... it is a property of the function P (in relation to the events A and B), and not of the events as such. [1975, vol. II, p. 259]

52. For an English translation see Kolmogorov [1956, pp. 8, 9].
53. See Bingham [1997] for Holgate's discussion of independence as it was considered by the Polish School.
54. See Shafer [1978, p. 345].
55. Kamlah [1987, p. 102] discusses Stumpf's concept of probability, the logical character of which is such that time plays no part in probability. For related arguments see Shafer [1983] and [1985]. In the former of these Shafer uses an explicit time scale and expectation of random quantities in the place of the rooted trees used earlier, and in the latter ([p. 262]) he argues that the change from $\Pr[A]$ to $\Pr[A|B] = \Pr[A \cap B]/\Pr[B]$ 'is defensible and justifiable only on the basis of a protocol that tells circumstances under which the information B will be acquired'.
56. See also Hailperin [1996, p. 135].
57. Similar accounts have been given by Fisher, Barnard, and Edwards; see Edwards [1978, p. 116] for references. Dale [1999, §3.4] provides a formulation of the results of this section in terms of a table of unit area whereas Edwards [op. cit., p. 117] notes that the corollary to Proposition 8 holds even if the table is not uniform. Timerding [1908, p. 48] replaces Bayes's 'table-and-balls' example by one involving balls and a chest-of-drawers.
58. Hogben writes 'Bayes's prior probabilities are assignable to hypotheses in a meaningful sense only in the domain of what von Mises speaks of (p. 138) an experiment carried out in two stages, i.e. when each of the hypotheses of the comprehensive set is referable to a real population at risk' [1957, p. 450], and see also his [1957, pp. 130, 141].
59. It is sometimes suggested that Proposition 9, framed in terms of 'table and balls thrown', provides the desired solution to Bayes's problem. I believe that Proposition 10 is the required result, and that the preceding quotation provides the link between this result and that for the 'unknown event' in Proposition 10. For further discussion see Dale [1982] and Edwards [1978, p. 117].
60. See Dale [1999, §9.6] and Hald [1998, §15.6] for details.
61. For a detailed history of the indifference principle from Laplace to Jeffreys see Hald [1998, §15.6]. On Edgeworth's 'seductive analogy'

between the principle of indifference and the method of clones in *Mathematical Psychics* see Mirowski [1994, p. 40].

62. See Edwards, Lindman, & Savage [1963] for a discussion of whether people do in fact learn from experience in the way that probabilities are revised by Bayes's Theorem. This work is commented upon in Gigerenzer et al. [1989, pp. 216–222].

63. See, for example, Boudot [1972, pp. 60–69], Jeffreys [1983, pp. 118, 123], and von Mises [1981, pp. 155–157].

64. This seems to imply exchangeability; see Dale [1999, §3.4].

65. See Hald [1998, p. 144].

66. See, in particular, Hacking [1965, chaps XII & XIII].

67. For a detailed examination of this function see Dutka [1981]; the term is apparently due to J.P.M. Binet (see Dutka [1991, p. 241]).

68. See Molina & Deming [1940, p. xii].

69. That Bayes's main result holds if one assumes only that the prior does not vanish 'in a neighbourhood of the central value' was shown by Watanabe, who proved the following limiting result.

> If an event whose probability p is unknown à priori, has been observed to have occurred r times in n independent trials made on the event, n being a very large number, the probability à posteriori that p lies in the neighbourhood of r/n is nearly equal to unity. [1933, p. 390]

This result of Watanabe's is perhaps more akin to what I have described as the Inverse Bernoulli Theorem [Dale, 1999, chap. 1] than to Bayes's Theorem.

In their analysis of *The Federalist* papers, Mosteller & Wallace state 'We hope to produce such strong statistical evidence as to overwhelm any moderate assessment of initial odds' [1984, p. 50], and again 'if the likelihood ratio is large enough, or small enough, the final probabilities (not odds) are changed only slightly by wide changes in the initial odds' [p. 57]. The overwhelming of initial opinion by incisive data was referred to by Savage [1962, p. 20] as *precise measurement*.

70. Dutka [1981, p. 18] notes that this result was proved in Montmort's book of 1714.

71. See Molina & Deming [1940, pp. xi–xii].

72. On the Rule of Succession see Boudot [1972, pp. 83–85], Hald [1998, §15.6], Venn [1886], and Zabell [1989].

73. For a discussion of the derivation of the posterior distribution by a series of applications of Bayes's result see O'Hagan [1994, §3.6].

74. See Zabell [1988], Hald [1998, §8.5], and Dale [1999, §4.6].
75. See Hald [1998, p. 146] for a comparison with Hume.
76. We are perhaps more used to thinking of Bayes's problem as one that is the inverse to Bernoulli's, and it is perhaps worth bearing in mind that by 'inverse problem' here Price means the latter.
77. In addition to the references cited in this and the following paragraphs, the reader may be referred to Cajori [1991, pp. 377–378], Jordan [1923], Keynes [1921, chaps 30 & 31], and van Rooijen [1942].
78. See Feller [1957, §VII.2].
79. See Dempster [1966, p. 356].
80. I am indebted to Mrs J. Currie of Edinburgh University Library for this information.
81. See Dale [1999, p. 535].

Chapter 8

1. The ideas of Laplace and de Moivre were generalized to give what we now call the de Moivre–Laplace Central Limit Theorem:

    Let $\{X_k\}$ be a sequence of mutually independent and identically distributed random variables with mean μ and variance σ^2, and let $S_n = \sum_1^n X_k$. Then

    $$\Pr[(S_n - n\mu)/\sigma\sqrt{n} < z] \to \Phi(z),$$

    where $\Phi(\cdot)$ denotes the standard Normal distribution function.

    Hald [1990b, §10.3] may be consulted for details of Laplace's work on this matter.

2. For an applied Bayesian problem see Mosteller & Wallace [1984].
3. See Hald [1990b, §3] for a derivation using the integral calculus.
4. For ease of printing I have given things like $\alpha : \beta$ as $\frac{\alpha}{\beta}$ here.
5. For $n = 5, 10$, and 100 the values of $2\Phi(\sqrt{n/2(n-1)}) - 1$ are 0.57, 0.544, and 0.522, respectively.
6. See also Hald [1990b, p. 147] for reference to Timerding on this approximation, and also Wishart [1927, p. 12].
7. See Hald [1990b, p. 147].

Chapter 9

1. One finds this referred to in Act III, Scene III, of Shakespeare's *Twelfth Night; or, What you Will*, where Malvolio says 'It did come to his hands, and commands shall be executed. I think, we do know the sweet Roman hand.' See also Walsh [1894, p. 979].
2. The Advertisement to the first volume includes the words

    > The character of the author of this work, so well known by his other learned and elaborate writings, would have been sufficient to have recommended it to the public, if he had thought proper to have printed it during his own life.

3. Other comments on style from Russell are: a moving and pathetical way of writing (Burnet on Richard Baxter (1615-1691)); elegant and chaste (Warton on Nicholas Rowe (1673–1718)); clear, simple and easily understandable (Cumberland on Oliver Goldsmith (1728–1774)); refined without false delicacy, correct without insipidity (Campbell on ditto); an air of inimitable ease and carelessness, united with a high degree of correctness (Hall on William Cowper (1731–1800)); delicate in diction and interesting in narrative (Anon on Lord Macaulay (1800–1859)).
4. That this is the correct interpretation of 'vulgar' here is shown by Hazlitt's further saying 'It is clear you cannot use a vulgar English word, if you never use a common English word at all' [1822, vol. II, p. 187].
5. This Lecture is amply illustrated with quotes from Cicero.
6. We have already mentioned Berkeley's use of this rhetorical device in Chapter 5.
7. Walsh [1894, p. 1037] notes that this saying is sometimes given as 'le style est de l'homme'.
8. This quotation is taken partly from Buffon's essay itself and partly from the version given in Nisard [18–, vol. 4, pp. 410, 411].
9. Hubbell, in his translation of Cicero in *Cicero: Brutus ..., Orator* [1962, p. 297], discusses the training of an orator under five heads: invention (inventio), arrangement (collocatio), diction and style (elocutio), delivery (actio), and memory (memoria).
10. See Jones [1963, pp. 5, 18].
11. Walsh [1894, p. 1056–1057] relates the traditional story as follows. On being reproved by his confessor for indulging in certain 'conjugal infidelities', Henry IV asked his cleric what his favourite dish was. 'Partridges', came the reply. Shortly afterwards the confessor was

imprisoned, and was given nothing but partridges for his meals, until he complained with loathing of this constant diet. Henry reminded him of his earlier expressed fondness for the bird, to which the confessor replied 'Mais toujours perdrix!' The King then explained that he was devoted to his Queen, 'Mais toujours perdrix!'

12. One might note, in passing, that Parker stated that 'Cicero did less to form style than Jerome' [1867, p. 8].
13. See the *Dictionary of National Biography* and *Bibliotheca Britannica* for further details.
14. Walsh [1894, p. 399] refers to this aphorism as 'that beautiful proverb of which Pythagoras is reputed the author'.
15. The translations from *Diogenes Laertius. Lives of Eminent Philosophers* are by R.D. Hicks.
16. See Grant [1884, vol. I, pp. 146-148, 191-192; vol. II, p. 317].
17. It might be nice to recall the old Scottish Students' song 'A Chequer'd Career' (sung to the eighteenth century air 'Oh, dear! What can the matter be?', and with lyrics by David Rorie), the first verse of which runs as follows.

> When I first was a *civis* I studied Humanity,
> *Cos* and *Sine* show'd me Life's utter inanity,
> Hegel and Kant prov'd that all things were vanity,
> All save a chequer'd career!

Chapter 10

1. A genitive seems to be needed here.
2. In a fair copy of this letter, passed on to me by S.M. Stigler, who in turn had received it from Churchill Eisenhart, J.D. Holland gave the reading of this word as 'perfectly'. It seems, however, from a similar occurrence in Bayes's Notebook (p. 51(c)R, 6↑), that it should be taken as 'particularly', and this I feel makes better sense.
3. These same definitions are indeed given in all editions of the *Essai philosophique sur les probabilités*.
4. See Maistrov [1974, p. 83] and Sheynin [1973a, p. 281].
5. Plackett [1958, p. 133] notes that Tycho Brahe also eliminated systematic errors by using the arithmetic mean.
6. See Sheynin [1973a] and [1973b].
7. A complete study of Simpson's tract would include discussion of the last problem considered there, viz. 'to determine the odds, that the mean of a given number of observations is nearer to the truth than one single observation, taken indifferently' [1757a, p. 74]. In his discussion of this problem Shoesmith [1985] notes that Simpson demonstrates here the connexion between what we now know as the probability density and the probability distribution functions.

8. Plackett notes that

> since the generating function for (ii) [i.e., the same set of errors with probabilities proportional to $1, 2, \ldots, v + 1, \ldots, 2, 1$, respectively] is the square of what it is for (i) [i.e., possible errors are $-v, \ldots, -2, -1, 0, 1, \ldots, v$ and equal probabilities are attached to them], Simpson's initial contribution amounted mainly to realizing the physical interpretation of a mathematical result. [1958, p. 133]

9. See Hald [1990a, p. 211].
10. See Kendall [1961, p. 1], Maistrov [1974, p. 82], Seal [1954, p. 62], and Sheynin [1971b, p. 251].
11. See also Plackett [1958, p. 134].
12. See Todhunter [1865, Arts 563–569] and Sheynin [1973a, pp. 282 et seq. & 292].
13. Emphasis added.
14. See Stigler [1999, p. 303] for a discussion.
15. This manuscript was drawn to the attention of the statistical community by R.W. Home [1974–1975], and the reader may be directed to this paper for details of other work on electricity that was published at that time.
16. Tweddle in fact writes 'It is impossible to investigate 18th century mathematics without encountering Leonhard Euler' [1988, p. 5].
17. Cavallo writes

> The earliest account we have of any known electrical effect, is by the famous ancient naturalist THEOPHRASTUS, who flourished about 300 years before the present era. He tells us that *amber* (whose Greek name is $\eta\lambda\varepsilon\kappa\tau\rho o\nu$, and from whence the name *Electricity* is derived) as well as the *lyncurium* [identified in a footnote as *Tourmalin*], has the property of attracting light bodies. This was all that was known of the subject, for about fifteen centuries after THEOPHRASTUS; in which long period we find no mention in history of any person having made any discoveries, nor even any experiments in this branch of philosophy; the science remaining quite in the dark till the time of WM. GILBERT, an English physician, whose work *de magnete*, which contains several electrical experiments, was published in the year 1600; and who, for his discoveries in this new and uncultivated field, may be justly deemed the father of the present Electricity. [1782, pp. xix–xx]

Roller [1953] has shown that Bacon was acquainted with Gilbert's *De magnete*. According to Singer [1814, p. 3], the first *treatise* on electricity was written by Hauksbee at the beginning of the eighteenth century.

18. Adams earlier gave the definition

> Amber, silk, jet, dry wood, and a variety of other substances, being excited, attract and repel light bodies, these are called *electrics*. Such substances, as metals, water, &c. the friction of which will not produce this power of attraction and repulsion, are called *non-electrics*.
> [1799, pp. 34–35]

19. As if it were not enough to have to cope with the ether, the electric fluid, the etherial fluid, and the unctuous effluvia, we find Benjamin Martin writing

> The *Electric Virtue* consists in a fine subtile Matter, omitted from some Sorts of Bodies under the Circumstances of Attrition, which Bodies are called ELECTRICS *per se*.
> [1746, p. 9]

(Martin also refers to the 'Electric Effluvia', 'this subtle Matter, or Spirit', and 'Electrical Virtue'.) Writing somewhat later, Singer has 'The *cause* of electrical phenomena is *material*, and possesses the properties of an elastic fluid' [1814, p. 55].

20. The page is actually numbered 'vii', but so is the preceding page.
21. Canton is the only other person to have a query identified by Priestley in this way.
22. Other works by Wilson: *An essay towards an explication of the phænomena of electricity, deduced from the Æther of Sir Isaac Newton, contained in three papers which were read before the Royal-Society*. London: C. Davis, 1746; *An essay towards an explication of the phænomena of electricity*. Dublin: 1747; *A Short View of Electricity*. London: C. Nourse, 1780; *A Treatise on Electricity* of 1750 (the first edition is just 'by B. W.'; the second of 1752 has the author's name on the title page): *A Letter from Mr. Benj. Wilson ... to Mr. Aepinus ... Read at the Royal Society December 23, 1763, and March 1764*. London, 1764; *Observations upon lightning, and the method of securing buildings from it's effects, in a letter to Sir Charles Frederick ... by B. Wilson ... and others*. London: for L. Davis, 1773; *A series of experiments on the subject of phosphori, and their prismatic colours; in which are discovered some new properties of light. Also, a new translation of two memoirs of the late J.B. Beccaria*. 2nd edition. With additions. London: for J. Nourse, 1776; *An account of*

experiments made at the Pantheon, on the nature and use of conductors: to which are added, some new experiments with the Leyden phial. Printed for J. Nourse, 1778; *Further observations upon lightning; together with some experiments . . . communicated to the Royal Society, and rejected.* London: for L. Davis, 1774.

23. The reference is to the *Philosophical Transactions* 51, Pt. 2, p. 899 et seq. Priestley also details an experiment, akin to one of Franklin's, by 'Dr. Darwin of Litchfield', read at the Royal Society on the 5th May, 1757. This Dr Darwin was Charles Darwin's grandfather Erasmus (1731–1802).

24. Writing of Franklin's letters to Peter Collinson, Priestley says

> Nothing was ever written upon the subject of electricity which was more generally read, and admired in all parts of Europe than these letters. [1767, p. 159]

25. The first passage in quotation marks is from the biography of Sir John Pringle in the *English Cyclopedia.*

The almost Lilliputian dispute between the 'round-endians' and the 'pointed-endians' related here is reported somewhat differently elsewhere in Mottelay's book. In a passage for the year 1769, Mottelay notes that St Paul's Cathedral was first provided with lightning conductors in 1769, and he then adds

> Dr. Tyndall, who mentions this fact (Notes of Lecture VI, March 11, 1875) likewise states that Wilson, who entertained a preference for blunt conductors as against the views of Franklin, Cavendish and Watson, so influenced King George III that the pointed conductors on Buckingham House were, during the year 1777, changed for others ending in round balls. [1922, p. 231]

In 1799 Adams noted that 'conductors terminated by sharp points are sometimes advantageous, and at other times prejudicial' [p. 339], and further

> The cases against which we wish principally to provide, are the explosions of extensive and highly electrified clouds; and here we have seen that blunted ends, as acting to a much smaller distance, are entitled to the preference. [1799, p. 340]

In the second appendix to the second edition of 1782 of his *Complete Treatise on Electricity* Cavallo recounts that a house belonging to the Board of Ordnance, at Purfleet, was struck by lightning on the 12th May 1777, although furnished with a conductor. Experiments as to the best form of lightning conductor were then carried out by Benjamin Wilson and Edward Nairne,

> the former giving constantly the preference to short conductors terminating in a ball, and the latter preferring long Conductors acutely pointed. [Cavallo, 1782, p. 456]

Elsewhere in this second edition of his book Cavallo preserved the text of the first edition, concluding that

> The upper end of the Conductor should be terminated in a pyramidal form, with the edges, as well as the point, very sharp. [1782, p. 80]

In 1779 Charles, Viscount Mahon, gave a detailed account of experiments designed to test the relative efficacies of round and pointed lightning conductors, and included a list of the necessary requisites in erecting conductors.

Wilson's work seemed to become less acceptable to, and less accepted by, the scientific community as the years progressed: in fact, we find him publishing, in 1774, *Further experiments upon lightning; together with some experiments ... communicated to the Royal Society, and rejected.* And in 1779 Charles, Viscount Mahon, in his *Principles of Electricity*, omitted all discussion of Wilson's electrical experiments, reserving criticism of these for a later work and

> [thinking] it improper, to confound an Explanation of new and important *Facts*, with a Refutation of Doctrines, which, had it not been for the various Doubts they have occasioned in the Minds of some Men, would, in the Author's Opinion, not have deserved much consideration. [1779, p. iv]

On the subject of natural phenomena, note the following.

> In the Shetland iles they [i.e., the northern lights] are called 'merry dancers,' and are the regular attendants of clear evenings, giving a diversity and cheerfulness to the long winter nights. [Singer, 1814, p. 256]

26. Singer notes that 'The separation of electrified bodies is usually ascribed to repulsion; an assumption quite hypothetic and unnecessary' [1814, p. 24].
27. In his comments on Newton's twenty-second question Roller writes 'Newton felt it necessary to question the electrical effluvium theory, which played a prominent role in early eighteenth-century English work in electricity' [1959, p. 183].

28. I am uncertain as to whether any distinction was in general observed (or preserved) at this time between the terms *gravitation* and *gravity*, though that some were careful in this regard is seen in the following passage from one of James Stirling's notebooks — recorded as Bundle 136 in the National Register of Archives (Scotland) survey of the Stirling of Garden papers (Survey No. 2362 (1982)).

> I call gravitation the force with which bodies would descend if the Earth did not turn around it's axis and I call gravity the force with which they would descend while the Earth does turn round it's axis. [1988, p. 117]

29. Although it is generally recorded that Stanhope died in 1786, I have seen his birthdate given variously as 1710, 1713, 1714, and 1717.
30. This is presumably the Sir John Pringle, M.D., whose name we have already come across in §10.5. Pringle and Murdoch were both elected Fellows of the Royal Society in 1745 (o.s.).

Chapter 11

1. For a sketch of the contents of part of the notebook see Holland [1962].
2. The will is given in Chapter 12 of this work.
3. Samuel Pepys used Shelton's shorthand in writing his diary (see Rouse Ball & Coxeter [1974, pp. 388–389]).
4. Note that 'atq$_3$' is a contraction of 'atque': similar contractions are used elsewhere.
5. Bayes is probably following Newton in adopting this notation. Newton also used s' and t' for the complementary sine and tangent (i.e., cosine and cotangent), a notation that D.T. Whiteside ascribes to Seth Ward — see Whiteside [1971, vol. IV, pp. 117 & 134]. Ward also used s' for sine (see Whiteside, op. cit., 199n).
6. On the history of the use of a rectangle to indicate integration see Cajori [1929, §622].
7. I am grateful to my friend and colleague Hugh Murrell whose skilful use of the computer package *Mathematica* showed the connexion between Bayes's Rule and Newton's Method.
8. Expressions for sums of powers may be found in Edwards [1987]: see Hall & Knight [1948, §562] for details of a method for finding the sum of an assigned power of the roots of an equation.
9. One feels that there ought to be a word like 'quindenic' to describe such an equation.

10. See Cajori [1927, p. 43]. A still more general result is given in Burnside & Panton [1928, vol. I, chap. IX]: consider the equation

$$x^n + a_1 x^{n-1} + a_2 x^{n-2} + \cdots + a_n = 0,$$

the first negative term in which is $-a_r x^{n-r}$. If $-a_k$ is the greatest (sic) negative coefficient, then $\sqrt[r]{a_k} + 1$ is a superior limit of the positive roots.

11. See Appendix 11.3.

12. One is tempted to write the left-hand side of this inequality in terms of the von Ettingshausen symbol, but the ratio of 'binomial coefficients' thus obtained perhaps casts doubt on the method of proof to be adopted.

13. For some reason, obscure to me, the preposition in the title of this work was changed to *on* in the second edition (printed for William Baynes, 54 Paternoster Row, and William Davis) of 1801.

14. There will be little harm in our regarding these fluxions as derivatives (though it is of course wrong to do so).

15. For a discussion of the separate derivations of this formula by Euler (who essentially used an equivalent to the Leibnizian differential calculus) and MacLaurin (who essentially used an equivalent to the Newtonian fluctional calculus) see Mills [1985] and Tweddle [1998]. Gould & Squire [1963] may be consulted for an alternative result, given by MacLaurin, to the Euler–MacLaurin sum formula, a result that, combined with the latter, provides useful upper and lower bounds for the desired sum.

16. Following Hardy [1949, §13.2] we define the Bernoulli numbers B_r by

$$\frac{t}{e^t - 1} = 1 - (t/2) + \sum_{n=1}^{\infty} (-1)^{n-1} B_n t^{2n}/(2n)!$$

The first few B_r are $B_1 = 1/6$, $B_2 = 1/30$, $B_3 = 1/42$, $B_4 = 1/30$, and $B_5 = 5/66$. Various formulae for the constant C in the Euler–MacLaurin sum formula are given in Hardy [op. cit., §13.13].

The Bernoulli functions $B_n(x)$ are defined by

$$\frac{t e^{xt}}{e^t - 1} = 1 + \sum_{n=1}^{\infty} B_n(x) t^n/n!.$$

On putting $x = 0$ here and comparing the resulting series with that given above for $t/(e^t - 1)$ one finds that

$$B_{2r}(0) = (-1)^{r-1} B_r; \quad B_{2r+1}(0) = 0 \quad (r > 0).$$

17. To show the equivalence of the series for $\log\left(\frac{z+1}{z}\right)^{z+1/2}$ given in (11.11), notice firstly that

$$\Delta^n(1/z) = (-1)^n n!/z(z+1)\ldots(z+n)$$

and

$$(1/z)^{(n)} = (-1)^n n!/z^{n+1}.$$

Thus

$$\frac{1}{(2r)!}\Delta(1/z)^{(2r-2)} = \frac{1}{(2r)(2r-1)}\Delta\left(1/z^{2r-1}\right).$$

This establishes the equivalence of the first two series in (11.11); that of the second and third series follows from the fact that, for any appropriately differentiable function $f(\cdot)$ of z, $(\Delta f(z))^{(n)} = \Delta f^{(n)}(z)$.

18. The English language is somewhat handicapped when dealing with the fluxionary calculus. The terms 'fluxion' and 'fluent' seem to lack an appropriate corresponding verb — 'fluctuate' has a well-defined sense, and in the absence of a verb such as 'fluxate' (which I do not feel strongly enough about to coin) one is reduced to the sort of circumlocution expressed here.

19. Compare Simpson [1823, vol. 1, §6].

20. The Stirling numbers of the first kind, $S_n^{(m)}$, have the generating function

$$x(x-1)\ldots(x-n+1) = \sum_{m=0}^{n} S_n^{(m)} x^m,$$

and those of the second kind, $\mathfrak{S}_n^{(m)}$, can be generated by

$$x^n = \sum_{m=0}^{n} \mathfrak{S}_n^{(m)} x(x-1)\ldots(x-n+1).$$

A closed form expression connecting the two is

$$S_n^{(m)} = \sum_{k=0}^{n-m}(-1)^k\binom{n-1+k}{n-m+k}\binom{2n-m}{n-m-k}\mathfrak{S}_{n-m+k}^{(k)}.$$

For further details see, for example, Abramowitz & Stegun [1965].

21. We have already noted, in Chapter 6, the rapidly convergent series given by Burnside [1916–1917] for $\log N!$: according to Dutka [1991, p. 244] another rapidly convergent series was given by Legendre in his *Exercices de calcul intégral* [1811, p. 299].

22. I have been unable to find this in Cajori's *History of Mathematical Notations*, though it might well be there. Bayes in fact writes [p. 73L]

If $\boxed{abcde}^{\,n}$ signify y^e sum of all y^e factors of y^e n^{th} dimension which can be formed of a, b, c, d, e

$$\frac{a^n - b^n}{a - b} = \boxed{ab}^{\,n-1}.$$

& in general

$$\frac{\boxed{abcde\&c}^{\,n} - \boxed{bcde\&cf}^{\,n}}{a - f} = \boxed{abcde\&cf}^{\,n-1}$$

It seems that this is a sort of convolution, in powers rather than subscripts.

23. See also Guicciardini [1989, pp. 41, 83].
24. See de Morgan [1852b]. For further details of the early use of basic terms in the fluxionary calculus see Guicciardini [1989, §3.2].
25. Hardy and Wright [1960, p. 316] attribute the 'three-square' theorem to Legendre [1798].
26. See Beatty [1955]. The *Dictionary of National Biography* records that Lord Stanhope, while at Cambridge, 'was attracted by the mathematical lectures of the blind professor, Nicholas Saunderson'.
27. Needham's Christian name is spelled variously with or without a final 'e': the spelling used here is as the name is given in the *Philosophical Transactions*. *Larousse du XXe Siècle* gives the first names as Jean Turbeville, and the later *Larousse grand Encyclopedique* gives John Turberville.
28. In a paper following this one by Needham, Le Monnier discusses the communication of electricity, considering the following questions.

> How is this electric Virtue to be communicated to such Bodies as have it not, and which are not capable of acquiring it by bare Friction only? How is the electric Matter propagated? And, lastly, in what Proportion is it distributed? [1746, p. 290]

29. This experiment is also recorded in Priestley [1767, p. 97].
30. Nollet opposed the theory of electricity proposed by Franklin; see Priestley [1767, p. 160].
31. In 1725 it was stated in Boyle's *Philosophical Works* that a cubic inch of water (at 60°) weighed 256 grains. In Aikin & Aikin [1807] this figure was given as 252.506 grains, and in Partington [1835] it appeared as 252.52 grains.

32. It is not clear whence Bayes's references to Troy and Avoirdupois measurements come; there was, however, a paper entitled 'A state of the English weights and measures of capacity, as they appear from the laws as well ancient as modern' published in No. 491 of the *Philosophical Transactions* in 1749, though it does not seem particularly relevant.

33. Since some of the musical terms used by Robert Smith in his *Harmonics, or the philosophy of musical sounds*, a work to which Bayes refers here, have different meanings from those used today, we give a brief discussion of some of Smith's terminology.

    Smith's 'perfect consonances' are what we would call 'intervals', and our 'tempered intervals' are what he refers to as 'imperfect consonances'. Further,

    > A Comma is the interval of two sounds whose single vibrations have the ratio of 81 to 80, and is the difference of the major and minor tones.*
    >
    > *For the ratio of 9 to 8 [i.e., the ratio of a major tone] diminished by the ratio of 10 to 9 [i.e., the ratio of a minor tone], is the ratio of 9×9 to 8×10, or of 81 to 80. [p. 14]

    And again

    > If no other primes but 1, 2, 3 were admitted to the composition of musical ratios, a system of sounds thence resulting could have no perfect thirds; nor any perfect consonance whose vibrations are in any ratio having the number 5, or any multiple of it, for either of its terms, as 5 to 4, 10 to 9, 16 to 15, &c: it being impossible for any powers and products of the given primes 1, 2, 3 to compose any other prime or multiple of it.
    >
    > The minor tones DE, GA being thus excluded, and major tones being put in their places, every perfect major III^d will be increased by a comma, as being the difference of the tones; and every hemitone and perfect minor 3^d will be as much diminished; because the 4^{ths} and v^{ths}, as CF and Fc, cG and GC, are perfect, whether 5 be admitted or not, as depending on the primes 1, 2, 3, only.
    >
    > These diminished hemitones being called Limmas, the octave is now divided into 5 major tones and 2 limmas ....
    > [pp. 31–32]

    Augustus de Morgan recorded the difficulty he had in reading Smith's *Harmonics* in a comic poem; see Ross [1994] for a discussion of the latter.

34. The description is from Kassler [1979, vol. II, p. 949].

35. Kepler's Problem is concerned with the determination of the true anomaly of a planet given the mean anomaly. It may also be viewed geometrically, as requiring the division, in a given ratio, of the area of a semicircle by a straight line drawn from a given point of the base of the semicircle.

The following description is from Armitage.

> Johannes Kepler had announced in 1609 that the planet Mars revolved in an ellipse, the Sun occupying one of the two foci of the curve. He went on to establish that the other planets, including the Earth, also describe elliptic orbits having the Sun as a common focus. Now ellipses differ in size and in shape; and the astronomer needs to be able to specify the length of what is called the major axis AA' of a planet's elliptic orbit, and the eccentricity ($CS:CA$) which measures the proportional displacement of the focus S from the centre C of the curve ... In determining these quantities, astronomers who strictly followed the ideas of Kepler had to take account of his second law of planetary motion, that the radius vector SP, joining the Sun S to a planet P, sweeps out equal areas in equal times. However, by the middle of the seventeenth century it had been found easier to substitute some other condition for Kepler's second law, as, for example, that a planet revolves with a uniform angular velocity about the vacant focus H of its elliptic orbit — the focus *not* occupied by the Sun. This is very nearly true when the orbit is of small eccentricity; and in 1656, the reputed year of Halley's birth, Seth Ward, an Oxford astronomer, had published a book on geometrical astronomy in which he adopted this simplifying assumption as the basis of his method for determining a planet's orbital elements. [1966, pp. 13–14]

36. In this quotation the ';' indicates an abbreviation: thus 'cujusq;' is, in full, 'cujusque'.
37. In the third edition of the *Principia* of 1726 this result is given as follows.

> Propositio XXVI. Problema VII. Invenire incrementum horarium areæ quam luna, radio ad terram ducto, in orbe circulari describit.

38. This corollary is given in the following form in the third edition of the *Principia*.

> Et, cum tempora periodica sint in ratione composita ex ratione radiorum directe, & ratione velocitatum inverse; vires

centripetae sunt in ratione composita ex ratione radiorum directe, & ratione duplicata temporum periodicorum inverse.

39. See Ward [1740, pp. 147–148].
40. If 'Greaves' is pronounced as 'graves' (and the 1974 edition of *Chambers Twentieth Century Dictionary* in fact gives these two words as paronymous, when meaning 'dregs of melted tallow'), then a pun is intended here!
41. On the expectations of bodies politic, and in particular the Jews, see Hartley [1749, Part II, chap. IV, Sect. II].
42. Bayes's evaluation of the incomplete beta-integral is commented on by Lidstone [1941, pp. 178–179], Molina & Deming [1940, pp. xi–xii], Sheynin [1969, p. 4], Timerding [1908, pp. 50–51], and Wishart [1927]. A detailed study of the incomplete beta-function is to be found in Dutka [1981], the contributions of Bayes and Price to the evaluation of the beta probability integral being discussed in Hald [1990b]. The beta *distribution* seems to be ascribable to Bayes (see Sheynin [1971a, p. 235]).

    There seems to be some difference of opinion as to what is meant by the term 'incomplete beta-function': Beyer [1968] uses it for $I_x(a,b) = B_x(a,b)/B(a,b)$, whereas Dutka [1981] and Jordan [1965] use it for $B_x(a,b)$. Dutka (loc. cit.) calls $I_x(a,b)$ the *incomplete beta-function ratio*.
43. Although Scott [1938] is probably the definitive treatise on Wallis's mathematical writings, its relative rarity would no doubt make the modern reader search somewhere else for information on Wallis: Edwards [1987, chap. 8] may be recommended for a discussion of Wallis's *Arithmetica infinitorum* of 1655, in particular for the treatment of Wallis's product.

Chapter 12

After I had chosen the quotation from Jerome K. Jerome that appears at the head of this chapter I came across Meller's *London Cemeteries*. There lists of tombstone symbols are given and illustrated, and by a happy coincidence the illustration for *skull*, a symbol denoting mortality, is taken from a tomb in Bunhill Fields!

1. See the 1983 City of London pamphlet *Bunhill Fields*.
2. This was not the first time that the pestilence had placed pressure on churchyards: in the second edition of his *Survey of London* of 1603,

John Stow relates that in 1348, the year of the beginning of the first great plague of the reign of Edward III,

> for want of room in churchyards to bury the dead of the city and of the suburbs, one John Corey, clerk, procured of Nicholas, prior of the Holy Trinity within Aldgate, one toft of ground near unto East Smithfield, for the burial of them that died, with condition that it might be called the churchyard of the Holy Trinity; which ground he caused, by the aid of divers devout citizens, to be inclosed with a wall of stone. [1912, pp. 113–114]

3. See also Bell [1951, p. 211].
4. I use the word 'Nonconformist' loosely here, to include dissenters; for the distinction between the two see Appendix 2.1.
5. In the second edition of his *Survey of London* of 1603, John Stow relates of an earlier plague 'In the mean season, a bitter plague fell among them, consuming in short time such a multitude that the quick were not sufficient to bury the dead' [1912, p. 9].
6. Bell is somewhat dismissive of Defoe's *A Journal of the Plague Year* of 1722, stating that he cannot regard it as 'anything other than an historical novel' [1951, p. ix]. G.A. Aitken, however, in his introduction to the 1908 *Everyman* edition of Defoe's *Journal*, finds the narrative to be 'generally accurate'. Wheatley [1891, I, p. 302] notes that 'it is a mistake to connect it [i.e., Bunhill Fields] with "the great pit in Finsbury" mentioned by Defoe in his *Memoirs of the Plague*'.
7. The data are from *London's Dreadful Visitation, or a collection of all the Bills of Mortality from Dec. 27, 1664, to Dec. 19, 1665*, by the Company of Parish Clerks of London, 1665, and from Graunt's *Observations on the Bills of Mortality*. 'There exist trifling discrepancies, due to their being presented in a condensed form' [Boghurst, 1894, p. xxi].
8. Lawrence Green [1947, chapter VI] relates that wine seemed to be a protection against bubonic plague when that pestilence ravaged Cape Town.
9. Landa, in his Notes to the 1969 edition of Defoe's *A Journal of the Plague Year*, mentions thirty-four other herbs, spices, barks, flowers, etc. used as 'preservatives'.
10. See Latham & Matthews [1983, vol. X, p. 331].
11. The weekly Bills of Mortality were drawn up by the Company of Parish Clerks.
12. In his *History of Speldhurst* MacKinnon notes that

> The passing bell is the tenor for adults, the fifth for those between 15 and 21 years, and the treble or third for infants and children under 15 years. [1930, p. 31]

13. Defoe writes in his *A Journal of the Plague Year* that these words were 'set close *over* the same cross'. Landa remarks in his Explanatory Notes to the 1969 edition of this work, that the practice of marking houses dates from 1518 in England, and somewhat earlier on the Continent.
14. As reprinted in Defoe's *A Journal of the Plague Year*. Bell [1951] notes that these orders were in fact issued in 1646 (see Defoe [1969, p. xxxviii]).
15. James Hadley was Parish Clerk of St Olave, Hart Street. This church, whose old gate was ornamented with stone skulls and cross-bones, the skulls having iron spikes a-top, was immortalized by Dickens as *Saint Ghastly Grim* in Chapter XXI of his *The Uncommercial Traveller* (see Holmes [1896, p. 78]). Besant relates that '[St Olave's] is one of the smallest of the City churches being a square of only 54 feet' [1910, p. 280].
16. See Bell [1951, pp. 224–225].
17. Reported in Landa's Notes to the 1969 edition of Defoe's *A Journal of the Plague Year*, p. 276.
18. This was not the only disparaging description of the Presbyterians: Jeremy remarks that

> In the eighteenth century wind-guards fixed on chimney-pots were called *Presbyterians*, in derisive allusion to the want of fixedness in the theological opinions of the Denomination of that name, who were charged with turning with every wind of doctrine. The comparison, however, was faulty. [1885, p. viii]

19. This was probably the Plague pit sited where Liverpool Street Station now stands. A similar curiosity was also exhibited in Defoe's *A Journal of the Plague Year*.
20. There were separate cemeteries for those of the Jewish and Roman Catholic persuasions. See Meller [1981, chap. 4] for further details.
21. Dickens, *The Uncommercial Traveller*, Chapter 9.
22. Holmes points out [1896, p. 187] that two things can be understood by the phrase 'private cemetery': either one belonging to a particular person, family, or institution, or one that is established as a private speculation.
23. See Meller [1981, p. 88].

24. The date given here is from the City of London's *Bunhill Fields* guide of 1983; Meller [1981, p. 88] gives the year as 1884, as does Black, who records 'The cemetery had already been closed, but her body was allowed to be buried in her family's vault' [1990, p. 102].
25. It appears from Light [1913] that 'north' means on the northern side of the main path through the cemetery.
26. Jones [1849] records that the original inscription on Richard Price's tomb read 'The Rev. Richard Price, D.D., F.R.S., who died 19th April, 1791, aged 68 years'. Richard was in fact interred in the tomb of his uncle, the Rev. Samuel Price, who died on the 21st of April, 1756, aged 80 years.
27. A more comprehensive list is given in Meller [1981, pp. 90–93]. The present tombstone to Susanna Wesley was erected in 1936; the present gates, on which the names of some of those buried in Bunhill Fields are listed, mistakenly carry the name of Samuel rather than Susanna.
28. We are, unfortunately, denied the exuberant 'Let joy be unconfined' expressed by Lord Byron in his *Childe Harold*, c. III, xxii.
29. Now the Family Record Centre.
30. The date of Anne Bayes's death is given on the present vault as the 7th of January: Rippon gives it as the 17th — perhaps more in keeping with the date in the Burial Register?
31. Note that Rippon also records here that the tomb was 'formerly a H.S. 64$\mathcal{EW}$ 51$\mathcal{NS}$ lately repaired.' (Presumably 'H.S.' stands for 'headstone'.)
32. The inscriptions commemorating Nathaniel Bayes and Joshua Cotton are given here in slightly smaller print to allow their fitting into one line, and hence to preserve the actual appearance of the wording on the vault. Notice too that the inscriptions are generally in capital letters.
33. That the entry is indeed wrong may be seen by referring to Bowen and Kitchen [1971].
34. It is extremely difficult to reconcile the inscriptions as given by Jones with those given by Rippon.
35. The *Oxford English Dictionary* gives *conuenant* as an alternative for 'covenant', among its uses being 'A particular clause of agreement contained in a deed; e.g. the ordinary covenants to pay rent, etc. in a lease'. The legal term *messuage* means 'a dwelling and offices with the adjoining lands appropriated to the household'.
36. The interpretation of the abbreviations 'Surr. pres^t.' is felicitously forwarded by the examination of Joshua Bayes's will, which gives them in full as 'Surrogate' and 'present'. 'N.P.' of course stands for 'Notary Public'.

37. The Bayes family wealth stemmed largely from the Sheffield cutlery days. Joshua Bayes was rich enough to lend £1,500 to Thomas Gibson and Henry Jacomb at 4% interest, the borrowers putting up, as collateral, a property in Wiltshire worth £41,000. It later transpired that this property was actually collateral they had received for money owed them, and Thomas and Samuel Bayes were later peripherally involved in a court case concerning this property as trustees of Joshua's estate on his death.
38. Sarah's grocer son by her marriage to Robert, John, proved to be rather a black sheep, being declared bankrupt in 1816.
39. The quotation is from C.W. Murphy, Dan Lipton, and John Neat's *She's a Lassie from Lancashire* of 1907, the chorus of which runs as follows.

>She's a lassie from Lancashire,
>Just a lassie from Lancashire,
>She's the lassie that I love dear,
>   Oh! so dear.
>Though she dresses in clogs and shawl,
>She's the prettiest of them all,
>None could be fairer nor rarer than Sarah,
>My lass from Lancashire.

40. Samuel Wildman was the son of Watkinson Wildman and Susannah Carpenter, Anne Bayes's sister.
41. Charles Darwin's views on choristers was somewhat different; in his autobiography he writes of his time in Cambridge:

> I used generally to go by myself to King's College, and I sometimes hired the chorister boys to sing in my rooms.
> [F. Darwin, 1887, vol. I, p. 49]

Nathaniel was not alone in desiring that his body be committed to the earth without song: according to Mrs Basil Holmes, Sir John Morden, the founder of Morden's College in Blackheath, instructed that he be interred in the chapel of the college 'without any pomp or singing boys, but decently' [1896, p. 260].

42. A free translation of the Archer inscription runs as follows.

>Here lie
>The mortal remains
>Of the Reverend John Archer,
>Truly a venerable man
>On account of the singular modesty of his life,
>The untouched probity of his character,
>His faithful worship of God,

594   Notes: chapter 12

His liberal benevolence towards all,
And his most faithful preaching
of evangelical truth.
(Near Tunbridge Wells
in the County of Kent)
Who, weighed down for a long time by the pains of the flesh,
Which he bravely and patiently bore,
Ardently aspired to the rest of eternal life,
Which he at last obtained with great joy,
A man of the rarest virtues
Who left us sadly repining
Born 15th March, A.D. 1675,
Died 23rd September, 1733.

Here also lies
Anne Archer,
His most faithful wife,
Died 14th July, 1750, in her 78th year.

The quotation *Quis desiderio sit pudor aut modus tam cari capitis* (note: 'cari', and not 'chari') given at the end of this inscription is from Horace's, *Odes*, Book I, Ode 24, line 1. A free translation runs

What shame or limit to our grief can there be
in the loss of so dear a man?

Bibliography

A Satyr on the New Tunbridge Wells. Being a Poetical Description of the Company's Behaviour & each other ... Occasion'd by a most stupid Pamphlet that was impos'd upon the Town, under the Title of Islington: Or, The *Humours of New Tunbridge Wells* ... 1733. London.

Abramowitz, M. & Stegun, I.A. (Eds) 1965. *Handbook of Mathematical Functions, with Formulas, Graphs, and Mathematical Tables.* New York: Dover Publications, Inc.

Adams, G. 1799. *An Essay on Electricity, explaining the principles of that useful science; and describing the instruments, contrived either to illustrate the theory, or render the practice entertaining.* 5th edition, with corrections and additions by William Jones. London: J. Dillon, and Co.

Addison, J., Steele, R., et al. 1907. *The Spectator.* 8 books in 4 vols. Edited by G. Gregory Smith. London: J.M. Dent & Co.; New York: E.P. Dutton & Co.

Aikin, A. & Aikin, C.R. 1807. *A Dictionary of Chemistry and Mineralogy.* London: J. & A. Arch.

Ainger, A. 1905. *Charles Lamb.* London: Macmillan & Co., Ltd.

Allen, T. 1760. From *The London Magazine, or, Gentleman's Monthly Intelligencer* 29: 613.

Altmann, A. 1973. *Moses Mendelssohn. A Biographical Study.* London: Routledge & Kegan Paul.

Amsinck, P. 1810. *Tunbridge Wells, and its Neighbourhood, illustrated by a Series of Etchings, and Historical Descriptions.* London: W. Miller.

Anon. $171\frac{4}{5}$. An account of the book entituled *Commercium Epistolicum Collinii & aliorum, De Analysi promota*; published by order of the Royal-Society, in relation to the dispute between Mr. Leibnitz and Dr. Keill, about the right of invention of the method of fluxions, by some call'd the differential method. *Philosophical Transactions* 29: 173–224. See Newton [$171\frac{4}{5}$].

Anon. 1731–1732. *A view of the Dissenting Intrest in London of the Presbyterian and Independent Denominations from the year 1695 to the 25 of December 1731. With a postscript of the present state of the Baptists.* (Copied by S. Palmer.) Housed in Dr Williams's Library.

Anon. 1742–1743. An account of the proportions of the English and French measures and weights, from the standards of the same, kept at the Royal Society. *Philosophical Transactions* 42: 185–188.

Anon. 1750. Extract on the National Debt from *The London Magazine, or, Gentleman's Monthly Intelligencer* 19: 150–151.

Arbuthnot, J. 1692. *Of the Laws of Chance, or, a Method of Calculation of the Hazards of Game.* London.

——— 1706. *A Sermon preached to the People at the Mercat Cross of Edinburgh, on the subject of the UNION in 1706, while the Act for Uniting the Two Kingdoms was depending before the Parliament there.* Reprinted in Arbuthnot [1770], vol. II, pp. 169–191.

——— *A Letter to the Reverend Mr. Dean Swift, occasioned by a Satire said to be written by him, entitled, a Dedication to a Great Man, concerning Dedications. Discovering, among other wonderful Secrets, what will be the present Posture of Affairs a Thousand Years hence. By a Sparkish pamphleteer of Button's Coffee-House.* Reprinted in Arbuthnot [1770], vol. II, pp. 115–125.

——— 1770. *Miscellaneous Works of the late Dr. Arbuthnot. With an Account of the Author's Life.* 2 vols. London.

Archer, J. 1720. *A Sermon at the Opening of the New-Chappel at Tunbridge-Wells. August 1, 1720. etc.* London: Richard Ford.

Archibald, R.C. 1926. A rare pamphlet of Moivre and some of his discoveries. *Isis* 8: 671–676 + 7p. facsimile.

Armitage, A. 1966. *Edmond Halley.* London: Thomas Nelson and Sons Ltd.

Artmann, B. 1998. The liberal arts. *The Mathematical Intelligencer* 20, No. 3: 40–41.

Bach, C.N. 1998. Review of Dawson [1997]. *The Mathematical Intelligencer* 20, No. 4: 61–64.

Bacon, F. 1886. *Bacon's Essays.* With Introduction, Annotations, Notes and Indexes by F. Storr & C.H. Gibson. 2nd edition. London: Rivingtons.

Bailey, L. & Bailey, B. 1970. *History of Non-conformity in Tunbridge Wells.* (n.p.).

Balguy, J. 1730. *Divine Rectitude: or, a brief Inquiry Concerning the Moral Perfections of the Deity; Particularly in respect of Creation and Providence.* London: printed for John Pemberton.

—— 1732. *A Letter to a Deist, concerning the Beauty and Excellency of Moral Virtue, and the Support and Improvement which it receives from the Christian Revelation. By a Country Clergyman.* 3rd edition. London: John Pemberton.

—— 1741. *An Essay on Redemption. Being the Second Part of Divine Rectitude. By the Author of the Former.* London: printed for J. and H. Pemberton.

Ball, W.W. Rouse & Coxeter, H.S.M. 1974. *Mathematical Recreations & Essays.* 12th edition. Toronto: University of Toronto Press.

Barbauld, A.L. 1804. *The Correspondence of Samuel Richardson, Author of Pamela, Clarissa, and Sir Charles Grandison. Selected from the Original Manuscripts, bequeathed by him to his family. To which are prefixed, a biographical account of that author, and observations on his writings.* 6 vols. London: printed for Richard Philips. Reprinted in 1966; New York: AMS Press, Inc.

Barnard, G.A. 1964. Article on Thomas Bayes. *Encyclopaedia Britannica,* 14th edition.

Bartholomew, D.J. 1988. Probability, statistics and theology. *Journal of the Royal Statistical Society, A* 151: 137–178.

Bartlett, M.S. 1962. *Essays on Probability and Statistics.* London: Methuen & Co. Ltd.

Barton, M. 1937. *Tunbridge Wells.* London: Faber & Faber Limited.

Beatty, F.M. 1955. The scientific work of the third Earl Stanhope. *Notes and Records of the Royal Society of London* 11: 202–221.

Bebbington, G. 1972. *Street Names of London.* London: B.T. Batsford.

Beckenbach, E.F. & Bellman, R. 1971. *Inequalities.* New York: Springer-Verlag.

Bell, W.G. 1951. *The Great Plague in London in 1665.* (Revised edition) London: The Bodley Head.

Bellhouse, D.R. 1992. Article in *The IMS Bulletin* 21, No. 3: 225–227.

―――― 1999. 2001: A Bayes Odyssey. *Chance* 12, No. 3: 48–50.

―――― 2002. On some recently discovered manuscripts of Thomas Bayes. *Historia Mathematica* 29: 383–394.

Berkeley, G. 1707. *Arithmetica absque Algebra aut Euclide demonstrata & Miscellanea Mathematica sive Cogitata Nonnulla de radicibus surdis, de æstu aeris, de cono æquilatero et cylindro eidem sphæræ circumscriptis, de ludo algebraico et parænetica quædam ad studium matheseos præsertim algebræ.* Published in one volume. London & Dublin.

―――― 1710. *A Treatise concerning the Principles of Human Knowledge. Part I. Wherein the chief causes of error and difficulty in the sciences, with the grounds of scepticism, atheism, and irreligion, are inquired into.* Dublin.

―――― 1734. *The Analyst; or, a Discourse Addressed to an Infidel Mathematician. Wherein it is Examined Whether the Object, Principles, and Inferences of the Modern Analysis are more Distinctly Conceived, or more Evidently Deduced, than Religious Mysteries and Points of Faith, by the Author of the Minute Philosopher.* London.

―――― 1735a. *A Defence of Free-Thinking in Mathematics. In Answer to a Pamphlet of Philalethes Cantabrigiensis, intituled,* Geometry no Friend to Infidelity, or a Defence of Sir Isaac Newton, and the British Mathematicians. *Also an Appendix Concerning Mr. Walton's* Vindication of the Principles of Fluxions Against the Objections Contained in *The Analyst. Wherein it is Attempted to put this Controversy in such a Light as that every Reader may be able to Judge thereof.* Dublin.

―――― 1735b. *Reasons for not Replying to Mr. Walton's Full Answer in a letter to P.T.P., By the Author of the Minute Philosopher.* Dublin: M. Rhames, for R. Gunne.

―――― 1744. *Siris: a Chain of Philosophical Reflexions and Inquiries concerning the Virtues of Tar-water and divers other subjects connected together and arising one from another.* Dublin.

―――― 1784. *The Works of George Berkeley, D.D. Late Bishop of Cloyne in Ireland. To which is added, An Account of his Life, and Several of his Letters to Thomas Prior, Esq. Dean Gervais, and Mr. Pope, &c. &c.* Two volumes. Dublin.

Berman, D. (Ed.) 1989. *George Berkeley. Eighteenth Century Responses.* 2 vols. New York: Garland Publishing, Inc.

Bernardo, J.M., Degroot, M.H., Lindley, D.V. & Smith, A.F.M. (Eds) 1988. *Bayesian Statistics 3. Proceedings of the Third Valencia International Meeting, June 1–5, 1987.* Oxford: Clarendon Press.

Bernoulli, D. 1777. (Published 1778.) Diiudicatio maxime probabilis plurium observationum discrepantium atque verisimillima inductio inde formanda. *Acta Academia Scientiarum Imperialis Petropolitanae*, pars prior 3–23. Reprinted in Bernoulli [1982], pp. 361–375. Translated in Kendall [1961].

—— 1982. *Die Werke von Daniel Bernoulli. Band 2: Analysis, Wahrscheinlichkeitsrechnung.* Basel: Birkhäuser Verlag.

Bernoulli, J. 1713. *Ars conjectandi.* Basileæ: Thurnisiorum, Fratrum. Reprinted in Bernoulli [1975].

—— 1975. *Die Werke von Jakob Bernoulli.* Band 3. Basel: Birkhäuser Verlag.

Bernoulli, N. 1709. *De usu artis conjectandi in jure.* Basel. Reprinted in J. Bernoulli [1975], pp. 287–326.

Besant, W. 1902. *London in the Eighteenth Century.* London: Adam & Charles Black.

—— 1910. *London City.* London: Adam and Charles Black.

Besant, W. & Mitton, G.E. 1903. *The Fascination of London. Holborn and Bloomsbury.* London: Adam and Charles Black.

Beyer, W.H. (Ed.) 1968. *CRC Handbook of Tables for Probability and Statistics.* Cleveland: The Chemical Rubber Company.

Bickerstaff, I. See *The Tatler*.

Bingham, N.H. 1997. Studies in the history of probability and statistics XLV. The late Philip Holgate's paper 'Independent functions: Probability and analysis in Poland between the Wars'. *Biometrika* 84: 159–173.

Bingham, N.H. & Kiesel, R. 1998. *Risk-Neutral Valuation. Pricing and Hedging of Financial Derivatives.* London: Springer-Verlag.

Black, M. 1933. *The Nature of Mathematics: A Critical Survey.* London: Kegan Paul, Trench, Trubner & Co., Ltd.

Black, S.E. 1990. *Bunhill Fields: the Great Dissenters' Burial Ground.* Provo, Utah: Religious Studies Center, Brigham Young University.

Blair, D.G. 1975. On purely probabilistic theories of scientific inference. *Philosophy of Science* 42: 242–249.

Blair, H. 1784. *Lectures on Rhetoric and Belles Lettres.* Philadelphia: printed and sold by Robert Aitken.

Blanchard, R. (Ed.) 1955. *The Englishman. A political journal by Richard Steele.* Oxford: Clarendon Press.

────── 1963. Richard Steele and the Secretary of the SPCK. pp. 287–295 in McKillop [1963].

Boghurst, W. 1666. *Loimographia: Or an Experimentall relation of the Plague, of what happened Remarkable in the last Plague in the City of London.* Ed. by Dr. J.F. Payne for the Epidemiological Society of London, 1894, and printed as *Loimographia. An Account of the Great Plague of London in the Year 1665.* London: Shaw and Sons.

Bogue, D. & Bennett, J. 1808–1812. *History of Dissenters, from the Revolution, in 1688, to the year 1808.* 4 vols (1808, 1809, 1810, 1812). London: printed for the authors.

Bond, J.D. 1921. The development of trigonometric methods down to the close of the XVth Century (with a general account of the methods of constructing tables of natural sines down to our days). *Isis* N° 11, vol. IV, 2: 295–323.

Boole, G. 1847. *The Mathematical Analysis of Logic, being an essay towards a calculus of deductive reasoning.* Cambridge: Macmillan, Barclay, & Macmillan.

────── 1877. *A Treatise on Differential Equations.* 4th edition. London: Macmillan and Co. Reprinted in 1959(?) as 5th edition; New York: Chelsea.

Booth, J. (Ed.) 1865. *Epigrams, ancient and modern: humorous, witty, satirical, moral, and panegyrical.* London: Longmans, Green, and Co.

Borel, É. 1909. Les probabilitès dènombrable et leur applications arithmètiques. *Rend. Circ. Mat. Palermo* 27: 247–270.

Boscovich, R. *De calculo probabilitatum que respondent diversis valoribus summe errorum post plures observationes, quarum singule possint esse erronee certa quadam quantitate.* Manuscript No. 62 from the Boscovich Archive, Department of Rare Books and Special Collections, University of California Library.

——— 1758. *Philosophiæ naturalis theoria redacta ad unicam legem virium in natura existentium.* Vienna.

——— 1763. *Theoria philosophiæ naturalis, redacta ad unicam legem virium in natura existentium ... Perpolita, et aucta, ac a ... mendis expurgata, etc.* Venetiis: Ex Typographia Remondiniana.

Boudot, M. 1972. *Logique inductive et probabilité.* Paris: Armand Colin.

Bowen, E. & Kitchin, T. c.1763–c.1828. *The Royal English Atlas. Eighteenth century county maps of England and Wales.* Reprinted in 1971, with an Introduction by J.B. Harley and Donald Hodson. Newton Abbot, Devon: David and Charles Reprints.

Bowley, A.L. 1926. *Elements of Statistics.* 5th edition. London: P.S. King & Son, Ltd.

Boyd, P. 1934. *Roll of the Draper's Company of London. Collected from the Company's Records and other sources.* Croydon: J.A. Gordon.

Boyer, C.B. 1959. *The History of the Calculus and its Conceptual Development (The Concepts of the Calculus).* New York: Dover Publications, Inc. Originally published by Hafner in 1949 as *The Concepts of the Calculus, A Critical and Historical Discussion of the Derivative and the Integral.*

Boyle, R. 1725. *The Philosophical Works of the Honourable Robert Boyle.* London: W. & J. Innys.

Bretherton, F.F. 1916. John Wesley's visits to Tunbridge Wells. *Proceedings of the Wesley Historical Society* 10: 197–199.

Brewer, E. Cobham 1978. *The Dictionary of Phrase and Fable.* Classic edition. New York: Avenel Books.

Britton, J. 1832. *Descriptive Sketches of Tunbridge Wells and the Calverley Estate; with brief notices of the picturesque scenery, seats, and antiquities in the vicinity. Embellished with Maps and Prints.* London: published by the author.

Broad, C.D. 1923. *Scientific Thought.* London: Routledge and Kegan Paul, Limited.

Bromwich, T.J.I'A. 1931. *An Introduction to the Theory of Infinite Series.* (Reprint of the 2nd revised edition of 1926.) London: Macmillan and Co., Limited.

Browne, T. 1658. *Hydriotaphia, Urne-buriall, or, A Discourse of the Sepulchrall Urnes lately found in Norfolk.* Reprinted in Volume 3 of Browne [1904].

——— 1904. *The Works of Sir Thomas Browne.* 3 volumes. Edited by Charles Sayle. London: Grant Richards.

Browne, W. 1714. *Christiani Hugenii Libellus de Ratiociniis in Ludo Aleæ. Or, the Value of all Chances in Games of Fortune; Cards, Dice, Wagers, Lotteries etc. Mathematically Demonstrated.* London.

Browning, A. (Ed.) 1953. *English Historical Documents 1660–1714.* vol. VIII of *English Historical Documents*, general editor D.C. Douglas. London: Eyre & Spottiswoode.

Bryant, A.W.M. 1960. *Restoration England.* London: Collins. (First published, by Longmans & Co., in 1934 as *The England of Charles II.*)

Buckingham, C.P. 1875. *Elements of the differential and integral calculus, by a new method, founded on the true system of Sir Isaac Newton, without the use of infinitesimals or limits.* Chicago: S.C. Griggs & Co. Revised edition 1880.

Buffon, G.L.L. *Discours sur le style.* Reprinted in Cahour [1854], pp. 411–421.

——— 1735. Solution de problèmes qui regardoient le jeu du franc carreau. *Histoire de l'Académie royale des Sciences Paris*: 43–45.

Bunhill Fields Burial Ground: proceedings in reference to its preservation. With inscriptions on the tombs. 1867. London: Hamilton, Adams, and Co.

Burnside, W. 1916–1917. (Published 1917.) A rapidly convergent series for $\log N!$. *The Messenger of Mathematics* 46: 157–159.

Burnside, W.S. & Panton, A.W. 1928. *The Theory of Equations: with an Introduction to the Theory of Binary Algebraic Forms.* 2 vols. 9th edition. Dublin: Hodges, Figgis, & Co.

Burr, T.B. 1766. *The History of Tunbridge Wells.* London.

Burton, D.M. 1980. *Elementary Number Theory.* Boston, London, Sydney, Toronto: Allyn & Bacon, Inc.

Burton, R. 1621. *The Anatomy of Melancholy.* Oxford.

Butler, J. 1736. *The Analogy of Religion, Natural and Revealed, to the Constitution and Course of Nature.* Reprinted, with an introduction by R. Bayne, in 1906; London: J.M. Dent & Sons, Ltd. Also printed (1840?) as *The Analogy of Religion, to the Constitution and Course of Nature: also, Fifteen Semons, on subjects chiefly ethical. With a*

Life of the Author, a Copious Analysis, Notes and Indexes. By Joseph Angus. London: The Religious Tract Society.

Cahour, A. (Ed.) 1854. *Chefs-d'œuvre d'éloquence française.* Paris: Julien, Lanier et c^e.

Cajori, F. 1917. Discussion of fluxions: from Berkeley to Woodhouse. *American Mathematical Monthly* 24: 145–154.

——— 1919. *A History of the Conceptions of Limits and Fluxions in Great Britain from Newton to Woodhouse.* Chicago & London: Open Court Publishing Company.

——— 1927. *An Introduction to the Modern Theory of Equations.* New York: Macmillan.

——— 1928–1929. *A History of Mathematical Notations.* Volume I (1928): *Notations in elementary mathematics.* Volume 2 (1929): *Notations mainly in higher mathematics.* Chicago: Open Court Publishing Company.

——— 1991. *A History of Mathematics.* 5th edition. New York: Chelsea.

——— 1993. *A History of Mathematical Notations.* Two volumes bound as one. New York: Dover Publications Inc.

Calamy, E. 1830. *An Historical Account of My Own Life, with some reflections on the times I have lived in. (1671–1731.).* Edited and illustrated with notes, historical and biographical, by John Towill Rutt. 2 vols. 2nd edition. London: Henry Colburn and Richard Bentley.

Cantor, M. 1880–1908. *Vorlesungen über Geschichte der Mathematik.* 4 vols. Leipzig: Teubner. Reprinted in 1965; New York: Johnson Reprint Corporation.

Carlisle, N. 1808. *A Topographical Dictionary of England.* 2 vols. London.

Carlyle, T. 1915a. *English and other Critical Essays.* London & Toronto: J.M. Dent & Sons Ltd.

——— 1915b. *Scottish & Other Miscellanies.* London & Toronto: J.M. Dent & Sons Ltd.

Carnap, R. 1962. *Logical Foundations of Probability.* 2nd edition. Chicago: University of Chicago Press.

Cavallo, T. 1782. *A Complete Treatise on Electricity, in theory and practice; with original experiments.* 2nd edition. London. 1st edition 1777.

Chuaqui, R. 1991. *Truth, Possibility and Probability. New Logical Foundations of Probability and Statistical Inference.* Amsterdam: North-Holland.

Cicero. 1962. *Brutus*, with an English translation by G.L. Hendrickson. *Orator*, with an English translation by H.M. Hubbell. London: William Heinemann Ltd.

Clarke, F.M. 1929. *Thomas Simpson and his Times.* New York: Columbia University Press.

Clarke, S. 1731. *Sermons on the following subjects ... Of God's disposing all Things to their proper Ends.* vol. VII, Sermon XIV. London: printed by W. Botham, for James and John Knapton.

Clay, J.W. (Ed.) 1895. *Familiæ Minorum Gentium*, vol. III. Volume 39 of *The Publications of the Harleian Society.* London.

Clero, J.P. 1988. *Thomas Bayes. Essai en vue de resoudre un probleme de la doctrine des chances.* With a preface by B. Bru. Cahiers d'Histoire et de Philosophie des Sciences, N. 18. Paris: Société Française d'Histoire des Sciences et des Techniques.

Clifford, J. 1826. *The Tunbridge Wells Guide or An account of the ancient and present State of that place. with a particular description of all the Towns, Villages, Antiquities, Natural Curiosities, Ancient and Modern Seats, Founderies, &c. within the circumference of sixteen Miles, with accurate Views of the principal Objects.* Tunbridge Wells: J. Clifford.

Cobbett, W. 1912. *Rural Rides in the Counties of Surrey, Kent, Sussex, Hampshire, Wiltshire, Gloucestershire, Herefordshire, Worcestershire, Somersetshire, Oxfordshire, Berkshire, Essex, Suffolk, Norfolk and Hertfordshire: With Economical and Political Observations relative to matters applicable to, and illustrated by, the State of those Counties respectively.* London: J.M. Dent & Sons Ltd.

Colbran, J. c.1852. *Colbran's Hand-Book and Visitor's Guide for Tunbridge Wells and its Neighbourhood.* Tunbridge Wells: John Colbran.

Coles, E. 1674. *The Newest, Plainest, and the Shortest Shorthand.* London.

Colligan, J.H. 1915. *Eighteenth Century Nonconformity.* London: Longmans, Green and Co.

——— 1923. Early Presbyterianism at Tunbridge Wells. *Journal of the Presbyterian Historical Society of England*, 2, part 4: 222–224.

The Concise Scots Dictionary. 1985. Editor-in-chief: M. Robinson. Aberdeen(?): Aberdeen University Press.

Condorcet, M.J.A.N. Caritat, le Marquis de. 1785. *Essai sur l'application de l'analyse à la probabilité des décisions rendues à la pluralité des voix.* Paris: de l'imprimerie royale. Reprinted in 1972; New York: Chelsea.

Cotes, R. *Canonotechnia, sive constructio tabularum per differentias.* Printed in Cotes [1722], pp. 35–71.

—— 1722. *Harmonia Mensurarum, sive analysis & synthesis per rationum & angulorum mensuras promotae: accedunt alia opuscula mathematica.* Ed. R. Smith. Cambridge.

Cournot, A.A. 1843. *Exposition de la théorie des chances et des probabilités.* Paris: Librairie de L. Hachette.

Creasy, J. (Compiler) 1964. *Index to the John Evans List of Dissenting Congregations and Ministers 1715–1729 in Dr. Williams's Library.* London: Dr. Williams's Trust.

Cumberland, R. 1672. *De legibus naturae disquisitio philosophica in qua earum forma, summa capita, ordo, promulgatio, & obligatio è rerum natura investigantur; quinetiam elementa philosophiæ Hobbianæ cùm moralis tum civilis, considerantur & refutantur.* London. Translated into English in 1750 by John Towers, under the title *A Philosophical Enquiry into the Laws of Nature: wherein the essence, the principal heads, the order, the publication, and the obligation of these laws are deduced from the nature of things. Wherein also, the principles of Mr. Hobbes's philosophy, both in a state of nature, and of civil society, are examined into, and confuted.* Dublin.

Dale, A.I. 1982. Bayes or Laplace? An examination of the origin and early applications of Bayes' theorem. *Archive for History of Exact Sciences* 27: 23–47.

—— 1986. A newly-discovered result of Thomas Bayes. *Archive for History of Exact Sciences* 35: 101–113.

—— 1988. On Bayes' theorem and the inverse Bernoulli theorem. *Historia Mathematica* 15: 348–360.

—— 1990. Thomas Bayes: some clues to his education. *Statistics & Probability Letters* 9: 289–290.

—— 1991. Thomas Bayes's work on infinite series. *Historia Mathematica* 18: 312–327.

―― 1995. *Pierre-Simon Laplace. Philosophical Essay on Probabilities.* New York: Springer-Verlag.

―― 1999. *A History of Inverse Probability from Thomas Bayes to Karl Pearson.* 2nd edition. New York: Springer-Verlag.

Dale, R.W. 1907. *A History of English Congregationalism.* Completed and edited by A.W.W. Dale. London: Hodder & Stoughton.

Dalzel, A. 1862. *History of the University of Edinburgh from its Foundation.* 2 vols. Edinburgh: Edmonston & Douglas.

Darwin, F. (Ed.) 1887. *The Life and Letters of Charles Darwin, including an autobiographical chapter.* 3 vols. London: John Murray.

Daston, L.J. 1988. *Classical Probability in the Enlightenment.* Princeton, N.J.: Princeton University Press.

David, F.N. 1962. *Games, Gods and Gambling. The origins and history of probability and statistical ideas from the earliest times to the Newtonian era.* London: Charles Griffin & Co. Ltd.

Davidson, F. (Ed.) 1954. *The New Bible Commentary.* 2nd edition. London: Inter-Varsity Fellowship.

Davies, C.M. 1875. *Mystic London: or, phases of occult life in the metropolis.* London: Tinsley Brothers.

Dawson, J.W., Jr. 1997. *Logical Dilemmas: the life and work of Kurt Gödel.* Wellesley, Mass.: A.K. Peters.

de Coste, H. 1649. *La Vie du R. P. Marin Mersenne, Theologien, Philosophe et Mathematicien de l'Ordre des Peres Minimes. Par F.H.D.C. Religieux du mesme Order.* Paris: Sebastien Cramoisy.

de Finetti, B. 1937. La Prevision: ses lois logiques, ses sources subjectives. *Annales de l'Institut Henri Poincaré* 7: 1–68. Translated as 'Foresight: its logical laws, its subjective sources' in Kyburg & Smokler [1964], pp. 93–158.

―― 1972. *Probability, Induction and Statistics: the art of guessing.* London, New York, Sydney, Toronto: John Wiley & Sons.

―― 1974 & 1975. *Theory of Probability. A critical introductory treatment.* 2 vols. London, New York, Sydney, Toronto: John Wiley & Sons.

de Moivre, A. 1714. Solutio generalis altera praecedentis Problematis, ope combinationum et serierum infinitarum. *Philosophical Transactions* 29: 145–158.

―――― 1725. *Annuities upon Lives: or, The Valuation of Annuities upon any Number of Lives; as also, of Reversions. To which is added, An Appendix concerning the Expectations of Life, and Probabilities of Survivorship.* London: F. Fayram, B. Motte and W. Pearson.

―――― 1756. *The Doctrine of Chances: or, a Method of Calculating the Probabilities of Events in Play.* 3rd edition. London: A. Millar. (1st edition 1718). Reprinted in 1967; New York: Chelsea.

de Morgan, A. 1838a. *An Essay on Probabilities and on their Application to Life Contingencies and Insurance Offices.* London: Longman, Orme, Brown, Green, & Longmans.

―――― 1838b. On a question in the theory of probabilities. *Transactions of the Cambridge Philosophical Society* 6: 423–430.

―――― 1852a. On the authorship of the 'Account of the Commercium Epistolicum' published in the Philosophical Transactions. *The London, Edinburgh and Dublin Philosophical Magazine and Journal of Science*, Series 4, 3: 440–444.

―――― 1852b. On the early history of infinitesimals in England. *The London, Edinburgh and Dublin Philosophical Magazine and Journal of Science*, Series 4, 4: 321–330.

―――― 1860. Rev. Thomas Bayes, *&c. Notes and Queries*, 2nd Series, 9: 9–10.

Defoe, D. 1908/1963. (First published 1722). *A Journal of the Plague Year: being Observations or Memorials of the most Remarkable Occurrences, as well as Publick as Private, which happened in London during the last Great Visitation in 1665.* London: J.M. Dent & Sons Ltd. Also published, with an Introduction and Notes by the Editor Louis Landa, by Oxford University Press, London, in 1969.

Deming, W. Edwards. 1940. Some remarks concerning Bayes' note on the use of certain divergent series. pp. xv–xvi in *Facsimiles of two papers by Bayes*. Washington D.C.: The Graduate School, The Department of Agriculture.

Dempster, A.P. 1966. New methods for reasoning towards posterior distributions based on sample data. *Annals of Mathematical Statistics* 37: 355–374.

Derham, W. 1798. *Physico-theology: or, a Demonstration of the Being and Attributes of God, from his Works of Creation. Being the Substance of Sixteen Discourses delivered in St. Mary-le-Bow Church, London, at the Hon. Mr. Boyle's Lectures, in the Years 1711 and 1712.* A new edition. 2 vols. London. Also published, in the same year, as *Physico-Theology; or a Demonstration of the Being and Attributes of God, from his Works of Creation: being the Substance of 16 Sermons preached at the Honble Mr. Boyle's Lectures, with large Notes and many curious Observations.* London. (1st edition 1713.)

Digby, K. 1661. *A Discourse Concerning the Vegetation of Plants. Spoken by Sir Kenelme Digby, at Gresham College, on the 23. of January, 1660. At a Meeting of the Society for promoting Philosophical Knowledge by Experiments.* London.

Dingle, R.B. 1973. *Asymptotic Expansions: Their Derivation and Interpretation.* London & New York: Academic Press.

Diogenes Laertius. 1925. *Lives of Eminent Philosophers.* With an English translation by R.D. Hicks. London: William Heinemann Ltd, & Cambridge, Mass.: Harvard University Press.

Ditton, H. 1706. *An Institution of Fluxions: Containing the First Principles, The Operations, With some of The Uses and Applications of that Admirable Method; According to the Scheme perfix'd to his Tract of* Quadratures *by (its First Inventor) the incomparable Sir* ISAAC NEWTON. London: by W. Botham, for James Knapton.

Doddridge, P. 1803. *The Works of the Rev. P. Doddridge, D.D.* 10 volumes. Leeds: E. Baines.

——— 1829–1831. *The Correspondence and Diary of Philip Doddridge, D.D. Illustrative of various particulars in his life hitherto unknown: with notices of many of his contemporaries; and a sketch of the ecclesiastical history of the times in which he lived.* Ed. J.D. Humphries. 5 vols (1829, 1829, 1830, 1830, 1831). London: Henry Colburn and Richard Bentley.

Dodson, J. 1753. On infinite series and logarithms. *Philosophical Transactions* 48: 273–284.

Dutka, J. 1981. The incomplete beta function — a historical profile. *Archive for History of Exact Sciences* 24: 11–29.

——— 1991. The early history of the factorial function. *Archive for History of Exact Sciences* 43: 225–249.

Eames, J. 1736. A brief account, by Mr. John Eames, F.R.S. of a work entitled, The Method of Fluxions and Infinite Series, with its Application to the Geometry of Curve Lines, by the inventor Sir Isaac

Newton, Kt. &c. Translated from the original, not yet made public. To which is subjoined a perpetual comment upon the whole, &c. by John Colson, M.A. & F.R.S. *Philosophical Transactions* 39: 320–328.

Edgeworth, F.Y. 1881. *Mathematical Psychics: An Essay on the Application of Mathematics to the Moral Sciences.* London: C. Kegan Paul & Co.

—— 1884. The philosophy of chance. *Mind* 9: 223–235.

—— 1887. *Metretike, or the Method of Measuring Probability and Utility.* London: Temple Company.

Edwards, A.W.F. 1972. *Likelihood. An account of the statistical concept of* likelihood *and its application to scientific inference.* Cambridge: Cambridge University Press. Reprinted in 1992, as *Likelihood. Expanded edition.* Baltimore and London: Johns Hopkins University Press.

—— 1974. A problem in the doctrine of chances. *Proceedings of Conference on Foundational Questions in Statistical Inference, Aarhus, May 7–12, 1973.* (Eds O. Barndorff-Nielsen, P. Blaesild, G. Schou.) Department of Theoretical Statistics, Institute of Mathematics, University of Aarhus. Memoirs No. 1: 41–60.

—— 1978. Commentary on the arguments of Thomas Bayes. *Scandinavian Journal of Statistics* 5: 116–118.

—— 1983. Pascal's problem: the "gambler's ruin". *International Statistical Review* 51: 73–79. Reprinted as Appendix 2 to Edwards [1987].

—— 1986. Is the reference in Hartley (1749) to Bayesian inference? *The American Statistician* 40: 109–110.

—— 1987. *Pascal's Arithmetical Triangle.* London: Charles Griffin & Co. Ltd., and New York: Oxford University Press.

Edwards, W., Lindman, H. & Savage, L.J. 1963. Bayesian statistical inference for psychological research. *Psychological Review* 70: 193–242.

Eisenhart, C. 1961. Boscovich and the combination of observations. pp. 200–212 in Whyte [1961]. Reprinted in Kendall & Plackett [1977], pp. 88–100.

Eizat, E. 1695. *Apollo Mathematicus, or the Art of curing Diseases by the Mathematicks, According to the Principles of $D^{R.}$ Pitcairn. A Work both Profitable and Pleasant; and never Published in English before. To which is subjoined, A Discourse of Certainty, according to the Principles of the same Author.* London.

Elwig, H. 1941. *A Biographical Dictionary of Notable People at Tunbridge Wells. 17th to 20th Century. Also a List of Local Place Names.* Tunbridge Wells(?).

Emerson, W. 1763. *The Method of Increments. Wherein The Principles are demonstrated; and The Practice thereof shewn in the Solution of Problems.* London: for J. Nourse.

—— 1776. *Miscellanies, or a Miscellaneous Treatise: containing Several Mathematical Subjects.* London: J. Nourse.

Encyclopædia Britannica; or, a Dictionary of Arts and Sciences, compiled upon a new plan. In which the different SCIENCES *and* ARTS *are digested into distinct Treatises or Systems; and The various* TECHNICAL TERMS, *&c. are explained as they occur in the order of the Alphabet*. 1771. Edinburgh.

Euler, L. 1748. *Introductio in Analysin Infinitorum.* Lausannæ. Reprinted in *Opera Omnia*, Series I, vols VIII & IX.

—— 1755. *Institutiones calculi differentialis cum ejus usu in analysi finitorum ac doctrina serierum.* Petropolitanae. Reprinted in *Opera Omnia*, Series I, vol. X.

—— 1760. De seriebus divergentibus. *Novi Commentarii Academiae Scientiarum Petropolitanae* 5 (1754/1755): 205–237. Reprinted in *Opera Omnia*, Series I, vol. XIV.

—— 1770a. *Vollständige Anleitung zur Algebra.* First published in Russian in 1768–69, and reprinted in *Opera Omnia*, Series I, vol. I.

—— 1770b. *Lettres a une princesse d'allemagne sur divers sujets de physique & de philosophie.* 3 vols. A Mietau et Leipsic. Chez Steidel et compagnie. Reprinted in *Opera Omnia*, Series III, vols XI & XII.

—— 1795. *Letters to a German Princess, on different subjects in physics and philosophy.* Translated from the French by Henry Hunter, D.D. 2 vols. London: printed for the translator.

—— 1911–. *Leonardi Euleri Opera Omnia.* Lipsiae et Berolini. Allgemeine Schweizerische Gesellschaft für gesammten Natuwissenschaften.

Evans, G.E. 1897. *Vestiges of Protestant Dissent: being lists of ministers, sacramental plate, registers, antiquities, and other matters pertaining to most of the churches (and a few others) included in the National Conference of Unitarian, Liberal Christian, Free Christian, Presbyterian, and other non-subscribing or kindred congregations.* Liverpool: F. & E. Gibbons.

Fairfield, S. 1983. *The Streets of London. A dictionary of the names and their origins.* London: Macmillan.

Farebrother, R.W. 1990. Studies in the history of probability and statistics XLII. Further details of contacts between Boscovich and Simpson in June 1760. *Biometrika* 77: 397–400.

―――― 1999. *Fitting Linear Relationships: A History of the Calculus of Observations 1750–1900.* New York: Springer-Verlag.

Farrar, F.W. (Ed.) 1867. *Essays on a Liberal Education.* London: Macmillan and Co. Reprinted in 1969; Westmead, Farnborough, Hants.: Gregg International Publishers Ltd.

Farthing, R. 1990. *Royal Tunbridge Wells. A Pictorial History.* Chichester: Phillimore & Co. Ltd.

Feller, W. 1957. *An Introduction to Probability Theory and Its Applications.* vol. 1. 2nd edition. New York, London, Sydney: John Wiley & Sons. (1st edition 1950; 3rd edition 1968.)

Fetzer, J.H. 1981. *Scientific Knowledge. Causation, Explanation, and Corroboration.* Dordrecht: D. Reidel.

Fine, T.L. 1973. *Theories of Probability: an examination of foundations.* New York: Academic Press.

Fisher, R.A. 1962. Some examples of Bayes' method of the experimental determination of probabilities *a priori. Journal of the Royal Statistical Society, B* 24: 118–124.

―――― 1995. *Statistical Methods, Experimental Design, and Scientific Inference.* A Re-issue of *Statistical Methods for Research Workers, The Design of Experiments,* and *Statistical Methods and Scientific Inference.* Edited by J.H. Bennett. Oxford: Oxford University Press. (First published in this edition in 1990.)

Force, J.E. 1985. *William Whiston: Honest Newtonian.* Cambridge: Cambridge University Press.

Foster, J. 1749, 1752. *Discourses on all the Principal Branches of Natural Religion and Social Virtue.* 2 vols. London.

Franklin, B. 1774. *Experiments and Observations on Electricity, made at Philadelphia in America. To which are added, Letters and Papers on Philosophical Subjects.* 5th edition. London: printed for F. Newbery.

Fraser, A.C. 1901. *The Works of George Berkeley D.D.; Formerly Bishop of Cloyne. Including his Posthumous Works.* 4 vols. Clarendon Press: Oxford.

Fraser, J.B. 1838. *Narrative of the Residence of the Persian Princes in London, in 1835 and 1836. With an account of their Journey from Persia, and subsequent adventures.* 2nd edition. London: Richard Bentley.

Freedman, D. 1999. From association to causation: some remarks on the history of statistics. *Statistical Science* 14: 243–258.

Freudenthal, H. 1980. Huygens' foundations of probability. *Historia Mathematica* 7: 113–117.

Fyfe, A. 1997. The reception of William Paley's *Natural Theology* in the University of Cambridge. *British Journal for the History of Science* 30: 321–335.

Gaspey, W. 1863. *Brackett's Descriptive Illustrated Hand Guide to Tunbridge Wells, and the Neighbouring Towns, Seats, and Villages; containing Historical Notices, Information for Visitors, And a Short Directory.* Tunbridge Wells: W. Brackett.

Geisser, S. 1988. The future of statistics in retrospect. pp. 147–158 in Bernardo et al. [1988].

Gibson, G.A. 1898. (Published 1899.) A review: with special reference to the *Analyst* controversy. *Proceedings of the Edinburgh Mathematical Society* 17: 9–32. (A review of M. Cantor's *Vorlesungen über Geschichte der Mathematik.* Dritter (Schluss) Band. Dritte Abteilung, 1727–1758.)

Gigerenzer, G., Swijtink, Z., Porter, T., Daston, L., Beatty, J. & Krüger, L. 1989. *The Empire of Chance. How probability changed science and everyday life.* Cambridge: Cambridge University Press.

Gilbert, W. 1600. *De magnete, magneticisque corporibus, et de magno magnete tellure; Physiologia nova, plurimus & argumentis, & experimentis demonstrata.* London: printed for Peter Short. Reprinted in 1967; Bruxelles: Culture et Civilisation.

Gillispie, C.C. 1970. *Dictionary of Scientific Biography.* vol. I. New York: Charles Scribner's Sons.

Godolphin, F.R.B. (Ed.) 1949. *The Latin Poets.* New York: Random House, Inc.

Goldstein, J.A. 1996. *An ending? Mathematics and Natural Philosophy Education at Liberal Dissenting Academies in the Aftermath of the Scientific Revolution 1689–1796.* Ph.D. dissertation, Temple University.

Goldstine, H.H. 1977. *A History of Numerical Analysis from the 16th through the 19th Century.* New York: Springer-Verlag.

Good, I.J. 1950. *Probability and the Weighing of Evidence.* London: Charles Griffin & Company Ltd.

—— 1983. *Good Thinking. The Foundations of Probability and Its Applications.* Minneapolis: University of Minnesota Press.

—— 1988. Bayes's red billiard ball is also a herring, and why Bayes withheld publication. *Journal of Statistical Computing and Simulation* 29: 335–340.

Gordon, A.G. 1902. *Early Nonconformity and Education. Address by Principal Gordon, M.A., at the Opening of the Session of the Unitarian Home Missionary College, Memorial Hall, Manchester, on 8th October 1902.* Manchester: H. Rawson & Co., Printers.

Gordon, T. 1751. *A Cordial for Low-Spirits: being a Collection of Valuable Tracts.* 2nd edition. London: R. Griffiths.

Gould, H.W. & Squire, W. 1963. MacLaurin's second formula and its generalization. *American Mathematical Monthly* 70: 44–52.

Grant, A. 1884. *The Story of the University of Edinburgh during its first three hundred years.* 2 vols. London: Longmans, Green, and Co.

Grant, J. 1880–1883. *Cassell's Old and New Edinburgh: Its History, its People, and its Places.* 3 vols. London, Paris & New York: Cassell, Petter, Galpin & Co.

Greaves, J. 1736a. *A Discourse of the Roman Foot and Denarius: From whence, as from two Principles, The Measures and Weights used by the Ancients may be deduced.* London: J. Brindley.

—— 1736b. *Pyramidographia: or, a description of the pyramids in Aegypt.* London: J. Brindley. Reprinted in Greaves [1737], vol. I.

—— 1737. *Miscellaneous Works.* 2 vols. London: Thomas Birch.

Green, L.G. 1947. *Tavern of the Seas.* Cape Town: Howard Timmins.

Grove, H. 1734. *Wisdom, the first Spring of Action in the Deity. A Discourse in which, Among other Things, the Absurdity of God's being acted by Natural Inclinations, and of an Unbounded Liberty, is shewn. The Moral Attributes of God are explain'd. The Origin of Evil is consider'd. The Fundamental Duties of Natural Religion are shewn to be reasonable; and several things, advanc'd by some late authors, and others, relating to these subjects, are freely examin'd.* London: printed for James, John, and Paul Knapton.

Guicciardini, N. 1989. *The development of Newtonian calculus in Britain 1700–1800.* Cambridge: Cambridge University Press.

Hacking, I. 1965. *Logic of Statistical Inference.* Cambridge: Cambridge University Press.

——— 1970. Bayes, Thomas. pp. 531-532 in Gillispie [1970].

——— 1975. *The Emergence of Probability: a philosophical study of early ideas about probability, induction and statistical inference.* Cambridge: Cambridge University Press.

Hailperin, T. 1996. *Sentential Probability Logic. Origins, Development, Current Status, and Technical Applications.* Bethlehem, Pa.: Lehigh University Press. London: Associated University Presses.

——— 2000. Probability semantics for quantifier logic. *Journal of Philosophical Logic* 29: 207–239.

Hald, A. 1984. Commentary on 'De Mensura Sortis.' *International Statistical Review* 52: 229–236.

——— 1990a. *A History of Probability and Statistics and Their Applications before 1750.* New York: John Wiley & Sons.

——— 1990b. Evaluations of the beta probability integral by Bayes and Price. *Archive for History of Exact Sciences* 41: 139–156.

——— 1998. *A History of Mathematical Statistics from 1750 to 1930.* New York: John Wiley & Sons, Inc.

Hall, A.R. & Tilling, L. (Eds) 1976. *The Correspondence of Isaac Newton.* vol. VI, 1713–1718. Cambridge: published for the Royal Society by Cambridge University Press.

Hall, H.S. & Knight, S.R. 1948. *Higher Algebra. A sequel to Elementary Algebra for Schools.* London: Macmillan and Co., Ltd.

Hamilton, A. 1859. *Memoirs of the Court of Charles the Second, by Count Grammont, with numerous additions and illustrations. Also: the personal history of Charles, including the King's own account of his escape and preservation after the Battle of Worcester, as dictated to Pepys. And the Boscobel Tracts; or, contemporary narratives of His Majesty's adventures, from the murder of his father to the Restoration.* Edited by Sir Walter Scott. London: Henry G. Bohn.

Harben, H.A. 1918. *A Dictionary of London, being notes topographical and historical relating to the streets and principal buildings in the City of London.* London: Herbert Jenkins Ltd.

Hardy, G.H. 1949. *Divergent Series.* Oxford: Clarendon Press. Reprinted in 1991; New York: Chelsea.

Hardy, G.H. & Wright, E.M. 1960. *An Introduction to the Theory of Numbers.* 4th edition. Oxford: Clarendon Press.

Harris, J. 1702. *A New Short Treatise of Algebra: with the Geometrical Construction of Equations, As far as the Fourth Power or Dimension. Together with a Specimen of the Nature and Algorithm of Fluxions.* London.

——— 1708, 1710. *Lexicon Technicum: Or, An Universal English Dictionary of Arts and Sciences: Explaining not only the Terms of Art, but the Arts Themselves.* 2 vols. 2nd edition. London.

Hartley, D. 1749. *Observations on Man, His Frame, His Duty, And His Expectations.* London: Richardson. Reprinted, as 6th edition, in 1834; London: Thomas Tegg and Son. Reprinted 1966; Gainesville, Fla.: Scholars' Facsimiles & Reprints.

Hastings, J. (Ed.) 1971. *Encyclopædia of Religion and Ethics.* 12 vols. Reprint of edition published 1908–1921. (Volume 9 was reprinted in 1967.) Edinburgh: T. & T. Clark.

Hazlitt, W. 1821, 1822. *Table-Talk; or, Original Essays.* 2 vols. London: John Warren.

——— 1826. Of persons one would wish to have seen. *New Monthly Magazine.* Reprinted pp. 523–539 in Hazlitt [1930].

——— 1930. *Selected Essays of William Hazlitt 1778 : 1830.* Edited by Geoffrey Keynes. London: Nonesuch Press.

Hicks, H. 1887. *History of the Bunhill Fields Burial Ground, with some of the Principal Inscriptions.* London: Charles Skipper and East, Printers.

Hill, G.B. (Ed.) 1934. *Boswell's Life of Johnson. Together with Boswell's Journal of a Tour to the Hebrides and Johnson's Diary of a Journey into North Wales.* Revised and enlarged edition by L.F. Powell. 6 vols. Oxford: Clarendon Press.

Hoadly, B. & Wilson, B. 1756. *Observations on a Series of Electrical Experiments.* London: printed for T. Payne. Second edition 1759.

Hodgson, J. 1736. *The Doctrine of Fluxions, founded on Sir Isaac Newton's Method, published by Himself in his Tract upon the Quadrature of Curves.* London.

Hogben, L. 1957. *Statistical Theory. The Relationship of Probability, Credibility and Error. An examination of the contemporary crisis in statistical theory from a behaviourist viewpoint.* London: George Allen & Unwin Ltd.

—— 1960. *Mathematics in the Making.* London: Macdonald & Co.

Holland, J.D. 1962. The Reverend Thomas Bayes, F.R.S. (1702–1761). *Journal of the Royal Statistical Society, A* 125: 451–461.

—— 1965. The Rev. Thomas Bayes. *Electronics and Power* 11: 238.

—— 1968. An eighteenth-century pioneer: Richard Price, D.D., F.R.S. (1723–1791). *Notes and Records of the Royal Society of London* 23: 43–64.

Holmes, Mrs B. 1896. *The London Burial Grounds. Notes on their History from the Earliest Times to the Present Day.* London: T. Fisher Unwin.

Home, R.W. 1974–1975. Some manuscripts on electrical and other subjects attributed to Thomas Bayes, F.R.S. *Notes and Records of the Royal Society of London* 29: 81–90.

Hooper, G. 1721. *An Inquiry into the State of the Ancient Measures, the Attick, the Roman, and especially the Jewish. With an Appendix, concerning our old English Money, and Measures of Content.* London: R. Knaplock.

Horn, D.B. & Ransome, M. (Eds) 1957. *English Historical Documents 1714–1783.* vol. X of *English Historical Documents*, general editor D.C. Douglas. London: Eyre & Spottiswoode.

Horne, H.H. 1916. *The Philosophy of Education. Being the foundations of education with the related natural and mental sciences.* New York: Macmillan.

Horwich, P. 1982. *Probability and Evidence.* Cambridge: Cambridge University Press.

Howson, C. & Urbach, P. 1989. *Scientific Reasoning: The Bayesian Approach.* La Salle, Ill.: Open Court Publishing Company.

Huber, P.J. 1972. Robust statistics: a review. *The Annals of Mathematical Statistics* 43: 1041–1067.

Hume, D. 1894. *Essays Literary, Moral and Political.* Sir John Lubbock's Hundred Books. London: George Routledge and Sons.

Hutcheson, F. 1725. *An Inquiry into the Original of our Ideas of Beauty and Virtue; In Two Treatises. In which the principles of the late Earl of* SHAFTESBURY *are Explain'd and Defended, against the Author of the Fable of the Bees; and the Ideas of Moral Good and Evil are establish'd, according to the Sentiments of the Antient Moralists. With an Attempt to introduce a Mathematical Calculation in Subjects of Morality.* London. Re-issued as Volume 1 of Hutcheson [1971].

——— 1971. *Collected Works of Francis Hutcheson.* 7 vols. Hildesheim: Georg Olms Verlagsbuchhandlung.

Huxley, T.H. 1889. The value of witness to the miraculous. Reprinted in Huxley [1909], pp. 160–191.

——— 1905. *Science and Education. Essays.* London: Macmillan and Co., Limited.

——— 1909. *Science and Christian Tradition.* London: Macmillan and Co., Ltd.

Huygens, C. 1657. *Tractatus de Ratiociniis in Aleæ Ludo**. Printed in van Schooten [1657].

Ineichen, R. 1996. *Würfel und Wahrscheinlichkeit. Stochastisches Denken in der Antike.* Heidelberg, Berlin, Oxford: Spektrum Akademischer Verlag.

James, T.S. 1867. *The History of Litigation & Legislation respecting Presbyterian Chapels and Charities in England and Ireland, between 1816 and 1849.* London: Hamilton Adams & Co.

---

*The title is thus given in one place, and has the order of the two last words switched in another.

Jaynes, E.T. 1979. Where do we stand on maximum entropy? pp. 15–118 in Levine & Tribus [1979].

Jeffrey, R.C. 1983. *The Logic of Decision.* 2nd edition. Chicago: University of Chicago Press.

Jeffreys, H. 1961/1983. *Theory of Probability.* 3rd edition. Oxford: Clarendon Press.

——— 1973. *Scientific Inference.* 3rd edition. Cambridge: Cambridge University Press.

Jeremy, W. D. 1885. *The Presbyterian Fund and Dr Williams's Trust: with biographical notes of the trustees, and some account of their academies, scholarships and schools.* London: Williams and Norgate.

Jesseph, D.M. 1993. *Berkeley's Philosophy of Mathematics.* Chicago: University of Chicago Press.

Johns, O. 1917. *Asphalt and Other Poems.* New York: Alfred A. Knopf.

Johnston, G.A. (Ed.) 1930. *Berkeley's Commonplace Book.* London: Faber & Faber Ltd.

Johnston, S.P. 1901. An unpublished essay by Berkeley. *Hermathena: a series of papers on literature, science, and philosophy, by members of Trinity College, Dublin* XI: 180–185. [This essay is now known as 'Of Infinites'.]

Jones, J.A. (Ed.) 1849. *Bunhill Memorials. Sacred Reminiscences of Three Hundred Ministers and other persons of note, who are Buried in Bunhill Fields, of every denomination. With the inscriptions on their tombs and gravestones, and other historical information respecting them, from Authentic Sources.* London: James Paul.

Jones, R.F. 1963. The Rhetoric of Science in England of the Mid-Seventeenth Century. pp. 5–24 in McKillop [1963].

Jones, W.B. & Thron, W.J. 1980. *Continued Fractions: Analytic Theory and Applications.* London, Amsterdam: Addison-Wesley.

Jordan, C. 1923. On the inversion of Bernoulli's theorem. *The London, Edinburgh, and Dublin Philosophical Magazine and Journal of Science, Series 6*, 45: 732–735.

——— 1965. *Calculus of Finite Differences.* 3rd edition. New York: Chelsea.

Jurin, J. 1734. *Geometry no Friend to Infidelity, or, a Defence of Sir Isaac Newton and the British Mathematicians, in a Letter to the Author of the Analyst. Wherein it is examined how far the Conduct of such Divines as intermix the Interest of Religion with their private Disputes and Passions, and allow neither Learning nor Reason to those they differ from, is of Honour or Service to Christianity, or agreeable to the Example of our Blessed Saviour and his Apostles. By Philalethes Cantabrigiensis.* London.

Kamlah, A. 1987. The decline of the Laplacian theory of probability: a study of Stumpf, von Kries, and Meinong. pp. 91–116 in Krüger et al. [1987].

Kassler, J.C. 1979. *The Science of Music in Britain, 1714–1830. A Catalogue of Writings, Lectures and Inventions.* 2 vols. New York and London: Garland.

Keckermann, B. 1617. *Systema compendiosum totius mathematices, hoc est geometriae, opticae, astronomiae, et geographiae publicis praelectionibus anno 1605, in celeberrimo gymnasio Dantiscano propositum.* Later edition 1661. Hanoviae, apud P. Antonium.

Kendall, M. & Plackett, R.L. 1977. *Studies in the History of Statistics and Probability.* vol. II. London: Charles Griffin & Company Ltd.

Kendall, M.G. 1961. Studies in the history of probability and statistics XI. Daniel Bernoulli on maximum likelihood. *Biometrika* 48: 1–2. (Translation, by C.G. Allen, of Bernoulli [1777], pp. 3–18.) Reprinted in Pearson & Kendall [1970], pp. 155–172.

——— 1963. Ronald Aylmer Fisher, 1890–1962. *Biometrika* 50: 1–15.

Keynes, J.M. 1921. *A Treatise on Probability.* London: Macmillan. Reprinted in 1973; London: Macmillan.

Kitcher, P. 1973. Fluxions, limits, and infinite littlenesse: A study of Newton's presentation of the calculus. *Isis* 64: 33–49.

Kjeldsen, T.H. 1993. The early history of the Moment Problem. *Historia Mathematica* 20: 19–44.

Kline, M. 1990. *Mathematical Thought from Ancient to Modern Times.* 3 vols. New York & Oxford: Oxford University Press.

Knipe, H.R. (Ed.) 1916. *Tunbridge Wells and Neighbourhood. A Chronicle of the Town from 1608 to 1915; and papers by various writers relating to the Geology, Plant and Animal Life, Archæology, and other matters of the District.* Tunbridge Wells: Pelton.

Knopp, K. 1928. *Theory and Application of Infinite Series.* London & Glasgow: Blackie & Sons. Second English edition of 1951 reprinted in 1990; New York: Dover Publications, Inc.

Köbel, J. 1536. *Geometrei, vonn küstlichen messen unnd absehen, allerhand höhe, fleche, ebene, weite unnd breyte, als thürn, kirchen, bäw, baum, velder unnd äcker, &c. mit künstlich zůbereiten Jacob stab, philosphischen speigel, schatten, unnd messrůten, durch schöne figuren und exemple.* Frankfurt.

Kolmogoroff, A. 1933. *Grundbegriffe der Wahrscheinlichkeitsrechnung.* Berlin: Julius Springer. Reprinted in 1977; Berlin: Springer-Verlag. Also reprinted, in translation, as Kolmogorov [1956].

Kolmogorov, A.N. 1956. *Foundations of the Theory of Probability.* Translation edited by N. Morrison. New York: Chelsea.

Körner, S. 1960. *The Philosophy of Mathematics: an introductory essay.* London: Hutchinson & Co. Ltd.

Kotz, S. & Johnson, N.L. (Eds) 1992. *Breakthroughs in Statistics.* 2 vols. New York: Springer-Verlag.

Krüger, L., Daston, L.J., & Heidelberger, M. (Eds) 1987. *The Probabilistic Revolution. Volume I: Ideas in History.* Cambridge, Mass.: MIT Press.

Kruskal, W.H. & Tanur, J.M. (Eds) 1978. *International Encyclopedia of Statistics.* vol. 1. New York: Free Press.

Kyburg, H.E., Jr. 1970. *Probability and Inductive Logic.* London: The Macmillan Company, Collier-Macmillan Limited.

Kyburg, H.E., Jr. & Smokler, H.E. (Eds) 1964. *Studies in Subjective Probability.* New York: John Wiley & Sons, Inc. 2nd edition (with some changes in papers reprinted), 1980; Huntington, New York: Robert E. Krieger Publishing Company.

Laboucheix, H. 1970. *Richard Price: théoricien de la révolution américaine; le philosophe et la sociologue; le pamphlétaire et l'orateur.* Paris: Didier. Translated by Sylvia and David Raphael as *Richard Price as moral philosopher and political theorist,* 1982. Oxford, at The Taylor Institution: The Voltaire Foundation.

Lamb, C. 1894(?). *Essays of Elia.* London: The Home Library Book Co. Geo. Newnes Ltd.

Lancaster, H.O. 1994. *Quantitative Methods in Biological and Medical Sciences. A Historical Essay.* New York: Springer-Verlag.

Landau, E. 1927. *Elementare Zahlentheorie.* Leipzig: S. Hirzel. Translated by J.E. Goodman as *Elementary Number Theory.* New York: Chelsea, 1966.

Landen, J. 1760. A new method of computing the sums of certain series. *Philosophical Transactions* 51: 553–565.

Laplace, P.-S. 1774. Mémoire sur la probabilité des causes par lés événements. *Mémoires de l'Académie royale des Sciences de Paris (Savants étrangers)* 6: 621–656. Reprinted in *Œuvres complètes de Laplace* 8: 27–65.

——— 1778. (Published 1781.) Mémoire sur les probabilités. *Mémoires de l'Académie royale des Sciences de Paris,* 227–332. Reprinted in *Œuvres complètes de Laplace* 9: 383–485.

——— 1814. *Essai philosophique sur les probabilités.* Paris: Ve Courcier.

——— 1820. *Théorie analytique des probabilités.* 3rd edition. Re-issued in two volumes in 1995, Paris: Jacques Gabay.

——— 1825. *Essai philosophique sur les probabilités.* 5th edition. Paris: Bachelier. Reprinted in 1986 with a preface by René Thom and a postscript by Bernard Bru, Paris: Christian Bourgois. Translated as Dale [1995].

——— 1878–1912. *Œuvres complètes de Laplace.* 14 vols. Paris: Gauthier-Villars.

Latham, R.C. & Matthews, W. 1971–1983. *The Diary of Samuel Pepys.* 11 vols. London: G. Bell & Sons, Ltd.; Bell & Hyman.

Le Monnier, L.G. 1746. Extract of a Memoir concerning the Communication of *Electricity*; read at the public Meeting of the *Royal Academy* of *Sciences* at *Paris, Nov. 12. 1746* by Monsieur *le Monnier* the younger, M.D. of that *Academy,* and F.R.S. communicated by the Author to the *President* of the *Royal Society. Philosophical Transactions* 44: 290–295.

Leader, J.D. 1879. *Extracts from the earliest Book of Accounts belonging to the Town Trustees of Sheffield, dating from 1566 to 1701, with explanatory notes.* Sheffield: J.D. Leader & Sons.

——— 1897. *The Records of the Burgery of Sheffield, commonly called the Town Trust, with Introduction and Notes.* London: Elliot Stock.

Lee, E. 1856. *The Southern Watering-Places: Hastings, St. Leonard's, Dover, and Tunbridge Wells.* London: J. Church.

Leeman, A.D. 1986. *Orationis Ratio. The stylistic theories and practice of the Roman orators historians and philosophers.* Amsterdam: Adolf M. Hakkert.

Legendre, A.M. 1798. *Essai sur la théorie des nombres.* Paris: Chez Duprat.

Levine, R.D. & Tribus, M. (Eds) 1979. *The Maximum Entropy Formalism.* Cambridge, Mass.: M.I.T. Press.

Lewis, C.T. 1937. *A Latin Dictionary for Schools.* Oxford: Clarendon Press.

Lewis, S. 1849. *A Topographical Dictionary of England.* 7th edition. 4 vols. London: S. Lewis & Co. (1st edition 1831.)

L'Hospital G.-F.-A. de. 1696. *Analyse des infiniment petits, pour l'intelligence des lignes courbes.* Paris.

Lidstone, G.J. 1941. Review of *Facsimiles of two papers by Bayes* (Molina and Deming, 1940). *The Mathematical Gazette* 25: 177–180.

Light, A.W. 1913. *Bunhill Fields. Written in honour and to the memory of the many saints of God whose bodies rest in this old London cemetery.* London: C.J. Farncombe & Sons, Ltd.

Lillywhite, B. 1963. *London Coffee Houses. A reference book of coffee houses of the seventeenth eighteenth and nineteenth centuries.* London: George Allen and Unwin Ltd.

Linnell, C.L.S. (Ed.) 1964. *The Diaries of Thomas Wilson, D.D. 1731–1737 and 1750.* London: S.P.C.K.

Locke, J. 189-?. *An Essay concerning Human Understanding.* Sir John Lubbock's Hundred Books. London: George Routledge and Sons.

——— 1964. *Some Thoughts concerning Education.* Abridged and edited with an introduction and commentary by F.W. Garforth. London: Heinemann.

Lockie, J. 1813. *Topography of London.* 2nd edition. London: printed for Sherwood, Neely, and Jones. Reprinted in 1994 by the London Topographical Society, Publication 148.

Lucas, J.R. 1970. *The Concept of Probability.* Oxford: Clarendon Press.

Luce, A.A. & Jessop, T.E. (Eds) 1948–1957. *The Works of George Berkeley Bishop of Cloyne.* 9 vols. London: Thomas Nelson and Sons Ltd.

Lynd, R. 1928. *The Pleasures of Ignorance.* 2nd edition. London: Methuen & Co. Ltd.

Macaulay, A. 1756. *Polygraphy or Short-hand made easy to the meanest capacity: being an universal character fitted to all languages: which may be learn'd by this book, without the help of a master.* 3rd edition. London.

McClintock, B. 1984. Translation of de Moivre's *De mensura sortis* as 'On the Measurement of Chance, or, on the Probability of Events in Games Depending Upon Fortuitous Chance'. *International Statistical Review* 52: 237–262.

M'Cormick, J. 1774. *State-papers and Letters, addressed to William Carstares.* Edinburgh.

Machin, J. 1738. The solution of Kepler's problem. *Philosophical Transactions* 40: 205–230.

McKillop, A.D. 1963. *Restoration and Eighteenth-Century Literature: Essays in Honor of Alan Dugald McKillop.* Chicago: University of Chicago Press.

MacKinnon, D.D. 1902. *History of Speldhurst.* Tunbridge Wells: H.G. Groves. 2nd edition, 1930 (revised by D. James).

McLachlan, H. 1931. *English Education under the Test Acts: being the history of the non-conformist academies 1662–1820.* Manchester: Manchester University Press.

MacLaurin, C. 1742. *A Treatise of Fluxions. In two books.* Edinburgh: T.W. & T. Ruddimans. 2nd edition 1801.

Mahon, Charles Viscount. 1779. *Principles of Electricity, containing Divers new Theorems and Experiments, together with An Analysis of the superior Advantages of high and pointed Conductors.* London: for P. Elmsley.

Maistrov, L.E. 1974. *Probability Theory: a historical sketch.* Translated and edited by S. Kotz. New York & London: Academic Press. Originally published in Russian in 1967.

Maitland, W. 1756. *The History of London from its foundation to the present time.* 2 vols. London.

Mallet, C.E. 1924. *A History of the University of Oxford.* 3 vols. London: Methuen & Co. Ltd.

Malone, E. (Ed.) 1807. *The Life of Samuel Johnson LL.D. With his Correspondence and Conversations. By James Boswell Esq.* Paterson's Universal Library of Standard Authors. Edinburgh: William Paterson.

Manuel, F. E. 1968. *A Portrait of Isaac Newton.* Cambridge, Mass.: The Belknap Press of Harvard University Press. (The chapter 'The Lad from Lincolnshire' is reprinted in Palter [1970].)

Marryat, H. & Broadbent, U. 1930. *The Romance of Hatton Garden.* London: James Cornish & Sons.

Marsaglia, G. & Marsaglia, J.C.W. 1990. A new derivation of Stirling's approximation to $n!$. *American Mathematical Monthly* 97: 826–829.

Martin, B. 1746. *An Essay on Electricity: being an Enquiry into the Nature, Cause and Properties thereof, on the Principles of Sir Isaac Newton's Theory of Vibrating Motion, Light and Fire; and the various Phænomena of forty-two Capital Experiments; with some Observations relative to the uses that may be made of this Wonderful Power of Nature.* Bath: printed for the Author.

Marvell, A. 1971. *The Rehearsal Transpros'd and The Rehearsal Transpros'd: The Second Part.* Ed. D.I.B. Smith. Oxford: Clarendon Press.

Matthews, A.G. 1934. *Calamy Revised. Being a revision of Edmund Calamy's Account of the ministers ejected and silenced, 1660–2.* Oxford: Clarendon Press.

Maxwell, G. 1975. Induction and Empiricism: A Bayesian-Frequentist Alternative. pp. 106–165 in Maxwell & Anderson [1975].

Maxwell, G. & Anderson, R.M., Jr. (Eds) 1975. *Induction, Probability, and Confirmation.* Volume VI of the Minnesota Studies in the Philosophy of Science. Minneapolis: University of Minnesota Press.

Maxwell, J. Clerk. 1879. *The Electrical Researches of the Honourable Henry Cavendish, F.R.S. Written between 1771 and 1781, edited from the original manuscripts in the possession of the Duke of Devonshire, K.G.* Cambridge: Cambridge University Press.

Medawar, P.B. 1969. *Induction and Intuition in Scientific Thought.* London: Methuen & Co. Ltd.

Meller, H. 1981. *London Cemeteries. An Illustrated Guide and Gazetteer.* Amersham, England: Avebury Publishing Company.

Melville, L. 1912. *Society at Royal Tunbridge Wells in the Eighteenth Century — and after.* London: Eveleigh Nash.

Mill, J.S. 1843. *A System of Logic, Ratiocinative and Inductive: Being a Connected View of the Principles of Evidence and the Methods of Scientific Investigation.* 8th edition 1872; London: Longmans, Green, Reader, and Dyer. Reprinted in 1892 (Sir John Lubbock's Hundred Books); London: George Routledge and Sons. Reprinted in 1961, Longmans. Also published as Volumes 7 & 8 in *Collected Works of John Stuart Mill*, ed. J.M. Robson, 1974; Toronto: University of Toronto Press, and London: Routledge and Kegan Paul.

Mills, S. 1985. The independent derivations by Leonhard Euler and Colin MacLaurin of the Euler-MacLaurin summation formula. *Archive for History of Exact Sciences* 33: 1–13.

Minute Books of the Body of Protestant Dissenting Ministers of the Three Denominations in and about the cities of London and Westminster. 3 vols. (11 July 1727–7 July 1761; 6 Oct. 1761–11 April 1797; 3 April 1798–3 April 1827).

Mirowski, P. (Ed.) 1994. *Edgeworth on Chance, Economic Hazard, and Statistics.* Lanham, Md.: Rowman & Littlefield Publishers Inc.

Molina, E.C. 1932. Expansion for Laplacian integrals in terms of incomplete gamma functions. *Bell System Technical Journal* 11: 563–575.

Molina, E.C. & Deming, W. Edwards. 1940. *Facsimiles of Two Papers by Bayes.* (i) An Essay towards solving a Problem in the Doctrine of Chances, with Richard Price's Foreword and Discussion; *Phil. Trans. Royal Soc.*, pp. 370–418, 1763. With a Commentary by Edward C. Molina. (ii) A Letter on Asymptotic Series from Bayes to John Canton; pp. 269–271 of the same volume. With a Commentary by W. Edwards Deming. Washington D.C.: The Graduate School, The Department of Agriculture.

Monro, C.J. 1874. Note on the inversion of Bernoulli's Theorem in probabilities. *Proceedings of the London Mathematical Society* 5: 74–78. (Errata, pp. 145–146.)

Montmort, P.R. de 1713. *Essay d'analyse sur les jeux de hazard.* 2nd edition. Paris: J. Quillau. Reprinted, as 3rd edition, in 1980; New York: Chelsea.

Moore, D.S. 1998. Statistics among the liberal arts. *Journal of the American Statistical Association* 93: 1253–1259.

Morgan, W. 1815. *Memoirs of the Life of The Rev. Richard Price, D.D. F.R.S.* London: R. Hunter.

Morris, C.N. 1988. Approximating posterior distributions and moments. pp. 327–344 in Bernardo et al. [1988].

Mosteller, F. & Wallace, D.L. 1984. *Applied Bayesian and Classical Inference. The Case of* The Federalist *Papers.* 2nd edition of *Inference and Disputed Authorship: The Federalist*, 1964. New York: Springer-Verlag.

Mottelay, P.F. 1922. *Bibliographical History of Electricity and Magnetism chronologically arranged. Researches into the domain of the early sciences, especially from the period of the revival of scholasticism, with biographical and other accounts of the most distinguished natural philosophers throughout the middle ages.* London: Charles Griffin & Company Ltd.

Murray, F.H. 1930. Note on a scholium of Bayes. *Bulletin of the American Mathematical Society* 36: 129–132.

Nagel E. 1939. *Principles of the Theory of Probability.* International Encyclopedia of Unified Science, vol. 1, number 6. Chicago: University of Chicago Press.

Needham, T. 1746. Extract of a letter from Mr. Turbervill Needham to Martin Folkes, Esq., Pr.R.S. concerning some new electrical experiments lately made at Paris. *Philosophical Transactions* 44: 247–263.

Nevill, R. 1911. *London Clubs their history and treasures.* London: Chatto & Windus.

Newell, D.J. 1984. Present position and potential developments: some personal views. Medical statistics. *Journal of the Royal Statistical Society, A* 147: 186–197.

Newton, I. 1687. *Philosophiæ Naturalis Principia Mathematica.* London. Reprinted in 1965; Bruxelles: Culture et Civilisation. Third edition of 1726 translated into English by Andrew Motte in 1729. The translations revised, and supplied with an historical and explanatory appendix, by Florian Cajori, 1934. Berkeley: University of California Press.

——— 1704a. *Tractatus de quadratura curvarum.* pp. 170–211 in Newton [1704b].

——— 1704b. *Opticks: or, a Treatise of the Reflexions, Refractions, Inflexions and Colours of Light. Also two Treatises of the Species and Magnitude of Curvilinear Figures.* London.

——— 1707. *Arithmetica Universalis; sive de compositione et resolutione arithmetica liber.* London.

——— 1710. Quadrature of Curves. Article in Harris [1710]. Reprinted in Whiteside [1964].

——— 171$\frac{4}{5}$. An account of the book entituled *Commercium Epistolicum Collinii & aliorum, De Analysi promota*; published by order of the *Royal-Society*, in relation to the Dispute between Mr. *Leibnitz* and Dr. *Keill*, about the Right of Invention of the Method of *Fluxions*, by some call'd *the Differential Method. Philosophical Transactions* 29: 173–224.

——— 1736. *The Method of Fluxions and Infinite Series: with its Application to the Geometry of Curve-Lines. By the Inventor Sir Isaac Newton, K^t. Late President of the Royal Society. Translated from the Author's Latin Original not yet made publick. To which is subjoin'd, A Perpetual Comment upon the whole Work, Consisting of Annotations, Illustrations, and Supplements. In order to make this Treatise a Compleat Institution for the use of Learners.* Translation and notes by John Colson. London.

——— 1737. *A Treatise of the Method of Fluxions and Infinite Series, With its Application to the Geometry of Curve Lines.* London: printed for T. Woodman & J. Millan. (A pirated version of Newton [1736].)

——— *A Dissertation upon the Sacred Cubit of the Jews and the Cubits of the several Nations; in which, from the Dimensions of the greatest Egyptian Pyramid, as taken by Mr. John Greaves, the antient Cubit of Memphis is determined. Translated from the Latin of Sir Isaac Newton, not yet published.* Printed in Greaves [1737], vol. II.

Nightingale, B. 1890–1893. *Lancashire Nonconformity; or, sketches, historical & descriptive, of the Congregational and Old Presbyterian Churches in the County.* 6 vols. Manchester: John Heywood.

Nisard, D. 189-. *Histoire de la Littérature Française.* 18th edition. 4 vols. Paris; Librairie de Firmin-Didot et Cie.

O'Donnell, T. 1936. *History of Life Assurance in Its Formative Years. Compiled from approved sources.* Chicago: American Conservation Company.

Ogborn, M.E. 1962. *Equitable Assurances. The Story of Life Assurance in the Experience of the Equitable Life Assurance Society 1762–1962.* London: George Allen and Unwin Ltd.

O'Hagan, A. 1994. *Kendall's Advanced Theory of Statistics. Volume 2B. Bayesian Inference.* London: Edward Arnold.

Oldham, J. 1960. *Poems of John Oldham.* With an introduction by Bonamy Dobrée. London: Centaur Press Ltd.

Onely, R. 1771. *A General Account of Tunbridge Wells, and its Environs: historical and descriptive.* London: G. Pearch.

Orton, J. 1766. *Memoirs of the Life, Character and Writings of the Late Reverend Philip Doddridge, D.D. of Northampton.* Salop.

Oxford Latin Dictionary. 1983. Edited by P.G.W. Glare. Oxford: Clarendon.

Page, N. (Ed.) 1987. *Dr Johnson. Interviews and Recollections.* Houndmills, Basingstoke, Hampshire, and London: Macmillan.

Paley, W. 1785. *The Principles of Moral and Political Philosophy.* Reprinted in Volume IV of Paley [1825].

——— 1802. *Natural Theology, or Evidences of the Existence and Attributes of the Deity collected from the Appearances of Nature.* Reprinted in Volume V of Paley [1825].

——— 1825. *The Works of William Paley, D.D. With additional sermons, etc. etc. and a corrected account of the life and writings of the author, by the Rev. Edmund Paley, A.M.* 7 volumes. London: printed for C. and J. Rivington, et al.

Palter, R. (Ed.) 1970. *The* Annus Mirabilis *of Sir Isaac Newton 1666–1966.* Cambridge, Mass. and London, England: M.I.T. Press.

Parker, C.S. 1867. On the history of classical education. pp. 1–80 in Farrar [1867].

Parker, I. 1914. *Dissenting Academies in England, Their Rise and Progress and their Place among the Educational Systems of the Country.* Cambridge: Cambridge University Press.

Parker, S. 1669. *A Discourse of Ecclesiastical Politie, wherein the authority of the Civil Magistrate over the Consciences of Subjects in matters of External Religion is asserted; the Mischiefs and Inconveniences of Toleration are represented, and all Pretenses pleaded in behalf of Liberty of Conscience are fully answered.* London.

Partington, C.F. 1835. *The British Cyclopædia of the Arts and Sciences.* London: Orr & Smith.

Pascal, B. 1665. *Traité du triangle arithmétique*. Paris: Desprez.

—— 1735. *Les Provinciales, ou Lettres ecrites par Louis de Montalte a un provincial de ses amis, Et aux RR.PP. Jesuites sur la Morale & la Politique de ces Peres. Avec les notes de Guillaume Wendrock.* (Nouvelle edition), 3 volumes. Amsterdam: J. Fr. Bernard.

Passler, D.L. 1971. *Time, Form, and Style in Boswell's Life of Johnson.* New Haven and London: Yale University Press.

Patin, J.M. 1989. A very short proof of Stirling's formula. *American Mathematical Monthly* 96: 41–42.

Payne, J.F. 1894. See Boghurst [1894].

Pearce, L. — *Free Churches of Tunbridge Wells.* (n.p.).

Pearson, E.S. & Kendall, M.G. (Eds) 1970. *Studies in the History of Statistics and Probability.* vol. I. London: Charles Griffin & Company Ltd.

Pearson, K. 1900. *The Grammar of Science.* Second edition. London: Adam and Charles Black.

—— (Ed.) 1934. *Tables of the Incomplete Beta-Function.* Printed at the University Press, Cambridge, for the Biometrika Trustees.

—— 1978. *The History of Statistics in the 17th and 18th Centuries, against the changing background of intellectual, scientific and religious thought. Lectures given at University College London during the academic sessions 1921–1933.* Ed. E.S. Pearson. London: Charles Griffin & Co., Ltd.

Pemberton, H. 1728. *A View of Sir Isaac Newton's Philosophy.* London: printed by S. Palmer.

—— 1741. Dr. Pemberton's Animadversions on some Particulars relating to himself in Dr. Jurin's Letter to ____ ____, Esq; in Answer to Mr. Robins's Full Confutation, *&c.* With an Explanation of the first Lemma in Sir Isaac Newton's Principia. Art. VI, January 1741, pp. 68–78 in *The History of the Works of the Learned. Giving A General View of the State of Learning throughout EUROPE, and containing an impartial Account and accurate Abstracts of the most valuable Books publish'd in Great-Britain and Foreign Parts. Interspers'd with Dissertations on several curious and entertaining Subjects, critical Reflections, and Memoirs of the most eminent Writers in all Branches of polite Literature.* vol. IX. London.

Perkins, J. 1697. *Mr Perkins's Letter to Mr. Cornwell, And other Ministers at Tunbridge-Wells, Who denied him the Use of the Pulpit there. And have not Answered the Letter, as desired.* London: J. Bradford.

Phippen, J. (Ed.) 1840. *Colbran's New Guide for Tunbridge Wells.* London: Bailey & Co.

Pike, G.H. 1870. *Ancient Meeting-houses; or, Memorial Pictures of Nonconformity in old London.* London: S.W. Partridge & Co.

Plackett, R.L. 1958. Studies in the history of probability and statistics VII. The principle of the arithmetic mean. *Biometrika* 45: 130–135.

Playfair, J. 1857. *Elements of Geometry, containing the first six books of Euclid, with a supplement on the quadrature of the circle and the geometry of solids. To which are added, elements of plane and sphericale trigonometry.* Philadelphia: J.B. Lippincott.

Poems on Affairs of State, from the Reign of K. James I. to the Year 1703. Written by the Greatest Wits of the Age. 1716. vol. II. London.

Poincaré, H. 1903. *La Science et l'Hypothèse.* Paris: Ernest Flammarion.

Poisson, S.-D. 1830. Mémoire sur la proportion des naissances des filles et des garçons. *Mémoires de l'Académie des Sciences, Paris* 9: 239–308.

———— 1837. *Recherches sur la probabilité des jugements en matière criminelle et en matière civile, précédées des règles générales du cacul des probabilités.* Paris: Bachelier.

Pope, A. 1872. *The Works of Alexander Pope.* vol. VIII. London: John Murray.

Popper, K.R. 1968. *The Logic of Scientific Discovery.* Revised edition. London: Hutchinson & Co. Ltd.

Press, S.J. 1989. *Bayesian Statistics: Principles, Models, and Applications.* New York: John Wiley & Sons.

Price, R. 1787/1974. *A Review of the Principal Questions in Morals. Particularly Those respecting the Origin of our Ideas of Virtue, its Nature, Relation to the Deity, Obligation, Subject-matter, and Sanctions. The third edition corrected, and enlarged by an appendix, containing additional notes, and a dissertation on the being and attributes of the Deity.* London: printed for T. Cadell. Edited (1974) by D.D. Raphael. Oxford: Clarendon.

Priestley, J. 1767. *The History and Present State of Electricity, with Original Experiments.* London: printed for J. Dodsley.

Quetelet, L.A.J. 1835. *Sur l'Homme et le Développement de ses Facultés, ou Essai de physique sociale.* 2 vols. Paris: Bachelier. Translated as Quetelet [1842].

────── 1842. *A Treatise on Man and the Development of his Faculties.* Edinburgh: William and Robert Chambers. Reprinted in 1969; Gainesville, Fla.: Scholars' Facsimiles & Reprints.

Ramsey, F.P. 1965. *The Foundations of Mathematics and other Logical Essays.* Edited by R.B. Braithwaite. Totowa, N.J.: Littlefield, Adams & Co.

Rashed, R. (Ed.) 1988. *Sciences à l'Époque de la Révolution française: recherches historiques.* Paris: Librairie scientifique et technique Albert Blanchard.

Rawlings, G.B. 1926. *The Streets of London: their history and associations.* London: Geoffrey Bles.

The Record of the Royal Society of London for the Promotion of Natural Knowledge. 1940. Fourth edition. London: printed for The Royal Society by Morrison & Gibb Ltd., Edinburgh.

Reichenbach, H. 1971. *The Theory of Probability. An Inquiry into the Logical and Mathematical Foundations of the Calculus of Probability.* 2nd edition. Berkeley: University of California Press.

Rigaud, S.J. 1844. *A Defence of Halley against the Charge of Religious Infidelity.* Oxford: The Ashmolean Society.

Rippon, J. 1803. *Proposals for printing by subscription (On a beautiful Paper, and in a handsome Style.) The History of Bunhill Fields Burying Ground, An Estate in the Possession of the City.* Printed as an attachment to *A Discourse on the Origin and Progress of the Society for promoting Religious Knowledge among the Poor, From it's Commencement in 1750 to the Year 1802.* London.

Robins, B. 1735. *A Discourse Concerning the Nature and Certainty of Sir Isaac Newton's Methods of Fluxions, and of Prime and Ultimate Ratios.* London: W. Innys. Reprinted in Robins [1761], vol. II.

────── 1761. *Mathematical Tracts.* 2 vols. London: James Wilson.

Robinson, E.F. 1893. *The Early History of Coffee Houses in England. With some account of the first use of coffee and a bibliography of the subject.* London: Kegan Paul, Trench, Trübner & Co., Ltd.

Rockwell, J.E. 1893. *Shorthand Instruction and Practice*. Washington D.C.: Government Printing Office.

Rogers, P. (Ed.) 1978. *The Eighteenth Century*. The Context of English Literature. London: Methuen & Co. Ltd.

Roller, D.D. 1953. Did Bacon know Gilbert's De Magnete? *Isis* 44: 10–13.

—— 1959. *The De Magnete of William Gilbert*. Amsterdam: Menno Hertzberger.

Ross, S. 1994. De Morgan tussles with Smith's *Harmonics* in a comic poem. *British Journal for the History of Science* 27: 467–471.

Rothstein, E. & Weinbrot, H.D. 1976. The *Vicar of Wakefield*, Mr. Wilmont, and the 'Whistonean Controversy'. *Philological Quarterly* 55: 225–240.

Rousseau, G.S. 1978. Science. pp. 153–207 in Rogers [1978].

Rowzee, L. 1670. *The Queen's Wells: That is, a Treatise of the Nature and Vertues of Tunbridge Water. Together with an Enumeration of the chiefest Diseases, which it is good for, and against which, it may be used, and the Manner and order of taking it*. London. Printed in London in 1746 in The Harleian Miscellany: or, a Collection of Scarce, Curious; and Entertaining Pamphlets and Tracts, As well in Manuscript as in Print, Found in the late Earl of Oxford's Library.

Russell, B.A.W. 1961. *History of Western Philosophy and its Connection with political and social circumstances from the earliest times to the present day*. London: George Allen & Unwin Ltd.

Russell, W. Clark. 187-? *The Book of Authors. A collection of criticisms, ana, môts, personal descriptions, etc. etc. etc.* (The 'Chandos Classics.') London: Frederick Warne and Co.

Rutherforth, T. 1744. *An Essay on the Nature and Obligations of Virtue*. Cambridge.

Saglio, E. 1919. *Dictionnaire des antiquités grecques et romaines d'après les textes et les monuments*. vol. 5 (T-Z). Paris: Librairie Hachette.

Sainte-Beuve, C.A. 1953–1955. *Port-Royal*. Texte présenté et annoté par Maxime Leroy. 'Bibliothèque de la Pléiade'. Paris: Gallimard.

Salmon, W.C. 1967. *The Foundations of Scientific Inference*. Pittsburgh: University of Pittsburgh Press.

Savage, L.J. 1960. Reading note on Bayes' theorem. First printed as the Appendix to the first (1991) edition of Dale [1999].

—— 1961. The Foundations of Statistics Reconsidered. *Proceedings of the Fourth [1960] Berkeley Symposium on Mathematical Statistics and Probability*, vol. I: 575–586. Reprinted in Kyburg & Smokler [1964], pp. 173–188, and in Savage [1981], pp. 296–307.

—— et al. 1962. *The Foundations of Statistical Inference. A Discussion.* Second Impression. London: Methuen & Co. Ltd.

—— 1981. *The Writings of Leonard Jimmie Savage — A Memorial Selection.* Washington D.C.: The American Statistical Association and The Institute of Mathematical Statistics.

Say, S. 1745. *Poems on Several Occasions; and Two Critical Essays, viz. The First, on the Harmony, Variety, and Power of Numbers, whether in Prose or Verse. The Second, On the Numbers of Paradise Lost.* London.

Scott, J., Green, C., Falconer, W. & Meader, J. 1765. *A General Dictionary of Arts and Sciences: or, A Complete System of Literature.* London: printed for S. Crowder.

Scott, J.F. 1938. *The Mathematical Works of John Wallis, D.D., F.R.S. (1616–1703).* London: Taylor and Francis, Ltd.

Seal, H.L. 1954. A budget of paradoxes. *Journal of the Institute of Actuaries Students' Society* 13: 60–65. Reprinted in Kendall & Plackett [1977], pp. 24–29.

—— 1978. Bayes, Thomas. pp. 7–9 in Kruskal & Tanur [1978].

Selby-Bigge, L.A. (Ed.) 1897. *British Moralists, being selections from writers principally of the eighteenth century.* 2 vols. Oxford: Clarendon.

Shafer, G. 1976. *A Mathematical Theory of Evidence.* Princeton, N.J.: Princeton University Press.

—— 1978. Non-additive probabilities in the work of Bernoulli and Lambert. *Archive for History of Exact Sciences* 19: 309–370.

—— 1982. Bayes's two arguments for the rule of conditioning. *Annals of Statistics* 10: 1075–1089.

—— 1983. A subjective interpretation of conditional probability. *Journal of Philosophical Logic* 12: 453–466.

―――― 1985. Conditional probability. *International Statistical Review* 53: 261–277.

Shelton, T. 1641. *Tachygraphy the most exact and compendius methode of short and swift writing that hath ever yet beene published by any*. Cambridge. Republished as *Tachygraphia: sive, Exactissima & compendiosissima breviter scribendi methodus*. London: Tho. Creake, 1660.

―――― 1654. *Zeiglographia: or, a new art of short-writing never before published. More easie, exact, short and speedie than any heretofore*. London.

Sherwin, H. 1705. *Mathematical Tables, contrived after a most comprehensive method*. London. (There are several later editions.)

Sheynin, O.B. 1968. Studies in the history of probability and statistics XXI. On the early history of the law of large numbers. *Biometrika* 55: 459–467. Reprinted in Pearson & Kendall [1970], pp. 231–239.

―――― 1969. On the work of Thomas Bayes in probability theory. [In Russian.] *The Academy of Sciences of the USSR, The Institute of History of Natural Sciences and Technology. Proceedings of the 12th Conference of Doctorands and Junior Research Workers. Section of History of Mathematics and Mechanics*. Moscow. (n.p.).

―――― 1971a. Studies in the history of probability and statistics XXV. On the history of some statistical laws of distribution. *Biometrika* 58: 234–236. Reprinted in Kendall & Plackett [1977], pp. 328–330.

―――― 1971b. J.H. Lambert's work on probability. *Archive for History of Exact Sciences* 7: 244–256.

―――― 1973a. Finite random sums (a historical essay). *Archive for History of Exact Sciences* 9: 275–305.

―――― 1973b. R.J. Boscovich's work on probability. *Archive for History of Exact Sciences* 9: 306–324.

Shiokawa, T. 1977. *Pascal et les miracles*. Paris: Librairie A.-G. Nizet.

Shoesmith, E. 1985. Thomas Simpson and the arithmetic mean. *Historia Mathematica* 12: 352–355.

Sidgwick, H. 1867. The theory of classical education. pp. 81–143 in Farrar [1867].

Simpson, T. 1740. *The Nature and Laws of Chance*. London.

——— 1742. *The Doctrine of Annuities and Reversions, Deduced from General and Evident Principles: With useful Tables, shewing the Values of Single and Joint Lives, etc. at different Rates of Interest*. London: John Nourse.

——— 1748. On the fluents of multinomials, and series affected by radical signs, which do not begin to converge till after the second term. *Philosophical Transactions* 45: 328–335.

——— 1751. A general method for exhibiting the value of an algebraic expression involving several radical quantities in an infinite series: wherein Sir Isaac Newton's theorem for involving a binomial, with another of the same author, relating to the roots of equations, are demonstrated. *Philosophical Transactions* 47: 20–27.

——— 1755. A Letter to the Right Honourable *George* Earl of *Macclesfield*, President of the *Royal Society*, on the *Advantage* of taking the Mean of a Number of Observations, in practical Astronomy. *Philosophical Transactions* 49: 82–93.

——— 1757a. An Attempt to shew the Advantage arising by Taking the Mean of a Number of Observations, in practical Astronomy. pp. 64–75 in Simpson [1757b].

——— 1757b. *Miscellaneous Tracts on Some curious, and very interesting Subjects in Mechanics, Physical-Astronomy, and Speculative Mathematics; wherein the Precession of the Equinox, the Nutation of the Earth's Axis, and the Motion of the Moon in her Orbit, are determined*. London: printed for J. Nourse.

——— 1758. The invention of a general method for determining the sum of every 2d, 3d, 4th, or 5th, &c. term of a series, taken in order; the sum of the whole series being known. *Philosophical Transactions* 50: 757–769.

——— 1762. *A Treatise of Algebra. Wherein The Principles are Demonstrated And Applied. In many useful and interesting Enquiries, and in the Resolution of a great Variety of Problems of different kinds. To which is added, The Geometrical Construction of a great Number of Linear and Plane Problems, With the Method of resolving the same Numerically*. 3rd edition. Revised. London: J. Nourse.

——— 1823. *The Doctrine and Application of Fluxions. Containing (besides what is common on the subject) a number of new improvements in the theory: and the solution of a variety of new and very interesting problems, in different branches of the mathematics*. A new edition, carefully revised, and adapted, by copious appendixes, to the present *advanced state of science. By a graduate of the University of Cambridge.* 2 vols. London: J. Collingwood, and G. & W. B. Whittaker.

Singer, G.J. 1814. *Elements of Electricity and Electro-Chemistry.* London: printed for Longman, Hurst, Rees, Orme, and Brown.

Smith, G.C. 1980. Thomas Bayes and fluxions. *Historia Mathematica* 7: 379–388.

Smith, R. 1738. *A Compleat System of Opticks. In Four Books, viz. A Popular, A Mathematical, a Mechanical, and a Philosophical Treatise. To which are added Remarks upon the Whole.* Cambridge.

—— 1749. *Harmonics, or the philosophy of musical sounds.* Cambridge: J. Bentham. Reprinted in 1966; New York: Da Capo Press.

Speidell, J. 1616. *A geometricall extraction; or, A compendiovs collection of the chiefe and choyse problems, collected out of the best, and latest writers. Wherevnto is added, about 30. problemes of the authors inuention, being for the most part, performed by a better and briefer way, then by any former writer.* Later edition 1657. London.

Sprange, J. 1786/1780/1797. *The Tunbridge Wells Guide; or An Account of the ancient and present State of that Place, To which is Added a particular Description of the Towns and Villages, Gentlemens Seats, Remains of Antiquity, Founderies, &c. &c. within the Circumference of Sixteen Miles.* (n.p.).

—— 1817. *The Tunbridge Wells Guide, or An Account of the ancient & present State of that Place with a particular description of all the Towns, Villages, Antiquities, Natural Curiosities, Ancient & Modern Seats, Founderies, &c. within the Circumference of sixteen Miles, with accurate Views of the principal Objects.* Tunbridge Wells: J. Sprange.

Sprat, T. 1667. *The History of the Royal Society of London, for the Improving of Natural Knowledge.* London.

Stevenson, B. 1958(?) *Stevenson's Book of Quotations. Classical and Modern.* 9th edition. London: Cassell.

Stewart, L. 1999. Other centres of calculation, or, where the Royal Society didn't count: commerce, coffee-houses and natural philosophy in early modern London. *British Journal for the History of Science* 32: 133–153.

Stigler, S.M. 1982. Thomas Bayes's Bayesian inference. *Journal of the Royal Statistical Society, A* 145: 250–258.

——— 1983. Who discovered Bayes's theorem? *The American Statistician* 37: 290–296.

——— 1984. Studies in the history of probability and statistics XL. Boscovich, Simpson and a 1760 manuscript note on fitting a linear relationship. *Biometrika* 71: 615–620.

——— 1986. *The History of Statistics. The Measurement of Uncertainty before 1900*. Cambridge, Mass., and London, England: the Belknap Press of Harvard University Press.

——— 1999. *Statistics on the Table: The History of Statistical Concepts and Methods*. Cambridge, Mass.: Harvard University Press.

Stirling, J. 1730/1753/1764. *Methodus Differentialis: sive Tractatus de Summatione et Interpolatione Serierum Infinitarum*. London. Translated into English in 1749.

——— 1735. Of the figure of the earth, and the variation of gravity on the surface. *Philosophical Transactions* 39: 98–105.

Stock, J. 1776. *An Account of the Life of George Berkeley, D.D. Late Bishop of Cloyne in Ireland. With Notes, Containing Strictures Upon his Works.* 2nd edition. London: John Murray. Reprinted in Berman [1989], pp. 5–85.

Stone, E. 1730. *The Method of Fluxions both Direct and Inverse. The Former being a Translation from the Celebrated Marquis De L'Hospital's Analyse des Infiniments Petits: and the Latter Supply'd by the Translator.* 2 vols. London: for W. Innys.

Stow, J. 1603. *A Survay of London. Conteyning the Originall, Antiquity, Increase, Moderne estate, and description of that City, writtten in the yeare 1598, by John Stow Citizen of London. Since by the same Author increased, with diuers rare notes of Antiquity, and published in the yeare, 1603. Also an Apologie (or defence) against the opinion of some men, concerning that Citie, the greatnesse thereof. With an Appendix, containing in Latine Libellum de situ & nobilitate Londinii: Written by William Fitzstephen, in the raigne of Henry the second.* Reprinted in 1912; London: J.M. Dent & Sons Ltd.

Strange, C.H. 1949. *Nonconformity in Tunbridge Wells*. Tunbridge Wells: The Courier Printing and Publishing Co. Ltd.

Strode, T. 1678. *A Short Treatise of the Combinations, Elections, Permutations & Composition of Quantities. Illustrated by Several Examples, with a new Speculation of the Differences of the Powers of Numbers.* London.

Strype, J. 1720. *A Survey of the Cities of London and Westminster: containing the Original, Antiquity, Increase, Modern Estate and Government of those Cities. Written at first in the Year MDXCVIII. By John Stow, Citizen and Native of London.* 2 volumes. In six books. London.

Suetonius. (Gaius Suetonius Tranquillus) 1962. *The Twelve Caesars.* Translated by Robert Graves. Harmondsworth, Middlesex: Penguin Books.

Swijtink, Z.G. 1987. The objectification of observation: measurement and statistical methods in the nineteenth century. pp. 261–285 in Krüger et al. [1987].

The Tatler, with notes and illustrations. 1831. 3 vols. Edinburgh: Thomas Nelson, and Peter Brown.

Taylor, B. 1717. A treatise on infinite series; part first. By Peter Remund de Monmort, F.R.S. To which is added an appendix, in which several parts are treated in a different way. *Philosophical Transactions* 30: 633–689.

Taylor, E.G.R. 1966. *The Mathematical Practitioners of Hanoverian England 1714–1840.* Cambridge: Cambridge University Press.

Thomas, D.O. 1977. *The Honest Mind. The Thought and Work of Richard Price.* Oxford: Clarendon.

Thomson, J. Radford 1871. *Pelton's (Late Brackett's) Illustrated Guide to Tunbridge Wells, and the Neighbouring Seats, Towns, and Villages: comprising a History of the Town, Information useful to Visitors, and Chapters on the Botany and Geology of the Neighbourhood; with three maps and numerous engravings.* 5th edition. Tunbridge Wells: Richard Pelton.

Thorburn, W.M. 1918. The myth of Occam's razor. *Mind* 27: 345–353.

Thornbury, W. 1887–1893. *Old and New London. A Narrative of Its History, Its People, and Its Places. The City, Ancient and Modern.* A new edition, carefully revised and corrected. vol. II. London: Cassell, Petter, Galpin & Co.

Tierney, L. & Kadane, J.B. 1986. Accurate approximations for posterior moments and marginal densities. *Journal of the American Statistical Association* 81: 82–86.

Timerding, H.E. (Ed.) 1908. *Versuch zur Lösung eines Problems der Wahrscheinlichkeitsrechnung von Thomas Bayes.* Ostwald's Klassiker der Exakten Wissenschaften, Nr. 169. Leipzig: Wilhelm Engelmann.

Timpson, T. 1859. *Church History of Kent: from the earliest period to MDCCCLVIII.* London: Ward & Co.

Todhunter, I. 1865. *A History of the Mathematical Theory of Probability from the time of Pascal to that of Laplace.* Cambridge.

Tunbrigalia. or, the Tunbridge Miscellany, For the years, 1737, 1738, 1739. Being A Curious Collection of Miscellany Poems, &c. Exhibited upon the Walks at Tunbridge Wells, in the last Years. By a Society of Gentlemen and Ladies. 1740. London.

Turner, G.L. (Ed.) 1911. *Original Records of Early Nonconformity under Persecution and Indulgence.* 3 vols. London & Leipsic: T. Fisher Unwin.

Tweddle, I. 1988. *James Stirling 'This about series and such things'.* Edinburgh: Scottish University Press.

——— 1998. The prickly genius — Colin MacLaurin (1698–1746). *The Mathematical Gazette* 82: 373–378.

Tweedie, C. 1922. *James Stirling: a sketch of his life and works, along with his scientific correspondence.* Oxford: Clarendon.

Urwick, W. 1884. *Nonconformity in Herts. Being Lectures upon the Nonconforming Worthies of St. Albans, and Memorials of Puritanism and Nonconformity in all the Parishes of the County of Hertford.* London: Hazell, Watson, and Viney, Ltd.

Uspensky, J.V. 1937. *Introduction to Mathematical Probability.* New York: McGraw-Hill.

van Rooijen, J.P. 1942. De waarschijnlijkheidsrekening en het theorema van Bayes. *Euclides (Groningen)* 18: 177–199.

van Schooten, F. 1657. *Exercitationum Mathematicarum, Liber V.* Amsterdam: Ex Officina Johannis Elsevirii.

Velikovsky, I. 1957. *Worlds in Collision.* London: Victor Gollancz Ltd.

Venn, J. 1866. *The Logic of Chance.* London: Macmillan. Reprinted as the 4th edition 1962 (unaltered reprint of the 3rd edition of 1888); New York: Chelsea.

Venn, J. & Venn, J.A. 1922. *Alumni Cantabrigiensis. A biographical list of all known students, graduates and holders of office at the University of Cambridge, from the earliest times to 1900.* Part I. From the earliest times to 1751. Cambridge: Cambridge University Press.

von Braunmühl, J.A. 1900. *Vorlesungen über Geschichte der Trigonometrie.* Leipzig.

von Mises, R. 1942. On the correct use of Bayes' formula. *Annals of Mathematical Statistics* 13: 156–165.

——— 1981. *Probability, Statistics and Truth.* New York: Dover Publications, Inc.

von Wright, G.H. 1960. *A Treatise on Induction and Probability.* Paterson, N.J.: Littlefield, Adams & Co.

Walker, K. (Ed.) 1984. *The Poems of John Wilmot, Earl of Rochester.* Oxford: published for the Shakespeare Head Press by Basil Blackwell.

Waller, J.F. (Ed.) 1865. *The Imperial Dictionary of Universal Biography.* London, Glasgow and Edinburgh: Wm. MacKenzie.

Walley, P. 1991. *Statistical Reasoning with Imprecise Probabilities.* London: Chapman and Hall.

Wallis, J. 1655. *Arithmetica infinitorum.* pp. 355–478 in Wallis [1657].

——— 1657. *Opera mathematicorum pars prima.* Oxford.

Wallis, R.V. & Wallis, P.J. 1986. *Biobibliography of British Mathematics and its Applications. Part II. 1701–1760.* Newcastle upon Tyne: Project for Historical Biobibliography.

Walmsley, P. 1990. *The Rhetoric of Berkeley's Philosophy.* Cambridge: Cambridge University Press.

Walsh, W.S. 1894. *Handy-book of Literary Curiosities.* London: Gibbings & Company, Ltd.

Walton, J. 1735. *A Vindication of Sir Isaac Newton's Principles of Fluxions, against the Objections contained in the Analyst.* Dublin: S. Powell and London: J. Roberts.

Warburton, W. 1755. *The Divine Legation of Moses demonstrated, on the principles of a religious Deist, from the omission of the doctrine of a future state of reward and punishment in the Jewish dispensation.* In nine books. 4th edition. (1st edition 1738.) London: J. & P. Knapton.

Ward, J. 1740. *The Lives of the Professors of Gresham College: To which is prefixed The Life of the Founder, Sir Thomas Gresham. With An Appendix, consisting of Orations, Lectures, and Letters, written by the Professors, with other Papers serving to illustrate the Lives.* London:

printed for John Moore. Reprinted in 1967; New York and London: Johnson Reprint Corporation.

——— 1759. *A System of Oratory, Delivered in a Course of Lectures Publicly read at Gresham College, London: To which is prefixed An Inaugural Oration, Spoken in Latin, before the Commencement of the Lectures, according to the usual Custom.* 2 vols. London.

Warnock, G.J. 1953. *Berkeley.* London: Penguin Books.

Watanabe, M. 1933. On the à posteriori probability. *Tôhoku Mathematical Journal* 38: 390–396.

Watson, D. 1792. *The Works of Horace, translated into English prose, as near as the propriety of the two languages will admit. Together with the original Latin, from the best editions. Wherein the words of the Latin text are ranged in their grammatical order; the ellipses carefully supplied; the observations of the most valuable commentators both ancient and modern, represented; and the author's design and beautiful descriptions fully set forth in a key annexed to each poem; with notes geographical and historical; also the various readings of Dr. Bentley. The whole adapted to the capacities of youth at school, as well as of private gentlemen.* A new edition, by W. Crakelt. London: printed for T. Longman et al.

Watson, W. 1748. Some further inquiries into the nature and properties of electricity. *Philosophical Transactions* 45: 93–120.

Weatherford, R. 1982. *Philosophical Foundations of Probability Theory.* London: Routledge & Kegan Paul.

Weinstock, R. 1982. Dismantling a centuries-old myth: Newton's *Principia* and inverse-square orbits. *American Journal of Physics* 50: 610–617.

——— 1998. Newton's *Principia* and inverse-square orbits in a resisting medium: a spiral of twisted logic. *Historia Mathematica* 25: 281–289.

Westfall, R.S. 1973. Newton and the fudge factor. *Science* 179: 751–758.

Wheatley, H.B. 1891. *London Past and Present. Its History, Associations, and Traditions.* London: John Murray.

Whiston, W. 1749. *Memoirs of the Life and Writings of Mr. William Whiston. Containing Memoirs of Several of his Friends also.* 2nd edition, corrected, 1753. London: J. Whiston & B. White.

Whitaker, W.B. 1940. *The Eighteenth Century English Sunday. A Study of Sunday Observance from 1677 to 1837.* London: The Epworth Press.

Whitam, J.H. 1948. *An Account of the Company of Cutlers in Hallamshire.* (n.p.).

White, J.G. 1901. *The Churches & Chapels of Old London, with A Short Account of those who have Ministered in Them.* London: printed for private circulation.

White, J.T. 1880. *Latin-English Dictionary.* London: Longmans, Green, and Co.

Whitehead, A.N. 1967. *Adventures of Ideas.* New York: Free Press. First published 1933, Macmillan.

Whiteside, D.T. 1964. *The Mathematical Works of Isaac Newton.* vol. I. Assembled with an Introduction by Dr. Derek T. Whiteside. The Sources of Science, No. 3. New York and London: Johnson Reprint Corporation.

——— 1971. *The Mathematical Papers of Isaac Newton.* With the assistance in publication of M.A. Hoskin. vol. IV. Cambridge: Cambridge University Press.

Whittaker, E.T. 1910. *A History of the Theories of Aether and Electricity from the age of Descartes to the close of the nineteenth century.* London: Longmans, Green, and Co.

Whittaker, E.T. & Watson, G.N. 1902. *A Course of Modern Analysis. An introduction to the general theory of infinite processes and of analytic functions; with an account of the principal transcendental functions.* 4th edition, reprinted, 1973. Cambridge: Cambridge University Press.

Whittle, P. 1976. *Probability.* 2nd edition. New York: John Wiley & Sons. (1st edition 1970.) 3rd edition of 1992: *Probability via Expectation.* New York: Springer-Verlag.

Whyte, L.L. (Ed.) 1961. *Roger Joseph Boscovich, S.J., F.R.S., 1711–1787; Studies of his Life and Work on the 250th Anniversary of his Birth.* London: George Allen & Unwin Ltd.

Wilson, B. 1752. *A Treatise on Electricity.* 2nd edition. London.

Wilson, J.H. 1967. *The Court Wits of the Restoration. An Introduction.* London: Frank Cass & Co. Ltd.

Wilson, W. 1814. *The History and Antiquities of Dissenting Churches and Meeting Houses, in London, Westminster, and Southwark; including the lives of their ministers, from the rise of non-conformity to the present time. With an appendix on the origin, progress, and present state of Christianity in Britain.* 4 vols. London.

Wise, M.E. 1950. The incomplete beta function as a contour integral and a quickly converging series for its inverse. *Biometrika* 37: 208–218.

Wishart, J. 1927. On the approximate quadrature of certain skew curves, with an account of the researches of Thomas Bayes. *Biometrika* 19: 1–38.

Wollaston, W. 1726. *The Religion of Nature Delineated.* London.

Woodhouse, R. 1803. *The Principles of Analytical Calculation.* Cambridge.

Yates, F. 1962. Sir Ronald Aylmer Fisher (1890–1962). *Revue de l'Institut International de Statistique* 30: 280–282.

Young, T. 1807. *A Course of Lectures on Natural Philosophy and the Mechanical Arts.* 2 vols. London: printed for Joseph Johnson.

Zabell, S.L. 1988. Buffon, Price, and Laplace: scientific attribution in the 18th century. *Archive for History of Exact Sciences* 39: 173–181.

——— 1989. The rule of succession. *Erkenntnis* 31: 283–321.

Index

Abilities, 146
Abramowitz, Milton, 585, 595
Absolute value, 221
Absurdity, 111, 133, 143
Abu'l-Wefa, 496
Academy:
    Attercliffe, 44
    Coward's, 536
    Dissenting, 41, 42
    Doddridge's, 45, 381
    Fund, 41
    Hoxton, 42
    John Jennings's, 53
    Market Harborough, 44
    Newington Green, 44
    Tenter Alley, 535
    Training at, 44
Act:
    Conventicle, 528
    Corporation, 528
    Five Mile, 9, 14, 67, 506, 528
    of Toleration, 67, 528
    of Uniformity, 8, 9, 14, 15, 36, 67, 528
    of Union, 48
    Schism, 67, 68
Actions, 132
    beneficent, 113, 114, 144, 148, 168
    benevolent, 116
    indifferent, 118
    morality of, 146
    reasonable, 113
Adams, George, 399, 579, 581, 595
Addison, Joseph, 26, 91, 188, 190, 534, 547, 595
Addition formulae, 431
Adeney, Walter F., 36
Æther, 395, 401–403, 579

Affection, bubonic, 551
Affections, 116, 117, 130, 133, 148, 552
Agent, moral, 127
Agricola, James, 35
Aikin, Arthur, 586, 595
Aikin, Charles R., 586, 595
Ainger, Alfred, 43, 595
Ainsworth, Robert, 382
Aitken, G.A., 590
Algebra:
    negative quantity in, 215
    rule in, 216
Alice, 301
Allen, Thomas, 423, 479, 595
Alsop, Vincent, 16
Alsted, Johann Heinrich, 54
Altmann, Alexander, 552, 595
Amber, 400
Amis, Kingsley, 526
Amoz, 141
Amsinck, Paul, 70, 544, 595
Anabaptists, 34, 35, 245
Anacoinosis, 377
Analysis, 561, 562
    non-standard, 223, 249
Analyst, The, 185, 219
Anarchism, 35
Anaxagoras, 407
Anderson, Robert M., 624
Andrews, Mordecai, 94, 548
Angels, 151
    faculties of, 144
    happiness of, 133
Angus, Joseph, 551, 552
Animals, beauty of, 155
Annesley, Samuel, 15, 509
Antinomianism, 23, 35
Antiquity, aspects of, 373

Antithesis, 377
Aphorism:
	Aristotle's, 259
	Boyle's, 42
	Pythagoras's, 577
Apostolius, Michael, 567
Apparatus, wonderful, 154
Appetites, 552
Arbuthnot, John, 28, 48, 81, 85, 300, 301, 526, 538, 596
Archer, Anne, 525, 594
Archer, John, 68, 70, 74, 544, 593, 596
	memorial to, 524
Archibald, Raymond C., 255, 596
Arminianism, 34
Arminians, 25, 33, 36
Arminius, James, 34
Armitage, Angus, 587, 596
Art, works of, 157
Artmann, Benno, 536, 596
Arts, liberal, 39, 41, 536
Ashurst, 53
Aspect, melancholy, 111, 143
Astragali, 546
Astronomical observations, 385
Astronomy, practical, 386
Atmospheres, 395–397
Augustinian Friars, 19

Bach, Craig, 6, 596
Bacon, Francis, 43, 381, 579, 596
Bailey, Brian, 33, 68, 94, 544, 548, 597
Bailey, Lorna, 33, 68, 94, 544, 548, 597
Baker, Thomas, 94
Baldwin's Gardens, 18, 532
Baldwin, Richard, 532
Balguy, John, 102, 551, 555, 597
Ball, Walter W. Rouse, 583, 597
Balls, billiard, 315
Baptism, 31, 34
	infant, 26, 34
Baptists, 64
Barbauld, Anna L., 77, 331, 546, 597
Barnard, George A., 335, 536, 597
Barrington, Sophia, *see* Bayes, Sophia

Barrow, Isaac, 193
Bartholomew, David J., 550, 597
Bartlett, Maurice S., 306, 597
Barton, Margaret, 77, 79, 82, 87, 544, 597
Bashford, Henry H., 544
Batson, Edmund, 18, 532
Baxter, Richard, 33, 576
Baxterians, 25, 33, 36
Bayes's *Essay*, 148
	definitions, 273
	errata, 297
	the Appendix, 289
	the postulate, 278, 312, 320
	the problem, 270, 273, 299
	the Rules, 286–288, 323, 330, 336, 343
		uses and defects of, 296
	the Scholium, 282, 318–320
Bayes's Rule, *see* Bayes's Theorem
Bayes's Theorem, 261, 264, 583, 569, 570
	inverse to Bernoulli's, 331
	reverse use of, 320, 329
Bayes, Alice, 8
Bayes, Anne I, 12, 30, 511, 592
Bayes, Anne II, 30, 32 *see* West, Ann(e)
Bayes, Bezalel, 527
Bayes, Elizabeth, 11, 12
Bayes, Frances, 8
Bayes, Hugh, 7
Bayes, John I, 8, 511
Bayes, John II, 12, 31, 511
	will of, 522
Bayes, John III, 30, 31
Bayes, Joshua I, 11
Bayes, Joshua II, 12, 15, 18, 19, 24, 25, 30–33, 46, 52, 69, 140, 382, 507, 509–511
	will of, 516
Bayes, Martha, 12
Bayes, Mary, 30, 32, 511, 517
	will of, 522
Bayes, Nathaniel, 30–32, 511, 517, 518
	will of, 522
Bayes, Rebecca I, 527

Bayes, Rebecca II, 527
Bayes, Rebecca III, 30, 32
    see Cotton, Rebecca
Bayes, Richard I, 7, 8
Bayes, Richard II, 8
Bayes, Robert, 527
Bayes, Rose I, 7
Bayes, Rose II, 7
Bayes, Ruth, 12
Bayes, Samuel I, 8–10
Bayes, Samuel II, 30, 31, 511, 513, 517, 518, 522
Bayes, Samuel III, 527
Bayes, Sarah I, 8
Bayes, Sarah II, 11, 529
Bayes, Sophia, 12, 517
Bayes, Theodosia, 511, 513
Bayes, Thomas, *presque partout*
Bayes–Laplace scheme, 267
Bayes/Cotton vault, 509, 510, 512, 515, 516
    inscriptions on, 512
Bayesian Wills, 516
Bayne, R., 550, 553
Beatty, F.M., 586, 597
Beatty, John, 612
Beauty, 110, 111, 119, 120, 122–126, 130, 134, 135, 137, 138, 143, 151, 154, 155, 157, 165, 166
Bebbington, Gillian, 597
Beckenbach, Edwin F., 455, 597
Beings:
    innocent, 168
    malicious, 118
    rational, 117
    sensible, 117, 168
    sentient, 177
    solitary, 117, 118
Belchier, John, 92
Bell, passing, 590
Bell, Walter G., 499, 503–505, 532, 589, 590, 597
Bellhouse, David R., 92, 382, 384, 412, 418, 535, 597
Bellman, Richard, 455, 597
Beneficence, 115, 180
Benevolence, 102, 115, 116, 118, 128, 137, 138, 148–150, 161, 164, 166, 173, 175–177, 180, 550, 552, 553
    divine, 145, 149
    human, 146
    instinctive, 150
    natural, 169, 170
    rational, 150
Benevolent, *see* Benevolence
Benignity, 123, 174
Bennett, James, 12, 39, 42, 43, 530, 531, 538, 600
Bennett, Joseph, 15, 531, 532
Bentham, Jeremy, 376
Berkeley, George, 185, 237, 238, 249, 557, 561, 562, 598
Berman, David, 598
Bernardo, Jose M., 599
Bernoulli functions, 584
Bernoulli numbers, 446, 447, 568, 584
Bernoulli trials, 313
Bernoulli's inversion problem, 332
Bernoulli's Theorem, 331
    inverse, 575
Bernoulli, Daniel, 305, 394, 599
Bernoulli, Jakob, 261, 300, 313, 331, 393, 572, 599
Bernoulli, Johann, 191
Bernoulli, Nicholas, 303, 393, 599
Bernoulli–Poisson Theorem, 265
Besant, Walter, 19, 532, 546, 591, 599
Beta probability integral, 589
Beta-function, 353
    incomplete, 253, 268, 321–323, 333, 368, 589
Beta-integral, incomplete, 261, 323, 589
Bethesda, pool of, 77
Beyer, William H., 589, 599
Bickerstaff, Isaac, 51, 52, 80, 90, 599
Bigots, 215, 245, 567
Bills of Mortality, 502, 503
Binet, Jacques P.M., 574
Bing, Frederick M., 316
Bingham, Nicholas H., 263, 561, 573, 599

Binomial:
    skew, 325
    symmetric, 325
Biographer, 6
Biography, 6, 37
Black, Max, 562, 599
Black, Susan E., 591, 599
Blair, David G., 322, 600
Blair, Hugh, 37, 94, 376, 378, 600
Blake, William, 509
Blanchard, Rae, 534, 543, 600
Blessings, communication of, 108, 140
Blinkard, 567
Boffin, Nicodemus, xi
Boghurst, William, 19, 502, 590, 600
Bogue, David, 12, 39, 42, 43, 530, 531, 538, 600
Bond, J.D., 496, 600
Boole, George, 322, 600
Booth, John, 538, 600
Booth, Mr, 94
Booth, Thomas, 12
Borel, Émile, 573, 600
Boscovich, Ruggiero, x, 392–394, 526, 600
Boudot, Maurice, 266, 574, 575, 601
Bounty, sagacious, 156
Bovingdon, 18
Bowen, Emanuel, 592, 601
Bowley, Arthur L., 322, 601
Box Lane, 18
Boyd, Percival, 601
Boyer, Carl B., 559, 601
Boyle, Robert, 381, 586, 601
Boys, singing, 522, 593
Bradshaw, Ebenezer, 15, 532
Brahe, Tycho, 578
Bramhall, John, 97
Bretherton, Francis F., 95, 601
Brewer, E. Cobham, 539, 601
Bridges, George, 29
Briggs, Henry, 454
Brighton, 60, 79
Britton, John, 63, 64, 76, 519, 601
Broad, Charlie D., 548, 601
Broadbent, Una, 624

Bromwich, Thomas J.I'A., 256, 431, 568, 601
Browne, Thomas, 187, 376, 382, 519, 601
Browne, William, 303, 602
Browning, Andrew, 10, 67, 68, 505, 528, 529, 543, 602
Bru, Bernard, 258
Bryant, Arthur W.M., 29, 542, 602
Buckingham, Catherinus P., 453, 602
Buckingham, Duke of, 29, 38, 96, 548
Buffon, George L.L., 303, 378, 461, 463, 577, 602
Bunce, an agèd man, 94
Bunhill Fields Burial Ground, 499, 591
    burials in, 510
    consecrated?, 501, 507
    monuments, 512
Bunyan, John, 376, 509
Buris, Mr, 18
Burnet, Gilbert, 67
Burns, Robert, 38
Burnside, William S., 257, 583, 585, 602
Burr, Thomas B., 64, 65, 76, 540, 602
Burton, David M., 457, 602
Burton, Robert, 51, 245, 378, 602
Butcher, Edmund, 46
Butler, Joseph, 550, 551, 553, 602
Butler, Samuel, 90
Buze, Samuel, 10
Byron, George G., 592

Cahour, Arsène, 603
Cajori, Florian, 240, 241, 249, 258, 439, 555, 557, 559–561, 563–565, 567, 568, 575, 583, 585, 603
Calamities, 108, 140
Calamy, Edmund I, 15, 24, 25, 30, 49–51, 60, 68, 382, 529–533, 538, 540, 544, 549, 603
Calamy, Edmund II, 47, 52, 537

Calculus, 562, 563
    differential, 185, 225, 250, 564
    fluxionary, 182, 184–186, 250, 561, 586
    infinitesimal, 560, 566
Calendar, 535
Calvin, John, 33
Calvinism, 34
Calvinists, 25, 33, 36
Canton, John, 255, 261, 336, 384, 386, 392, 412, 580
Cantor, Moritz B., 559, 560, 603
Career, chequer'd, 577
Carlisle, Nicholas, 529, 532, 603
Carlyle, Thomas, 38, 533, 603
Carnap, Rudolf, 263, 317, 603
Carp, Augustus, 77, 544
Carpenter, Anne, see Bayes, Anne I
Carpenter, Nathaniel, 47, 382, 537
Carstaires, see Carstares
Carstares, William, 48–52, 538
Carthusians, 464
Catechism, Westminster, 531
Cauchy, Augustin-Louis, 225
Cause, intelligent, 271
Causes, 338
    probability of, 265
    problem of, 570
    stable, 271
Cavallo, Tiberius, 406, 579, 581, 603
Cavendish, Henry, 555, 581
Celerities, 221
Celestial mechanics, §11.7 passim
Cemetery:
    common, 501
    private, 501, 507, 508, 591
Central Limit Theorem, 575
Cervantes Saavedra, Miguel de, 556
Chalmers, Thomas, 550
Chance, Decima, 514, 516
Chance, 387, 388
    as used by Bayes, 271
Chances, 388
    Doctrine of, 270
Chandler, Samuel, 535
Change, continuous, 221
Changes, synchronal, 200, 202, 205, 209
Chaos, 122, 123, 154

Chapel of King Charles the Martyr, 66
Chapman, Alice, see Bayes, Alice
Character:
    base, 160
    cruel, 114
    villainous, 129, 160
Cheese, xi
Cheops, pyramid of, §11.8.1 passim
Chocolate, 534
Chocolate-house, 29
Choice, 148, 155
Cholera epidemics, 508
Christ, 103, 106, 132, 163, 174, 374
Christians, 194, 220
Chuaqui, Rolando, 263, 604
Church Book, Independents', 94
Churchmen, political, 505
Cicero, Marcus Tullius, 373, 375, 377, 378, 381, 604
Clapham, 32
Clarendon, Earl of, 542
Claret, 538
Clark, Samuel, 527
Clarke, Frances M., 182, 526, 604
Clarke, Samuel, 121, 153, 193, 604
Clay, John W., 7, 8, 11, 527, 604
Clegg, James, 15
Clero, Jean P., 258, 312, 604
Clifford, J., 544, 604
Clones, method of, 574
Clow, William M., 35, 533
Cobbett, William, 79, 545, 604
Code:
    Clarendon, 528
    Laudian, 530
Coffee, 533
Coffee-house, 10, 26, 29
    Button's, 27, 28
    Child's, 26, 27
    Conversation, 26, 30
    Daniel's, 26
    Garraway's, 27
    Giles's, 534
    Hamlin's, 25, 26, 28
    Jonathan's, 27
    Marine, 184
    North's, 25, 26, 28, 33
    Slaughter's, 27, 534

St James's, 28
The Chapter, 26
The Grecian, 27, 28
White's, 28
Will's, 27, 28
Colbran, John, 74, 604
Coles, Elisha, 421, 422, 486, 604
Collett, 545
Collier, Theodosia, 31
Colligan, James Hay, 95, 531, 604
Collinson, Peter, 580
Colson, John, 556
Comma, 587
Condorcet, Marie J.A.N. Caritat, Marquis de, 326, 605
Conduct, divine, 110, 111, 114, 118, 119, 148, 169
Conductors:
   electrical, 404, 405
   lightning, 580
Confusion, 115, 119, 125, 151, 154
Congregational Fund Board, 42, 46, 47
Congregationalists,
   see Independents
Consonances:
   imperfect, 586
   perfect, 586
Constitution, 155
Continued fractions, 458
Conventicle, 543
Copper, Mathias, 70, 83, 544
Cornish, John, 18, 25
Cost:
   of coffee, 10, 26
   of snuff, 10
   of tobacco, 10
Cotes, Roger, 423, 452, 468, 498, 561, 605
Cotton, Bayes, 522
Cotton, Benjamin, 524
Cotton, Decima, see Chance, Decima
Cotton, Joshua, 511, 517, 518
Cotton, Rebecca, 511, 517
   see Bayes, Rebecca III
Cotton, Samuel, 524
Cotton, Thomas, 32, 511, 517
   will of, 524

Cotton, Thomas Bayes, 516
Cotton, William, 524
Cournot, Antoine A., 322, 605
Covenanters, Scottish, 36
Cowper, William, 576
Coxeter, Harold S.M., 583, 597
Craufurd, Thomas, 538
Creasy, John, 530, 605
Creation, 112, 121, 125, 135, 142, 151, 156, 165, 180, 553
Creatures:
   intelligent, 114, 137, 144
   rational, 114, 127, 137, 145
   welfare of, 143
Creed, Athanasian, 83, 84
Crimes, 158
Crisp, Tobias, 23
Cromwell, Thomas, 32
Cudworth, Ralph, 260
Cumberland, Richard, 304, 468, 572, 605
Currie, Jo, 575
Curtis, William A., 531
Cutlers, Company of, 8, 11, 527

Dale, Andrew I., 47, 265, 298, 307, 313, 320, 326, 329, 331, 444, 526, 552, 569, 574, 575, 605
Dale, Robert W., 10, 42, 46, 528, 530, 533, 606
Dalzel, Andrew, 52, 606
Dame Agnes the Cleere's Well, 499
Darwin, Charles, 580, 593
Darwin, Erasmus, 580
Darwin, Francis, 526, 606
Daston, Lorraine J., 306, 571, 606, 612, 620
Daventry, 14
David, Florence N., 546, 572, 573, 606
Davidson, Francis, 551, 553, 606
Davies, Charles M., 569, 606
Davis, Moll, 58
Dawson, John W., Jr, 6, 606
de Coste, Hilarion, 606
de Finetti, Bruno, 262, 266, 306, 309, 317, 570, 571, 573, 606

de la Rose, Elizabeth, 517, 530
de la Rose, John, 12, 44, 530
de Moivre's rules, 330
de Moivre, Abraham, 27, 253, 255, 270, 296, 299, 303–305, 324, 325, 368, 393, 418, 428, 534, 568, 572, 575, 606
de Moivre–Laplace limit theorem, 332
de Morgan, Augustus, 219, 223, 334, 453, 556–558, 571, 586, 587, 607
de Quincy, Thomas, 567
Decision, 264
Decrements, 239
    synchronal, 237
Defoe, Daniel, 44, 502, 503, 505, 507, 509, 543, 590, 607
Deformity, 119, 125, 151
Degrees of belief, 310
Degroot, Morris H., 599
Deity, existence of, 271
Demerit, 127, 130, 159, 161
Deming, William Edwards, 255, 258, 574, 575, 589, 607, 625
Demosthenes, 377
Dempster, Arthur P., 333, 575, 607
Densham, 42
Density function, beta, 321
Dependence:
    Bayes's definition, 307
    de Moivre's definition, 308
    Emerson's definition, 309
Derham, William, 42, 89, 101, 607
Derrick, Samuel, 545
Desirability, 263
Dickens, Charles, 77, 95, 370, 508, 526, 528, 591
Difference:
    absolute, 390, 563
    assignable, 210
Differential method, 451, 453, §11.5.6 *passim*
Digby, Kenelm, 41, 608
Dingle, Robert B., 257, 608
Diogenes Laertius, 382, 608
Diophantine equations, 457, 467
Disbelief, Mill and, 547

Discipline, mental, 187
Disobedient, the, 130, 161
Disposition, *see* Propension
    benevolent, 115, 116, 148
    good, 116, 130–132, 161, 170
    inward, 132
    virtuous, 133
Dissenters, 36, 42, 44, 68
Dissenting:
    academies, 12
    schools, 12
Dissimilitude, 156
Diston, Thomas, 11
Distribution:
    χ^2, 368
    beta, 319, 320, 325, 332, 353, 589
    beta posterior, 362
    binomial, 268, 277, 310
    continuous, 393
    incomplete negative binomial, 323
    Normal, 325
        expansion of, 337, 362
    posterior, 316, 322, 575
    prior, 329
        swamping effect of, 322
    uniform, 319, 320
Ditton, Humphrey, 66, 68, 74, 83, 84, 101, 219, 543, 546, 561, 608
Doctrine, fluxionary, 190
Doddridge, Philip, 44, 53, 171, 174, 181, 527, 530, 537, 553, 555, 608
Dodson, James, 253, 608
Dowding, 84, 87, 546
Drummond, Colin, 53, 99
Dryden, John, 96, 548
Dudley, Lord North, 58
Dutka, Jacques, 321, 368, 574, 575, 585, 589, 608
Dux Bucks, *see* Buckingham, Duke of

Eaglesham, Mr, 544
Eames, John, 41–43, 253, 509, 536, 537, 608

Edgeworth, Francis Y., 89, 145, 302, 322, 548, 552, 572, 574, 609
Edinburgh University, 46, 537
   college records, 99
   library rules, 55
   Regents, 48, 383
   theological faculty, 57
Edinburgh, College of, 375
Education:
   artificial, 40
   for the ministry, 44
   ingenuous, 40
   liberal, 39–41, 43
   mathematical, 248
   natural, 40
Edwards, Anthony W.F., 262, 298, 305, 317, 318, 331, 574, 583, 589, 609
Edwards, Ward, 574, 609
Eeles, Henry, 400
Effects, 338
Effluvia, unctuous, 401, 579
Effluvium, equine, 19
Eisenhart, Churchill, 578, 609
Eizat, Edward, 54, 316, 328, 540, 609
Elasticity, 395
Election, 26
Electric, (n) 400, 401
   virtue, 579
Electricity, 399, §11.6.1 *passim*, 579
   negative, 400
   notes on, 384
   positive, 400
Electrics, 579
Electrification, plus and minus, 404
Ellicott, John, 65, 518
Elwig, H., 83, 544, 610
Emerson, William, 222, 309, 610
Energy, 554
England, Church of, 101
Enon Chapel, 508
Enthusiast, 215, 245, 567
Epigram, 405
Episcopalians, 36
Epitrope, 377

Equations, solution of, §11.5.4 *passim*
Equitable Life Assurance Society, 259, 420
Equity, divine, 133
Erasmus, Desiderius, 381
Error, chaff of, 6
Errors:
   distribution, x, 393
   independent, 393
   observational, 384–386
   symmetric distribution of, 392
   systematic, 578
   theory of, 6, 394
Esteem, 127, 132, 159, 175
Ether, *see* Æther
Ethics, 550
Euclid's algorithm, 457
Euclidean:
   demonstrations, 315
   propositions, 299
Euler constant, 257
Euler, Leonhard, 255, 256, 399, 402, 450, 459, 579, 584, 610
Euler–MacLaurin sum formula, 444–446, 584
Evanescent, 224, 238, 240, 250
Evans, George E., 610
Evans, John, 69
Events:
   future, 271, 325, 329
   independent, 276
   subsequent, 274, 275, 312
   timing of, 311
Evil, 135, 170
   judicial, 158
   penal, 158
Exchangeability, 309, 574
Expectation, 300–302, 304, 310, 387
   additivity of, 306
   of benefit, 306, 314
   of events, 305
   of sums, 305
   value of, 301, 304
Experience:
   learning from, 574
   uniform, 292
Experiment, two-stage, 314, 574
Experiments, repeated, 270

Faces, human, 126
Fact:
    future, 271
    past, 271
Factorial, descending, 354
Fairfield, Sheila, 611
Faith, 194
Faithful, the, 130
Falconer, William, 633
Fanatics, 506
Fancourt, Samuel, 102
Fantoccini, 96, 548
Farebrother, Richard W., 526, 546, 611
Farrar, Frederic W., 611
Farthing, Roger, 58, 64, 65, 69, 70, 74, 85, 519, 520, 542, 544–546, 611
Fatality, 173
Favour, 127, 131–133, 159, 161
Felicity, 114, 117, 119, 129, 133, 134, 151, 161
Feller, William, 575, 611
Fellow-commoner, 528
Ferguson, David, 548
Fergusson, *see* Ferguson
Fermat, Pierre de, 259
Fetzer, James H., 267, 611
Fiducial approach, 320
Fielding, Henry, 529
Fine, Terrence L., 309, 310, 611
Finney, William, 518
Finsbury Estate, 499
Fire:
    chemical, 401
    electrical, 398, 404
    elementary, 401
Fisher, Ronald A., 258, 263, 266, 307, 313, 317, 320, 569–571, 611
Fitness, 143
    and happiness, 112
    moral, 106, 107, 109, 115, 140, 159, 554
    natural, 554
    of things, 109, 110, 112, 143
FitzGerald, Edward, 181
Flowing quantity, *see* Fluent
Fluent(s), 197, 209, 213, 221, 565

Fluid:
    ætherial, 395
    electric, 400, 401, 579
    electrical, 395–397, 400, 406, 409
    etherial, 401, 402, 406, 579
Fluxions, 197, 221, 560, 564, 566
    as velocities, 209, 213
    decreasing, 203
    doctrine of, 193, 195, 201
    increasing, 202
    method of, 193–196
    mystery of, 194, 219
    notion of, 196
    permanent, 200, 202
    principles of, 195
    proposed treatise on, 209, 237
    second, etc., 201, 202, 214
    theory of, 182
Folkes, Martin, 459
Forbush, William B., 528
Force, James E., 84, 85, 546, 611
Foster, James, 32, 185, 535, 611
Frankland, Richard, 12, 530
Franklin, Benjamin, 400, 404, 405, 411, 580, 581, 586, 611
Frant, 58, 63
Fraser, Alexander C., 249, 557–559, 612
Fraser, James B., 88, 612
Frederick the Great, 567
Freedman, David, 571, 612
Freedom, 148
Freethinker, 185
Free-will, 26
Frequency, in Bayes's definitions, 313
Freudenthal, Hans, 300, 301, 572, 612
Friend, ingenious, 331, 428
Friends, Society of, 74
Fur, 85
Fyfe, Aileen, 551, 612

Galton, Francis, 258
Gambler's Ruin problem, 305
Gambles, equivalent, 302
Garforth, Francis W., 40
Garth, Samuel, 188, 190

654   Index

Gaskoin, Charles J.B., 36
Gaspey, William, 60, 63, 64, 78, 612
Geisser, Seymour, 320, 612
Generating functions, 393
Geometry, §11.5.3 *passim*
George, Earl of Macclesfield, 386
Giant Despair, 239
Gibbon, Edward, x, 376
Gibson, George A., 186, 187, 190, 225, 241, 556, 557, 559, 560, 564, 566, 612
Gigerenzer, Gerd, 574, 612
Gilbert, William, 579, 612
Gillies, Donald A., 550
Gillispie, Charles C., 612
Gimcrack, Nicholas, 90
Gloucester, Duke of, 60
God, 106
    actions of, 138
    acts of, 112, 143
    attributes of, 102, 103, 105, 107–109, 111, 129, 138, 140, 141, 164, 180, 550, 551
    benevolence of, 112–117, 143, 172
    ends of, 110–112, 119, 126, 137, 142
    happiness of, 113
    perfections of, 105, 107, 109–111, 115, 116, 120, 135, 140, 141, 150, 152, 156
    promise of, 132, 163
    wisdom of, 113
    works of, 107, 110–112, 114, 119, 124, 142, 154
Godolphin, Francis R.B., 555, 612
Gödel, Kurt, 6
Goldsmith, Oliver, 529, 576
Goldstein, Joel A., 530, 613
Goldstine, Herman H., 613
Good, Irving J., ix, 261, 262, 320, 571, 613
Good, natural, 152
Goodman, Nelson, 266
Goodness, 106–108, 113–115, 119, 127–140, 145, 149, 156, 158, 160, 162, 164, 167
    divine, 121
    infinite, 114, 145, 167
    moral, 115, 140, 148, 152
Goodrich, Mr, 527
Gordon, Alexander G., 12, 14, 44, 53, 67, 613
Gordon, Thomas, 89, 613
Gospel, 131
Gough, 70, 96
Gould, H.W., 584, 613
Government, 151, 158
    divine, 128, 169
    ends & intentions of, 126
    moral, 159
Grammont, Comte Philibert de, 60, 542, 548
Grant, Alexander, 39, 49, 50, 57, 375, 537, 538, 577, 613
Grant, James, 48, 538, 613
Gratitude, 118
Graunt, John, 590
Gravitation, 407, 409, 582
Gravity, 395, 408, 582
Gray's Inn Lane, 22
Greaves, John, 475, 480–483, 485, 588, 613
    on the pyramid of Cheops, §11.8.1 *passim*
    on weights & measures, §11.8.2 *passim*
Greek, 39, 381
    and Latin authors, 373
Green, Charles, 633
Green, Lawrence G., 590, 613
Gregory of St Vincent, 556
Gregory the Great, 41
Gregory, James, 46, 53, 219
Grendon St Mary, 9
Gresham College, 370
Grove, Henry, 102, 167, 180, 553–555, 614
Guicciardini, Niccolò, 557, 561, 586, 614
Guide, blind, 217
Gulliver, Lemuel, 547
Gwyn, Nell, 58, 548

Hacking, Ian M., 302, 320, 536, 574, 614

Hadley, James, 504, 591
Hailperin, Theodore, 268, 573, 614
Haines, Joseph, 73
Hald, Anders, 266, 298, 313–316, 318, 324, 325, 328, 353, 358, 362, 364, 366, 367, 369, 569, 571, 574–576, 578, 589, 614
Hall:
    Founders', 96
    Pinners', 23
    Salters', 23, 25, 537
Hall, A. Rupert, 557, 566, 614
Hall, Henry S., 583, 614
Halley, Edmond, 189, 559, 588
Hamilton, Anthony, 60, 61, 100, 542, 615
Hamilton, William, 99
Happiness, 108, 110–115, 117–119, 122, 124, 126, 131–134, 137, 141, 142, 144, 149–151, 159, 162, 163, 166, 168, 172, 176, 553
    communication of, 173
    greatest, 180
    perfect, 161
    production of, 165
Harben, Henry A., 533, 615
Hardy, Godfrey H., 255, 567, 584, 586, 615
Harmony, 111, 125, 134, 143, 165, §11.6.4 *passim*
Harris, John, 183, 184, 240, 565, 566, 615
Hartley, David, 76, 175, 179, 331, 428, 550, 554, 588, 615
Hastings, James, 67, 528, 531, 533, 543, 544, 615
Hatred, 108
Hauksbee, Francis I, 28, 579
Hauksbee, Francis II, 28
Hawkins, Sir John, 471
Hazlitt, William, 29, 180, 376, 577, 615
Hearne, Thomas, 503
Heat, 123
Heathen, impure doctrines of, 374
Hecuba, 526
Heidelberger, Michael, 620

Henry, Matthew, 25, 533
Heresy, 36
Heretics, burning of, 540
Herring, red, 261
Hezekiah, 109, 141, 551
Hicks, H., 499, 577, 615
Hill, George B., 616
Hill, Joseph, 15, 532
Hoadly, Benjamin, 84, 384, 404, 411, 616
Hodge, Nathaniel, 505
Hodgson, James, 453, 616
Hogben, Lancelot, 261, 267, 299, 300, 315, 316, 574, 616
Holgate, Philip, 561, 573, 599
Holiness, 553
Holland, John D., ix, 421, 530, 536, 553, 578, 583, 616
Holland, particularities of, 88
Holmes, Mrs Basil, 499, 501, 508, 591, 593, 616
Holmes, Sherlock, 569
Holt, 403
Home, John, 538
Home, R.W., 403, 409, 411, 578, 616
Hood, Thomas, 504
Hook, *see* Hooke
Hooke, Robert, 407
Hooper, George, 482, 616
Horace (Quintus Horatius Flaccus), 527, 544, 555, 594
Horn, David B., 536, 537, 616
Horne, Herman H., 40, 536, 616
Horsley, John, 46, 47, 52, 56, 99, 537
Horsley, Samuel, 542
Horwich, Paul, 266, 617
Howes, Francis, 555
Howsley, Frances, *see* Bayes, Frances
Howson, Colin, 265, 617
Hoyle, John, 421, 518, 522
Hubbell, Harry M., 577, 604
Huber, Peter J., 617
Hugenius, 485
Hughes, Dr, 376
Hughes, Peg, 58
Hughes, William, 46
Humanitarians, 509

Humanity studies, 374, 382
Hume, David, 82, 88, 266, 267, 571, 617
Humpty Dumpty, 301
Hungerford, Margaret, 557
Hunter, Adam M., 543
Huntingdon, Lady, 76
Hurd, Richard, 376
Hurst, James, 7
Hutcheson, Francis, 102, 123, 146, 154, 155, 180, 388, 554, 617
Huxley, Thomas, 39, 490, 536, 617
Huygens, Christiaan, 300, 302–304, 617
Hypobole, 377
Hypothesis:
  shifting of the, 565
  simplest, 550
Hypothetico-deductive method, 266, 571

Ignorance, Bayes's criterion of, 315
Ill-nature, 108
Inclinations, 132, 170, 172
Increments, 239
  synchronal, 237
Independence:
  and the Polish school, 573
  Bayes's definition, 307
  comparative, 309
  de Moivre's definition, 308
  Emerson's definition, 309
  statistical, 310
  stochastic, 309, 573
Independents, 23, 73, 94, 95
Indians, and Tunbridge Wells, 547
Indifference, divine, 115
Induction, 265, 571
  insufficient, 338
Ineichen, Robert, 546, 573, 617
Inference, 264
  Bayesian, 266, 569
  causal, 571
  scientific, 263, 569
  statistical, 263, 317, 569
Infidelity, 23
Infinitesimals, 215, 564, 566
  theory of, 183

Ingen-housz, Jan, 405
Innocence, 168
Instinct, perverse, 156
Instrument:
  goodness of, 390
  imperfect, 386, 392
Integral limit theorem, 324
Integration:
  Daniell, 307
  Lebesgue, 307
Integrity, 131
Intelligence, mere, 113, 144, 168
Interest, 156
Invariance principle, 316
Inverse Principle of Maximum Likelihood, 264
Inverse Principle of Maximum Probability, 263
Inversion theorems:
  Bayes's, 264
  Bernoulli's, 331, 333
Isaiah, 141, 551
Islington, Wells at, 549
Isocrates, 377

Jacob, a Jew, 26
James, Thomas S., 69, 617
Jarret, Edward, 94
Jaynes, Edwin T., 332, 618
Jeffery, John, 520
Jeffery, Richard I, 520
Jeffery, Richard II, 518
Jeffery, Richard III, 518, 519
Jeffery, Robert, 520
Jeffery, Sarah, 518, 519
Jeffrey, John, 65, 518
Jeffrey, Richard C., 263, 618
Jeffrey, Sarah, 65, 518, 519
Jeffreys, Harold, ix, 263, 306, 313, 550, 569–571, 573, 574, 618
Jenkins, the Rev. Mr, 94
Jenner, Robert, 94
Jeremy, Walter D., 42, 46–48, 260, 530, 591, 618
Jerome, Jerome K., 589
Jesseph, Douglas M., 188–190, 219, 222, 249, 557–559, 561–563, 565, 567, 618

Jessop, Thomas E., 247, 558, 567, 622
Jesuits, 39
Jesus, 35
Johns, Orrick, 6, 96, 618
Johnson, Ben, 375
Johnson, Dr, 87
Johnson, James, 87
Johnson, Mrs, 85
Johnson, Norman L., 258, 620
Johnson, Samuel, 44, 85, 376, 546
Johnson, Thomas, 553
Johnson, William, 70, 73, 74, 95, 259
  memorial to, 524
Johnston, George A., 558, 618
Johnston, Swift P., 558, 618
Jolie, Timothy, 44
Jones, John A., 516, 548, 591, 592, 618
Jones, Richard F., 380, 577, 618
Jones, William B., 459, 618
Jordan, 64
Jordan, Charles, 575, 589, 618
Jordan, Mr, 69
Jourdan's Lane, 65
Judgments, 108, 140
Jurin, James, 188, 240, 241, 244, 559, 560, 619
Justice, 106–109, 111, 133, 140, 142, 145, 161, 163, 180

Kadane, Joseph B., 638
Kamlah, Andreas, 306, 573, 619
Kassler, Jamie C., 468, 471, 587, 619
Katabaptists, *see* Anabaptists
Keckermann, Bartholomaeus, 54, 619
Keill, John, 566
Kemp, John, 382
Kendall, Maurice G., 263, 569, 578, 619, 629
Kentish, Thomas, 16
Kepler's laws, 475
Kepler's problem, 432
Kepler, Johannes, 432, 587
Keynes, John M., 315, 316, 575, 619
Khayyám, Omar, 181
Kiesel, Rüdiger, 263, 599
Killigrew, Harry, 29

Kilpatrick, Thomas B., 102, 179
Kindness, 131
Kine, Pharoah's, 73
King, Mrs, 527
King, William, 15, 531, 532
King, William, D.D., 102
King:
  Charles I, 58
  Charles II, 8, 28, 29, 58, 96, 544
  Edward III, 589
  Edward VI, 33
  George I, 36
  George II, 85, 530
  George III, 399, 405, 581
  George IV, 60
  James II, 58, 542
  James VI, 537
  of Siam, 88, 547
  William II, 36
  William III, 24, 48, 49, 530
Kirby Hall, 532
Kitchen, Thomas, 592, 601
Kitcher, Philip, 182, 619
Kjeldsen, Tinne H., 459, 619
Kline, Morris, 434, 619
Knight, Samuel R., 583, 614
Knipe, Henry R., 78, 546, 619
Knopp, Konrad, 459, 567, 620
Knowledge, perfect, 167
Knox, John, 57
Köbel, Jacob, 394, 620
Körner, Stephan, 562, 620
Kolmogoroff, Andrei N., 309, 620
Kolmogorov, Andrei N., 573
Kotz, Samuel, 258, 620
Krüger, Lorenz, 612, 620
Kruskal, William H., 620
Kyburg, Henry E., Jr, 262, 263, 317, 322, 570, 620

L'Hospital, Guillaume-F.-A. de, 622
La Bruyère, Jean de, 259
Laboucheix, Henri, 260, 522, 620
Lafaye, Georges, 383
Lagrange, Joseph Louis de, 393
Lamb, Charles, 43, 184, 376, 620
Lancaster, Henry O., 621
Landa, Louis, 590
Landau, Edmund, 456, 621

Landen, John, 253, 621
Language, Jewish, 153
Laplace, Pierre-Simon, 261, 313, 315, 317, 321, 325, 332–334, 393, 526, 571, 574, 575, 621
Large numbers:
    first law of, 265
    inverse law of, 264
    second law of, 264
Latham, Robert C., 26, 502, 507, 529, 590, 621
Latin, 39, 53, 381
    language, purity of, 372
Law(s):
    divine, 108, 141
    fixed, 271
    general, 114
    simplest, 550
    statistical, 265
Le Monnier, Louis G., 459, 461, 462, 464, 586, 621
Leader, John D., 8, 621
Learning from experience, 317
Leather Lane, 19, 23, 24
Leather Lane Chapel, 18, 19, 69, 507, 522
Lee, Edwin, 63, 622
Leeman, Anton D., 381, 622
Legendre, Adrien-Marie, 567, 568, 585, 622
Leges aleariae, 546
Leibniz, Gottfried Wilhelm, 183, 225, 245, 562, 566
Leibniz:
    differential method of, 213
    method of, 245
    method of differences, 215
Lemma, Berkeley's, 240
Leslie, Charles, 540
Lettres Provinciales, 487
Levine, Raphael D., 622
Lewis, Charlton T., 375, 622
Lewis, Martin, 33
Lewis, Samuel, 9, 17, 528, 529, 622
Lidstone, G.J., 258, 589, 622
Light, Alfred W., 591, 622
Lillywhite, B., 28, 622

Limits, 249
    and ultimate ratios, 564, 565
    doctrine of, 560
    theory of, 190, 223, 231, 245, 563
Limma, 587
Lindley, Dennis V., 307, 599
Lindman, Harold, 574, 609
Linnell, Charles L.S., 63, 66, 82, 87, 546, 622
Lipton, Dan, 593
Little Kirby-Street, 19, 84
Little Saint Helens, 15
Locke, John, 40, 88, 622
Lockie, John, 622
Logan, *see* Loggan
Logarithm, 454
    of $(2z-1)!!$, 254
    of $\sqrt{2\pi}$, 254
    of $z!$, 254
Loggan, Thomas, 85
Logic, 373, 381
    rules of, 215, 217
Longitude:
    determination of, 83
    problem, 546
Lorenz, Ludwig V., 316
Lotteries, 293, 294, 329
    conditional, 307
    fair, 302
Louise Marie de Gonzaga, Queen of Poland, 489
Love, 118
Lucas, John R., 264, 622
Luce, Arthur A., 247, 558, 567, 622
Luther, Martin, 35
Luxuriance, exhibition of, 400
Lynd, Robert, 526, 623

Macaulay, Aulay, 623
Macaulay, Thomas B., 376, 576
McClintock, Bruce, 572, 623
M'Cormick, Joseph, 49, 538, 623
M'Glothlin, William J., 34, 544
Machin, John, 432, 623
Mackie, Charles, 99
McKillop, Alan D., 623
MacKinnon, Donald H.O.D., 31, 53, 81, 540, 590, 623

McLachlan, Herbert, 42, 623
MacLaurin, Colin, 187, 418, 444, 446, 453, 559, 560, 584, 623
Maddox, Isaac, 47, 52, 55, 56, 99, 537, 542
Madox, *see* Maddox
Magnitudes:
    evanescent, 183
    nascent, 183
Mahon, Charles, Viscount, 581, 623
Maistrov, Leonid E., 578, 623
Maitland, William, 499, 623
Malevolence, 176
Malice, 108
Mallet, Charles E., 530, 623
Malone, Edmond, 85, 546, 547, 624
Malone, sweet Molly, 60
Manuel, Frank E., 37, 535, 624
Marryat, Howard, 624
Marsaglia, George, 568, 624
Marsaglia, John C.W., 568, 624
Martin, Benjamin, 579, 624
Marvell, Andrew, 97, 98, 549, 624
Mason, J.I., 421
Mason, John, 537
Mathematician:
    business of, 196, 249
    infidel, 187, 188, 190, 193
Mathematicians:
    disputes among, 217
    Greek, 307
    Islamic, 307
Mathematics, 373, 381
    and scepticism, 44
    most general use of, 218
Mathews, 64
Matter:
    electric, 400, 401
    insensible, 138, 167
Matthews, Arnold G., 10, 527, 528, 624
Matthews, William, 26, 502, 507, 529, 590, 621
Mauldon, Jean, 65, 548
Maxim:
    absurd, 216
    Buffon's, 378
    in arithmetic, 216
    shocking, 216
Maxwell, Grover, 322, 624
Maxwell, James Clerk, 555, 624
May, Mr, 33
Meader, James, 633
Mean:
    arithmetic, 394
    golden, 33
    of several observations, 385–387, 389, 390
Mearns, Hughes, 248
Measures, 481, 482
    English, 482
    French, 482
Medawar, Peter B., 571, 624
Medium, subtile, 401
Meller, Hugh, 499, 501, 589, 591, 624
Melvil, Thomas, 102
Melville, Lewis, 545, 624
Mendelssohn, Moses, 552
Mercer, Robert, 31
Merchant Taylors' School, 43
Merchant, English Turkey, 26
Merchants' Lecture, 23, 24
Mercy, 111, 129, 142, 174
Merit, 127, 130, 131, 133, 148, 158, 161, 162
    divine, 116
    moral, 115, 130–132, 161
Merry dancers, 582
Mersennus, 485
Mess-Johns, 51, 539
Method of false position, 459
Methodism, 96
    Arminian, 76
    Calvinistic, 76
    Primitive, 74
Methods:
    Bayesian, 569
    unreliable, 254
Middleton, Conyers, 559
Mikado, The, 552
Mill, John S., 322, 547, 625
Mills, Stella, 584, 625
Minion, xi
Mirowski, Philip, 574, 625
Mischief, 110
Misery, 108, 113–115, 118, 132, 150, 168, 170, 176

Mitton, Geraldine E., 19, 532, 546, 599
Modesty, 259, 260, 555, 569
Molina, Edward C., 258, 368, 574, 575, 589, 625
Moment generating function, 320
Moments, 183, 184, 213, 238, 243, 566
  via proportions, 244
Monro, Cecil J., 333, 625
Montagu, Elizabeth, 250, 546
Montmort, Pierre Remond de, 575, 625
Moore, David S., 536, 625
Moral evil, 146
Moral good, 146
Morality, 118, 150, 550
Morden, John, 593
Morgan, Alexander, 99
Morgan, William, 259, 550, 625
Morris, Carl N., 626
Morris, Christopher, 502, 507
Mortality, Bills of, 590
Moses, law of, 35
Mosteller, Frederick, 575, 576, 626
Motion, muscular, 407
Motte, Andrew, 565, 567
Mottelay, Paul F., 580, 626
Mount Ephraim, 58, 63, 64, 76
Mount Sion, 58, 63, 64, 76
Mount Sion Chapel, 68, 73, 74, 94–96, 544
Murdoch, Patrick, 413, 415
Murphy, C.W., 593
Murray, Forest H., 320, 626
Murrell, Hugh, 583
Muschenbroeck, see Musschenbroek
Musschenbroek, Pieter van, 462
Mysteries, profane, 374
Mythology, 374

Nagel, Ernest, 569, 626
Nairne, Edward, 581
Nascent, 224, 238, 240, 244, 251
Nash, Richard, 78, 79, 85, 545
National debt, estimate of, 486
Nature, ignorance of, 292
Neat, John, 593
Necessity, 155
  moral, 156
Needham, Turbervill, 459, 586, 626
Nero, 546
Nevill, Ralph, 626
Newbold, Sarah, see Bayes, Sarah I
Newell, D.J., 268, 626
Newington Green, 15
Newman, Henry, 27
Newton's method, 437, 583
Newton, Isaac, 43, 182, 183, 191, 193, 195, 197–199, 201, 205, 209, 210, 213, 223, 225, 237, 239, 250, 259, 268, 395, 401, 404, 407, 408, 474, 476, 479, 482, 555, 557, 559, 562, 564, 582, 583, 626
  on cubits, §11.8.3 passim
Nightingale, Benjamin, 627
Nightingale, Florence, 268
Nihilarians, 185
Nisard, Jean-Marie-Napoléon-Désiré, 378, 577, 627
Nollet, Jean Antoine, 464, 586
Non-abjurors, 36
Non-electrics, 579
Nonconformists, 36, 48, 67, 76, 509, 590
Nonconformity, 23, 28, 36
  English, 505
Nonjurors, 26, 36, 534
Normal probability integral, 366, 368, 369
Notation, Stanhope's, 451, 459
Numbers, §11.5.7 passim

O'Donnell, Terence, 627
O'Hagan, Anthony, 329, 569, 575, 628
Obedience, 118, 131
Observations, calculus of, 526
Occam's razor, 102, 550
Odds, 387, 388
Ogborn, Maurice E., 420, 421, 535, 536, 627
Okill, 64
Oldfield, Joshua, 50
Oldham, John, 51, 539, 540, 628
Oliver, Elizabeth Howell, 509

Onely, Richard, 53, 60, 70, 81, 154, 381, 540, 628
Optics, §11.6.3 *passim*
Order, 110, 111, 115, 119, 120, 122, 124–126, 134, 135, 137, 143, 151, 154, 157, 165, 172
    destruction of, 144
Ordination, lay, 26
Organs, imperfect, 386
Orton, Job, 44, 628
Oughtred, William, 563
Ounce, Roman, 464
Oxymoron, 377

Page, Norman, 547, 628
Pain, 174
Paley, William, 101, 143, 145, 550, 551, 628
Palmer, Samuel, 25
Palter, Robert, 628
Panton, Arthur W., 583, 602
Parabole, 377
Paradox:
    amazing, 398
    grue, 266, 571
Parasites, xi
Parker, Charles S., 39, 41, 44, 46, 53, 531, 536, 540, 577, 628
Parker, Irene, 14, 530, 628
Parker, Samuel, 22, 97, 628
Paroemia, Scottish, 89
Parramour, Richard, 8
Partial exchangeability, 309
Partington, Charles F., 586, 628
Partridges, 577
Pascal, Blaise, 306, 310, 487, 629
Passions, 113, 552
Passler, David L., 527, 629
Patience, 111, 142
Patin, J.M., 568, 629
Patrick, Simon, 121, 153, 552
Payne, Joseph F., 502, 629
Pearce, L., 544
Pearson, Egon S., 629
Pearson, Karl, 101, 322, 368, 490, 529, 536, 555, 629
Pearson, Sarah, *see* Bayes, Sarah II

Pemberton, Henry, 32, 535, 556, 560, 629
Pensioner, 528
Pepys, Samuel, 29, 502–504, 506, 507, 583
Perfections, 122, 124, 154
    moral, 106, 107, 111, 158, 177
    natural, 116
    objective, 122, 124
Périer, Marguerite, 488, 489
Perkins, Joseph, 95, 548, 630
Perpole Lane, 22
Perrault, 407
Philosophers:
    experimental, 381
    mechanical, 381
Philosophy:
    desideratum in, 338
    experimental, 269, 271
    moral, 550
    natural, 101, §11.6 *passim*
Phippen, James, 73, 87, 630
Pike, Godfrey H., 23, 630
Pitcairn, *see* Pitcairne
Pitcairne, Archibald, 54, 540, 571
Plackett, Robin L., 578, 619, 630
Plague, 499, 589
    and fleas, 507
    and Nonconformist clergy, 505
    and the Royal Society, 505
    and toads, 506
    and tobacco, 503
    and vinegar, 503
    nurses, 505
    the Great, 19
Planet, 155
Platt, Frederic, 34
Playfair, John, 497, 630
Pleasure, 552
    occasion of, 156
Pleasures, Cyprian, 19
Pliny (Gaius Plinius Secundus), 400
Poincaré, Henri, 391, 568, 630
Poisson, Siméon-Denis, 333, 334, 571, 630
Pólya–Eggenberger urn model, 267
Polydore Virgil, 245

Pope, Alexander, 32, 90, 183, 184, 630
Pope, Michael, 46, 522
Popper, Karl R., 262, 550, 630
Popularity in a preacher, 94
Pores, 401, 402
Porson, Richard, 375
Porter, Theodore, 612
Powlet, William, 19
Predestination, 35, 153
Presbyterian Fund, 46, 47
Presbyterianism, 95
Presbyterians, 23, 25, 33, 64, 66, 73
    disparaging remarks, 591
Prescription, 490
Press, S. James, 630
Previsions, 306, 307, 573
Price, Richard, 101, 140, 179, 255, 259, 260, 266, 298, 324, 325, 328, 329, 336, 403, 421, 509, 514, 518, 535, 550, 553, 591, 630
Price, Samuel, 591
Price, Sarah, 509
Priestley, Joseph, 403, 580, 586, 630
Princes:
    Indian, 88
    Persian, 88
Princesses:
    Anne, 58, 60
    German, 399
    Mary, 58
Principle:
    of indifference, 315–317, 574
    of insufficient reason, *see* principle of indifference
    of non-sufficient reason, *see* principle of indifference
Pringle, John, 405, 580
Priscian, 400
Probability density function, 578
Probability distribution function, 578
Probability, 388, §11.5.1 *passim*, 491
    additivity of, 307, 311
    Arbuthnot's definition, 303
    as used by Bayes, 271
    Bayes's definition, 300, 303
    chance of, 298
    de Moivre's definition, 302
    inverse, 265, 316, 570
    maximum, 265
    prior, 263, 316, 322, 574
    theory, validity of, 561
Probation, state of, 129–134, 161–163
Prolepsis, 377, 558
Propension, physical, 115, 148
Propensities, 148
Proportion, 120, 122, 124–126, 148, 151, 154, 157
Protestants, 194, 220
Proverb, Greek, 373, 382
Providence, 110, 112, 121, 125, 137, 143, 157, 180
Psychology:
    introspective, 552
    observational, 552
Pun, sophisticated, 299
Punishment, 108, 126–129, 132, 153, 158, 160–162, 554
Puritanism, 33, 67
Pyramid of Cheops, 480, 482

Quadrature, 183
Qualifications, virtuous, 130, 161
Quantity:
    determinate, 235
    flowing, 198
    fluxion of, 205
    nascent, 198, 200
    negative, 221
    permanent, 198, 225, 232
    standing, *see* Quantity, determinate
    uniformly flowing, 200
    vanishing, 198, 210, 223
Queen:
    Anne, 26, 48, 49, 51, 67, 68, 533
        death of, 33, 68
    Caroline, 82, 84
    Catherine, 58, 76, 544, 548
    Elizabeth, 8, 33
    Henrietta Maria, 58
    Katherine de Braganza, *see* Catherine
Quetelet, Lambert A.J., 29, 631
Quin, James, 545

Ramsey, Frank P., 302, 306, 631
Ransome, Mary, 536, 537, 616
Raphael, David D., 259, 260, 267, 552
Raphael, Sylvia, 260
Rashed, Rashdi, 569, 631
Rathmell, 12, 14
Ratio, 126, 157
  prime, 183, 190, 199, 202–204, 206, 565
  ultimate, 178, 183, 190, 198, 199, 203, 204, 206, 223, 563, 564, 565
Rawlings, Gertrude B., 631
Reade, Charles, 536
Reason, 143, 156, 158
Reasoning:
  analogical, 270
  inductive, 270
Rectitude, 111, 132, 140–143, 149, 159, 162, 164, 171, 173–175, 180, 553
  divine, 110, 111, 116, 127–129, 149, 158, 161, 169
  moral, 106, 107, 109–111, 115, 118, 141, 142, 150
  negative, 169
  of delight, 169
  of judgment, 169
  of will, 169
Redemption, 174
Regression, absolute deviations, x, 393
Rehearsal, The, 38, 97, 98
Reichenbach, Hans, 267, 570, 631
Religion:
  interests of, 193
  mysteries of, 194, 219
  natural, 101
  revealed, 101
Repentance, 139, 160
Reprobation, 26
Respectability, 533
Restoration, 89
  wits of the, 28, 91
Revelation, 143, 158
Revell, Lionell, 11
Revenge, 108
Revolution, American, 404

Rewards, 108, 128, 129, 131, 132, 162
Reynolds, Thomas, 15, 531, 532
Rheims, Jackdaw of, 82
Richardson, Samuel, 76, 77, 83, 331
Riemann zeta function, 257
Rigaud, Stephen J., 189, 190, 631
Righteous, rewards of the, 161, 162
Righteousness, 108
Rippon, John I, 66, 510, 512–514, 516, 592, 631
Rippon, John II, 510
Rising of the sun, 292, 327–329
Rites, impious, 374
Robins, Benjamin, 190, 224, 241, 250, 474, 556, 557, 560, 563, 565, 631
Robinson, Edward F., 631
Rochester, Earl of, 60, 61, 97, 549
Rockwell, Julius E., 422, 486, 632
Rogers, Pat, 528, 632
Rogers, Tom, 503
Roller, Duane H. Du B., 579, 582, 632
Roman hand, 375, 576
Romanism, 23
Rome, Church of, 101
Rorie, David, 577
Ross, Sydney, 587, 632
Rothstein, E., 632
Rousseau, George S., 91, 632
Rowe, Nicholas, 576
Rowzee, Ludowick, 58, 60, 70, 77, 632
Royal Society, 380, 404
  Fellows of, 90–92
  Statutes of, 92
Rule of Succession, 571, 575
  Laplace's, 266, 326
Russell, Bertrand A.W., 526, 632
Russell, W. Clark, 376, 576, 632
Rusty, 374
Rutherforth, Thomas, 554, 632
Rutt, John T., 531

Saglio, Edmond, 383, 632
Sainte-Beuve, Charles-Augustin, 490, 632

Salmon, Wesley C., 266, 322, 571, 632
Samples, statistical, 265
Sampling, 570
Sancroft, William, 505
Sanctity, 159
  odour of, 19
Sankey, 529
Saunderson, Nicholas, 586
Savage, Leonard J., 265, 307, 317, 570, 574, 609, 633
Saw, an old, 373
Say, Samuel, 32, 633
Schism, 36
School:
  British, 74
  National, 74
Science, human, 194, 220
Scotland, mediæval degree system in, 375
Scott, James, 633
Scott, Joseph F., 589, 633
Scott, Walter, 533
Scottish *merk*, 56
Scripture, 151
Seal, Hilary L., 568, 578, 633
Seal, Thomas, 69
Seale, Philip, 69
Searchers of the dead, 503, 504
Selby-Bigge, Lewis A., 550, 633
Self-interest, 179, 554
Self-love, 118, 150, 552
Senna treacle, 87
Sensation, 407
Sentences, figures of, 376
Series, §11.5.5 *passim*
  asymptotic, 255, 256
  divergent, 253, 256
  for arcsin z, 429
  infinite, 253, 384
  interpolation of, 452
  semi-convergent, 255
Servant, Greek, 26
Shadwell, Thomas, 58, 90
Shafer, Glenn, 263, 311, 312, 571, 573, 633
Sheffield, 7, 8, 11
Sheffield, John, 18, 532
Shelton, Thomas, 421, 583, 634

Sheridan, Richard B., 526
Sherwin, Henry, 454, 634
Sheynin, Oscar B., 325, 353, 578, 589, 634
Shiokawa, Tetsuya, 489, 634
Shoesmith, Eddie, 393, 578, 634
Shorthand verse, 486
Shorthand, §11.3 *passim*
  as used by Bayes, 421
  Aulay Macaulay's, 486
  Coles's, 421
  Rich's, 44
  Shelton's, 421
Sidgwick, Henry, 40, 634
Simpson, Thomas, x, 182, 253, 303, 384–392, 412, 578, 585, 634
Sin, 126–129, 139, 158, 167, 174
  degree of, 126, 127
  original, 35
Singer, George J., 579, 582, 636
Sizar, 528
Skewness, 337, 357
Smith, Adrian F.M., 599
Smith, G.C., 222, 227, 231, 562, 565, 636
Smith, Robert, 82, 466, 468, 469, 471, 474, 546, 559, 586, 636
Smith, Skinner, 47, 52, 99, 370, 382, 537
Smith, Sydney, 259, 376, 378
Smokler, Howard E., 570, 620
Snake, in the grass, 540
Snow-hill, 32
Snuff, invisible, 508
Socinianism, 23, 35, 533
Solinus, 400
Some, Mr, 527
Southey, Robert, 376, 510
Sozzini, Fausto P., 35
Sozzini, Lelio F.M., 35
Speidell, John, 55, 636
Speldhurst, 53, 58, 63, 81, 95, 540
Sprange, J., 64, 67, 77, 79, 636
Sprat, Thomas, 380, 636
Squire, William, 584, 613
St Bartholomew, Feast of, 8, 528
St Bartholomew's Day, 83

St Olave, Hart Street, 591
St Thomas's, Southwark, 18
Stamford, 530
Stanhope, Philip, Earl, 412, 582, 586
Steele, Richard, 26, 27, 31, 91, 595
Stegun, Irene A., 585, 595
Sterrett, J. MacBride, 35
Stevenson, Burton, 567, 636
Stewart, Larry, 546, 636
Stigler, Stephen M., 28, 54, 260, 261, 305, 316, 320, 331, 429, 526, 527, 540, 572, 578, 636
Stirling numbers, 450, 585
Stirling, James, 84, 255, 443, 450, 453, 477, 478, 568, 582, 637
Stirling–de Moivre formula, 255, 257, 332
Stock, Joseph, 186, 188, 559, 637
Stoke Newington, 535
Stone, Edmund, 453, 637
Stones, heap of, 125
Stott, David, 66
Stow, John, 501, 532, 534, 589, 637
Strange, Charles H., 548, 637
Streatfield, George, 535
Stretton, Mr, 16
Strode, Thomas, 572, 637
Strype, John, 512, 638
Style, 372, 373, 375, 376
    elegant, 379
    florid, 379
    neat, 379
    obscure, 379
    plain, 379
Stylus, 375
Suckling, John, 185, 534
Süßmilch, Johann Peter, 101
Suetonius (Gaius Suetonius Tranquillus), 526, 546, 638
Sunday observance, 66
Supererogation, 180
Swift, Jonathan, 28, 67, 90
Swijtink, Zeno G., 569, 612, 638
Sydserf, Thomas, 28, 534
Syllogism, 246

Sylvester, Matthew, 16, 529
Symmetry, 125, 157

Tangency, 183
Tanur, Judith M., 620
Tarugo's Wiles, 28
Taylor, Brook, 253, 638
Taylor, Christopher, 18, 25, 52
Taylor, Eva G.R., 44, 84, 638
Taylor, Jeremy, 376
Taylor, Martha, *see* Bayes, Martha
Taylor, Sarah, 520
Temper, 130, 132
Temple, William, 376
Temptation, 113, 130, 131, 133, 134
Tenter Alley, 41, 42, 370
Terms, horrid, 216
Tessera, 383
Thackeray, William M., 376
Thales, 400
Theophrastus, 400, 579
Theorems, universal, 156
Thermodynamics, first law of, 315
Things, nature of, 148
Thomas, David O., 41, 42, 259, 535, 536, 554, 638
Thompson, Josiah, 533
Thomson, George, 506
Thomson, J. Radford, 66, 74, 638
Thorburn, W.M., 550, 638
Thornbury, George Walter, 30, 638
Thron, Wolfgang J., 459, 618
Tiberius (Tiberius Claudius Nero), 526
Tibullus, Albius, 544
Tierney, Luke, 638
Tilling, Laura, 557, 566, 614
Tillotson, John, 376
Timerding, Heinrich C.F.E., 258, 261, 313, 325, 353, 366–368, 574, 576, 589, 638
Timpson, Thomas, 69, 94, 639
Tindal, Matthew, 175, 550
Tobacco, and the Plague, 503
Todhunter, Isaac, 333, 334, 569, 571, 578, 638
Toleration, 34
Tonbridge, 58
Tong, William, 537

Trencher-chaplains, 51
Trial, state of, 131, 133, 134, 162
Triangle, largest possible, 136, 166
Triangles:
    rectilinear, 432
    spherical, 432
Tribus, Myron, 622
Trigonometric:
    formulae, 431
    methods, 496
Trigonometry, §11.5.2 *passim*
Truth, *see* Veracity, 132, 162
    wheat of, 6
Truths, singular, 156
Tunbridge, 63
Tunbridge Wells, 505, 545
    circulating library, 77
    coffee at, 58
    coffee-house, 77
    company at, 60, 77, 80–82, 85
    East Indian visitors to, 87
    games of chance at, 85
    lodging houses, 65
    Master of Ceremonies, 78, 79
    Nonconformity in, 66
    rheumatism at, 76
    rules for company at, 79
    season, 70
    situation of, 58
    tobacco at, 58
Turner, George L., 10, 639
Tweddle, Ian, 579, 584, 639
Tweedie, Charles, 255, 450, 568, 639
Tyndall, John, 499, 501

Unfaithful, the, 130, 161
Uniformity, 155
    amidst variety, 123, 156
Unitarian, 35
Universe:
    bodies of, 155
    most happy, 137, 138, 166
    order of, 148
    regularity of, 155
    uniform, 137
University:
    Cambridge, 14
    Oxford, 14
Urbach, Peter, 265, 617

Urwick, William, 639
Usefulness, 122
Uspensky, James V., 321, 639
Utility:
    expected, 263, 572
    theory, 302

Valiant-for-truth, 239
Value, 305
van der Wærden, Bartel L., 301, 572
van Rooijen, Jozephus P., 575, 639
van Schooten, Francis, 572, 639
Variety, 155
    beautiful, 138
Velikovsky, Immanuel, 83, 551, 639
Velocities, 221
    instantaneous, 221
Velocity, idea of, 201
Vengeance, divine, 158
Venn, John, 528, 571, 575, 639
Venn, John A., 528, 639
Veracity, 106, 109, 111, 142, 145, 180
Vice, 127, 128, 139, 158, 160, 164, 174
Vices of the day, 66
Villiers, George, *see* Buckingham, Duke of
Vincent, Benjamin, 528
Vineyard, and labourers, 163
Virgil (Publius Vergilius Maro), 540, 544
Virtue, 108, 122, 123, 127, 128, 130, 133, 134, 139, 155, 158, 160, 163, 164, 167, 180, 553
Virtues, acquired, 131, 132, 161
Virtuosi, 89
Visitors:
    to London, 88
    to Tunbridge Wells, 87
von Braunmühl, Johann A., 496, 640
von Ettingshausen symbol, 583
von Mises, Richard, 261, 264, 322, 569, 574, 640
von Wright, Georg H., 258, 263, 265, 568, 570, 571, 640
Vulgar, 577
Vulgate, The, 536

Wadsworth, J., 44
Wager, 303
Wakefield, Vicar of, 85
Walker, Keith, 542, 549, 640
Wallace, David L., 575, 576, 626
Wallace, Lew, 557
Waller, John F., 536, 640
Walley, Peter, 306, 307, 570, 571, 573, 640
Wallis's product, 256, 443, 458, 589
Wallis, John, 563, 589, 640
Wallis, Peter J., 32, 537, 546, 561, 640
Wallis, Ruth V., 32, 537, 546, 561, 640
Walmsley, Peter, 558, 567, 640
Walsh, William S., 576, 577, 640
Walton, John, 220, 238, 244, 563, 640
Warburton's syllogism, 487
Warburton, William, 487, 640
Ward's hypothesis, 475
Ward, John I, 370
Ward, John II, 31, 41, 370, 480, 536, 559, 588, 640
Ward, Seth, 475, 583, 588
Warnock, Geoffrey J., 561, 641
Warwick, Guy, 522
Watanabe, Magoichirô, 329, 574, 641
Watering-places:
    and barrenness, 58, 60, 76
    description of, 79
Waters, chalybeate, 58
Watson, David, 641
Watson, George N., 562, 642
Watson, William, 401, 581, 641
Watts, 32
Watts, Isaac, 32, 46, 509
Weakness, 160
Weatherford, Roy, 322, 641
Weights and measures, §11.8.2
    Avoirdupois, 465
    Troy, 465
Weinbrot, Howard D., 632
Weinstock, Robert, 191, 560, 641
Wen, 79, 546
Wendrock, Guillaume, 487
Wesley, Charles, 509

Wesley, John, 35, 76, 509
Wesley, Samuel, 46, 509, 592
Wesley, Susanna, 509, 592
Wesleyans, see Methodism
West, Ann(e), 511, 517
    see Bayes, Anne II
West, Gilbert, 250
West, Thomas, 32, 511
    will of, 522
Westfall, Richard S., 560, 641
Wheatley, Henry B., 32, 499, 509, 532, 590, 641
Whewell, William, 43
Whiston, William, 27, 28, 70, 82, 84, 101, 546, 641
Whitaker, Wilfred B., 66, 641
Whitam, J.H., 527, 642
White Horse Inn, 22
White Knight, 301
White, James G., 531, 642
White, John T., 375, 642
Whitehead, Alfred N., 392, 642
Whitely, W.T., 531
Whiteside, Derek T., 556, 642
Whittaker, Edmund T., 562, 642
Whittle, Peter, 310, 573, 642
Whyte, Lancelot L., 526, 642
Wicked, 132, 162
Wickedness, 127, 128, 139, 158
Wickens, a draper, 546
Wildman, Aunt, 518
Wildman, Samuel, 518, 522
Williams, Daniel, 16, 23, 50
Wilmot, John, see Rochester, Earl of
Wilson, Benjamin, 384, 404, 409, 411, 580, 581, 616, 642
Wilson, James, 556
Wilson, John H., 28, 78, 532, 545, 546, 642
Wilson, Thomas, 61, 63, 66, 82, 87, 546
Wilson, Walter, 12, 89, 532, 536, 642
Winchelsea, Earl of, 532
Wisdom, 114, 123, 131, 136, 143, 145, 171–173
    divine, 111
    infinite, 114, 120, 133, 136, 145
Wise, M.E., 323, 368, 643

Wishart, John, 367, 368, 576, 589, 643
Wollaston, William, 554, 643
Woodhouse, Robert, 186, 240, 565, 643
Wordsworth, Elias, 12, 518
Worgan, John, 471
Workmanship, waste of, 154
World:
    infinite and eternal, 166
    regular, 123
Worm, xi
    homage of a, 152
Worth, *see* Value
Wretch, 160
Wright, Edward M., 586, 615
Wycherley, William, 29

Yates, Frank, 569, 570, 643
Young, Thomas, 401, 643

Zabell, Sandy L., 575, 643

Sources and Studies in the History of Mathematics and Physical Sciences

Continued from page ii

C.C. Heyde/E. Seneta
I.J. Bienaymé: Statistical Theory Anticipated

J.P. Hogendijk
Ibn Al-Haytham's *Completion of the Conics*

J. Høyrup
Length, Widths, Surfaces: A Portrait of Old Babylonian Alegbra and Its Kin

A. Jones (Ed.)
Pappus of Alexandria, Book 7 of the *Collection*

E. Kheirandish
The Arabic Version of Euclid's *Optics***,** Volumes I and II

J. Lützen
Joseph Liouville 1809–1882: Master of Pure and Applied Mathematics

J. Lützen
The Prehistory of the Theory of Distributions

G.H. Moore
Zermelo's Axiom of Choice

O. Neugebauer
A History of Ancient Mathematical Astronomy

O. Neugebauer
Astronomical Cuneiform Texts

F.J. Ragep
Naṣīr al-Dīn al-Ṭūsī's *Memoir on Astronomy*
(al-Tadhkira fī ᶜilm al-hay'a)

B.A. Rosenfeld
A History of Non-Euclidean Geometry

J. Sesiano
Books IV to VII of Diophantus' *Arithmetica***: In the Arabic Translation Attributed to Qusṭā ibn Lūqā**

L.E. Sigler
Fibonacci's *Liber Abaci***: A Translation into Modern English of Leonardo Pisano's Book of Calculation**

Sources and Studies in the History of Mathematics and Physical Sciences

Continued from the previous page

B. Stephenson
Kepler's Physical Astronomy

N.M. Swerdlow/O. Neugebauer
Mathematical Astronomy in Copernicus's De Revolutionibus

G.J. Toomer (Ed.)
Appolonius *Conics* Books V to VII: The Arabic Translation of the Lost Greek Original in the Version of the Banū Mūsā, Edited, with English Translation and Commentary by G.J. Toomer

G.J. Toomer (Ed.)
Diocles on Burning Mirrors: The Arabic Translation of the Lost Greek Original, Edited, with English Translation and Commentary by G.J. Toomer

C. Truesdell
The Tragicomical History of Thermodynamics, 1822–1854

K. von Meyenn/A. Hermann/V.F. Weisskopf (Eds.)
Wolfgang Pauli: Scientific Correspondence II: 1930–1939

K. von Meyenn (Ed.)
Wolfgang Pauli: Scientific Correspondence III: 1940–1949

K. von Meyenn (Ed.)
Wolfgang Pauli: Scientific Correspondence IV, Part I: 1950–1952

K. von Meyenn (Ed.)
Wolfgang Pauli: Scientific Correspondence IV, Part II: 1953–1954